Handbook of Chemical Biology

Handbook of Chemical Biology

Edited by
Oliver Stone

Willford Press
www.willfordpress.com

Published by Willford Press,
118-35 Queens Blvd., Suite 400,
Forest Hills, NY 11375, USA

ISBN: 978-1-68285-406-8

Cataloging-in-Publication Data

Handbook of chemical biology / edited by Oliver Stone.
p. cm.
Includes bibliographical references and index.
ISBN 978-1-68285-406-8
1. Biochemistry. 2. Chemistry. I. Stone, Oliver.
QH345 .H36 2018
572--dc23

For information on all Willford Press publications
visit our website at www.willfordpress.com

Willford Press

Printed in China.

Contents

Preface

This book has been a concerted effort by a group of academicians, researchers and scientists, who have contributed their research works for the realization of the book. This book has materialized in the wake of emerging advancements and innovations in this field. Therefore, the need of the hour was to compile all the required researches and disseminate the knowledge to a broad spectrum of people comprising of students, researchers and specialists of the field.

Chemical biology is a rapidly growing field which combines traditional chemistry with modern concepts of biology. The concepts related to the subject of chemical biology are proteomics, combinatorial chemistry, glycobiology, molecular sensing, etc. This book provides comprehensive insights into the field of chemical biology. It includes contributions of experts and scientists which will provide innovative insights into this field. It will benefit students and researchers alike. As this field is emerging at a rapid pace, the contents of this book will help the readers understand the modern concepts and applications of the subject.

At the end of the preface, I would like to thank the authors for their brilliant chapters and the publisher for guiding us all-through the making of the book till its final stage. Also, I would like to thank my family for providing the support and encouragement throughout my academic career and research projects.

Editor

1

Dysregulation of transition metal ion homeostasis is the molecular basis for cadmium toxicity in *Streptococcus pneumoniae*

Stephanie L. Begg[1], Bart A. Eijkelkamp[1], Zhenyao Luo[2,3,4], Rafael M. Couñago[2,3,4], Jacqueline R. Morey[1], Megan J. Maher[5], Cheryl-lynn Y. Ong[2,3], Alastair G. McEwan[2,3], Bostjan Kobe[2,3,4], Megan L. O'Mara[2,6], James C. Paton[1] & Christopher A. McDevitt[1]

Cadmium is a transition metal ion that is highly toxic in biological systems. Although relatively rare in the Earth's crust, anthropogenic release of cadmium since industrialization has increased biogeochemical cycling and the abundance of the ion in the biosphere. Despite this, the molecular basis of its toxicity remains unclear. Here we combine metal-accumulation assays, high-resolution structural data and biochemical analyses to show that cadmium toxicity, in *Streptococcus pneumoniae*, occurs via perturbation of first row transition metal ion homeostasis. We show that cadmium uptake reduces the millimolar cellular accumulation of manganese and zinc, and thereby increases sensitivity to oxidative stress. Despite this, high cellular concentrations of cadmium (~17 mM) are tolerated, with negligible impact on growth or sensitivity to oxidative stress, when manganese and glutathione are abundant. Collectively, this work provides insight into the molecular basis of cadmium toxicity in prokaryotes, and the connection between cadmium accumulation and oxidative stress.

[1] Research Centre for Infectious Diseases, School of Biological Sciences, The University of Adelaide, Adelaide, South Australia 4072, Australia. [2] School of Chemistry and Molecular Biosciences, University of Queensland, Brisbane, Queensland 4072, Australia. [3] Australian Infectious Diseases Research Centre, University of Queensland, Brisbane, Queensland 4072, Australia. [4] Institute for Molecular Bioscience, University of Queensland, Brisbane, Queensland 4072, Australia. [5] La Trobe Institute for Molecular Science, La Trobe University, Melbourne, Victoria 3086, Australia. [6] School of Mathematics and Physics, University of Queensland, Brisbane, Queensland 4072, Australia. Correspondence and requests for materials should be addressed to C.A.M. (email: christopher.mcdevitt@adelaide.edu.au).

Global cadmium (Cd) production has risen by >1,000-fold since the beginning of the twentieth century to ~20,000 tons per year[1]. Correspondingly, anthropogenic release of Cd into the atmosphere now significantly outstrips natural fluxes and predominantly occurs via non-ferrous ore processing, combustion of fossil fuels, and manufacturing, use and disposal of Cd-containing products[1]. As Cd cannot be degraded, its flux into marine and terrestrial ecosystems has increased the risk of exposure as it accumulates in the food chain, and it is estimated that humans ingest 30 μg of Cd every day[1,2]. Cd, which occurs as the divalent cation in the natural environment, is acutely toxic to all forms of life, although there are exceptions to this rule for organisms that have evolved in environments with extremely low zinc abundance[3]. Cd toxicity in humans and microbes is, at a cellular level, closely associated with oxidative stress, despite the inability of Cd^{2+} to directly produce reactive oxygen species[4]. As a consequence, understanding the molecular basis for how Cd^{2+} causes toxicity is a crucial issue.

Cd^{2+} accumulation in microbes and humans primarily occurs by uptake via divalent metal transporters, such as manganese (Mn^{2+}) transporters, although the molecular basis for this process remains unknown[5,6]. We investigated Cd^{2+} toxicity in the Gram-positive bacterium *Streptococcus pneumoniae* (pneumococcus) as it has only a single Mn^{2+}-specific uptake pathway. The pneumococcus acquires Mn^{2+} via the PsaBCA permease, which comprises the ATP-binding cassette (ABC) transporter, PsaBC, and a cell-surface solute-binding protein (SBP), PsaA. The Psa permease is essential for both Mn^{2+} uptake and *in vivo* virulence[7–9]. Similar to other ABC permeases, the SBP PsaA defines the specificity of the uptake pathway. However, our recent studies showed that the functional specificity of the Psa permease in Mn^{2+} acquisition arose not from the specificity of Mn^{2+} binding by PsaA, but instead resulted from the concerted action of protein metal ion coordination chemistry and structural rearrangements that prevented release of other divalent transition metal ions, such as Zn^{2+}, into the permease, despite their ability to bind to PsaA[8]. Thus, although it is conceivable that Cd^{2+} may interact with the Psa permease, whether it causes injury to the cell by non-productively competing for metal ion uptake or by accumulation in the cytosol remains unclear.

Here we use a combination of *in vitro*, *in vivo* and structural analyses to investigate how Cd^{2+} impacts transition metal homeostasis in *S. pneumoniae*. We show that *S. pneumoniae* accumulates millimolar concentrations of transition metal ions, and that Cd^{2+} dysregulates metal ion homeostasis by perturbing the Mn^{2+} and Zn^{2+} uptake and efflux pathways, although via strikingly different mechanisms. Manganese accumulation is most affected, and this arises from Cd^{2+} competing with Mn^{2+} ions for uptake via the Psa permease and the concomitant upregulation of the Mn^{2+}-efflux pathway. High-resolution structural analyses reveal that although the Mn^{2+}-specific Psa permease uses coordination chemistry and structural features to prevent the mistranslocation of other ions, such as Zn^{2+}, the larger ionic radius of Cd^{2+} enables this metal to circumvent these regulatory mechanisms. The increased sensitivity to oxidative stress associated with Cd^{2+} toxicity is due to impaired Mn^{2+} acquisition and not directly due to the intracellular accumulation of Cd^{2+} ions, which are buffered on cellular glutathione. Collectively, these findings highlight the challenge that biological systems encounter when non-physiological elements enter the biosphere. This study provides novel insights into the intracellular concentrations of transition metal ions in a Gram-positive cell, the mechanisms associated with their homeostasis, and how Cd^{2+} dysregulates these processes.

Results

S. pneumoniae accumulates high concentrations of metal ions. We first sought to determine the cellular quotient of first row transition metal ions in *S. pneumoniae*. Inductively coupled plasma–mass spectrometry (ICP–MS) data were combined with scanning electron microscopy analyses and colony-forming unit (c.f.u.) counts to derive the intracellular metal ion accumulation data in molar units. In this context, the intracellular metal ion concentration corresponds to the total accumulation of transition metal ions in the cytoplasmic compartment of *S. pneumoniae* and represents all cytoplasmic metal ions, from those bound in tight metal–protein complexes to those occurring as labile pools of exchangeable ions. Although the total number of metal ions per cell was similar to that previously reported for *Escherichia coli*[10], the pneumococcus has a smaller mean (±s.d.) cell volume, 0.113±0.04 fl ($n=44$; Table 1), and only a single cellular compartment, which results in the total cellular metal ion concentrations being in the millimolar range (Fig. 1a). The Irving–Williams stability series is the near-universal order of metal–protein-binding affinity, which arises from the imperfect steric selection of metal ions that occurs within protein–metal-binding sites. In order of increasing atomic number this can be represented as $Mn^{2+} < Fe^{2+} < Co^{2+} < Ni^{2+} < Cu^{2+} > Zn^{2+}$ (ref. 11). Our data also revealed that the total cellular quotient of Mn^{2+} (8.6±0.5 mM) was higher than the accumulation of other more competitive ions, such as Zn^{2+} (6.8±0.4 mM). Hence, in *S. pneumoniae*, the accumulation of Mn^{2+} to higher intracellular concentrations may be one factor that aids in increasing its bioavailability, relative to more competitive ions, such as Zn^{2+}, and thereby aid in its acquisition by proteins that require Mn^{2+}. We then proceeded to ascertain the impact of Cd^{2+} on the cellular accumulation of transition metal ions.

Cd dysregulates transition metal homeostasis. Titrations of Cd^{2+} into a cation-defined medium (CDM), with a constant concentration of Mn^{2+} (1 μM), showed that pneumococcal growth was perturbed (Fig. 1b and Supplementary Fig. 1a), but that this growth defect could essentially be abrogated by the addition of a stoichiometric concentration of Mn^{2+} to the Cd^{2+}-supplemented CDM (Fig. 1c and Supplementary Fig. 1b). *S. pneumoniae* growth in the presence of 30 μM Cd^{2+} reduced the cellular concentrations of Mn^{2+} and Zn^{2+}. Manganese accumulation was most affected, with a reduction of ~90% (Fig. 1a; 1 μM Mn^{2+}: 30 μM Cd^{2+}, $P<0.0001$ (one-way analysis of variance (ANOVA))), despite no significant effect on cell volume (Table 1; 1 μM Mn^{2+}: 30 μM Cd^{2+}, $P=0.376$ (one-way ANOVA)). Supplementation with Mn^{2+} restored pneumococcal Mn^{2+} accumulation to ~60% of the unchallenged levels (Fig. 1a and Supplementary Table 1; 30 μM Mn^{2+}: 30 μM Cd^{2+}, $P<0.0001$ (one-way ANOVA)). Zinc also showed a significant decrease in accumulation in the presence of Cd^{2+} (Fig. 1a; 1 μM Mn^{2+}: 30 μM Cd^{2+}, $P<0.0001$; 30 μM Mn^{2+}: 30 μM Cd^{2+}, $P<0.0001$ (one-way ANOVA)). Cd^{2+} accumulation increased in the presence of 30 μM Cd^{2+}, with or without Mn^{2+} (Fig. 1a). The degree to which Cd^{2+} was accumulated was more than that

Table 1 | *S. pneumoniae* cell volume parameters.

Strain	Cell volume (fl)*
D39 1 μM Mn^{2+}	0.113±0.04
D39 1 μM Mn^{2+}: 30 μM Cd^{2+}	0.124±0.07
D39 30 μM Mn^{2+}: 30 μM Cd^{2+}	0.127±0.07

*The data correspond to the mean ± s.d. from >40 independent measurements.

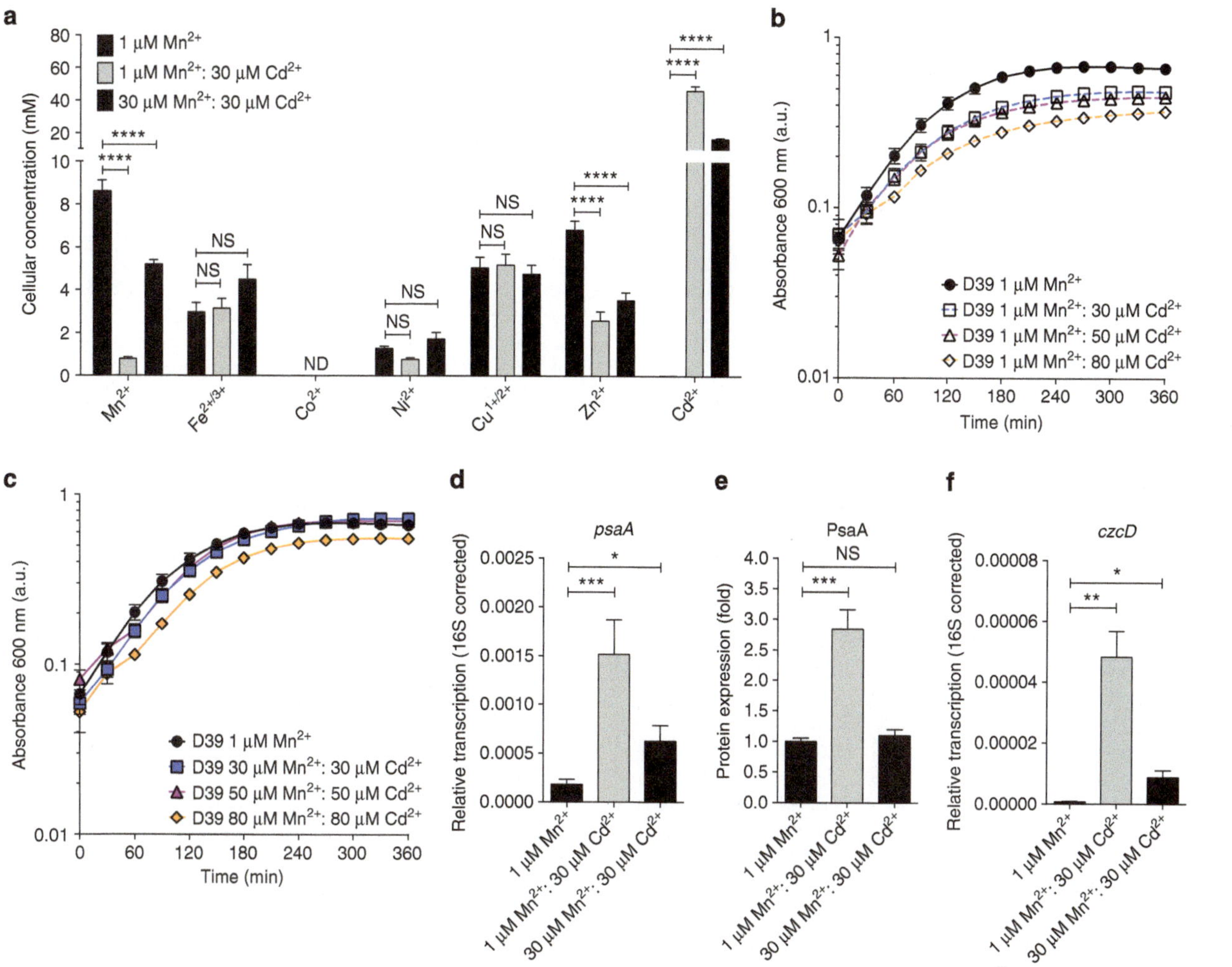

Figure 1. | Impact of Cd^{2+} on *S. pneumoniae* growth and first row transition metal homeostasis. (**a**) Total cellular metal ion accumulation of *S. pneumoniae* represented as the mean (± s.e.m.) concentration of ions per cell (determined as c.f.u.) of known cell volume (determined by scanning electron microscopy (SEM)) grown in the specified conditions. The values represent the total cellular metal ion quota, that is, bound, buffered and unbound metal ions, and are from at least three independent biological experiments. Metal ions assessed were Mn^{2+}, $Fe^{2+/3+}$, Co^{2+}, Ni^{2+}, $Cu^{1+/2+}$, Zn^{2+} and Cd^{2+}. The statistical significance of the differences in the mean metal concentrations was determined by a one-way ANOVA (ND, not detected; NS, not significant and ****$P<0.0001$). (**b,c**) *S. pneumoniae* (D39) grown in CDM supplemented with metal ions as indicated. The data correspond to mean (± s.e.m.) absorbance at 600 nm measurements from three independent biological experiments. Errors bars, where not visible, are overlapped by the representative symbols. (**d**) Relative transcription, corrected to 16S rRNA, of *psaA* by *S. pneumoniae* when grown in CDM supplemented with 1 μM Mn^{2+} (black), 1 μM Mn^{2+}: 30 μM Cd^{2+} (light grey) and 30 μM Mn^{2+}: 30 μM Cd^{2+} (dark grey). (**e**) Quantitative immunoblot data of PsaA (Supplementary Fig. 2), from equivalent total protein loadings of whole-cell extracts, from *S. pneumoniae* when grown in CDM supplemented with 1 μM Mn^{2+} (black), 1 μM Mn^{2+}: 30 μM Cd^{2+} (light grey) and 30 μM Mn^{2+}: 30 μM Cd^{2+} (dark grey). (**f**) Relative transcription, corrected to 16S rRNA, of *czcD* by *S. pneumoniae* when grown in CDM supplemented with 1 μM Mn^{2+} (black), 1 μM Mn^{2+}: 30 μM Cd^{2+} (light grey) and 30 μM Mn^{2+}: 30 μM Cd^{2+} (dark grey). The data correspond to the mean (± s.e.m.) of three independent biological experiments. The statistical significance of the differences in the mean data was determined by two-tailed unpaired *t*-tests (*$P<0.05$, **$P<0.01$ and ***$P<0.001$).

of any other transition metal ion in the pneumococcus, despite having no physiological role. At 1 μM Mn^{2+}: 30 μM Cd^{2+} in the growth medium, intracellular Cd^{2+} concentration was 46.4 ± 2.9 mM. Cd^{2+} accumulation was reduced by nearly 70% on the addition of 30 μM Mn^{2+} to the growth medium (16.5 ± 0.7 mM Cd^{2+}), but cellular accumulation of the ion still remained more than that of the other first row transition metal ions.

We then analysed the impact of Cd^{2+} on transition metal ion homeostatic pathways. Transcriptional analyses revealed that *psaA* transcription increased by 8.4-fold ($P=0.0007$ (two-tailed unpaired *t*-test)) and PsaA expression by 2.8 ± 0.3-fold ($P<0.0001$ (two-tailed unpaired *t*-test)) when grown in the presence of 30 μM Cd^{2+} (Fig. 1d,e), consistent with the Mn^{2+}-responsive transcriptional regulator, PsaR, sensing the Mn^{2+} depletion[7,12]. Transcription of *psaA* remained elevated during growth supplemented with 30 μM Mn^{2+} albeit to a lesser, but still significant, extent (3.5-fold, $P=0.0135$ (two-tailed unpaired *t*-test)), although PsaA expression did not appear to be significantly increased (Fig. 1d,e). Zinc accumulation in *S. pneumoniae* also occurs via an ABC permease, the Adc permease[9,13,14]. Transcriptional analysis of the Zn^{2+}-responsive regulator, *adcR*[15], and the Zn^{2+}-recruiting SBPs associated with the Adc permease, *adcA* and *adcAII* (refs 13,14,16), did not show significant changes in the presence of Cd^{2+} (Supplementary Fig. 3). Despite this, the Zn^{2+} accumulation levels (2.6 ± 0.4 mM) were comparable to strains without a functional Adc permease, Δ*adcCBA* (2.7 ± 0.3 mM) and Δ*adcA*/Δ*adcAII* (2.5 ± 0.3 mM).

The perturbation of Zn^{2+} acquisition was not due to competition for the Adc permease as the dominant Zn^{2+}-recruiting SBP, AdcA, showed very poor interaction with Cd^{2+} by comparison with Zn^{2+} (Supplementary Fig. 3). Taken together, these findings implicate Cd^{2+} as being able to exert an effect on the Zn^{2+}-responsive transcriptional regulators. *S. pneumoniae* uses two Zn^{2+}-responsive transcriptional regulators: AdcR, which regulates the Zn^{2+}-uptake machinery described above, and SczA, which regulates transcription of the cation diffusion facilitator efflux protein CzcD[15,17]. Consequently, we examined whether CzcD was also affected by Cd^{2+} abundance. Growth of *S. pneumoniae* in CDM with 1 μM Mn^{2+}: 30 μM Cd^{2+} increased *czcD* transcription by ~70-fold ($P = 0.0011$ (two-tailed unpaired *t*-test); Fig. 1f), while supplementation with Mn^{2+} (30 μM Mn^{2+}: 30 μM Cd^{2+}) reduced this to only ~12.7-fold more than cells grown in unsupplemented CDM (Fig. 1f). Although CzcD from *Cupriavidus metallidurans* is protective against Co^{2+}, Zn^{2+} and Cd^{2+} ions[18], there is no evidence to suggest that the *S. pneumoniae* orthologue has similar poly-specificity. Consequently, we examined an isogenic Δ*czcD* mutant strain to ascertain whether pneumococcal *czcD* contributed to Cd^{2+} resistance. The *S. pneumoniae* Δ*czcD* strain was hypersensitive to extracellular Zn^{2+}, consistent with a role in Zn^{2+} homeostasis (Fig. 2). Furthermore, the Cd^{2+}-induced efflux of Zn^{2+} observed in wild-type *S. pneumoniae* was abolished in the Δ*czcD* strain (Δ*czcD* 1 μM Mn^{2+}, 4.4 ± 0.3 mM Zn^{2+}; Δ*czcD* 1 μM Mn^{2+}: 30 μM Cd^{2+} 4.2 ± 0.3 mM Zn^{2+}; $P = 0.78$ (two-tailed unpaired *t*-test)). However, although Cd^{2+} accumulation in the Δ*czcD* strain showed a mild elevation of ~1.4-fold compared with the wild-type, the growth phenotypes of the Δ*czcD* strain were indistinguishable from the wild-type strain and also showed no increase in susceptibility to extracellular Cd^{2+} (Fig. 2 and Supplementary Fig. 4). Taken together, these data indicate that, despite the upregulation of *S. pneumoniae czcD* in response to Cd^{2+} exposure, CzcD is primarily a Zn^{2+} efflux transporter and appears to serve, at best, only a minor protective role against Cd^{2+}.

In addition to *czcD*, the pneumococcal genome encodes four other putative metal ion efflux pathways. These are as follows: spd1384 (*mntE*), a paralogue of CzcD involved in Mn^{2+} efflux[19]; spd1438 (*cadD*), a membrane protein of unknown function associated with Cd^{2+} tolerance[20]; two P-type ATPases, spd0635 (*copA*), which is responsible for Cu^{1+} efflux[21]; and spd1927, which has homology to *pmtA*, a PerR-regulated gene in *S. pyogenes* associated with Zn^{2+} resistance[22]. The *mntE* gene has previously been reported to be constitutively expressed and not responsive to cellular Mn^{2+} abundance[19]. Here, we observed that extracellular Cd^{2+} induced a 2.1-fold increase in *mntE* transcription ($P = 0.0211$ (two-tailed unpaired *t*-test); Supplementary Fig. 5a). On supplementation with 30 μM Mn^{2+}, *mntE* transcription was reduced to non-challenged levels ($P = 0.3665$ (two-tailed unpaired *t*-test) Supplementary Fig. 5a), suggesting that intracellular Cd^{2+} concentrations modulated *mntE* expression via an unknown metal-responsive regulator. We then analysed an isogenic Δ*mntE* strain, which lacks the primary Mn^{2+}-efflux pathway[19], for sensitivity to Cd^{2+}. Consistent with Cd^{2+} also perturbing Mn^{2+} homeostasis via MntE, the Δ*mntE* strain retained a higher cellular Mn^{2+} concentration (2.53 ± 0.1 mM) than the wild-type (0.78 ± 0.07 mM) ($P < 0.0001$ (two-tailed unpaired *t*-test)), when grown in 1 μM Mn^{2+}: 30 μM Cd^{2+}, and showed greater resistance to extracellular Cd^{2+} (Fig. 2). The *cadD* gene showed a similar trend to *czcD* and *mntE* with a ~3.6-fold increase in transcription when challenged with 30 μM Cd^{2+} ($P < 0.0001$ (two-tailed unpaired *t*-test); Supplementary Fig. 5b), followed by abrogation when supplemented with 30 μM Mn^{2+}.

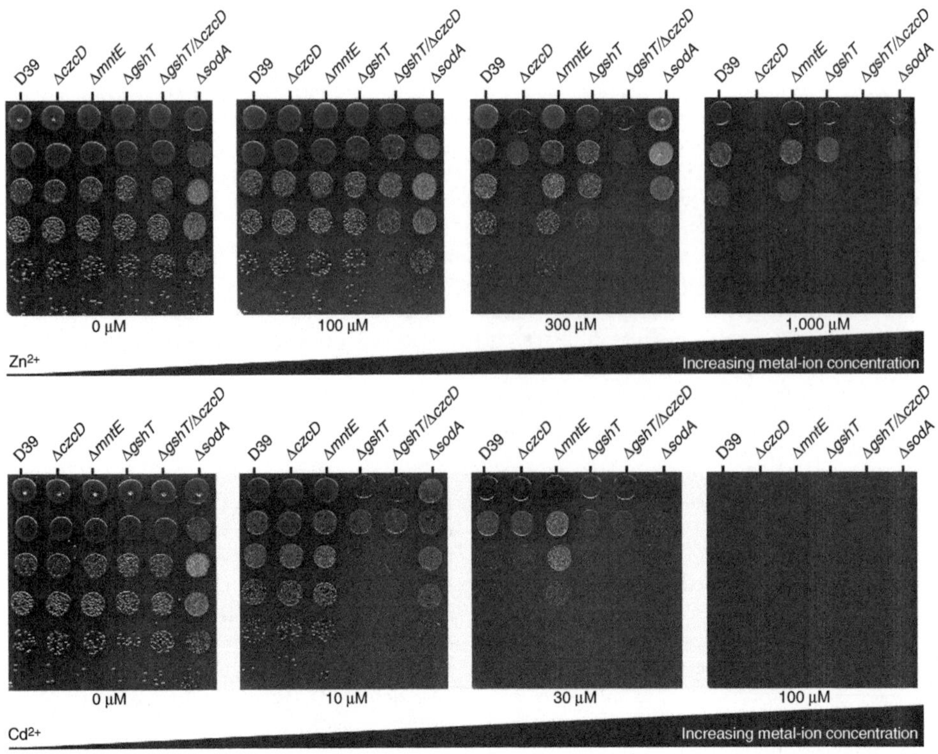

Figure 2 | Phenotypic impact of metal ions on *S. pneumoniae* growth. Drop test analysis of *S. pneumoniae* wild-type (D39), Δ*czcD*, Δ*mntE*, Δ*gshT*, Δ*gshT*/Δ*czcD* and Δ*sodA* on blood agar (BA) supplemented with: 0, 100, 300 and 1,000 μM $ZnSO_4$ or 0, 10, 30 and 100 μM $CdCl_2$. Cells were grown and adjusted to A_{600} of 0.4 and were serially diluted. Drops (5 μl) were spotted onto the BA plate, starting from the 10^0 dilution (top) down to the 10^{-5} dilution (bottom).

However, the P-type ATPases, *copA* and spd1927, were transcriptionally unresponsive to Cd^{2+} (Supplementary Fig. 5c,d), suggesting that they were unlikely to contribute to Cd^{2+} management. Collectively, these data show that Cd^{2+} accumulation was associated with transcriptional activation of several metal ion membrane protein transporters. However, of those pathways with known efflux function, that is, CzcD and MntE, Cd^{2+} ions subverted their primary functional roles and, instead, further dysregulated transition metal ion homeostasis in *S. pneumoniae* via amplification of Zn^{2+} and Mn^{2+} efflux, respectively. The mechanism(s) of Cd^{2+} efflux in *S. pneumoniae*, if any, remains to be identified.

Cd^{2+} competes with Mn^{2+} for the Psa permease. To elucidate the hitherto unexplained molecular basis for Cd^{2+} uptake via Mn^{2+} transporters, we examined the interaction of Cd^{2+} with the Mn^{2+}-recruiting SBP PsaA. Although the physiological role of PsaA is in recruiting Mn^{2+}, the metal-binding site of the protein has previously been shown to interact with other divalent transition metal ions[8]. However, coordination chemistry at the metal-binding site and structural dynamics prevent release of these other ions, and thereby maintain the functional role of the Psa permease in Mn^{2+} uptake[8]. Here the interaction of Cd^{2+} with PsaA was analysed by determining the structure of wild-type PsaA with Cd^{2+} at 2.0 Å resolution (Fig. 3a and Supplementary Fig. 6). The PsaA–Cd^{2+} complex revealed that Cd^{2+} was coordinated by the metal-binding site residues His67, His139, Glu205 and Asp280 in a structure that was highly similar (root-mean-square deviation (RMSD) < 0.3 Å over 242 Cα atoms) to our previously reported Mn^{2+}- and Zn^{2+}-bound structures[23,24]. Nevertheless, small re-orientations of the side chains of the metal-coordinating residues allowed the bulkier Cd^{2+} ion (ionic radius of 95 pm for Cd^{2+} versus 84 pm for Mn^{2+}) to be accommodated into the protein–metal-binding site (Fig. 3b and Supplementary Table 2). In the PsaA–Cd^{2+} structure, the distances between metal-coordinating atoms and the metal ion were increased by ~0.2 Å compared with those for PsaA-Zn^{2+} (Supplementary Table 2). Similar increases in metal–protein distances have been observed for the Cd^{2+}-thiolate-binding sites of Cd^{2+}-

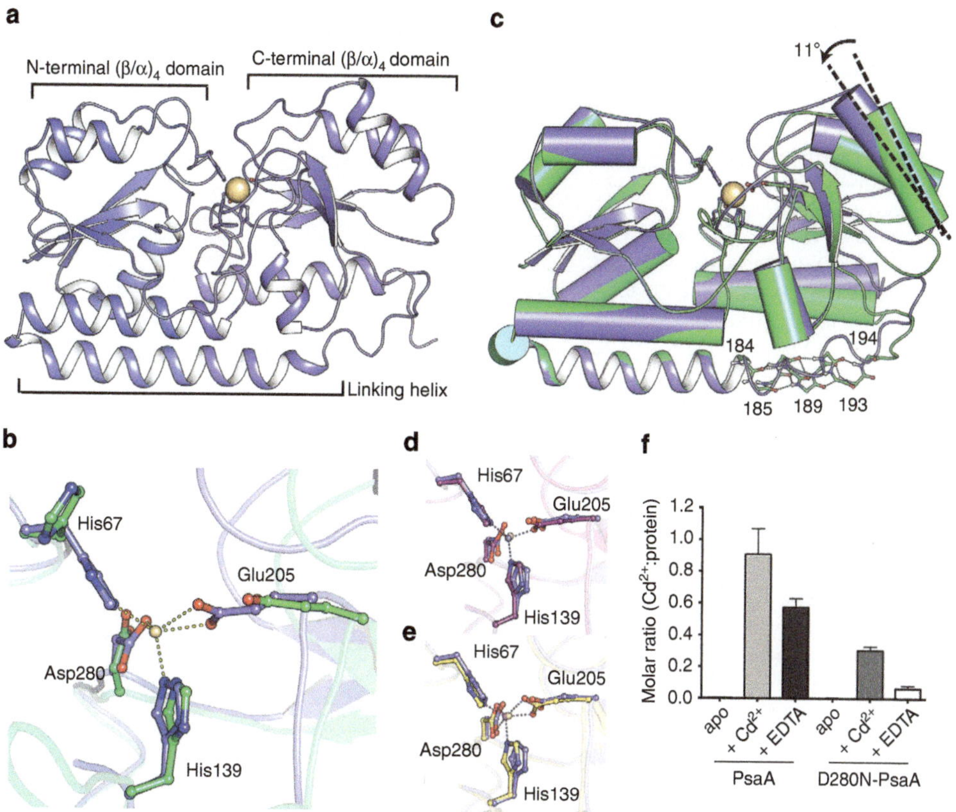

Figure 3 | Structure of Cd^{2+}-bound PsaA. (**a**) Cartoon illustration of Cd^{2+}-bound wild-type PsaA in closed conformation (PDB accession code 4UTP). The structure consists of N-terminal and C-terminal $(\beta/\alpha)_4$ domains with a domain-linking helix. The metal-binding site is situated between the two domains, in which a Cd^{2+} ion (pale-yellow sphere) is bound. (**b**) The metal-binding site of wild-type PsaA in the presence (light blue, PDB accession code 4UTP) and absence (green, PDB accession code 3ZK7) of Cd^{2+} (pale-yellow sphere), with the metal coordination shown as dashed lines. (**c**) Superposition of Cd^{2+}-bound (light blue, PDB accession code 4UTP) and apo (green, PDB accession code 3ZK7) wild-type PsaA structures (Cα atoms, residues 32–163, were used for the superposition). The linking helix is shown in cartoon representation, with other helices shown as cylinders. The crossing angle of the C-terminal $(\beta/\alpha)_4$ domains between the structures is marked. The bound Cd^{2+} ion is shown as a pale-yellow sphere. In the apo structure, backbone atoms in the flexible region of the linking helix (residues 184–194) are shown in stick representation, with hydrogen bonds between backbone atoms as dashed lines. (**d**) Comparison of the metal-binding sites between Cd^{2+}-bound (light blue, PDB accession code 4UTP) and Zn^{2+}-bound (purple, PDB accession code 1PSZ) wild-type PsaA structures. The metal ions are shown as spheres (Cd^{2+}, pale yellow; Zn^{2+}, grey) and their coordination as dashed lines. (**e**) Comparison of the metal-binding sites between Cd^{2+}-bound (light blue, PDB accession code 4UTP) and Mn^{2+}-bound (yellow, PDB accession code 3ZTT) wild-type PsaA structures. The metal ions are shown as spheres (Cd^{2+}, pale yellow; Mn^{2+}, blue) and their coordination as dashed lines. In **a–e**, the metal-coordinating residues are shown in stick representation. (**f**) *In vitro* metal binding of Cd^{2+} to wild-type or D280N variant apo-PsaA analysed by ICP-MS. Data correspond to the mean (± s.d.) molar ratio of Cd^{2+} to the PsaA isoform for apo-PsaA (black), Cd^{2+}-PsaA (light grey), EDTA-Cd^{2+}-PsaA (dark grey). apo-D280N-PsaA (black), Cd^{2+}-D280N-PsaA (grey) and EDTA-Cd^{2+}-D280N-PsaA (white).

substituted metallothionein and rubredoxin proteins, compared with the Zn^{2+}-bound equivalents[25,26]. Although Cd^{2+} and Zn^{2+} often display similar coordination spheres in proteins, the longer bond distances observed for Cd^{2+} usually result in an increase in the coordination number. Accordingly, the coordination geometry in the PsaA–Cd^{2+} structure is an intermediate between a square-pyramid and trigonal-bipyramid ($n = 5$), in contrast to the 4- and 6-coordinate sites observed for the Zn^{2+}- and Mn^{2+}-bound PsaA structures, respectively (Fig. 3d,e and Supplementary Table 3). A comparison of the observed coordination of PsaA–Cd^{2+} with other metal-binding sites of metalloprotein structures in the PDB, composed of mixed N/O ligating atoms, revealed that Cd^{2+} is found in a variety of coordination geometries with coordination numbers ranging from 5 to 7 (Supplementary Table 4). Nevertheless, the only bona fide protein–metal-binding site for Cd^{2+} is in carbonic anhydrase from *Thalassiosira weissflogii*, which has been shown to be a cambialistic enzyme with Zn^{2+} in a tetrahedral geometry and Cd^{2+} in a trigonal-bipyramidal geometry[3,27,28]. Thus, the coordination geometry found in PsaA–Cd^{2+} conforms to what has been observed for the only known naturally occurring Cd^{2+}-binding protein, and for other protein structures deposited in the PDB. We further examined the interaction of metal-free (apo) PsaA with Cd^{2+} and observed that apo-PsaA bound Cd^{2+} in a 1:1 molar ratio (Fig. 3f), with a K_D of 5.6 ± 1 nM (Supplementary Fig. 7). As the cognate ligand, Mn^{2+}, has a K_D of 3.3 ± 1 nM for PsaA, the relatively small (approximately twofold) difference in the K_Ds is consistent not only with the competitive effect on Mn^{2+} acquisition but also the continued, albeit reduced, Cd^{2+} accumulation during growth in the presence of equimolar concentrations of both ions (30 μM Mn^{2+}: 30 μM Cd^{2+}). Collectively, these findings are consistent with the observed impact of Cd^{2+} on Mn^{2+} accumulation in the wild-type and mutant *S. pneumoniae* strains and indicate that it is a competitive effect mediated via PsaA. However, they also indicate that interaction of Cd^{2+} with PsaA is distinct from the previously characterized Mn^{2+} and Zn^{2+} interactions.

PsaA is permissive for Cd^{2+} binding and release. PsaA uses a 'spring-hammer'-like mechanism for metal ion binding, in which the dynamics of the metal ion-loaded protein are dictated by a combination of metal ion geometry and the distortion of a helix linking the N- and C-terminal lobes of the protein[8]. We further analysed the metal-binding site coordination geometry by use of an Asp280 variant isoform (PsaA-D280N), as it would be incapable of facilitating the trigonal-bipyramidal coordination observed in wild-type PsaA[8]. The structure of PsaA-D280N bound to Cd^{2+}, refined to 1.7 Å resolution, revealed a partially solvated Cd ion loosely interacting with residue Glu205 (Fig. 4a,b). The C-terminal lobe of PsaA showed minor movement (~1.0° relative to the open, metal-free conformation), but the inability of Cd^{2+} to ligate to residue Asp280 prevented further conformational changes. Consistent with these observations, PsaA-D280N was greatly impaired in Cd^{2+} binding (Fig. 3f).

We then assessed the distortion of the linking helix by monitoring the number of disrupted backbone hydrogen bonds in its flexible region (residues 184–194) in wild-type PsaA–Cd^{2+} (Figs 3c and 4a, and Supplementary Table 5)[8]. Both the hinge-bending angle between the N- and C-terminal lobes and the number of main-chain hydrogen bonds disrupted in the residue range 184–194 corresponded more closely to the Mn^{2+}-bound structure than to the 'locked' Zn^{2+}-structure. This indicated that Cd^{2+} binding was a reversible process (Supplementary Table 5), consistent with the observation that Cd^{2+} could be extracted

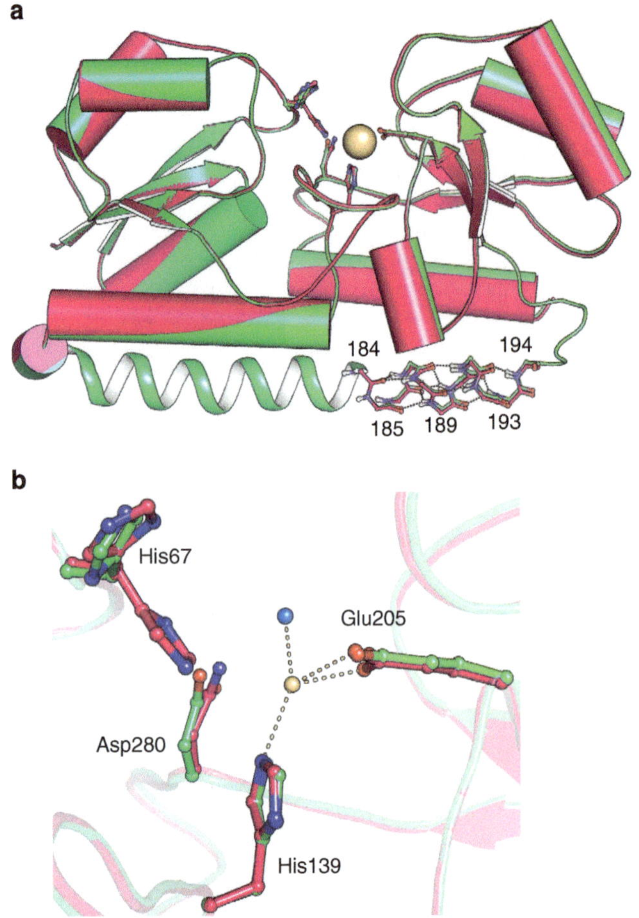

Figure 4 | Structure of Cd^{2+}-bound PsaA-D280N in an open conformation. (**a**) Superposition of Cd^{2+}-bound PsaA-D280N (magenta, PDB accession code 4UTO) and apo wild-type (green, PDB accession code 3ZK7) PsaA structures (Cα atoms, residues 32–163, were used for the superposition). The bound Cd^{2+} ion is shown as a pale-yellow sphere. The linking helix is shown in cartoon representation (up to residue 183) and backbone atoms in the flexible region of the linking helix (residues 184–194) are shown in stick representation, with hydrogen bonds between backbone atoms as dashed lines. (**b**) Close-up view of the metal-binding sites between the superimposed Cd^{2+}-bound PsaA-D280N (magenta, PDB accession code 4UTO) and apo wild-type (green, PDB accession code 3ZK7) PsaA structures. The bound Cd^{2+} ion is shown as a pale-yellow sphere and its coordination with the protein residues is shown as dashed lines. The water molecule coordinating the metal is shown as a light-blue sphere. In **a**,**b**, the metal-coordinating residues are shown in stick representation.

(~40%) from the PsaA–Cd^{2+} complex by using the chelating agent EDTA (Fig. 3f). Thus, these findings provide further support for the conclusion that both Mn^{2+} and Cd^{2+} compete for transport via the Psa permease, consistent with the cellular accumulation data, which shows that Mn^{2+} uptake occurs at the expense of Cd^{2+} import and vice versa.

Reduced Mn^{2+} levels result in sensitivity to oxidative stress. We then sought to assess how Cd^{2+} affected oxidative stress management in *S. pneumoniae*. Manganese has a prominent role in regulating the expression of superoxide dismutase (*sodA*), where it also serves as a cofactor[29–31]. Transcription of *sodA* was decreased under Cd^{2+}-induced Mn^{2+} starvation, but this was restored when supplemented with Mn^{2+} (Fig. 5a), and direct measurement of SodA activity from *S. pneumoniae* showed a similar trend (Fig. 5b). The impact of Cd^{2+} on oxidative stress response was ascertained using paraquat, which causes oxidative damage by promoting a futile redox cycle in the cytoplasm. We observed that although growth in the presence of Cd^{2+} resulted in a significant decrease in survival (Fig. 5c, $P = 0.0087$ (two-tailed unpaired *t*-test)), the increased sensitivity was not due to Cd^{2+}, but instead was a result of the decreased Mn^{2+} accumulation, as *S. pneumoniae* grown in 30 μM Mn^{2+}: 30 μM Cd^{2+} exhibited wild-type levels of survival (Fig. 5c). Hence, the perturbation of Mn^{2+} homeostasis by Cd^{2+} heightens the sensitivity to oxidative stress, while Cd^{2+} could be accumulated to high intracellular concentrations with no apparent deleterious effect. This ability to accumulate Cd^{2+} without direct toxicity indicated that intracellular buffering of the ion was crucial.

Reduced thiol groups on small peptides, primarily glutathione in *S. pneumoniae*, have been implicated in having an essential role in intracellular transition metal ion management (Fe^{2+} and Zn^{2+}) and in ameliorating Cd^{2+} toxicity[32]. *S. pneumoniae* is incapable of *de novo* glutathione synthesis and acquires glutathione via a high-affinity ABC permease, the Gsh permease[32]. Here we investigated the contribution of glutathione to management of transition metal ion stress. In *S. pneumoniae*, the total glutathione content, in response to Cd^{2+} exposure, increased by approximately twofold to a mean (± s.e.m.) pneumococcal cell concentration of 19.0 ± 0.9 mM (Fig. 5d). Glutathione abundance is crucial for Cd^{2+} management, as a mutant strain incapable of glutathione acquisition, *S. pneumoniae* Δ*gshT*, was hypersensitive to Cd^{2+} stress (1 μM Mn^{2+}: 30 μM Cd^{2+}; Figs 2 and 5e) and, in contrast to the wild type, could not be rescued by addition of Mn^{2+} (Fig. 5e; 30 μM Mn^{2+}: 30 μM Cd^{2+}). By contrast, the Δ*sodA* strain, which we have previously shown to be hypersensitive to oxidative stress challenge[30,33], demonstrated a wild-type growth

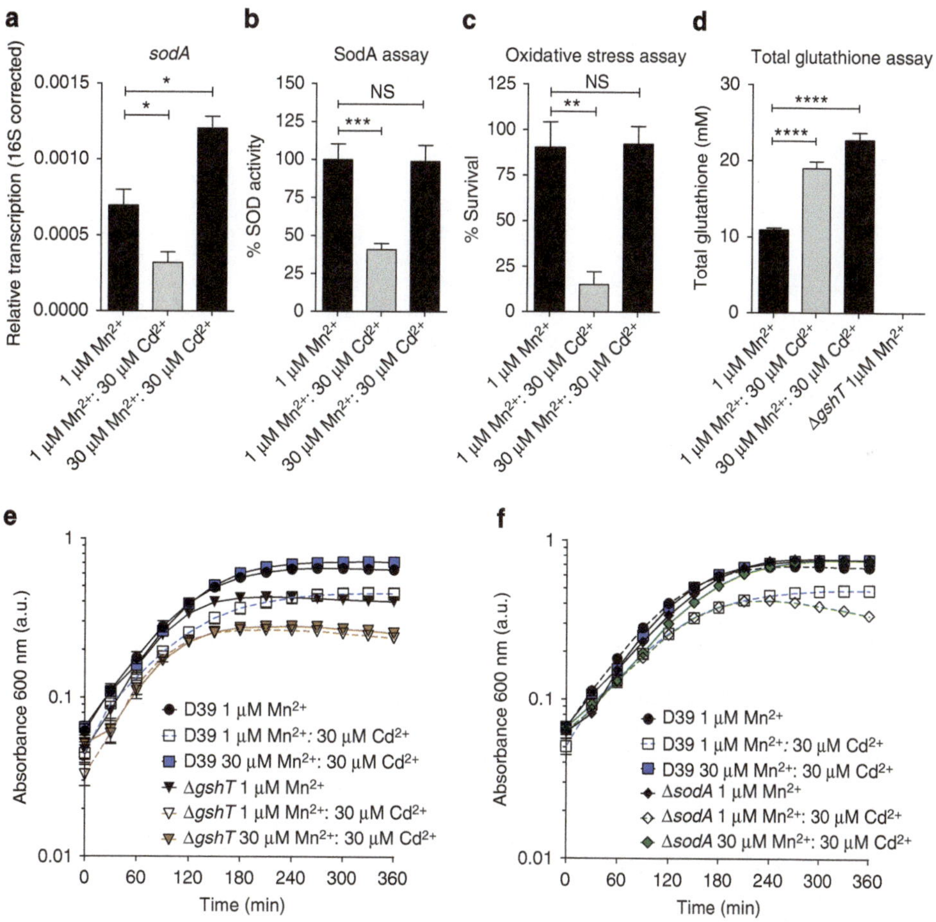

Figure 5 | The impact of Cd^{2+} on sensitivity to oxidative stress. (**a**) Relative transcription, corrected to 16S rRNA, of *sodA* by *S. pneumoniae* when grown in CDM supplemented with 1 μM Mn^{2+} (black), 1 μM Mn^{2+}: 30 μM Cd^{2+} (light grey) and 30 μM Mn^{2+}: 30 μM Cd^{2+} (dark grey). The data correspond to the mean (± s.e.m.) of three independent biological experiments. (**b**) SodA activity assay *S. pneumoniae* when grown in CDM supplemented with 1 μM Mn^{2+} (black), 1 μM Mn^{2+}: 30 μM Cd^{2+} (light grey) and 30 μM Mn^{2+}: 30 μM Cd^{2+} (dark grey). Activity was calculated as a percentage of SodA activity observed for *S. pneumoniae* cells grown in 1 μM Mn^{2+}. The data correspond to the mean (± s.e.m.) of three independent biological experiments. (**c**) Paraquat killing of *S. pneumoniae* grown in CDM supplemented with 1 μM Mn^{2+} (black), 1 μM Mn^{2+}: 30 μM Cd^{2+} (light grey) and 30 μM Mn^{2+}: 30 μM Cd^{2+} (dark grey). Survival was calculated as a percentage of colonies at 30 min compared with cells not challenged with paraquat. The data correspond to the mean (± s.e.m.) of three independent biological experiments. (**d**) Total glutathione determination per mean cell volume for *S. pneumoniae* grown in CDM supplemented with 1 μM Mn^{2+} (black), 1 μM Mn^{2+}: 30 μM Cd^{2+} (light grey) and 30 μM Mn^{2+}: 30 μM Cd^{2+} (dark grey). The statistical significance of the differences in the mean data for **a–d** was determined by two-tailed unpaired *t*-tests (NS, not significant; $*P<0.05$, $**P<0.01$, $***P<0.001$, $****P<0.0001$). (**e**,**f**) *S. pneumoniae* Δ*gshT* (**e**) and Δ*sodA* (**f**) grown in CDM supplemented with metal ions as indicated. The data correspond to mean (± s.e.m.) absorbance 600 nm measurements from three independent biological experiments. Errors bars, where not visible, are overlapped by the representative symbols.

phenotype in the presence of Cd^{2+}, with no apparent increase in sensitivity to metal ion stress (Figs 2 and 5f). Taken together, these findings indicate that glutathione serves a major role in Cd^{2+} buffering, whereas SodA appears to be dispensable, consistent with the lack of redox activity of Cd^{2+} ions.

To ascertain whether this was also the case for Zn^{2+}, which is also presumed to be buffered by glutathione, we examined the response of the Δ*gshT* strain to Zn^{2+} exposure. Intriguingly, while the Δ*gshT* strain was insensitive to subinhibitory concentrations of Zn^{2+}, similar to the Δ*czcD* strain, the combination mutant (Δ*gshT*/Δ*czcD*) showed significant attenuation of growth (100 µM Zn^{2+}; Fig. 2). At higher concentrations of Zn^{2+}, the growth of the Δ*gshT*/Δ*czcD* strain corresponded with the hypersensitive phenotype of the Δ*czcD* strain, while the Δ*gshT* strain showed a very mild impact, relative to wild type (300 µM Zn^{2+}; Fig. 2). Taken together, these data show that intracellular management of Zn^{2+} occurs via the concerted actions of both glutathione and Zn^{2+} efflux, wherein the activation and efflux via CzcD is crucial to prevent Zn^{2+} intoxication. By contrast, management of Cd^{2+}, which has no major efflux pathway, is predominantly dependent on cellular glutathione to prevent mismetallation of other proteins. Concordantly, when Cd^{2+} accumulation surpasses the buffering capacity of the glutathione pool (1 µM Mn^{2+}: 30 µM Cd^{2+}), or in the absence of intracellular glutathione (Δ*gshT*), this leads to Cd^{2+} toxicity.

Discussion

Anthropogenic-facilitated entry of non-physiological elements, such as Cd^{2+}, into the biosphere presents unique cellular challenges for biological systems. The prior absence of these elements from the biosphere obviated any selective pressure towards evolving mechanisms for managing their impact on biological processes. As a consequence, many non-physiological elements rapidly accumulate in the food chain leading to significant toxicity in higher organisms[1,2]. We observed that Cd^{2+} was toxic to *S. pneumoniae* and that this arose from acute dysregulation of the transition metal ion homeostatic mechanisms. Elucidation of the total cellular quotient of metal ions in *S. pneumoniae* not only showed the impact of Cd^{2+} but also revealed novel aspects of cellular transition metal homeostasis. *S. pneumoniae* is a Gram-positive organism, comprising only a single cellular compartment; thus the concentration of transition metal ions accurately reflects their total intracellular accumulation and encompasses metal ions occurring in labile exchangeable pools and those bound weakly or tightly to proteins, peptides, nucleic acids and other molecules. Determination of the cellular quotient of metal ions in wild-type *S. pneumoniae* and mutant variants revealed the range over which cellular variation of transition metal ion concentrations is permissible. Manganese accumulation shows significant plasticity, with cells being viable at concentrations as low as ~3% of typical accumulation levels[23,30]. By contrast, Zn^{2+} accumulation is regulated in a much narrower window, with its minimal cellular quotient being between 20 and 40% of typical accumulation. Intriguingly, the higher total cellular quotient of Mn^{2+} ions suggests that the concentration 'set-points' for cellular accumulation of transition metal ions may also contribute to ensuring efficient acquisition of metal cofactors by proteins. Manganese, which has been reported to be buffered by molecules such as citrate, phosphate and histidines residues from proteins in the cytoplasm[34,35], has fewer reported cellular roles. Hence, its presence at such high cellular concentrations suggests that in *S. pneumoniae* it could occur predominantly as a labile pool of cytoplasmic-buffered Mn^{2+}. By contrast Zn^{2+}, which is reported in numerous studies to be highly utilized in intracellular proteins[36,37], has been predicted to have a significantly smaller labile pool[10,36]. Comprehensive intracellular metal ion speciation studies have been limited by a combination of technical and methodological challenges that remain to be surmounted. However, the data here allow us to speculate that, in *S. pneumoniae*, the potentially larger labile pool of Mn^{2+} ions, relative to Zn^{2+} ions, could be one mechanism that aids in Mn^{2+}-dependent proteins correctly acquiring their cognate metal cofactor. The millimolar cellular quotient of Zn^{2+} ions in *S. pneumoniae* may also provide a plausible basis for the unexplained lower sensitivity of the zinc-responsive transcriptional regulator AdcR to Zn^{2+} ions, when compared with the homoplastic regulator, Zur, from *E. coli*[10,15]. In *E. coli*, Zn^{2+} has been shown to accumulate to an overall (cytoplasmic plus periplasmic) cellular concentration of ~0.2 mM and Zur is highly sensitive to Zn^{2+} abundance, sensing Zn^{2+} ions in the femtomolar range[10]. By contrast, AdcR is 2–3 orders of magnitude less sensitive, suggesting that a higher cellular 'set-point' for sensing of Zn^{2+} ions in *S. pneumoniae* may exist[15].

In the context of exposure to Cd^{2+}, the impact on transition metal ion homeostasis corresponds to specific dysregulation of Mn^{2+} and Zn^{2+} accumulation, but via different mechanisms. Manganese accumulation was disrupted by direct competition of Cd^{2+} for the Psa permease and the upregulation of the Mn^{2+}-efflux pathway MntE. These results also suggest that PsaR, the transcriptional regulator of the Psa permease, is not functionally activated by Cd^{2+} and thus permits upregulation of *psaBCA*. In this way, the interaction of Cd^{2+} with PsaR may mimic the binding of Zn^{2+} rather than Mn^{2+}, which occurs in a manner that does not permit repression of *psaBCA* transcription[12]. Although competition for the Psa permease was consistent with expectations, based on the inability of the SBP to be specific for solely Mn^{2+} ions, the mistranslocation of Cd^{2+} ions via the permease highlights an evolutionary limitation in managing extracellular stress by non-physiological elements. Specific acquisition of Mn^{2+} via the Psa permease arises from a combination of mechanisms that prevent the transport of Zn^{2+} ions, which are more abundant in the native niche of *S. pneumoniae*. However, the larger ionic radius and 'strained' coordination of the Cd^{2+} ion at the metal-binding site, similar to Mn^{2+} (Supplementary Table 3), resulted in an easily destabilized metal–protein complex, such that PsaA was incapable of preventing release of Cd^{2+} ions into the permease. This underscores the challenge encountered by metalloregulatory mechanisms when interacting with elements that have only recently entered the biosphere. In PsaA, the regulatory mechanisms that have evolved to prevent mistranslocation of Zn^{2+} ions are not capable of preventing the translocation of Cd^{2+} ions. By contrast with Mn^{2+}, the Cd^{2+}-induced disruption of Zn^{2+} homeostasis occurred via upregulation of *czcD*, the Zn^{2+}-efflux pathway, and the apparent inability of the pneumococcus to transcriptionally activate additional Zn^{2+}-recruiting proteins (that is, AdcAII (ref. 13)) to compensate for the depletion of cellular Zn^{2+} ions. Hence, Cd^{2+} accumulation activated the coordinated pneumococcal physiological response to Zn^{2+} intoxication or overload[13,15,17,23], resulting in the futile export and depletion of cellular Zn^{2+}. Our findings suggest that Cd^{2+} accumulation results in the mismetallation of the Zn^{2+}-responsive transcriptional regulators, AdcR and SczA[13,15,17]. Whether this is due to Cd^{2+} directly interacting with the Zn^{2+}-binding sites of intracellular metal sensors[38–40] or via Cd^{2+}-mediated displacement of Zn^{2+} from thiol buffering sites[41], which results in the non-physiological activation of the Zn^{2+}-responsive transcriptional regulators, remains to be determined. Irrespective of this fact, transcriptional activation of *czcD* is often construed as a mechanism to facilitate Cd^{2+}

efflux based on studies of the archetypal protein from *C. metallidurans*[18]. However, in *S. pneumoniae*, similar to *S. pyogenes*[42], CzcD is primarily a Zn^{2+}-efflux pathway and does not provide a major protective role against Cd^{2+} ions. As a consequence, in *S. pneumoniae*, the Cd^{2+}-induced upregulation of *czcD* appears to preferentially deplete intracellular Zn^{2+}, despite the greater abundance of Cd^{2+} ions.

Cd^{2+} accumulation is associated with oxidative stress, despite the redox-inert nature of the ion[4]. The data here show that, in *S. pneumoniae*, the increased sensitivity to oxidative stress arises from the loss of *sodA* expression and activity due to Cd^{2+}-induced Mn^{2+} depletion. However, the loss of SodA was an indirect by-product of Cd^{2+} accumulation and, importantly, not associated with the ability of the pneumococcus to tolerate Cd^{2+} ions. This was demonstrated by the Δ*sodA* strain, which showed wild-type growth phenotypes in the presence of extracellular Cd^{2+}, and in wild-type *S. pneumoniae*, which retained the ability to accumulate high cellular concentrations of Cd^{2+} without an increase in sensitivity to oxidative stress, when supplemented with Mn^{2+}. However, SodA still has a major role in the *in vitro* and *in vivo* fitness of *S. pneumoniae* owing to a lack of other mechanisms to detoxify superoxide[23,30,33]. Here, in the presence of Cd^{2+}, the loss of SodA also occurs in conjunction with dysregulation of Mn^{2+} and Zn^{2+} homeostatic mechanisms and the accumulation of very high concentrations of Cd^{2+}. Hence, it is the combination of these factors that compromises pneumococcal viability in the presence of Cd^{2+}.

Intracellular accumulation of Cd^{2+} in *S. pneumoniae* was crucially dependent on glutathione, which acts as a low-affinity metal ion buffer for certain transition metal ions (Fe^{2+}, Zn^{2+} and Cd^{2+}), as phenotypic growth perturbations were seen when Cd^{2+} accumulation exceeded the total cellular glutathione pool (19–23 mM). Although it cannot be completely discounted that glutathione contributes to pneumococcal oxidative stress management[32], the hypersensitivity of the Δ*gshT* strain to extracellular Cd^{2+}, in contrast to the Δ*sodA* strain, strongly implicates the major role for cellular glutathione is in metal ion buffering. The observed hypersensitivity of the Δ*gshT*/Δ*czcD* strain to Zn^{2+} stress further underscores how intracellular metal ion homeostasis relies on the contributions of both cellular glutathione and the efflux machinery[13,43]. The results are also consistent with the central role of low-molecular-weight thiols for metal ion homeostasis in other microorganisms. Bacillithiol has recently been shown to have a crucial role in the buffering of Zn^{2+} ions in *Bacillus subtilis*[41], while it also serves an undefined, but protective role in Cd^{2+} stress for *B. subtilis* and *Staphylococcus aureus*[44]. Although *S. pneumoniae* and other microorganisms lack a bacillithiol biosynthetic pathway, the role of low-molecular-weight species for metal ion homeostasis is emerging as a crucial mechanism. Overall, our findings here highlight the role of glutathione as an intracellular metal ion buffer, in preference to oxidative stress management. This finding has broad ramifications for the assumed role(s) of cellular glutathione, and related thiol derivatives, in prokaryotes and potentially eukaryotes[43,41].

In conclusion, Cd^{2+} accumulation in the food chain and its toxicity are crucial, but largely overlooked concerns. Our findings reveal the molecular basis by which Cd^{2+} is mistranslocated via a Mn^{2+} transporter and provides new insights into how Cd^{2+} dysregulates intracellular transition metal homeostasis. Disruption of multiple essential metal ion homeostatic mechanisms is highly damaging to biological organisms, such as *S. pneumoniae*[14], and in this way Cd^{2+} acts to severely dysregulate both Mn^{2+} and Zn^{2+} homeostasis (Supplementary Fig. 8). By revealing the remarkably high (millimolar) cellular quotients of transition metal ions and glutathione, this work also shows how Cd^{2+}, despite its inability to directly generate reactive oxygen species, is associated with oxidative stress. This work provides a new understanding of the mechanisms by which Cd^{2+} enters cells and causes toxicity, and thereby, opens the way to identifying new routes towards developing specific therapeutic agents capable of preventing Cd^{2+} toxicity.

Methods

Growth experiments and whole-cell assays. The *S. pneumoniae* D39 Δ*sodA*, Δ*czcD*, Δ*gshT*, Δ*mntE*, Δ*adcCBA* and Δ*adcA*/Δ*adcAII* strains have been generated previously[13,32]. The Δ*mntE* and Δ*gshT*/Δ*czcD* strains were generated as described previously[13], using primer sequences listed in Supplementary Table 6. *S. pneumoniae* D39, Δ*sodA*, Δ*czcD*, Δ*mntE*, Δ*gshT*, Δ*gshT*/Δ*czcD*, Δ*adcCBA* and Δ*adcA*/Δ*adcAII* were grown in CDM, which corresponded to the C + Y media without transition metal ion supplementation[23]. The base ion content of the CDM was ascertained by ICP–MS on an Agilent 7500cx ICP–MS (Adelaide Microscopy, University of Adelaide)[23]. Growth experiments were conducted in CDM supplemented with 1 μM $MnSO_4$ and concentrations of 10, 30, 50 or 80 μM of $CdCl_2$ or $MnSO_4$, as specified. All chemicals used in this study were purchased from Sigma-Aldrich, unless otherwise specified. For bacterial growth experiments, an inoculum was prepared from overnight grown culture on a blood-agar (BA) plate and resuspended in CDM to an absorbance at 600 nm (A_{600}) of 1.0. The inoculum was then diluted in 200 μl of CDM in a 96-well flat-bottom plate (Greiner Bio One) to a final A_{600} of 0.05 and sealed with gas-permeable seal (Breathe-Easy, Diversified Biotech). The plate was then incubated in a FLUOStar Omega spectrophotometer with an ACU gas controller (BMG Labtech) at 310 K in 5% CO_2. A_{600} readings were recorded using the well-scan function every 30 min. Data from at least six independent growth experiments was averaged to ascertain the effect of $CdCl_2$ on bacterial growth. Bacterial density at A_{600} was measured as the c.f.u. per ml determined by serial dilutions on BA plates.

Bacterial growth for ICP–MS, paraquat killing and glutathione assays used identical growth parameters to the microplate experiments, with $MnSO_4$ and $CdCl_2$ supplementation as specified. For ICP–MS, 50 ml of culture was grown to A_{600} = 0.3–0.4, harvested and prepared for analysis by ICP–MS, as described previously[23]. For the paraquat killing assays, 1 ml of A_{600} = 0.3 culture was incubated for 30 min with 10 mM paraquat and then serially diluted and plated on blood agar. Total glutathione assays were performed using 1 ml of A_{600} = 0.3 culture with the Promega GSH-Glo assay kit according to the manufacturer's instructions (Promega, USA). To examine the effect of Cd^{2+} stress on cellular SodA activity, *S. pneumoniae* D39 was grown to an A_{600} of 0.3 as for the ICP–MS experiments in CDM supplemented with 1 μM $MnSO_4$, 30 μM $CdCl_2$:1 μM $MnSO_4$ or 30 μM $CdCl_2$:30 μM $MnSO_4$. The cells were washed three times in PBS before lysis by sonication and intact cells and other insoluble material was removed by centrifugation at 277 K for 30 min at 14,000 *g*. The assay was performed using a SOD activity kit (Calbiochem) and the data were normalized to the total protein concentration of the cell-free extract, as determined using the Dc Bio-Rad protein determination assay, followed by normalization (in percentage) to the average SOD activity measured for the untreated cells (1 μM $MnSO_4$).

Metal sensitivity assays were also performed using a *S. pneumoniae* drop test. Bacteria were grown in CDM supplemented with 1 μM $MnSO_4$ and grown until the A_{600} reached 0.3–0.4. Cells were then serially diluted 10-fold up to a 10^{-5} dilution, and 5 μl of each dilution were spotted on BA plates supplemented with varying concentrations of $ZnSO_4$ (0, 100, 300 and 1,000 μM) or $CdCl_2$ (0, 10, 30 and 100 μM). Plates were photographically documented following overnight incubation at 37 °C, 5% CO_2.

RNA extraction and RT-PCR analysis. Pneumococci were grown as for the ICP–MS analyses, then 500 μl of the culture was mixed with 1 ml of RNA Protect (Qiagen). RNA was extracted and purified using an RNeasy Protect Bacteria Mini Kit (Qiagen) after enzymatic lysis using lysozyme and mutanolysin, all according to the manufacturer's instructions. DNase I treatment (Roche) was performed followed by quantitative reverse transcription PCR using SuperScript III (Invitrogen) with a Roche LC480 Real-Time Cycler. The transcription levels of genes analysed were normalized to those obtained for 16S rRNA. Primer sequences are in Supplementary Table 6.

Cell volume and concentration determination. Scanning electron microscopy was used to determine the dimensions of the pneumococcal cell. Bacteria were grown, as described above, harvested and then prepared for, and analysed by a Philips XL30 FEG scanning electron microscope as described in ref. 45. Cell measurements were obtained using instrument software. The cell dimensions were used to calculate volume assuming an ellipsoid:

$$V = \frac{4}{3}\pi r_1 r_2 r_3 \qquad (1)$$

where r_1, r_2, and r_3 are the radii of the ellipsoidal cell in three dimensions determined by scanning electron microscopy (SEM). The total quotient of transition metal ions was then derived using cell volume (V, litres), where the total cell density is known (c.f.u.), and concentration of metal ions in a sample of known volume and known number of cells (M, moles).

$$C = \frac{(M/_{c.f.u.})}{V} \quad (2)$$

The derived concentration represents mean molarity of metal ion per cell of mean dimensions.

Expression and purification of apo-PsaA. Recombinant PsaA was expressed in *E. coli* LEMO21(DE3) from the pCAMcLIC01-PsaA construct[8]. The dodecahistidine tag was removed from affinity-purified PsaA by enzymatic digestion by the human rhinovirus 3C protease and the protein purified further on a HisTrap HP column. Apo-PsaA was prepared by dialyzing the dodecahistidine tag-cleaved protein in a 20-kDa molecular-weight-cutoff membrane (Slide-A-Lyzer, Pierce) against 4 l of sodium acetate buffer, pH 4.0, with 20 mM EDTA at 300 K. The sample was then dialysed against 4 l of 20 mM Tris-HCl, pH 7.2, and 100 mM NaCl, at 277 K and centrifuged at 18,000 *g* for 10 min to remove any insoluble material. Protein samples were analysed for metal content by heating 5 μM protein at 370 K for 15 min in 3.5% HNO_3 and the metal ion content was measured by ICP–MS.

Protein assays. Protein concentration determination was performed using the Dc Bio-Rad protein determination assay. Metal-loading assays were performed on purified apo-PsaA (30 μM) by mixing with 300 μM $CdCl_2$ in a total volume of 2 ml in the assay buffer (20 mM MOPS, pH 7.2, and 100 mM NaCl) for 60 min at 277 K. The sample was desalted on a PD10 column (GE Healthcare) into the assay buffer and the protein concentration was quantified. Protein was then either kept for ICP–MS analysis or mixed with 3 mM EDTA in a total volume of 2 ml for 60 min at 277 K. Samples were then desalted on a PD10 column as before. Solutions (5–10 μM) of control, metal-loaded and EDTA-treated protein were prepared in 3.5% HNO_3 and boiled for 15 min at 370 K. Samples were then cooled and centrifuged for 20 min at 14,000 *g*. The supernatant was then analysed by ICP–MS and the protein-to-metal ratio was determined.

Immunoblot analyses of PsaA expression levels. Wild-type and mutant *S. pneumoniae* were grown under the same conditions as for ICP–MS. After reaching an A_{600} of 0.4, cells were incubated with 0.1% sodium deoxycholate at 310 K for 60 min to induce lysis. Protein concentrations were determined and 10 μg of total protein was loaded into each lane. After electrophoretic separation by SDS–PAGE, proteins were transferred to a nitrocellulose membrane using the iBlot (Life Technologies) system. The blots were incubated with murine anti-PsaA serum (1:2,000; ref. 23), followed by anti-mouse IRDye 800 (LI-COR; 1:50,000), and were scanned using an Odyssey infrared imaging system (LI-COR). Band intensities were measured using the manufacturer's application software and the results correspond to the mean (± s.e.m.) of two independent biological experiments.

Determination of K_D for PsaA with Cd^{2+}. Excitation–emission spectra were determined on a FLUOStar Omega (BMG Labtech) at 301 K using black half-volume 384-well microtitre plates (Greiner Bio One). All experiments were performed in 20 mM MOPS (pH 6.7) and 50 mM NaCl with FluoZin-3 (Life Technologies) at a concentration of 50 nM. Deionized water and buffers were prepared and treated with Chelex-100 (Sigma-Aldrich) to avoid metal contamination. Filters used for FluoZin-3 were excitation (485 ± 10 nm) and emission (520 ± 5 nm). To determine the dissociation constant between a metal (X) and FluoZin-3 (F), we considered the following equilibria:

$$F + X \overset{K_{D.X}}{\leftrightarrow} F.X \quad (3)$$

where, for a metal that increases the fluorescence of the probe by more than ~10%, the following equation, which is an exact analytical relationship derived from the mass action equation for the formation of a 1:1 complex between probe and metal, was used to estimate the dissociation constant, $K_{D.X}$

$$\frac{f - f_{min}}{f_{max} - f} = \frac{[X]}{K_{D.X}} \quad (4)$$

where f is the measured fluorescence intensity in the presence of metal, f_{max} is the fluorescence in the presence of saturating metal and f_{min} is the fluorescence in the absence of metal. In all cases, a low concentration (50 nM) of probe was used and we assumed that the free-metal concentration was equal to the added metal concentration. The mean ± s.e.m. ($n = 6$) K_D determined for FluoZin-3 with $CdCl_2$ in the buffer system used in this study was determined to be 250 ± 35 nM. Competition by PsaA for Cd^{2+} binding was assessed by monitoring the decrease in the fluorescence of 50 nM FluoZin-3-Cd^{2+} in response to increasing apo-PsaA concentrations and analysed using $\log_{10}$[inhibitor] versus response model, with the experimentally derived K_D for FluoZin-3 with Cd^{2+}, in GraphPad prism to determine the K_D value for Cd^{2+} binding by PsaA.

Protein crystallization and structure determination. Cd-bound PsaA-D280N protein crystals were obtained in 12.5% (w/v) polyethylene glycol (PEG) 1000, 12.5% (w/v) PEG 3350, 12.5% (v/v) MPD, 0.1 M Trizma-Bicine, pH 8.7, and 0.01 M $CdCl_2$ using the hanging drop vapour diffusion method[8]. Cd^{2+}-bound wild-type PsaA crystals were grown in 26–36% PEG 400, 0.1 M NaCl, 0.1 M Trizma-HCl, pH 8.0, and 0.1 M $CdCl_2$ also using vapour diffusion. Before data collection, the crystals were flash-cooled by rapid immersion in liquid nitrogen. The diffraction data were collected on a single crystal at the Australian Synchrotron (MX2 beamline). Cd^{2+}-bound D280N and wild-type PsaA structures were determined by molecular replacement with Phaser[46] using the Mn^{2+}-bound D280N (PDB accession code 3ZKA) and the Zn^{2+}-bound (PDB accession code 1PSZ) PsaA crystal structures, respectively, as phasing models followed by automatic model building in Phenix.AutoBuild[47]. The structures were refined iteratively using Phenix.Refine and Refmac5 manual model fitting in Coot[48]. Both structures had two molecules (chains A and B) in each asymmetric unit and chain A molecules were used in subsequent structural analyses. Structural analyses (superpositions, metal ion coordination and N-/C-terminal domain-crossing angles) were performed in Chimera[49]. Secondary structure elements were assigned by DSSP (version 2.1.0; ref. 50). The WHAT-IF[51] and PISA[52] web services were used to identify hydrogen bonds and salt bridges. Data processing and structure refinement statistics can be found in Supplementary Table 7.

References

1. Nriagu, J. O. & Pacyna, J. M. Quantitative assessment of worldwide contamination of air, water and soils by trace metals. *Nature* **333,** 134–139 (1988).
2. Joseph, P. Mechanisms of cadmium carcinogenesis. *Toxicol. Appl. Pharmacol.* **238,** 272–279 (2009).
3. Lane, T. W. *et al.* A cadmium enzyme from a marine diatom. *Nature* **435,** 42 (2005).
4. Cuypers, A. *et al.* Cadmium stress: an oxidative challenge. *Biometals* **23,** 927–940 (2010).
5. Archibald, F. S. & Duong, M. N. Manganese acquisition by *Lactobacillus plantarum*. *J. Bacteriol.* **158,** 1–8 (1984).
6. Hubert, N. & Hentze, M. W. Previously uncharacterized isoforms of divalent metal transporter (DMT)-1: implications for regulation and cellular function. *Proc. Natl Acad. Sci. USA* **99,** 12345–12350 (2002).
7. McAllister, L. J. *et al.* Molecular analysis of the *psa* permease complex of *Streptococcus pneumoniae*. *Mol. Microbiol.* **53,** 889–901 (2004).
8. Couñago, R. M. *et al.* Imperfect coordination chemistry facilitates metal ion release in the Psa permease. *Nat. Chem. Biol.* **10,** 35–41 (2014).
9. Dintilhac, A. & Claverys, J. P. The *adc* locus, which affects competence for genetic transformation in *Streptococcus pneumoniae*, encodes an ABC transporter with a putative lipoprotein homologous to a family of streptococcal adhesins. *Res. Microbiol.* **148,** 119–131 (1997).
10. Outten, C. E. & O'Halloran, T. V. Femtomolar sensitivity of metalloregulatory proteins controlling zinc homeostasis. *Science* **292,** 2488–2492 (2001).
11. Irving, H. & Williams, R. J. P. The stability of transition-metal complexes. *J. Chem. Soc.* 3192–3210 (1953).
12. Lisher, J. P., Higgins, K. A., Maroney, M. J. & Giedroc, D. P. Physical characterization of the manganese-sensing pneumococcal surface antigen repressor from *Streptococcus pneumoniae*. *Biochemistry* **52,** 7689–7701 (2013).
13. Plumptre, C. D. *et al.* AdcA and AdcAII employ distinct zinc acquisition mechanisms and contribute additively to zinc homeostasis in *Streptococcus pneumoniae*. *Mol. Microbiol.* **91,** 834–851 (2014).
14. Dintilhac, A., Alloing, G., Granadel, C. & Claverys, J. P. Competence and virulence of *Streptococcus pneumoniae*: Adc and PsaA mutants exhibit a requirement for Zn and Mn resulting from inactivation of putative ABC metal permeases. *Mol. Microbiol.* **25,** 727–739 (1997).
15. Reyes-Caballero, H. *et al.* The metalloregulatory zinc site in *Streptococcus pneumoniae* AdcR, a zinc-activated MarR family repressor. *J. Mol. Biol.* **403,** 197–216 (2010).
16. Loisel, E. *et al.* AdcAII, a new pneumococcal Zn-binding protein homologous with ABC transporters: biochemical and structural analysis. *J. Mol. Biol.* **381,** 594–606 (2008).
17. Kloosterman, T. G., van der Kooi-Pol, M. M., Bijlsma, J. J. & Kuipers, O. P. The novel transcriptional regulator SczA mediates protection against Zn^{2+} stress by activation of the Zn^{2+}-resistance gene *czcD* in *Streptococcus pneumoniae*. *Mol. Microbiol.* **65,** 1049–1063 (2007).
18. Anton, A., Grosse, C., Reissmann, J., Pribyl, T. & Nies, D. H. CzcD is a heavy metal ion transporter involved in regulation of heavy metal resistance in *Ralstonia* sp. strain CH34. *J. Bacteriol.* **181,** 6876–6881 (1999).
19. Rosch, J. W., Gao, G., Ridout, G., Wang, Y. D. & Tuomanen, E. I. Role of the manganese efflux system *mntE* for signalling and pathogenesis in *Streptococcus pneumoniae*. *Mol. Microbiol.* **72,** 12–25 (2009).
20. Crupper, S. S., Worrell, V., Stewart, G. C. & Iandolo, J. J. Cloning and expression of *cadD*, a new cadmium resistance gene of *Staphylococcus aureus*. *J. Bacteriol.* **181,** 4071–4075 (1999).
21. Fu, Y. *et al.* A new structural paradigm in copper resistance in *Streptococcus pneumoniae*. *Nat. Chem. Biol.* **9,** 177–183 (2013).

22. Brenot, A., Weston, B. F. & Caparon, M. G. A PerR-regulated metal transporter (PmtA) is an interface between oxidative stress and metal homeostasis in *Streptococcus pyogenes*. *Mol. Microbiol.* **63,** 1185–1196 (2007).
23. McDevitt, C. A. *et al.* A molecular mechanism for bacterial susceptibility to zinc. *PLoS Pathog.* **7,** e1002357 (2011).
24. Lawrence, M. C. *et al.* The crystal structure of pneumococcal surface antigen PsaA reveals a metal-binding site and a novel structure for a putative ABC-type binding protein. *Structure* **6,** 1553–1561 (1998).
25. Braun, W. *et al.* Comparison of the NMR solution structure and the x-ray crystal structure of rat metallothionein-2. *Proc. Natl Acad. Sci. USA* **89,** 10124–10128 (1992).
26. Graham, S. C., Maher, M. J., Simmons, W. H., Freeman, H. C. & Guss, J. M. Structure of *Escherichia coli* aminopeptidase P in complex with the inhibitor apstatin. *Acta Crystallogr. D Biol. Crystallogr.* **60,** 1770–1779 (2004).
27. Price, N. M. & Morel, F. M. M. Cadmium and cobalt substitution for zinc in a marine diatom. *Nature* **344,** 658–660 (1990).
28. Xu, Y., Feng, L., Jeffrey, P. D., Shi, Y. & Morel, F. M. Structure and metal exchange in the cadmium carbonic anhydrase of marine diatoms. *Nature* **452,** 56–61 (2008).
29. Ogunniyi, A. D. *et al.* Central role of manganese in regulation of stress responses, physiology, and metabolism in *Streptococcus pneumoniae*. *J. Bacteriol.* **192,** 4489–4497 (2010).
30. Eijkelkamp, B. A. *et al.* Extracellular zinc competitively inhibits manganese uptake and compromises oxidative stress management in *Streptococcus pneumoniae*. *PLoS ONE* **9,** e89427 (2014).
31. Yesilkaya, H. *et al.* Role of manganese-containing superoxide dismutase in oxidative stress and virulence of *Streptococcus pneumoniae*. *Infect. Immun.* **68,** 2819–2826 (2000).
32. Potter, A. J., Trappetti, C. & Paton, J. C. *Streptococcus pneumoniae* uses glutathione to defend against oxidative stress and metal ion toxicity. *J. Bacteriol.* **194,** 6248–6254 (2012).
33. Tseng, H. J., McEwan, A. G., Paton, J. C. & Jennings, M. P. Virulence of *Streptococcus pneumoniae*: PsaA mutants are hypersensitive to oxidative stress. *Infect. Immun.* **70,** 1635–1639 (2002).
34. Hider, R. C. & Kong, X. L. Glutathione: a key component of the cytoplasmic labile iron pool. *Biometals* **24,** 1179–1187 (2011).
35. Tabares, L. C. & Un, S. *In situ* determination of manganese(II) speciation in *Deinococcus radiodurans* by high magnetic field EPR: detection of high levels of Mn(II) bound to proteins. *J. Biol. Chem.* **288,** 5050–5055 (2013).
36. Sun, X. *et al.* Putative copper- and zinc-binding motifs in *Streptococcus pneumoniae* identified by immobilized metal affinity chromatography and mass spectrometry. *Proteomics* **11,** 3288–3298 (2011).
37. Andreini, C., Bertini, I., Cavallaro, G., Holliday, G. L. & Thornton, J. M. Metal ions in biological catalysis: from enzyme databases to general principles. *J. Biol. Inorg. Chem.* **13,** 1205–1218 (2008).
38. Ralston, D. M. & O'Halloran, T. V. Ultrasensitivity and heavy-metal selectivity of the allosterically modulated MerR transcription complex. *Proc. Natl Acad. Sci. USA* **87,** 3846–3850 (1990).
39. Guerra, A. J. & Giedroc, D. P. Metal site occupancy and allosteric switching in bacterial metal sensor proteins. *Arch. Biochem. Biophys.* **519,** 210–222 (2012).
40. Sun, Y., Wong, M. D. & Rosen, B. P. Role of cysteinyl residues in sensing Pb(II), Cd(II), and Zn(II) by the plasmid pI258 CadC repressor. *J. Biol. Chem.* **276,** 14955–14960 (2001).
41. Ma, Z. *et al.* Bacillithiol is a major buffer of the labile zinc pool in *Bacillus subtilis*. *Mol. Microbiol.* **94,** 756–770 (2014).
42. Ong, C. L., Gillen, C. M., Barnett, T. C., Walker, M. J. & McEwan, A. G. An antimicrobial role for zinc in innate immune defense against group A *Streptococcus*. *J. Infect. Dis.* **209,** 1500–1508 (2014).
43. Helbig, K., Bleuel, C., Krauss, G. J. & Nies, D. H. Glutathione and transition-metal homeostasis in *Escherichia coli*. *J. Bacteriol.* **190,** 5431–5438 (2008).
44. Rajkarnikar, A. *et al.* Analysis of mutants disrupted in bacillithiol metabolism in *Staphylococcus aureus*. *Biochem. Biophys. Res. Commun.* **436,** 128–133 (2013).
45. Fairbrother, L. *et al.* Effect of the cyanide-producing bacterium *Chromobacterium violaceum* on ultraflat Au surfaces. *Chem. Geol.* **265,** 313–320 (2009).
46. McCoy, A. J. Solving structures of protein complexes by molecular replacement with Phaser. *Acta Crystallogr. D Biol. Crystallogr.* **63,** 32–41 (2007).
47. Terwilliger, T. C. *et al.* Iterative model building, structure refinement and density modification with the PHENIX AutoBuild wizard. *Acta Crystallogr. D Biol. Crystallogr.* **64,** 61–69 (2008).
48. Emsley, P., Lohkamp, B., Scott, W. G. & Cowtan, K. Features and development of Coot. *Acta Crystallogr. D Biol. Crystallogr.* **66,** 486–501 (2010).
49. Pettersen, E. F. *et al.* UCSF Chimera--a visualization system for exploratory research and analysis. *J. Comput. Chem.* **25,** 1605–1612 (2004).
50. Kabsch, W. & Sander, C. Dictionary of protein secondary structure: pattern recognition of hydrogen-bonded and geometrical features. *Biopolymers* **22,** 2577–2637 (1983).
51. Hekkelman, M. L. *et al.* WIWS: a protein structure bioinformatics Web service collection. *Nucleic Acids Res.* **38,** W719–W723 (2010).
52. Krissinel, E. & Henrick, K. Inference of macromolecular assemblies from crystalline state. *J. Mol. Biol.* **372,** 774–797 (2007).

Acknowledgements

We acknowledge the use of the Australian Synchrotron (Victoria, Australia) and the University of Queensland Remote-Operation X-Ray (UQ-ROCX) Diffraction Facility. This work was supported by the Australian Research Council (ARC) grant DP120103957 to C.A.M. and J.C.P., and the National Health and Medical Research Council (NHMRC) Project grants 1022240, to C.A.M., and 1080784, to C.A.M. and M.J.M., and Program grants 565526, to A.G.M., B.K. and J.C.P., and 1071659, to B.K. and J.C.P. M.L.O. holds an ARC DECRA (DE1211550), B.K. is a NHMRC Senior Research Fellow (1003325) and J.C.P. is a NHMRC Senior Principal Research Fellow (1043070). We thank the beamline staff at the Australian Synchrotron for their support and Dr C. Adolphe for discussions.

Author contributions

S.L.B., B.A.E. and C.A.M. designed and executed all the biochemical studies. C.-I.Y.O. generated the mutant strains. Z.L., R.M.C. and M.J.M. designed and executed the crystallographic experiments. S.L.B., R.M.C., M.J.M., M.L.O. and C.A.M. drafted the manuscript. All authors contributed to the design, analysis, discussion of the research and writing of the final manuscript.

Additional information

Accession codes: Coordinates and structure factors for Cd^{2+}-bound PsaA-D280N, partially occluded state, and for Cd^{2+}-bound wild-type PsaA, closed state, have been deposited in the RCSB Protein Data Bank under accession codes 4uto and 4utp, respectively.

Competing financial interests: The authors declare no competing financial interests.

2

Spontaneous formation and base pairing of plausible prebiotic nucleotides in water

Brian J. Cafferty[1,2], David M. Fialho[1,2], Jaheda Khanam[1,2], Ramanarayanan Krishnamurthy[2,3] & Nicholas V. Hud[1,2]

The RNA World hypothesis presupposes that abiotic reactions originally produced nucleotides, the monomers of RNA and universal constituents of metabolism. However, compatible prebiotic reactions for the synthesis of complementary (that is, base pairing) nucleotides and mechanisms for their mutual selection within a complex chemical environment have not been reported. Here we show that two plausible prebiotic heterocycles, melamine and barbituric acid, form glycosidic linkages with ribose and ribose-5-phosphate in water to produce nucleosides and nucleotides in good yields. Even without purification, these nucleotides base pair in aqueous solution to create linear supramolecular assemblies containing thousands of ordered nucleotides. Nucleotide anomerization and supramolecular assemblies favour the biologically relevant β-anomer form of these ribonucleotides, revealing abiotic mechanisms by which nucleotide structure and configuration could have been originally favoured. These findings indicate that nucleotide formation and selection may have been robust processes on the prebiotic Earth, if other nucleobases preceded those of extant life.

[1] School of Chemistry and Biochemistry, Georgia Institute of Technology, 901 Atlantic Drive, Atlanta, Georgia 30332, USA. [2] NSF-NASA Center for Chemical Evolution, Atlanta, Georgia 30332, USA. [3] Department of Chemistry, The Scripps Research Institute, La Jolla, California 92037, USA. Correspondence and requests for materials should be addressed to N.V.H. (email: hud@chemistry.gatech.edu).

The current role of mononucleotides and RNA polymers in numerous cellular functions gave rise to the long-standing hypothesis that these molecules were involved early in the emergence of life: the RNA World hypothesis[1]. Supporting this hypothesis, model prebiotic reactions and analyses of carbonaceous meteorites provide evidence that the canonical nucleobases of RNA (adenine, guanine, cytosine, uracil) were likely present on the prebiotic Earth[2–4]. In addition, progress has been made towards finding abiotic routes to ribose and related sugars from simple molecules (for example, formaldehyde, glyoxylate), as well as mechanisms for ribose phosphorylation and ribose selection from complex sugar mixtures[5–8]. Nevertheless, despite decades of effort, the chemical origin of nucleosides and nucleotides (that is, nucleobases glycosylated with ribose and phosphorylated ribose) remains an unsolved problem[9,10]. In the 1970s, Orgel and co-workers showed that adenosine (the nucleoside of adenine) can be formed in 1–5% yield if a solution of adenine and ribose is dried and heated[11,12], but no comparable reactions have been demonstrated for the other three canonical nucleosides. Frustrated by what became known as *The Nucleoside Problem*, Orgel proposed an alternative pathway for prebiotic pyrimidine nucleoside and nucleotide synthesis that bypassed glycosidic bond formation, a pathway in which the cytosine nucleobase is built stepwise on a sugar scaffold[13]. This approach was furthered by Sutherland and co-workers, who developed a synthetic route to cytidine and uridine nucleotides starting from glycolaldehyde and glyceraldehyde[13,14]. However, the necessity to temporally separate specific reagents and reaction steps caused some to question the relevance of this synthesis to the origin of RNA[15,16]. Sutherland and co-workers have since proposed a spatially separated geochemical scenario that includes an ordered delivery of reagents on the prebiotic Earth that would coincide with the sequential steps of their pyrimidine nucleotide synthesis, a scenario that is initiated by meteorite impacts[17].

The persistent challenge of finding a simple, robust and plausible prebiotic route to the canonical nucleosides—juxtaposed with the exquisite functionality of RNA—have caused many researches to consider RNA a product of chemical and/or biological evolution[18,19]. Inspired by the possibility that RNA evolved from a *proto*-RNA with alternative nucleobases that more easily formed nucleosides, Miller and co-workers demonstrated that urazole (a triazole analog of uracil) is efficiently glycosylated by ribose in water[20]. Subsequent demonstrations of nucleoside formation with different plausible prebiotic heterocycles[12,21,22] suggest that other nucleosides may have been common on the prebiotic Earth. While encouraging, a model prebiotic reaction has yet to be reported that produces two extant nucleosides that form a Watson–Crick base pair or two noncanonical nucleosides that form a similar base pair—a property used by extant life for information transfer and, arguably, essential for the emergence of RNA-based life.

As part of our search for the possible components of proto-RNA, we evaluated a series of heterocycles for the ability to be spontaneously glycosylated by ribose and to selectively assemble in water[23,24]. In this work, we report the glycosylation reactions and nucleotide assembly/mutual selection of two candidate proto-nucleobases, barbituric acid (BA) and melamine. These heterocycles are well suited to function as the recognition units of a primitive informational system, as BA and melamine: (i) are produced in the same model prebiotic reaction[25]; (ii) can form Watson–Crick-like base pairs with each other[26]; (iii) have H-bond donor and acceptor groups that are complementary with uracil (for melamine) and adenine (for BA), making these heterocycles 'forward compatible' for base pairing with two extant nucleobases[9]; (iv) possess chemical properties that indicate favourable glycosylation by ribose, specifically at the C5 position of BA and the exocyclic amines of melamine. Additional reasons have been previously put forth for why BA and melamine should be considered potential ancestral nucleobases of RNA[27,28].

Here, we show that glycosylation of melamine and BA by ribose-5-phosphate (R5P) occurs spontaneously in water to produce nucleotides in yields of up to 55% and 82%, respectively. When combined, the nucleotides form supramolecular assemblies with Watson–Crick-like base pairs, even within the crude reaction mixtures. These assemblies are shown to preferentially incorporate and increase the fraction of the β-anomer of the melamine nucleotide over the α-anomer. These findings demonstrate prebiotically plausible mechanisms for the selection of nucleotides in both nucleobase and sugar structure.

Results

Spontaneous formation of nucleosides and nucleotides. Given previous reports that phosphorylated sugars can be produced in model prebiotic reactions[7], we were motivated to explore the potential for BA and melamine to be glycosylated by R5P, which could represent a model prebiotic route to nucleotides that can form base pairs that are similar to those formed by the canonical nucleobases (Fig. 1a). Nucleotides spontaneously form when BA is mixed in water with one equivalent of R5P at 20 °C (without the need for drying). The reaction is surprisingly efficient with BA + R5P conjugates exceeding 80% in the unpurified (crude) reaction mixture after 24 h and with significant yields for reactions performed over the range of pH 3–11 (Fig. 1b and Supplementary Fig. 1). Reactions between melamine and R5P are also productive, forming melamine + R5P conjugates at yields ranging from 33 to 55% when the reaction was carried out for 24 h at 65 °C (Fig. 1c). Glycosylation of melamine with R5P was observed from pH 3 to 9 and at 20 °C (Supplementary Figs 2 and 3). Glycosylation of BA and melamine by (unphosphorylated) ribose was also found to occur spontaneously in water, producing four nucleoside isomers for each reaction (Supplementary Figs 4 and 5). The robustness (for example, wide pH and temperature range) and good yields for both nucleoside and nucleotide formation with BA and melamine are noteworthy among model prebiotic reactions, especially considering that none of the four canonical nucleobases form nucleosides in detectable yields when heated with R5P in water (Supplementary Fig. 6). On the contrary, the canonical nucleosides and nucleotides of RNA are thermodynamically disfavoured (but kinetically stable) in water[29].

To characterize the nucleotides formed from the BA reactions with R5P, the products formed in water after 24 h at 20 °C were isolated by column chromatography for further analysis. Two-dimensional (2D) NMR spectroscopy confirmed that the BA + R5P conjugates are C-nucleotides, with a C–C glycosidic bond between ribose and BA (5-ribofuranosyl-C-barbiturate-5′-monophosphate, C-BMP; Fig. 2b and Supplementary Fig. 7). One-dimensional (1D) rotating frame NOE (ROE) spectroscopy confirmed that the β-anomer is preferentially formed in a 67:33 ratio over the α-anomer (Fig. 2c). Similar to our previous report that reactions between 2,4,6-triaminopyrimidine and ribose yield the C-linked β-ribofuranoside as the major product[10], BA reactions with R5P again demonstrate that the biologically significant β-anomeric sugar form can be preferentially selected through nucleosidation with an alternative nucleobase. Intermediate products suggest that the BA nucleotide-formation reaction proceeds through a Knoevenagel condensation (Supplementary Fig. 8a). We note that β-C-BMP shares a close structural relationship with the C-nucleotide pseudouridine, the most common posttranscriptional modification of RNA in biology[30].

We next isolated and characterized the melamine + R5P conjugates formed from a reaction between melamine and R5P

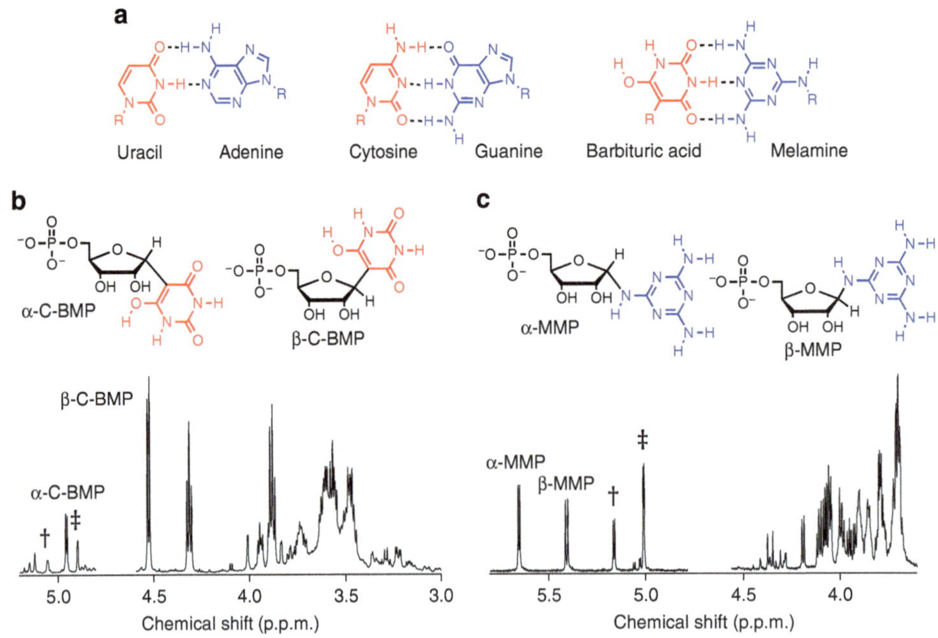

Figure 1 | Spontaneous formation of nucleotides by barbituric acid (BA) and melamine in water. (**a**) Chemical structures of the four canonical nucleobases of RNA shown in their Watson-Crick base pairs, and BA with melamine in an analogous Watson-Crick-like base pair. The R group on all nucleobases, which is H for the free nucleobases, indicates the position of ribose attachment through a glycosidic bond on the canonical bases and for melamine and BA in the current work. (**b**) Chemical structures of the two C-nucleotide anomers of BA-ribosyl-monophosphate (C-BMP) and ^{1}H NMR spectrum of a BA + R5P crude reaction mixture revealing the formation of α-C-BMP and β-C-BMP. (**c**) Chemical structures of the two anomers of melamine-ribosyl-monophosphate (MMP) and the ^{1}H NMR spectrum of a melamine + R5P crude reaction mixture revealing the formation of α-MMP and β-MMP. The anomeric proton resonances for each nucleotide are labelled, and those for R5P are marked with † indicating α-R5P, and ‡ indicating β-R5P. Relative integrated intensities of the nucleotide anomeric resonances show that, for these two reactions, the total C-BMP yield was 82%, and the total MMP yield was 55%. The HOD peaks have been removed from the NMR spectra for clarity. See the Methods for reaction details.

that was performed at 65 °C over 24 h. 2D NMR analysis of the melamine + R5P conjugates confirmed that glycosylation of melamine occurs at an exocyclic amine (Fig. 3b and Supplementary Fig. 9). Glycosylation proceeds through a reversible Schiff base intermediate (Supplementary Fig. 8b), which is partially stabilized by ring closure, to form two *N*-nucleotides (N-ribofuranosyl-melamine-5′-monophosphate (MMP)). 1D ROE analysis confirms that the α- and β-anomers of MMP equilibrate to approximately equal amounts in aqueous solution (Fig. 3c). The stability of the glycosidic bond of exocyclic *N*-linked triaminotriazine nucleosides in water had previously been shown to be quite poor, with total hydrolysis occurring on the order of minutes[27]. The hydrolytic stability of a 5 mM solution of MMP was evaluated to determine the rate of hydrolysis at pH 5 and 5 °C. Under these conditions, the hydrolysis rate was determined to be 3×10^{-6} min^{-1}, with a half-life of 6 months (Supplementary Fig. 10). Remarkably, the dissociation constant between melamine and R5P was found to be 3.7 mM. For comparison, the work of Miller and co-workers[20] revealed that the K_d for the ribonucleoside of urazole is 700 mM at 5 °C, whereas the K_d for the ribonucleoside of uracil is estimated to be around 700 M. Thus, the equilibrium of nucleoside formation between melamine and ribose could be several orders of magnitude more favourable than that of the extant nucleobases.

MMP and C-BMP form supramolecular assemblies. We next tested the ability of MMP and C-BMP to exhibit base pairing in aqueous solution. Base pairing is *not* exhibited by the canonical mononucleotides[31], but such a property would have been advantageous for the prebiotic mutual selection and co-localization of base-pairing nucleotides, particularly if proto-biopolymers emerged from a complex 'prebiotic soup'[32,33]. The crude products of melamine and BA reactions with R5P were mixed and the mixtures were analysed by circular dichroism (CD) spectroscopy as an initial test of whether melamine and BA nucleotides can form chiral assemblies even without purification. To maximize assembly of the two heterocycles, solution pH was adjusted to between pH 4 and 5. This pH range was chosen due to pK_a considerations of the parent heterocycles (BA pK_a = 4; melamine pK_a = 5), which would be expected to minimize the difference in relative ionization between the bases[34]. This mixture of nucleotides, side products and unreacted starting materials from the two reactions (50 mM each in total BA and melamine) exhibits a substantial CD signal, whereas separately the crude products of each reaction exhibit either no CD signal (the melamine-R5P reaction) or a much lower CD signal and with a different wavelength profile (the BA-R5P reaction; Fig. 4a). The loss of the CD signal upon heating the mixture of the crude reactions above 30 °C, and return of signal upon cooling to 5 °C (Supplementary Fig. 11a) indicates the reversible formation of non-covalent assemblies.

The assemblies formed upon combining the crude reaction mixtures were imaged by atomic force microscopy (AFM), which revealed linear supramolecular polymers with diameters of ca 2 nm (Fig. 4b). These structures are fully consistent with the presence of stacked hydrogen-bonded hexads with paired BA and melamine bases (Fig. 4f)[23], assemblies that have been observed previously with analogous molecules that also form hexads[10,35]. The length of these supramolecular polymers (typically > 1 μm) indicates that tens of thousands of heterocycles are paired within

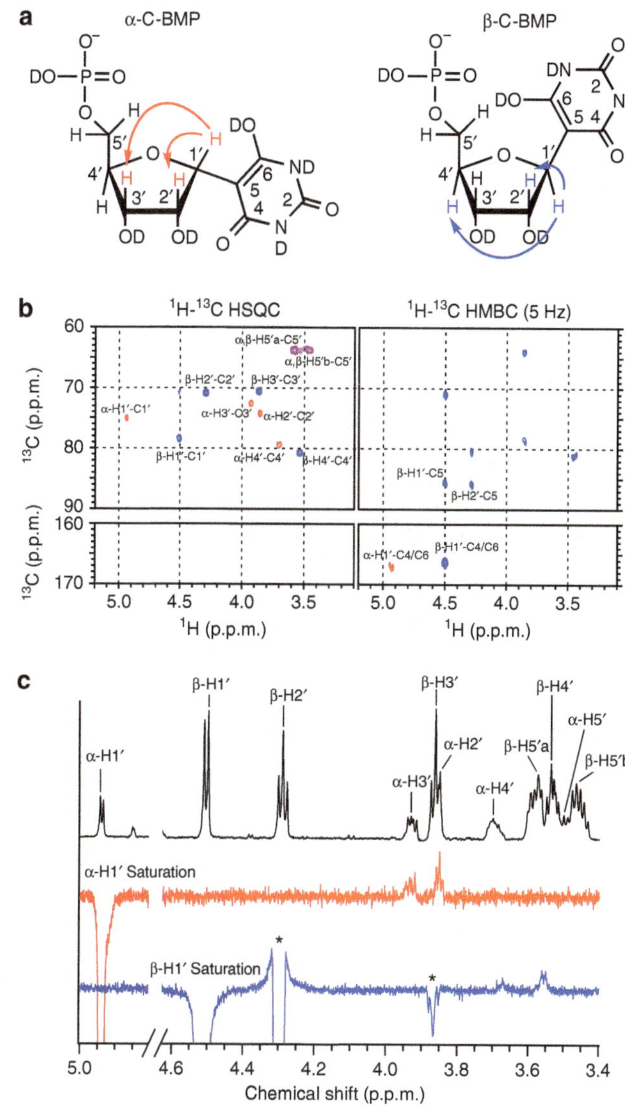

Figure 2 | NMR characterization of C-BMP nucleotides. (**a**) Chemical structure of α-C-BMP and β-C-BMP with arrows indicating through-space proton-proton magnetization transfer (ROE) as shown in **c**. (**b**) Heteronuclear single-quantum correlation (HSQC) and heteronuclear multiple bond correlation (HMBC) spectra showing ^{1}H-^{13}C couplings for α-C-BMP and β-C-BMP. ^{1}H-^{13}C correlation observed between H1′ and C5 of β-C-BMP and H1′-C4/C6 of α-C-BMP and β-C-BMP in the HMBC as well as the absence of H5-C5 correlations in the HSQC support the C-nucleoside assignment. α-Anomer cross peaks are shown in red, β-anomer cross peaks in blue and overlapping cross peaks in purple. ^{1}H Correlation spectroscopy (COSY) spectrum of a mixture of C-BMP nucleotides is provided in the Supplementary Information. (**c**) ^{1}H NMR and 1D ROE spectra of a solution containing a 1:2 ratio of α-C-BMP to β-C-BMP. (Top) ^{1}H NMR spectrum with resonance assignments as indicated in **a**. (Middle) Irradiation of the H1′ of α-C-BMP results in through space magnetization transfer to the H2′ and H3′ of α-C-BMP. (Bottom) Irradiation of the H1′ of β-C-BMP results in through space magnetization transfer to the H4′ of β-C-BMP. * Indicates TOCSY transfer from β-H1′ to β-H2′ and β-H3′.

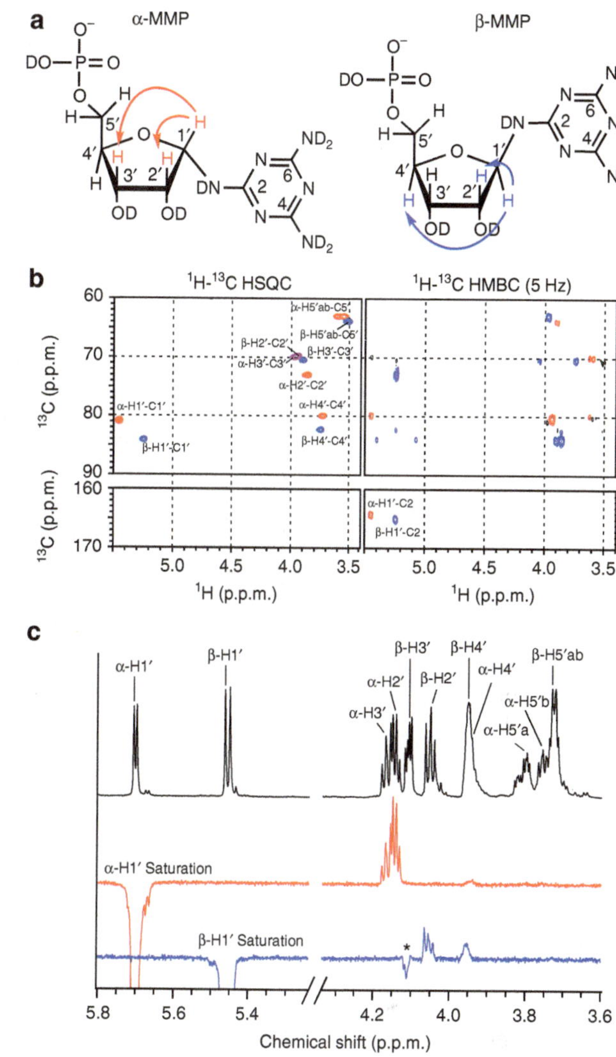

Figure 3 | NMR characterization of MMP nucleotides. (**a**) Chemical structure of α-MMP and β-MMP with arrows indicating through-space proton-proton magnetization transfer as shown in **c**. (**b**) HSQC and HMBC spectra showing ^{1}H-^{13}C couplings for α-MMP and β-MMP. ^{1}H-^{13}C correlation observed in the HMBC (H1′-C5) of α-MMP and β-MMP show coupling between ribose and melamine. α-Anomer cross peaks are shown in red, β-anomer cross peaks in blue and overlapping cross peaks in purple. ^{1}H COSY spectrum of a mixture of MMP nucleotides is provided in the Supplementary Information. (**c**) ^{1}H NMR and 1D ROE spectra of a solution containing an approximately equal concentration of α-MMP to β-MMP. (Top) ^{1}H NMR spectrum with resonance assignments as indicated in **a**. (Middle) Irradiation of the H1′ of α-MMP results in through space magnetization transfer to the H3′ of α-MMP. (Bottom) Irradiation of the H1′ of β-MMP results in through space magnetization transfer to the H2′ and H4′ of β-MMP. * Indicates TOCSY transfer from β-H1′ to β-H3′.

a single assembly. When solutions containing purified MMP and C-BMP were combined and analysed by AFM, supramolecular polymers with a diameter of 2 nm were also observed (Fig. 4c and Supplementary Fig. 11b). As can be seen in Fig. 4b,c, the assemblies formed from purified MMP and BMP are noticeably shorter than those formed upon combining the crude reaction mixtures. This observation is consistent with length being limited by greater peripheral charge[36], an effect that is expected to be greater for supramolecular polymers formed from the purified nucleotides than it is for assemblies containing both nucleotides and the parent heterocycles (that is, those present in the crude reaction mixtures).

C-BMP and MMP nucleotides also form water-soluble supramolecular assemblies when mixed with free melamine and free BA, respectively. These C-BMP-melamine and MMP-BA assemblies were visualized by AFM (Fig. 4d,e and Supplementary Fig. 11c,d), revealing 2 nm fibres that are again indicative of

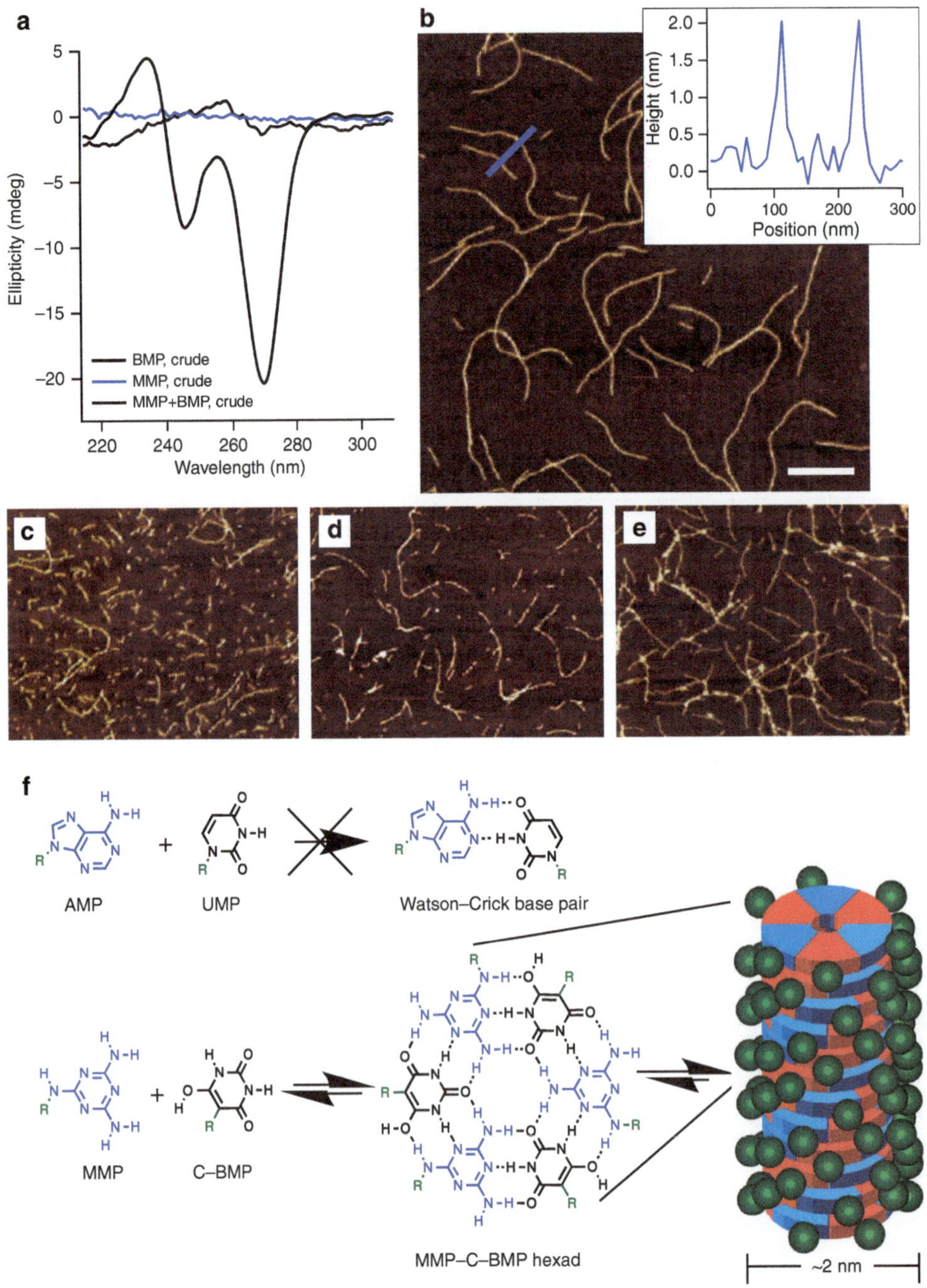

Figure 4 | Nucleotides assemble into supramolecular polymers. (**a**) CD spectra of melamine-R5P and BA-R5P crude reaction mixtures, separate and combined at 5 °C. Black curve is mixture of melamine-R5P and BA-R5P crude reaction mixtures; red curve is BA-R5P crude reaction mixture; blue curve is melamine-R5P crude reaction mixture. Ultraviolet spectra associated with CD spectra are provided in Supplementary Fig. 11. (**b–e**) AFM topographic images of (**b**) mixture of melamine-R5P and BA-R5P crude reaction mixtures, (**c**) mixture of purified C-BMP and MMP, (**d**) purified C-BMP mixed with melamine, (**e**) purified MMP mixed with BA. Inset in **b** shows height measurements of blue line in image. Scale bar in **b** is 300 nm, and all AFM images are shown at the same magnification. (**f**) Chemical structures of MMP and C-BMP and their association into 2 nm wide stacked hexads. The green R and green spheres indicate R5P. The similarity of AMP to MMP and UMP to C-BMP, and the inability of these canonical nucleosides to assemble in aqueous solution is also illustrated. All solutions contained 1 M NaCl.

stacked hexad assemblies. Although free BA and free melamine form insoluble precipitates when mixed in aqueous solution (Supplementary Fig. 12), the steric bulk and charge provided by conjugation with R5P on one nucleobase favours the formation of the water soluble, linear assemblies of stacked hexads[35].

Supramolecular assemblies preferentially incorporate β-MMP. ^{1}H-NMR spectroscopy was used to further characterize C-BMP-melamine and MMP-BA assemblies (Supplementary Figs 13 and 14). Nucleotides incorporated into these supramolecular assemblies exhibit extreme ^{1}H NMR line broadening (to baseline), which render them invisible to solution state NMR spectroscopy. In contrast, free nucleotides that exist in equilibrium with the assemblies exhibit virtually no change in ^{1}H resonance chemical shift or line width. Quantitative analysis of ^{1}H resonance intensity can be used to determine the fraction of free nucleotides, and subtraction of these values from the known total concentration of nucleotides in a sample reveals the concentration of assembled nucleotides[35]. Analysis of ^1H spectra of solutions containing MMP + BA, 50 mM in nucleotide and heterocycle, showed temperature-dependent assembly of nucleotides from 5 to 40 °C (Fig. 5a,b and Supplementary Fig. 14). Unexpectedly, the preferential incorporation for β-MMP over α-MMP was observed at all temperatures where assemblies are present. In particular, a twofold preference for β-MMP incorporation is observed at 5 °C, and only incorporation of β-MMP at 20 °C (Fig. 5b).

As the β-anomer is selectively incorporated into the assemblies over the α-anomer, and anomerization of MMP occurs in the solution (see the Methods for details), we next explored if supramolecular assembly will affect the anomeric ratio of MMP. As noted above, at equilibrium, a solution of MMP contains a mixture of α- and β-anomers (45% α and 55% β), however, when 50 mM MMP was incubated with 50 mM BA at 5 °C, we observe the conversion of α-MMP to β-MMP, a conversion that reached a maximum of 63% β-MMP after 8 days (Fig. 5c). This conversion was not observed when BA was omitted or when MMP was below the minimal concentration required for MMP-BA assembly. Therefore, we have found that the assemblies formed with BA preferentially select/stabilize the β-anomer of MMP, and as a result, with anomerization, the assemblies enrich β-MMP (Fig. 5d). These observations indicate that solutions containing MMP and BA assemblies will change overtime in both molecular (that is, enrichment of β-MMP) and supramolecular composition (that is, more MMP and BA assembled).

Discussion

The data presented here demonstrate the efficient single-step syntheses of complementary nucleosides and nucleotides, starting with the plausible proto-nucleobases melamine and BA and ribose or R5P. Although R5P forms an exocyclic N-glycosidic bond with melamine (MMP) and R5P forms a C-nucleotide with BA (C-BMP), both of these nucleotides favour their β anomers. Specifically, for the free nucleotides in solution, the β and α anomers of C-BMP exist at equilibrium in a 67:33 ratio, and for MMP the β-anomer is favoured in a 55:44 ratio over the α-anomer. Perhaps coincidentally, extant life uses the β-anomeric form of ribonucleotides, and our observations indicate that this form may have been enriched on the early Earth for β-ribonucleotides for which glycosidic bond anomerization is under equilibrium control. This nucleotide structural preference is apparently not limited to the nucleotides of melamine and BA, as previous studies have also revealed that nucleosides formed by drying and heating ribose with two other pyrimidines, 2-pyrimidinone[21,37] and 2,4,6-triaminopyrimidine[23], also favour the β-anomer.

Unlike the mononucleotides of extant RNA, we observe that MMP and C-BMP will pair as monomers in aqueous solution, with each other and with their unmodified pairing partners (that is, MMP with C-BMP, MMP with BA, and C-BMP with melamine), producing, in all cases, supramolecular structures that indicate the highly efficient stacking of H-bonded hexads that are themselves composed of Watson–Crick-like base pairs.

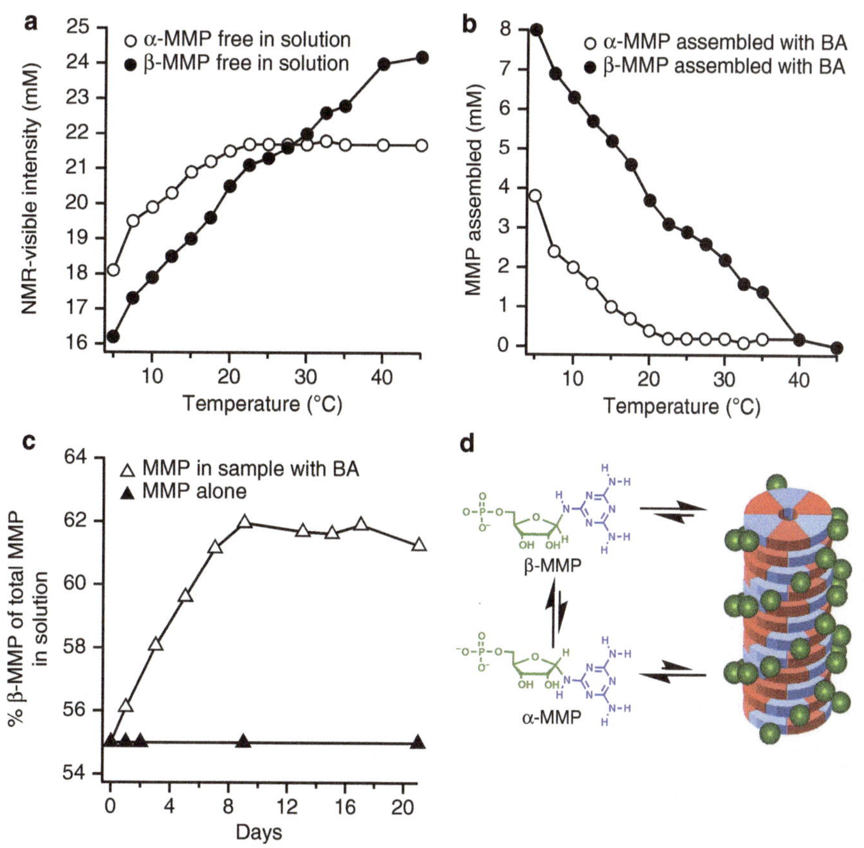

Figure 5 | MMP assembly and anomerization in the presence of BA. (**a**) Plot of NMR visible resonance intensity of unassembled α-MMP and β-MMP as a function of temperature in a solution containing 50 mM MMP and BA. (**b**) Plot showing fraction of both MMP anomers assembled at various temperatures (plot generated by subtracting data shown in **a** from measured total concentration of α-MMP and β-MMP in each sample). (**c**) Plot showing the change in anomeric ratio (by percent) of β-MMP as a function of time in solutions containing both MMP and BA, or MMP alone. Samples were maintained at 5 °C during the experiment and diluted just prior to analysis to disassemble MMP in order to enable quantification of the total MMP in solution by NMR. (**d**) Schematic showing preferential assembly (stacked hexads) and anomerization. All solutions contained 0.3 M NaCl.

This pairing and stacking is sufficiently robust to drive assembly in the presence of the side products and unreacted starting materials of the crude nucleotide reactions. Based on these observations, and because C-BMP and MMP structurally resemble two nucleotides found in life today (UMP and AMP, respectively) and have been reported to pair with extant, complementary nucleobases[38,39], it is tempting to speculate that these heterocycles could represent ancestral nucleotides of the contemporary genetic polymers. In particular, the ability for C-BMP and MMP to form *noncovalent* supramolecular assemblies could have facilitated the prebiotic localization, organization and subsequent linking of these (or similar) nucleotides into *covalent* polymers that were then capable of storing and transferring information (for example, by templating the formation of sequence-specific assemblies for the polymerization of additional monomers[40]).

The ability of C-BMP and MMP to form supramolecular assemblies might have also facilitated the emergence of early RNA-like polymers by selecting nucleotides with sugars (or earlier trifunctional linkers[19]) that were structurally compatible with the assemblies and their subsequent coupling into covalent polymers. In the present study, we have, for practical reasons, used D-ribose and D-R5P for our nucleoside and nucleotide reactions with melamine and BA, but L-ribose or L-R5P would exhibit equivalent reactivity with these two heterocycles. Nevertheless, it has been often postulated that a racemic mixture of nucleotides would have inhibited the prebiotic synthesis of RNA polymers[41], and so the question of how the present system might address this challenge deserves some discussion. Although we have not shown chiral nucleotide selection, in the current study we have demonstrated that the β-anomer of MMP is enriched in supramolecular assemblies over the α-anomer of MMP, and this selection leads to a detectable increase in the ratio of the β-anomer over the α-anomer of MMP in the entire solution (presumably due to anomerization and selective stabilization by the assembly). As a recent example of the ability of supramolecular polymers to promote local chiral resolution, Aida and co-workers demonstrated that racemic solutions of chiral macrocycles self-sort into homochiral supramolecular polymers[42]. It is therefore possible that supramolecular assemblies, formed by nucleotides with different sugars, including different anomers and enantiomers, could have been selectively enriched in individual supramolecular assemblies before polymerization. Current investigations of this possibility are actively being pursued in our laboratory.

Methods

Materials. Melamine and BA were purchased from Acros Organic, D-R5P disodium salt and D-ribose were purchased from Sigma-Aldrich. All chemicals were used as received.

Synthesis of C-BMP. BA (2.5 mmol) and R5P (2.5 mmol) were dissolved in 5 ml of H_2O and the pH was adjusted to 9 with NaOH (unless otherwise noted). The solution was stirred for 24 h at 20 °C, at which time a clear, pale yellow solution was present. This solution is referred to as the crude BA-R5P reaction mixture. To isolate C-BMP, the crude reaction mixture was loaded onto a gravity column containing QAE Sephadex A-25 anion exchange media and eluted with a gradient of NH_4HCO_3 buffer (pH 9.4) from 50 mM to 0.5 M. The fractions containing product were lyophilized, redissolved in water and pooled. When the reaction was performed at pH 9, yield of C-BMP was 82% (α-C-BMP 22%, β-C-BMP 60%); ^{1}H NMR of the crude reaction mixture is shown in Fig. 1b. HRMS (*m/z*): $[M]^-$ calculated for $C_9H_{12}N_2O_{10}P^-$, 339.0235; found, 339.0243. Ultraviolet-visible: (50 mM NaH_2PO_4, pH 7) λ_{max} = 260 nm; ε_{260} = 23,000 mol l^{-1} cm^{-1}.

α-C-BMP. ^{1}H NMR (500 MHz, D_2O): δ 4.94 (d, *J* = 4.2 Hz, H1′), 3.85 (dd, *J* = 4.2, X Hz, H2′), 3.93 (dd, *J* = 4.1, 3.5 Hz, H3′), 3.70 (m, H4′), 3.59 (m, H5′a), 3.46 (m, H5′b); ^{13}C NMR (126 MHz, D_2O): δ 167.1 (C4/C6), 152.7 (C2), 85.5 (C5), 79.5 (d, *J* = 8.2 Hz, C4′), 75.1 (C1′), 74.3 (C2′), 72.7 (C3′), 63.6 (d, *J* = 4.4 Hz, C5′).

β-C-BMP. ^{1}H NMR (500 MHz, D_2O): δ 4.51 (d, *J* = 5.7 Hz, H1′), 4.29 (t, *J* = 5.7 Hz, H2′), 3.86 (t, *J* = 6.2 Hz, H3′), 3.54 (m, H4′), 3.58 (m, H5′a); 3.46 (m, H5′b); ^{13}C NMR (126 MHz, D_2O): δ 166.4 (C4/C6), 153.1 (C2), 85.7 (C5), 80.8 (d, *J* = 7.9 Hz, C4′), 78.6 (C1′), 70.9 (C2′), 70.5 (C3′), 63.8 (d, 4.6 Hz C5′).

Synthesis of MMP. Melamine (1 mmol) and R5P (1 mmol) were dissolved in 5 ml of H_2O and the pH was adjusted with HCl to pH 5, unless otherwise noted. The solution was stirred at 65 °C for 24 h, at which point a clear, light brown solution was present. This solution is referred to as the melamine-R5P crude reaction mixture. To isolate MMP, the crude reaction mixture was loaded onto a gravity column containing SP Sephadex C-50 cation-exchange media and eluted with NH_4OAc buffer, 50 mM, pH 4.3. The fractions containing product were lyophilized, redissolved in water and pooled. When the reaction was performed at pH 5, yield of MMP was 55% (α-MMP 26.4%, β-MMP 28.6%); ^{1}H NMR of the crude reaction mixture is shown in Fig. 1c. HRMS (*m/z*): $[M]^-$ calculated for $C_8H_{14}N_6O_7P^-$, 337.0667; found, 337.0659. Ultraviolet-visible: (50 mM NaH_2PO_4, pH 7) λ_{max} = 235 nm; ε_{235} = 5,500 mol l^{-1} cm^{-1}.

α-MMP. ^{1}H NMR (500 MHz, D_2O): δ 5.46 (d, *J* = 4.1 Hz, H1′), 3.97 (dd, *J* = 4.7, 1.9 Hz, H3′), 3.94(dd, *J* = 4.7, 4.1 H2′), 3.74 (m, H4′), 3.64 (m, H5′a), 3.55 (m, H5′b); ^{13}C NMR (126 MHz, D_2O): δ 164.4 (C2), 164.2 (C4/C6), 81.0 (C1′), 79.5 (d, *J* = 8.2 Hz, C4′), 70.0 (C3′), 69.9 (C2′), 63.4 (d, *J* = 5.3, C5′).

β-MMP. ^{1}H NMR (500 MHz, D_2O): δ 5.25 (d, *J* = 6.6 Hz, H1′), 3.89 (dd, *J* = 5.4, 3.2 Hz, H3′), 3.85 (dd, *J* = 6.6, 5.4 Hz, H2′), 3.75 (m, H4′), 3.49 (m, H5′a,b); ^{13}C NMR (126 MHz, D_2O): δ 165.1 (C2), 164.9 (C4/C6), 84.2 (C1′), 82.3 (d, *J* = 8.0 Hz, C4′), 73.1 (C2′), 70.5 (C3′), 64.2 (d, 4.5 Hz, C5′).

General assembly protocol. All solutions were 50 mM in each heterocycle. Final concentration of solutions containing crude reaction mixtures of either nucleotide was 50 mM in total heterocycle (that is, both unmodified and glycosylated). Upon combining melamine and BA nucleotides and/or parent heterocycles, the pH of solution was adjusted with HCl or NaOH to pH 4.5–5. This pH range was chosen as it would be expected to maximize assembly of the heterocycles due to pK_a considerations (pK_a of protonated melamine is 5 and the pK_a of BA is 4)[34]. All solutions contained either 0.3 or 1.0 M NaCl as noted. Anomerization experiments were performed with MMP that was first incubated at 5 °C for 24 h to enable the anomeric ratio to equilibrate under the conditions tested. Variable temperature NMR experiments were performed on samples containing MMP and BA that were mixed and stored at 5 °C for 15 h before analysis to ensure that the association between MMP and BA had reached equilibrium. Stock solutions of nucleotides and crude reaction mixtures were stored at −20 °C.

Spectroscopic analysis of the supramolecular assemblies. CD analysis was carried out on a Jasco J-810 CD spectrometer equipped with a six-cell Quantum Northwest peltier temperature controller. Strain-free 0.01 mm demountable cells from Starna were used for all CD analysis. ^{1}H NMR spectra were collected on a Bruker DRX-500 NMR and were the sum of 32 transients. All NMR samples were D_2O exchanged and lyophilized before analysis in D_2O with an internal standard of 3-(trimethylsilyl)-2,2′,3,3′-tetradeuteropropionic acid at 1.11 mM. Variable temperature NMR experiments were performed by gradually heating the solutions in order to disassemble supramolecular assemblies present, or slow cooling to enable reassembly, followed by short incubation at the desired temperature to allow for the assemblies to reach equilibrium. To analyse anomerization of MMP over time, a sample of MMP and BA (50 mM each) was stored at 5 °C, and at specified time points, samples were removed and diluted with D_2O to 5 mM in MMP total (a concentration below the minimal assembly concentration) and analysed by NMR at 20 °C

Atomic force microscopy. Imaging was performed with a Nanoscope IIIa (Digital Instruments) in tapping mode, using Si tips (MicroMash, 16 N m^{-1}). A freshly cleaved mica substrate was pre-activated by incubation with a solution of 20 mM $MgCl_2$ that was then rinsed with water and dried under N_2 (*g*). A 3-μl sample (initially stored on ice) was deposited on the mica substrate and spread with N_2 (*g*) followed by the addition of 50 μl of cold water to remove sample that was not adsorbed to the mica surface and quickly dried again with N_2 (*g*).

Analytical HPLC. Analysis of nucleotide samples was performed by running a linear gradient of 90% acetonitrile/10% water to 50% acetonitrile/H_2O for 7 min, followed by isocratic 90% aqueous acetonitrile for 6 min on a Waters XBridge Amide (150 × 2.5 × 3.5 μm^3) column. Analytical HPLC analysis of nucleoside samples was performed by running a linear gradient of 85% acetonitrile/15% water to 40% acetonitrile/60% water over 25 min on a Waters XBridge Amide (150 × 2.5 × 3.5 μm^3) column.

References

1. Joyce, G. F. The antiquity of RNA-based evolution. *Nature* **418,** 214–221 (2002).
2. Callahan, M. P. *et al.* Carbonaceous meteorites contain a wide range of extraterrestrial nucleobases. *Proc. Natl Acad. Sci. USA* **108,** 13995–13998 (2011).

3. Barks, H. L. *et al.* Guanine, adenine, and hypoxanthine production in UV-irradiated formamide solutions: Relaxation of the requirements for prebiotic purine nucleobase formation. *ChemBioChem.* **11,** 1240–1243 (2010).
4. Hayatsu, R. Orgueil Meteorite: organic nitrogen contents. *Science* **146,** 1291–1293 (1964).
5. Ricardo, A., Carrigan, M. A., Olcott, A. N. & Benner, S. A. Borate minerals stabilize ribose. *Science* **303,** 196–196 (2004).
6. Springsteen, G. & Joyce, G. F. Selective derivatization and sequestration of ribose from a prebiotic mix. *J. Am. Chem. Soc.* **126,** 9578–9583 (2004).
7. Krishnamurthy, R., Guntha, S. & Eschenmoser, A. Regioselective alpha-phosphorylation of aldoses in aqueous solution. *Angew. Chem. Int. Ed. Engl.* **39,** 2281–2285 (2000).
8. Ponnamperuma, C. & Mack, R. Nucleotide synthesis under possible primitive Earth conditions. *Science* **148,** 1221–1223 (1965).
9. Orgel, L. E. Prebiotic chemistry and the origin of the RNA world. *Crit. Rev. Biochem. Mol. Biol.* **39,** 99–123 (2004).
10. Cafferty, B. J. & Hud, N. V. Abiotic synthesis of RNA in water: a common goal of prebiotic chemistry and bottom-up synthetic biology. *Curr. Opin. Chem. Biol.* **22,** 146–157 (2014).
11. Fuller, W. D., Sanchez, R. A. & Orgel, L. E. Studies in prebiotic synthesis. VI. Synthesis of purine nucleosides. *J. Mol. Biol.* **67,** 25–33 (1972).
12. Fuller, W. D., Sanchez, R. A. & Orgel, L. E. Studies in prebiotic synthesis: VII. Solid-state synthesis of purine nucleosides. *J. Mol. Evol.* **1,** 249–257 (1972).
13. Sanchez, R. A. & Orgel, L. E. Studies in prebiotic synthesis. V. Synthesis and photoanomerization of pyrimidine nucleosides. *J. Mol. Biol.* **47,** 531–543 (1970).
14. Powner, M. W., Gerland, B. & Sutherland, J. D. Synthesis of activated pyrimidine ribonucleotides in prebiotically plausible conditions. *Nature* **459,** 239–242 (2009).
15. Benner, S. A., Kim, H.-J. & Carrigan, M. A. Asphalt, water, and the prebiotic synthesis of ribose, ribonucleosides, and RNA. *Acc. Chem. Res.* **45,** 2025–2034 (2012).
16. Eschenmoser, A. Etiology of potentially primordial biomolecular structures: From vitamin B12 to the nucleic acids and an inquiry into the chemistry of life's origin: A retrospective. *Angew. Chem. Int. Ed. Engl.* **50,** 12412–12472 (2011).
17. Patel, B. H., Percivalle, C., Ritson, D. J., Duffy, C. D. & Sutherland, J. D. Common origins of RNA, protein and lipid precursors in a cyanosulfidic protometabolism. *Nat. Chem.* **7,** 301–307 (2015).
18. Joyce, G. F., Schwartz, A. W., Miller, S. L. & Orgel, L. E. The case for an ancestral genetic system involving simple analogs of the nucleotides. *Proc. Natl Acad. Sci. USA* **84,** 4398–4402 (1987).
19. Hud, N. V., Cafferty, B. J., Krishnamurthy, R. & Williams, L. D. The origin of RNA and 'My Grandfather's Axe'. *Chem. Biol.* **20,** 466–474 (2013).
20. Kolb, V. M., Dworkin, J. P. & Miller, S. L. Alternative bases in the RNA world: the prebiotic synthesis of urazole and its ribosides. *J. Mol. Evol.* **38,** 549–557 (1994).
21. Bean, H. D. *et al.* Formation of a β-pyrimidine nucleoside by a free pyrimidine base and ribose in a plausible prebiotic reaction. *J. Am. Chem. Soc.* **129,** 9556–9557 (2007).
22. Kim, H. J. & Benner, S. A. Prebiotic glycosylation of uracil with electron-donating substituents. *Astrobiology* **15,** 301–306 (2015).
23. Chen, M. C. *et al.* Spontaneous prebiotic formation of a β-ribofuranoside that self-assembles with a complementary heterocycle. *J. Am. Chem. Soc.* **136,** 5640–5646 (2014).
24. Cafferty, B. J. & Hud, N. V. Was a pyrimidine-pyrimidine base pair the ancestor of Watson-Crick base pairs? Insights from a systematic approach to the origin of RNA. *Israel J. Chem.* **55,** 891–905 (2015).
25. Menor-Salvan, C., Ruiz-Bermejo, D. M., Guzman, M. I., Osuna-Esteban, S. & Veintemillas-Verdaguer, S. Synthesis of pyrimidines and triazines in ice: Implications for the prebiotic chemistry of nucleobases. *Chem. Eur. J* **15,** 4411–4418 (2009).
26. Whitesides, G. M. *et al.* Noncovalent synthesis: using physical-organic chemistry to make aggregates. *Acc. Chem. Res.* **28,** 37–44 (1995).
27. Hysell, M., Siegel, J. S. & Tor, Y. Synthesis and stability of exocyclic triazine nucleosides. *Org. Biomol. Chem.* **3,** 2946–2952 (2005).
28. van Vliet, M. J., Visscher, J. & Schwartz, A. W. Hydrogen-bonding in the template-directed oligomerization of a pyrimidine nucleotide analog. *J. Mol. Evol.* **41,** 257–261 (1995).
29. Rios, A. C., Yu, H. T. & Tor, Y. Hydrolytic fitness of *N*-glycosyl bonds: comparing the deglycosylation kinetics of modified, alternative, and native nucleosides. *J. Phys. Org. Chem.* 173–180 (2014).
30. Hamma, T. & Ferre-D'Amare, A. R. Pseudouridine synthases. *Chem. Biol.* **13,** 1125–1135 (2006).
31. Ts'o, P. O. P. in *Basic Principles in Nucleic Acid Chemistry* Vol. 1 (ed. Ts'o, P. O. P.) 453–584 (Academic Press, 1974).
32. Engelhart, A. E. & Hud, N. V. Primitive genetic polymers. *Cold Spring Harb. Perspect. Biol.* **2** doi: 10.1101/cshperspect.a002196 (2010).
33. Krishnamurthy, R. On the emergence of RNA. *Israel J. Chem.* **55,** 837–850 (2015).
34. Cafferty, B. J., Avirah, R. R., Schuster, G. B. & Hud, N. V. Ultra-sensitive pH control of supramolecular polymers and hydrogels: pK(a) matching of biomimetic monomers. *Chem. Sci.* **5,** 4681–4686 (2014).
35. Cafferty, B. J. *et al.* Efficient self-assembly in water of long noncovalent polymers by nucleobase analogues. *J. Am. Chem. Soc.* **135,** 2447–2450 (2013).
36. Besenius, P. *et al.* Controlling the growth and shape of chiral supramolecular polymers in water. *Proc. Natl Acad. Sci. USA* **107,** 17888–17893 (2010).
37. Sheng, Y., Bean, H. D., Mamajanova, I., Hud, N. V. & Leszczynski, J. A comprehensive investigation of the energetics of pyrimidine nucleoside formation in a model prebiotic reaction. *J. Am. Chem. Soc.* **131,** 16088–16095 (2009).
38. Xia, X., Piao, X. & Bong, D. Bifacial peptide nucleic acid as an allosteric switch for aptamer and ribozyme function. *J. Am. Chem. Soc.* **136,** 7265–7268 (2014).
39. Voet, D. Crystal and molecular structure of intermolecular complex 9-ethyladenine-5,5-diethylbarbituric acid. *J. Am. Chem. Soc.* **94,** 8213–8222 (1972).
40. Hud, N. V. & Anet, F. A. L. Intercalation-mediated synthesis and replication: a new approach to the origin of life. *J. Theor. Biol.* **205,** 543–562 (2000).
41. Joyce, G. *et al.* Chiral selection in poly(C)-directed synthesis of oligo(G). *Nature* **310,** 602–604 (1984).
42. Sato, K., Itoh, Y. & Aida, T. Homochiral supramolecular polymerization of bowl-shaped chiral macrocycles in solution. *Chem. Sci.* **5,** 136–140 (2014).

Acknowledgements

We thank J. Forsythe for accurate mass analysis, L.A. Bottomley for use of AFM, L.D. Williams and G.B. Schuster for discussions. This work was supported by the NSF and the NASA Astrobiology Program under the NSF Center for Chemical Evolution (CHE-1504217).

Author contributions

B.J.C., R.K. and N.V.H. designed the research, D.M.F. and J.K. synthesized and purified MMP and BMP, B.J.C. and D.M.F. developed and characterized the nucleotide reactions, B.J.C. performed the NMR, CD and AFM analysis, all authors analysed and discussed the results, B.J.C., R.K. and N.V.H. wrote the paper.

Additional information

Competing financial interests: The authors declare no competing financial interests.

3

Modulating protein activity using tethered ligands with mutually exclusive binding sites

Alberto Schena[1,2,3,*], Rudolf Griss[1,2,3,*] & Kai Johnsson[1,2,3]

The possibility to design proteins whose activities can be switched on and off by unrelated effector molecules would enable applications in various research areas, ranging from biosensing to synthetic biology. We describe here a general method to modulate the activity of a protein in response to the concentration of a specific effector. The approach is based on synthetic ligands that possess two mutually exclusive binding sites, one for the protein of interest and one for the effector. Tethering such a ligand to the protein of interest results in an intramolecular ligand-protein interaction that can be disrupted through the presence of the effector. Specifically, we introduce a luciferase controlled by another protein, a human carbonic anhydrase whose activity can be controlled by proteins or small molecules *in vitro* and on living cells, and novel fluorescent and bioluminescent biosensors.

[1]École Polytechnique Fédérale de Lausanne, Institute of Chemical Sciences and Engineering, Avenue Forel 2, EPFL SB ISIC LIP BCH-4303, CH-1015 Lausanne, Switzerland. [2]École Polytechnique Fédérale de Lausanne, Institute of Bioengineering, CH-1015 Lausanne, Switzerland. [3]National Centre of Competence in Research in Chemical Biology, CH-1015 Lausanne, Switzerland. *These authors contributed equally to this work. Correspondence and requests for materials should be addressed to K.J. (email: kai.johnsson@epfl.ch).

Allosteric proteins act as molecular switches in which binding of a molecule to a site different from the active site changes the conformation of the protein and its underlying activity. Such proteins are fundamental for the regulation of most natural signalling processes. Designing new allosteric proteins is a formidable test for our understanding of protein function and such designer proteins can find applications in synthetic biology and biosensing[1–4]. For example, the most popular fluorescence-based biosensor for live-cell imaging is based on an engineered fluorescent protein allosterically regulated by calcium ions[5].

The design of novel allosteric proteins is usually based on the insertion of a pre-existing allosteric protein domain into another protein; binding of the allosteric modulator then changes the conformation of the allosteric domain and of the protein in which it is inserted[6–8]. However, the generality of the approach is limited as it relies on pre-existing allosteric proteins and modulators. In addition, the identification of an appropriate insertion site is difficult. Alternative approaches for regulating protein activity have been developed. For example, enzymatic activities have been regulated by tethering an inhibitor to an enzyme through a single-stranded oligonucleotide; binding of a complementary DNA sequence prevents an intramolecular inhibition of the enzyme[9,10]. However, in its present form the approach is limited to DNA as an effector molecule. In another approach, the activity of an enzyme is regulated by introducing new binding sites in close proximity to its active site[11]. This principle is the basis for the enzyme multiplied immunoassay technique—the first homogeneous immunoassays—which are still commonly used in diagnostics today[12,13]. However, not every enzyme lends itself to such modifications of its active site. In short, alternative design principles for the generation of proteins that can exist in two different states that are energetically similar but differ in activity are needed.

We introduce here a conceptually novel approach to regulate protein activity. As with allosteric proteins, the activity of the reporter is modulated by an external effector. However, the modulation is not based on a conformational change within the reporter protein but on the steric displacement of a ligand in a larger semisynthetic protein construct. This is achieved by generating synthetic ligands that possess two mutually exclusive binding sites. We demonstrate the potential of the approach by generating a novel bioluminogenic protein as well as bioluminescent and fluorescent biosensors for applications *in vitro* and in live cells.

Results

Effector-modulated reporters. We have previously introduced fluorescent and bioluminescent sensor proteins that are based on the competition of a tethered ligand with an analyte for a common binding site[14–18]. The binding and unbinding of the tethered ligand leads to a change in fluorescence resonance energy transfer (FRET) or bioluminescence resonance energy transfer (BRET) efficiency and can be used as readout for the concentration of the competing analyte. We speculated that adding a second synthetic ligand that binds to its target in a mutually exclusive manner with respect to the first tethered ligand could allow the modulation of protein activity by an unrelated effector. In this approach, the effector binds to one of the two synthetic ligands, making the other one unavailable for interactions with the protein to which the synthetic ligands are tethered. The task of modulating protein activities is thus reduced to either the modification of an existing ligand–protein interaction or the design of a new one and opens up the possibility to transform regular proteins into effector-modulated reporters resembling allosteric proteins. In the following, we will abbreviate our approach as chemical ligand-associated steric hindrance (CLASH).

Control of a luciferase by an exogenous effector. Controlling the light-emitting properties of luciferases through molecules of interest is an attractive approach to generate powerful biosensors. Luciferases allosterically controlled by either calcium ions or cyclic AMP have been generated by inserting naturally occurring allosteric protein domains[19,20], but to the best of our knowledge no luciferases have been designed that are regulated by other proteins. We thus exploited CLASH to transform the luciferase from *Renilla reniformis* into one that is activated by an effector protein. For first proof-of-principle experiments, we chose the protein streptavidin as the effector, since its natural ligand biotin can be readily synthetically derivatized. We synthesized a dual ligand containing the luciferase inhibitor coelenteramide[21] and biotin in such close proximity that simultaneous binding to the luciferase and to streptavidin should be disfavoured (Fig. 1a,b). Using SNAP-tag technology[22], we chemically linked the dual ligand to the *R. reniformis* luciferase, generating CLASH-Strep/Luc (Fig. 1c). In the absence of streptavidin, the tethered coelenteramide inhibits the luciferase, resulting in a dark state. On addition of streptavidin, the light intensity increased in a concentration-dependent manner: binding of streptavidin to the biotin moiety of the synthetic ligand forces coelenteramide to dissociate from the luciferase, leading to an active luciferase. The light intensity increased by 1,700% in the presence of streptavidin, permitting streptavidin detection by naked eye (Fig. 1d,e). The marked increase in light emission in response to an analyte makes the approach an attractive starting point for generating biosensors for various proteins. This then would require the replacement of biotin in our synthetic ligand with an appropriate binder to the protein of interest.

Control of human carbonic anhydrase. To demonstrate the generality of CLASH, we applied it to regulate the activity of the enzyme human carbonic anhydrase II (HCA). Furthermore, we wanted to translate the activity modulation into a signal compatible with biosensing. We synthesized a dual ligand in which the HCA inhibitor benzenesulfonamide and a biotin moiety are connected in a way that binding to their respective protein receptors would be mutually exclusive (Fig. 2a–c) and used again streptavidin as the effector. The choice of the linker composition and length in the labelling molecule was based on the crystal structure of the receptor proteins. To couple HCA activity switching to a readout compatible with biosensing, we expressed HCA as a fusion protein with the luciferase NanoLuc, a 30-proline linker (PP30) and SNAP-tag, following the design principles of LUCIDs—our previously described class of BRET-based biosensors[18]. The SNAP-PP30-NLuc-HCA fusion protein was labelled with the synthetic ligand also containing the fluorophore Cy3 for the BRET readout. The resulting semisynthetic protein CLASH-Strep/HCA (Fig. 2b) is expected to show an increase in HCA activity and display a change in BRET efficiency between NanoLuc and Cy3 when the tethered ligand is displaced from HCA through binding of the effector streptavidin. The interaction between streptavidin and biotin has a K_d in the femtomolar range; since this is much lower than the concentration of CLASH-Strept/HCA in the assay, the transition of the titration curve occurs around the concentration of the CLASH construct.

To assess the control of HCA activity in CLASH-Strep/HCA, we performed a standard HCA activity assay based on the catalysis of a chromogenic reaction[23]. In the absence of streptavidin, the construct showed low HCA activity, but on

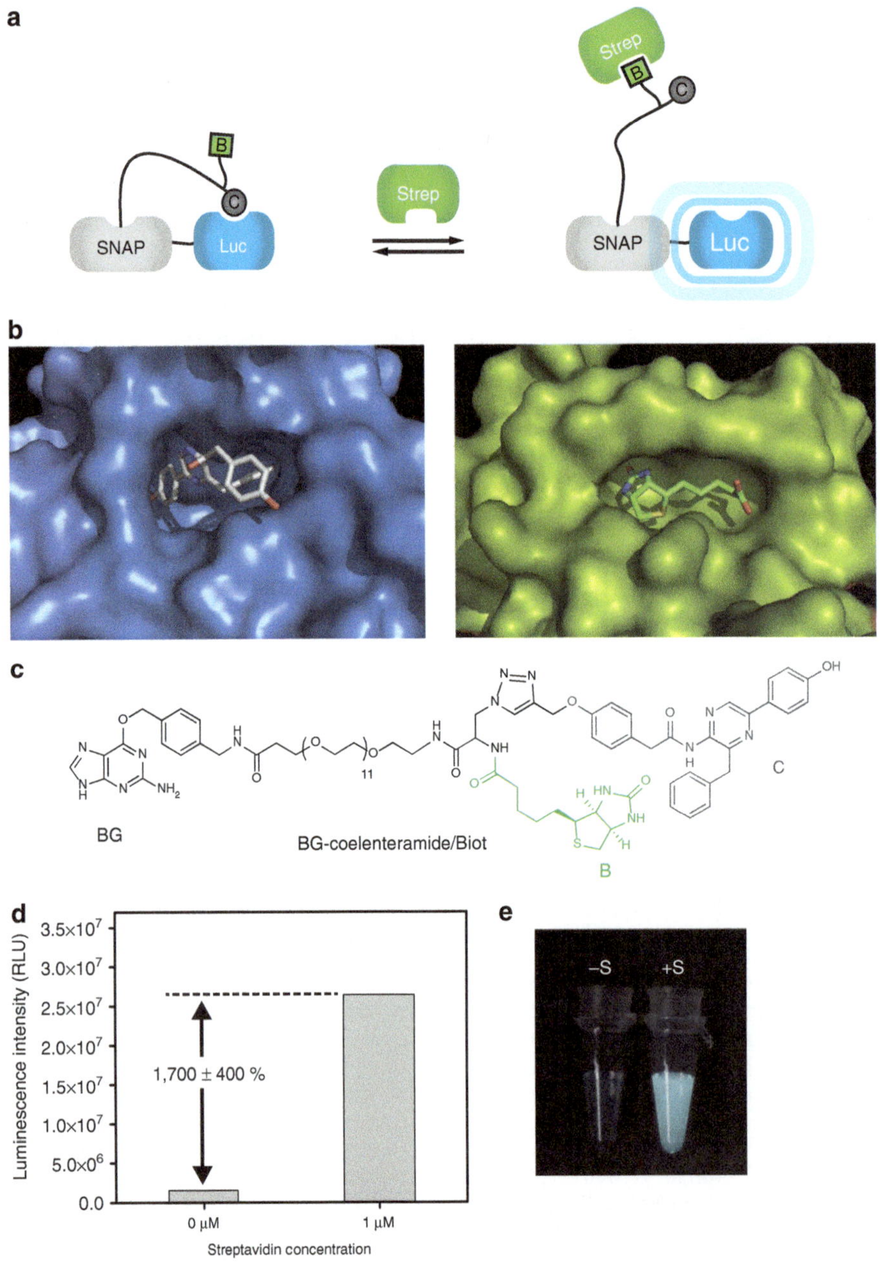

Figure 1 | Control of a luciferase. (**a**) Schematic principle of the regulation of a luciferase by an exogenous effector: the dual ligand coelenteramide/biotin can bind to luciferase (Luc) or streptavidin (Strep) separately but not to both at the same time. Binding of streptavidin to the secondary ligand biotin (B) makes the tethered coelenteramide (C) unbind from the luciferase active site, leading to a change in bioluminescence. (**b**) Crystal structures of the active sites of *Renilla* luciferase with the inhibitor coelenteramide (blue, PDB ID 2PSJ) and of streptavidin bound to biotin (green, PDB ID 3RY2). (**c**) Structure of the dual ligand labelling molecule. (**d**) Luminescence intensity of 100 µl of 25 nM CLASH-Strep/Luc mixed with 2.5 µg ml^{-1} of the substrate coelenterazine in absence and in presence of 1 µM streptavidin. (**e**) Picture of the same solutions in transparent tubes, taken with a Canon 600D camera.

addition of saturating concentrations of streptavidin, the catalytic activity was restored (Fig. 2d). This is in agreement with an inhibition of the HCA activity through the intramolecular benzenesulfonamide in the absence of streptavidin, while on binding of streptavidin to the dual ligand, the benzenesulfonamide is displaced from HCA and the hydrolytic activity is restored. We also tested CLASH-Strep/HCA for BRET efficiency changes at increasing concentrations of streptavidin. We measured an overall BRET ratio change of 640 ± 50% (Fig. 2e,f). The same ratio change was measured when opening CLASH-Strep/HCA by addition of high concentrations of the HCA inhibitor benzenesulfonamide (Supplementary Fig. 1), confirming that the binding of streptavidin to biotin indeed leads to the unbinding of benzenesulfonamide from HCA. When the synthetic ligand BG-Cy3-SA that lacks the biotin moiety was used to label the protein, no response to streptavidin was observed (Fig. 2f).

To demonstrate the modularity of our approach, we took the same HCA fusion protein as above, but used the enzyme dihydrofolate reductase (DHFR) as the effector. Towards this end, we labelled SNAP-PP30-NLuc-HCA with a synthetic ligand that consists of the DHFR-ligand trimethoprim and benzenesulfonamide, generating CLASH-DHFR/HCA (Fig. 3a,b). In the absence of DHFR, the benzenesulfonamide occupies the

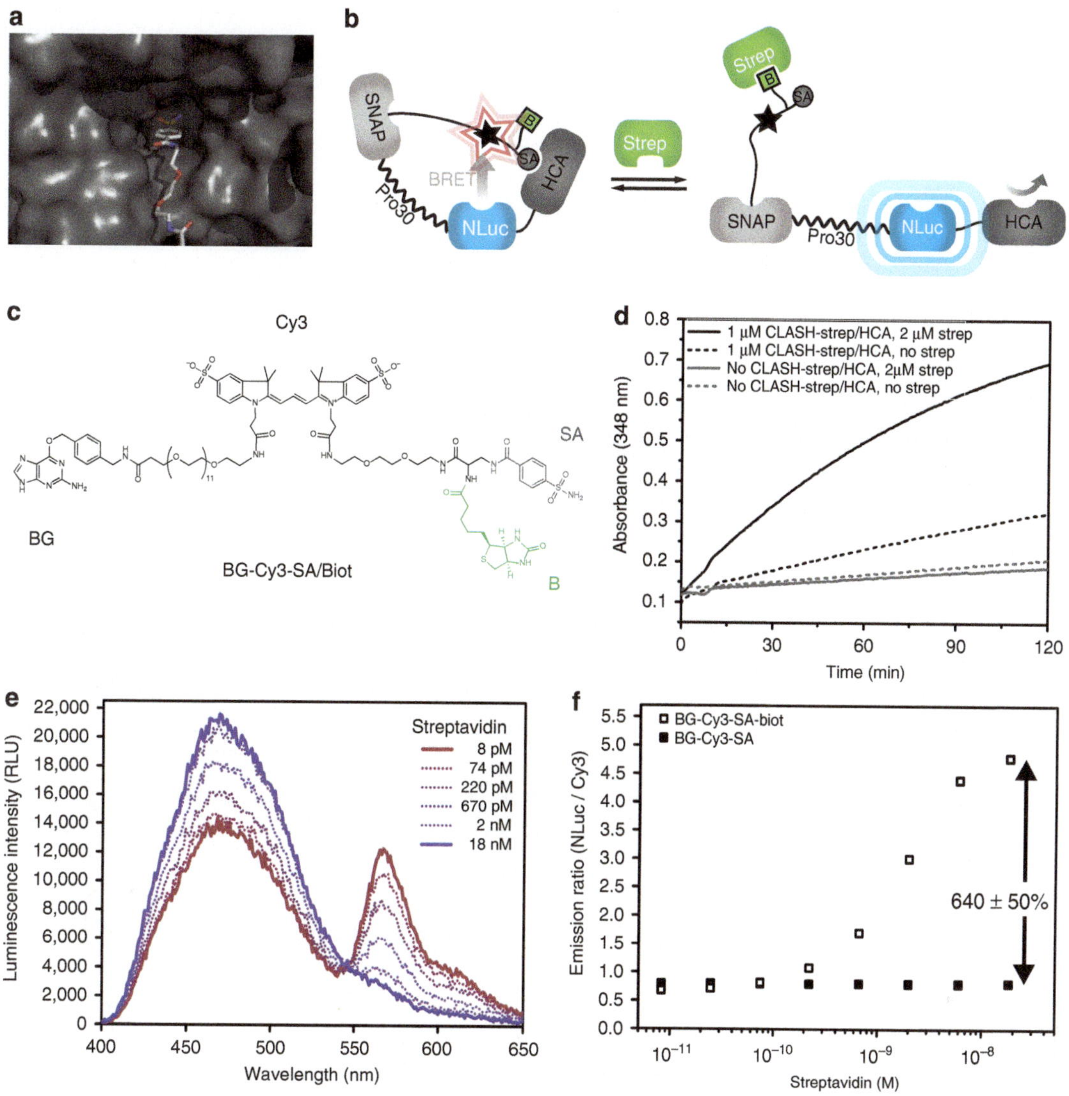

Figure 2 | Control of HCA by streptavidin. (**a**) Crystal structure of the active site of HCA bound to a benzenesulfonamide (PDB ID 1CNW). (**b**) Modulation of HCA activity in CLASH-Strep/HCA. A fusion protein of SNAP-tag, a 30-proline linker, NanoLuc Luciferase (NLuc) and HCA and is labelled with a synthetic molecule containing a fluorophore (red star) and the two intramolecular ligands benzenesulfonamide (SA) and biotin (B). Binding of streptavidin (Strep) to the tethered B displaces SA from HCA, leading both to a conformational change that can be seen by a decrease in BRET efficiency, and to an increase of enzymatic activity. (**c**) Chemical structure of the labelling compound. (**d**) Enzymatic activity of HCA and followed by measuring absorbance at 348 nm. Streptavidin (Strep) acts as an effector of the hydrolytic activity of HCA: in the absence of Strep, SA binds to and inhibits HCA (black dashed line), while on addition of saturation concentrations of Strep, the catalytic activity increases (black solid line). (**e**) Emission spectra of 10 nM CLASH-Strep/HCA at increasing streptavidin concentrations. (**f**) Titration of 10 nM CLASH-Strep/HCA with streptavidin. As a control, the synthetic ligand BG-Cy3-SA lacking the tethered biotin is used.

active site of HCA and CLASH-DHFR/HCA is in a closed conformation, as shown by the high BRET efficiency (Fig. 3c). In the presence of sufficient concentrations of DHFR, the binding of the tethered sulfonamide to the active site of HCA is prevented and CLASH-DHFR/HCA adopts an open conformation, as demonstrated by its low BRET efficiency.

By mixing CLASH-DHFR/HCA with a defined concentration of free DHFR, CLASH-DHFR/HCA can be easily transformed into a biosensor for DHFR ligands such as the anticancer agent methotrexate (MTX), commonly monitored in hospital patients[24]. DHFR brings the sensor molecule into its open conformation and can be displaced by MTX. If the concentration of MTX is lower than the concentration of free DHFR, the drug is quantitatively bound to DHFR, while above this threshold the excess of free MTX disrupts the interaction between DHFR and tethered trimethoprim, thereby bringing CLASH-DHFR/HCA to its high-BRET state. At the stage where all free DHFR is bound to MTX, only a small relative increase in MTX concentration is necessary to displace the DHFR bound to the sensor molecule, giving a sharp transition. The transition can be easily seen with the naked eye and can be shifted by simply changing the concentration of free DHFR (Fig. 3d; Supplementary Fig. 2).

While a similar system could also be obtained using the previously introduced DHFR-based LUCID for MTX[18], the present approach opens up the possibility to construct ratiometric BRET sensors with binding proteins that are geometrically not well suited for LUCID sensors. For example, an antibody can replace the free DHFR and one of the two binding sites of our synthetic ligand can be the antigen. The plethora of commercially available antibodies would allow transforming

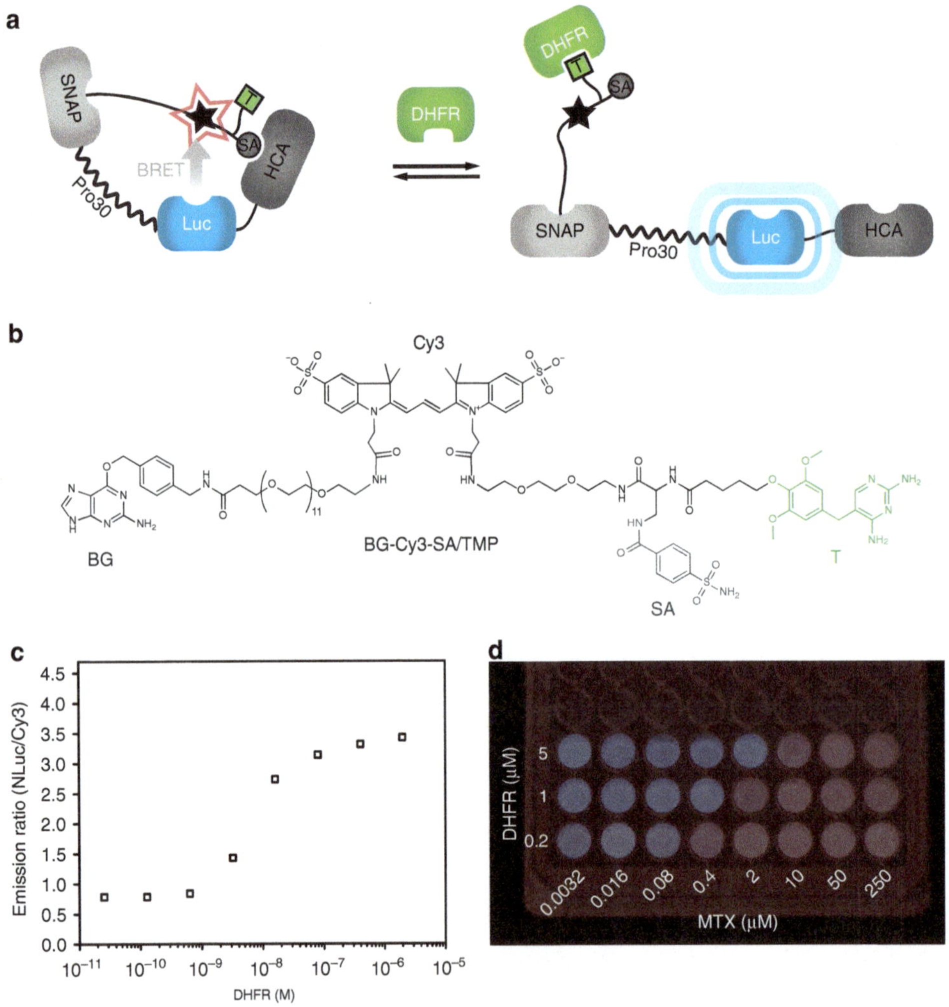

Figure 3 | Modulation of HCA by DHFR and DHFR ligands. Cartoon of CLASH-DHFR/HCA for control of HCA with DHFR: DHFR acts as an effector by binding to the tethered trimethoprim (T) and leading to the unbinding of benzenesulfonamide (SA) from HCA. (**b**) Chemical structure of the labelling molecule. (**c**) The opening of the biosensor can be detected by the increase in emission ratio on addition of DHFR. (**d**) Titrations with the anticancer agent methotrexate (MTX) at three different free DHFR concentrations show that the colour of the luminescence emission changes at the threshold concentration. Picture taken with a Canon 600D camera. The colour balance was modified to give a clear difference between the blue and red colours.

CLASH-antigen/HCA into biosensors for a multitude of analytes. We demonstrated this principle by titrating CLASH-Strep/HCA with anti-biotin monoclonal antibody. In accordance with the CLASH principle, at increasing concentrations of the antibody the BRET efficiency decreases (Supplementary Fig. 3).

Use of CLASH in live-cell imaging. Labelling of SNAP-tag fusion proteins can be achieved *in vitro* and *in vivo*, including whole animals[25]. While the low permeability of the synthetic ligands we describe here would make intracellular applications difficult, the selectivity of the labelling reaction opens up the possibility to generate effector-regulated proteins on the surface of living cells, either to control protein activities or for sensing applications. To demonstrate this possibility, we aimed at displaying a CLASH-AChE/HCA on HEK293 cells in which the activity of HCA is regulated by compounds targeting acetylcholine esterase (AChE). Controlling the activity of an enzyme such as HCA that is part of a family of closely related enzymes is difficult since specific inhibitors are rarely available. Achieving control with an unrelated compound that does not interact with any other member of such family is of interest for mechanistic studies. As compounds targeting AChE are used as drugs, pesticides or nerve agents, our CLASH-AChE/HCA could in addition be used for biosensing applications. We have previously reported a FRET-based biosensor for detecting such compounds, ACh-SNIFIT[26], which however suffered from relatively small ratio changes[16].

To create CLASH-AChE/HCA, AChE was fused to the N-terminus of an HCA-based SNIFIT (Fig. 4a). We targeted the construct to the outer cell membrane of transiently transfected HEK293 cells. The cells were labelled with a dual ligand containing benzenesulfonamide as primary ligand and edrophonium, an inhibitor of AChE[26], as secondary ligand (Fig. 4b). In the absence of free AChE inhibitors, the stronger interaction of edrophonium with AChE prevents the binding of the primary ligand to HCA, keeping the sensor in its low-FRET state. When the AChE inhibitor tacrine, a drug used in the past to treat Alzheimer's disease, was perfused on living cells expressing CLASH-AChE/HCA, tacrine displaced the secondary ligand from AChE, allowing the benzenesulfonamide to bind HCA and leading to high FRET efficiency of the sensor (Fig. 4c-d). Importantly, the switching between the two states was fully reversible, and the overall measured FRET ratio change was

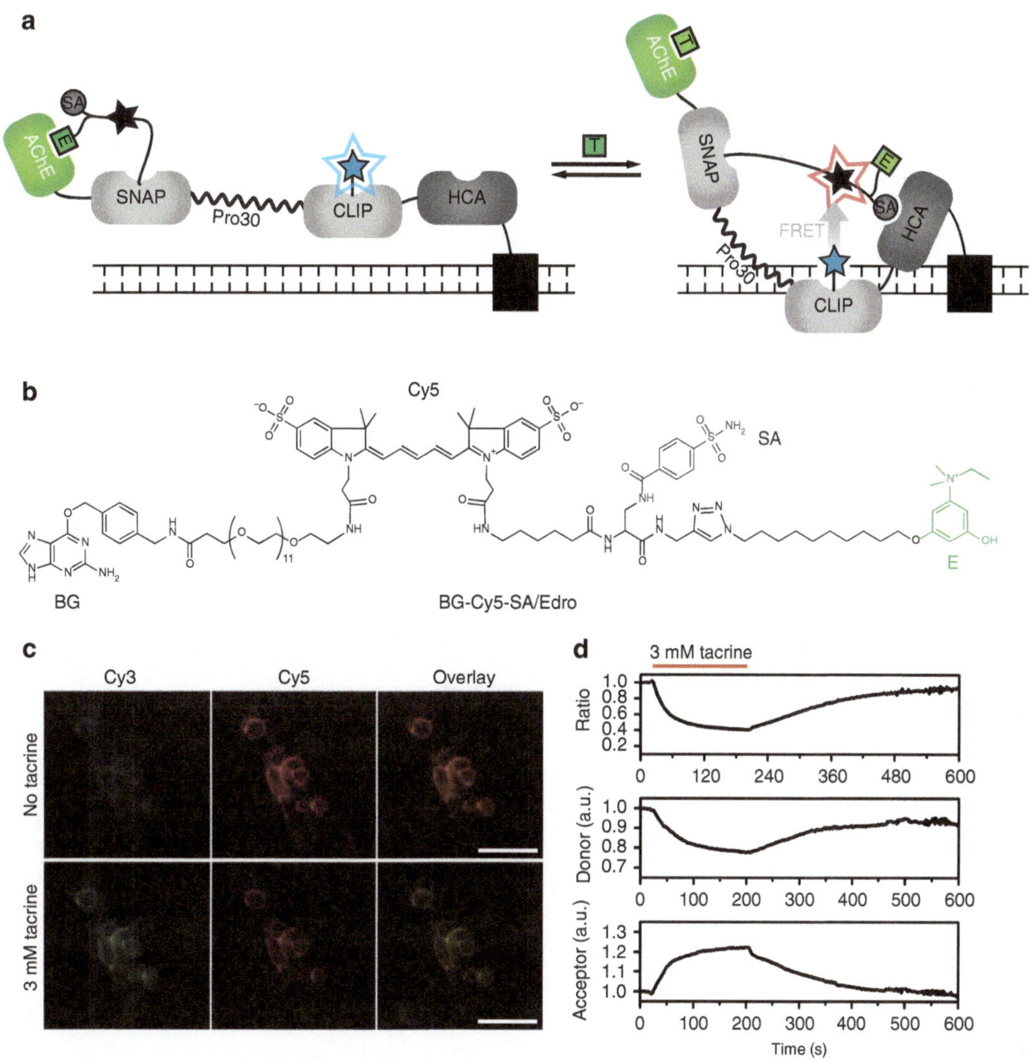

Figure 4 | Modulation of HCA by tacrine on living cells. (**a**) Cartoon of CLASH-AChE/HCA. The acetylcholine esterase (AChE) inhibitor tacrine (T) displaces edrophonium (E) from AChE and benzenesulfonamide (SA) can bind to HCA inducing an increase in FRET efficiency. (**b**) Structure of the labelling compound. (**c**) Fluorescence micrographs of HEK293 cells displaying CLASH-AChE/HCA on their surface. In the absence of tacrine Cy3 emission is more intense, while on addition tacrine Cy5 emission increases; scale bars, 50 μm. (**d**) Changes observed in FRET ratio, donor (Cy3) and acceptor (Cy5) emission on perfusion of HEK293 cells displaying CLASH-AChE/HCA on their surface with 3 mM tacrine.

twofold higher than the original ACh-SNIFIT[26]. In CLASH-AChE/HCA, the relative strength of the two synthetic ligands is such that HCA is active in the absence of the effector targeting AChE and inhibited by its addition (see also Supplementary Discussion for a more detailed discussion of the underlying design principles). Choosing a stronger HCA ligand, it should be equally possible to generate a variant, namely CLASH-HCA/AChE, in which AChE activity is controlled by small benzenesulfonamides. As AChE is a key enzyme for neurotransmission both at the central and at the peripheral nervous systems, such a CLASH-HCA/AChE could find applications as a research tool in neurobiology. For example, inactivating its AChE activity by the addition of a biorthogonal small molecule, for example, a sulfonamide, could help in elucidating the role of cholinergic pathways in neurobiology.

Discussion

In conclusion, CLASH represents a general concept to control protein function in response to an unrelated effector. The approach allowed us to transform a regular luciferase in a bioluminogenic probe responding with marked increase in emitted light on addition of a regulator. We applied the same principle to control the activity of the enzyme HCA by three completely different ligands, two proteins and a small molecule. The possibility to modulate protein activities on the surface of living cells opens up applications beyond *in vitro* assays. While the feasibility of our approach was demonstrated for a number of unrelated proteins, we would like to reiterate that it is restricted to proteins for which ligands with mutually exclusive binding sites exist or can be designed. Finally, this work highlights how the combination of synthetic chemistry and protein engineering can be exploited to create proteins with entirely new functionalities.

Methods

Synthesis of the labelling compounds. Detailed procedures for the synthesis of all compounds and SNAP-tag substrates, their characterization, the titration in multiwell plates and the imaging experiments are given in the Supplementary Methods.

Cloning and protein expression. All sensor constructs were obtained from the pET51b(+)-based construct SNAP-PP30-NLuc-HCA described in ref. 18 or from pDisplay-based construct SNAP-PP30-CLIP-HCA[16] using standard cloning techniques. The amino-acid sequences of the described proteins can be found in Supplementary Note 1. The proteins were expressed in the *Escherichia coli* strain Rosetta-gami (DE3). Bacterial cultures in LB or 2 × YT medium were grown at 37 °C to an $OD_{600\,nm}$ of 0.8, at which point the temperature was lowered to 16 °C, and 0.5 mM isopropyl β-D-thiogalactopyranoside was added. After 16 h, the cells were harvested by centrifugation and lysed by sonication. The cell extracts were cleared by centrifugation and purified in two steps using Ni-NTA (Qiagen) and Strep-Tactin (IBA) according to the suppliers' instructions. The proteins were then transferred into HEPES buffer (50 mM HEPES, pH 7.2, 50 mM NaCl) using Amicon Ultra-0.5 Ultracel-30K centrifugal filters (Millipore). Solutions with proteins containing HCA were supplemented with a twofold molar excess of $ZnCl_2$.

SNAP-tag labelling. The sensor proteins were diluted to 1 µM in HEPES buffer containing 4 µM of the corresponding SNAP-tag substrate. After incubation at room temperature for 1 h, the labelled sensor constructs were purified using Amicon Ultra-0.5 Ultracel-30K centrifugal filters (Millipore), washing three times with 0.4 ml HEPES buffer.

Luciferase activity assay. The labelled protein SNAP-hRLuc was diluted to a concentration of 50 nM in HEPES buffer with or without 2 µM streptavidin (New England Biolabs). One hundred microlitres were added to the wells of a white nonbinding 96-well plate (Greiner Bio-One). The luciferase activity was then measured on an EnVision Multilabel Reader (PerkinElmer): 100 µl of 5 mg l^{-1} benzyl coelenterazine (NanoLight Technology) was added to the wells using the system's injector and the luminescence intensity was measured after 5 s. Photographs were taken with a Canon 600D digital camera.

HCA activity assay. HCA hydrolytic activity assays were performed using the standard *p*-nitrophenyl acetate assay[23]: 1 µM of the protein was mixed with 1 mM *p*-nitrophenyl acetate in HEPES buffer. The rates of the HCA catalysed reactions were measured by following the absorbance at 348 nm characteristic of the reaction product using a SPECTRAmax 340 spectrophotometer (Molecular Devices).

BRET assays. The labelled proteins were diluted to concentrations of 20 nM in 50 µl HEPES buffer containing 0.5 mg ml^{-1} bovine serum albumin and spiked with known concentrations of analyte in white nonbinding 96-well plates (Greiner Bio-One). After incubation at room temperature for 15–30 min, 50 µl Nano-Glo Luciferase Assay Substrate (Promega) diluted 50-fold in Nano-Glo Luciferase Assay Buffer (Promega) was added. Bioluminescence was measured on an EnVision Multilabel Reader (PerkinElmer). The signal was collected using an emission filter for Umbelliferone (wavelength: 460 nm and bandwidth: 25 nm) to record NanoLuc emission and a filter for Cy3 (wavelength: 595 nm and bandwidth: 60 nm) to record Cy3 emission. Emission spectra were measured on an Infinite M1000 spectrofluorometer (Tecan) with a step size of 1 nm, a bandwidth of 10 nm and an integration time of 100 ms. Assays involving DHFR were performed in the presence of 50 or 100 µM NADPH. Biotin monoclonal antibody (BTN.4) was purchased from Fischer Scientific (THERMO SCIENTIFIC PIERCE AB MA5-11251).

Live-cell FRET imaging. HEK293 cells were grown on polylysine-coated glass coverslips (Ø 15 mm) in DMEM Glutamax medium (Lonza) supplemented with 10% fetal bovine serum (Lonza) and transiently transfected with the appropriate plasmid using Lipofectamine (Invitrogen) according to the manufacturer's protocol. Twenty-four hours after transfection, HEK293 cells were labelled with a solution of 2 µM of the corresponding BG derivative and 10 µM of the corresponding O2-benzylcytosine derivative in Hank's buffered salt solution complemented with 10 mg ml^{-1} bovine serum albumin for 10 min at room temperature. After being labelled, cells were washed four times with Hank's buffered salt solution. Glass coverslips with labelled HEK293 cells were transferred to a Warner imaging chamber (RC-20). Perfusion of the chamber was performed gravity fed at a flow rate of 0.5 ml min^{-1}. Time-course experiments of sensor imaging were performed using a Leica LAS AF 7000 widefield microscope equipped with a 40 × plan Apochromat 1.25 numerical aperture oil immersion objective lens. A xenon arc lamp was used for imaging of the HEK293 cells. For each frame, the two channels (donor and FRET) were measured consecutively with an interval of 30 ms between the two emission channels. The following filter sets were used for the FRET ratio imaging: for Cy3/Cy5, excitation at 530 nm (bandwidth 35 nm), emission at 580 nm (bandwidth 40 nm) (Cy3) and at 700 nm (bandwidth 72 nm) (Cy5). If not indicated otherwise, the image size was 293 × 293 µM, and an average of five cells per image was analysed for the intensity ratio plots.

References

1. Buskirk, A. R. & Liu, D. R. Creating small-molecule-dependent switches to modulate biological functions. *Chem. Biol.* **12,** 151–161 (2005).
2. Dueber, J. E., Yeh, B. J., Bhattacharyya, R. P. & Lim, W. A. Rewiring cell signaling: the logic and plasticity of eukaryotic protein circuitry. *Curr. Opin. Struct. Biol.* **14,** 690–699 (2004).
3. Golynskiy, M. V., Koay, M. S., Vinkenborg, J. L. & Merkx, M. Engineering protein switches: sensors, regulators, and spare parts for biology and biotechnology. *Chembiochem* **12,** 353–361 (2011).
4. Lemke, E. A. & Schultz, C. Principles for designing fluorescent sensors and reporters. *Nat. Chem. Biol* **7,** 480–483 (2011).
5. Miyawaki, A. *et al.* Fluorescent indicators for Ca^{2+} based on green fluorescent proteins and calmodulin. *Nature* **388,** 882–887 (1997).
6. Ostermeier, M. Engineering allosteric protein switches by domain insertion. *Protein Eng. Des. Sel.* **18,** 359–364 (2005).
7. Ostermeier, M. Designing switchable enzymes. *Curr. Opin. Chem. Biol.* **19,** 442–448 (2009).
8. Karginov, A. V., Ding, F., Kota, P., Dokholyan, N. V. & Hahn, K. M. Engineered allosteric activation of kinases in living cells. *Nat. Biotechnol.* **28,** 743–747 (2010).
9. Saghatelian, A., Guckian, K. M., Thayer, D. A. & Ghadiri, M. R. DNA detection and signal amplification via an engineered allosteric enzyme. *J. Am. Chem. Soc.* **125,** 344–345 (2003).
10. Gianneschi, N. C. & Ghadiri, R. M. Design of molecular logic devices based on a programmable dna-regulated semisynthetic enzyme. *Angew. Chem. Int. Ed.* **46,** 3955–3958 (2007).
11. Legendre, D., Soumillion, P. & Fastrez, J. Engineering a regulatable enzyme for homogeneous immunoassays. *Nat. Biotechnol.* **17,** 67–72 (1999).
12. Rubenstein, K. E., Schneider, R. S. & Ullman, E. F. "Homogeneous" enzyme immunoassay. A new immunochemical technique. *Biochem. Biophys. Res. Commun.* **47,** 846–851 (1972).
13. Wu, A. H. B. A selected history and future of immunoassay development and applications in clinical chemistry. *Clin. Chim. Acta* **369,** 119–124 (2006).
14. Brun, M. A., Tan, K. T., Nakata, E., Hinner, M. J. & Johnsson, K. Semisynthetic fluorescent sensor proteins based on self-labeling protein tags. *J. Am. Chem. Soc.* **131,** 5873–5884 (2009).
15. Reymond, L. *et al.* Visualizing biochemical activities in living cells through chemistry. *Chimia. (Aarau).* **65,** 868–871 (2011).
16. Brun, M. A. *et al.* Semisynthesis of fluorescent metabolite sensors on cell surfaces. *J. Am. Chem. Soc.* **133,** 16235–16242 (2011).
17. Masharina, A., Reymond, L., Maurel, D., Umezawa, K. & Johnsson, K. A fluorescent sensor for GABA and synthetic GABA(B) receptor ligands. *J. Am. Chem. Soc.* **134,** 19026–19034 (2012).
18. Griss, R. *et al.* Bioluminescent sensor proteins for point-of-care therapeutic drug monitoring. *Nat. Chem. Biol.* **10,** 598–603 (2014).
19. Saito, K. *et al.* Luminescent proteins for high-speed single-cell and whole-body imaging. *Nat. Commun.* **3,** 1262 (2012).
20. Contag, C. H. & Bachmann, M. H. Advances in *in vivo* bioluminescence imaging of gene expression. *Annu. Rev. Biomed. Eng.* **4,** 235–260 (2002).
21. Matthews, J. C., Hori, K. & Cormier, M. J. Substrate and substrate analogue binding properties of Renilla luciferase. *Biochemistry* **16,** 5217–5220 (1977).
22. Keppler, A. *et al.* A general method for the covalent labelling of fusion proteins with small molecules in vivo. *Nat. Biotechnol.* **21,** 86–89 (2003).
23. Verpoorte, J. A., Mehta, S. & Edsall, J. T. Esterase activities of human carbonic anhydrases B and C. *J. Biol. Chem.* **242,** 4221–4229 (1967).
24. Schulz, M. & Schmoldt, A. Therapeutic and toxic blood concentrations of more than 800 drugs and other xenobiotics. *Pharmazie* **58,** 447–474 (2003).
25. Bojkowska, K. *et al.* Liver-specific ablation of Krüppel-associated box-associated protein 1 in mice leads to male-predominant hepatosteatosis and development of liver adenoma. *Hepatology* **56,** 1279–1290 (2012).
26. Schena, A. & Johnsson, K. Sensing acetylcholine and anticholinesterase compounds. *Angew. Chem. Int. Ed.* **53,** 1302–1305 (2014).

Acknowledgements

This work was supported by the Swiss National Science Foundation and the National Centre of Competence in Research (NCCR) in Chemical Biology, Lausanne, Switzerland.

Author contributions

A.S., R.G. and K.J. designed the experiments; A.S. and R.G. performed the experiments; and A.S. performed the chemical synthesis and characterization of the synthetic ligands. All of the authors contributed to the writing of the manuscript.

Additional information

Competing financial interests: The authors declare competing financial interests. A.S., R.G. and K.J. have filed the following patent application: Patent Cooperation Treaty No. PCT/EP2013/073078 Title 'SENSOR MOLECULES AND USES THEREOF'.

A biosynthetic model of cytochrome *c* oxidase as an electrocatalyst for oxygen reduction

Sohini Mukherjee[1], Arnab Mukherjee[2], Ambika Bhagi-Damodaran[2], Manjistha Mukherjee[1], Yi Lu[2] & Abhishek Dey[1]

Creating an artificial functional mimic of the mitochondrial enzyme cytochrome *c* oxidase (C*c*O) has been a long-term goal of the scientific community as such a mimic will not only add to our fundamental understanding of how C*c*O works but may also pave the way for efficient electrocatalysts for oxygen reduction in hydrogen/oxygen fuel cells. Here we develop an electrocatalyst for reducing oxygen to water under ambient conditions. We use site-directed mutants of myoglobin, where both the distal Cu and the redox-active tyrosine residue present in C*c*O are modelled. *In situ* Raman spectroscopy shows that this catalyst features very fast electron transfer rates, facile oxygen binding and O-O bond lysis. An electron transfer shunt from the electrode circumvents the slow dissociation of a ferric hydroxide species, which slows down native C*c*O (bovine 500 s^{-1}), allowing electrocatalytic oxygen reduction rates of 5,000 s^{-1} for these biosynthetic models.

[1] Department of Inorganic Chemistry, Indian Association for the Cultivation of Science, 2A&2B Raja SC Mullick Road, Jadavpur Kolkata 700032, India.
[2] Department of Chemistry, University of Illinois at Urbana-Champaign, Champaign, Illinois 61801, USA. Correspondence and requests for materials should be addressed to A.D. (email: icad@iacs.res.in).

Mimicking the sophistication of naturally occurring enzymes has been a long-term goal of the scientific community. An artificial analogue that can perform equally well as its natural predecessor will not only provide deeper understanding of the native enzymes, but also enable the development of efficient artificial catalysts. For several decades now chemists have embarked on this daunting pursuit of emulating the efficiency and selectivity of naturally occurring enzymes and several important milestones have been achieved. Efforts from synthetic inorganic chemists have resulted in synthetic models of myoglobin (Mb), galactose oxidase, tyrosinase, cytochrome P450 and cytochrome *c* oxidase (*Cc*O)[1–5]. Alternatively, there has been fervent pursuit of biochemical constructs inspired by natural metalloenzymes. A series of binuclear non-haem iron, cytochrome *c*, haem oxidases and iron–sulfur enzyme models have resulted from such efforts[4,6–12]. While none of the synthetic or biochemical models reported so far could match the reactivity exhibited by their natural counterparts, fundamental insights regarding the structure–function correlations of several metalloenzymes have been gained in the process[7,13–15]. In addition, key information about the secondary coordination sphere interactions present in the protein-active site, which play a dominating role in determining the electronic structure and reactivity of these metalloenzymes, have been identified[16,17].

In a biosynthetic approach, stable naturally occurring proteins have been used as scaffolds for creating mimics of several metalloenzymes, such as hydrogenases which are involved in the reversible generation of H_2 from water, haem proteins participating in electron transfer and O_2-binding, non-haem iron and copper enzymes active in small molecule activation, and even novel enzymes containing non-native cofactors[18–24]. For example, using this approach, biosynthetic models that structurally and functionally mimic *Cc*O and nitric oxide reductase have been reported[7,25]. Despite decades of focused effort, however, biosynthetic models with catalytic efficiencies approaching those of the naturally occurring metalloenzymes have remained elusive[26–28]. In this report, we communicate a biosynthetic model of *Cc*O bearing the distal Cu_B and a tyrosine residue that is kinetically more competent in reducing O_2 electrochemically than any known synthetic analogue, as well as native *Cc*O itself.

X-ray crystallography of Mb and its mutant have revealed that its two propionate side chains project out of the protein surface into the solvent (Fig. 1)[7]. Taking advantage of this structural feature, we have previously developed an electrocatalytic O_2 reduction system where the native haem cofactor in Mb is replaced by a modified hemin cofactor bearing an alkyne group (Hemin-yne, Fig. 2) so that electrons can be injected directly into the haem from a gold electrode to facilitate O_2 reduction[29]. This method resulted in a Mb-functionalized electrode bearing 2.15×10^{-12} mol per cm^2 of protein, which was characterized using several microscopic and spectroscopic techniques[29]. Over the last few years, a biosynthetic model of *Cc*O has been reported in which two distal residues of Mb (L29 and F43) have been mutated to His, which along with the native His64, form a Cu-binding site, mimicking the distal Cu_B-binding site present in *Cc*O (Cu_BMb)[30]. Furthermore, in an attempt to mimic the conserved Tyr 244 residue in the *Cc*O-active site, a G65Y mutant of Cu_BMb (G65YCu_BMb) containing redox-active tyrosine residue in the distal site and a variant where a tyrosine residue was crosslinked to the active site histidines were also created (Fig. 1)[31].

Herein we report the electrocatalytic properties of the G65YCu_BMb (higher synthetic yields than the tyrosine cross-linked variant) immobilized on an Au electrode using the method developed for WT Mb[29].

Results

Electrode characterization by SERRS. Surface-enhanced resonance Raman spectroscopy (SERRS) data (Fig. 3a) of the electrodes bearing the G65YCu_BMb protein with and without Cu_B show the oxidation and spin state marker ν_4, ν_3, ν_2 and ν_{10} bands at 1,375 cm^{-1}, 1,493 cm^{-1},1,585 cm^{-1} and 1,641 cm^{-1}, respectively. The ν_4, ν_3 and ν_2 values are consistent with the presence of a five-coordinated high spin haem in the active site on these electrodes bearing the biochemical constructs of *Cc*O[32]. Also the bands at 1,504 cm^{-1} and ν_4 at 1,641 cm^{-1} suggest the presence of a mixture of six-coordinate low spin species, which likely has H_2O as the axial ligand. The positions of these bands in the G65YCu_BMb and their relative intensities are different from Hemin-yne (Supplementary Fig. 1)[29].

Electrode characterization by X-ray photoelectron spectroscopy. X-ray photoelectron spectroscopic (XPS) data of a G65YCu_BMb-bound Au electrode clearly indicate the presence of Fe, Cu, C, N and O elements (Supplementary Fig. 2, Supplementary Table 1). The $3p_{3/2}$, $2p_{3/2}$ and $2p_{1/2}$ binding energy peak for the Fe^{III} of haem group appear at 56.5 eV, 709.4 eV and 722.4 eV, respectively[33,34]. The $2p_{3/2}$ and $2p_{1/2}$ binding energy peak for Cu^{II} in the distal site appear at 931.7 eV and 951.8 eV, respectively[35]. The N_{1s}

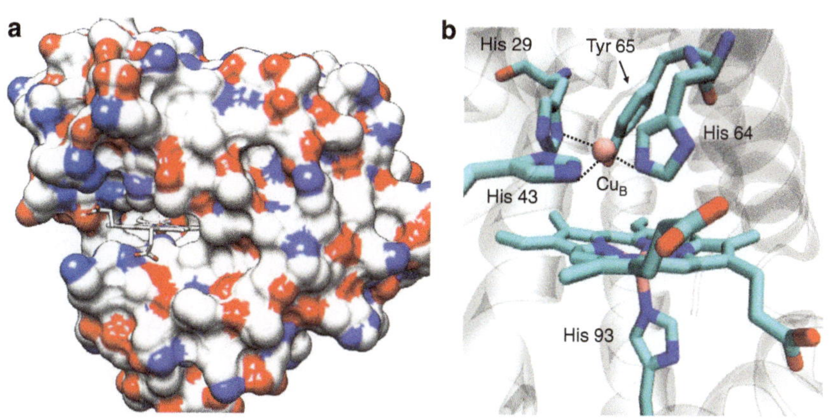

Figure 1 | Crystal structure. Crystal structure of a Mb-based biosynthetic model of *Cc*O, F33Y-Cu_BMb; pdb id: 4FWY. (**a**) The haem cofactor is in a cleft on the molecule protein surface, (colour coded according to the charge of the residues), with the propionate groups exposed to the solvent. (**b**) The computer model of G65YCu_BMb showing its catalytic centre containing the distal Cu_B bound to histidines and a tyrosine 65.

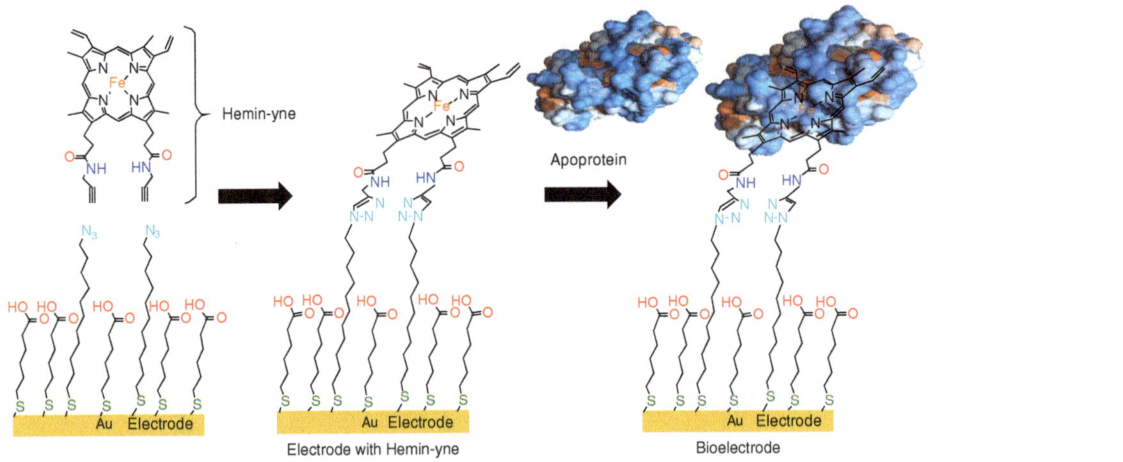

Figure 2 | Construction of the electrode bearing the biosynthetic model. Reconstitution of apoprotein *in situ* with Hemin-yne groups that are covalently attached to mixed self-assembled monolayers of thiols on an Au electrode. The modified hemin is indicated as Hemin-yne.

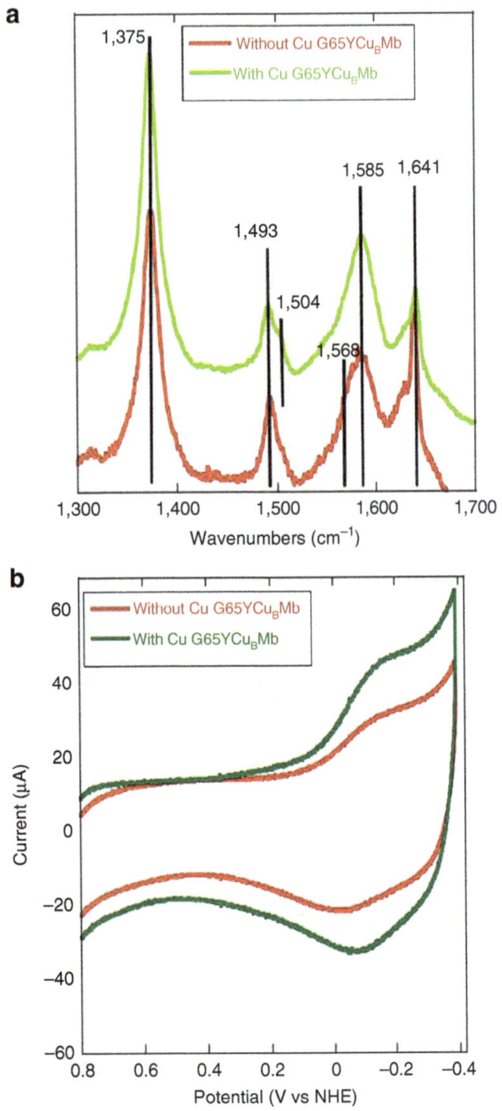

Figure 3 | SERRS and CV data for the electrode fabricated with G65YCu$_B$Mb mutant. (**a**) SERRS spectra of G65YCu$_B$Mb with (green) and without (red) the distal Cu^{2+} in air-saturated 100 mM phosphate buffer (pH 7) solution. (**b**) Anaerobic CV of G65YCu$_B$Mb without Cu$_B$ (red) and after Cu$_B$ binding (green). 2 V s^{-1} scan rate, in degassed, pH 7, 100 mM phosphate buffer using a Pt counter electrode.

peak is broad (Supplementary Fig. 2), as it contains several components due to the presence of amide, haem pyrroles and the triazole groups (resulting from the covalent attachment of Hemin-yne) on the surface[33]. Similarly, the C_{1s} peak (Supplementary Fig. 2) contains contributions from different types of C atoms (aromatic, aliphatic, haem and so on) on these protein-modified surfaces[36].

Electrode characterization by CV. Cyclic voltammetry (CV) of G65YCu$_B$Mb with and without the distal Cu$_B$ immobilized onto the electrodes in degassed buffer show the haem $Fe^{3+/2+}$ mid-point reduction potential ($E^{1/2}$) at −97 mV and −57.5 mV, respectively (Fig. 3b). The peak separation between the cathodic and the anodic peak for both the cases is ~70 mV (ref. 37). Hemin-yne displays the $Fe^{3+/2+}$ reduction potential at −70.0 mV in the absence of a protein and −135.0 mV when bound to wild-type apo Mb[29]. In the case of the G65YCu$_B$Mb protein-bound electrodes, the $Cu^{2+/+}$ process overlaps with the Hemin-yne $Fe^{3+/2+}$ process, resulting in approximately twice the area under these CV peaks relative to the G65YCu$_B$Mb-bearing electrodes prior to Cu$_B$ loading. The integrated area under these CV features in the absence of Cu^{2+} indicates that there are $2.55 \pm 0.05 \times 10^{-12}$ mol of protein per cm^2 of the surface. The ratios of the integrated area under the CV features of G65YCu$_B$Mb functionalized before and after loading the Cu$_B$ is ~1:2 (Table 1, fourth column) which is consistent with the expected 1:1 stoichiometry (that is, every G65YCu$_B$Mb binds one Hemin-yne and one Cu^{2+} ion). Note that the $E^{1/2}$ values of the hemin and Cu$_B$ measured for these electrodes are slightly different from those estimated from potentiometric titration in solution[7,19]. This is likely due to the interfacial microenvironment of the -COOH-terminated Self Assembled Monolayer (SAM) which is known to shift the apparent formal potentials of redox-active species in its vicinity[38]. Thus the *in situ* reconstitution of the protein with Hemin-yne on the electrode is evident from the SERRS, XPS (Supplementary Fig. 2) and CV data (Supplementary Fig. 3, Supplementary Fig. 4). The presence of the Cu^{2+} at Cu$_B$ site on the electrode is indicated by XPS and CV data. Taken together, these data indicate the assembly of the G65YCu$_B$Mb, biosynthetic model of C*c*O covalently attached to the electrode via the linkage between the Hemin-yne and the azide terminated thiols created using click reaction (Fig. 2).

O_2 reduction reactivity of the electrode. In linear sweep voltammetry experiments performed in aerated buffers, large electrocatalytic O_2 reduction currents are observed by the G65YCu$_B$Mb (with and without Cu^{2+}) bearing bio-electrodes at pH 7 at room temperature, as the applied potential is lowered below +100 mV versus NHE (Fig. 4a, Supplementary Fig. 5). Thus as the potential of the electrode is lowered such that the iron in these proteins is reduced to Fe^{II}, an electrocatalytic O_2 reduction current is observed. It is important to note that the potential of O_2 reduction reaction (ORR) (E_{ORR}) is −263 mV, which is more negative than the $E^{1/2}$(−97 mV), suggesting that the potential determining step of ORR is not the reduction of resting Fe^{III} to Fe^{II} but the reduction of a different species with −166 mV more negative potential (Fig. 4a inset). A more negative E_{ORR} relative to $E^{1/2}$ is mechanistically significant (*vide infra*).

In these active sites, O_2 may be reduced by $4e^-$ and $4H^+$ to H_2O, or by fewer electrons to produce partially reduced oxygen species (PROS) like O_2^- and H_2O_2. The extent of $4e^-$ reduction and the second order rate constant (k_{ORR}) of the ORR can be determined using rotating disc electrochemistry (RDE) where the catalytic O_2 reduction current increases with increasing rotation rates (Fig. 4a,b) following the Kouteky–Levich (K–L) equation (equation (1))[39].

$$i_{cat}^{-1} = i_K(E)^{-1} + i_L^{-1} \quad (1)$$

where, $i_K(E)$ is the potential-dependent kinetic current and i_L is the Levich current. i_L is expressed as

$$i_L = 0.62nFA[O_2](D_{O2})^{2/3}\omega^{1/2}\nu^{-1/6} \quad (2)$$

where n is the number of electrons transferred to the substrate, A is the macroscopic area of the disc (0.096 cm^2), $[O_2]$ is the concentration of O_2 in an air-saturated buffer (0.26 mM) at 25 °C,

Table 1 | $E^{1/2}$ and coverage.

Protein	Metal-binding sites	$E^{1/2}$ (mV)	Integrated coverage (mol cm^{-2})
Hemin-yne reconstituted myoglobin[29]	Fe	−135.0 ± 5.0	$2.15 \pm 0.05 \times 10^{-12}$
Hemin-yne reconstituted G65YCu$_B$Mb	Fe	−57.5 ± 5.0	$2.55 \pm 0.05 \times 10^{-12}$
	Cu$_B$, Fe	−97.0 ± 5.0	$4.65 \pm 0.05 \times 10^{-12}$

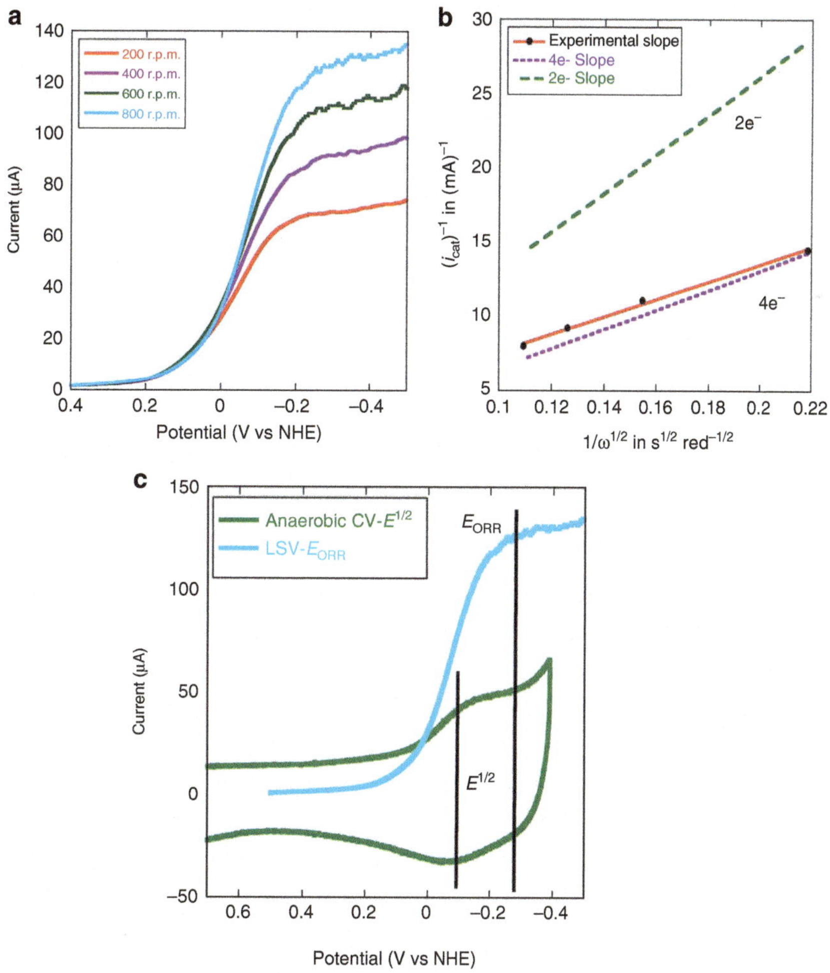

Figure 4 | RDE data for electrode modified with G65YCu$_B$Mb with Cu^{2+}. (**a**) Linear sweep voltammogram of G65YCu$_B$Mb with Cu^{2+} in air-saturated pH 7, 100 mM phosphate buffer solution at 100 mV s^{-1} scan rate, potentials are scaled relative to NHE and a Pt counter electrode is used. Data are collected at different rotation speeds (200 r.p.m.—red, 400 r.p.m.—purple, 600 r.p.m.—green, 800 r.p.m.—blue). (**b**) Plot of i_{cat}^{-1} for G65YCu$_B$Mb-bearing bioelectrode at −300 mV potential and at multiple rotation rates, with the inverse square root of the angular rotation rate ($\omega^{-1/2}$) (**c**) Difference between the potential for O_2 reduction (E_{ORR}) and the midpoint reduction potential of $Fe^{3+/2+}$ redox couple ($E^{1/2}$).

D_{O2} is the diffusion coefficient of O_2 ($1.8 \times 10^{-5}\,cm^2\,s^{-1}$) at 25 °C, ω is the angular velocity of the disc and v is the kinematic viscosity of the solution ($0.009\,cm^2\,s^{-1}$) at 25 °C (ref. 40).

Plot of i_{cat}^{-1} at multiple rotation rates with the inverse square root of the angular rotation rate (i_{cat}^{-1}) for G65YCu$_B$Mb (with Cu^{2+}) (Fig. 4b) is linear. The slope of K–L plot is expressed as $1/[n\{0.62FA(D_{O2})^{2/3}v^{-1/6}\}]$, which can be used to experimentally estimate the value of n where n is the number of electrons donated to the substrate, that is, O_2. The slope obtained from the experimental data for G65YCu$_B$Mb (Fig. 4b) is close to the theoretical slope (Fig. 4b, dotted purple line) expected for a $4e^-$ process and very different from the slope for a $2e^-$ process (Fig. 4b, dotted green line). Thus the G65YCu$_B$Mb bioelectrode predominantly catalyses a $4e^-/4H^+$ reduction of O_2 to H_2O at pH 7.

The intercept of the K–L plot is the inverse of the kinetic current ($i_K(E)^{-1}$), where $i_K(E)$ is expressed as[41]

$$i_K(E) = k_{ORR}nFA[O_2]\Gamma_{cat} \quad (3)$$

where, n is the number of electrons, A is the geometric surface area, $[O_2]$ is the bulk concentration of O_2, Γ_{cat} is the surface coverage of the catalyst (obtained from the integration of the anaerobic CV data) and k_{ORR} is the second order rate constant for O_2 reduction estimated at -300 mV. At this potential, in an oxygenated buffer, the G65YCu$_B$Mb catalyst is involved in substrate diffusion-limited ORR. Using this equation (equation (3)) and the experimentally obtained $i_K(E)$ at -300 mV, the second order rate constant for O_2 reduction G65YCu$_B$Mb is evaluated to be $1.98 \times 10^7\,M^{-1}\,s^{-1}$ (Table 2). The pseudo first order rate can be determined (Table 2) from the second order rate by taking into account the substrate, O_2 concentration under these experimental conditions to be 0.26 mM (Supplementary Fig. 6). The catalytic ORR rate by G65YCu$_B$Mb surpasses those reported for the best artificial synthetic analogues (Table 2).

The G65YCu$_B$Mb biosynthetic Mb scaffold-based bio-electrode for O_2 reduction is remarkably stable. Monolayers bearing covalently attached O_2-reducing electrocatalysts reported so far have never been stable enough to allow these dynamic electrochemical experiments to determine the kinetic parameters (k_{ORR}, number of electrons and so on). Enzymes like laccases, directly attached to chemically modified graphite electrodes, were found to be stable enough to be investigated with these hydrodynamic techniques[27]. The failure to perform these experiments has been attributed to degradation of the catalyst during RDE experiment, that is, very small turnover numbers presumably due to the production of PROS during ORR. Rotating ring disc electrochemistry (RRDE) shows formation of only ~6% PROS by the G65YCu$_B$Mb (Supplementary Fig. 7) during ORR, indicating that it reduces 96% of O_2 to H_2O consistent with the RDE data. During the RDE experiments (Fig. 4a), the G65YCu$_B$Mb-functionalized electrodes bearing 10^{-12} mol of the catalyst reduced $1.8 \pm 0.3 \times 10^{-8}$ mol of O_2 ($7 \pm 1 \times 10^{-3}$ C total charge and $4e^-$ per O_2 molecule) yielding a turnover number of at least 10^4. The role of PROS in degrading the catalyst is established by the fact that the electrolytic current (at -0.3 V) remains stable in the presence of 50 μM catalase in solution (Supplementary Fig. 8).

Table 2 | k_{ORR} of different ORR catalysts.

ORR catalysts	Metals	k_{ORR}	PROS (%)
G65YCu$_B$Mb	Cu$_B$, Haem	$1.98 \times 10^7\,M^{-1}s^{-1}$ or $5{,}148\,s^{-1}$	$\sim 6 \pm 1$
Synthetic model[60]	Cu$_B$, Haem	$1.2 \times 10^5\,M^{-1}s^{-1}$	$\sim 10 \pm 1$

ORR, O_2 reduction reaction; PROS, partially reduced oxygen species.
k_{ORR} determined at -0.4 V versus NHE; PROS determined at -0.12 V versus NHE in both cases.

Discussion

To understand the facile and selective O_2 reduction catalysed by the G65YCu$_B$Mb biochemical model, the recently developed SERRS-RDE technique is employed[42]. In this technique, the rR spectra of the catalyst (that is, G65YCu$_B$Mb) bearing electrode is collected while the system is involved in steady state O_2 reduction and the species accumulated in the steady state can be identified. For any species to accumulate in steady state, its rate of formation has to be greater than its rate of decay. Thus, while the species preceding the rate-determining step (rds) will accumulate at steady state, the accumulation of a species in steady state does not immediately imply its decay as the rds. In the absence of O_2, a high spin ferrous species is formed, characterized by a ν_4 and ν_3 vibrations at $1{,}357\,cm^{-1}$ and $1{,}473\,cm^{-1}$ (Fig. 5a, cyan), respectively, when a cathodic potential of -0.4 V is applied signifying reduction of the resting ferric state (Fig. 5a, red) to the active ferrous state at these potentials. However, when the same reducing potential is applied in an oxygenated buffer the SERRS-RDE data (Supplementary Fig. 9) clearly show the presence of different species during electrocatalytic ORR, which leads to broadening of the ν_4, ν_3 and ν_2 regions (Fig. 5a, green) relative to the oxidized and reduced states (Fig. 5a, red and cyan). In particular, the ν_3 and ν_2 vibrations discernibly shift to higher energies as indicated by clear increase in intensities at $1{,}508\,cm^{-1}$ and $1{,}591\,cm^{-1}$ (Fig. 4b), suggesting the accumulation of $Fe^{IV}{=}O$ species during steady state ORR[43]. Signals from high spin ferrous, resting high spin ferric, low spin ferric and ferryl species with ν_3 at $1{,}473\,cm^{-1}$ (Fig. 5a, cyan; and Fig. 5b, brown), $1{,}493\,cm^{-1}$ (Fig. 5a, red and Fig. 5b, dashed green), $1{,}504\,cm^{-1}$ (Fig. 5a, cyan and Fig. 5b, cyan) and $1{,}508\,cm^{-1}$ (Fig. 5a, green and Fig. 5b, green), respectively, could be convoluted by fitting the ν_3 region of the spectrum. The lack of significant signal from the high spin ferrous species (weak ν_3 at $1{,}473\,cm^{-1}$) suggests that O_2 binding to these species is facile in the steady state. These mutants use the basic design of Mb which has a very fast O_2-binding rate ($10^7\,M^{-1}s^{-1}$) (refs 44,45). This rate is indeed ~10 times faster than O_2 binding to the haem a$_3$ site of CcO (ref. 46). Similarly, the very weak intensity of the high spin (HS) Fe^{III} species indicates that the ET to Fe^{III} resting state is very facile at these potentials as may be expected due to direct attachment of the Hemin-yne to the electrode. The significant intensity of ferryl species (ν_2 at $1{,}591\,cm^{-1}$ and ν_3 at $1{,}508\,cm^{-1}$) entails the O–O bond cleavage leading to its formation to be faster than its decay via reduction under steady state. Thus the reduction of the resting Fe^{III} state, O_2 binding to Fe^{II} are facile in G65YCu$_B$Mb under the reaction conditions. The low spin ferric species accumulated during steady state ORR could be a dioxygen adduct or peroxide adduct similar to those observed in native CcO and its model systems[46–51]. The low frequency region shows $^{18}O_2$-sensitive bands suggestive of the formation of a low spin ferric peroxide and $Fe(IV){=}O$ (Supplementary Fig. 10).

If one were to conceive of a Gedanken steady state turnover experiment with CcO where the electron transfer to the active site is very fast (that is, in the hypothetical situation where ET from Cyt *c* to CcO is not the rds) as the ET from the electrode to the active site is very fast due to direct attachment of the later to the former, the species that would accumulate during turnover, based on the Babkock–Wikström mechanism (Fig. 6, the parameters of native CcO are indicated in purple and G65YCu$_B$Mb are

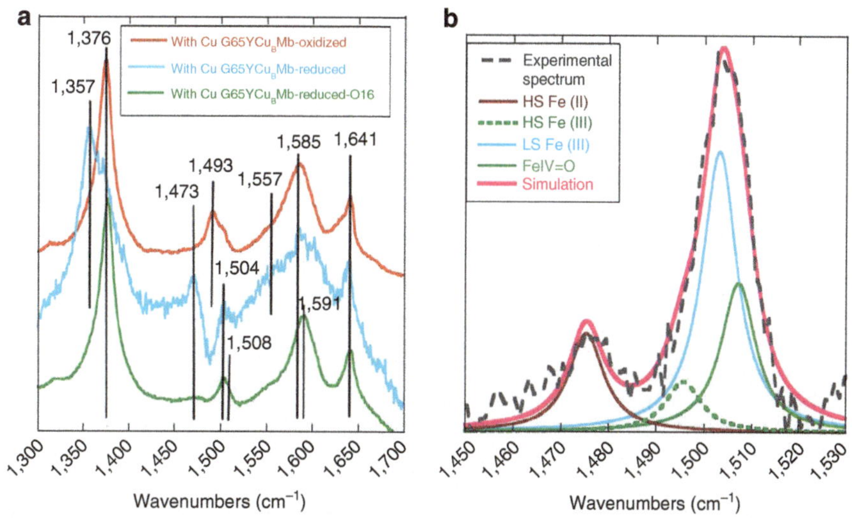

Figure 5 | SERRS-RDE data of G65YCu$_B$Mb-bearing electrode. (**a**) SERRS-RDE data of G65YCu$_B$Mb-bearing electrode at oxidized (applied potential was 0 V with respect to Ag/AgCl reference electrode), reduced (applied potential was − 0.4 V with respect to Ag/AgCl reference electrode) state and in the presence of O_2 (O16) saturated 100 mM pH 7 phosphate buffer and (**b**) Components of the rR spectrum determined by simulating the spectra of G65YCu$_B$Mb-bearing electrode in the presence of O_2 (O16) saturated 100 mM pH 7 phosphate buffer indicated in green in **a**.

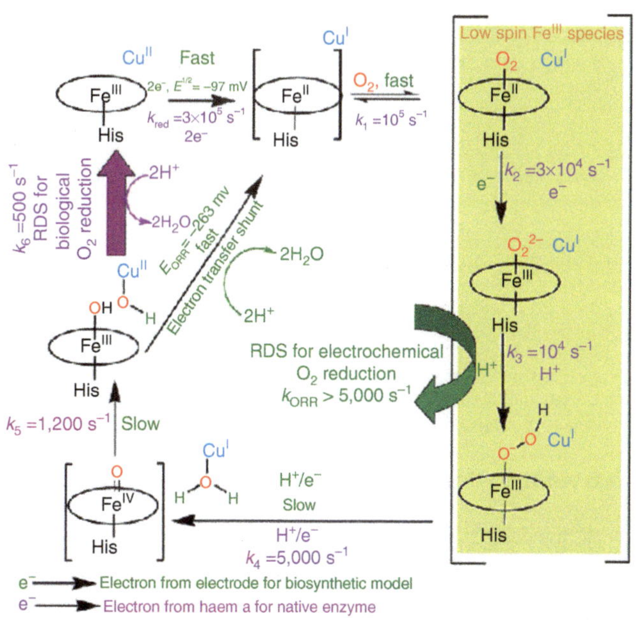

Figure 6 | Mechanisms. Comparison between the mechanism of native C*c*O (ref. 46) in solution and the biosynthetic G65YCu$_B$Mb model on electrode.

indicated in green[52] and the arrows in black indicate general route for O_2 reduction by C*c*O followed in both the native system, as well as in the G65YCu$_B$Mb-immobilized electrode) are the Fe^{II},Fe^{II}-O_2, Fe^{III}-O_2^{2-}, Fe^{III}-OOH, $Fe^{IV}=O$ and Fe^{III}-OH species as the rates of formation of these species are greater than their rates of decay[53,54]. Out of these, the Fe^{II}-O_2, Fe^{III}-O_2^{2-} and Fe^{III}-OOH species will have Raman signatures of low spin haem (Fig. 6), Fe^{III}-OH will have rR signature of high spin haem and the $Fe^{IV}=O$ will have signatures unique to haem ferryl species[46,54,55]. The SERRS-RDE data show the presence of species having signatures of low spin Fe^{III} and $Fe^{IV}=O$. While the later can originate from only a single species, the former can indicate the presence of any of the three species or a combination of them. The lack of significant high spin signal indicates that the biosynthetic model circumvents accumulation of Fe^{III}-OH and resting Fe^{III} species in the steady state by facile ET. The overall rate-limiting step of native C*c*O in solution is the dissociation of hydroxide of the Fe^{III}-OH end product of O_2 reduction from haem to generate the active ferric resting form and has a first order rate constant of 500 s^{-1} (refs 46,54–56). This dissociation is required during turnover as the potential of this hydroxide-bound form is likely to be more negative that the five-coordinate resting oxidized site (which will be regenerated after hydroxide dissociation) and will not be reduced by haem a. While the $E^{1/2}$ of a haem a_3 Fe^{III}-OH species cannot be determined with confidence due to strong interaction potential and co-operativity between the haem a and haem a_3 sites, the potential of formate and azide-bound high spin haem a_3 site (analogous to hydroxide) is ~130 to 200 mV more negative than the resting ferric site[57,58]. The potential determining step of ORR (defined as the electron transfer (ET) step in catalysis having lowest potential) by the G65YCu$_B$Mb is 166 mV more negative than the $E^{1/2}$ for the resting high spin ferric state and is likely to be the reduction of the Fe^{III}-OH species. Thus the bioelectrode can circumvent the kinetic barrier associated with the dissociation of the hydroxide by directly reducing it to ferrous at 166 mV lower potential. This direct electron transfer to the ferric hydroxide species, which is an intermediate in the catalytic cycle of C*c*O, circumventing a slow step in catalysis, is an electron transfer shunt analogous to peroxide shunt in cytochrome P450, which overcomes the rate-determining O_2 activation step of the native enzyme[59]. In the mass transfer controlled region of the catalytic current, all ET steps are facile and steps like O_2 binding, protonation and O–O bond cleavage can be rds at these potential. The lack of HS Fe^{II} accumulation in the SERRS-RDE indicates that O_2 binding to Fe^{II} is very fast and not the rds. The O–O bond cleavage in the Babkock–Wikström mechanism involves ET to the active site and not the rds as well. Thus rds of ORR by these biosynthetic models is likely to be the protonation of the Fe^{III}-O_2^- species with a first order rate of 5,000 s^{-1} (Fig. 6).

In an air-saturated buffer (0.26 mM O_2), the pseudo first order rate constant of ORR by G65YCu$_B$Mb is determined to be ~5–6 × 10^3 s^{-1} (k_{ORR} [O_2]).The highest second order O_2 reduction rate reported for any synthetic mimic of C*c*O is 1.2 × 10^5 $M^{-1}s^{-1}$; that too on a multilayer having 1,000 times more catalyst than the G65YCu$_B$Mb electrodes[60]. The second order rate constant of G65YCu$_B$Mb is $10^7\ M^{-1}\ s^{-1}$ which is, thus, 2 orders of magnitude higher than best synthetic haem/Cu-based O_2 reduction electrocatalyst. Thus the selectivity and kinetic rate of the G65YCu$_B$Mb-bearing electrode surpasses those reported for smaller synthetic analogues and illustrates the advantages of using a biochemical scaffold over a synthetic scaffold. Although the pseudo first order rate constant of the G65YCu$_B$Mb is 10 times faster than the rate of native *Cc*O in solution, such a comparison is vulnerable to differences in reaction conditions (for example, G65YCu$_B$Mb is water soluble but *Cc*O exists in membranes). Alternatively, erstwhile efforts resulting in electrodes bearing native *Cc*O in a manner similar to these bio-electrodes show extremely sluggish O_2 reduction[60–62]. This is due to improper alignment of this membrane-bound protein on the electrode, which precludes efficient electron transfer to the active site[63–65]. However, the direct attachment of haem to the electrode utilizing its solvent-exposed propionate groups (that is, a short circuit) enables fast electron transfer to the active site[29]. This is further supported by the fact that when ethynylferrocene (Fc) is attached to the same surface ~25 mV peak separation is observed even at $5\ V\ s^{-1}$ (Supplementary Fig. 11)[38], suggesting that the ET is indeed fast. As a result when a *Cc*O-functionalized SAM-covered Au electrode produces <1 μA electrochemical O_2 reduction current at −300 mV, this bio-electrode produce ~100 μA current at similar potentials[66].

Finally, the G65YCu$_B$Mb mutant has residues in the distal pocket that can help both electron and proton transfer during O_2 reduction (Y65 in G65YCu$_B$Mb). In *Cc*O, the involvement of Tyr 244 residues in proton/electron transfer during O_2 reduction is now widely accepted[67,68]. Previous biochemical and structural studies on these mutants had indeed indicated the close proximity of this residue to the distal site[7,19,30]. An analogous biochemical model without the Y65 residue, Cu$_B$Mb, is not as stable as the G65YCu$_B$Mb as the former degrades rapidly during the RDE experiments (Supplementary Fig. 12). In summary, a electron transfer shunt which circumvents the rate-determining dissociation of a ferric hydroxide species by directly reducing it at slightly negative potential, fast O_2 binding, fast electron transfer to the active site and the presence of a protective Y65 residue in a biochemical model of *Cc*O results in O_2 reduction activity 100 times faster than the best synthetic models, order of magnitude faster than *Cc*O immobilized on electrode and follows a mechanism comparable to that of native *Cc*O in solution.

Methods

Materials. 1-Azidoundecane-11-thiol and Hemin-yne were synthesized following the reported procedure[29,69]. 6-Mercaptohexanoic acid was purchased from Sigma Aldrich. Di-sodium hydrogenphosphate dihydrate (Na_2HPO_4. $2H_2O$) was purchased from Merck. 2, 6-lutidine was purchased from Avra Synthesis Pvt. Ltd. These chemicals were used without further purification. Au wafers were purchased from Platypus Technologies (1,000 Å of Au on 50 Å of Ti adhesion layer on top of a Si(III) surface). Transparent Au wafers (100 Å of Au on 10 Å of Ti) were purchased from Phasis, Switzerland. Au and Ag discs for the RRDE and SERRS experiments, respectively, were purchased from Pine Instruments, USA. The Mb mutants were prepared as reported in the literature[7,30]. Analysis of the components of the rR spectrum was done by using Lorenztian line shape of peak fit software.

Instrumentation. All electrochemical experiments were performed using a CH Instruments (model CHI710D Electrochemical Analyzer). Bipotentiostat, reference electrode and Teflon plate material evaluating cell (ALS, Japan; http://www.als-japan.com/1398.html) were purchased from CH Instruments. The RRDE set-up from Pine Research Instrumentation (E6 series ChangeDisk tips with AFE6M rotor) was used to obtain the RRDE data. The mutant Mb-functionalized or SAM-covered Au surface (disc of $0.1\ cm^2$ area for RDE, RRDE and wafer of $0.45\ cm^2$ area for CV) was always used as the working electrode. The XPS data were collected in a Omicron (model: 1712-62-11) spectrometer using a high-resolution monochromatic Al-Kα source at 1,486.7 eV under 15 kV voltage and 10 mA current maintaining a base pressure of 5×10^{-10} mbar. The binding energies were calibrated to the Ag $3d_{5/2}$ peak at 368.2 eV. The resonance Raman experiments were done in the Kr^+ Laser (Sabre Innova, Model—SBRC-DBW-K) purchased from Coherent, and the data were collected using the Spectrograph (Model—Trivista 555) from Princeton Instruments.

Formation of mixed SAM and covalent attachment of Hemin-yne to it. Mixed self-assembled monolayer of 1-azidoundecan-11-thiol and 6-mercaptohexanoic acid was formed on immersing the properly cleaned Au wafers or disks into the deposition solution containing 1-azidoundecan-11-thiol and 6-mercaptohexanoic acid in 10 ml of ethanol in a desired ratio (typically 1:49). The total thiol concentration of these deposition solutions were always maintained at 1 mM. On this SAM Hemin-yne was covalently attached using 'Click' reaction[29].

Reconstitution of Apo-G65YCu$_B$Mb mutants to G65YCu$_B$Mb. For all the experiments on heterogeneous SAM surfaces, the Hemin-yne modified -COOH SAM surfaces were incubated with a 20 μM apoprotein (Apo-G65YCu$_B$Mb) solution for 2 h. The supernatant solution was drained and the surface was cleaned with water. The presence of Cu^{2+} in the non-haem-binding site is confirmed by electron paramagnetic resonance (EPR) spectroscopy (Supplementary Fig. 13). The immobilization of the mutant is further confirmed by the absorption spectra and SERRS of the surface fabricated with those mutants (Supplementary Fig. 14, Supplementary Fig. 15).

Cyclic voltammetry. The CV was performed using Au wafers sandwiched between two Teflon blocks of the Plate material evaluating cell. All electrochemical experiments were done in pH 7 phosphate buffer containing potassium hexafluorophosphate. Anaerobic cyclic voltammetric experiments were done by using degassed buffer (three cycles of freeze–pump–thaw). Ag/AgCl reference electrode and Pt counter electrode were used throughout all the electrochemical experiments except the case of anaerobic experiments where only Ag wire was used as the reference electrode.

The peak areas were estimated by integrating the anodic/cathodic peak of the anaerobic CV of the mutant Mb-functionalized SAM-covered Au surface, using the data acquisition software itself. A line collinear with the background is used to subtract the background. The estimated area has been further confirmed by subtracting the background current of a SAM-functionalized electrode (Supplementary Fig. 11) bearing ferrocene. Both these approaches provide the same estimate.

To ensure that the SAM surface is stable during the electrocatalytic investigations, disc bearing just the SAM was subjected to several rotations (200–1,000 r.p.m.) and its capacitive current was found not to change, indicating that the SAM is retained on the electrode during these dynamic electrochemistry experiments (Supplementary Fig. 16). SAM can also be damaged when the protein atop the SAM degrades during ORR due to the reactive oxygen species produced. When an unstable electrocatalyst (Hemin-yne) decayed there was a steady loss of ORR current, indicating degradation of the active site but the capacitive currents of the SAM were unaltered (Supplementary Fig. 17).

SERRS and SERRS-RDE. The excitation wavelength used in the Resonance Raman experiments was 406.7 nm and the power applied to the sample was 10–15 mW. The spectrograph was calibrated against naphthalene. The Ag surfaces were roughened before SERRS experiments following literature protocols[70]. The SERRS-RDE set-up is described in ref. 42. The data for the oxidized state was obtained by holding the potential of the disc at 0 mV versus NHE, and the data during steady state ORR was obtained by holding the disc at −400 mV versus NHE and the disc was rotated at 300 r.p.m. Normally data were acquired over a period of 300 s.

References

1. Collman, J. P. *et al.* A cytochrome c oxidase model catalyzes oxygen to water reduction under rate-limiting electron flux. *Science* **315**, 1565–1568 (2007).
2. Mirica, L. M. *et al.* Tyrosinase reactivity in a model complex: an alternative hydroxylation mechanism. *Science* **308**, 1890–1892 (2005).
3. Wang, Y., DuBois, J. L., Hedman, B., Hodgson, K. O. & Stack, T. D. Catalytic galactose oxidase models: biomimetic Cu(II)-phenoxyl-radical reactivity. *Science* **279**, 537–540 (1998).
4. Beinert, H., Holm, R. H. & MÃ¼nck, E. Iron-sulfur clusters: nature's modular, multipurpose structures. *Science* **277**, 653–659 (1997).
5. Cole, A. P., Root, D. E., Mukherjee, P., Solomon, E. I. & Stack, T. D. P. A trinuclear intermediate in the copper-mediated reduction of O2: four electrons from three coppers. *Science (Washington, D. C.)* **273**, 1848–1850 (1996).
6. Robertson, D. E. *et al.* Design and synthesis of multi-heme proteins. *Nature (London)* **368**, 425–432 (1994).
7. Yeung, N. *et al.* Rational design of a structural and functional nitric oxide reductase. *Nature* **462**, 1079–1082 (2009).
8. Tard, C. *et al.* Synthesis of the H-cluster framework of iron-only hydrogenase. *Nature* **433**, 610–613 (2005).

9. Aboelella, N. W. *et al.* Effects of thioether substituents on the O_2 reactivity of β-diketiminate-Cu(I) complexes: probing the role of the methionine ligand in copper monooxygenases. *J. Am. Chem. Soc.* **128,** 3445–3458 (2006).
10. Shin, H., Lee, D.-H., Kang, C. & Karlin, K. D. Electrocatalytic four-electron reductions of O_2 to H_2O with cytochrome c oxidase model compounds. *Electrochim. Acta* **48,** 4077–4082 (2003).
11. Kim, E. *et al.* Superoxo, μ-peroxo, and μ-oxo complexes from heme/O_2 and heme-Cu/O_2 reactivity: Copper ligand influences in cytochrome c oxidase models. *Proc. Natl Acad. Sci. USA* **100,** 3623–3628 (2003).
12. Tolman, W. B. & Que, J. L. Sterically hindered benzoates: a synthetic strategy for modeling dioxygen activation at diiron active sites in proteins. *J. Chem. Soc. Dalton Trans.* **31,** 653–660 (2002).
13. Osyczka, A., Moser, C. C., Daldal, F. & Dutton, P. L. Reversible redox energy coupling in electron transfer chains. *Nature* **427,** 607–612 (2004).
14. Koder, R. L. *et al.* Design and engineering of an O2 transport protein. *Nature* **458,** 305–309 (2009).
15. Page, C. C., Moser, C. C., Chen, X. & Dutton, P. L. Natural engineering principles of electron tunneling in biological oxidation-reduction. *Nature* **402,** 47–52 (1999).
16. Tani, F., Matsu-ura, M., Nakayama, S. & Naruta, Y. Synthetic models for the active site of cytochrome P450. *Coord. Chem. Rev.* **226,** 219–226 (2002).
17. Dey, A. *et al.* Solvent tuning of electrochemical potentials in the active sites of HiPIP versus ferredoxin. *Science* **318,** 1464–1468 (2007).
18. Astrow, M. L., Peacock, A. F. A., Stuckey, J. A. & Pecoraro, V. L. Hydrolytic catalysis and structural stabilization in a designed metalloprotein. *Nat. Chem.* **4,** 118–123 (2012).
19. Lin, Y.-W. *et al.* Roles of glutamates and metal ions in a rationally designed nitric oxide reductase based on myoglobin. *Proc. Natl Acad. Sci. USA.* **107,** 8581–8586 (2010).
20. Esselborn, J. *et al.* Spontaneous activation of [FeFe]-hydrogenases by an inorganic [2Fe] active site mimic. *Nat. Chem. Biol.* **9,** 607–609 (2013).
21. Fontecave, M. & Artero, V. Bioinspired catalysis at the crossroads between biology and chemistry: A remarkable example of an electrocatalytic material mimicking hydrogenases. *C. R. Chim.* **14,** 362–371 (2011).
22. Lu, Y. & Valentine, J. S. Engineering metal-binding sites in proteins. *Curr. Opin. Struct. Biol.* **7,** 495–500 (1997).
23. Cracknell, J. A. & Blanford, C. F. Developing the mechanism of dioxygen reduction catalyzed by multicopper oxidases using protein film electrochemistry. *Chem. Sci.* **3,** 1567–1581 (2012).
24. dos Santos, L., Climent, V., Blanford, C. F. & Armstrong, F. A. Mechanistic studies of the 'blue' Cu enzyme, bilirubin oxidase, as a highly efficient electrocatalyst for the oxygen reduction reaction. *Phys. Chem. Chem. Phys.* **12,** 13962–13974 (2010).
25. Sigman, J. A., Kim, H. K., Zhao, X., Carey, J. R. & Lu, Y. The role of copper and protons in heme-copper oxidases: kinetic study of an engineered heme-copper center in myoglobin. *Proc. Natl Acad. Sci. USA.* **100,** 3629–3634 (2003).
26. Bhagi-Damodaran, A., Petrik, I. D., Marshall, N. M., Robinson, H. & Lu, Y. Systematic tuning of heme redox potentials and its effects on O2 reduction rates in a designed oxidase in myoglobin. *J. Am. Chem. Soc.* **136,** 11882–11885 (2014).
27. Blanford, C. F., Foster, C. E., Heath, R. S. & Armstrong, F. A. Efficient electrocatalytic oxygen reduction by the 'blue' copper oxidase, laccase, directly attached to chemically modified carbons. *Faraday Discuss.* **140,** 319–335 (2009).
28. Blanford, C. F., Heath, R. S. & Armstrong, F. A. A stable electrode for high-potential, electrocatalytic O_2 reduction based on rational attachment of a blue copper oxidase to a graphite surface. *Chem. Commun.* **43,** 1710–1712 (2007).
29. Mukherjee, S., Sengupta, K., Das, M. R., Jana, S. S. & Dey, A. Site-specific covalent attachment of heme proteins on self-assembled monolayers. *J. Biol. Inorg. Chem.* **17,** 1009–1023 (2012).
30. Sigman, J. A., Kwok, B. C. & Lu, Y. From myoglobin to heme-copper oxidase: design and engineering of a CuB center into sperm whale myoglobin. *J. Am. Chem. Soc.* **122,** 8192–8196 (2000).
31. Miner, K. D. *et al.* A designed functional metalloenzyme that reduces O2 to H2O with over one thousand turnovers. *Angew. Chem. Int. Ed.* **51,** 5589–5592 (2012).
32. Hu, S., Smith, K. M. & Spiro, T. G. Assignment of protoheme resonance Raman spectrum by heme labeling in myoglobin. *J. Am. Chem. Soc.* **118,** 12638–12646 (1996).
33. Bandyopadhyay, S., Mukherjee, S. & Dey, A. Modular synthesis, spectroscopic characterization and in situ functionalization using "click" chemistry of azide terminated amide containing self-assembled monolayers. *RSC Adv.* **3,** 17174–17187 (2013).
34. Guo, L.-H., McLendon, G., Razafitrimo, H. & Gao, Y. Photo-active and electro-active protein films prepared by reconstitution with metalloporphyrins self-assembled on gold. *J. Mater. Chem.* **6,** 369–374 (1996).
35. Ghodselahi, T., Vesaghi, M. A., Shafiekhani, A., Baghizadeh, A. & Lameii, M. XPS study of the Cu@Cu2O core-shell nanoparticles. *Appl. Surf. Sci.* **255,** 2730–2734 (2008).
36. Schmitt, S. K., Murphy, W. L. & Gopalan, P. Crosslinked PEG mats for peptide immobilization and stem cell adhesion. *J. Mater. Chem. B* **1,** 1349–1360 (2013).
37. Boulatov, R., Collman, J. P., Shiryaeva, I. M. & Sunderland, C. J. Functional analogues of the dioxygen reduction site in cytochrome oxidase: mechanistic aspects and possible effects of Cu_B. *J. Am. Chem. Soc.* **124,** 11923–11935 (2002).
38. Mukherjee, S., Bandyopadhyay, S. & Dey, A. Tuning the apparent formal potential of covalently attached ferrocene using SAM bearing ionizable COOH groups. *Electrochim. Acta* **108,** 624–633 (2013).
39. Bard, A. J. & Faulkner, L. R. *Electrochemical Methods: Fundamentals and Applications* 2nd edn (Wiley, 1980).
40. McCrory, C. C. L. *et al.* Electrocatalytic O2 reduction by covalently immobilized mononuclear copper(I) complexes: evidence for a binuclear Cu2O2 intermediate. *J. Am. Chem. Soc.* **133,** 3696–3699 (2011).
41. Shi, C. & Anson, F. C. Potential-dependence of the reduction of dioxygen as catalysed by tetraruthenated cobalt tetrapyridylporphyrin. *Electrochim. Acta* **39,** 1613–1619 (1994).
42. Sengupta, K., Chatterjee, S., Samanta, S. & Dey, A. Direct observation of intermediates formed during steady-state electrocatalytic O_2 reduction by iron porphyrins. *Proc. Natl Acad. Sci. USA* **110,** 8431–8436 (2013).
43. Oertling, W. A., Kean, R. T., Wever, R. & Babcock, G. T. Factors affecting the iron-oxygen vibrations of ferrous oxy and ferryl oxo heme proteins and model compounds. *Inorg. Chem.* **29,** 2633–2645 (1990).
44. Sono, M., Smith, P. D., McCray, J. A. & Asakura, T. Kinetic and equilibrium studies of the reactions of heme-substituted horse heart myoglobins with oxygen and carbon monoxide. *J. Biol. Chem.* **251,** 1418–1426 (1976).
45. Nagao, S. *et al.* Structural and oxygen binding properties of dimeric horse myoglobin. *Dalton Trans.* **41,** 11378–11385 (2012).
46. Ferguson-Miller, S. & Babcock, G. T. Heme/copper terminal oxidases. *Chem. Rev.* **96,** 2889–2908 (1996).
47. Iwata, S., Ostermeier, C., Ludwig, B. & Michel, H. Structure at 2.8 A resolution of cytochrome c oxidase from Paracoccus denitrificans. *Nature* **376,** 660–669 (1995).
48. Collman, J. P. & Decreau, R. A. Functional biomimetic models for the active site in the respiratory enzyme cytochrome c oxidase. *Chem. Commun.* **44,** 5065–5076 (2008).
49. Garcia-Bosch, I. *et al.* A "naked" Fe^{III}-(O_2^{2-})-Cu^{II} species allows for structural and spectroscopic tuning of low-spin heme-peroxo-Cu complexes. *J. Am. Chem. Soc.* **137,** 1032–1035 (2015).
50. Kim, E., Chufan, E. E., Kamaraj, K. & Karlin, K. D. Synthetic models for heme-copper oxidases. *Chem. Rev.* **104,** 1077–1134 (2004).
51. Kim, E. *et al.* Dioxygen reactivity of copper and heme-copper complexes possessing an imidazole-phenol cross-link. *Inorg. Chem.* **44,** 1238–1247 (2005).
52. Babcock, G. T. & Wikstrom, M. Oxygen activation and the conservation of energy in cell respiration. *Nature* **356,** 301–309 (1992).
53. Varotsis, C., Zhang, Y., Appelman, E. H. & Babcock, G. T. Resolution of the reaction sequence during the reduction of oxygen by cytochrome oxidase. *Proc. Natl Acad. Sci. USA* **90,** 237–241 (1993).
54. Han, S., Takahashi, S. & Rousseau, D. L. Time dependence of the catalytic intermediates in cytochrome c oxidase. *J. Biol. Chem.* **275,** 1910–1919 (2000).
55. Han, S., Ching, Y.-C. & Rousseau, D. L. Ferryl and hydroxy intermediates in the reaction of oxygen with reduced cytochrome c oxidase. *Nature* **348,** 89–90 (1990).
56. Rousseau, D. L. Bioenergetics: two phases of proton translocation. *Nature* **400,** 412–413 (1999).
57. Kojima, N. & Palmer, G. Further characterization of the potentiometric behavior of cytochrome oxidase. Cytochrome a stays low spin during oxidation and reduction. *J. Biol. Chem.* **258,** 14908–14913 (1983).
58. Namslauer, A., Branden, M. & Brzezinski, P. The rate of internal heme-heme electron transfer in cytochrome c oxidase. *Biochemistry* **41,** 10369–10374 (2002).
59. Denisov, I. G., Makris, T. M., Sligar, S. G. & Schlichting, I. Structure and chemistry of cytochrome P450. *Chem. Rev.* **105,** 2253–2278 (2005).
60. Boulatov, R., Collman, J. P., Shiryaeva, I. M. & Sunderland, C. J. Functional analogues of the dioxygen reduction site in cytochrome oxidase:mechanistic aspects and possible effects of CuB. *J. Am. Chem. Soc.* **124,** 11923–11935 (2002).
61. Friedrich, M. G. *et al.* In situ monitoring of the catalytic activity of cytochrome c oxidase in a biomimetic architecture. *Biophys. J.* **95,** 1500–1510 (2008).
62. Haas, A. S. *et al.* Cytochrome c and cytochrome c oxidase: monolayer assemblies and catalysis. *J. Phys. Chem. B* **105,** 11351–11362 (2001).
63. Burgess, J. D., Rhoten, M. C. & Hawkridge, F. M. Cytochrome c oxidase immobilized in stable supported lipid bilayer membranes. *Langmuir* **14,** 2467–2475 (1998).
64. Naumann, R. *et al.* The peptide-tethered lipid membrane as a biomimetic system to incorporate cytochrome c oxidase in a functionally active form. *Biosens. Bioelectron.* **14,** 651–662 (1999).
65. Burgess, J. D., Jones, V. W., Porter, M. D., Rhoten, M. C. & Hawkridge, F. M. Scanning force microscopy images of cytochrome c oxidase immobilized in an electrode-supported lipid bilayer membrane. *Langmuir* **14,** 6628–6631 (1998).

66. Su, L., Kelly, J. & Hawkridge, F. M. Electroreduction of O_2 on cytochrome c oxidase modified electrode for biofuel cell. *ECS Trans.* **2,** 1-6 (2007).
67. Yoshikawa, S., Muramoto, K. & Shinzawa-Itoh, K. Proton-pumping mechanism of cytochrome c oxidase. *Annu. Rev. Biophys.* **40,** 205–223 (2011).
68. Morgan, J. E., Verkhovsky, M. I., Palmer, G. & Wikström, M. Role of the PR intermediate in the reaction of cytochrome c oxidase with O2. *Biochemistry* **40,** 6882–6892 (2001).
69. Collman, J. P., Devaraj, N. K., Eberspacher, T. P. A. & Chidsey, C. E. D. Mixed azide-terminated monolayers: a platform for modifying electrode surfaces. *Langmuir* **22,** 2457–2464 (2006).
70. Arūnas Bulovas, Z. T. *et al.* Double-layered Ag/Au electrode for SERS spectroscopy: preparation and application for adsorption studies of chromophoric compounds. *Chemija* **18,** 9–15 (2007).

Acknowledgements

This research is sponsored by Department of Science and Technology, India and US National Institute of Health (GM062211).

Author contributions

S.M. acquired all the data, analysed the data and wrote the paper. A.M. and A.B. purified and provided the mutant proteins and helped write the paper. M.M. helped S.M. in acquiring some data. Y.L. provided the proteins, helped write the paper and developed the science. A.D. conceived the study, analysed the data, developed the science and wrote the paper

Additional information

Competing financial interests: The authors declare no competing financial interests.

5

Target engagement and drug residence time can be observed in living cells with BRET

Matthew B. Robers[1], Melanie L. Dart[1], Carolyn C. Woodroofe[2,†], Chad A. Zimprich[1], Thomas A. Kirkland[2], Thomas Machleidt[1], Kevin R. Kupcho[1], Sergiy Levin[2], James R. Hartnett[1], Kristopher Zimmerman[1], Andrew L. Niles[1], Rachel Friedman Ohana[1], Danette L. Daniels[1], Michael Slater[1], Monika G. Wood[1], Mei Cong[1], Yi-Qiang Cheng[3] & Keith V. Wood[1]

The therapeutic action of drugs is predicated on their physical engagement with cellular targets. Here we describe a broadly applicable method using bioluminescence resonance energy transfer (BRET) to reveal the binding characteristics of a drug with selected targets within intact cells. Cell-permeable fluorescent tracers are used in a competitive binding format to quantify drug engagement with the target proteins fused to Nanoluc luciferase. The approach enabled us to profile isozyme-specific engagement and binding kinetics for a panel of histone deacetylase (HDAC) inhibitors. Our analysis was directed particularly to the clinically approved prodrug FK228 (Istodax/Romidepsin) because of its unique and largely unexplained mechanism of sustained intracellular action. Analysis of the binding kinetics by BRET revealed remarkably long intracellular residence times for FK228 at HDAC1, explaining the protracted intracellular behaviour of this prodrug. Our results demonstrate a novel application of BRET for assessing target engagement within the complex milieu of the intracellular environment.

[1] Promega Corporation, 2800 Woods Hollow Road, Fitchburg, Wisconsin 53711, USA. [2] Promega Biosciences Incorporated, 277 Granada Drive, San Luis Obispo, California 93401, USA. [3] UNT System College of Pharmacy, Department of Pharmaceutical Sciences, University of North Texas Health Science Center, Fort Worth, Texas, USA. † Present address: National Institutes of Health, 9800 Medical Center Drive, Rockville MD 20850, USA. Correspondence and requests for materials should be addressed to M.B.R. (email: Matt.robers@promega.com).

Deciphering how small molecule modulators bind their intracellular targets is fundamental to understanding pharmacological mechanism. In addition to the specificity and affinity of target engagement, binding dynamics under non-equilibrium conditions may also underlie the therapeutic potential of new drug candidates[1–3]. These parameters are routinely assessed through biochemical means, which may fail to adequately mimic the complexity of the intracellular environment. Proteins reside in structurally intricate settings within the cells and typically function as components of extended molecular complexes, and thus they may exhibit significantly different behaviours than they would as isolated polypeptides[4–7]. It is not surprising that biochemical analysis of target engagement often fails to correlate with compound potency measured by cellular phenotype. Preferably, correlations between binding interactions and physiological outcomes should be made within a common physiological context. For this reason, the pharmaceutical industry has directed increased efforts towards assessing target engagement within intact cells[8–10].

While quantitation of compound binding to purified proteins or surface receptors (in particular G-protein coupled receptors) is well established[11–13], similar analysis for intracellular targets has been more difficult. Indirect approaches are often used instead, relying on deconvolution of cellular responses to infer target engagement[14]. For example, expression profiling may be used as an indicator of altered target activity in response to agonists or antagonists. However, compounds typically bind to multiple targets within cells, where only a few are mechanistically associated with the relevant phenotype. Unambiguously resolving the molecular targets of compounds within complex pathways and establishing that a cellular response serves as an adequate proxy for physical binding by the compound can be challenging.

More recently, various qualitative approaches based on ligand-induced protein stabilization have been used to characterize target engagement[9,10,15,16]. Such methods can be limited by the incremental stability imparted by compound binding relative to the inherent stability of the intracellular target. Consequently, these methods are prone to false negative results as many targets fail to exhibit measurable stabilization upon ligand binding[17]. For some of these techniques, elevated temperatures are required for the analysis, and thus may not represent physiological conditions for compound binding. Importantly, these methods are limited to end point analysis, complicating the application of such methods for measurements of binding kinetics or compound residence time.

Assessments of target engagement are especially challenging for prodrug inhibitors that require intracellular activation for maximal potency[18–20]. Mechanistic studies for such prodrug inhibitors may not be adequately represented in a biochemical framework, and may require analysis in cells to be physiologically meaningful. For example, the clinically approved histone deacetylase (HDAC) prodrug FK228 (depsipeptide, romidepsin, Istodax) as well as the related natural product thailandepsin A (TDP-A) utilize a unique mechanism that require intracellular reduction to achieve maximal potency[18,19,21]. It has been recently demonstrated that pulse-treatment of cells with FK228 results in highly potent and persistent inhibition of pan-HDAC activity[22–24]. Although various alternate intracellular mechanisms have been proposed for this observation[24], it has not been determined whether the sustained potency of FK228 is mechanistically associated with the intracellular residence time at HDAC isozymes. Biophysical methods compatible with living cells are therefore needed to interrogate target engagement and residence time for this compound class.

Bioluminescence resonance energy transfer (BRET) can reveal real-time molecular interactions within intact cells without cell lysis or non-physiological temperatures[25]. Energy transfer techniques such as BRET or fluorescence resonance energy transfer (FRET) are well established for quantifying intracellular protein–protein interactions within cells; however, BRET is often preferred owing to increased detection sensitivity[26–28]. While both energy transfer techniques have been utilized to measure compound binding to extracellular or lysate-derived analytes[12,13,29,30], neither has been successfully applied to the interrogation of target engagement and compound residence time within intact cells.

In contrast to previous applications of energy transfer, the approach presented here utilizes live cells expressing an intracellular target protein genetically fused to NanoLuc luciferase and a cell-permeable fluorescent tracer derived from a suitable drug or tool compound. BRET is achieved inside intact cells by reversible binding of the fluorescent tracer to the intracellular target. The binding characteristics of an interacting compound are revealed by its ability to compete with the tracer and thus influence the production of BRET. The general applicability of this approach for intracellular proteins is supported by our ability to quantify intracellular engagement by inhibitors of HDACs, bromodomains (BRDs) and kinases. Our analysis focused on target engagement at HDACs because this new capability should enable the biophysical characterization of target engagement and residence time for a diverse set of HDAC inhibitors, including the prodrugs FK228 and TDP-A. The BRET approach enabled a mechanistic interrogation on the enhanced potency and sustained efficacy of FK228, allowing us to directly measure both target engagement and residence time against target HDACs within intact cells. Using BRET, we were able to determine that the prolonged phenotypic effects of FK228 and TDP-A prodrugs in non-equilibrium conditions are due to the remarkably stable complexes formed between these molecules and HDAC protein within cells.

Results

BRET enables target engagement analysis for HDACs. We used a small luciferase (NanoLuc/Nluc, 19 kDa) as a BRET donor, which has been recently demonstrated to produce extraordinarily intense and stable luminescence with a relatively narrow spectral distribution[31]. For BRET tracers, we applied Non-Chloro-TOM dye (NCT) as a BRET acceptor, which demonstrated adequate cell-permeability and provided significant spectral resolution from Nluc[25]. Detection of tracer binding by energy transfer should be broadly applicable to intracellular proteins so long as the tracer is permeable to the cells and the proteins can be suitably tagged with Nluc. Intracellular target engagement is measured by competitive displacement of the tracer resulting in a loss of BRET (Fig. 1a).

HDACs served as the primary model system, where a single broad-coverage tracer could enable analysis of target engagement over multiple classes of HDACs. To elucidate inhibitor interactions across classes I and IIb HDACs, a broad-coverage target engagement assay was developed using a fluorescent tracer derived from SAHA coupled to NCT dye (Fig. 1b). Intracellular binding of the SAHA-NCT tracer to selected HDAC isozymes was detected by measuring the bioluminescence emission from a population of cells expressing a genetic fusion between the isozyme and Nluc. The putative targets of SAHA[32]; classes I and IIb HDACs, were included in the study to reveal isozyme-specific differences in inhibitor affinity. As HDACs coordinate binding of hydroxamic acid-based inhibitors such as SAHA via two highly conserved active-site His residues[33,34], a binding-deficient construct encoding Nluc-HDAC6 CD2 (H610A/H611A) was included in the analysis as a negative control. Furthermore, class IIa HDACs (HDACs 4, 5, 7 and 9) and class IV HDAC

(HDAC11) were also included as controls because of their expectedly low affinity to the SAHA-NCT tracer[32]. Microplate measurements of HeLa cells transiently expressing the genetic fusions confirmed both specific and concentration-dependent BRET only for classes I and IIb HDACs (Fig. 1c, Supplementary Fig. 1). To determine specific BRET signal, BRET data were background-corrected using a molar excess of unlabelled SAHA (Supplementary Fig. 1a). Both N- and C- terminal fusions were evaluated, with the orientation producing the largest BRET ratio for each isozyme shown in Fig. 1c. HDACs are known to tolerate reporter tags or purification tags larger than Nluc[35], and we found that tethering of Nluc did not significantly alter HDAC activity, using HDAC2 as a an example (Supplementary Fig. 1c).

The highest tracer affinity was evident for HDACs 1, 2, 6 and 10, with lower affinity observed for HDACs 3 and 8 (Fig. 1c). While the presence of the fluorescent moiety on SAHA impacted affinity of the tracer, the observed rank-order engagement by the SAHA-NCT tracer was in general agreement with the relative affinities reported for unmodified SAHA[32] (Supplementary Table 1). Because of the high luminescence intensity of Nluc interaction of the SAHA-NCT tracer could be detected at expression levels comparable to the endogenous protein, as demonstrated for HDAC1 in multiple cell types (Supplementary Fig. 1d). These results support that HDAC-Nluc fusions do not require overexpression to detect target engagement analysis via BRET.

In contrast to other members of the class I/IIb HDAC family, the HDAC6 isotype possesses two putative catalytic domains (CDs) as determined by homology analysis[36–38]. To determine whether the SAHA-NCT tracer binds one of the domains preferentially, we generated Nluc fusions with segregated CDs of HDAC6 and queried tracer binding in cells. BRET measurements demonstrated engagement of the tracer with the C-terminal CD (CD2) (Fig. 1c), whereas negligible BRET signal was observed at CD1 (Supplementary Fig. 1b). It has been reported that CD1 may lack deacetylase activity, and our results suggest that altered binding specificity may contribute to this[38]. As expected, the HDAC6 CD2 (H610A/H611A) mutant engaged SAHA-NCT poorly as determined via BRET (Fig. 1c).

To verify that BRET was arising from within living cells rather than from cellular debris that may be present in the culture medium, we imaged the luminescence emission from individual cells expressing Nluc fusions of HDAC6 or HDAC10. The addition of the SAHA-NCT tracer (1 μM) resulted in observable BRET clearly located within HeLa cells transiently expressing Nluc fusions of HDAC6 or HDAC10 (Fig. 1d, Supplementary Fig. 1e). Challenge with a molar excess of unmodified SAHA (10 μM) resulted in attenuation of the BRET signal, indicating that the BRET signal was generated by a specific and reversible interaction of the tracer with the Nluc fusion protein. BRET signal was observed exclusively in the cytoplasm of HeLa cells expressing Nluc-HDAC6, whereas HDAC10-Nluc expressing cells generated BRET in both the nucleus and cytosol. This is consistent with the expected localization patterns of both HDAC6 and HDAC10, and further demonstrates that the BRET signal is imposed by the placement of the luciferase even though the SAHA-NCT tracer is expected to bind multiple intracellular HDACs[36].

The ability to displace the tracer by an unlabelled compound allowed the relative binding efficiency by HDAC inhibitors to be determined for multiple intracellular isozymes. Unlabelled SAHA effectively attenuated the BRET signal in a concentration-dependent manner when added to cells in combination with the SAHA-NCT tracer (Fig. 1e). The relative binding efficiency of SAHA for different HDAC isozymes is apparent by the concentration required to displace a defined amount of the tracer. Using a fixed concentration (1 μM) of SAHA-NCT for competitive displacement, relative binding efficiencies were estimated by the Cheng-Prusoff equation[39] (see Supplementary Table 1 for observed and adjusted values). By this method, the greatest binding was observed with Nluc fusions of HDAC1, 2, 6 and 10, whereas HDACs 3 and 8 showed lower relative binding, as expected.[32] Although SAHA and the SAHA-NCT tracer share the same binding motif, this approach should be applicable to any molecule able to competitively displace the tracer from the target.

Profiling engagement and potency of HDAC inhibitors in HeLa. To demonstrate the utility of the BRET method to profile intracellular target engagement across HDACs, we used a structurally diverse set of inhibitors that included both hydroxamate- and non-hydroxamate-based compounds. Hydroxamate-based inhibitors were represented by SAHA, trichostatin A (TSA), ACY1215, M344 and panobinostat. Mocetinostat is an aminobenzamide-based inhibitor, and FK228 is a bicyclic depsipeptide prodrug. All six of the tested inhibitors engaged HDACs 1, 2 and 3, with FK228 demonstrating the highest engagement for these isozymes (Fig. 1f). Negligible binding of FK228 and mocetinostat were observed for HDAC6, HDAC8 and HDAC10 (Fig. 1f). Rank-order binding of the compounds for HDACs 1 and 2 were in agreement, likely explained by the close sequence homology between these proteins[40]. Other HDAC isozymes had more varied selectivity profiles, providing selectivity signatures that were largely in agreement with previous reports, albeit with some exceptions[32,40]. The bicyclic depsipeptide inhibitors of HDACs (including FK228 and the recently described TDP-A) are notable in that they rely on a unique mechanism of prodrug activation

Figure 1 | Measuring target engagement at intracellular HDACs. (**a**) Illustration of intracellular target engagement assay. A permeable fluorescent tracer binds in dynamic equilibrium to an intracellular target protein fused to Nluc, resulting in BRET. Introduction of compounds that bind the same target cause the tracer to be displaced, resulting in a decrease in BRET. (**b**) The HDAC tracer was derived from the adduct of a broad-coverage HDAC inhibitor (SAHA, shown in black) and the NCT dye (shown in red). (**c**) Introduction of SAHA-NCT to cell medium resulted in specific and concentration-dependent intracellular BRET with Nluc fusion of various HDACs (classes I and IIb) expressed in HeLa cells, as measured using a microplate luminometer. A control HDAC6 CD2 construct encoding a binding-deficient mutant (H610A/H611A) showed weakened engagement with the tracer. BRET values at each tracer concentration were background-corrected by parallel measurements made in the presence of an excess of unmodified SAHA (20 μM) (as described in Methods and Supplementary Fig. 1a). Apparent tracer affinities were estimated using equation (1) and reported in Supplementary Table 1. Data are mean ± s.e.m. of three independent experiments. (**d**) Luminescence images of HeLa cells expressing HDAC6 or HDAC10 fused to Nluc, shown with false colouring to represent light emission from 420–500 nm (donor) or above 590 nm (acceptor). Images were acquired in the presence of SAHA-NCT tracer alone (1 μM), or in combination with excess unlabelled SAHA (10 μM). Signal suppression caused by excess SAHA indicates specificity of the BRET signal. (**e**) Concentration-dependent attenuation of BRET from intracellular HDAC fusions with titration of SAHA in the presence of 1 μM SAHA-NCT tracer. Data were collected on a microplate luminometer and are mean ± s.e.m. of three independent experiments. See Supplementary Table 1 and equations (2) and (3) for estimation of apparent affinities of SAHA to HDACs. (**f**) BRET measurements showing the relative affinity of HDAC inhibitors in HeLa cells (1 μM SAHA-NCT). See Supplementary Table 2 for relative compound affinities to individual HDACs. Data are mean ± s.d. of four data points.

which requires intracellular reduction of a disulfide bond[18,19,41]. The ability to observe target engagement within living cells thus allows characterization of inhibitor mechanism for these compounds within the physiological environment necessary for prodrug conversion.

It is well established that upregulated HDAC 1 and 2 activities inhibit apoptosis and mediate proliferation in cervical cancer cell lineages such as HeLa[40,42,43]. To verify that our intracellular target engagement profiles measured by BRET accurately correlate with antiproliferation, we determined potencies for the

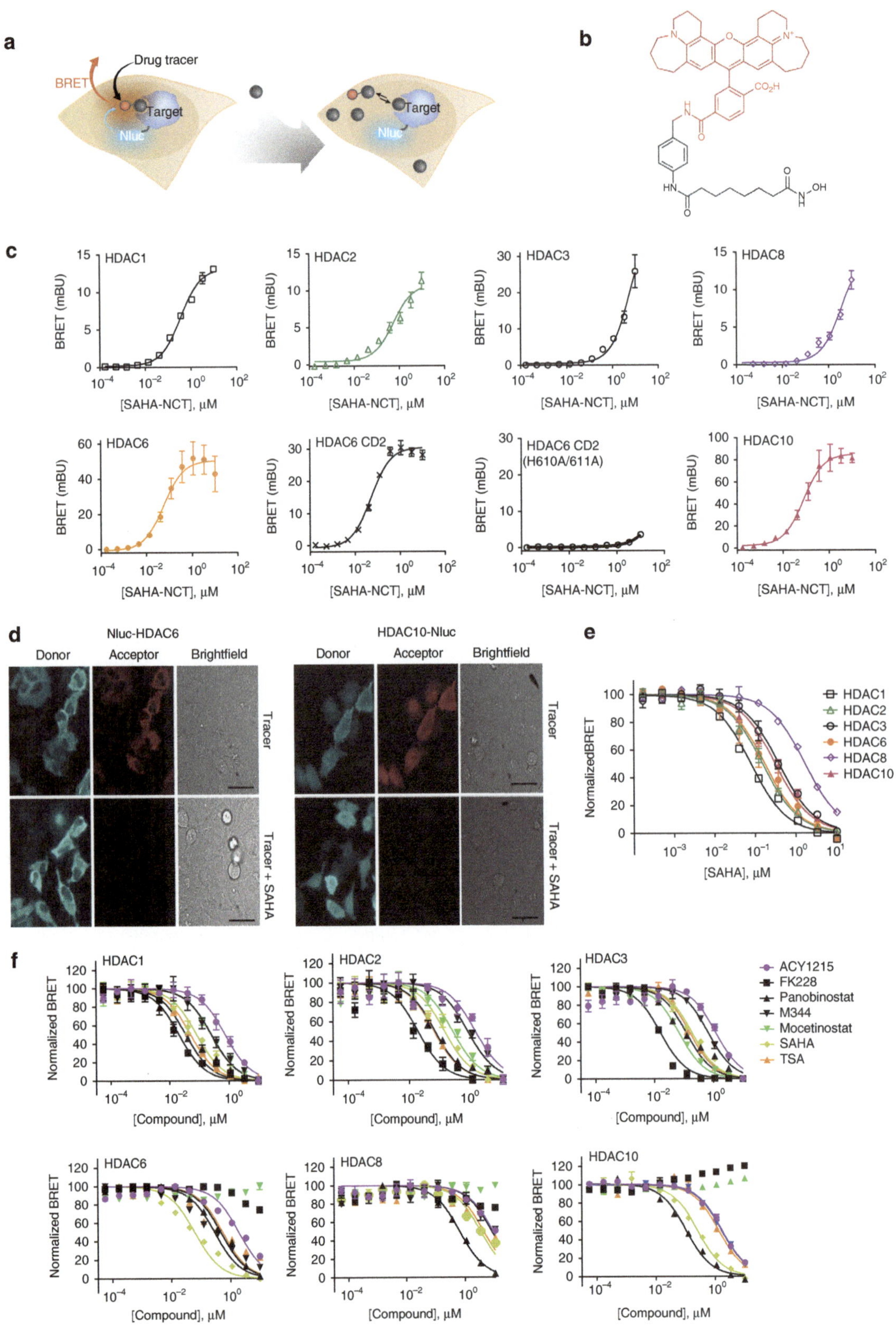

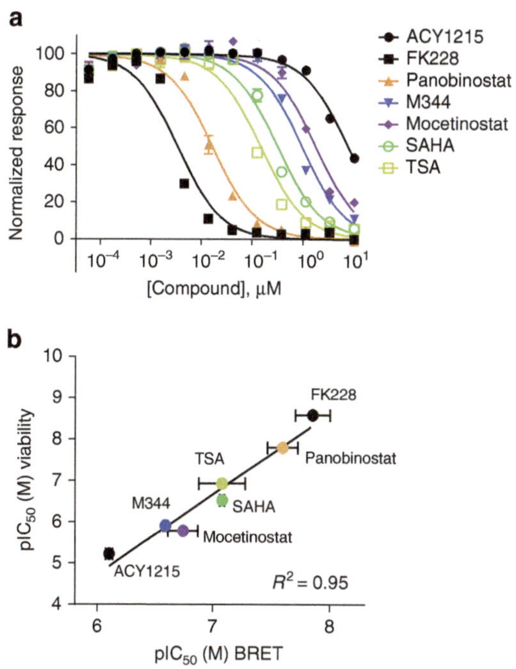

Figure 2 | Correlation of phenotypic potency with target engagement to intracellular HDAC isozymes. (**a**) 48 h treatment with HDAC inhibitors results in antiproliferative effects in HeLa cells, as measured by intracellular ATP levels. Data are the mean of three independent experiments ± s.e.m.. (**b**) Antiproliferative potency of HDAC inhibitors, as determined by cellular ATP levels in HeLa cells, is strongly correlated with target engagement to HDAC1. See Supplementary Fig. 2b and Supplementary Table 2 for the comparative correlation with other HDAC isozymes. Data are the mean of 3 independent experiments ± s.d.

panel of inhibitors using assays of apoptosis (caspase 3/7 activation) and cell proliferation (cellular ATP). A range of potencies was observed with similar rank-order produced by either method (Fig. 2a, Supplementary Fig. 2a). As expected[40], target engagement was found to strongly correlate with antiproliferation for HDAC1 and HDAC2, yielding $R^2 = 0.95$ and 0.88 respectively (Fig. 2b, Supplementary Fig. 2b and Supplementary Table 2). Unsurprisingly, the correlations with less relevant HDACs were relatively poor (Supplementary Fig. 2b, Supplementary Table 2). As HDAC1 and HDAC2 are the dominant HDACs mediating cell proliferation in many cancer cell lines, these correlations reinforce the validity of the target engagement measurements provided by BRET[40,42,43].

BRET reveals long residence times for FK228 and TDP-A. FK228 and related TDP-A require the reducing environment within cells to activate their sulfhydryl groups and achieve maximal potency[18] (Fig. 3a). Although there has been speculation regarding the possible causes for protracted efficacy of FK228 in pulse-treatment experiments[22,24], the precise mechanism has not been determined. We investigated this for FK228 and TDP-A by analysing temporal changes in the BRET measurements to characterize kinetic aspects of target engagement. As a prerequisite, it was necessary to establish that SAHA-NCT demonstrates rapid binding kinetics to ensure that dynamic interaction of the tracer with the intracellular target remains close to equilibrium over the time interval needed for engagement by the inhibitors[44]. As SAHA has fast binding kinetics, the SAHA-NCT tracer should be suitable for kinetic analysis of relatively slow-binding HDAC inhibitors in a competitive format[45].

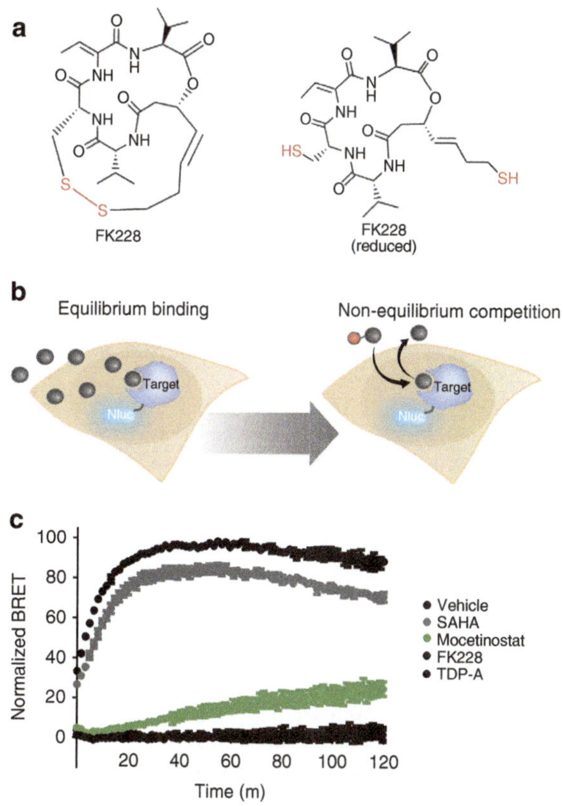

Figure 3 | Measuring the intracellular residence time of HDAC inhibitors at HDAC1. (**a**) Structure of the inhibitor FK228 in the prodrug form (left) and the form activated by intracellular reduction (right). (**b**) Illustration of assay method for measuring intracellular residence time using BRET. Live cells expressing the target protein fused to Nluc are equilibrated with a near-saturating concentration of compound. The cells are then washed to remove unbound compound and treated with a near-saturating concentration of a tracer. The profile of compound with slow dissociation kinetics from the target impedes tracer binding, which slows production of the BRET signal. (**c**) Residence time analysis by BRET reveals remarkably slow dissociation rates for FK228 (red) and TDP-A (blue), compared with mocetinostat (green), SAHA (grey) or vehicle/DMSO control (black). Data are normalized to maximum signal (vehicle/DMSO-treated) versus full-occupancy control (10 μM SAHA, no washout). The kinetic traces for FK228 and TDP-A are nearly indistinguishable from the full occupancy control (10 μM SAHA, no washout). Data are mean ± s.e.m.. of four independent experiments.

Addition of a 1 μM solution of tracer to cells expressing HDAC1-Nluc resulted in a rapid increase in BRET, reaching a plateau within ~20 min of tracer addition (Supplementary Fig. 3a). This rate is at least partially impacted by permeability across the plasma membrane, as addition of digitonin as a permeabilizing agent resulted in near-instantaneous BRET between the tracer and HDAC1-Nluc (Supplementary Fig. 3e).

Unlike hydroxamate-based inhibitors such as SAHA, it as been previously reported that prodrug inhibitors in the FK228 family possess a slow but protracted activity in cancer cell lineages[23]. Using the BRET approach, we confirmed that fast binding kinetics are observed for unmodified SAHA in HeLa cells (Supplementary Fig. 3b). In contrast to SAHA, FK228 and TDP-A equilibrated slowly, displaying an IC_{50} value that dropped continuously over the 180 min timecourse. The slow-binding kinetics of the prodrug inhibitors in cells were corroborated in an enzymatic assay using purified HDAC1 and chemically reduced FK228 or TDP-A (Supplementary Fig. 3c). Our results therefore support previous observations regarding the delayed effect of FK228 inside cancer cells.

Because of the reported functional persistence of FK228 activity, it has been suggested that these inhibitors may accumulate within cells, impact HDAC expression levels, or release slowly from their targets[22,24]. To directly assess the possibility of slow release from intracellular HDAC protein, the BRET method was reconfigured to measure the relative rates of compound dissociation from HDAC1. HeLa cells expressing HDAC1-Nluc were first equilibrated with a saturating concentration of compound. The aminobenzamide inhibitor mocetinostat was included as a control, as it has been demonstrated that inhibitors in this class possess slow dissociation kinetics with HDAC1 (refs 45–47). Cells were treated with FK228 at 100 nM, TDP-A at 10 nM, mocetinostat at 10 μM, or SAHA at 10 μM, followed by ligand removal (including an additional wash for FK228 and TDP-A) and immediate challenge with a saturating concentration of SAHA-NCT tracer. Increased BRET is indicative of tracer binding to the HDAC, which cannot occur until after the compound has dissociated (Fig. 3b, see Supplementary Fig. 3d for a detailed schematic representation of the assay). This analysis was also performed in lysed cells to address the proposed possibility that the reduced FK228 and TDP-A could accumulate and become trapped within cells, leading to residual target engagement after compound removal[22,24]. In live cells, dissociation of SAHA from HDAC1 was rapid and beyond the detection limits of the assay, whereas the reference compound mocetinostat displayed a slower dissociation rate as expected (Fig. 3c). The intracellular dissociation rate for FK228 and TDP-A were slower than the control compound mocetinostat at HDAC1, with a residence time profile similar to fully bound HDAC1 (samples without 10 μM SAHA removal, Fig. 3c). Results were corroborated in lysed cells, suggesting that intracellular trapping of FK228 is insufficient to explain the observed kinetics of dissociation (Supplementary Fig. 3e). These results indicate a remarkably slow dissociation rate (long residence time) for FK228 and TDP-A from HDAC1, and suggest that the strong binary complex between the inhibitors and HDAC protein are the cause of the prolonged phenotypic effect.

BRET for target engagement at alternate target classes. In addition to HDACs, we evaluated the BRET method for the BET family of BRDs and the lymphocyte specific protein tyrosine kinase (LCK) using BRET tracers derived from known inhibitors of these target proteins (Supplementary Fig. 4a, Supplementary Methods). An NCT tracer derived from iBET-762 engaged full-length Nluc-BRD4 proteins (Fig. 4a, Supplementary Figure 4b), as well as other members of the BET family of BRDs (Supplementary Fig. 4c). Target engagement profiles at BRD4 were then determined in competitive displacement formats with a panel of known BRD4 inhibitors (Fig. 4b) as well as segregated domains of BRD4 (Supplementary Fig. 4d). Applicability to intracellular kinases was then demonstrated at LCK using an NCT tracer derived from BIBF-1120 (Supplementary Fig. 4e,f). For both BRDs and LCK, rank-order affinities of various reference compounds observed using BRET were consistent with previous reports[48–50] Taken together, these results indicate that energy transfer to cell-permeable tracer may provide a broadly applicable means for evaluating target engagement over key intracellular target classes.

Discussion

This report describes the first method to directly assess target engagement and binding kinetics within intact mammalian cells under physiological conditions. Compound binding is detected by BRET, a biophysical process which provides significant advantages over alternative methods that rely on more complex biological proxies, operate under non-physiological or disruptive conditions, or are unsuitable for real-time analysis of compound binding and dissociation[10,15]. The simplified workflow we describe also eliminates the laborious liquid transfer steps and excessive hands-on time required of many alternative methodologies, particularly those utilizing thermal denaturation. Compared with biochemical methods, detection of intracellular target engagement provides an assessment of binding efficiency in an environment more representative of the cellular milieu where it would occur in tissues. The intracellular measurements also circumvent the need for isolated enzymes, which may be difficult to obtain by heterologous expression with sufficient purity and in a structurally appropriate form. This is the case with HDACs, for example, which generally function in protein complexes[51].

The ratiometric nature of energy transfer measurements mitigates potential assay interferences, such as variations in cell number or expression level between samples[26,27,52]. BRET was chosen as the preferred energy transfer system over FRET due to the increased detection sensitivity it provides in microplate formats[26–28,30,52,53]. NanoLuc, in particular, provides excellent sensitivity as a BRET donor because of its small and stable structure, exceptionally bright luminescence, and relatively narrow emission spectrum[31]. Corroborating the signal strength and sensitivity of this method within live cells, energy transfer between Nluc and the NCT tracers produced sufficient BRET signal to directly image target engagement at the single-cell level using bioluminescent imaging microscopy. Using this BRET technique, it may be therefore possible to study target engagement distribution over a cell population.

The binding profiles for the tracers described in this report provide evidence that the red-emitting fluorophore, NCT, can be

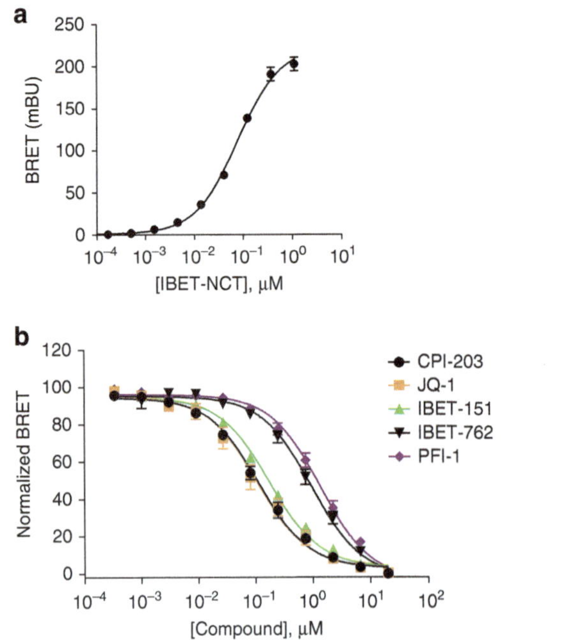

Figure 4 | Measuring intracellular target engagement at bromodomain BRD4 using BRET. (**a**) Specific and dose-dependent BRET was observed with Nluc-fusions to full-length BRD4 expressed in HeLa. BRET values at each tracer concentration were background-corrected by parallel measurements made in the presence of an excess of unmodified compound (20 μM) (described in Methods). Data are mean ± s.d. of four data points. (**b**) Use of BRET to measure rank-order affinity of BRD inhibitor panel against BRD4 within live HeLa cells (1 μM IBET-NCT held constant). Data are mean ± s.e.m. of four independent experiments.

coupled onto structurally diverse chemical probes to provide cell-permeable BRET tracers for a variety of target classes. For tracer derivatization, we utilized probes with known mechanism of action and putative target engagement profiles. Probe molecules with less established structure–activity relationships may represent additional challenge for NCT modification. In the context of phenotypic screening (when knowledge of target selectivity for a lead compound is unknown), it may be useful to explore the suitability of the BRET technique for target identification. As an extension of the principle demonstrated here, it may be possible to apply an NCT derivative of a lead compound against a diverse panel of Nluc fusion constructs as a screen for potential intracellular targets. Such an approach could serve as a complement to more traditional mass spectrometry-based methods for target identification.

Although the fluorescent moiety of the tracer may impart altered target affinity compared with the unlabelled parent compound, our results support general applicability of the method so long as the tracer is permeable to cells, specific BRET can be measured over a titratable concentration range, and that competitive displacement can reveal accurate target engagement profiles for unlabelled compounds. However, use of BRET tracers for target engagement studies may represent a limitation for mechanistic studies on allosteric ligands that occupy binding sites distinct from that of the tracer. Ligands that engage the target but fail to compete with the tracer would therefore represent false negatives. While this may represent a potential liability, it has been well-documented that allosteric ligands may bind in a mutually exclusive manner with orthosteric (active-site) tracers[54]. Future studies are warranted to determine if this BRET technique could enable target engagement studies in this context. Non-tracer-based techniques (such as those utilizing thermal stability or thermal shift) are prone to a distinct set of false negatives, as high affinity ligands may not impart a measurable increase in stability for certain target proteins[17]. As this BRET technique is independent of target stability or denaturation steps, it may be a useful complement to stability-based assessments of target engagement.

By using a broad-spectrum tracer for analysis of class I/IIb HDACs, we successfully demonstrated that a single tracer could enable a target engagement study across multiple classes of HDACs. Although our engagement results were largely in agreement with literature[32,40], small discrepancies could be attributed to the behaviour of isolated proteins versus the intracellular targets interrogated in our analysis. Indeed, there is a lack of consensus in the literature regarding the profiles of our compound set against purified HDACs[4,32,40], possibly explained by the diversity and variability of available biochemical HDAC assay components.

To evaluate the validity of our HDAC target engagement results, we performed the first systematic correlation analysis between intracellular isozyme affinity and antiproliferative potency in a common cellular context. Target engagement measured by BRET for HDAC1 and HDAC2 for the compound panel showed a strong correlation with both apoptosis and proliferation of HeLa cells. The involvement of HDACs 1 and 2 in these phenotypic outcomes is consistent with proposed models for their roles in tumorigenesis in cervical cancer cells such as HeLa. This relationship is not surprising, as HDAC1 and HDAC2 share 87% sequence similarity and form intracellular protein complexes which coordinate the transition from G1 to S-phase in the cell cycle[42,43]. They are also commonly upregulated in tumour cells, including HeLa. Consequently, it is not surprising that pharmacological inhibition of these enzymes should inhibit proliferation in this cell model[40]. While a strong correlation was observed for HDACs 1 and 2 (and to a lesser extent HDAC3), the relatively poor correlation observed at HDACs 6, 8 and 10 support that these targets are not directly implicated in the antiproliferative effects of HDAC inhibitors in this cell model. Our observed profiles for target engagement therefore corroborate the proposed link between HDAC1 and HDAC2 activities in tumour cell proliferation and inhibition of apoptosis.

Although assessments of target occupancy are often assessed under equilibrium (using affinity values such as K_d, IC_{50}), this may not be adequate for predicting target occupancy *in vivo*, where equilibrium conditions may not apply[1]. Under the dosing dynamics that are representative of *in vivo* conditions, dissociation rate, or residence time, may better characterize the ligand-receptor complex[1–3]. Furthermore, intracellular residence time may be impacted by intracellular target densitites, suggesting that binding analysis using purified analytes may not be adequate to predict drug residence time *in vivo*[7]. However, no methods currently exist to directly measure intracellular target engagement under non-equilibrium conditions. To enable a biophysical assessment of this process as it occurs inside intact cells, we explored the feasibility for using intracellular BRET to observe the kinetics of drug dissociation for a select group of HDAC inhibitors.

To mechanistically characterize the persistent effects of FK228 following inhibitor washout experiments, we configured the BRET assay to directly monitor dissociation of the prodrug inhibitors at HDAC1 inside cells. In this configuration, extremely slow apparent dissociation kinetics (long residence time) were observed for these HDAC inhibitors compared with SAHA. Moreover, the prodrug inhibitors showed even longer residence time than mocetinostat, a compound with reportedly slow dissociation rates at HDACs[45–47].

Consistent with our residence time analysis, it has been well-documented that pulse-treatment of cells with FK228 results in enhanced and persistent inhibition of intracellular HDAC activity compared with SAHA[22,24]. This enhanced duration of HDAC inhibition was concomitant with increased HIV RNA expression and virion release from infected T-cells, indicating that FK228 should be further assessed for treatment of latent viral reservoir in HIV-infected patients[22]. In the previous studies, it was not determined whether the sustained efficacy of FK228 resulted from long residence time at HDACs versus alternate mechanisms (potentially involving HDAC expression levels, cellular accumulation of the reduced drug, inhibition of HDAC complexes and so forth). For prodrug molecules such as FK228, acellular assay formats are unsuitable to explore these mechanisms. The intracellular BRET technique was able to interrogate the dynamics of FK228 target engagement because of its compatibility with equilibrium and non-equilibrium binding analysis. Results presented here enabled a mechanistic interrogation into the enhanced phenotypic effect of FK228 under such pulse-treatment conditions, and indicate that HDAC occupancy is enhanced for this class of prodrug molecules under open system conditions. In a broader context, optimizing a lead compound for increased intracellular residence time at primary targets, while decreasing residence time at collateral targets, could mitigate potential drug side effects and increase safety profiles. The ability to configure the BRET assay to interrogate binding kinetics could be used to guide medicinal chemistry efforts during the workflow of lead optimization.

In summary, we have found that BRET can provide quantitative analysis of intracellular target engagement over a diverse set of target classes. This approach can be used to assess target engagement and binding kinetics in a simplified, homogeneous assay format. We have demonstrated that intracellular binding can be correlated with phenotypic potency, potentially allowing for primary drug targets to be differentiated from targets engaged collaterally. The use of BRET for interrogating compound binding to intracellular targets should therefore advance capabilities in drug design and mechanistic analysis.

Methods

Expression plasmid construction. To produce Nluc fusions with the targets interrogated in this study, pF31K Nluc [CMV/Neo] and pF32K [CMV/Neo] were used to place Nluc at the N-terminus or the C-terminus of the target protein (respectively) using the manufacturer's protocol (Promega). For N-terminal tagging of Nluc to the target protein in pF31K, the resulting fusion encoded a flexible Gly-Ser-Ser-Gly-Ala-Ile-Ala linker connecting Nluc with the target. For C-terminal tagging of Nluc to the target protein in pF32K, the resulting fusion encoded a flexible Val-Ser-Leu-Gly-Ser-Ser-Gly linker connecting the target protein with Nluc. Complementary DNAs encoded the following target proteins, and were a 100% match to their respective NCBI reference sequence identifiers; BRD2 (NP005095), BRD3 (AB383723.1), BRD4 (NP490597), BRDT (NP001229739), Kat2b (NP003875), HDAC1 (NP004955), HDAC2 (NP001518), HDAC3 (NP003874), HDAC4 (NP006028), HDAC5 (NP005465), HDAC6 (NP006035), HDAC7 (NP056216), HDAC8 (NP060956), HDAC9 (NP848510), HDAC10 (NP114408), HDAC11 (NP079103) and LCK (NP005347). To generate Nluc fusions with segregated domains of HDAC6, ORFs encoding amino acid residues 74-455 and residues 479-845 were used for HDAC6/CD1 and HDAC6/CD2, respectively. Nluc-HDAC6 CD2 was mutated at H610A/H611A (residues represented from full-lenth HDAC6) to generate a binding-deficient mutant protein. To generate Nluc fusions with segregated domains of BRD4, ORFs encoding amino acids 44-168 and residues 333-460 were used for BRD4/BD1 and BRD4/BD2, respectively.

Cell transfection and BRET measurements under equilibrium conditions. To lower intracellular expression levels of the reporter fusion, Nluc/target fusion constructs into carrier DNA (pGEM3ZF-, Promega) at a mass ratio of 1:10 (mass/mass), before forming FuGENE HD complexes according to the manufacturer's protocol (Promega). DNA:FuGENE complexes were formed at a ratio of 1:3 (μg DNA per μl FuGENE). One part of the transfection complexes was then mixed with 20 parts (v/v) of HeLa cells (ATCC) suspended at a density of 2×10^5 in DMEM (Gibco) + 10% fetal bovine serum (FBS) (GE Healthcare), followed by incubation in a humidified, 37 °C/5% CO_2 incubator for 20 h. Cells were trypsinized and resuspended in Opti-MEM without phenol red (Life Technologies). Cells were then seeded into white, nonbinding surface plates (Corning) at a density of 2×10^4 cells/well. All chemical inhibitors were prepared as concentrated stock solutions in dimethylsulphoxide (DMSO) (Sigma-Aldrich). TDP-A was prepared by bacterial fermentation from the Cheng group, as described[19,41]. Remaining chemical inhibitors were purchased from Selleck Chemicals, with the exception of iBET, iBET-151, PFI-1 and CPI-203 (Xcessbio) and JQ-1 (EMD Biosciences). For generating tracer isotherms against targets expressed in cells, serially diluted tracer was added to cells in the presence or absence of 20 μM competing unlabelled compound (with the exception of BIBF-1120, for which Dasatinib was used). Cells were then equilibrated for 2 h before BRET measurements. For determining unlabelled compound isotherms, BRET tracers were added to the cells at fixed concentrations before test compound addition. A final concentration of 1 μM was used for SAHA-NCT and iBET-NCT. Experiments with BIBF1120-NCT utilized 3 μM. Serially diluted test compounds were then added to the cells and allowed to equilibrate for 2 h before BRET measurements. To measure BRET, NanoBRET NanoGlo Substrate-(Promega) was added, and filtered luminescence was measure on a BMG LABTECH Clariostar luminometer equipped with 450 nm BP filter (donor) and 610 nm LP filter (acceptor), using 0.5 s integration time with gain settings of 2,800 and 3,500, respectively. Background-corrected BRET ratios were determined by subtracting the BRET ratios of samples with excess competing ligand (20 μM) from the BRET ratios in the absence of competing ligand. Milli-BRET units (mBU) are the BRET values × 1,000. Apparent tracer affinity values were determined as described previously[13] using curve fits in GraphPad Prism with the equation (equation (1)):

$$Y = Bmax \times X/(Kd + X). \tag{1}$$

Competitive displacement data were then graphed with GraphPad Prism software using a three-parameter curve fit with the following equation (equation (2));

$$Y = Bottom + (Top - Bottom)/(1 + 10^{((X - LogIC50))}) \tag{2}$$

Normalized data were generated by assigning 100% to the theoretical maximum of the three-parameter curve fit and 0% for the theoretical minimum value of the three-parameter curve fit. Apparent affinity values for each compound were calculated from observed IC_{50} values according to the Cheng-Prusoff equationref. 39 (equation (3)):

$$K_i = \frac{IC_{50}}{1 + \frac{[L]}{K_D}} \tag{3}$$

where $[L]$ is the concentration of fluorescent ligand in μM and K_D is the apparent intracellular K_D of fluorescent ligand in μM. The apparent intracellular K_D values were calculated from the tracer saturation binding experiments described above.

Kinetic analysis of target engagement via BRET. For kinetic analysis of the equilibration rates of SAHA, FK228 and TDP-A, cells transfected with Nluc-HDAC fusions as described above were pre-equilibrated for 2 h with 1 μM SAHA-NCT tracer prior addition of test compound. NanoBRET NanoGlo Substrate was added to the samples and equilibrated at room temperature for 30 min before BRET measurements. Immediately after addition of the unlabelled test compound, kinetic BRET analysis was performed at room temperature over the time indicated on a Thermo Varioskan Luminometer equipped with 450 nm BP filter (donor) and 610 nm LP filter (acceptor), using 0.5 s integration times.

Kinetic analysis of ligand dissociation via BRET. For direct analysis of relative compound dissociation rates via BRET, 2×10^6 cells transfected with HDAC1-Nluc (in 10 ml of Opti-MEM) were aliquoted into 15 ml conical tubes and first pre-equilibrated with DMSO (vehicle), SAHA (10 μM), mocetinostat (10 μM), FK228 (100 nM) or TDP-A (10 nM) for 3 h at 37 °C/5% CO_2. Final concentration of DMSO was 0.1%. Cells treated with SAHA or mocetinostat were pelleted and resuspended in 10 mL of Opti-MEM in the presence of NanoBRET NanoGlo Substrate and 1 μM SAHA-NCT tracer immediately before BRET measurements. FK228 and TDP-A were subjected to an additional centrifugation and 10 ml wash step (with 5 min incubation) before BRET measurements, to ensure adequate removal of inhibitor from the cell medium. In addition to a zero-occupancy (DMSO/vehicle-treated) sample for determination of $BRET_{100}$, a full-occupancy control sample ($BRET_0$) was included, wherein 10 μM unmodified SAHA was left incubating on the cells (no washout) during the kinetic BRET analysis. To ensure that FK228 and TDP-A were not trapped within intact cells during the residence time analysis, a separate experiment was performed with digitonin added as lytic reagent (50 $\mu g\,ml^{-1}$) added at the time of 1 μM SAHA-NCT tracer addition. BRET data were normalized to 100% signal from the positive control (defined as the maximum BRET value observed from cells treated with 1uM SAHA-NCT in the absence of unlabelled competing compound) and 0% signal (defined as the BRET value observed from cells treated with 1 uM SAHA-NCT + 10 μM unmodified SAHA).

BRET imaging. HeLa cells were transfected with Nluc/target fusion constructs as described previously (see above). Cells suspended at a density of 2×10^5 cells per ml in DMEM (Gibco) + 10% FBS (Hyclone) were plated into 35 mm tissue culture treated imaging dishes (ibidi) at 2 ml per dish and incubated 18–24 h at 37 °C/5% CO_2. Media was removed from imaging dishes via gentle aspiration and replaced with 2 ml warm Opti-MEM without phenol red (Gibco) in the presence or absence of tracer +/− competing cold compound. Tracers and cold compound pairs used include 1 μM SAHA-NCT +/− 10 μM SAHA, 3 μM IBET-NCT +/− 10 μM IBET and 3 μM BIBF-1120-NCT +/− 10 μM Nilotinib. Cells were equilibrated for 2 h at 37 °C + 5% CO_2 before addition of NanoBRET NanoGlo Substrate. Images were captured on an Olympus LV200 microscope equipped with an environmental stage and a Hamamatsu ImagEM EMCCD camera. All images were acquired using a × 100/1.4 UPLanSApo objective and Olympus cellSens software. To image BRET events, images of donor and acceptor emission were acquired sequentially using a 460/80 bandpass filter and a 590 nm long-pass filter, respectively.

Apoptosis and cell proliferation assays. To measure induction of apoptosis, HeLa cells were seeded into 96-well plates at a density of 10^4 cells per well in DMEM + 10% FBS. Serially diluted test compounds were added to the cells and incubated for 20 h. After incubation, an equal volume of Caspase-Glo (3/7) reagent (Promega) was added to the cells, and quantified on a GloMax-Multi Microplate Luminometer (Promega) according to the manufacturer's protocol. To measure inhibition of proliferation, HeLa cells were seeded into 96-well plates at a density of 2.5×10^3 cells per well in DMEM + 10% FBS. Serially diluted test compounds were added as described above and incubated 48 h. After incubation, an equal volume of CellTiter-Glo Luminescent Cell Viability Assay (Promega) was added to the cells, and luminescence was quantified according to the manufacturer's protocol. Data were then graphed with GraphPad Prism software as described above for the BRET target engagement experiments.

Kinetic analysis of compound inhibition on purified HDAC1. Full-length human recombinant HDAC 1 was purchased from SignalChem. HDAC 1 enzymatic activity was measured using HDAC-Glo I/II Assay (Promega)[55]. To determine real-time IC_{50} values, 90 μl of 1/2,400 diluted HDAC 1 in HDAC-Glo I/II assay buffer was added to a opaque white, sterile, TC-treated 96-well assay plate (Corning). To enzyme, 100 μl of HDAC-Glo I/II detection reagent was added (190 μl total volume per well). The assay plate was mixed by a brief plate shake and the HDAC reaction was allowed to initiate and come to a steady-state luminescent signal by incubating for 20 min at room temperature. During this incubation, a serial dilution of SAHA, FK228, TDP-A was performed at a 200 × concentration in 100% DMSO in a separate master 96-well plate. A 10 µl aliquot of this master titration series was added to HDAC-Glo I/II assay buffer to generate a 20 × concentrated intermediate dilution containing 2.5 mM DTT for the prodrug inhibitors. Once the previous 20 min incubation was complete, 10 µl replicates ($n = 2$ for SAHA and $n = 3$ for FK228 and TDP-A) from this master intermediate titration series of compounds were transferred to the opaque white assay plate for a final total volume of 200 µl. The plate was then placed in a POLARstar OPTIMA (BMG LABTECH, Ortenberg, Germany) and luminescence was measured every 5 min for 120 min. For each time point, the raw luminescent signal for each test well of the appropriate compound was normalized to the no inhibition control well average (to calculate % inhibition). For each compound, the % inhibition was graphed versus inhibitor concentration for each time point to determine IC_{50}. Data were then graphed with GraphPad Prism software as described above for the BRET target engagement experiments.

Analysis of expression level of HDAC-NLuc fusions. HEK293 or HeLa cells were transfected with DNA constructs encoding HDAC1-NLuc at 0.8 µg ml^{-1} using PEI as transfection reagent as described previously[56]. To reduce expression levels the DNA was undiluted, diluted 1:10 and diluted 1:100 into a promoterless carrier DNA plasmid to generate a final total DNA concentration of 0.8 µg ml^{-1}. Control cells were transfected with the promoterless carrier DNA plasmid only. Twenty-four hours after transfection the cells were collected using Cellstriper (Corning), washed with PBS and then lysed using detergent lysis buffer (mammalian lysis buffer from Promega) supplemented with 1:50 dilution of RQ1 DNase (Promega) and 1 × RQ1 DNAase buffer and 1 × protease inhibitor cocktail (Promega). 5% of each cell lysate was analysed by SDS-PAGE and electro-transferred onto a PVDF membrane (Life Technologies). The membrane was blocked for 1 h with 5% BSA (Promega) in TBS buffer and probed overnight at 4 °C with the primary antibody in TBS supplemented with 0.1% Tween-20 (TBST). After three washes in TBST, the membrane was incubated with a secondary HRP - conjugated antibody (Jackson laboratories) in TBST for 1 h, washed five times with TBST and one time with TBS. The immune-stained proteins were detected using enhanced chemiluminescent (ECL) reagent (Promega) and detected on the LAS400 imager (GE Healthcare). Antibody source: anti-HDAC1 (Abcam; ab46985), anti-HDAC2 (Abcam; ab51832).

Activity of HDAC2 expressed in cell free expression system. HDAC2-NLuc, HDAC2 and NLuc were expressed in the S30 T7 High-Yield Protein expression system (Promega) as recommended by the manufacturer. A total of 40 µl of each expression reaction (in triplicates) were tested for HDAC2 activity using HDAC-Glo 2 assay (Promega). Luminescence was measured on a GloMAX luminometer (Promega).

References

1. Copeland, R. A., Pompliano, D. L. & Meek, T. D. Drug-target residence time and its implications for lead optimization. *Nat. Rev. Drug. Discov.* **5,** 730–739 (2006).
2. Lu, H. & Tonge, P. J. Drug-target residence time: critical information for lead optimization. *Curr. Opin. Chem. Biol.* **14,** 467–474 (2010).
3. Tummino, P. J. & Copeland, R. A. Residence time of receptor-ligand complexes and its effect on biological function. *Biochemistry* **47,** 5481–5492 (2008).
4. Bantscheff, M. *et al.* Chemoproteomics profiling of HDAC inhibitors reveals selective targeting of HDAC complexes. *Nat. Biotechnol.* **29,** 255–265 (2011).
5. Becher, I. *et al.* Chemoproteomics reveals time-dependent binding of histone deacetylase inhibitors to endogenous repressor complexes. *ACS Chem. Biol.* **9,** 1736–1746 (2014).
6. Vauquelin, G. & Charlton, S. J. Exploring avidity: understanding the potential gains in functional affinity and target residence time of bivalent and heterobivalent ligands. *Br. J. Pharmacol.* **168,** 1771–1785 (2013).
7. Vauquelin, G. Rebinding: or why drugs may act longer *in vivo* than expected from their *in vitro* target residence time. *Expert Opin. Drug Discov.* **5,** 927–941 (2010).
8. Savitski, M. M. *et al.* Proteomics. Tracking cancer drugs in living cells by thermal profiling of the proteome. *Science* **346,** 1255784 (2014).
9. Jafari, R. *et al.* The cellular thermal shift assay for evaluating drug target interactions in cells. *Nat. Protoc.* **9,** 2100–2122 (2014).
10. Martinez Molina, D. *et al.* Monitoring drug target engagement in cells and tissues using the cellular thermal shift assay. *Science* **341,** 84–87 (2013).
11. Zwier, J. M. *et al.* A fluorescent ligand-binding alternative using Tag-lite(R) technology. *J. Biomol. Screen.* **15,** 1248–1259 (2010).
12. Lebakken, C. S. *et al.* Development and applications of a broad-coverage, TR-FRET-based kinase binding assay platform. *J. Biomol. Screen.* **14,** 924–935 (2009).
13. Stoddart, L. A. *et al.* Application of BRET to monitor ligand binding to GPCRs. *Nat. Methods.* **12,** 661–663 (2015).
14. Simon, G. M., Niphakis, M. J. & Cravatt, B. F. Determining target engagement in living systems. *Nat. Chem. Biol.* **9,** 200–205 (2013).
15. Moreau, M. J., Morin, I. & Schaeffer, P. M. Quantitative determination of protein stability and ligand binding using a green fluorescent protein reporter system. *Mol. Biosyst.* **6,** 1285–1292 (2010).
16. Taipale, M. *et al.* Chaperones as thermodynamic sensors of drug-target interactions reveal kinase inhibitor specificities in living cells. *Nat. Biotechnol.* **31,** 630–637 (2013).
17. Savitski, M. M. *et al.* Tracking cancer drugs in living cells by thermal profiling of the proteome. *Science* **346,** 1255784 (2014).
18. Furumai, R. *et al.* FK228 (depsipeptide) as a natural prodrug that inhibits class I histone deacetylases. *Cancer Res.* **62,** 4916–4921 (2002).
19. Wang, C. *et al.* Thailandepsins: bacterial products with potent histone deacetylase inhibitory activities and broad-spectrum antiproliferative activities. *J. Nat. Prod.* **74,** 2031–2038 (2011).
20. Giang, I., Boland, E. L. & Poon, G. M. Prodrug applications for targeted cancer therapy. *AAPS J.* **16,** 899–913 (2014).
21. Wilson, A. J., Cheng, Y. Q. & Khabele, D. Thailandepsins are new small molecule class I HDAC inhibitors with potent cytotoxic activity in ovarian cancer cells: a preclinical study of epigenetic ovarian cancer therapy. *J. Ovarian Res.* **5,** 12 (2012).
22. Wei, D. G. *et al.* Histone deacetylase inhibitor romidepsin induces HIV expression in CD4 T cells from patients on suppressive antiretroviral therapy at concentrations achieved by clinical dosing. *PLoS Pathog.* **10,** e1004071 (2014).
23. Crabb, S. J. *et al.* Characterisation of the *in vitro* activity of the depsipeptide histone deacetylase inhibitor spiruchostatin A. *Biochem. Pharmacol.* **76,** 463–475 (2008).
24. Ito, T. *et al.* Real-time imaging of histone H4K12-specific acetylation determines the modes of action of histone deacetylase and bromodomain inhibitors. *Chem. Biol.* **18,** 495–507 (2011).
25. Machleidt, T. *et al.* NanoBRET-A novel BRET platform for the analysis of protein-protein interactions. *ACS Chem. Biol.* **10,** 1797–1804 (2015).
26. Pfleger, K. D. & Eidne, K. A. Illuminating insights into protein-protein interactions using bioluminescence resonance energy transfer (BRET). *Nat. Methods* **3,** 165–174 (2006).
27. Pfleger, K. D., Seeber, R. M. & Eidne, K. A. Bioluminescence resonance energy transfer (BRET) for the real-time detection of protein-protein interactions. *Nat. Protoc.* **1,** 337–345 (2006).
28. Dacres, H., Dumancic, M. M., Horne, I. & Trowell, S. C. Direct comparison of bioluminescence-based resonance energy transfer methods for monitoring of proteolytic cleavage. *Anal. Biochem.* **385,** 194–202 (2009).
29. Friedman Ohana, R. *et al.* Deciphering the cellular targets of bioactive compounds using a chloroalkane capture tag. *ACS Chem. Biol.* **10,** 2316–2324 (2015).
30. Griss, R. *et al.* Bioluminescent sensor proteins for point-of-care therapeutic drug monitoring. *Nat. Chem. Biol.* **10,** 598–603 (2014).
31. Hall, M. P. *et al.* Engineered luciferase reporter from a deep sea shrimp utilizing a novel imidazopyrazinone substrate. *ACS Chem. Biol.* **7,** 1848–1857 (2012).
32. Bradner, J. E. *et al.* Chemical phylogenetics of histone deacetylases. *Nat. Chem. Biol.* **6,** 238–243 (2010).
33. Finnin, M. S. *et al.* Structures of a histone deacetylase homologue bound to the TSA and SAHA inhibitors. *Nature* **401,** 188–193 (1999).
34. Wang, D. F., Helquist, P., Wiech, N. L. & Wiest, O. Toward selective histone deacetylase inhibitor design: homology modeling, docking studies, and molecular dynamics simulations of human class I histone deacetylases. *J. Med. Chem.* **48,** 6936–6947 (2005).
35. Miller, K. M. *et al.* Human HDAC1 and HDAC2 function in the DNA-damage response to promote DNA nonhomologous end-joining. *Nat. Struct. Mol. Biol.* **17,** 1144–1151 (2010).
36. Yao, Y. L. & Yang, W. M. Beyond histone and deacetylase: an overview of cytoplasmic histone deacetylases and their nonhistone substrates. *J. Biomed. Biotechnol.* **2011,** 146493 (2011).
37. Zhang, Y., Gilquin, B., Khochbin, S. & Matthias, P. Two catalytic domains are required for protein deacetylation. *J. Biol. Chem.* **281,** 2401–2404 (2006).
38. Zou, H., Wu, Y., Navre, M. & Sang, B. C. Characterization of the two catalytic domains in histone deacetylase 6. *Biochem. Biophys. Res. Commun.* **341,** 45–50 (2006).
39. Cheng, Y. & Prusoff, W. H. Relationship between the inhibition constant (K1) and the concentration of inhibitor which causes 50 per cent inhibition (I50) of an enzymatic reaction. *Biochem. Pharmacol.* **22,** 3099–3108 (1973).
40. Witt, O., Deubzer, H. E., Milde, T. & Oehme, I. HDAC family: what are the cancer relevant targets? *Cancer. Lett.* **277,** 8–21 (2009).
41. Wang, C., Flemming, C. J. & Cheng, Y. Q. Discovery and activity profiling of thailandepsins A through F, potent histone deacetylase inhibitors, from E264. *MedChemComm* **3,** 976–981 (2012).
42. Wilting, R. H. *et al.* Overlapping functions of Hdac1 and Hdac2 in cell cycle regulation and haematopoiesis. *EMBO J.* **29,** 2586–2597 (2010).

43. Yamaguchi, T. *et al.* Histone deacetylases 1 and 2 act in concert to promote the G1-to-S progression. *Genes Dev.* **24,** 455–469 (2010).
44. Neumann, L., von Konig, K. & Ullmann, D. HTS reporter displacement assay for fragment screening and fragment evolution toward leads with optimized binding kinetics, binding selectivity, and thermodynamic signature. *Methods Enzymol.* **493,** 299–320 (2011).
45. Di Micco, S. *et al.* Structural basis for the design and synthesis of selective HDAC inhibitors. *Bioorg. Med. Chem.* **21,** 3795–3807 (2013).
46. Fournel, M. *et al.* MGCD0103, a novel isotype-selective histone deacetylase inhibitor, has broad spectrum antitumor activity *in vitro* and *in vivo*. *Mol. Cancer. Ther.* **7,** 759–768 (2008).
47. Bonfils, C. *et al.* Evaluation of the pharmacodynamic effects of MGCD0103 from preclinical models to human using a novel HDAC enzyme assay. *Clin. Cancer Res.* **14,** 3441–3449 (2008).
48. Schulze, J. *et al.* Cell-based protein stabilization assays for the detection of interactions between small-molecule inhibitors and BRD4. *J Biomol. Screen.* **20,** 180–189 (2014).
49. Davis, M. I. *et al.* Comprehensive analysis of kinase inhibitor selectivity. *Nat. Biotechnol.* **29,** 1046–1051 (2011).
50. Karaman, M. W. *et al.* A quantitative analysis of kinase inhibitor selectivity. *Nat. Biotechnol.* **26,** 127–132 (2008).
51. Guenther, M. G., Barak, O. & Lazar, M. A. The SMRT and N-CoR corepressors are activating cofactors for histone deacetylase 3. *Mol. Cell. Biol.* **21,** 6091–6101 (2001).
52. Couturier, C. & Deprez, B. Setting Up a bioluminescence resonance energy transfer high throughput screening assay to search for protein/ protein interaction inhibitors in mammalian cells. *Front. Endocrinol.* **3,** 100 (2012).
53. Dacres, H., Dumancic, M. M., Horne, I. & Trowell, S. C. Direct comparison of fluorescence- and bioluminescence-based resonance energy transfer methods for real-time monitoring of thrombin-catalysed proteolytic cleavage. *Biosens. Bioelectron.* **24,** 1164–1170 (2009).
54. Lebakken, C. S., Reichling, L. J., Ellefson, J. M. & Riddle, S. M. Detection of allosteric kinase inhibitors by displacement of active site probes. *J Biomol. Screen.* **17,** 813–821 (2012).
55. Halley, F. *et al.* A bioluminogenic HDAC activity assay: validation and screening. *J Biomol. Screen.* **16,** 1227–1235 (2011).
56. Ohana, R. F. *et al.* HaloTag-based purification of functional human kinases from mammalian cells. *Protein Expr. Purif.* **76,** 154–164 (2011).

Acknowledgements

We thank Frank Fan, Jennifer Wilkinson, Cesear Corona and Poncho Meisenheimer (Promega) for their thoughtful input on this manuscript.

Author contributions

M.B.R. and K.V.W. wrote the paper and designed the experiments. All authors contributed to the design and execution of the experiments.

Additional information

Competing financial interests: Besides Yi-Qiang Cheng, all authors are employees of Promega Corporation.

Metal ion-directed dynamic splicing of DNA through global conformational change by intramolecular complexation

Toshihiro Ihara[1,2], Hiroyuki Ohura[1], Chisato Shirahama[1], Tomohiro Furuzono[1], Hiroshi Shimada[1], Hirotaka Matsuura[1] & Yusuke Kitamura[1,2]

Chemically engineered DNAs—in which global conformation can be modulated in response to specific stimuli—could be allosteric functional DNAs themselves or work as a modulator of the functional nucleic acids such as DNAzymes and aptamers. Here, we show that two terpyridines built in the DNA backbone form a stable intramolecular 1:2 complex, $[M(terpy)_2]^{2+}$, with divalent transition metal ions. Upon complexation, the DNA conjugates adopt a Ω-shape structure, in which two distal sequences located outside the terpyridines connect with each other to form a continuous segment with a specific structure or sequence. Such a DNA structure is globally controlled by local metal complexation events that can be rationally designed based on general coordination chemistry. This method is regarded as metal ion-directed dynamic sequence edition or DNA splicing. DNAzymes with peroxidase-like activity can thus be regulated by several transition metal ions through sequence edition techniques based on the Ω-motif.

[1]Department of Applied Chemistry and Biochemistry, Graduate School of Science and Technology, Kumamoto University, 2-39-1 Kurokami, Chuo-ku, Kumamoto 860-8555, Japan. [2]CREST, Japan Science and Technology Agency, 7 Gobancho, Chiyoda-ku, Tokyo 102-0076, Japan. Correspondence and requests for materials should be addressed to T.I. email: (toshi@chem.kumamoto-u.ac.jp).

Recently, several research groups have reported on the fabrication of specific DNA structures using metal ion-directed techniques[1–21]. For example, two thymines and cytosines coordinate to Hg^{2+} and Ag^{+}, respectively, to form a M-DNA (metal ion DNA complex) duplex through the formation of metal-centred nucleobase dimers[7,8,12–15]. A silver ion can be accommodated in the nucleobase triplet CGC of a DNA triplex by substitution of the N3 proton of cytosine on the third strand, which stabilizes the triplex structure even under neutral pH[10,11]. The knowledge of the specific coordination of nucleobases to these soft metal ions enabled us to prepare stable duplexes or triplexes by designing the sequences. On the other hand, several DNA analogues carrying metal chelators instead of natural nucleobases have been reported. Chelators, such as catechol, phenylenediamine, aminophenol and hydroxypyridone, selectively coordinate to certain transition metal ions to form stable M-DNA duplexes or triplexes[4–6]. The DNA conjugates with chelator molecules on their backbone have also been reported to form unique structures[9,16–21].

In the past decade, a number of functional nucleic acids, such as ribozymes, DNAzymes and aptamers, have been isolated using techniques of molecular evolution[22,23]. These nucleic acids have been applied to analytical systems for various biomolecules as functional modules for analyte recognition and signal amplification[24–27]. Prescribed higher-ordered structures are critically important for these nucleic acids to express their intrinsic functions. Therefore, techniques that control formation of their structures should be useful to modulate their functions. If the chemical groups, which respond to certain stimuli by changing their structure, are incorporated into the backbone of the functional DNAs or their complementary sequences, we could regulate their functions by controlling conformational changes in the DNAs. This strategy is general in the sense that it targets the higher-ordered structure of functional nucleic acids rather than each of their particular functions themselves. The techniques for drastic changes in DNA conformation could also be applied to regulate gene expression by certain stimuli such as decoy, antisense and RNAi strategies.

Herein, we synthesized 2,2′:6′,2″-terpyridine (terpy)-based amidite reagents. Terpy units can be incorporated into any positions in the DNA backbone with these amidite reagents using automated DNA synthesis. Terpy forms a stable complex with Fe^{2+} and some other divalent transition metal ions[28,29]. We expected that the stiffness and conformation of the DNA backbone could be controlled through complexation with certain metal ions. Then, it would enable us to regulate the activity of functional nucleic acids. We have already reported the effect of metal ions on the stability of DNA structures consisting of the conjugates with one terpy (synthesized using **1** mentioned below). The terpy incorporated in the stem moiety of the duplex stabilized the structure by complexation, which confers entropic benefit on the DNA. In contrast, the structure of bimolecular triplexes carrying a terpy unit in its loop moiety significantly destabilizes on complexation with divalent transition metal ions[30]. This can be explained by the change in the length of the terpy unit in the loop of the triplex by metal coordination; the metal-bound terpy connecting two pyrimidine strands is too short to cap the base triplet in the structure of the DNA triplex.

In the present study, two terpy units are incorporated into the backbone of a DNA strand to form $terpy_2$–DNA conjugates, in which two terpys are located at distant positions from each other with an intervening sequence. It is expected that transition metal ions, such as Fe^{2+}, forms a 1:2 complex with the two terpys in a $terpy_2$–DNA conjugate. To form such an intramolecular complex, the backbone of $terpy_2$–DNA conjugates make adopt an Ω-shaped conformation, where the two distal sequences outside both terpys in the sequence are directly connected with each other to form a continuous stretch of the specific structure or sequence. Such global conformational control of DNA would be regarded as metal ion-directed sequence edition or reversible splicing of DNA. We perform allosteric regulation of catalytic activity of DNAzyme through metal ion-directed sequence edition of DNA.

Results

DNA preparation and system design. In this study, we aimed to regulate the function of DNAzyme with peroxidase-like activity[31,32] by conformational control of the $terpy_2$–DNA conjugate. To incorporate terpy units into the backbone of the DNA, terpy-based amidite reagents were prepared. We synthesized two amidite reagents **1** and **2** (Fig. 1a) according to the schemes shown in Supplementary Information. Although both of them have a 2,2′:6′,2″-terpyridine structure with two ethylene linker chains connecting to phosphoramidite and dimethoxytrityl-protected hydroxyl groups, the substitution position of the linkers are different. **1** and **2** tether the two linker chains from 6,6″- and 5,5″-positions of terpy, respectively.

The sequences of the DNAs used in this study are shown in Fig. 1b. The four $terpy_2$–DNA conjugates, $\mathbf{tp_2hp1}$, $\mathbf{tp_2hp2}$, $\mathbf{tp_2tmp1}$ and $\mathbf{tp_2tmp2}$, were prepared using **1** and **2** according to the standard phosphoramidite method shown in Supplementary Information. The last letter in the names of the $terpy_2$–DNA conjugates, '1' or '2', stands for the amidite reagents **1** and **2** built in their sequences, respectively. We designed two systems for regulation of oxidation reactions by DNAzyme. In one system, we used the modified DNAzyme, $\mathbf{tp_2hp1}$ and $\mathbf{tp_2hp2}$, which can adopt a hairpin form by intramolecular metal complexation. The sequence of DNAzyme was split into two parts[33], and each half was attached to both ends of a part of the complimentary sequence of the H-*ras* gene[34] through terpys to form $\mathbf{tp_2hp1}$ and $\mathbf{tp_2hp2}$. The catalytic activity of $\mathbf{tp_2hp1}$ and $\mathbf{tp_2hp2}$ were expected to be restored when forming a hairpin conformation induced by intramolecular 1:2 metal complexation (Fig. 1c). Reconstruction of an integrated active DNAzyme on the allosteric template, $\mathbf{tp_2tmp1}$ and $\mathbf{tp_2tmp2}$, was expected in another system. The same DNAzyme mentioned above was also split into two parts, and then the short binding arms were attached to the end of these parts to form a pair of the half DNAzymes[33], **zymR** and **zymL**. The outside two sequences in $\mathbf{tp_2tmp1}$ and $\mathbf{tp_2tmp2}$ were designed to be complementary to the binding arms of **zymR** and **zymL**. Therefore, $\mathbf{tp_2tmp1}$ and $\mathbf{tp_2tmp2}$ are expected to work as an effective template when both outside sequences come in proximity with each other by forming an Ω-shaped conformation through 1:2-type intramolecular metal complexation (Fig. 1d). The loop sequences between the two terpys of $\mathbf{tp_2tmp1}$ and $\mathbf{tp_2tmp2}$ are designed to be complementary to miR-34 (microRNA)[35].

Preliminary studies of complexation. Complexation of the $terpy_2$–DNA conjugates, $\mathbf{tp_2tmp1}$ and $\mathbf{tp_2tmp2}$, with metal ions were investigated by ultraviolet–visible (vis) titrations. Four divalent transition metal ions, Fe^{2+}, Ni^{2+}, Cu^{2+} and Zn^{2+}, were used in this study (Fig. 2). The aliquots of transition metal ions were added into the conjugate solutions, and their ultraviolet–vis spectrum was recorded after each addition. The spectral changes and the plots of absorbance around the peak with Fe^{2+} titration are shown in Fig. 2a,b. The complexation behaviour of the two conjugates was completely different. $\mathbf{tp_2tmp1}$ showed stoichiometric complexation with Fe^{2+}. The absorbance increased monotonously and saturated at the point of $[Fe^{2+}]/[\mathbf{tp_2tmp1}] = 2.0$ (Fig. 2a), indicating that each of the two terpys on a $\mathbf{tp_2tmp1}$ bound a Fe^{2+} to form two $[Fe(terpy)]^{2+}$ on a **tp_2tmp1**. In contrast, the stoichiometry of **tp_2tmp2** complexation was clearly shown to be 1.0 ($=[Fe^{2+}]/[$**tp_2tmp2**$]$) (Fig. 2b), indicating that the terpy units formed $[Fe(terpy)_2]^{2+}$ on **tp_2tmp2**. Further change was not observed in the spectra even with more additional Fe^{2+}. The characteristic band was observed at 530 nm. This is attributed to the metal-to-ligand charge-transfer band of the 1:2 complex, $[Fe(terpy)_2]^{2+}$ (refs 28,29). Interestingly, this meant that the complex $[Fe(terpy)_2]^{2+}$ did not form on **tp_2tmp1** but exclusively formed on **tp_2tmp2**. All of the results observed in the ultraviolet–vis titrations were reproduced in the presence of the complementary DNA, **ctmp**, which is the tandem sequence complementary to the two distal sequences outside of the two terpys.

The only difference between **tp_2tmp1** and **tp_2tmp2** is the tethering positions of the terpys connected to DNA backbones. We calculated the structures of both complexes using the density functional theory (DFT) method (B3LYP/6-31G*). The optimized structures of $[Fe(terpy)_2]^{2+}$ are shown in Fig. 3. The structure of Fig. 3a is the complex consisting of a Fe^{2+} and two 6,6″-tethered terpys and Fig. 3b is of a Fe^{2+} and two 5,5″-tethered terpys, each of which is the complex expected to form on **tp_2tmp1** (or **tp_2tmp1/ctmp**) and **tp_2tmp2** (or **tp_2tmp2/ctmp**), respectively. In the structure of $[Fe(terpy)_2]^{2+}$ with 6,6″-tethered terpys, the linker chains on a terpy almost touches the central pyridine ring of another terpy in the complex. This steric hindrance should make the complex unstable or impossible to form. In contrast, this hindrance was not observed in the complex consisting of 5,5″-tethered terpys. This would explain the difference in complexation behaviour between **tp_2tmp1** and **tp_2tmp2** observed in ultraviolet–vis titration.

The stoichiometry of Ni^{2+}–**tp_2tmp2** complexation was almost the same with Fe^{2+} mentioned above (Fig. 2c, spectral change: see Supplementary Fig. 1). That is, regardless of the feeding ratio, $[M^{2+}]/[$**tp_2tmp2**$]$, Fe^{2+} and Ni^{2+} seem to form a rather stable complex, $[M(terpy)_2]^{2+}$, on **tp_2tmp2**. Meanwhile, Cu^{2+} (Fig. 2d, spectral change: see Supplementary Fig. 2) and Zn^{2+} (see Supplementary Fig. 3) showed a multiphasic unique spectral change. For both titrations, two apparent inflection points were observed at $[M^{2+}]/[$**tp_2tmp2**$]$ ratios of 1.0 and 2.0, indicating that the two types of complexes with different stoichiometry were successively formed with the addition of the metal ions. 1:2 complex, $[M(terpy)_2]^{2+}$, formed on **tp_2tmp2** at a lower feeding ratio, which changed to a 1:1 complex, $[M(terpy)]^{2+}$ upon further addition of metal ions. The difference in the stoichiometry of the complexes of the four transition metal ions could be explained by the relative magnitude of successive binding constants, K_1 and K_2; Fe^{2+}: $\log K_1 = 7.1$, $\log K_2 = 13.8$; Ni^{2+}: 10.7, 11.2; Zn^{2+}: 6.7, 5.2; Cu^{2+}: 13.0, not available (that is, K_2 is much smaller than K_1). Whereas the magnitudes of K_1 are larger than K_2 for Fe^{2+} and Ni^{2+}, the opposite is true for Cu^{2+} and Zn^{2+} (refs 28,36).

To identify the complex, an equimolar mixture of terpy$_2$–DNA and Fe^{2+} was subjected to matrix-assisted laser desorption ionization time-of-flight mass spectrometry. The short terpy$_2$–DNA conjugate (18-mer DNA conjugate carrying two terpy units) was prepared for this study using the terpy amidite **2**. Using the conjugate with a lower molecular weight, we can obtain accurate and precise *m/z* values of the complexes, even for multimeric DNAs. The result are shown in Supplementary Fig. 4. The only complex observed was $[$terpy$_2$–DNA $+ 54]^-$ for the

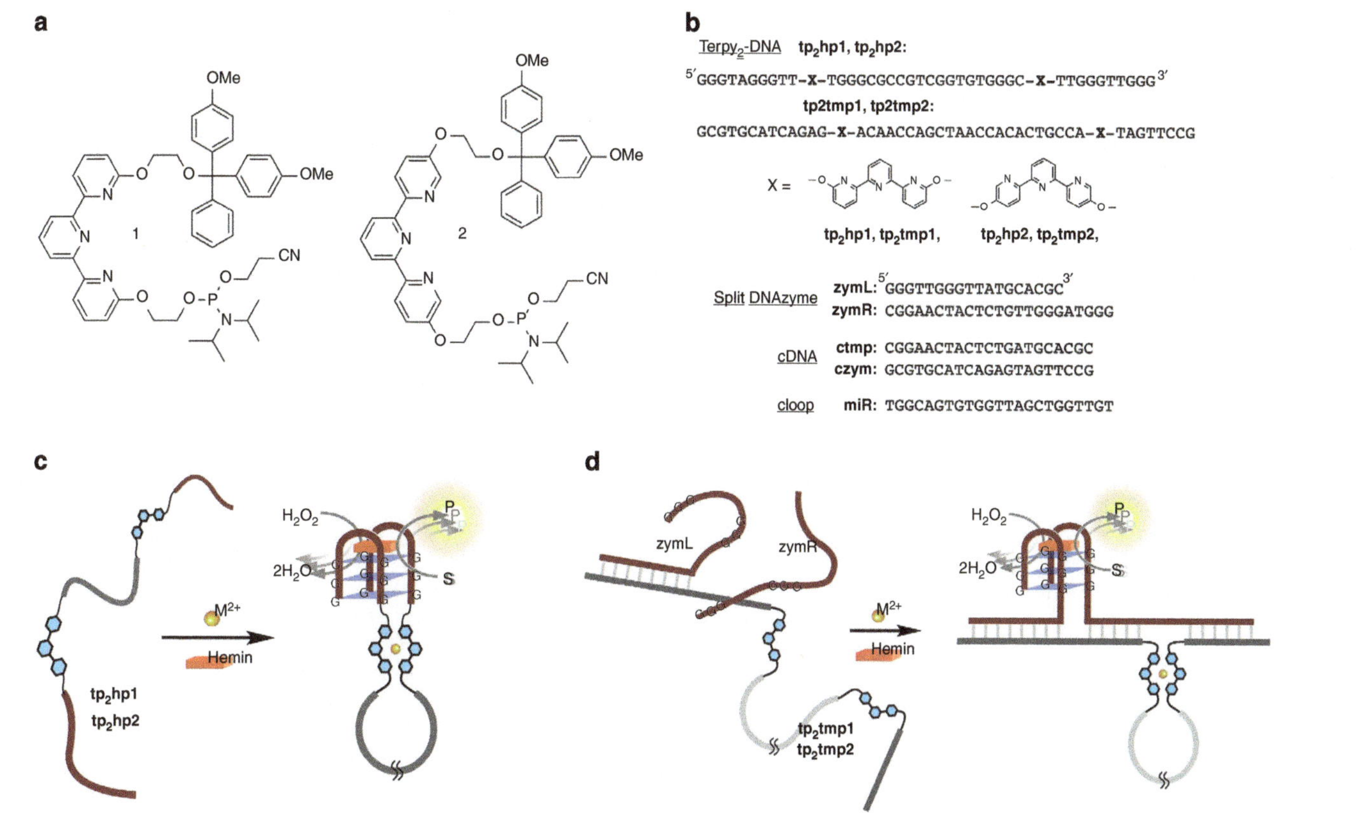

Figure 1 | DNA structures, sequences and constructions of allosteric DNAzymes. (**a**) Structures of terpyridine-based amidite reagents, **1** and **2**. (**b**) Sequences of terpy$_2$-DNA conjugates and DNAs used in this study. (**c**) Working principle of metal ion-dependent DNAzyme, **tp_2hp1** and **tp_2hp2**. The modified DNAzyme adopts an active hairpin structure by intramolecular metal complexation. (**d**) Allosteric regulation of split DNAzyme activity by metal ion-directed dynamic sequence edition of the template, **tp_2tmp1** and **tp_2tmp2**.

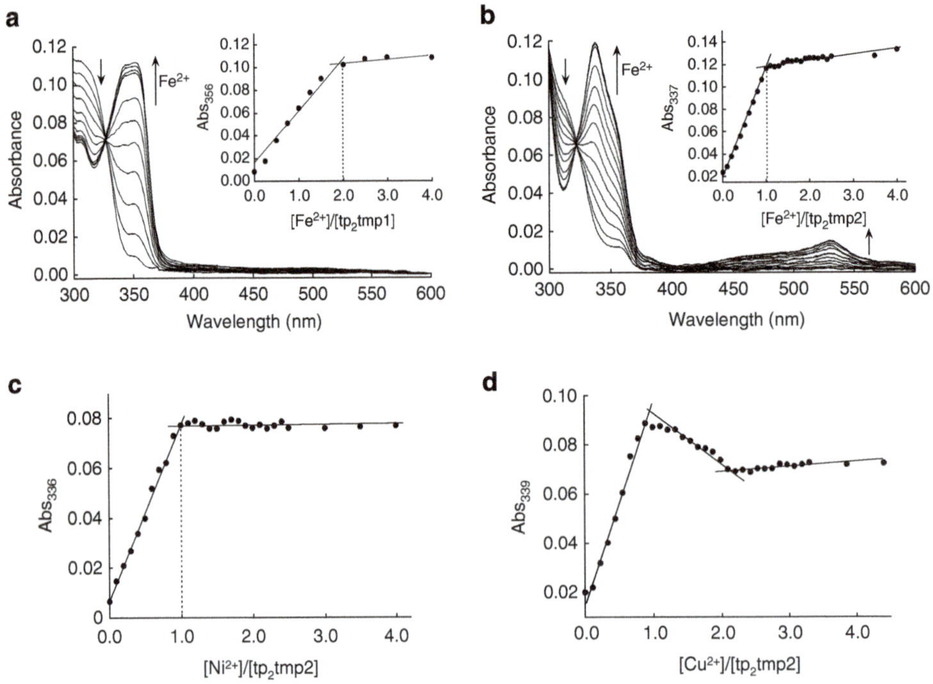

Figure 2 | Preliminary spectrophotometric studies for metal complexation of tp$_2$tmp1 and tp$_2$tmp2 with divalent transition metal ions. (**a**) Spectral change of **tp$_2$hp1** in ultraviolet-vis titration of Fe^{2+}. Inset: the absorbance at 356 nm was plotted against the feeding ratio, [Fe^{2+}]/[**tp$_2$hp1**]. (**b**) Spectral change of **tp$_2$hp2** in ultraviolet-vis titration of Fe^{2+}. Inset: the absorbance at 337 nm was plotted against the feeding ratio, [Fe^{2+}]/[**tp$_2$hp2**]. (**c**) The result of Ni^{2+} titration. The absorbance at 336 nm was plotted against the feeding ratio, [Ni^{2+}]/[**tp$_2$hp2**]. (**d**) The result of Cu^{2+} titration. The absorbance at 339 nm was plotted against the feeding ratio, [Cu^{2+}]/[**tp$_2$hp2**]. The titrations were performed with 2.0 μM DNA in HEPES buffer solution (10 mM, pH 7.0) containing 500 mM NaCl at 25 °C.

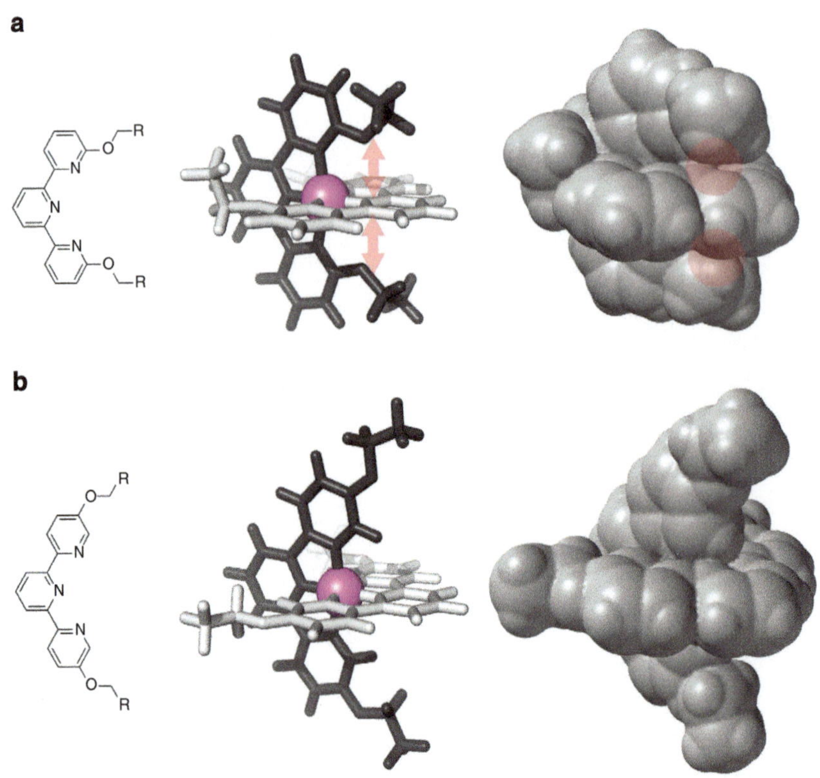

Figure 3 | Optimized structures of [Fe(terpy)$_2$]$^{2+}$ with 6,6″-substituted and 5,5″-substituted terpyridines. Both structures were optimized with DFT calculation (B3LYP/6-31G*). (**a**) Optimized structure of [Fe(terpy)$_2$]$^{2+}$ with 6,6″-terpyridines, which is proposed to form on **tp$_2$hp1** and **tp$_2$tmp1**. (**b**) Optimized structure of [Fe(terpy)$_2$]$^{2+}$ with 5,5″-terpyridines, which is proposed to form on **tp$_2$hp2** and **tp$_2$tmp2**. The two structures shown in the centre and the right-hand side are different representations of the same structures, stick model and Van der Waals surface, respectively.

equimolar mixture. The *m/z* of 54 corresponds to the mass difference between Fe^{2+} and two protons. No multimeric DNA complexes were observed. This result shows that the conjugate, terpy$_2$-DNA, does not form intermolecular multimeric complexes, but intramolecularly forms the complex $[Fe(terpy)_2]^{2+}$.

Metallo-regulation of thermal stability of the duplexes. The melting behaviour was studied for the duplexes consisting of the terpy$_2$-DNA conjugates and the complementary DNA, **ctmp**. The ultraviolet melting curves of the duplexes, **tp$_2$tmp1/ctmp** and **tp$_2$tmp2/ctmp**, recorded in the presence of the divalent transition metal ions are shown in Fig. 4. The melting curves of **tp$_2$tmp1/ctmp** were scarcely affected by any transition metal ions (Fig. 4a). In contrast, the duplex structure of **tp$_2$tmp2/ctmp** was apparently stabilized by the addition of an equimolar amount of transition metal ions (Fig. 4b). In addition, the metal ions appeared to change the shape of the melting curves of **tp$_2$tmp2/ctmp** to be more cooperative, indicating that the two sequences outside of the terpys cooperatively dissociated in a narrow temperature range in the presence of the metal ions.

The highly cooperative melting behaviour also suggests that $[Fe(terpy)_2]^{2+}$ observed in ultraviolet–vis titration for **tp$_2$tmp2** was an intramolecular complex formed on **tp$_2$tmp2•**Fe^{2+}. That is, **tp$_2$tmp2** formed an Ω-shaped conformation through intramolecular 1:2 complexation and, concomitantly, the distal two segments located outside of two terpys came close in proximity with each other to form a continuous stretch of the complementary sequence for **ctmp** (Fig. 4c). The duplexes would be stabilized by entropic benefit through metal ion-directed global conformational change of DNA. This could be regarded as dynamic sequence edition. Meanwhile, the complex formation does not affect the global conformation of **tp$_2$tmp1**, because each of the terpys binds Fe^{2+} independently, which is the reason the metal ions scarcely stabilize the duplex **tp$_2$tmp1/ctmp**.

The melting temperature of **czym/ctmp** was 38 °C under the same conditions. Although the Ω-shaped conformation of **tp$_2$tmp2•**Fe^{2+} gave the same sequence with **czym**, the duplex stability of **tp$_2$tmp2•**Fe^{2+}**/ctmp** did not come up to **czym/ctmp**. The structure of **tp$_2$tmp2•**Fe^{2+} could be improved for its structural continuity. Comparable results were commonly observed for chemical ligation products with non-natural structures at the connecting points[37]. It could be improved by optimization of the linker chain structure of terpy units in terpy$_2$-DNA.

At the point of equimolar ratio ($[M^{2+}]/[\mathbf{tp_2tmp2}] = 1$), all transition metal ions used in the melting study stabilized the duplex **tp$_2$tmp2/ctmp** to a similar extent. However, the dependences of the stabilization effects on the feeding ratio were different for each of the metal ions. For Fe^{2+} and Ni^{2+}, the duplex remained stable at $[M^{2+}]/[\mathbf{tp_2tmp2}] = 2$ and 3. In contrast, the duplex was destabilized at higher feeding ratios for Cu^{2+} and Zn^{2+} (Fig. 4d). The results essentially coincide with those obtained in the ultraviolet–vis titration. The duplex stability of **tp$_2$tmp2/ctmp** was maintained even in the presence of the excess amount of Fe^{2+} and Ni^{2+}, because the Ω-shaped conformation of **tp$_2$tmp2** would be preserved due to the binding properties of the metal ions with terpy, $K_1 < K_2$. Meanwhile, for Cu^{2+} and Zn^{2+}, the global conformation of **tp$_2$tmp2** would be changed from Ω- to a linear-form accompanying the transition of the complex types formed on **tp$_2$tmp2** from $[M(terpy)_2]^{2+}$ (on **tp$_2$tmp2•**M^{2+}) to $2[M(terpy)]^{2+}$ (on **tp$_2$tmp2•**$2M^{2+}$) with increasing amounts of the ions due to their magnitudes of the two successive binding constants with terpy, $K_1 > K_2$ (ref. 36). Especially, the binding of Cu^{2+} with endocyclic nitrogens of nucleobases is strong[38,39]. This would explain the significant inhibition of the duplex formation by Cu^{2+} at higher $[Cu^{2+}]/[\mathbf{tp_2tmp2}]$ ratios.

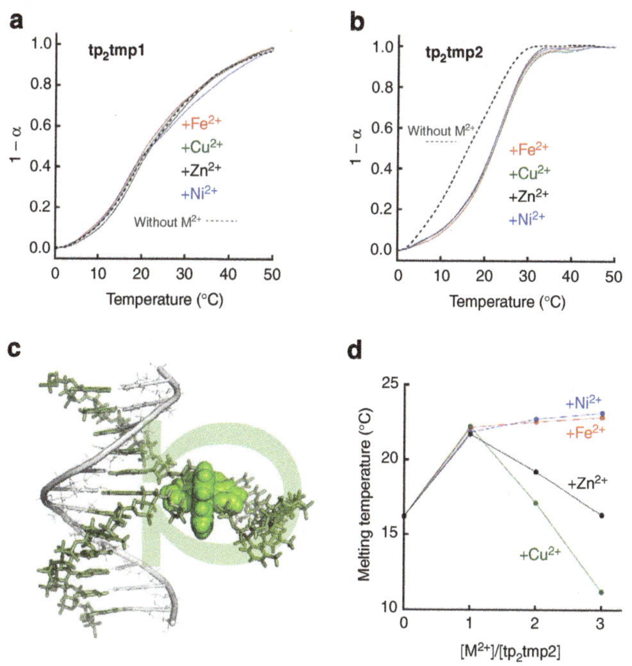

Figure 4 | Effects of the transition metal ions on melting behaviour of the duplex, terpy$_2$-DNA/ctmp. (**a**) Ultraviolet melting curves of **tp$_2$tmp1/ctmp** in the presence of divalent transition metal ions. The dotted curve was obtained in the absence of any transition metals. Red, green, black and blue curves show the melting curves obtained in the presence of Fe^{2+}, Cu^{2+}, Zn^{2+} and Ni^{2+}, respectively. (**b**) Ultraviolet melting curves of **tp$_2$tmp2/ctmp** in the presence of divalent transition metal ions. Assignment of the curves are the same with **a**. Melting experiments were carried out for 1.0 μM DNAs and the metal ions in HEPES buffer solution (10 mM, pH 7.0). (**c**) One of the possible structures of the Ω-shape DNA duplex, **tp$_2$tmp2•**Fe^{2+}**/ctmp**. The model was geometry-optimized by AMBER* force field with GB/SA (generalized Born/surface area) solvent model using MacroModel version 9.5. (**d**) The effect of the feeding ratio of the metal ions on the melting temperatures of the duplex, **tp$_2$tmp2/ctmp**.

Allosteric regulation of DNAzyme activity. We constructed the two systems to regulate the activity of DNAzyme by specific coordination of metal ions. The active form of the DNAzyme is a G-quadruplex (G4) structure that requires a hemin as a cofactor[31,32]. The catalytic activity was monitored by oxidation of ABTS (2,2′-Azino-bis(3-ethylbenzothiazoline-6-sulphonic acid), which shows green in colour after oxidation[40].

Figure 5a shows the time course of the reaction profiles for the conjugates, **tp$_2$hp1** or **tp$_2$hp2**, in the presence of an equivalent amount of Fe^{2+}. The activity of **tp$_2$hp2** was apparently restored by Fe^{2+}. Such activation was not observed for **tp$_2$hp1**. As expected from the preliminary studies mentioned above, only **tp$_2$hp2** seemed to adopt a hairpin structure by intramolecular $[Fe(terpy)_2]^{2+}$ formation and restored the active form of DNAzyme (Fig. 1c). The conjugate **tp$_2$hp2** could be regarded as a semi-artificial metallo-DNAzyme. However, the activity level of **tp$_2$hp2•**Fe^{2+} was very little compared with the system of **tp$_2$tmp2**. The G4-structure formed on **tp$_2$hp2•**Fe^{2+} might not be ideal as a catalyst. This could be improved by fine-tuning of the structures around terpys.

Another allosteric system was also tested to control the activity of DNAzyme. Here **tp$_2$tmp2** was used as the tunable template to activate the split DNAzyme (**zymR** and **zymL**) as shown in

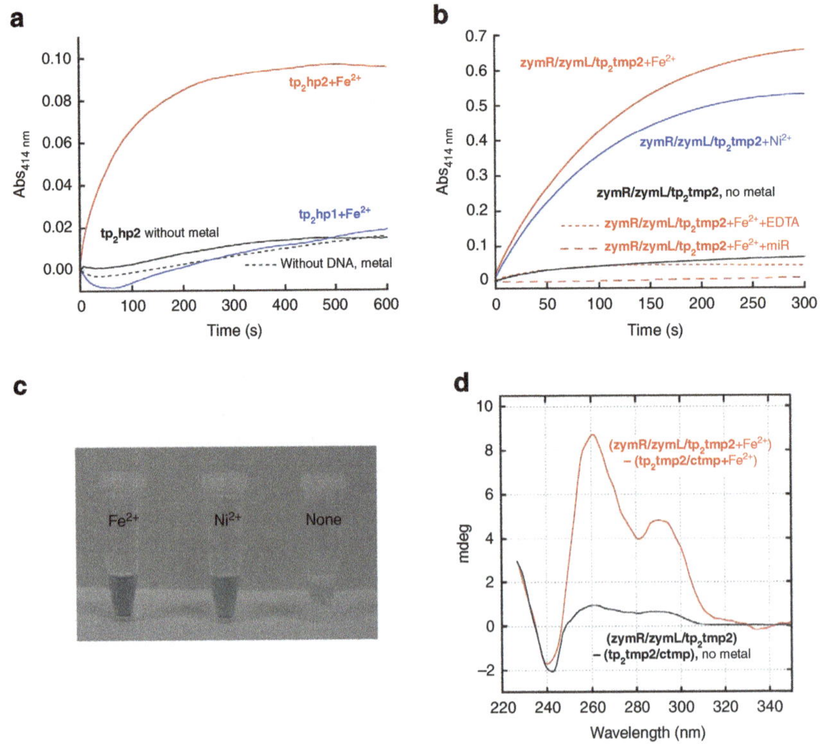

Figure 5 | Metal ion-directed regulation of DNAzyme activity. (**a**) Time courses of the ABTS oxidation by **tp$_2$hp1** and **tp$_2$hp2** in the presence of Fe^{2+}. The dotted curve was obtained without the conjugates in the absence of any transition metals. Solid curves show the reaction profiles obtained in the presence of the DNA conjugates. Blue: **tp$_2$hp1** + Fe^{2+}; red: **tp$_2$hp2** + Fe^{2+}, black: **tp$_2$hp2**, no metal ions. (**b**) Time courses of the ABTS oxidation by the split DNAzyme with **tp$_2$tmp2** in the presence of Fe^{2+} and Ni^{2+} and the effects of reaction inhibitors. Red: **zymR/zymL/tp$_2$tmp2** + Fe^{2+}; blue: **zymR/zymL/tp$_2$tmp2** + Ni^{2+}; black: **zymR/zymL/tp$_2$tmp2**, no metal ions; red dotted: **zymR/zymL/tp$_2$tmp2** + Fe^{2+} + EDTA; red broken: **zymR/zymL/tp$_2$tmp2** + Fe^{2+} + cDNA to the loop region (**miR**). Typical reaction profiles are shown in **a** and **b**. The errors in repetitive reactions were within 10% (s.d., $n = 3$). DNA complexes (all DNA components: **zymR**, **zymL**, **tp$_2$hp1**, **tp$_2$hp2**, **tp$_2$tmp1** and **tp$_2$tmp2**), 0.40 μM; hemin, 2.0 μM; MCl_2, 0.40 μM; H_2O_2, 2.0 mM; ABTS, 100 μM; HEPES, 14 mM (pH 7.0); NaCl, 700 mM; KCl, 140 mM. (**c**) Images of reaction solutions shown in **b**. (**d**) Differential circular dichroism (CD) spectra of the ternary complex, **zymR/zymL/tp$_2$tmp2**. Red: (**zymR/zymL/tp$_2$tmp2** + Fe^{2+})—(**tp$_2$tmp2/ctmp** + Fe^{2+}), black: (**zymR/zymL/tp$_2$tmp2**)—(**tp$_2$tmp2/ctmp**). CD spectra were measured for 3.0 μM DNAs and the metal ions in HEPES buffer solution (20 mM, pH 7.0) containing 1,000 mM NaCl, 200 mM KCl at 15 °C.

Fig. 1d. The time course of the reaction profiles is shown in Fig. 5b. Equivalent concentrations of Fe^{2+} and Ni^{2+} restored the activity of the split DNAzyme in the presence of **tp$_2$tmp2**. The effect of metal ions could be observed by naked eyes (Fig. 5c). The regained activity level was comparable to that of the positive control, **zymR/zymL/czym** (see Supplementary Fig. 6). Cu^{2+} and Zn^{2+} also showed moderate effect in restoration of split DNAzyme activity (see Supplementary Fig. 5). As we expected, global conformation of **tp$_2$tmp2** was fixed to a Ω-shape by intramolecular formation of $[M(terpy)_2]^{2+}$. Then, **tp$_2$tmp2•**M^{2+} worked as an effective template to reconstruct the integrated active form of DNAzyme.

The activity was almost completely suppressed by a large excess amount of EDTA and an equimolar amount of the single-stranded 23-mer DNA (**miR**) with an miR-34 sequence, which is complementary to the loop sequence (Fig. 5b). The results of these inhibition experiments confirmed our scheme for the regulation of DNAzyme activity. EDTA removed Fe^{2+} from the complex as a masking reagent. The single-stranded DNA with the miR-34 sequence straightened the loop region of **tp$_2$tmp2** by hybridization and made it impossible to form $[Fe(terpy)_2]^{2+}$.

The ternary DNA complex **zymR/zymL/tp$_2$tmp2** was subjected to circular dichroism (CD) spectroscopy. Figure 5d shows the differential CD spectra. To show only the contribution of the DNAzyme moiety, the CD spectrum of **tp$_2$tmp2/ctmp** (duplex + loop) was subtracted from those of **zymR/zymL/tp$_2$tmp2** (reconstituted DNAzyme + duplex + loop) obtained in the absence and in the presence of Fe^{2+}. The characteristic CD band for the mixed-type G4-structure (or the mixture of parallel and antiparallel folded species) was induced by the addition of Fe^{2+} (refs 41,42). This suggests that Fe^{2+} activated the DNAzyme by induction of a valid higher-ordered structure through sequence edition of the template, **tp$_2$tmp2**.

Discussion

Our aim in this study was to induce a drastic conformational change of global DNA (secondary) structure and its application for the allosteric regulation of the function of nucleic acids by certain stimuli. Although several stimuli-responsive DNA conjugates that change their local conformation have been reported[43–45], the systems of allosteric control of the global DNA structures are quite limited[9,17,46,47]. Even more, as far as we know, there is no precedent of conformational modulation of DNA aiming at sequence edition. The formation of a Ω-shape structure by intramolecular interaction between two specific points on a sequence could be regarded as stimuli-responsive artificial splicing. That is, the distal two segments are directly connected with each other by clipping out the intervening loop sequence.

In this study, the specific coordination of terpys to transition metal ions was chosen as a connecting reaction to show the availability of a Ω-motif for regulation of DNA function. Actually, while the phosphates in DNA backbone and nucleobases can coordinate to the transition metal ions[39], the ions, especially Fe^{2+} and Ni^{2+}, selectively bind two terpys to form a Ω-shaped structure under experimental conditions. The common preference in coordination chemistry of terpy was retained even for complexation on DNA. Therefore, the Ω-shape structure formation could be designed based on general knowledge in coordination chemistry. The kind and concentration of the metal ions, the structure of the terpy and the design of the conjugate architecture could be effective factors to control the system. Especially, it was interesting that local structural differences (just as ≈1.4 Å) in terpy units, such as substitution positions, were critically important to define the global structure.

We succeeded in the regulation of DNAzyme activity using the Ω-motif. Judging from the results of ultraviolet–vis titration, ultraviolet melting, CD spectroscopy, modelling analysis by DFT and ABTS oxidation and inhibition experiments, the activity of DNAzyme was regulated basically according to the proposed scheme shown in Fig. 1. However, the difference in melting temperatures of the duplex **tp$_2$tmp2/ctmp** observed in the presence and absence of Fe^{2+} might not be enough to solely explain the sharp response in restoration of DNAzyme activity. At least, the CD spectra showed that Fe^{2+} induced the active G4-structure of DNAzyme. Fe^{2+} could work as an essential factor not only to assemble the split DNAzymes but also to induce the active G4-structure through complexation.

To construct the system for artificial (non-enzymatic or chemical) splicing of DNA, we can choose appropriate clipping reactions from many bioorthogonal chemistries including click reactions[48,49] and photochemical reactions[37,44–46,50]. The techniques could be applied for signal amplification in bioanalyses through DNA circuits based on autonomous succession of strand exchange[26,51]. The regulation of cell function through the control of gene expression would also be very compatible and a promising application.

Methods

Synthesis. Preparation[30] of the amidite reagents, **1** and **2**, and DNAs are described in the Supplementary Information.

Ultraviolet-vis titration. The aliquots (typically, 0.40 μl)[16,52,53] of a 0.50 mM stock solution of MCl_2 ($M = Fe^{2+}, Ni^{2+}, Cu^{2+}$ or Zn^{2+}) were titrated into a 1.0 ml solution of 2.0 μM DNA or DNA complex (**tp$_2$tmp1**, **tp$_2$tmp2**, **tp$_2$tmp1/ctmp** or **tp$_2$tmp2/ctmp**, in 10 mM HEPES (pH 7.0), 500 mM NaCl) in a cuvette, and then the solution was mixed well. The ultraviolet–vis spectra were measured using a spectrophotometer V-650 (Jasco) at 25 °C, after 30 min incubation following each addition.

Ultraviolet melting. Thermal denaturation experiments were performed[9,54,55] in HEPES buffer solution (10 mM, pH 7.0) on a UV-1650 (Shimadzu) ultraviolet–vis spectrophotometer equipped with a Peltier thermal controller. Concentrations of each DNA strand and MCl_2 ($M = Fe^{2+}, Ni^{2+}, Cu^{2+}$ or Zn^{2+}) were 1.0 μM. Prior to the beginning of each melt, the samples were degassed at 85 °C for 5 min and then annealed by slowly cooling to 0 °C. In the denaturation experiments, the solutions were heated at a rate of 0.5 deg min^{-1} after equilibration for 20 min at 0 °C.

The temperature that gave the maximum first derivative of the melting curve was used as the melting temperature, a measure of the thermal stability of the duplexes.

Circular dichroism measurement. CD[56,57] spectra were measured in HEPES buffer solution (20 mM, pH 7.0) containing 1.0 M NaCl and 200 mM KCl on a J-725 Spectropolarimeter (Jasco) equipped with a Peltier thermal controller, at 15 °C. Concentrations of all DNA strands and $FeCl_2$ were 3.0 μM. Prior to the beginning of each melt, the samples were degassed at 85 °C for 5 min and then annealed by slowly cooling to 15 °C.

To only show the contribution of the DNAzyme moiety, the CD spectrum of **tp$_2$tmp2/ctmp** (duplex + loop) was subtracted from those of **zymR/zymL/tp$_2$tmp2** (reconstituted DNAzyme + duplex + loop) obtained in the absence and in the presence of Fe^{2+}.

Molecular modelling. Both complex ($[Fe(terpy)_2]^{2+}$) structures were calculated by the DFT method (B3LYP/6-31G*) using Spartan 10 (Wavefunction Inc., USA). The molecular models shown in Fig. 4c were constructed with MacroModel (ver. 9.5, Schrödinger Inc., USA). The initial structure was built as a standard double helix using the Maestro (Schrödinger Inc.) fragment library. The optimization was performed with an AMBER* force field with generalized Born/surface area treatment of solvation.

Allosteric catalytic reactions by split DNAzymes. Into the mixed solution of DNA and hemin,[40,58,59] was added the solution containing MCl_2 ($M = Fe^{2+}, Ni^{2+}, Cu^{2+}$ or Zn^{2+}), H_2O_2 and ABTS to prepare 1.0 ml of the reaction solution in a cuvette. The final concentrations of each of the components were as follows: DNA complexes (all DNA components: **zymR**, **zymL**, **czym**, **tp$_2$hp1**, **tp$_2$hp2**, **tp$_2$tmp1** and **tp$_2$tmp2**), 0.40 μM; hemin, 2.0 μM; MCl_2, 0.40 μM; H_2O_2, 2.0 mM; ABTS, 100 μM; HEPES, 14 mM (pH 7.0); NaCl, 700 mM; KCl, 140 mM. The progress of ABTS oxidation was monitored at 15 °C by the change in absorbance at 414 nm. The time courses of absorbance change were monitored using a spectrophotometer V-650 (Jasco) equipped with a Peltier thermal controller.

References

1. Clever, G. H., Kaul, C. & Carell, T. DNA–metal base pair. *Angew. Chem. Int. Ed.* **46**, 6226–6236 (2007).
2. Scharf, P. & Müller, J. Nucleic acids with metal-mediated base pairs and their applications. *ChemPlusChem* **78**, 20–34 (2013).
3. Biver, T. Stabilisation of non-canonical structures of nucleic acids by metal ions and small molecules. *Coord. Chem. Rev.* **257**, 2765–2783 (2013).
4. Tanaka, K. *et al.* Programmable self-assembly of metal ions inside artificial DNA duplex. *Nat. Nanotechnol.* **1**, 190–194 (2006).
5. Tanaka, K., Tengeiji, A., Kato, T., Toyama, N. & Shionoya, M. A discrete self-assembled metal array in artificial DNA. *Science* **299**, 1212–1213 (2003).
6. Takezawa, Y., Maeda, W., Tanaka, K. & Shionoya, M. Discrete self-assembly of iron(III) ions inside triple-stranded artificial DNA. *Angew. Chem. Int. Ed.* **48**, 1081–1084 (2009).
7. Ono, T. *et al.* Synthesis of covalently linked parallel and antiparallel DNA duplexes containing the metal-mediated base pairs T–Hg(II)–T and C–Ag(I)–C. *Chem. Commun.* **47**, 1542–1544 (2011).
8. Ono, A., Torigoe, H., Tanaka, Y. & Okamoto, I. Binding of metal ions by pyrimidine base pair in DNA duplex. *Chem. Soc. Rev.* **40**, 5855–5866 (2011).
9. Ihara, T., Takeda, Y. & Jyo, A. Metal ion-directed cooperative triple helix formation of glutamic acid–oligonucleotide conjugate. *J. Am. Chem. Soc.* **123**, 1772–1773 (2001).
10. Ihara, T., Ishii, T., Araki, N., Wilson, A. W. & Jyo, A. Silver ion unusually stabilizes the structure of a parallel-motif DNA triplex. *J. Am. Chem. Soc.* **131**, 3826–3827 (2009).
11. Urata, S., Miyahata, T., Matsuura, H., Kitamura, Y. & Ihara, T. Alteration of DNAzyme activity by silver ion. *Chem. Lett.* **43**, 1020–1022 (2014).
12. Urata, H., Yamaguchi, E., Funai, T., Matsumura, Y. & Wada, S. Incorporation of thymine nucleotides by DNA polymerase through T–Hg^{II}–T base pairing. *Angew. Chem. Int. Ed.* **49**, 6516–6519 (2010).
13. Park, K. S. & Park, H. G. Technological applications arising from the interaction of DNA bases with metal ions. *Curr. Opin. Biotechnol.* **28**, 17–24 (2014).
14. Guo, W. *et al.* Reversible Ag^+-crosslinked DNA hydrogels. *Chem. Commun.* **50**, 4065–4068 (2014).
15. He, X. & Ma, N. A general strategy for label-free sensitive DNA detection based on quantum dot doping. *Anal. Chem.* **86**, 3676–3681 (2014).
16. Shimada, H., Sakurai, T., Kitamura, Y., Matsuura, H. & Ihara, T. Metallo-regulation of the bimolecular triplex formation of peptide nucleic acid. *Dalton Trans.* **42**, 16006–16013 (2013).
17. Göritz, M. & Krämer, R. Allosteric control of oligonucleotide hybridization by metal ion-induced cyclization. *J. Am. Chem. Soc.* **127**, 18016–18017 (2005).
18. Graf, N., Göritz, M. & Krämer, R. A metal-ion-releasing probe for DNA detection by catalytic signal amplification. *Angew. Chem. Int. Ed.* **45**, 4013–4015 (2006).
19. Choi, J. S. *et al.* Synthesis of DNA triangles with vertexes of bis(terpyridine)iron(II) complexes. *J. Am. Chem. Soc.* **126**, 8606–8607 (2004).
20. Yang, H. & Sleiman, H. F. Templated synthesis of highly stable, electroactive, and dynamic metal-DNA branched junctions. *Angew. Chem. Int. Ed.* **47**, 2443–2446 (2008).
21. Bianké, G. & Häner, R. A metal-coordinating DNA hairpin mimic. *Chembiochem* **5**, 1063–1068 (2004).
22. Ellington, A. D. & Szostak, J. W. In vitro selection of RNA molecules that bind specific ligands. *Nature* **346**, 818–822 (1990).
23. Breaker, R. R. & Joyce, G. F. A DNA enzyme that cleaves RNA. *Chem. Biol.* **1**, 223–229 (1994).

24. Liu, J., Cao, Z. & Lu, Y. Functional nucleic acid sensors. *Chem. Rev.* **109,** 1948–1998 (2009).
25. Vinkenborg, J. L., Karnowski, N. & Famulok, M. Aptamers for allosteric regulation. *Nat. Chem. Biol.* **7,** 519–527 (2011).
26. Zhang, H., Li, F., Dever, B., Li, X.-F. & Le, X. C. DNA-mediated homogeneous binding assays for nucleic acids and proteins. *Chem. Rev.* **113,** 2812–2841 (2013).
27. Li, Y. & Lu, Y. (eds) *Functional Nucleic Acids for Analytical Applications* (Springer, 2009).
28. Schubert, U. S., Hofmeier, H. & Newkome, G. R. *Modern Terpyridine Chemistry* (Wiley, 2006).
29. Schwarz, G., Hasslauer, I. & Kurth, D. G. From terpyridine-based assemblies to metallo-supramolecular polyelectrolytes (MEPEs). *Adv. Colloid. Interface Sci.* **207,** 107–120 (2014).
30. Ihara, T., Sato, Y., Shimada, H. & Jyo, A. Metalloregulation of triple helix formation by control of the loop conformation. *Nucleos. Nucleot. Nucleic Acids* **27,** 1084–1096 (2008).
31. Li, Y. & Sen, D. Toward an efficient DNAzyme. *Biochemistry* **36,** 5589–5599 (1997).
32. Travascio, P., Li, Y. & Sen, D. DNA-enhanced peroxidase activity of a DNA aptamer-hemin complex. *Chem. Biol.* **5,** 505–517 (1998).
33. Kolpashchikov, D. M. Split DNA enzyme for visual single nucleotide polymorphism typing. *J. Am. Chem. Soc.* **130,** 2934–2935 (2008).
34. Reddy, E. P., Reynolds, R. K., Santos, E. & Barbacid, M. A point mutation is responsible for the acquisition of transforming properties by the T24 human bladder carcinoma oncogene. *Nature* **300,** 149–152 (1982).
35. Hau, A., Ceppi, P. & Peter, M. E. CD95 is part of a Let-7/p53/miR-34 regulatory network. *PLoS ONE* **7,** e49636 (2012).
36. Bjerrum, J., Schwarzenbach, G. & Sillén, L. G. *Compiled. Stability constants of Metal-Ion Complexes, with Solubility Products of Inorganic Substances* (The Chemical Society, 1957).
37. Ihara, T., Fujii, T., Mukae, M., Kitamura, Y. & Jyo, A. Photochemical ligation of DNA conjugates through anthracene cyclodimer formation and its fidelity to the template sequences. *J. Am. Chem. Soc.* **126,** 8880–8881 (2004).
38. Eichhorn, G. & Shin, Y. A. Interaction of metal ion with polynucleotides and related compounds. XII. The relative effect of various metal ions on DNA helicity. *J. Am. Chem. Soc.* **90,** 7323–7328 (1968).
39. Hud, N. V. (ed.) *Nucleic Acid–Metal Ion Interaction* (RSC Publishing, 2009).
40. Collins, P., Dobson, A. D. & Field, J. A. Reduction of the 2,2′-Azinobis (3-ethylbenzthiazoline-6-sulfonate) cation radical by physiological organic acids in the absence and presence of manganese. *Appl. Environ. Microbiol.* **64,** 2026–2031 (1998).
41. Ambrus, A. *et al.* Human telomeric sequence forms a hybrid-type intramolecular G-quadruplex structure with mixed parallel/antiparallel strands in potassium solution. *Nucleic Acids Res.* **34,** 2723–2735 (2006).
42. Masiero, S. *et al.* A non-empirical chromophoric interpretation of CD spectra of DNA G-quadruplex structures. *Org. Biomol. Chem.* **8,** 2683–2692 (2010).
43. Mori, K., Kodama, T., Baba, T. & Obika, S. Bridged nucleic acid conjugates at 6'-thiol: synthesis, hybridization properties and nuclease resistances. *Org. Biomol. Chem.* **9,** 5272–5279 (2011).
44. Nishioka, H., Liang, X., Kato, T. & Asanuma, H. A photo-fueled DNA nanodevices that contains two different photoswitches. *Angew. Chem. Int. Ed.* **51,** 1165–1168 (2012).
45. Rotaru, A. & Mokhir, A. Nucleic acid binders activated by light of selectable wavelength. *Angew. Chem. Int. Ed.* **46,** 6180–6183 (2007).
46. Arslan, P., Jyo, A. & Ihara, T. Reversible circularization of an anthracene-modified DNA conjugate through bimolecular triplex formation and its analytical application. *Org. Biomol. Chem.* **8,** 4843–4848 (2010).
47. Ogino, M. & Fujimoto, K. Photochemical synthesis of R-shaped DNA toward DNA recombination and processing in vitro. *Angew. Chem. Int. Ed.* **45,** 7223–7226 (2006).
48. Lim, R. K. V. & Lin, Q. Bioorthogonal chemistry: recent progress and future directions. *Chem. Commun.* **46,** 1589–1600 (2010).
49. Gramlich, P. M. E., Wirges, C. T., Manetto, A. & Carell, T. Postsynthetic DNA modification through the copper-catalyzed azide–alkyne cycloaddition reaction. *Angew. Chem. Int. Ed.* **47,** 8350–8358 (2008).
50. Mukae, M., Ihara, T., Tabara, M. & Jyo, A. Anthracene–DNA conjugates as building blocks of designed DNA structures constructed by photochemical reactions. *Org. Biomol. Chem.* **7,** 1349–1354 (2009).
51. Chen, X., Briggs, N., McLain, J. R. & Ellington, A. D. Stacking nonenzymatic circuits for high signal gain. *Proc. Natl Acad. Sci. USA* **110,** 5386–5391 (2013).
52. Ihara, T., Sueda, S., Inenaga, A., Fukuda, R. & Takagi, M. Synthetic DNA ligands conjugated with metal binding moiety. Regulation of the interaction with DNA by metal ions and the ligand effect on metal assisted DNA cleaving. *Supramol. Chem.* **8,** 93–111 (1997).
53. Ihara, T. *et al.* Metal ion-directed cooperative DNA binding of small molecules. *J. Inorg. Biochem.* **100,** 1744–1754 (2006).
54. Marky, L. A. & Breslauer, K. J. Calculating thermodynamic data for transition of any molecularity from equilibrium melting curve. *Biopolymers* **26,** 1601–1620 (1987).
55. Kitamura, Y. *et al.* Template-directed formation of luminescent lanthanide complexes: versatile tools for colorimetric identification of single nucleotide polymorphism. *J. Inorg. Biochem.* **102,** 1921–1931 (2008).
56. Cheng, X., Liu, X., Bing, T., Cao, Z. & Shangguan, D. General peroxidase activity of G-quadruplex–hemin complexes and its application in ligand screening. *Biochemistry* **48,** 7817–7823 (2009).
57. Li, C. *et al.* Backbone modification promotes peroxidase activity of G-quadruplex-based DNAzyme. *Chem. Commun.* **48,** 8347–8349 (2012).
58. Nakayama, S. & Sintim, H. O. Colorimetric split G-quadruplex probes for nucleic acid sensing: improving reconstituted DNAzyme's catalytic efficiency via probe remodeling. *J. Am. Chem. Soc.* **131,** 10320–10333 (2009).
59. Kong, D.-M., Yang, W., Wu, J., Li, C. X. & Shen, H.-X. Structure-function study of peroxidase-like G-quadruplex-hemin complexes. *Analyst* **135,** 321–326 (2010).

Acknowledgements

We thank Ms M. Nishi for the assistance in the measurements of mass spectra for amidite and conjugates identification. We also thank Professor H. Ihara and Professor M. Takafuji for their support in measuring CD spectra. This work was supported by the Grant-in-Aid for Scientific Research on Innovative Areas (Coordination Programming, Area 2107) (no. 24108734 to T.I.), Scientific Research (B) (no. 24350040 to T.I.) from the Ministry of Education, Culture, Sports, Science and Technology, Japan; and the Health Labor Sciences Research Grant (Research on Publicly Essential Drugs and Medical Devices) (to T.I.) from the Ministry of Health and Labor, Japan.

Author contribution

T.I. conceived the project, designed the systems and wrote the paper. H.O. synthesized the DNA conjugates and performed the DNAzyme activity measurements. C.S., T.F. and H.S. synthesized the amidites and DNA conjugates and conducted the ultraviolet measurements. H.M. and Y.K. analysed and discussed the data.

Additional information

Competing financial interests: The authors declare no competing financial interests.

7

Selective inhibition of the kinase DYRK1A by targeting its folding process

Isao Kii[1,2], Yuto Sumida[3,†], Toshiyasu Goto[4], Rie Sonamoto[1], Yukiko Okuno[1], Suguru Yoshida[3], Tomoe Kato-Sumida[3,†], Yuka Koike[1,2], Minako Abe[5], Yosuke Nonaka[5], Teikichi Ikura[5], Nobutoshi Ito[5], Hiroshi Shibuya[4], Takamitsu Hosoya[3] & Masatoshi Hagiwara[1]

Autophosphorylation of amino-acid residues is part of the folding process of various protein kinases. Conventional chemical screening of mature kinases has missed inhibitors that selectively interfere with the folding process. Here we report a cell-based assay that evaluates inhibition of a kinase at a transitional state during the folding process and identify a folding intermediate-selective inhibitor of dual-specificity tyrosine-phosphorylation-regulated kinase 1A (DYRK1A), which we refer to as FINDY. FINDY suppresses intramolecular autophosphorylation of Ser97 in DYRK1A in cultured cells, leading to its degradation, but does not inhibit substrate phosphorylation catalysed by the mature kinase. FINDY also suppresses Ser97 autophosphorylation of recombinant DYRK1A, suggesting direct inhibition, and shows high selectivity for DYRK1A over other DYRK family members. In addition, FINDY rescues DYRK1A-induced developmental malformations in *Xenopus laevis* embryos. Our study demonstrates that transitional folding intermediates of protein kinases can be targeted by small molecules, and paves the way for developing novel types of kinase inhibitors.

[1] Department of Anatomy and Developmental Biology, Graduate School of Medicine, Kyoto University, Yoshida-Konoe-cho, Kyoto 606-8501, Japan. [2] Pathophysiological and Health Science Team, Imaging Application Group, Division of Bio-Function Dynamics Imaging, RIKEN Center for Life Science Technologies, 6-7-3 Minatojima-minamimachi, Chuo-ku, Kobe 650-0047, Japan. [3] Laboratory of Chemical Bioscience, Institute of Biomaterials and Bioengineering, Tokyo Medical and Dental University, 2-3-10 Kanda-Surugadai, Tokyo 101-0062, Japan. [4] Department of Molecular Cell Biology, Medical Research Institute, Tokyo Medical and Dental University, 1-5-45, Yushima, Tokyo 113-8510, Japan. [5] Department of Structural Biology, Graduate School of Medical and Dental Sciences, Medical Research Institute, Tokyo Medical and Dental University, 1-5-45, Yushima, Tokyo 113-8510, Japan. † Present address: Chemical Biology Team, Imaging Chemistry Group, Division of Bio-Function Dynamics Imaging, RIKEN Center for Life Science Technologies, 6-7-3 Minatojima-minamimachi, Chuo-ku, Kobe 650-0047, Japan. Correspondence and requests for materials should be addressed to I.K. (email: isao.kii@riken.jp) or to M.H. (email: hagiwara.masatoshi.8c@kyoto-u.ac.jp).

Dysregulation of protein kinase activity has been implicated in pathological conditions, such as neurological disorders and tumorigenesis. The protein kinase family represents an attractive target for drug development[1,2]. Kinase inhibitors that target ATP-binding pockets sometimes cause adverse side effects by suppressing unintended kinases, because the sequence and structure of the pockets are well-conserved[3]. Innovative ideas are therefore necessary in drug discovery to obtain a highly selective inhibitor of the target kinase.

DYRK1A, dual-specificity tyrosine-phosphorylation-regulated kinase 1A, is a mammalian orthologue of *Drosophila minibrain* and is essential for brain development[4]. The physiological importance of *DYRK1A* has been suggested by its proposed relationship with various symptoms of Down syndrome (DS)[5–7]. *DYRK1A* resides within the obligate trisomic region of human chromosome 21 and the extra copy of the *DYRK1A* gene in people with DS causes a 1.5-fold increased expression of the dual-specificity tyrosine-phosphorylation-regulated kinase 1A (DYRK1A) protein[8]. The excessive DYRK1A activity is not only pivotal in causing the characteristic facial features[9,10] and congenital heart defects[9] of DS, but is also associated with early-onset of Alzheimer's disease[11,12]. The hypothesis that the elevated activity of DYRK1A contributes to these neurological disorders has stimulated an interest in DYRK1A as a potential target for therapeutic drugs[4,12,13]. In addition, inhibition of DYRK1A increases pancreatic β-cell proliferation, suggesting therapeutic promise for diabetes therapy[14,15]. To repress the excessive activity of DYRK1A, we had previously developed a synthetic small molecule, INDY, which potently suppresses the kinase activity of DYRK1A. INDY also inhibits other DYRK family members as well as Cdc2-like kinases[16].

DYRKs and glycogen synthase kinase 3β (GSK3β) autophosphorylate their own tyrosine residue in their transitional state and phosphorylate serine or threonine residues on their substrates after maturation[17–21]. Lochhead *et al.* postulated the existence of a transitional intermediate of these kinases that has biochemical properties distinct from the mature state[18,19]. Interestingly, the intermediate showed different sensitivity to chemical inhibitors from the mature kinase. For example, the transitional intermediate of *Drosophila* DYRK2 was inhibited by Purvalanol A, but not by 4,5,6,7-tetrabromo-1H-benzotriazole (TBB), whereas the mature kinase was inhibited by both[18]. So far, selective inhibitors of the transitional intermediate have not yet been identified, because the focus of conventional drug screening has been on the mature forms of the kinases[18,22].

Isolation of transitional intermediates is considered to be quite difficult because of their low population and short lifetimes[23]. Therefore, we developed a cell-based assay, named SPHINKS (substrate phosphorylation by sequential induction of kinase and substrate), to evaluate intermediate-selective inhibitors of DYRK1A. Through examination of our synthetic chemical library, we identified a small compound, referred to as FINDY (folding intermediate-selective inhibitor of DYRK1A), which interfered with the folding process of DYRK1A, but did not inhibit the substrate phosphorylation catalysed by the mature kinase. FINDY specifically suppressed autophosphorylation of Ser97 in DYRK1A, resulting in its degradation by proteasomes. Furthermore, we found that FINDY did not affect the kinase activity of the other members of the DYRK family, indicating the possibility that transitional intermediate-selective kinase inhibitors distinguish between the conserved family members.

Results

Design of the cell-based assay targeting the intermediate. To evaluate intermediate-selective inhibitors of DYRK1A, we developed the SPHINKS assay, enabling us to evaluate kinase inhibition at the transitional state during the folding process (Fig. 1a). We first established a HEK293 cell line with dual-inducible expression of DYRK1A and TAU, a well-characterized substrate of DYRK1A[24,25]. Expression of FLAG-tagged DYRK1A (FLAG-DYRK1A) was controlled by the *tet* operator, and was initiated by treatment with doxycycline (Fig. 1b, lanes 4–6). TAU was expressed in a fused form with the destabilization domain FKBP12 (DD), which causes degradation of DD fusion proteins in the absence of a small molecule, Shield-1 (ref. 26). Treatment with Shield-1 for 2 h stabilized DD-TAU in a dose-dependent manner (Fig. 1b, lanes 2, 3, 5 and 6). Phosphorylation of the stabilized DD-TAU at Thr212 was enhanced 4 h after the administration of doxycycline (Fig. 1b, lanes 5 and 6; and Supplementary Fig. 1), indicating that DYRK1A is produced from the doxycycline-induced transcript within 4 h.

To distinguish kinase inhibition at the transitional state from that after maturation, we set up two types of time courses. Small molecules were administered either before the induction by doxycycline (time course: Tc 1) or after the accumulation of mature DYRK1A (Tc 2; Fig. 1a). Following additional incubation with Shield-1 for 2 h, total cell lysates were harvested, then subjected to western blot. If the compound selectively inhibits the transitional intermediate, but not the mature DYRK1A, it should suppress TAU phosphorylation in Tc 1, but not in Tc 2, as illustrated in Fig. 1a.

Identification of FINDY. Using the SPHINKS assay, we examined our chemical library and found a small molecule that suppressed TAU phosphorylation in Tc 1, but not in Tc 2 (Fig. 1c). This small molecule was referred to as FINDY (Fig. 1d). In contrast, the canonical DYRK1A inhibitor INDY suppressed TAU phosphorylation in both Tc 1 and 2 (Fig. 1c).

FINDY is structurally similar to an ATP-competitive inhibitor, RD0392 (Fig. 1e and Supplementary Fig. 2), which we identified through *in vitro* screening using recombinant DYRK1A. RD0392 bears a five-membered heterocyclic ring, rhodanine. This moiety has been reported as a substructure of Pan Assay Interference Compounds[27], which are compounds that have frequently arisen as hits in high-throughput screening. Conversely, Mendgen *et al.* noted that small molecules bearing rhodanine are potentially valuable and should not be regarded as problematic or promiscuous binders *per se*[28]. Therefore, we synthesized structural derivatives of RD0392, one of which was FINDY.

In contrast to the dose-dependent inhibition by RD0392 and INDY in the *in vitro* assay, FINDY did not exert any observable inhibitory effect on the recombinant DYRK1A (Fig. 1f). This was supported by a computational docking simulation study, which showed that RD0392 was able to fit to the ATP-binding pocket of mature DYRK1A (Supplementary Fig. 3a,b), whereas FINDY could not (Supplementary Fig. 3c,d). FINDY decreased the amount of FLAG-DYRK1A in Tc 1, whereas DD-TAU and glyceraldehyde 3-phosphate dehydrogenase (GAPDH) were unaffected. In contrast, the amount of FLAG-DYRK1A was unaltered in Tc 2 (Fig. 1c). These results suggest that FINDY selectively impairs the stability or translation of DYRK1A, leading to the suppression of TAU phosphorylation in Tc 1.

To determine whether the stability or the translation is affected by FINDY, we constructed a translation reporter system using ribosome skipping mediated by a 2A peptide. The 2A peptide inhibits normal peptide bond formation without affecting translation[29], which enables the bicistronic expression of FLAG-DYRK1A and haemagglutinin (HA)-tagged enhanced green fluorescent protein (EGFP) (HA-EGFP) under the control of a doxycycline-driven promoter (Fig. 2a). When administered with doxycycline,

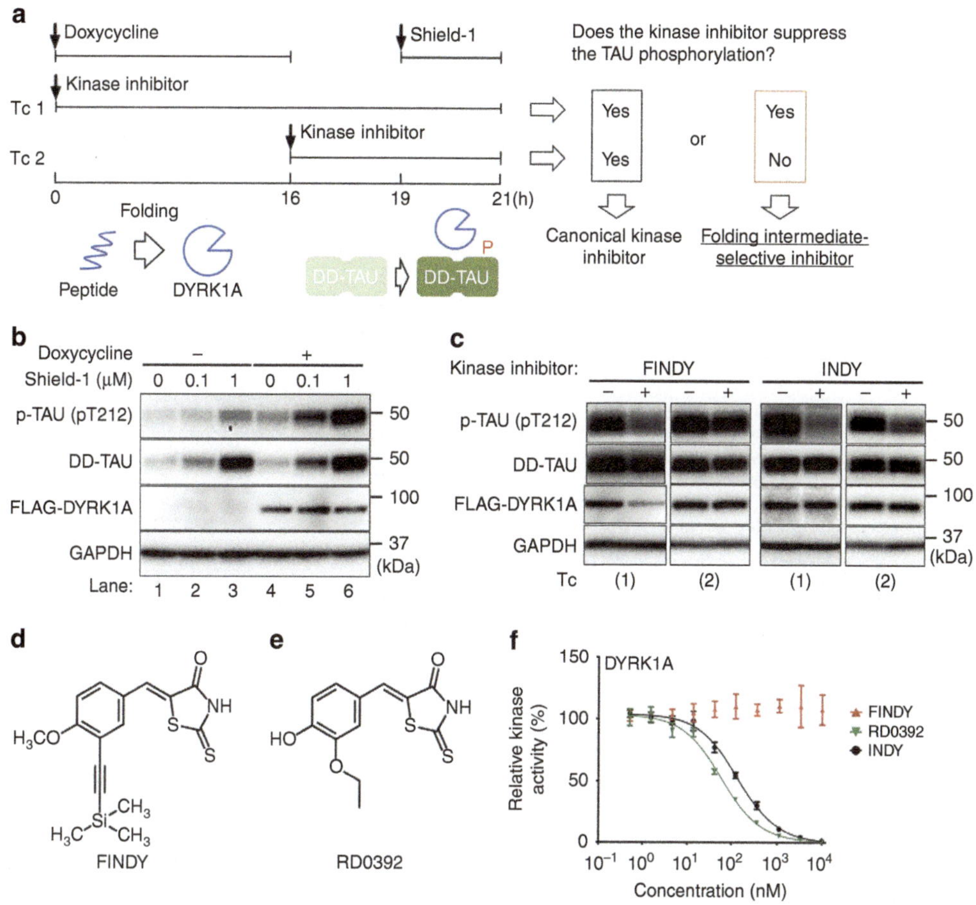

Figure 1 | Cell-based assay to evaluate a transitional intermediate-selective inhibitor of DYRK1A. (**a**) Schematic diagram of the SPHINKS assay. Doxycycline induces DYRK1A expression. Subsequently, Shield-1 stabilizes TAU fused with the destabilization domain of FKBP12 (DD-TAU). DYRK1A phosphorylates DD-TAU over the defined time period (19–21 h). Small molecules from our chemical library were added at the indicated points in Tc 1 and 2. Canonical kinase inhibitors suppress TAU phosphorylation in both Tc 1 and 2. Intermediate-selective inhibitors of DYRK1A should suppress TAU phosphorylation in Tc 1, but not in Tc 2. (**b**) Inducible expression of FLAG-DYRK1A and DD-TAU in HEK293 cells. Doxycycline induced FLAG-DYRK1A expression (lanes 4–6) and Shield-1 stabilized DD-TAU (lanes 2, 3, 5, 6). FLAG-DYRK1A predominantly phosphorylated Thr212 of TAU (p-TAU; lanes 5 and 6). p-TAU, TAU, FLAG and GAPDH were detected by western blot, using their corresponding antibodies. Representative data from the triplicate experiments are shown. (**c**) Identification of FINDY as the intermediate-selective inhibitor of DYRK1A. TAU phosphorylation was suppressed by FINDY (10 μM) in Tc 1, but not in Tc 2. In contrast, the canonical DYRK1A inhibitor INDY (10 μM) suppressed TAU phosphorylation in both Tc 1 and 2. Representative data from the triplicate experiments are shown. (**d**) Structure of FINDY. (**e**) Structure of RD0392, a canonical ATP-competitive inhibitor of DYRK1A. (**f**) FINDY did not inhibit the *in vitro* kinase activity of DYRK1A. Recombinant DYRK1A was incubated with the peptide substrate DYRKtide in the presence of FINDY, RD0392 or INDY. RD0392 and INDY inhibited the kinase activity with IC_{50} values of 60.2 and 139 nM, respectively. Representative dose-response curves with Hill slopes are shown. The results are presented as means ± s.d. ($n = 4$).

FINDY decreased the amount of FLAG-DYRK1A in a dose-dependent manner (EC_{50} value: 2.2 μM), without affecting HA-EGFP (Fig. 2b). FINDY also decreased the amount of endogenous DYRK1A in a primary culture of cortical neurons (Fig. 2c). On the other hand, administration of RD0392 did not decrease the amount of FLAG-DYRK1A (Fig. 2d). Next, we examined the effect of FINDY on the degradation of FLAG-DYRK1A in the presence of a translation inhibitor, cycloheximide. Cells were pre-treated with doxycycline for 16 h, then incubated with cycloheximide and FINDY. Under this condition, FINDY did not induce the degradation of FLAG-DYRK1A compared with the vehicle treatment (Fig. 2e), indicating that FINDY destabilizes newly translated FLAG-DYRK1A.

FINDY suppresses autophosphorylation on Ser97 of DYRK1A. Catalytic activity of several protein kinases is regulated by autophosphorylation to stabilize the active conformation[30]. To investigate whether autophosphorylation is involved in the stabilization of DYRK1A, we prepared a substituted mutant with Arg (K188R) in place of Lys188. Lys188 in DYRK1A interacts directly with ATP; the substitution thus renders the kinase inactive, leading to a 'kinase-dead' mutant, which lacks autophosphorylation ability[31]. In our study, the K188R mutant protein of FLAG-DYRK1A was barely detected with western blot, whereas EGFP from the bicistronic cassette was detected (Fig. 3a, K188R). This suggests that the abrogation of autophosphorylation destabilizes the DYRK1A protein. To identify the residues required for the stabilization of DYRK1A, we mutated potential autophosphorylation sites. Ser97 and Tyr111 in the non-catalytic amino-terminal domain of DYRK1A can be autophosphorylated[31]. In addition, Tyr319 or Tyr321 in the activation loop are required to retain catalytic activity[31–33], and the autophosphorylation of these residues has also been reported[21]. Similarly, the autophosphorylation of Ser520, which corresponds to Ser529 in the DYRK1A variant used in the present

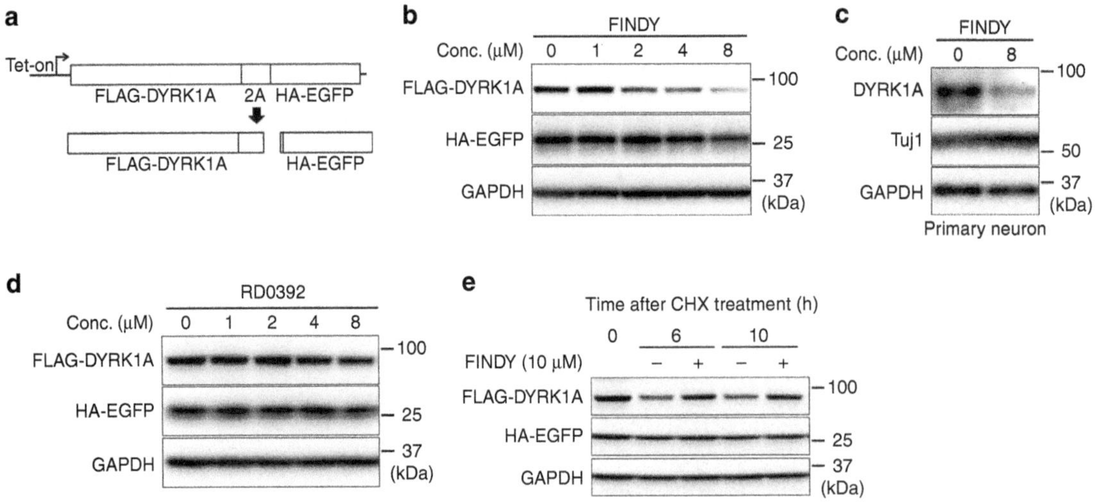

Figure 2 | FINDY destabilizes DYRK1A during the folding process. (**a**) Schematic diagram of the 2A peptide-mediated bicistronic expression of FLAG-DYRK1A and HA-EGFP. Expression is controlled by the *tet* operator. Effects of FINDY (**b**) or RD0392 (**d**) on destabilization of DYRK1A. FINDY (**b**) or RD0392 (**e**) was added with doxycycline to HEK293 cells harbouring the bicistronic expression vector, which was stably integrated in the genome. After incubation for 5 h, total cell lysates were subjected to SDS-PAGE followed by western blot analysis using the corresponding antibodies against FLAG, HA and GAPDH. Representative data from the pentaplicate experiments are shown. (**c**) The amount of endogenous DYRK1A was decreased by FINDY. Primary cortical neurons were cultured in the presence of FINDY for 3 days. DYRK1A, Tuj1 (neuron-specific class III β-tubulin) and GAPDH were detected by western blot analysis using their corresponding antibodies. Representative data from the triplicate experiments are shown. (**e**) FINDY did not induce the degradation of mature DYRK1A. The cells were treated with doxycycline for 16 h, then incubated with cycloheximide (CHX) with or without FINDY for the indicated time. Total cell lysates were subjected to SDS-PAGE followed by western blot analysis. Representative data from the triplicate experiments are shown. Conc., concentration.

study, is required for its binding to 14-3-3β (ref. 34). Therefore, we prepared mutants with these residues substituted. The substituted mutant of S97A and the double mutant of Y321F/Y319F were degraded, whereas the other mutants remained (Fig. 3a), indicating that Ser97 and one of the tyrosine residues in the activation loop are required for the DYRK1A stabilization.

Ser97 and its surrounding amino-acid sequence in DYRK1A are highly conserved among various species (Fig. 3b), suggesting their functional importance. To detect phosphorylated Ser97 (p-Ser97), we generated a polyclonal antibody against p-Ser97, and transiently transfected 293T cells with the expression vectors of FLAG-DYRK1A and its mutants, in which the mutants were detectable, probably because the overexpressed proteins were beyond the capacity of proteasomal degradation. This antibody detected the intact DYRK1A (wild type, WT), but barely identified the kinase-dead (K188R) and S97A mutants in the western blot analysis (Fig. 3c). In addition, competition experiments using phospho- and non-phospho-peptides clearly confirmed the antibody's specific recognition ability towards p-Ser97 (Supplementary Fig. 4). We then examined whether FINDY suppresses autophosphorylation. In the presence of FINDY, the amount of p-Ser97 significantly decreased, whereas signals for phosphorylated tyrosine residues (p-Tyr) were unaffected (Fig. 3d). In addition, mass spectrometry analysis of the trypsin-digested peptides of FLAG-DYRK1A showed that FINDY did not suppress the phosphorylation of Tyr321 in the activation loop (Supplementary Fig. 5). To confirm that the FINDY-mediated inhibition of Ser97 autophosphorylation is involved in degradation of DYRK1A, we examined whether FINDY destabilized the DYRK1A mutants of K188R and S97A, which were not able to catalyse Ser97 autophosphorylation. FINDY did not decrease the level of these mutants (Supplementary Fig. 6). Thus, FINDY acts as an inhibitor of the autophosphorylation and does not induce DYRK1A degradation by an independent mechanism. These results thus suggest that FINDY destabilizes DYRK1A by inhibiting Ser97 autophosphorylation.

To further determine whether Ser97 is autophosphorylated by an inter- or intramolecular mechanism, we generated HEK293 cells that co-expressed the kinase-dead K188R mutant (FLAG-DYRK1A K188R) and an intact DYRK1A fused with a DD tag (DD-FLAG-DYRK1A). If Ser97 autophosphorylation were mediated by an intermolecular mechanism, Ser97 of the kinase-dead K188R mutant would be phosphorylated when expressed in the presence of DD-FLAG-DYRK1A. However, we found that the overexpressed DD-FLAG-DYRK1A did not enhance the phosphorylation of Ser97 in the K188R mutant (Supplementary Fig. 7a). In addition, DD-FLAG-DYRK1A did not stabilize the co-expressed K188R protein (Supplementary Fig. 7b). These results suggest that the Ser97 autophosphorylation of DYRK1A is catalysed in an intramolecular manner. Furthermore, treatment with cantharidic acid, a protein phosphatase inhibitor, increased the amount of p-Ser97 signal in FLAG-DYRK1A (Supplementary Fig. 7c), suggesting that p-Ser97 is dephosphorylated by a phosphatase inhibited by cantharidic acid. Interestingly, the FLAG-DYRK1A protein, which was purified from HEK293 cells, did not significantly catalyse Ser97 autophosphorylation (Supplementary Fig. 7d,e). Consistently, FINDY did not decrease the levels of p-Ser97 in HEK293 cells when cells were treated with FINDY after the accumulation of mature DYRK1A (Fig. 3e). Taken together, these results suggest that Ser97 autophosphorylation is catalysed before maturation and that p-Ser97 is susceptible to dephosphorylation after maturation.

FINDY inhibits incorporation of biotin-ATP. We examined whether FINDY inhibited incorporation of ATP into the pocket of DYRK1A. An ATP analogue, the biotinylated acyl phosphate of ATP, which fits to the ATP-binding pocket of kinases, irreversibly reacts with the ATP-orienting lysine and transfers biotin to this lysine[35]. We applied this method in designing an experiment, which is illustrated in Fig. 4a. If FINDY occupies the ATP-binding pocket of the purified kinase, it would inhibit

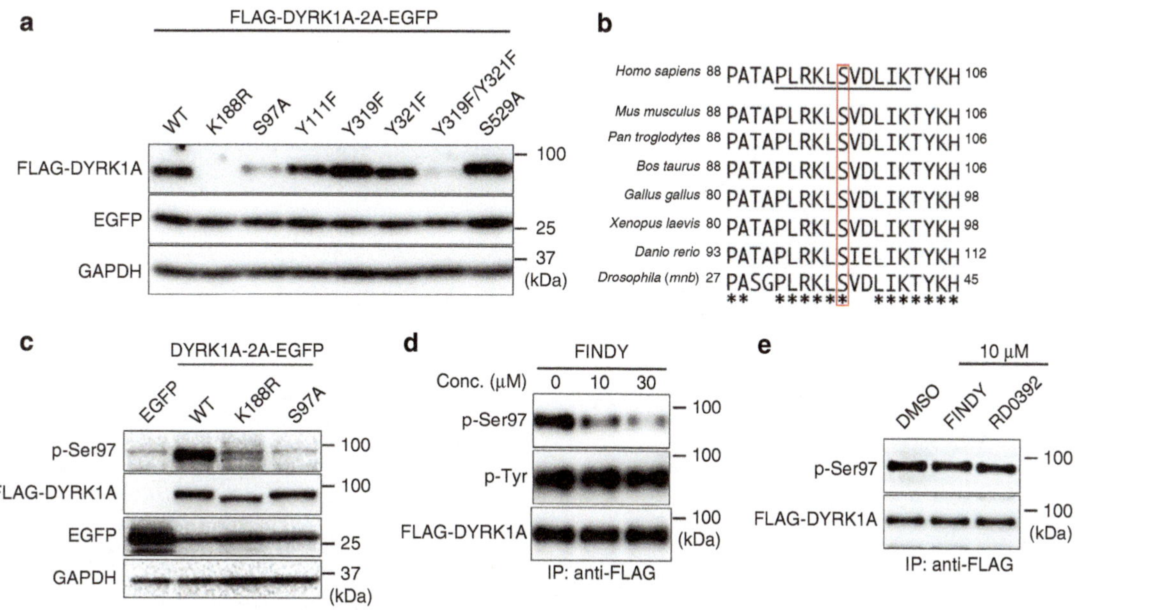

Figure 3 | FINDY suppresses autophosphorylation of Ser97 in DYRK1A. (**a**) The bicistronic expression of intact DYRK1A (WT) and the substituted mutants of Lys188 to Arg (K188R), Ser97 to Ala (S97A), Tyr111 to Phe (Y111F), Tyr319 to Phe (Y319F), Tyr321 to Phe (Y321F), Tyr319 and Tyr321 to Phe (Y319F/Y321F) and Ser529 to Ala (S529A) with EGFP in HEK293 cells. The expression vectors were stably integrated in the genome of the cells. Each protein was detected with the corresponding antibody against FLAG, EGFP or GAPDH. Representative data from the quadruplicate experiments are shown. (**b**) The alignment of Ser97 and its surrounding amino-acid sequences from *DYRK1A* of the indicated species. The conserved serine residues corresponding to Ser97 of *Homo sapiens DYRK1A* are highlighted by the red rectangular box. Underlined portions indicate the phospho-peptide used in the immunization. (**c**) The polyclonal antibody against phosphorylated Ser97 (p-Ser97) specifically detected FLAG-DYRK1A (WT), but not the kinase-dead (K188R) or S97A mutants. The cells were transiently transfected with the expression vectors of the indicated constructs and the proteins were overexpressed. Total cell lysates were subjected to SDS-PAGE followed by western blot analysis. Representative data from the triplicate experiments are shown. (**d**) FINDY suppressed Ser97 autophosphorylation. FLAG-DYRK1A was produced in HEK293 cells incubated with epoxomicin and FINDY for 5 h, then purified with anti-FLAG antibody-conjugated beads. Each protein was detected with the corresponding antibodies against p-Ser97, phospho-Tyr (p-Tyr) or FLAG. Representative data from the triplicate experiments are shown. (**e**) FINDY did not suppress Ser97 autophosphorylation in Tc 2. FLAG-DYRK1A was produced in HEK293 cells for 16 h in the presence of doxycycline, after which cells were treated with 10 μM of FINDY or RD0392 for 5 h. Total cell lysates were subjected to immunoprecipitation (IP) with anti-FLAG antibody-conjugated beads. Each protein was detected with the corresponding antibodies against p-Ser97 and FLAG. Representative data from the duplicate experiments are shown. Conc., concentration.

the binding of the biotin-ATP analogue to this region. FLAG-DYRK1A produced in the presence or absence of the small molecules was immunopurified from the cells. The biotinylation of the immunopurified FLAG-DYRK1A produced in the presence of FINDY was significantly suppressed compared with the vehicle control, whereas that of FLAG-DYRK1A produced with either INDY or RD0392 was not (Fig. 4b). This indicates that FINDY inhibits the incorporation of biotin-ATP into the DYRK1A intermediate during the folding process. As a positive control, treatment of the immunopurified FLAG-DYRK1A with 100 μM ATP before incubation with the biotin-ATP analogue blocked biotinylation (Fig. 4b).

Ser97 autophosphorylation in *in vitro* translation. To confirm the effect of FINDY on Ser97 autophosphorylation of DYRK1A in a simpler assay than living cells, we employed an *Escherichia coli*-based coupled *in vitro* transcription-translation system. The PURExpress system is reconstituted from recombinant proteins and purified ribosomes. The signal of p-Ser97 of DYRK1A was detected when expressed in the PURExpress system (Fig. 5a), indicating a possibility that Ser97 autophosphorylation does not require mammalian proteins. The band of p-Ser97 migrated slower than the major band of the FLAG-DYRK1A protein (Fig. 5a), indicating that a small portion of DYRK1A is autophosphorylated in this experiment. This suggests a low probability of the Ser97 autophosphorylation in the *in vitro* translation. We then performed the *in vitro* translation in the presence of FINDY or RD0392. FINDY and RD0392 suppressed Ser97 autophosphorylation in similar dosage ranges (Fig. 5b–d). In addition, FINDY did not inhibit substrate phosphorylation by the DYRK1A protein produced in the PURExpress system (Fig. 5e). In contrast, RD0392 inhibited substrate phosphorylation in a dose-dependent manner (Fig. 5e). We also examined tyrosine autophosphorylation of DYRK1A. In the PURExpress system, no significant effects of FINDY or RD0392 were observed on the signal from phosphorylated tyrosine (Supplementary Fig. 8). These results indicate that FINDY suppresses Ser97 autophosphorylation during the *in vitro* expression of DYRK1A.

We also confirmed whether the mature DYRK1A catalysed Ser97 autophosphorylation or not using the FLAG-DYRK1A protein produced in the PURExpress system. Immunopurified FLAG-DYRK1A bound on resin was reacted with lambda protein phosphatase, then allowed to autophosphorylate in the presence of ATP for 2 h. Consistent with the results using FLAG-DYRK1A purified from HEK293 cells (Supplementary Fig. 7d,e), the purified DYRK1A protein did not catalyse Ser97 autophosphorylation (Fig. 5f). Thus, Ser97 autophosphorylation is a one-off event during the translation or folding process. These results demonstrate that FINDY suppressed Ser97 autophosphorylation before the maturation of DYRK1A.

Ser97 autophosphorylation of recombinant DYRK1A proteins. To determine whether FINDY directly inhibits Ser97 autophosphorylation, we developed an *in vitro* autophosphorylation assay

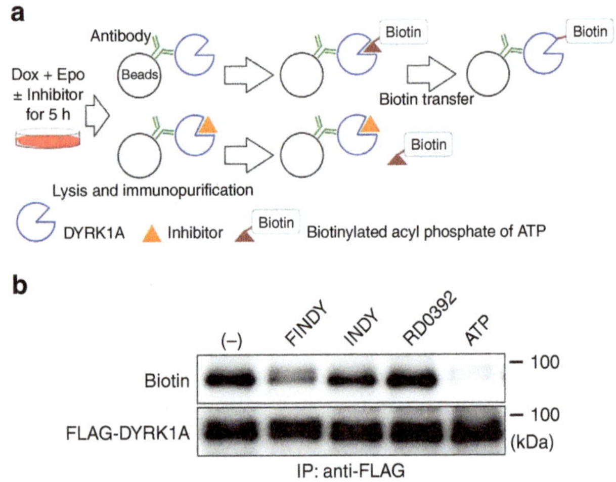

Figure 4 | FINDY affects the folding of DYRK1A, targeting the ATP-binding pocket. (**a**) Schematic diagram of the biotin-transfer experiment. Expression of FLAG-DYRK1A is induced by doxycycline (Dox) with epoxomicin (Epo) and kinase inhibitors (10 μM). The resulting FLAG-DYRK1A proteins are immunopurified and reacted with the biotinylated acyl phosphate of ATP (20 μM). If the inhibitor occupies the ATP-binding pocket, then biotin transfer is prevented. (**b**) FINDY competes with the biotin-ATP analogue for the DYRK1A intermediate. FINDY, INDY and RD0392 were used as described in **a**. As a positive control for the competition experiment, the FLAG-DYRK1A protein was pre-treated with ATP (100 μM) before the reaction with the biotin-ATP analogue. The labelled proteins were blotted with streptavidin and anti-FLAG antibody. Representative data from the triplicate experiments are shown. IP, immunoprecipitation.

using a recombinant DYRK1A protein. DYRK1A fused with a GST-tag at the amino-terminus and with a Twin-Strep-tag (TS) at the carboxy-terminus (GST-DYRK1A-TS) was produced in *E. coli* cells at 6 °C to suppress autophosphorylation, as we found that the p-Ser97 signal decreased in cells cultured at low temperature (Supplementary Fig. 9a). GST-DYRK1A-TS was purified using tandem affinity purification with *Strep*-Tactin Sepharose followed by GSH-conjugating resin (Supplementary Fig. 9b), then subjected to an *in vitro* reaction with ATP (100 μM). Under these conditions, the p-Ser97 signal did not increase, whereas the protein band was broadly shifted (Fig. 6a,b). Therefore, we wondered if the incubation at 6 °C suppressed prior autophosphorylation of the tyrosine residues (Tyr319 and Tyr321) in the activation loop, which are required for Ser97 autophosphorylation (Supplementary Fig. 9c). To verify this possibility, we generated pseudo-phosphorylation mutants, in which these tyrosine residues were substituted for phosphomimetic aspartic or glutamic acid residues (Y319D, Y319E, Y321D and Y321E), then examined their facilitation of Ser97 autophosphorylation. The Y319D and Y319E mutants were purified (Supplementary Fig. 9d), whereas the Y321D and Y321E mutants formed aggregates. As expected, the Y319D and Y319E mutations facilitated Ser97 autophosphorylation of recombinant GST-DYRK1A-TS (Fig. 6a,b). Furthermore, the Y319E mutant catalysed substrate phosphorylation at an equivalent level to the WT kinase (Supplementary Fig. 9e).

We then examined the effect of FINDY on *in vitro* Ser97 autophosphorylation using the Y319E mutant. FINDY inhibited Ser97 autophosphorylation with an IC_{50} value of 110 nM, but did not suppress tyrosine phosphorylation (Fig. 6c,d), which is consistent with results in Fig. 3d. The FINDY-mediated inhibition of Ser97 autophosphorylation decreased due to competition with excess ATP (1 mM; Fig. 6d and Supplementary Fig. 9e). In contrast, the canonical inhibitor RD0392 suppressed both Ser97 and tyrosine phosphorylation in the Y319E mutant (Fig. 6e,f), with IC_{50} values of 205 and 514 nM, respectively. These results demonstrate that FINDY directly and selectively inhibits Ser97 autophosphorylation.

In addition, we examined whether or not FINDY inhibited substrate phosphorylation by these recombinant proteins. The *in vitro* kinase assay showed that FINDY did not inhibit substrate phosphorylation by the WT and Y319E proteins in concentrations of up to 16 μM (Fig. 6g). In contrast, RD0392 inhibited substrate phosphorylation by the WT and Y319E proteins with IC_{50} values of 325 and 813 nM, respectively (Fig. 6g).

FINDY strengthens interaction between DYRK1A and CDC37. Our recent study indicated that mutations affecting autophosphorylation strengthen the interaction between DYRK1A and the kinase-specific co-chaperone CDC37 (ref. 36). CDC37 assists in the folding of target kinases in cooperation with HSP90 (refs 37,38). Taipale *et al.* have demonstrated that CDC37/HSP90 is associated with thermodynamically unstable kinases[39] and that CDC37 fused with luciferase acts as a thermodynamic sensor for kinase structures[40].

To investigate whether the mutation of Ser97 affects the thermodynamic stability of DYRK1A, we examined the interaction between the S97A mutant and CDC37 fused with luciferase nanoKAZ, as described in our recent study[36], with some modifications (see the Methods for details). The S97A, Y321F, Y319F/Y321F and K188R mutations significantly strengthened the interaction of DYRK1A with CDC37-nanoKAZ (Fig. 7a). The S97A mutant interacted with CDC37-nanoKAZ 2.2-fold more strongly than intact DYRK1A (Fig. 7a). The Y321F, Y319F/Y321F and K188R mutants interacted 5.2-, 6.2- and 54-fold more strongly than intact DYRK1A, respectively (Fig. 7a). Interestingly, the Y321F mutation increased the interaction (Fig. 7a), but did not destabilize the DYRK1A protein in HEK293 cells (Fig. 3a). Thus, the increased interaction with CDC37 is not always correlated with the degradation of DYRK1A mutants. Therefore, another functional role of Ser97 may be involved in the degradation of DYRK1A.

We next investigated whether FINDY decreases the thermodynamic stability of DYRK1A. Treatment with FINDY significantly strengthened CDC37-nanoKAZ interactions in a dose-dependent manner, whereas treatment with RD0392 weakened the interactions (Fig. 7b), suggesting that FINDY induces thermodynamic destabilization of DYRK1A. In addition, we examined the effect of FINDY on the S97A mutant. FINDY also increased the CDC37 interaction with the S97A mutant, but it tended to be less than that with intact DYRK1A (Supplementary Fig. 10). These results suggest that FINDY affected a part of the folding process in a mechanism independent of the Ser97 autophosphorylation.

Selective inhibition of DYRK1A by FINDY. To determine the selectivity of FINDY, potential inhibitory activity was assessed against a panel of 275 kinases using 10 μM of FINDY. Only five other kinases (GSK3β, MARK4, PIM1, PIM3, PLK3) were inhibited by over 75% and none of these showed over 85% inhibition (Supplementary Table 1). In addition, FINDY did not exhibit any inhibitory effect on DYRK1B and DYRK2 in an *in vitro* kinase assay, whereas RD0392 inhibited both kinases in a dose-dependent manner (Fig. 8a,b). DYRK1B, the member of the DYRK family closest to DYRK1A, possesses a conserved serine residue similar to Ser97 of DYRK1A (Supplementary Fig. 11a), the mutation of which also leads to destabilization of DYRK1B (Supplementary Fig. 11b). Therefore, we investigated whether

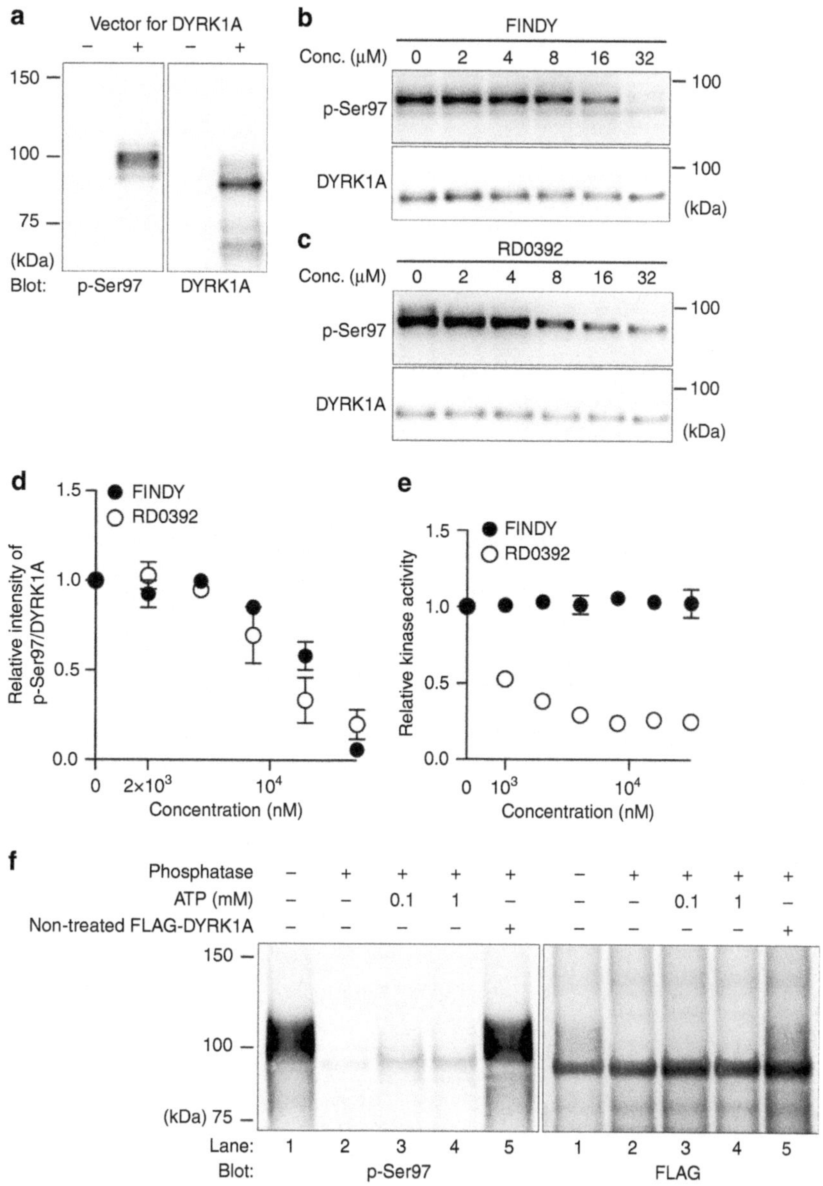

Figure 5 | FINDY suppresses Ser97 autophosphorylation of DYRK1A in an *in vitro* transcription-translation system. (**a**) Expression of DYRK1A in a cell-free *E. coli*-based coupled transcription-translation system. Representative data from the duplicate experiments are shown. (**b**,**c**) *In vitro* expression of DYRK1A in the presence of FINDY (**b**) or RD0392 (**c**). Representative data from the triplicate experiments are shown. (**d**) Relative activities of Ser97 autophosphorylation in the *in vitro* expression of DYRK1A in the presence of FINDY or RD0392. The band intensities in **b**,**c** were quantified. The graph shows the means ± s.d. ($n = 3$). (**e**) DYRK1A produced in the *in vitro* system was incubated with the peptide substrate DYRKtide in the presence of FINDY or RD0392 (up to 32 μM). The results are presented as means ± s.d. ($n = 3$). (**f**) DYRK1A produced in the *in vitro* system was purified with anti-FLAG antibody-conjugated beads, then reacted with lambda protein phosphatase for 2 h. After washout of phosphatase, the dephosphorylated DYRK1A was allowed to autophosphorylate in the presence of ATP and sodium orthovanadate (1 mM) for 2 h. To check the residual phosphatase activity, the non-treated FLAG-DYRK1A protein was added into the autophosphorylation reaction (lane 5). Representative data from the duplicate experiments are shown. Conc., concentration.

FINDY destabilizes DYRK1B and DYRK2 in cells expressing FLAG-DYRK1B-2A-EGFP and FLAG-DYRK2-2A-EGFP, respectively. Neither the expression of FLAG-DYRK1B nor FLAG-DYRK2 was affected by FINDY (Fig. 8c,d). These results suggest that the transitional intermediate of DYRK1A has a unique structural property distinct from that of DYRK1B and DYRK2. Moreover, FINDY decreased the amount of endogenous DYRK1A in HEK293 cells, whereas no alteration in the endogenous expression of the other 13 kinases was observed (Fig. 8e), indicating that FINDY selectively destabilizes the transitional intermediate of DYRK1A.

We have previously reported that microinjection of *DYRK1A* mRNA into *Xenopus laevis* embryos causes the malformation of neural tissues, and this system has been applied to assess the efficacy of kinase inhibitors *in vivo*[16]. To validate the subtype-selectivity of FINDY in the context of living organisms, we investigated whether FINDY rescues the developmental malformation of *Xenopus* embryos induced by the overexpression of DYRK1A or DYRK1B. *DYRK1A* or *DYRK1B* mRNA was injected into two dorsal blastomeres at the eight-cell stage, resulting in malformation of the eye and the head in stage 40 tadpoles (Fig. 9a). Control β-*globin* mRNA did not cause any

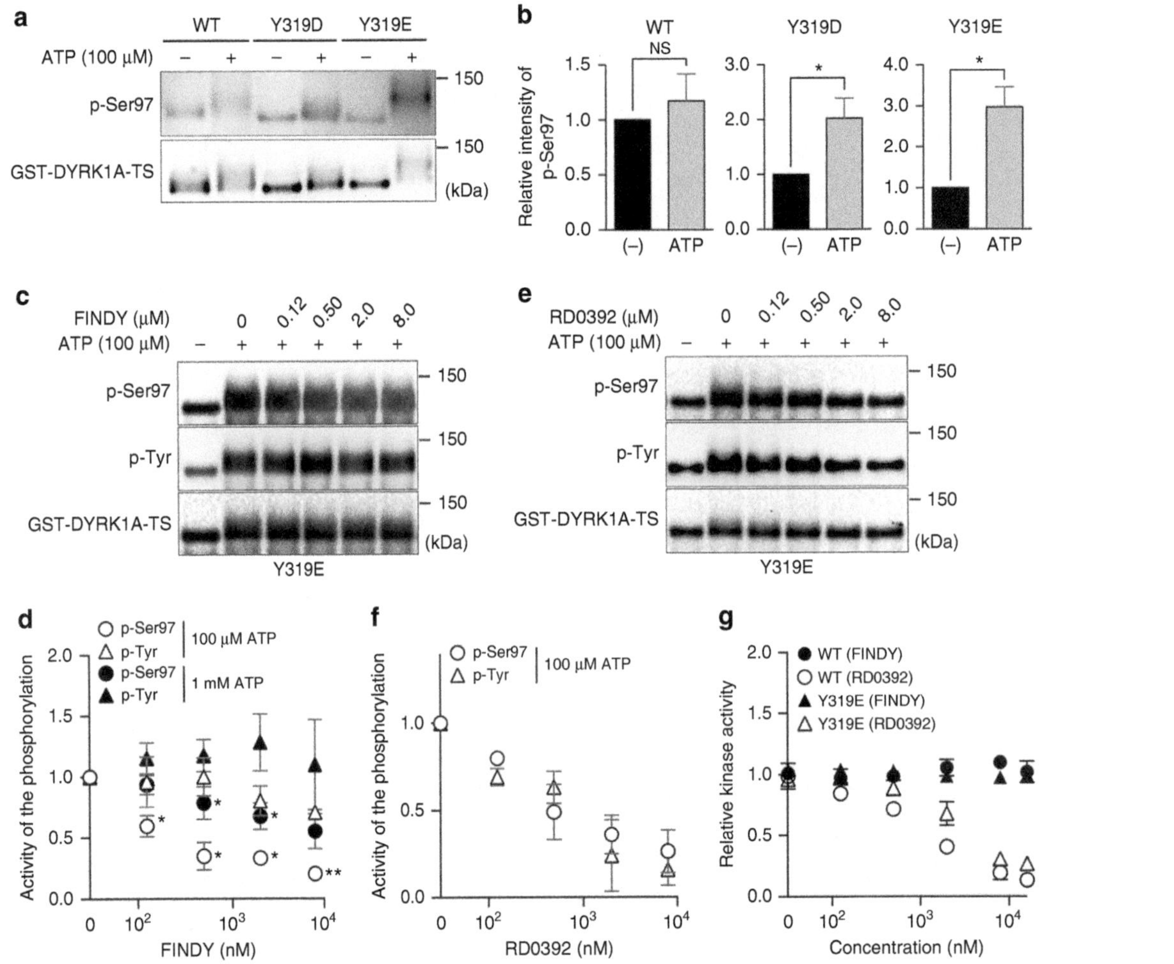

Figure 6 | FINDY inhibits Ser97 autophosphorylation of recombinant DYRK1A protein. (**a**) Recombinant GST-DYRK1A-TS protein (WT) and its pseudo-phosphorylation mutants (Y319D and Y319E), produced at 6 °C in *E. coli*, were reacted with ATP (100 μM), then analysed by SDS-PAGE followed by western blot analysis. Representative data from the quadruplicate experiments are shown. (**b**) The band intensities in **a** were quantified. The graph shows the means ± s.d. ($n = 4$). Statistical significance was calculated compared with the value without ATP. NS indicates not significant. $^{*}P < 0.05$. Statistical analysis was performed with the Mann-Whitney *U*-test. (**c**) The Y319E mutant was reacted with ATP (100 μM) in the presence of FINDY. Representative data from the triplicate experiments are shown. (**d**) Relative activities of Ser97 and tyrosine phosphorylation in the presence of ATP (100 μM and 1 mM) and FINDY, which were calculated from the ATP-dependent increase in the intensities of p-Ser97 and p-Tyr (see the Methods for details). The graph represents means ± s.d. ($n = 3$). Statistical significance was calculated compared with the value of the relative activity of p-Tyr at the same concentration of FINDY. $^{*}P < 0.05$ and $^{**}P < 0.01$. Data were analysed with the paired *t*-test. (**e**) RD0392 inhibited the phosphorylation of Ser97 and tyrosine residues. The Y319E mutant was reacted with ATP (100 μM) in the presence of RD0392. Representative data from the triplicate experiments are shown. (**f**) Relative activities of Ser97 and tyrosine phosphorylation in the presence of ATP (100 μM) and RD0392, which were calculated from the ATP-dependent increase in the intensities of p-Ser97 and p-Tyr (see the Methods for details). The graph represents means ± s.d. ($n = 3$). (**g**) The recombinant DYRK1A proteins were incubated with the peptide substrate DYRKtide in the presence of FINDY or RD0392 (up to 16 μM). The results are presented as means ± s.d. ($n = 3$).

morphological alterations compared with uninjected embryos (Fig. 9a). Administration of proINDY, which is an acetylated prodrug with enhanced cell membrane permeability hydrolysed in the cell to form INDY[16], repressed the malformation induced by both DYRK1A and DYRK1B (Fig. 9a). FINDY rescued the malformation induced by DYRK1A, but not that induced by DYRK1B, demonstrating marked subtype-selectivity (Fig. 9a). Embryos injected with *DYRK1A* mRNA were almost completely rescued with the administration of either FINDY or proINDY (Fig. 9b). Only 10.9% of the tadpoles developed normally with FINDY when injected with *DYRK1B* mRNA, whereas 89.1% were normal in the presence of proINDY (Fig. 9b). In the embryos injected with either *DYRK1A* or *DYRK1B* mRNA, decreased expression of the pan-neural marker *NCAM*, forebrain marker *BF-1*, eye markers *Pax-6* and *Rx-1*, fore-midbrain marker *Otx-2*, mid-hindbrain marker *En-2* and hindbrain marker *Krox-20* were observed (Fig. 9c). FINDY restored the expression of these markers only in DYRK1A-overexpressing tadpoles, whereas proINDY increased the expression of the markers in both DYRK1A- and DYRK1B-overexpressing embryos (Fig. 9c). These results indicate that FINDY rescues the neurological defects induced by DYRK1A in a highly specific manner.

Discussion

This study provides methods for evaluating chemical compounds that selectively inhibit the transitional intermediates of target kinases. We utilized these methods to identify and evaluate FINDY. Based on the following evidence, we demonstrate that FINDY selectively inhibits a folding intermediate of DYRK1A during its folding process, as illustrated in Fig. 10. First, FINDY suppresses the substrate phosphorylation by DYRK1A only when

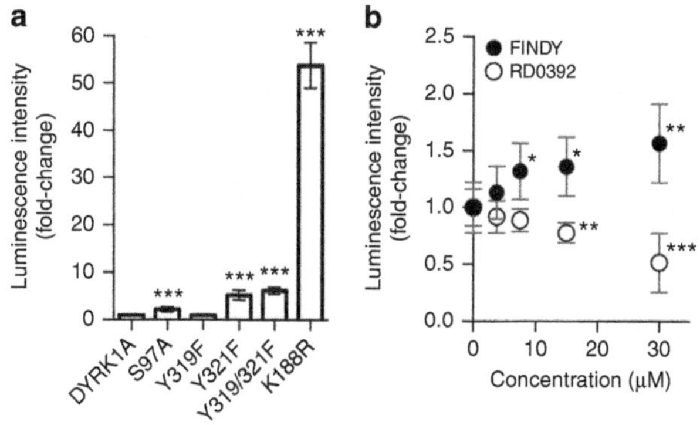

Figure 7 | FINDY strengthens the interaction between DYRK1A and the co-chaperone CDC37. (**a**) 293T cells stably expressing CDC37-nanoKAZ were transiently transfected with expression vectors of 3xFLAG-DYRK1A, as were mutants of S97A, Y319F, Y321F, Y319F/Y321F and K188R. Total cell lysates were subjected to the CDC37-nanoKAZ interaction assay (see the Methods for details). Luminescence intensities are shown as fold-changes relative to that of DYRK1A, normalized to the amount of bound protein on a 96-well plate. The graphs represent means ± s.d. ($n = 7$). Statistical significance was calculated compared with the value of the intact DYRK1A. ***$P < 0.001$. Statistical analysis was performed with the Mann-Whitney *U*-test. (**b**) 293T cells stably expressing CDC37-nanoKAZ were transiently transfected with an expression vector of 3xFLAG-DYRK1A. Twenty-four hours after the transfection, cells were treated with FINDY and RD0392, and incubated for an additional 24 h. Luminescence intensities are shown as fold-changes relative to that at 0 μM, normalized to the amount of 3xFLAG-DYRK1A bound on a 96-well plate. Points on the graph are means ± s.d. ($n = 8$). Statistical significance was calculated compared with the value at 0 μM. *$P < 0.05$, **$P < 0.005$ and ***$P < 0.001$. Statistical analysis was performed with the Mann-Whitney *U*-test.

it is administered before the induced expression (Fig. 1c). Second, neither in the cell-based nor in the *in vitro* assays is FINDY able to inhibit the substrate phosphorylation by the mature DYRK1A (Figs 1c,f, 5e and 6g). Third, FINDY suppressed Ser97 autophosphorylation in cells (Fig. 3d), in the *in vitro* translation (Fig. 5), and in the *in vitro* autophosphorylation assay with the recombinant protein (Fig. 6). The Ser97 autophosphorylation was a one-off intramolecular process that was not catalysed by the mature DYRK1A (Fig. 5f and Supplementary Fig. 7d,e). Furthermore, the CDC37-nanoKAZ interaction assay revealed that FINDY induces the thermodynamic destabilization of DYRK1A (Fig. 7b), suggesting that FINDY affects the folding state of DYRK1A.

Our study demonstrates that FINDY directly suppresses Ser97 autophosphorylation, as discussed below. First, FINDY suppressed Ser97 autophosphorylation in the *in vitro* translation system (Fig. 5), which contained only defined factors, but did not contain mammalian proteins. Second, FINDY suppressed the *in vitro* Ser97 autophosphorylation of the recombinant phosphomimetic mutant (Y319E) of DYRK1A, whereas tyrosine autophosphorylation was not suppressed by FINDY (Fig. 6 and Supplementary Fig. 5). These results are consistent with the results from the cellular assays (Fig. 3d). Several lines of evidence demonstrated that the mature DYRK1A was not able to catalyse Ser97 autophosphorylation (Figs 5f and 6a,b and Supplementary Fig. 7d,e). Thus, the kinase form that catalyses Ser97 autophosphorylation is not the mature form. Therefore, the recombinant Y319E probably reflects an aspect of the folding intermediate. Third, FINDY blocked the incorporation of biotin-ATP into DYRK1A (Fig. 4), indicating two possibilities. One is that FINDY occupies the ATP-binding pocket. The other is that FINDY affects the biotin-ATP incorporation by binding to an allosteric site, which may affect the ATP binding to the pocket. FINDY possesses the same polar groups of RD0392 that targets the ATP-binding pocket of DYRK1A (Supplementary Figs 2 and 3). Thus, the former scenario is most likely. The selectivity of FINDY for the Ser97 and tyrosine autophosphorylation suggests that the transitional state targeted by FINDY is distinct to that of tyrosine autophosphorylation. Further structural studies are necessary to conclude this discussion.

The phosphomimetics experiment suggested that the phosphorylated tyrosine residue in the activation loop is required for the subsequent Ser97 autophosphorylation (Fig. 6). Although intramolecular tyrosine autophosphorylation in the activation loop has been suggested to be a one-off inceptive event to achieve full activity during the co-translational folding process of *Drosophila* DYRKs[17,19,41], it is controversial whether the autophosphorylation of these tyrosine residues is essential for the catalytic activity of mammalian DYRK1A[21,33]. The double mutation of Y321F/Y319F in the activation loop destabilized FLAG-DYRK1A, whereas a single mutation gave a much smaller effect (Fig. 3a). Our study indicates that the autophosphorylation of at least one of these tyrosine residues is required for Ser97 autophosphorylation, suggesting that Ser97 autophosphorylation acts as a quality control mechanism. It remains elusive how Ser97 autophosphorylation regulates the stability of DYRK1A and why Ser97 autophosphorylation is a one-off process. In comparison to the *in vitro* translation (Fig. 5a), we did not observe the faster migrating band corresponding to the Ser97 non-phosphorylated DYRK1A in the cultured mammalian cells, which may be subjected to degradation during the folding process. The actual amount of the Ser97 autophosphorylation during the folding process in mammalian cells also remains unclear.

Although RD0392 inhibited the Ser97 autophosphorylation in the *in vitro* translation experiment (Fig. 5c,d) and in the autophosphorylation assay using the recombinant DYRK1A protein (Fig. 6e,f), RD0392 did not decrease the FLAG-DYRK1A protein in the cultured cells (Fig. 2d). This discrepancy is possibly due to low cell membrane permeability of RD0392. FINDY has a hydrophobic trimethylsilyl group that may serve to enhance the cell membrane permeability compared with that of RD0392. RD0392 inhibited the Tyr autophosphorylation of the recombinant DYRK1A protein (Fig. 6e,f). In contrast, RD0392 did not inhibit the Tyr autophosphorylation in the *in vitro* translation (Supplementary Fig. 8c). This indicates the possibility that RD0392 is capable of inhibiting post-translational Tyr autophosphorylation, but co-translational Tyr autophosphorylation is resistant to RD0392.

An interesting point in the biotin-ATP experiment is that INDY and RD0392 failed to inhibit biotinylation, whereas FINDY, on the other hand, did not (Fig. 4b). The most likely explanation is that FINDY has a slower rate of dissociation (K_{off}) from the intermediate of DYRK1A than the K_{off}s that INDY and RD0392 have to the mature form. This slow K_{off} for FINDY may be due to thermodynamic destabilization by FINDY, which is distinct from the effects of INDY and RD0392. FINDY uniquely strengthened the interaction between DYRK1A and CDC37-nanoKAZ (Fig. 7b). In contrast, RD0392 and INDY weakened the interaction (Fig. 7b)[36], which is in good agreement with previous reports that canonical kinase inhibitors weaken CDC37 interactions[39,40,42]. A previous study has demonstrated that CDC37 binding competes with ATP for the pocket of its client kinase[42], suggesting that the enhanced binding of CDC37 plays a role in the suppression of the biotinylation. Furthermore, the thermodynamic destabilization caused by FINDY may also cause hydrophobic aggregation of the DYRK1A protein, which masks the pocket and prevents the incorporation of ATP. Thus, we

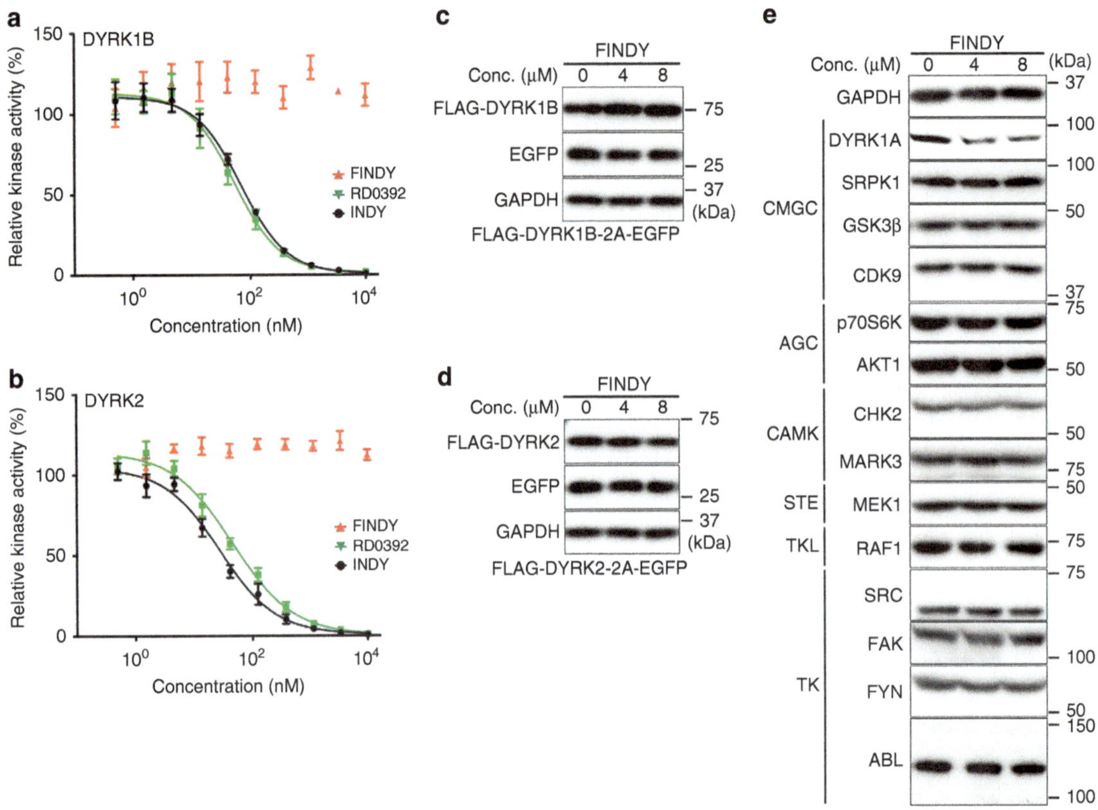

Figure 8 | FINDY selectively inhibits DYRK1A. (**a**,**b**) Recombinant DYRK1B (**a**) and DYRK2 (**b**) were incubated with the peptide substrate DYRKtide in the presence of FINDY, RD0392 or INDY. RD0392 and INDY inhibit the kinase activity with IC_{50} values of 53.3 and 69.2 nM, respectively, for DYRK1B, and with IC_{50} values of 45.6 and 27.7 nM, respectively, for DYRK2. The results are presented as means ± s.d. ($n = 4$). (**c**,**d**) Bicistronic expression of FLAG-DYRK1B (**c**) and FLAG-DYRK2 (**d**) with EGFP in HEK293 cells. FINDY administered at the indicated concentrations (Conc.) did not affect the expression of FLAG-DYRK1B and FLAG-DYRK2. Representative data from the triplicate experiments are shown. (**e**) Intact HEK293 cells were cultured in the presence of FINDY for 3 days. The indicated kinases and GAPDH were detected by western blot analysis using the corresponding antibodies. The categories of the indicated kinases in the kinome are shown. Representative data from the duplicate experiments are shown.

consider that thermodynamic destabilization by FINDY may result in strong competition with the biotin-ATP.

The SPHINKS assay consists of cells expressing DYRK1A and its substrate fused with a DD tag and the detection of TAU phosphorylation. This assay was developed to evaluate the effect of chemical compounds on the folding process of DYRK1A; however, it is not suitable for high-throughput screening of large-scale chemical libraries. Thus, the SPHINKS assay should be utilized as a second or third screening assay for a small number of chemical compounds that were roughly selected from the library by a robust high-throughput assay. The *in vitro* translation and autophosphorylation assays would be better methods for developing high-throughput assays. Some small molecules may show apparent selectivity in the SPHINKS assay because of slower actions, such as low solubility, pro-drug activation or slow uptake into cells. Thus, the properties of small molecules should be carefully examined.

This study demonstrates that targeting transitional intermediates is a promising strategy for the development of novel kinase inhibitors. Intramolecular autophosphorylation in the activation loop has been demonstrated not only in DYRK1A, but also in other kinases, such as DYRK2 (ref. 19), GSK3β (ref. 18), ERKs[43] and the Parkinson's disease-associated kinase LRRK2 (refs 44,45). In addition, autophosphorylation outside the activation loop has been reported in C-RAF[46] and protein kinase A (PKA)[47]. We believe that intermediate-selective inhibitors for these kinases can be identified. The close structural similarity between the intermediate-selective inhibitors and the canonical inhibitors raises the possibility that canonical ATP-competitive kinase inhibitors can be converted to intermediate-selective inhibitors with only a slight modification of the common chemical scaffold (Supplementary Figs 12 and 13). Numerous compounds targeting various kinases have been rejected in *in silico* or *in vitro* drug screening, but some could potentially inhibit a transitional intermediate of the target kinase. The methods developed in this study can be utilized to reassess these compounds and provide opportunities to rediscover selective kinase inhibitors.

The dynamic state of the transitional intermediate may provide conformational diversity for the inhibitor-binding pocket. It is possible to purify the metastable intermediate of the DYRK1A protein produced in the presence of FINDY, as we have preliminarily tried in Figs 4 and 5. Further studies using biophysical techniques are necessary in order to analyse the intermediate state targeted by FINDY. Resolving the intermediate states of kinases would expand the chemical options available for the design of selective inhibitors.

Methods

Materials. Small molecules were prepared using standard synthetic procedures, as described in Supplementary Methods. INDY and proINDY were prepared as described previously[16]. These compounds were dissolved in dimethylsulphoxide (DMSO) at 10 mM as a stock solution. Calyculin A (Cell Signaling Technology), okadaic acid (Tocris Bioscience) and cantharidic acid (Abcam) were purchased. The polyclonal antibody against p-Ser97 was produced by immunizing rabbits with a synthetic phospho-peptide, C-PEG-PLRKLS*VDLIK, which was coupled to KLH via the free cysteine, with S* representing phosphoserine. The specific antibody was purified by phospho-peptide-conjugating resin followed by a subtraction step on a

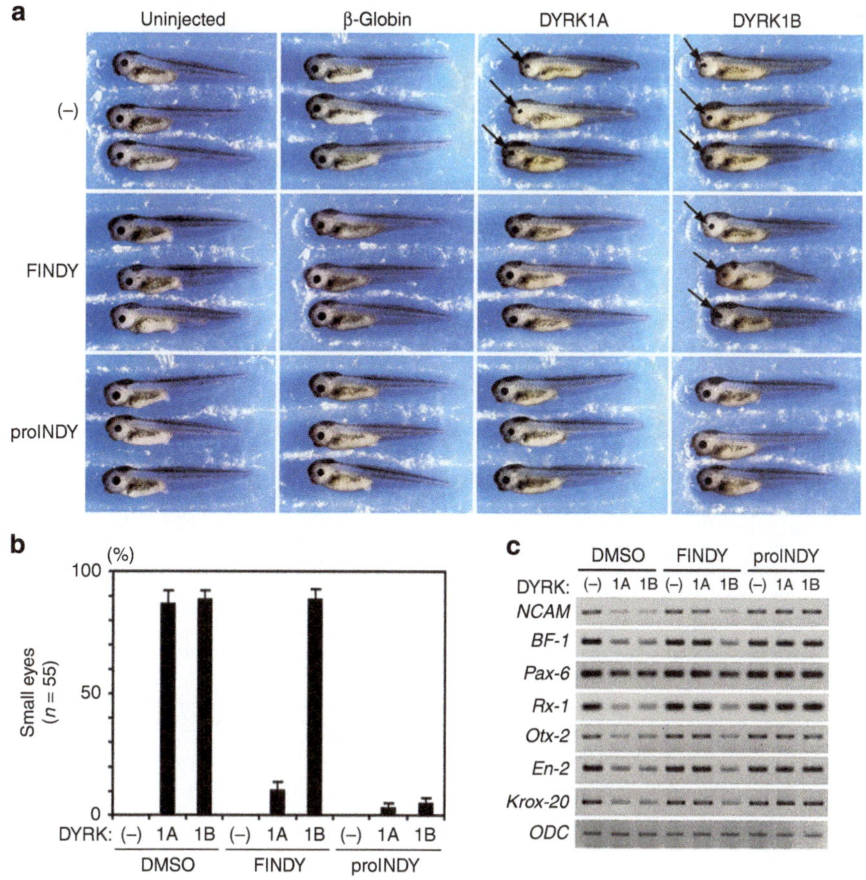

Figure 9 | FINDY selectively represses excessive DYRK1A activity in an animal model. (**a**) FINDY rescued the developmental malformation of *Xenopus laevis* embryos as induced by the overexpression of DYRK1A, but did not rescue that induced by DYRK1B. proINDY was effective in both cases. Embryos were treated with the vehicle alone (–), FINDY (2.5 μM) or proINDY (2.5 μM), then two dorsal blastomeres were injected with 750 pg of *β-globin*, *DYRK1A* or *DYRK1B* mRNA at the eight-cell stage. Uninjected embryos were used as controls. Development of the embryos was allowed to proceed to stage 40 in the dark. Representative images are shown. Arrows indicate small eyes. (**b**) Summary of the rescue rates of small eyes as induced by the overexpression of *DYRK1A* or *DYRK1B*. The results are the means ± s.e.m. of triplicate experiments ($n = 55$). (**c**) RT-PCR analysis of the neural marker genes in the head regions of the embryos at stage 25. Each of the marker genes indicates a specific region: *NCAM* for the pan-neural region, *Brain Factor-1* (*BF-1*) for the forebrain, *Pax-6* and *Rx-1* for the eye, *Otx-2* for the fore-midbrain, *Engrailed-2* (*En-2*) for the mid-hindbrain and *Krox-20* for the hindbrain. *Ornithine decarboxylase* (*ODC*) is an internal control.

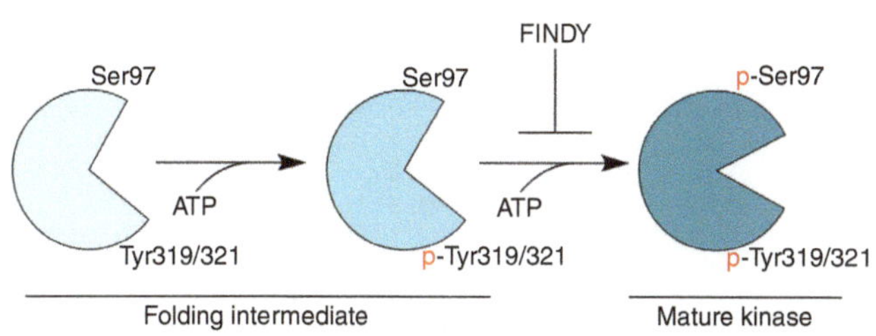

Figure 10 | Model of the folding intermediate-selective inhibition by FINDY. In the folding process of DYRK1A, a folding intermediate autophosphorylates Tyr319/321 and subsequently Ser97 in an intramolecular manner, which prevents degradation of DYRK1A. Autophosphorylated DYRK1A takes on a mature conformation. FINDY interferes only with the folding intermediate by preventing the incorporation of ATP, leading to its degradation.

column containing non-phospho-peptide. The polyclonal antibody against p-Ser97 (1/2,000) and antibodies listed below were diluted with Can Gel Signal Immunoreaction Enhancer Solution (TOYOBO) in western blot analysis. Mouse monoclonal anti-DYKDDDDK (FLAG) tag (1/1,000, clone 1E6, 018-22381/Lot: CTR5949) and anti-GFP (1/1,000, mFX75, 012-22541/Lot: STK2113) antibodies were purchased from Wako Pure Chemical Industries, mouse monoclonal anti-HA (1/1,000, clone HA124, 06340-54/Lot: M2E8170) antibody from Nacalai Tesque, mouse monoclonal anti-GAPDH (1/10,000, clone 6C5, AM4300/Lot: 1311029) from Thermo Fisher Scientific, mouse monoclonal anti-TAU (1/3,000, clone TAU-5, 577801/Lot: D00100518) from Calbiochem (Merck), rabbit polyclonal antibody against pT212 of TAU (1/3,000, 44740G/Lot: 1392715A) from Life Technologies, mouse monoclonal anti-phosphotyrosine (1/3,000, 4G10 Platinum, 05-1050/Lot: 2567063) from Millipore, mouse monoclonal anti-GST-tag (1/3,000, clone GT5, M209-3/Lot: 001) from MBL, mouse monoclonal anti-SRPK1 (1/3,000, clone 12/SRPK1, 611072) from BD Transduction Laboratories (BD Biosciences), mouse monoclonal anti-CDK9 (1/3,000, clone D-7, sc-13130/Lot: c0613) from Santa Cruz Biotechnology and mouse monoclonal anti-neuronal class III β-Tubulin (1/3,000, clone TUJ1, MRB-435P/Lot: E10045BF) from Covance. Rabbit polyclonal anti-DYRK1A (1/3,000, #2771S/Lot: 1), p70S6K (1/3,000, #9202S/Lot: 8), CHK2 (1/3,000, #2662S/Lot: 5), MARK3 (1/3,000, #9311BC/Lot: 1), RAF1 (1/3,000, #9422BC/Lot: 5), FAK (1/3,000, #3285BC/Lot: 6), FYN (1/3,000, #4023BC/Lot: 2) and ABL (1/3,000, #2862BC/Lot: 11) antibodies were obtained from Cell Signaling Technology. Rabbit monoclonal anti-GSK3β (1/3,000, clone 27C10, #9315S/Lot: 4), AKT1 (1/3,000, clone C73H10, #2938BC/Lot: 2), MEK1 (1/3,000, clone 30C8, #9146BC/Lot: 1) and SRC (1/3,000, clone 32G6. #2123BC/Lot: 3) antibodies were purchased from Cell Signaling Technology. HRP-linked anti-rabbit (1/4,000, NA9340V/Lot: 9611780) and anti-mouse (1/4,000, ab5887/Lot: GR16899-9) IgG were purchased from GE Healthcare and Abcam, respectively. HRP-labelled streptavidin (1/2,000, 474-3000/Lot: 060065) was obtained from Kirkegaard & Perry Laboratories.

Vector construction. In brief, the bicistronic expression vectors for the DYRKs were constructed with pcDNA5/FRT/TO (Life Technologies). DD-TAU and DD-FLAG-DYRK1A were inserted into the pCAGIPuro vector. To prepare *in vitro* transcribed mRNAs for injection, DYRKs were inserted into the pCS2+ vector.

pCold (Takara Bio) and pET28a(+) vectors were used for the preparation of recombinant protein in *E. coli* cells. The pET28a(+) vector was also used for the *in vitro* translation experiments. PCR-amplified fragments of the human microtubule-associated protein TAU (0N4R) and FLAG-tagged DYRK1A were fused in-frame to the carboxyl-terminus of DD tag by the overlap-extension PCR method to generate DD-TAU and DD-FLAG-DYRK1A, respectively, as described previously[48], with some modifications. The combined fragments were inserted into the pCAGIPuro vector, an IRES-based bicistronic expression vector where the gene of interest and a puromycin-resistant gene are expressed from a single mRNA, which enables almost all of the cells selected with puromycin to express the gene product. A PCR-amplified fragment of FLAG-tagged DYRK1A was subcloned into pcDNA5/FRT/TO. For the bicistronic cassette using a 2A peptide, PCR-amplified fragments of FLAG-tagged DYRKs were fused in-frame to the amino-terminus of EGFP via the F2A peptide sequence by the overlap-extension PCR method. The combined fragments were inserted into pcDNA5/FRT/TO (Life Technologies). The site-directed mutagenesis of DYRK1A and DYRK1B was performed as described previously[49], with some modifications. The reconstituted vector sequences are available upon request.

Cell culture and transfection. Flp-In/T-REx HEK293 (Life Technologies) and intact HEK293 cells were maintained in low-glucose DMEM (Nacalai Tesque) supplemented with 10% fetal bovine serum (FBS; Nichirei Biosciences), 100 U ml^{-1} of penicillin and 100 μg ml^{-1} of streptomycin (Nacalai Tesque). Cells were transfected with plasmid DNAs using polyethylenimine MAX (Polysciences) as described previously[50], then selected with hygromycin B (Life Technologies) for the pcDNA5/FRT/TO vectors and/or with puromycin (Nacalai Tesque) for the pCAGIPuro vectors to establish stable cell lines.

SPHINKS assay. Our original synthetic chemical library used in this study consisted of newly synthesized structural derivatives of ATP-competitive inhibitors of DYRK1A, which had been identified by an *in vitro* kinase assay in our laboratory. Flp-In/T-REx HEK293 cells transfected with pcDNA5/FRT/TO-FLAG-DYRK1A and pCAGIPuro-DD-TAU were seeded on polyethylenimine-coated six-well plates with 2 ml of DMEM and 1% FBS. Cells were then cultured 24 h later in the presence of 1 μg ml^{-1} of doxycycline (Nacalai Tesque) for 16 h, then treated with 1 μM of Shield-1 (Clontech, Takara Bio) for 2 h. Small molecules from our chemical library were diluted in 1 ml of DMEM and 1% FBS, then administered into the cells at a final concentration of 10 μM (0.1% DMSO), as illustrated in Fig. 1a. Total cell lysates were prepared with SDS-urea buffer (50 mM Tris-HCl, pH 7.0, 1% SDS, 4 M urea, 10% glycerol, 20 mM dithiothreitol, 0.5 mg ml^{-1} BSA), then analysed with SDS–polyacrylamide gel electrophoresis (SDS–PAGE) followed by western blot analysis using antibodies against phospho-TAU (pT212) (1/3,000) and total TAU (1/3,000) proteins (Supplementary Fig. 14). Details are summarized in Supplementary Table 2.

Immunoprecipitation of FLAG-DYRK1A. HEK293 cells were cultured in the presence of 1 μg ml^{-1} of doxycycline, 1 μM of Shield-1, 1 μM of epoxomicin and a proteasome inhibitor (Peptide Institute), with or without 10 μM of small molecules for 5 h. The cells were lysed in 50 mM Tris (pH 7.4), 150 mM NaCl, 0.5 mM EDTA, 1% Empigen BB and protease/phosphatase inhibitor cocktails (Nacalai Tesque). The clarified cell lysates were processed with anti-FLAG (clone M2) beads (F2426, Sigma-Aldrich) and the bound proteins were then eluted with SDS-urea buffer, then analysed with SDS–PAGE followed by western blot analysis (Supplementary Fig. 14).

Biotin transfer experiment. Biotinylation of the lysine in the ATP-binding pocket was performed using the Pierce Kinase Enrichment Kit with an ATP Probe (Thermo Fisher Scientific), according to the manufacturer's instructions with some modifications. HEK293 cells were cultured with 1 μg ml^{-1} of doxycycline and 1 μM of epoxomicin, with or without 10 μM of small molecules for 5 h, then harvested and lysed in Pierce IP Lysis Buffer containing 0.5% Empigen BB. The clarified cell lysates were processed with anti-FLAG (M2) beads, then equilibrated with the kit's reaction buffer. The beads were subsequently reacted with 20 μM of the kit's ActivX Desthiobiotin-ATP probe for 30 min at room temperature. The bound proteins were eluted with SDS-urea buffer, then analysed with SDS–PAGE followed by western blot analysis (Supplementary Fig. 14).

***E. coli*-based *in vitro* translation assay.** The PURExpress kit (New England Biolabs) was employed. DYRK1A was expressed at 37 °C for 40 min in a 12.5-μl reaction mixture containing 30 ng of pET28a(+)-FLAG-DYRK1A. For inhibitor experiments, FINDY or RD0392 was added prior to the initiation of transcription. For immunoprecipitation, 100 μl of PBS containing 0.5% Empigen BB, 0.5 mg ml^{-1} BSA and 5 mM EDTA was added into the reaction mixture, then incubated with anti-FLAG (M2) beads at 4 °C for 6 h. The samples were prepared with an SDS-urea buffer, then analysed with SDS–PAGE followed by western blot (Supplementary Fig. 14). The band intensities were measured using the LAS-3000 (Fujifilm) and Multi Gauge software (Fujifilm).

For the phosphatase experiment, immunoprecipitated FLAG-DYRK1A bound on the beads was incubated in a total volume of 100 μl containing 1,600 U of lambda protein phosphatase for 2 h at 30 °C. After washing the beads with autophosphorylation reaction (AR) buffer (see below) containing 0.1 mg ml^{-1} of BSA (Sigma-Aldrich), the beads were incubated with the indicated concentrations of ATP and sodium orthovanadate (1 mM) for 2 h at 37 °C and prepared with the SDS-urea buffer. To check the residual phosphatase activity, non-treated immunoprecipitated FLAG-DYRK1A bound on the beads was added into the reaction.

For the *in vitro* kinase assay, 0.5 μl of the reaction mixture was incubated in a total volume of 25 μl of AR buffer containing 0.1 mg ml^{-1} of BSA, 100 μM of ATP, 50 μM of DYRKtide peptide (Anaspec) and the indicated concentrations of the small molecules for 2 h at 37 °C. The consumption of ATP was measured with ADP-Glo Kinase Assay kit (Promega Corporation).

***In vitro* autophosphorylation assays.** *E. coli* strain Rosetta(DE3)pLysS was transformed with the pCold-I-GST-DYRK1A-TS vector. The cells were cultured with IPTG (final 1 mM) for 24 h at 6 °C, then lysed with GET buffer (20 mM HEPES-KOH, pH 8.0, 150 mM NaCl, 1 mM EDTA, 2 mM dithiothreitol, 10% glycerol, 0.5% Empigen BB, 0.5% Triton X-100). Clarified cell lysates were loaded onto a *Strep*-Tactin Sepharose column (IBA) and the bound proteins were then eluted with GET buffer containing 2.5 mM desthiobiotin (IBA) and 1 mg ml^{-1} BSA. The eluates were subsequently applied to GST-Accept resin (Nacalai Tesque). The resin was washed extensively, first with GET buffer, then with AR buffer (5 mM MOPS, pH7.2, 2.5 mM β-glycerol-phosphate, 5 mM $MgCl_2$, 1 mM EGTA, 0.4 mM EDTA, 2 mM tris(2-carboxyethyl)phosphine). The GST-DYRK1A-TS (0.5 pmol) bound on the resin (bed vol. 2 μl) was suspended in 100 μl of AR buffer, mixed with 1 μl of a stock DMSO solution of FINDY or RD0392, then incubated at 37 °C for 30 min. We then added 5 μl of 2 mg ml^{-1} BSA to the suspension (final concentration of 0.1 mg ml^{-1}), followed by incubation at 37 °C for 30 min. We then added 5 μl of ATP solution to the suspension (final concentrations of 100 μM and 1 mM), which was subsequently incubated at 37 °C for 2 h. The bound proteins were eluted with SDS-urea buffer, then analysed with SDS–PAGE followed by western blot (Supplementary Fig. 14).

The band intensities were measured using the ChemiDoc MP Imaging System (Bio-Rad). In Fig. 6d,e, the relative activity of the phosphorylation was calculated from the ATP-dependent increase in the intensity of the p-Ser97 and p-Tyr signals by subtracting the band intensity in the absence of ATP from that in the presence of ATP.

For the *in vitro* kinase assay, 2 μl (bed vol.) of the resin was incubated in a total volume of 25 μl of AR buffer containing 0.1 mg ml^{-1} of BSA, 100 μM of ATP, 50 μM of DYRKtide peptide and the indicated concentrations of the small molecules, for 2 h at 37 °C. The consumption of ATP was measured with ADP-Glo Kinase Assay kit.

CDC37-nanoKAZ interaction assay. The CDC37-nanoKAZ interaction assay was performed as described previously[36], with some modifications. 293T cells stably expressing CDC37-nanoKAZ were transiently transfected with the expression vector 3xFLAG-DYRK1A, then cultured in a 96-well plate. Twenty-four hours after the transfection, FINDY and RD0392 were added to the wells at the indicated concentrations and the cells were cultured for an additional 24 h. The cells were lysed in ice-cold HENG buffer (50 mM HEPES-KOH, pH 7.9, 150 mM NaCl, 20 mM Na_2MoO_4, 2 mM EDTA, 5% glycerol, 0.5% Triton X-100) containing protease inhibitor cocktail (Nacalai Tesque) on ice. The cleared lysates were added to 96-well plates (OptiPlate-96 HB; Perkin Elmer) coated with antibody against FLAG peptide (clone M2, F3165/Lot: SLBH1191V) diluted in sodium bicarbonate buffer (pH 9.6), and incubated at 4 °C for 3 h. The wells were wash three times with ice-cold HENG buffer and then a luminescence assay was performed. The luminescence of the captured CDC37-nanoKAZ was measured using the Centro LB 960 Microplate Luminometer (Berthold Technologies). A reaction mixture containing *bis*-coelenterazine (final concentration of 0.1–1 μg in 100 μl) in PBS containing 0.02% Tween-20 and 20 mM EDTA was injected. The luminescence intensity was recorded at 0.1 s intervals for 10 s or 10 min. The maximum luminescence intensity (I_{max}), represented in relative luminescence units, was used in the study. The captured 3xFLAG-DYRK1A was also measured using HRP-conjugated anti-FLAG antibody (clone M2, A8592/Lot: SLBH1183V, Sigma-Aldrich) and TMD Super Sensitive One Component HRP Microwell Substrate (SurModics, Eden Prairie). The relative amount of protein bound on the well was estimated from the absorbance. The relative luminescence intensities were normalized to the amount of captured 3xFLAG-DYRK1A, and calculated as the fold-change relative to the value at 0 μM of the compounds.

Handling of *Xenopus laevis* embryos and reverse transcription (RT)-PCR analysis. *DYRK1A* and *DYRK1B* mRNA, which were synthesized from linearized vectors with the mMessage mMachine Kit (Life Technologies), were injected into two dorsal blastomeres at the eight-cell-stage for RT–PCR analysis and observation of the embryo phenotypes. Solutions containing proINDY or FINDY (2.5 μM each) were used to treat the embryos just after fertilization. The head regions of the injected embryos were dissected at the neural stage (stage 25) and the total RNA was isolated using TRIzol (Life Technologies). cDNA synthesis was carried out using Moloney murine leukaemia virus reverse transcriptase (Life Technologies). The sequences of the primer pairs were as follows[51,52]: NCAM, 5′-GCGGGTACCT TCTAATAGTCAC-3′ and 5′-GGCTTGGCTGTGGTTCTGAAGG-3′; Brain factor 1 (BF-1), 5′-TCAACAGCCTAATGCCTGAAGC-3′ and 5′-GCCGTCCACTTTCT TATCGTCG-3′; Pax-6, 5′-CAGAACATCTTTTACCCAGGA-3′ and 5′-ACTACT GCTAATGGGAATGTG-3′; Rx1, 5′-CCCCAACAGGAGCATTTAGAAGAC-3′ and 5′-AGGGCACTCATGGCAGAAGGTT-3′; Otx-2, 5′-GGATGGATTTGTTGC ACCAGTC-3′ and 5′-CACTCTCCGAGCTCACTTCTC-3′; En-2, 5′-CGGAATTC ATCAGGTCCGAGATC-3′ and 5′-GCGGATCCTTTGAAGTGGTCGCG-3″; Krox-20, 5′-AACCGCCCCAGTAAGACC-3′ and 5′-GTGTCAGCCTGTCCTGTT AG-3′; ornithine decarboxylase (ODC), 5′-GTCAATGATGGAGTGTATGGA TC-3′ and 5′-TCCATTCCGCTCTCCTGAGCAC-3′. No randomization was used, and no blinding was done.

Mass spectrometry. HEK293 cells were cultured in the presence of 1 μg ml^{-1} doxycycline and 1 μM epoxomicin, with or without 10 μM FINDY for 8 h. The cells were then harvested and lysed in 50 mM Tris (pH 7.4), 150 mM NaCl, 0.5 mM EDTA, 1% Empigen BB and protease/phosphatase inhibitor cocktails (Nacalai Tesque). The clarified cell lysates were processed with anti-FLAG (M2) beads (Sigma-Aldrich) and the bound proteins were then eluted with glycine buffer (pH 3.0), which was immediately neutralized with 1 M Tris (pH 8.0). After reductive alkylation, the samples were trypsinized, then processed using ZipTip pipette tips containing C18 (Millipore). The digested and purified peptides were analysed by LC-MS/MS on TripleTOF 5600 (AB SCIEX). Database searches were performed using ProteinPilot Software (AB SCIEX).

RI-*In vitro* kinase assay. The RI-*in vitro* kinase assay was performed as described previously[16,53], with some modifications. His-tagged human DYRK1A kinase domain (residues 127–485) was produced in *E. coli* then purified with Ni-NTA. GST-tagged full-length human DYRK1B recombinant proteins and His-tagged full-length human DYRK2 recombinant proteins were obtained from Life Technologies and Millipore, respectively. The kinase reaction was performed in a reaction mixture containing serially diluted inhibitors, 10 mM MOPS-KOH (pH 7.0), 10 mM magnesium acetate, 200 μM EDTA, 1 μM ATP, 0.1–0.4 μCi [γ-^{32}P]ATP, 5 μM DYRKtide and 0.1–1 μg recombinant kinase, for a final volume of 25 μl. The final concentration of DMSO was adjusted to 1%, regardless of the inhibitor concentration. The reaction mixture was incubated at 30 °C for 10 min and phosphoric acid (final 5%) was then added to stop the reaction. We then dispensed 25 μl of the reaction mixture onto P81, a phosphocellulose membrane (Whatman, GE Healthcare), and washed four times in 5% phosphoric acid. Cherenkov light from the incorporated ^{32}P was measured using a liquid scintillation counter. The kinase assay conditions, including the incubation period and the concentration of the kinases and substrates, were optimized to maintain linearity during the incubation. The net radioactivity was determined by subtracting the background count from the reaction mixture without kinase, and the data were expressed as a percentage of the control sample containing the solvent.

For the ATP kinetics, the same methodology was applied as above, except that the ATP concentration ranged from 1.25 to 40 μM. The amount of incorporated ^{32}P was calculated from the standard line. The K_m, K_i and V_{max} values were calculated with Prism5 software (GraphPad Software) using the competitive inhibition model.

Inhibitory profiling against 275 recombinant kinases. The effects of FINDY against 275 kinases, listed in Supplementary Table 1, at an inhibitor concentration of 10 μM, were tested using the QuickScout Screening Assist Mobility Shift Assay or an immobilized metal ion affinity-based fluorescence polarization screening express kit with ATP concentration at K_m or 1 mM (Carna Biosciences). Detailed information on the assay conditions is available on the website of Carna Biosciences (http://www.carnabio.com).

Docking simulation. We selected the crystal structure of the human DYRK1A/INDY complex (RCSB Protein Data Bank accession code 3ANQ)[16] as the starting model for the simulation. Docking simulations of RD0392 to DYRK1A were performed with CDOCKER and related modules in Discovery Studio 3.0 (Accelrys). As the same procedure failed to produce any reasonable binding model for FINDY, its complex with DYRK1A was built by manually modifying that for DYRK1A/RD0392 obtained above, maintaining the common moiety of the two compounds. The energy of the hypothetical DYRK1A/FINDY complex was then minimized with the same program suite.

Statistical analysis. Statistical analysis of the experimental data (Figs 6b and 7 and Supplementary Figs 7e, 9e and 10) was performed with the Mann–Whitney *U*-test. The data in Fig. 6d were analysed with the paired *t*-test. The results are shown as means ± s.d. with *P* values (*$P<0.05$, **$P<0.01$, ***$P<0.001$). The data were fitted to a four-parameter logistic curve (variable slope) for Hill slope determination, from which the IC_{50} and EC_{50} values were calculated using Prism 6.0 (GraphPad Software).

References

1. Chico, L. K., Van Eldik, L. J. & Watterson, D. M. Targeting protein kinases in central nervous system disorders. *Nat. Rev. Drug Discov.* **8,** 892–909 (2009).
2. Zhang, J., Yang, P. L. & Gray, N. S. Targeting cancer with small molecule kinase inhibitors. *Nat. Rev. Cancer* **9,** 28–39 (2009).
3. Karaman, M. W. *et al.* A quantitative analysis of kinase inhibitor selectivity. *Nature Biotechnol.* **26,** 127–132 (2008).
4. Tejedor, F. J. & Hammerle, B. MNB/DYRK1A as a multiple regulator of neuronal development. *FEBS J.* **278,** 223–235 (2011).
5. Altafaj, X. *et al.* Neurodevelopmental delay, motor abnormalities and cognitive deficits in transgenic mice overexpressing Dyrk1A (minibrain), a murine model of Down's syndrome. *Hum. Mol. Genet.* **10,** 1915–1923 (2001).
6. Malinge, S. *et al.* Increased dosage of the chromosome 21 ortholog Dyrk1a promotes megakaryoblastic leukemia in a murine model of Down syndrome. *J. Clin. Invest.* **122,** 948–962 (2012).
7. Laguna, A. *et al.* Triplication of DYRK1A causes retinal structural and functional alterations in Down syndrome. *Hum. Mol. Genet.* **22,** 2775–2784 (2013).
8. Dowjat, W. K. *et al.* Trisomy-driven overexpression of DYRK1A kinase in the brain of subjects with Down syndrome. *Neurosci. Lett.* **413,** 77–81 (2007).
9. Arron, J. R. *et al.* NFAT dysregulation by increased dosage of DSCR1 and DYRK1A on chromosome 21. *Nature* **441,** 595–600 (2006).
10. Gwack, Y. *et al.* A genome-wide *Drosophila* RNAi screen identifies DYRK-family kinases as regulators of NFAT. *Nature* **441,** 646–650 (2006).
11. Kimura, R. *et al.* The DYRK1A gene, encoded in chromosome 21 Down syndrome critical region, bridges between beta-amyloid production and tau phosphorylation in Alzheimer disease. *Hum. Mol. Genet.* **16,** 15–23 (2007).
12. Wegiel, J., Gong, C. X. & Hwang, Y. W. The role of DYRK1A in neurodegenerative diseases. *FEBS J.* **278,** 236–245 (2011).
13. Becker, W. & Sippl, W. Activation, regulation, and inhibition of DYRK1A. *FEBS J.* **278,** 246–256 (2011).
14. Wang, P. *et al.* A high-throughput chemical screen reveals that harmine-mediated inhibition of DYRK1A increases human pancreatic beta cell replication. *Nature Med.* **21,** 383–388 (2015).
15. Shen, W. *et al.* Inhibition of DYRK1A and GSK3B induces human beta-cell proliferation. *Nat. Commun.* **6,** 8372 (2015).
16. Ogawa, Y. *et al.* Development of a novel selective inhibitor of the Down syndrome-related kinase Dyrk1A. *Nat. Commun.* **1,** 86 (2010).
17. Kinstrie, R. *et al.* Characterization of a domain that transiently converts class 2 DYRKs into intramolecular tyrosine kinases. *Sci. Signal.* **3,** ra16 (2010).
18. Lochhead, P. A. *et al.* A chaperone-dependent GSK3beta transitional intermediate mediates activation-loop autophosphorylation. *Mol. Cell* **24,** 627–633 (2006).
19. Lochhead, P. A., Sibbet, G., Morrice, N. & Cleghon, V. Activation-loop autophosphorylation is mediated by a novel transitional intermediate form of DYRKs. *Cell* **121,** 925–936 (2005).
20. Walte, A. *et al.* Mechanism of dual specificity kinase activity of DYRK1A. *FEBS J.* **280,** 4495–4511 (2013).
21. Soundararajan, M. *et al.* Structures of Down syndrome kinases, DYRKs, reveal mechanisms of kinase activation and substrate recognition. *Structure* **21,** 986–996 (2013).
22. Lochhead, P. A. Protein kinase activation loop autophosphorylation in cis: overcoming a Catch-22 situation. *Sci. Signal.* **2,** pe4 (2009).
23. Daggett, V. & Fersht, A. The present view of the mechanism of protein folding. *Nat. Rev. Mol. Cell Biol.* **4,** 497–502 (2003).
24. Liu, F. *et al.* Overexpression of Dyrk1A contributes to neurofibrillary degeneration in Down syndrome. *FASEB J* **22,** 3224–3233 (2008).
25. Ryoo, S. R. *et al.* DYRK1A-mediated hyperphosphorylation of Tau. A functional link between Down syndrome and Alzheimer disease. *J. Biol. Chem.* **282,** 34850–34857 (2007).
26. Banaszynski, L. A., Chen, L. C., Maynard-Smith, L. A., Ooi, A. G. & Wandless, T. J. A rapid, reversible, and tunable method to regulate protein function in living cells using synthetic small molecules. *Cell* **126,** 995–1004 (2006).
27. Baell, J. B. & Holloway, G. A. New substructure filters for removal of pan assay interference compounds (PAINS) from screening libraries and for their exclusion in bioassays. *J. Med. Chem.* **53,** 2719–2740 (2010).
28. Mendgen, T., Steuer, C. & Klein, C. D. Privileged scaffolds or promiscuous binders: a comparative study on rhodanines and related heterocycles in medicinal chemistry. *J. Med. Chem.* **55,** 743–753 (2012).
29. Szymczak, A. L. *et al.* Correction of multi-gene deficiency *in vivo* using a single 'self-cleaving' 2A peptide-based retroviral vector. *Nature Biotechnol.* **22,** 589–594 (2004).
30. Edelman, A. M., Blumenthal, D. K. & Krebs, E. G. Protein serine/threonine kinases. *Annu. Rev. Biochem.* **56,** 567–613 (1987).
31. Himpel, S. *et al.* Identification of the autophosphorylation sites and characterization of their effects in the protein kinase DYRK1A. *Biochem. J.* **359,** 497–505 (2001).

32. Kentrup, H. *et al.* Dyrk, a dual specificity protein kinase with unique structural features whose activity is dependent on tyrosine residues between subdomains VII and VIII. *J. Biol. Chem.* **271,** 3488–3495 (1996).
33. Adayev, T. *et al.* Dual-specificity tyrosine phosphorylation-regulated kinase 1A does not require tyrosine phosphorylation for activity *in vitro*. *Biochemistry* **46,** 7614–7624 (2007).
34. Alvarez, M., Altafaj, X., Aranda, S. & de la Luna, S. DYRK1A autophosphorylation on serine residue 520 modulates its kinase activity via 14-3-3 binding. *Mol. Biol. Cell* **18,** 1167–1178 (2007).
35. Patricelli, M. P. *et al.* Functional interrogation of the kinome using nucleotide acyl phosphates. *Biochemistry* **46,** 350–358 (2007).
36. Sonamoto, R. *et al.* Identification of a DYRK1A inhibitor that induces degradation of the target kinase using co-chaperone CDC37 fused with Luciferase nanoKAZ. *Sci. Rep.* **5,** 12728 (2015).
37. Caplan, A. J., Mandal, A. K. & Theodoraki, M. A. Molecular chaperones and protein kinase quality control. *Trends Cell Biol.* **17,** 87–92 (2007).
38. Karnitz, L. M. & Felts, S. J. Cdc37 regulation of the kinome: when to hold 'em and when to fold 'em. *Sci. STKE* **2007,** pe22 (2007).
39. Taipale, M. *et al.* Quantitative analysis of HSP90-client interactions reveals principles of substrate recognition. *Cell* **150,** 987–1001 (2012).
40. Taipale, M. *et al.* Chaperones as thermodynamic sensors of drug-target interactions reveal kinase inhibitor specificities in living cells. *Nature Biotechnol.* **31,** 630–637 (2013).
41. Lochhead, P. A. *et al.* dDYRK2: a novel dual-specificity tyrosine-phosphorylation-regulated kinase in *Drosophila*. *Biochem. J.* **374,** 381–391 (2003).
42. Polier, S. *et al.* ATP-competitive inhibitors block protein kinase recruitment to the Hsp90-Cdc37 system. *Nat. Chem. Biol.* **9,** 307–312 (2013).
43. Crews, C. M., Alessandrini, A. A. & Erikson, R. L. Mouse Erk-1 gene product is a serine/threonine protein kinase that has the potential to phosphorylate tyrosine. *Proc. Natl Acad. Sci. USA* **88,** 8845–8849 (1991).
44. Greggio, E. *et al.* The Parkinson disease-associated leucine-rich repeat kinase 2 (LRRK2) is a dimer that undergoes intramolecular autophosphorylation. *J. Biol. Chem.* **283,** 16906–16914 (2008).
45. Kamikawaji, S., Ito, G. & Iwatsubo, T. Identification of the autophosphorylation sites of LRRK2. *Biochemistry* **48,** 10963–10975 (2009).
46. Noble, C. *et al.* CRAF autophosphorylation of serine 621 is required to prevent its proteasome-mediated degradation. *Mol. Cell* **31,** 862–872 (2008).
47. Keshwani, M. M. *et al.* Cotranslational cis-phosphorylation of the COOH-terminal tail is a key priming step in the maturation of cAMP-dependent protein kinase. *Proc. Natl Acad. Sci. USA* **109,** E1221–E1229 (2012).
48. Horton, R. M., Hunt, H. D., Ho, S. N., Pullen, J. K. & Pease, L. R. Engineering hybrid genes without the use of restriction enzymes: gene splicing by overlap extension. *Gene* **77,** 61–68 (1989).
49. Ho, S. N., Hunt, H. D., Horton, R. M., Pullen, J. K. & Pease, L. R. Site-directed mutagenesis by overlap extension using the polymerase chain reaction. *Gene* **77,** 51–59 (1989).
50. Thomas, M. *et al.* Full deacylation of polyethylenimine dramatically boosts its gene delivery efficiency and specificity to mouse lung. *Proc. Natl Acad. Sci. USA* **102,** 5679–5684 (2005).
51. Hyodo-Miura, J. *et al.* Involvement of NLK and Sox11 in neural induction in Xenopus development. *Genes Cells* **7,** 487–496 (2002).
52. Satoh, K. *et al.* Nemo-like kinase-myocyte enhancer factor 2A signaling regulates anterior formation in Xenopus development. *Mol. Cell. Biol.* **27,** 7623–7630 (2007).
53. Muraki, M. *et al.* Manipulation of alternative splicing by a newly developed inhibitor of Clks. *J. Biol. Chem.* **279,** 24246–24254 (2004).

Acknowledgements

We thank Dr Melissa Gammons (University of Bristol) and Dr Yoko Tanimura (Kyoto University) for their critical reading of the manuscript, Dr Satoshi Inouye (JNC Corporation, Yokohama, Japan) for providing the nanoKAZ gene and technical advice, Ayako Hosoya (Tokyo Medical and Dental University) for the HRMS analyses, and Kohei Araki and Dr Akiko Nakano-Kobayashi (Kyoto University) for their technical assistance. We also thank the Radioisotope Research Center of Kyoto University for instrumental support in the radioisotope experiments, and the laboratory members for their critical comments and discussion. This work was supported by Grants-in-Aid from the Japan Science and Technology Agency (J-AMP 10103930 to M.H., CREST 11103769 to M.H.), the Ministry of Education, Culture, Sports, Science and Technology (MEXT) of Japan (KAKENHI 24241076 to M.H., Innovative Cell Biology by Innovative Technology 10102928 to M.H. and T.H.) and the Platform for Drug Discovery, Informatics, and Structural Life Science of MEXT, Japan (M.H. and T.H.).

Author contributions

I.K. and M.H. designed the experiments and wrote the manuscript. Y.S. and T.H. supported the manuscript preparation. I.K. conceived and developed the cell-based assays. I.K. conceived the *in vitro* autophosphorylation assay, and I.K. and Y.K. developed the assay. I.K. performed the *in vitro* translation experiment. R.S. performed the CDC37 interaction assay. Y.O. performed the *in vitro* kinase assay. Y.S., S.Y., T.K.-S. and T.H. synthesized the small molecules. Y.N., M.A., T.I. and N.I. performed the molecular modelling studies and supported the preparation of the recombinant protein. T.G. and H.S. performed the rescue experiments using the *Xenopus* embryos.

Additional information

Competing financial interests: The authors declare no competing financial interests.

8

Integrative genomic mining for enzyme function to enable engineering of a non-natural biosynthetic pathway

Wai Shun Mak[1,*], Stephen Tran[2,*], Ryan Marcheschi[3], Steve Bertolani[1], James Thompson[4], David Baker[4], James C. Liao[3,5] & Justin B. Siegel[1,6,7]

The ability to biosynthetically produce chemicals beyond what is commonly found in Nature requires the discovery of novel enzyme function. Here we utilize two approaches to discover enzymes that enable specific production of longer-chain (C_5-C_8) alcohols from sugar. The first approach combines bioinformatics and molecular modelling to mine sequence databases, resulting in a diverse panel of enzymes capable of catalysing the targeted reaction. The median catalytic efficiency of the computationally selected enzymes is 75-fold greater than a panel of naively selected homologues. This integrative genomic mining approach establishes a unique avenue for enzyme function discovery in the rapidly expanding sequence databases. The second approach uses computational enzyme design to reprogramme specificity. Both approaches result in enzymes with >100-fold increase in specificity for the targeted reaction. When enzymes from either approach are integrated *in vivo*, longer-chain alcohol production increases over 10-fold and represents >95% of the total alcohol products.

[1] Department of Chemistry, University of California Davis, One Shields Avenue, Davis, California 95616, USA. [2] Department of Bioinformatics, University of California, Los Angeles, 172 Boyer Hall Box 951570, 611 Charles E. Young Drive East, Los Angeles, California 90095-1570, USA. [3] Department of Chemical and Biomolecular Engineering, University of California Los Angeles, 5531 Boelter Hall, Los Angeles, California 90095-1592, USA. [4] Department of Biochemistry, University of Washington, 1705 NE Pacific St Seattle, Washington 98195-7350, USA. [5] Institute for Genomics and Proteomics, University of California Los Angeles, 611 Young Drive East, Los Angeles, California 90095, USA. [6] Department of Biochemistry and Molecular Medicine, University of California Davis, 2700 Stockton Boulevard, Suite 2102, Sacramento, California 95817, USA. [7] Genome Center, University of California Davis, 451 Health Sciences Drive, Davis, California 95616, USA. * These authors contributed equally to this work. Correspondence and requests for materials should be addressed to J.C.L. (email: liaoj@ucla.edu) or to J.B.S. (email: jbsiegel@ucdavis.edu).

A central goal of the bioeconomy is to reduce our dependence on petroleum through next-generation biomanufacturing[1]. The USDA projects the industrial production of bio-based specialty chemicals to reach ~$340 billion USD by 2025, replacing half of our needs for these products from petroleum[2]. To accomplish this ambitious goal, non-natural biochemical pathways are needed. An example of one such a pathway is the 'synthetic recursive +1' carbon elongation pathway that has been developed to produce a variety of alcohol products in *Escherichia coli*[3,4]. The primary product from the current pathway is 1-butanol, with longer-chain alcohols (for example, pentanol, hexanol, heptanol and octanol) being either minor products of the pathway or not produced at all[3,4]. Yet, there is significant interest in producing longer-chain alcohols, given their use as specialty chemicals[5], as well as their greater energy density when used as liquid fuels or fuel additives. To produce longer-chain alcohols through the synthetic recursive +1 carbon elongation pathway, we demonstrate that modulating the specificity of a single enzyme in the pathway could redirect metabolic flux towards longer-chain alcohols.

In the current synthetic recursive +1 pathway, the enzymes LeuA, LeuB, LeuC and LeuD (collectively known as LeuABCD) are recruited to recursively elongate 2-ketobutyrate into long-chain 2-ketoacids[3,4]. These 2-ketoacids are then converted to aldehydes by ketoisovalerate decarboxylase[6] (KIVD) and subsequently to alcohols by alcohol dehydrogenase[3] (ADH; Fig. 1). In this engineered pathway, the primary determinant for the final product chain length is the number of cycles through LeuABCD[7].

Previous efforts to engineer the product specificity of this pathway have focused on broadening the specificity of LeuA and KIVD[3]. When the specificity of LeuA is broadened, additional cycles through the pathway are carried out, resulting in a mixture of C_2–C_6 linear and branched alcohols[3]. When the specificity of both LeuA and KIVD are broadened, C_2–C_8 products are produced with major products being butanol and pentanol[4]. To enhance longer-chain alcohol production, we focused on discovering a ketoacid decarboxylase specific for long-chain ketoacids. By replacing KIVD with an enzyme specific for long-chain ketoacids, the short-chain substrates were able to re-enter the +1 iteration cycle until their conversion to long-chain ketoacids, which are then decarboxylated and reduced into longer chain alcohols.

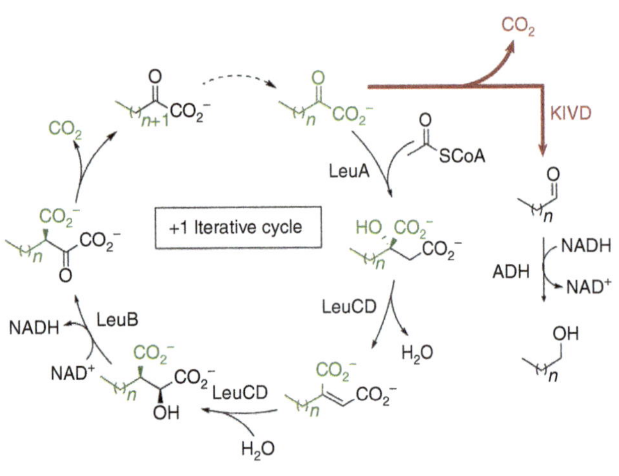

Figure 1 | The synthetic recursive +1 pathway. This pathway employs enzymes LeuABCD from leucine biosynthesis for carbon-chain elongation of 2-ketoacids. The enzyme KIVD performs the decarboxylation of 2-ketoacids and diverts carbon out from the +1 iterative cycle.

We employed two computationally directed approaches to discover a decarboxylase with the desired functional profile. The first is a new approach to mine the rapidly growing sequence databases. The majority of these proteins derived from the genomic sequencing of organisms (that is, genomic enzyme orthologues: GEOs) have not been experimentally characterized. Therefore, neither function nor specificity is known. Previous efforts to mine GEOs for function have relied on random sampling or sequence-based bioinformatics over an entire enzyme family[8–10]. These methods require a large number of genes to be produced and experimentally characterized. To enable a more efficient sampling method, we developed a novel computational pipeline that integrates molecular modelling and bioinformatics to forward predict GEOs' ability on carrying out the function of interest. Using this integrative genomic mining approach we identify a set of ketoacid decarboxylases diverse in sequence, but capable of utilizing 2-ketooctanoate (C8) as a substrate. In addition, we demonstrate that the median activity of GEOs selected using the integrative genomic mining approach is 75-fold greater than a set of naively selected proteins from the enzyme family. The second approach focuses on reprogramming the substrate specificity of KIVD by using computational enzyme design methods. Using the Rosetta Molecular Modeling Suite to direct the generation and screening of over 400 mutants results in the discovery of a mutant with >600-fold increase in specificity for C8 (ref. 11). Finally, the most specific and active enzyme from each approach is incorporated into the synthetic recursive +1 pathway to evaluate their ability to modulate *in vivo* alcohol production. Both enzymes result in a complete switch of the alcohol production profile towards longer chain alcohols, demonstrating the ability to rationally control biosynthetic product specificity for chemicals not commonly found in Nature.

Results

Integrative genomic mining for enzyme function. KIVD belongs to the TPP enzyme family that is composed of more than 17,000 sequences[12]. To identify GEOs with the desired function within this vast sequence space, we combined bioinformatics[13] and molecular modelling[14] with functional constraints to guide our exploration (Fig. 2). First, ketoacid decarboxylase GEOs were identified based on sequence similarity to the KIVD used in the current synthetic recursive +1 pathway. At the time the search was performed, 2,082 sequences were found in the non-redundant sequence database as significant matches. To broadly search KIVD sequence space, redundant sequences were filtered out with a sequence identity cutoff of 90% (ref. 15). Sequences derived from eukaryotic organisms were removed to increase the likelihood of producing soluble proteins in *E. coli* where the synthetic recursive +1 pathway has been implemented.

We assumed that all potential scaffolds with the desired function should be structurally homologous to KIVD, and further refined our sequence set accordingly. Since crystal structures for most of these predicted proteins are not available for analysis, we produced homology models to obtain a predicted ternary structure. Using Rosetta Comparative Modeling, one hundred models were generated for each of the 239 GEOs[14]. The lowest-energy model was selected as the representative for each GEO and evaluated for structural similarity to KIVD. We used the TMalign algorithm[16] to overlay the models with the native KIVD crystal structure, and only those with a TMalign score of >0.5 were kept. These bioinformatics and structural filters result in 239 GEOs as candidates for the targeted function (Supplementary Table 1). The pairwise identity of every sequence to each other was calculated and the mode of these values lies near 20% (Fig. 2).

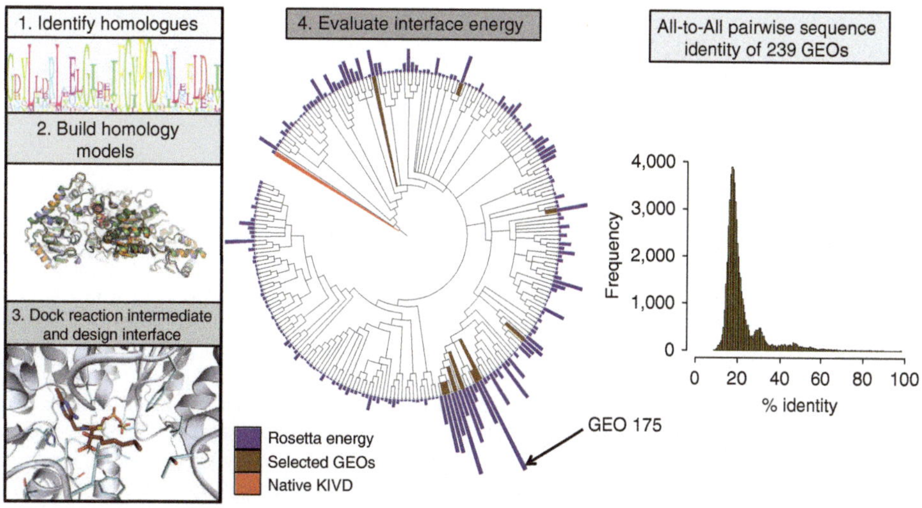

Figure 2 | The computationally directed genomic enzyme-mining pipeline. GEOs were identified based on amino-acid sequence homology to KIVD. Bioinformatics filters were used to identify a set of amino-acid sequences from genomic databases that are predicted to be KIVD homologues and likely to be decarboxylases. Once identified, homology models were built to obtain a predicted ternary structure of each GEO. Ligand docking and design simulations were subsequently run in the presence of our target ligand (for example, C8) to evaluate the potential protein-ligand interface energy. A phylogenetic tree for the 239 GEOs is depicted with a bar chart above each sequence. The bar height indicates the predicted protein-ligand interface energy; the higher the bar the lower the energy. Bar height is scaled linearly relative to the lowest protein-ligand interface energy. Ten GEOs (brown) were selected for experimental characterization. The pairwise sequence identity of all 239 GEOs to each other has a mode of 20%, indicating the high level of sequence diversity within this set of sequences.

Thus, the selected set of 239 GEOs represents a diverse sampling of sequence space for this fold family.

To further enrich the set for GEOs likely to function on C8, we docked a modelled reaction intermediate of the C8 substrate within the predicted active site (Fig. 2). The lowest-energy model for each GEO was used as a template for docking a modelled C8 intermediate with functional constraints, which ensured a productive geometric orientation between the predicted catalytic residues in the protein and intermediate. The C8 reaction intermediate used for docking is based on the synthetic thiamine pyrophosphate (TPP) analogue from a crystal structure of KIVD (PDB ID: 2VBG), which mimics the initial nucleophilic attack by the thiazolium[17]. Docking calculations were carried out using Rosetta Design in which both the identity and conformation of non-catalytic residues within the active site were allowed to change, and the backbone was allowed to move during minimization. Although the C8 intermediate may be docked without the introduction of mutations, the design was enabled during the simulations to allow for the possibility of introducing conservative mutations that significantly reduce interface energy. After docking and design simulations, the calculated interface energy was used to evaluate each GEO's ability to accommodate C8.

A phylogenetic tree based on sequence similarity was built for the 239 GEOs in which the lowest protein–ligand interface energy from the Rosetta Design simulation for each GEO are depicted (Fig. 2). The bar height above each GEO leaf is proportional to the lowest interface energy from all 239 GEOs, the higher the bar the lower the energy. On the basis of the calculated energies it is immediately apparent that a cluster of GEOs distant in sequence to KIVD (~15% sequence identity) is predicted to be capable of utilizing C8. The enzyme with the lowest energy, GEO 175, is a predicted protein in the genome database with no known function. The enzyme with the closest amino-acid sequence identity to GEO 175 (~50% sequence identity) for which significant biophysical characterization has been performed is benzoylformate decarboxylase from bacteria *Pseudomonas putida* (ppBFD)[18]. A structural analysis of the active site of ppBFD and GEO 175 reveals that the active sites are only 50% identical in sequence (Supplementary Fig. 1), and the activity of ppBFD on C8 has not been previously studied.

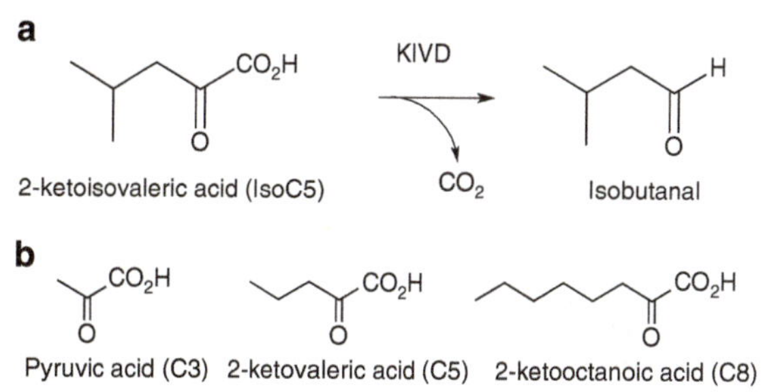

Figure 3 | Characterized reaction specificities. (**a**) KIVD utilizes various ketoacids as substrates to produce the corresponding aldehyde. (**b**) Ketoacids used for *in vitro* kinetic constant characterization in addition to 2-ketoisovaleric acid.

From the design simulations, 10 GEOs were chosen for experimental characterization based on the predicted protein–ligand interface energy as well as sequence diversity (Fig. 2 and Supplementary Table 1). Before ordering these GEOs, each mutation introduced through the Rosetta Design simulations was manually analysed. Mutations introduced during the design simulations were reverted if they were not found to significantly enhance the calculated Rosetta interface energy. Synthetic genes encoding each enzyme was obtained, expressed in *E. coli* and purified for *in vitro* kinetic characterization using a panel of 2-ketoacid substrates of different chain lengths including C8, C5, C3 and the natural substrate for KIVD, ketoisovaleric acid (Fig. 3). Out of the 10 selected GEOs, 6 (GEO 41, 74, 110, 175, 179 and 195) express and are able to be purified in a soluble form using our standardized method. Of these six, three exhibit detectable activities on at least one of the four ketoacids used for kinetic characterization.

Among the active GEOs, the enzyme with the highest efficiency on C8 is the computationally top ranked GEO 175, with a k_{cat}/K_M of 17,000 $M^{-1}s^{-1}$ (Table 1 and Fig. 4). This is only twofold lower than the efficiency of native KIVD on C8. However, GEO 175's catalytic efficiency on C8 is 33,000-fold higher than that on C3 and 354-fold higher than that on C5. When assayed under the same conditions, native KIVD's catalytic efficiency on C8 is 762-fold higher than that on C3 and 3.4-fold higher than that on C5. This corresponds to over a 100-fold improvement in specificity (C8 versus C5) relative to the native KIVD.

To evaluate whether the integrative genomic mining approach enriches the mining for C8 activity, an additional set of previously characterized ketoacid decarboxylases diverse in sequence were experimentally characterized in an equivalent manner. Five enzymes were selected (PDB: 1OVM, 2VBI, 3FZN, 1ZPD and 1OZF) and all were found to have detectable activity on at least one of the four ketoacids used for GEO characterization. The median C8 activity from the GEOs selected using the integrative genomic mining approach is 75-fold higher than the naively picked decarboxylases (Table 1 and Fig. 4). This indicates that the integrative genomic mining approach is effective in identifying functional enzymes that perform the desired reaction on C8.

The amino-acid sequence for GEO 175 is not predicted to require any mutations to be capable of utilizing the C8 substrate, and therefore represents the native enzyme from *Streptomyces sp. C*. Comparing the active-site pocket of GEO 175 with native KIVD (16.8% identity to each other) reveals significant shape differences (Fig. 5). On the basis of molecular modelling, GEO 175 is predicted to have an extended active site that is significantly more open to solvent relative to native KIVD. This results in the predicted binding mode for C8 to be extended in GEO 175 as opposed to a 'wrapped' conformation in KIVD. In GEO175, the third through fifth carbons of the C8 ketoacid ligand are predicted to occupy a solvent-exposed, open pocket with limited molecular interactions. However, the sixth through eighth carbons are predicted to interact with a narrow and hydrophobic pocket. Contrary to GEO 175, KIVD has an enclosed hydrophobic binding pocket and is predicted to make molecular interactions throughout the ketoacid alkyl chain (Figs 5 and 6).

On the basis of these observations we engineered a mutant of GEO 175 containing two amino-acid changes (L376T and T240S) predicted to remove the direct molecular interactions with seventh and eighth carbons of the ketoacid alkyl chain, while maintaining overall protein stability. On kinetic characterization, we found that the mutant protein's catalytic efficiency on C8 is 15-fold lower than GEO 175, but the decrease in efficiency for C5, C3 and isoC5 assayed under the same conditions is 2-fold or less (Table 1 and Fig. 6). The ability to rationally modulate activity

Table 1 | Enzyme kinetic constants of native, designed and discovered ketoacid decarboxylases.

$k_{cat}/K_M (M^{-1}s^{-1})$	C3	C5	C8	iso C5	
GEO179	0.9 ± 0.02	100 ± 15	1,200 ± 130	41 ± 1.2	Discovered
GEO195	2.4 ± 0.1	200 ± 14	1,400 ± 160	8 ± 0.3	Discovered
GEO175	0.51 ± 0.02	48 ± 3.7	17,000 ± 2,700	3.3 ± 0.2	Discovered
GEO175 L376T_T240S	0.27 ± 0.02	30 ± 7.2	1,100 ± 160	1.6 ± 0.1	Designed
1OVM	32 ± 0.5	2,100 ± 420	80,000 ± 4,300	1,300 ± 130	Native
2VBI	5,700 ± 1,400	52 ± 1.8	14 ± 0.8	1.8 ± 0.1	Native
3FZN	5.4 ± 0.1	1,700 ± 110	350 ± 30	110 ± 10	Native
1ZPD	8,200 ± 545	140 ± 5.6	19 ± 2	0.33 ± 0.01	Native
1OZF	n.d.	0.53 ± 0.04	n.d.	17 ± 0.9	Native
Native KIVD	42 ± 0.6	9,500 ± 470	32,000 ± 5,500	14,000 ± 1,100	Native
KIVD_VLV	0.71 ± 0.07	1.3 ± 0.2	2,800 ± 860	0.24 ± 0.02	Designed
k_{cat} (s^{-1})					
GEO179	n.d.	0.47 ± 0.03	0.32 ± 0.01	0.39 ± 0.01	
GEO195	n.d.	0.49 ± 0.01	0.56 ± 0.02	0.072 ± 0.001	
GEO175	n.d.	0.97 ± 0.04	10.1 ± 0.6	n.d.	
GEO175 L376T_T240S	n.d.	0.28 ± 0.04	3.4 ± 0.2	n.d.	
1OVM	0.2 ± 0.01	1 ± 0.1	1.7 ± 0.1	7.8 ± 0.4	
2VBI	25 ± 2.6	n.d.	0.051 ± 0.01	0.025 ± 0.001	
3FZN	n.d.	4.7 ± 0.1	1.3 ± 0.1	0.52 ± 0.02	
1ZPD	46.7 ± 1.4	0.95 ± 0.02	0.021 ± 0.001	n.d.	
1OZF	n.d.	n.d.	n.d.	0.03 ± 0.001	
Native KIVD	n.d.	14.3 ± 0.2	7 ± 0.3	61 ± 2.1	
KIVD_VLV	n.d.	0.013 ± 0.001	0.5 ± 0.03	n.d.	
K_M (mM)					
GEO179	n.d.	4.6 ± 0.6	0.27 ± 0.03	9.4 ± 0.2	
GEO195	n.d.	2.5 ± 0.2	0.4 ± 0.04	9.1 ± 0.3	
GEO175	n.d.	20 ± 1.3	0.58 ± 0.09	n.d.	
GEO175 L376T_T240S	n.d.	10 ± 2	3.1 ± 0.4	n.d.	
1OVM	7.5 ± 0.1	0.5 ± 0.1	0.021 ± 0.001	6.1 ± 0.6	
2VBI	4.3 ± 1.0	n.d.	3.6 ± 0.2	14 ± 0.5	
3FZN	n.d.	2.7 ± 0.2	3.8 ± 0.3	4.9 ± 0.3	
1ZPD	5.7 ± 0.3	7.1 ± 0.3	1.2 ± 0.1	n.d.	
1OZF	n.d.	n.d.	n.d.	1.8 ± 0.1	
Native KIVD	n.d.	1.5 ± 0.1	0.21 ± 0.04	4.5 ± 0.3	
KIVD_VLV	n.d.	10 ± 1	0.18 ± 0.05	n.d.	

The kinetic constants were determined, as described in Methods, against 2-ketooctanoate (C8), 2-ketoisovalerate (isoC5) and pyruvate (C3). The curves consisted of at least five points. The limit of detection under the conditions tested was 0.2 $M^{-1}s^{-1}$, kinetic constants beyond our detection limit are labeled as n.d. (not determined). Michaelis-Menten curve fits are shown in Supplementary Figs 3–6. Errors are shown as ± one s.d.

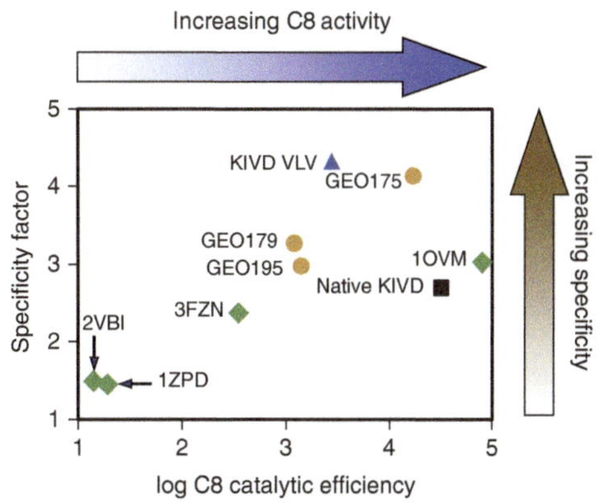

Figure 4 | Catalytic efficiency and specificity of characterized ketoacid decarboxylases. Specificity factor is calculated as $\frac{\log C8}{\log C5} + \frac{\log C8}{\log C3}$ where the log-scaled catalytic efficiencies for C8 relative to C5 and C3 are compared. The three active GEOs and the naively selected set of decarboxylases are represented in gold (circles) and green (diamonds), respectively. Native KIVD and KIVD_VLV are depicted in black (square) and blue (triangle), respectively. 1OZF is not included in this figure since activity on C8 is below our limit of detection. The genomic mining method and computational active-site redesign approach both produced enzymes with enhanced specificity relative to native KIVD.

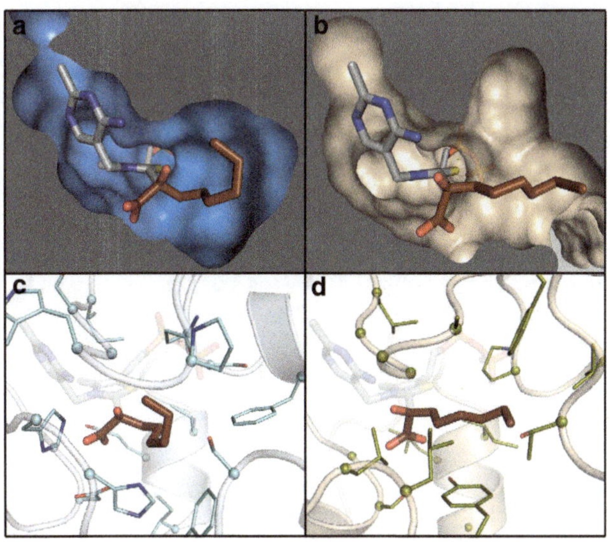

Figure 5 | Molecular model of the C8 intermediate docked into KIVD and GEO 175. Cross-sections of the space-filled active site are represented for KIVD (**a**) and GEO 175 (**b**). The C8 ketoacid and TPP ligand are shown in brown and grey, respectively. The relative orientation of the C8 ligand to the TPP cofactor represents one conformation of the reaction intermediate used in the simulations; additional details are provided in Methods. A detailed view of the amino acids within the active site are illustrated for KIVD (**c**) and GEO 175 (**d**). Residues within 5 Å of C8 ligand are shown in sticks and their corresponding C-alpha are shown in spheres. Figure was generated using PyMol v1.7.4 (ref. 34).

based on structure supports the accuracy of the GEO 175 molecular model and the hypothesis that the specificity of GEO 175 can be attributed to a combination of hydrophobic interactions with the sixth through eighth carbons of the alkyl chain, while lacking interactions with the third through fifth carbons (Fig. 6).

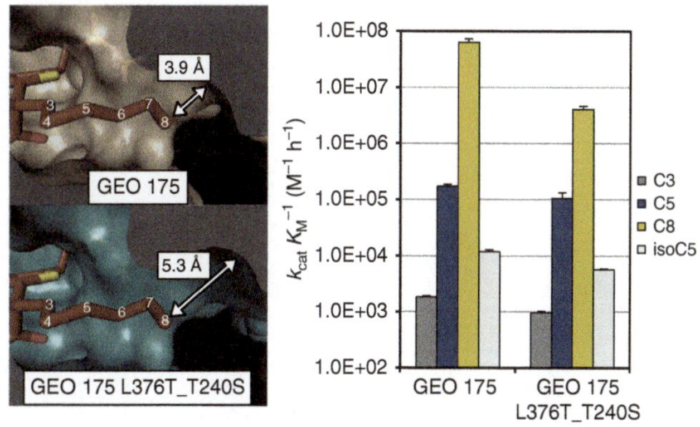

Figure 6 | Kinetic characterization of GEO 175 and GEO 175 L376T_T240S. Kinetic constants were measured as described in Methods. The substrate 2-ketooctanoate carbon chain numbering referred to in the text is numbered in white. According to the molecular model of GEO 175, the double mutation L376T_T240S is predicted to recede the pocket by 1.4 Å and remove interactions between the binding pocket and carbon 8 of the ketoacid alkyl chain. This mutant was observed to decrease catalytic efficiency on C8, but has a negligible effect on shorter-chain substrates. Errors are shown as ± one s.d.

Computationally directed KIVD-active-site redesign. In our second approach to improve the specificity of KIVD, we used Rosetta Design methods to explore the potential active-site sequence space that would accommodate 2-ketooctanoate as a substrate[11]. Design simulations were carried out with the same reaction intermediate and functional constraints utilized in the genomic mining pipeline. Ten residues in the active site were targeted for mutagenesis. At each site, one of eleven relatively hydrophobic amino acids (Val, Leu, Ile, Met, Phe, His, Gly, Ala, Thr, Tyr and Trp) was systematically introduced, and the remaining nine sites were allowed to be redesigned. The identities of amino acids at all other residues were kept constant (Fig. 7). The sequence profile from the 50 lowest-energy designs was used to guide the construction of a small library of roughly 400 KIVD mutants from the original combinatorial space of over 10^{11} possible active-site mutations.

Since KIVD can already efficiently utilize C8 as a substrate, enzyme specificity is of primary importance. Therefore, each mutant was screened against a panel of 2-ketoacid substrates to assess the specificity and activity of each mutant (Supplementary Fig. 2). This assay was performed using a high-throughput pH-dependent colorimetric assay of enzyme activity in crude cell extracts. Mutations that increased specificity for C8 were then combined to produce combinatorial mutants and screened for specificity with the colorimetric assays. The mutant that exhibits the highest specificity and activity is G402V/M538L/F542V (KIVD_VLV). Kinetic constant characterization on purified enzymes shows that the engineered enzyme KIVD_VLV has a 600-fold improvement in specificity between C8 and C5, in terms of catalytic efficiency, relative to the native KIVD (Table 1).

***In vivo* testing of longer-chain alcohol production.** The measured specificity and activity for each enzyme was used to select two candidates for further characterization of their ability to modulate alcohol production *in vivo*. The two enzymes with the highest activity and specificity (Fig. 4), KIVD_VLV and GEO 175, were tested for their ability to reprogramme the synthetic recursive +1 pathway to produce longer-alcohol products *in vivo*. The alcohol-producing biochemical pathway was adapted from a previous study used to synthesize alcohols of various

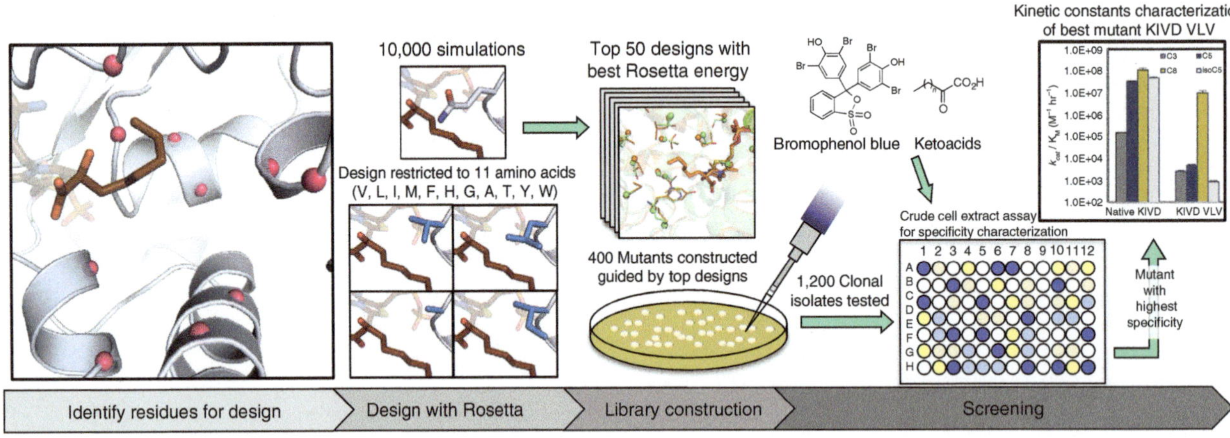

Figure 7 | Screening process for identifying an engineered KIVD with altered substrate specificity. The design process started with identifying active-site residues of KIVD (depicted in red). These amino acids were allowed to either remain native or sample any of 11 relatively hydrophobic amino acids. A total of 10,000 design simulations were run and the amino acids identified in the 50 lowest-energy designs were used to guide construction of a small library of roughly 400 KIVD mutants from the original combinatorial space of 10^{11} possible active-site mutations. Overall, 1,200 clonal isolates were screened for activity and specificity. The KIVD_VLV mutant was selected and subsequently purified for *in vitro* kinetic constant characterization.

lengths[3,4]. Starting from glucose, the aforementioned +1 pathway, involving LeuABCD, recursively elongates 2-ketoacids, starting with 2-ketobutyrate, into long-chained 2-ketoacids. These 2-ketoacids are then converted to aldehydes by ketoacid decarboxylases (native KIVD, KIVD_VLV and GEO 175) and subsequently to alcohols by an ADH. The carbon flux through this alcohol-synthesizing biochemical pathway is enhanced by overexpressing thrABC, ilvA, LeuABCD and ADH6 on inducible plasmids transformed into *E. coli*[3]. The effects of overexpression were, furthermore, supported and maintained by knocking out the threonine exporter enzyme RhtA and the primary DNA recombination degradation enzyme RecA. The enzyme AdhE was also knocked out, which helps raise longer-chain alcohol production by eliminating a competing pathway where acetyl-CoA is directly converted to ethanol[19].

Consistent with previous reports, the primary products (82% of total alcohol production) with native KIVD are short alcohols (C_2–C_4; Fig. 8 and Supplementary Table 2). The KIVD_VLV triple mutant completely switches the product profile of this pathway to predominantly longer-chain ($\geq C_5$) alcohols at 728 mg l^{-1}. Hexanol (C_6) is the major alcohol detected at a titre of 341 mg l^{-1} (47%), with titres of 269 mg l^{-1} (37%) heptanol (C_7), and 118 mg l^{-1} pentanol (16%) also observed. GEO 175 also switches the product profile of the pathway so that longer-chain alcohols are the primary products and produced at a level of 522 mg l^{-1}. For GEO175, the product profile is further shifted, resulting in heptanol (C_7) as the major product at a titre of 274 mg l^{-1} (50%), with hexanol production at 160 mg l^{-1} (29%) and 88 mg l^{-1} pentanol (16%). Similar to the results for KIVD_VLV, no ethanol or propanol production is observed, and only 19 mg l^{-1} butanol is produced. In addition, 10 mg l^{-1} of octanol was also observed, a product not observed when either native KIVD or KIVD_VLV were tested under equivalent conditions. The *in vivo* butanol and octanol titres for GEO175 compared with KIVD_VLV are consistent with the *in vitro* enzyme kinetics in which GEO175 has a significantly higher activity on C8, but lower specificity relative to C5, than KIVD_VLV. Total longer-chain alcohol titres for both GEO 175 and KIVD_VLV are increased significantly in comparison with native KIVD, with a >10-fold increase in heptanol production in both cases, and a >95% yield of longer-chain alcohols.

Replacing KIVD with GEO 175 or KIVD_VLV results in a significant increase in longer-chain alcohol production; however,

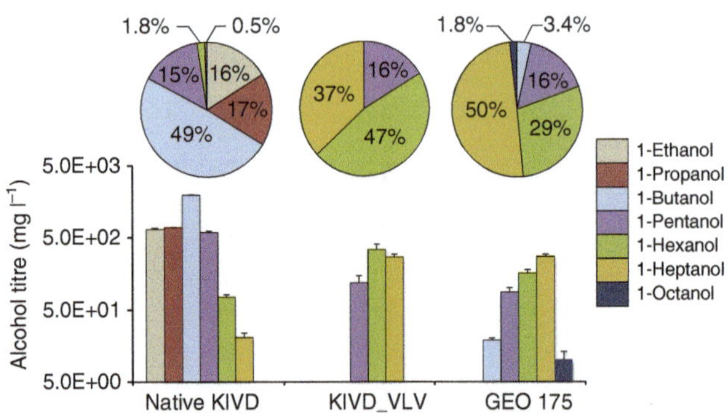

Figure 8 | *In vivo* alcohol production of the synthetic recursive +1 pathway with native KIVD, KIVD_VLV and GEO 175. Cells were incubated for 40 h in microaerobic conditions in a defined media as described in Methods. Each assay was performed in triplicate and titres reported if all three samples had observed product production above the limit of quantitation (5 mg l^{-1}) (Supplementary Table 2). Error bars show one s.d over the 3 biological replicates.

the overall alcohol production drops ~10-fold from the 4.0 g l^{-1} obtained with KIVD. We hypothesized that a potential reason for this could be due to the toxicity of longer-chain alcohols. To explore the potential of toxicity being a limiting factor in alcohol production, *E. coli* growth was monitored over a 7-h incubation with 10–1,000 mg l^{-1} C_5–C_8 alcohols supplemented at the beginning of growth (Fig. 9). Longer-chain alcohols are observed to be toxic as the supplemented concentration of C_6–C_8 alcohols reaches 250 mg l^{-1}. The final optical density (OD) of *E. coli* cultures drops from 1.9 to ~1.6 for C_6 and C_7 alcohols, and 1.0 for C_8. At 500 mg l^{-1} the OD drops to 1.0 and 0.1 for C_6 and C_7 alcohols, respectively. No significant growth is observed for C_8. At 1,000 mg l^{-1} no significant growth is observed for heptanol, and only an OD of 0.2 is achieved in the presence of hexanol. However, growth up to an OD of 1.5 is observed for pentanol. Significant toxicity in the range of 250–1,000 mg l^{-1} for longer-chain alcohols is consistent with the level of longer-chain alcohol titre, roughly 500 mg l^{-1}, produced in the engineered *E. coli* strains. This result indicates that to achieve higher total alcohol titre with this pathway, a strain of *E. coli* with high tolerance towards longer-chain alcohols is likely necessary[20,21]. However,

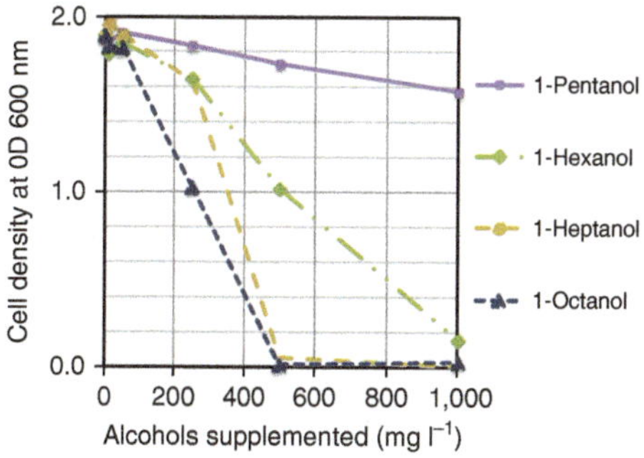

Figure 9 | Toxicity as a function of alcohol chain length. The optical density of the engineered *E. coli* strain after a 7-h growth was measured as a function of the alcohol concentration added to growth media at the beginning of incubation.

similar issues with product toxicity have been addressed through the design of continuous extraction methods and could be applied to the system here to increase levels of longer-chain alcohols titres[22].

Discussion

In this study we introduce an integrative genomic mining approach to enable the discovery of enzymes for a targeted function from sequence databases. The function of interest here was an enzyme with high activity for the decarboxylation of long-chain ketoacids to modulate the carbon flux of a synthetic pathway and increase titres of longer alcohols. The integrative genomic mining approach led to the discovery of a previously uncharacterized protein that we demonstrate has the targeted functional properties, and only required experimental characterization of 10 new proteins. In parallel, a more traditional computationally directed library-screening approach was utilized to reengineer the activity of a well-established enzyme. This required two successive rounds of screening over 400 mutants against a panel of substrates to identify an enzyme with the desired functional properties. When evaluated for their ability to modulate the carbon flux *in vivo*, both are able to completely shift the product profile towards longer-chain alcohols. However, the genomic mining approach provides a viable alternative to often expensive and laborious enzyme-engineering efforts that require screening libraries of mutants. This new approach for mining protein databases for enzymes of novel function stands to become an increasingly powerful approach as the methods for protein modelling improve and the size of genomic databases grow.

The enzyme GEO 175 represents the product of a general and rapid approach for obtaining enzymes with a desired function from the rapidly growing sequence databases. By combining bioinformatics and molecular modelling, this approach enables the identification of proteins likely to carry out a targeted function regardless of their native or putatively annotated activity. This overcomes issues with misannotation of protein function[23] or biased assumptions based on the closest characterized protein in the amino-acid sequence. The enzyme identified in this study, GEO175, is a clear illustration of this as it is roughly 15% identical to KIVD and its closest characterized sequence homologue (ppBFD, roughly 50% sequence identity) has a significantly different functional profile.

While GEO175 and KIVD_VLV both increased longer-chain alcohol titre >10-fold, we identified that toxicity needs to be addressed to further increase total longer alcohol titre. The titre of longer-chain alcohols produced is on the order of $0.5\,g\,l^{-1}$, which we then demonstrate is highly toxic for the cell lines used in this study. For industrial applications, bio-alcohol production often requires titres on the order of over tens of grams per litre (ref. 24). To avoid the need of a continuous extraction fermentation system, future efforts for increasing longer-chain alcohol production should focus on engineering or finding *E. coli* strains that are tolerant to this level of product formation. Efforts to integrate pumps and reengineer the cell wall to be resistant to alcohols could potentially mitigate toxicity effects and concurrently enable higher titres of longer alcohols to be produced through this pathway[20,21].

In addition to toxicity we observed that KIVD_VLV exhibits substrate inhibition kinetics above 2 mM ($0.3\,g\,l^{-1}$) of C8. This could be another potential cause of total alcohol titre loss observed *in vivo*. Redesigning KIVD_VLV to alleviate inhibition would be important for alcohol production at an industrially relevant level with this variant. However, the commensurate specificity and sixfold greater catalytic efficiency possessed by GEO_175 (relative to KIVD_VLV) make it a prime candidate for future industrial strain development for longer-chain alcohol production.

In summary, a new genomic mining approach and computationally directed library design efforts were both successfully implemented to obtain enzymes that enabled specific production of longer-chain alcohols *in vivo*. These results demonstrate the modularity of the synthetic +1 recursive pathway and provided a clear path forward to engineer the industrial-level production of longer alcohols. Finally, the integrative genomic mining approach introduced here is highly general, and with the rapid growth of sequence databases it has the potential to revolutionize the development and discovery of enzyme catalysts.

Methods

Integrative genomic mining. To obtain the GEO sequences, the native KIVD sequence was input for a homologous sequence search using HMMER3's* online server[13]. The resulting sequences were filtered using the CD-HIT* online server with a 90% identity cutoff[15,25] (Supplementary Figs 7 and 8). A homology model of each sequence was made using Rosetta Comparative Modeling[14]. At this point, the intermediate was placed into the active site, and 1,000 simulations were run to relax the intermediate according to the constraints. For each model, the lowest 100 in overall protein energy models were selected and then from that subset the lowest protein–ligand interface was chosen as the energy for the GEO.

An intermediate of the C8 product-yielding reaction was modelled using Spartan* (ref. 26). This intermediate is modelled based on the synthetic TPP analogue from a crystal structure of KIVD (PDB ID: 2VBG), which mimics the initial nucleophilic attack by the thiazolium. The six-carbon alkyl chain and the acid were added to the structure of the TPP analogue within Spartan. Different conformations of the alkyl chain were included in the modelling and a conformational library was made using OpenEye Omega* (ref. 27). For the enzyme design of KIVD, this intermediate was placed into the active site using distance and angle constraints. Rosetta Enzyme Design was run with default settings, an example of which is provided in the Rosetta Molecular Modeling Suite demos.

From the design simulations for each GEO, the lowest protein–ligand interface Rosetta energy was used to select a tractable number of GEOs for experimental characterization. During these simulations, any amino acids with a C_α within 8 Å of the active site could be mutated to any of the 20 amino acids. Ten GEOs of significant interest were chosen with the following criteria: five GEOs were chosen because they had the lowest predicted energies; the second five were chosen with the purpose of maximizing sequence space diversity. The 234 GEOs (less the five lowest in energy) were filtered with a sequence identity cutoff of 40%, and the five sequences from this filtered list with the lowest energy were picked as the second five GEOs in our final list. Each model was evaluated in the Foldit interface, and mutations made during the design simulations were reverted to the native amino acid if not predicted to improve the interface energy by more than two Rosetta energy units.

Phylogenetic tree of GEOs was generated with the Geneious software* using a Muscle sequence alignment[28]*. The resulting tree was visualized using iTOL online tool in the circular tree mode and rooted at native KIVD[29,30].

Synthetic genes coding for each GEO were synthesized as a DNA String by Life Technologies or the Joint Genome Institute. Genes were codon-optimized for *E. coli*, and the DNA as well as amino-acid sequence provided in Supplementary Tables 3 and 4. The string was cloned into the pET-29b(+) plasmid vector using the Gibson assembly between the NdeI and XhoI restriction sites that added a C-terminal 6x-His tag in-frame to the gene.

*Default settings were used unless otherwise specified.

KIVD active-site redesign. In the design simulations, 10 residues in the proposed active-site pocket were allowed to either remain native or sample any of 11 relatively hydrophobic amino acids (Val, Leu, Ile, Met, Phe, His, Gly, Ala, Thr, Tyr and Trp). The identities of amino acids at all other positions were kept constant. Residues within 12 Å of the ligand were allowed to undergo conformational sampling during simulation. A total of 10,000 design simulations were run, from which the 50 designs lowest in ligand–protein interface energy, and non-redundant in terms of sequence, were selected to represent the potential sequence space predicted to accommodate the C8 substrate. The profile was used to construct a small library of ~400 KIVD mutants. Each amino acid in the library was sampled as a single mutation, with the exception of residues that were within five residues from one another. These were sampled in a combinatorial manner as both single and double mutants to evaluate synergistic effects, given their proximity to one another. A complete list of oligos and the amino acids allowed for each site are provided in the Supplementary Tables 5 and 6.

Construction and selection of KIVD libraries. All *E. coli* strains and plasmids used are listed in Supplementary Table 7. Partially degenerate oligonucleotides (Supplementary Table 6) were ordered from Integrated DNA Technologies (San Diego, CA) and were used to generate libraries of *kivd* mutants (with N-terminal histidine tags) using PCR. Library DNA was purified, inserted into the PCR-amplified pQE9 vector by isothermal Gibson assembly[31] and transformed into XL1-Blue cells. Individual ampicillin-resistant colonies were picked and grown in 96-well blocks at 37 °C overnight. The number of colonies picked was three times the theoretical library size to ensure ~95% probability of all possible mutation combinations occurring. Glycerol was added to 25% (w/v) and libraries were stored at −80 °C until enzyme activity and specificity were assayed. KIVD mutant libraries were screened using a pH-monitored enzyme assay (method explained below) to measure the rate of H^+ consumption resulting from the decarboxylation of 2-ketoacids.

KIVD mutant libraries were grown overnight in culture blocks, diluted 1:100 into fresh media (Luria Broth), grown at 37 °C for 3 h to OD_{600} ~0.6, induced with 0.1 mM isopropyl-b-D-thiogalactoside (IPTG; GoldBio, Saint Louis, MO) and grown for two additional hours at 37 °C. Cells from the well blocks were transferred to 96-well assay plates (Costar, Corning, NY) and cell density was measured at 600 nm. An equal volume of permeabilization solution (8.7 mM potassium phosphate, 43.4 mM KCl, 0.87 mM $MgSO_4$, pH 7.1 ± 0.1, 8.7% (v/v) chloroform, 0.0043% (w/v) SDS and 0.26% (v/v) 2-mercaptoethanol) was then added to break the cell membranes. Bromothymol blue, TPP (Sigma-Aldrich, St Louis, MO) and substrate were added to 0.008% (w/v), 0.5 and 10 mM, respectively. All substrates (2-ketobutyrate, 2-ketovalerate, 2-ketoisovalerate, 2-ketocaproate and 2-ketooctanoate; Sigma-Aldrich) were dissolved in MilliQ dH_2O and pH was adjusted to 7.1 ± 0.1. Absorbance was measured at 615 nm in a spectrophotometer (TEK Powerwave XS, BioTek, Winooski, VT). Data were acquired for 15 min at 30 °C. Cells containing pQE_hiskivd_wt and pQE9 were used as positive and negative controls, respectively. Enzyme activity data were corrected for cell density. Mutants chosen from this screening had a 25% higher activity than wild-type KIVD for any of the 2-ketoacids.

The library derived from the pH-coupled assay was further screened using a second enzyme kinetics assay. This assay involved a second enzyme, ADH from *S. Cerevisiae* (ADH6), which is a key enzyme in the alcohol production pathway in reducing the aldehyde produced by KIVD into an alcohol. ADH couples oxidation of NADPH to $NADP^+$ to the reduction of aldehyde into an alcohol. Stoichiometrically, conversion of 1 M of 2-ketoacid to n-alcohol depletes exactly 1 M of NADPH. Therefore, activity of KIVD could be measured directly by measuring depletion of NADPH through ultraviolet spectrophotometry.

Overnight cultures of XL1B with pQE9 containing KIVD mutations were grown at 37 °C in a 96-well block. Cultures were diluted 1:100 and then grown for 3 h at 37 °C to OD_{600} of 0.6. Cultures were induced with 0.1 mM IPTG (GoldBio) for 3 h at 37 °C. Cultures were then centrifuged, and pellets were lysed with BugBuster (Novagen, Madison, WI). Cell lysate (5 µl) from each culture and a buffer mix (175 µl) containing coenzyme 1.5 mM TPP (Sigma), 0.2 mM NADPH (Fisher Scientific, Waltham, MA), 0.045 U ADH6, 100 mM $NaPO_4$, 100 mM NaCl and 10 mM $MgCl_2$, pH of 7, were added together into a 96-well plate. In all, 20 µl of 100 mM substrate was added to dilute to make a final concentration of 10 mM. The substrates tested were IsoC5, C4, C5, C6, C8 and H_2O (Supplementary Fig. 2). Absorbances were measured with a plate reader (TEK Powerwave XS, BioTek) at 340 nm at 30 °C for 15 min. The protein concentration in each sample was determined with a BCA assay (Thermo Scientific, Waltham, MA) and used to normalize slope values. Candidate KIVD mutants were chosen for greater activity on long-chained 2-ketoacids and decreased activity on shorter 2-ketoacids. This was determined from graphing absorbance versus time for each substrate and qualitatively comparing slope values versus wild type (Supplementary Fig. 2).

Site-directed mutagenesis. Oligonucleotides (Supplementary Table 6) encoding specific *kivd* mutations were ordered from Integrated DNA Technologies (Coralville, Iowa) and were used to mutate *kivd* in pZE_LeuA*BCDKA6 and pQE_hiskivd_wt using PCR. Amplified DNA fragments were purified, inserted into either the PCR-amplified pZE vector containing *leuA**BCD* and *adh6* or the pQE vector by isothermal Gibson assembly, and transformed into XL1-Blue cells. Plasmid DNA was purified (Qiagen, Hilden, Germany) from overnight cultures of antibiotic-resistant colonies and the plasmid sequences were verified (Laragen, Culver City, CA).

Chromosomal gene knockout. Genes were removed from the ATCC 98082 *ΔrhtA* strain genome using P1 transduction from the Keio collection as previously described[32]. The aldehyde-ADH gene (*adhE*) was knocked out to eliminate ethanol production from acetyl-coA. In all strains, *recA* was knocked out to prevent recombination between the genome and plasmids, thereby stabilizing the transformants. Primers used to target *recA* for knockout (Supplementary Table 6) were designed based on the Keio collection (Genobase, http://ecoli.aist-nara.ac.jp/) and ordered from Integrated DNA Technologies.

Fermentation procedure and analysis. For *n*-alcohol production, strains of ST128 were transformed with pZS_thrO, pZAlac_ilvA$_{BS}$leuA and pZE_-LeuA*BCDK*A6 containing various *kivd* mutations. Fermentation conditions were adapted from those in previous lines of works[3,4], with the following changes: 20 ml of medium was used, with 100 µg ml^{-1} ampicillin, 50 µg µl^{-1} kanamycin and 100 µg ml^{-1} spectinomycin added. Cells were grown to an optical density at 600 nm of ~0.6, followed by induction with 0.1 mM IPTG. After fermentation, cells were centrifuged for 15 min at 4,000*g* and 4 °C. The supernatant was split into two fractions for analysis, 5 ml for short-chain alcohols (ethanol, 1-propanol and 1-butanol) and 15 ml for longer-chain alcohols (1-pentanol, 1-hexanol, 1-heptanol and 1-octanol). Longer-chain alcohols were extracted from the 15-ml fraction by 3 ml *n*-hexane before analysis. GC-FID analysis was performed as previously described[4].

Protein purification and enzymatic assay of KIVD. For the native KIVD and KIVD_VLV mutant, 2 ml overnight cultures of XL1B cells were transformed with pQE9 containing N-terminal his-tagged KIVD enzymes and grown in Terrific Broth (BP Biomedical, Cat# 3046-042) with 50 µg ml^{-1} of carbenicillin (Fisher Scientific, Cat# BP2648-5). For the GEOs, 2-ml overnight cultures of BLR cells were transformed with pet29b+ plasmid containing N-terminal his-tagged GEOs, and grown in Terrific Broth with 50 µg ml^{-1} of kanamycin (Fisher Scientific, Cat# BP906-5). These cultures were diluted 1:1,000 in 500 ml of Terrific Broth with 1 mM $MgSO_4$, 1% glucose and 50 µg ml^{-1} of corresponding antibiotics, and then grown at 37 °C for 24 h. Cultures were pelleted down at 4,700 r.p.m. for 10 min and resuspended in auto-induction media (LB broth, 1 mM $MgSO_4$, 0.1 mM TPP, 1 × NPS and 1 × 5052) for induction at 18 °C for 34 h. At the end of induction, cells were centrifuged (4,700 r.p.m., 4 °C, 20 min), supernatant was removed and cells were resuspended in 40 ml lysis buffer (100 mM Hepes, pH 7.5, 100 mM NaCl, 10% glycerol, 0.1 mM TPP, 1 mM $MgSO_4$, 10 mM Imidazole, 1 mM TCEP) and 1 mM phenylmethylsulphonyl fluoride and sonicated for 2 min. Lysed cells were centrifuged at 4,700 r.p.m. at 4 °C for 60 min to remove cell debris. Supernatant was loaded on gravity flow column with 700 µl of cobalt slurry (Fisher Scientific, CAT# PI-90091) washed with 10 ml of wash buffer (100 mM Hepes, pH 7.5, 100 mM NaCl, 10% glycerol, 0.1 mM TPP, 1 mM $MgSO_4$, 10 mM Imidazole and 1 mM TCEP). Cobalt bead bed was washed with 15 ml of wash buffer five times and proteins were eluted with 1,000 µl of elution buffer (100 mM Hepes, pH 7.5, 100 mM NaCl, 10% glycerol, 0.1 mM TPP, 1 mM $MgSO_4$, 200 mM Imidazole and 1 mM TCEP). Protein samples were immediately buffer-exchanged with spin concentrators (Satorius, CAT# VS0112) into storage buffer (100 mM Hepes, pH 7.5, 100 mM NaCl, 10% glycerol, 0.1 mM TPP, 1 mM $MgSO_4$ and 1 mM TCEP) and stored at 4 °C until kinetics characterization. Protein concentrations were determined using a Synergy H1 spectrophotometer (Biotek) by measuring absorbance at 280 nm using their calculated extinction coefficients with the ExPASy ProtParam Tool[33]. All other buffers and salts were purchased from Fisher Scientific, unless otherwise specified.

The k_{cat} and K_M values of selective KIVD mutants were measured for the substrates C_3, isoC_5 and C_8. All substrates were dissolved in MilliQ H_2O and pH was adjusted to 7.5 as necessary. Activity was measured at 0.005–10 mM substrates. The assay was performed in a 96-well half-area plate. Each reaction contains a final concentration of 0.5 mM NADH, 1 mM dithiothreitol, 0.1 mM TPP, 1 mM MgSO4, reaction buffer (100 mM Hepes, 100 mM NaCl, 10% glycerol, pH 7.5) and ADH (Sigma-Aldrich, A7011, 10 U ml^{-1} for C_3, C_5 and C_8 reactions, 500 U ml^{-1} for isoC_5 reactions). A wide range of ketoacid decarboxylase concentrations, 4.5 nM–15 µM, was used according to the activity of each enzyme towards different substrates to perform steady-state kinetics measurement over a period of an hour. Absorbance readings were taken every 1 min at OD_{340} at 21 °C for 60 min using the Synergy H1. Kinetic parameters (k_{cat} and K_M) were determined by fitting initial velocity versus substrate concentration data to the Michaelis–Menten equation.

Alcohol toxicity. Alcohol tolerance of the *in vivo* alcohol production strain was evaluated by supplementing the fermentation media with specified quantities of longer-chain alcohols, and measuring growth over 7 h. Both the *E. coli* strain (ATCC 98082 pZS_thrO, pZAlac_ilvA_LeuA and pZE12LeuA*BCDKA6_KIVD_wt) and fermentation media (1X M9 metals + 1X trace metal mix + 0.5% yeast extract + 2% glucose + antibiotics) and conditions are the same as described above (Fermentation procedure and analysis). Cells were grown in media without IPTG induction to a starting $OD_{600} \sim 0.02$ and then supplemented with either 1-petanol, 1-hexanol, 1-heptanol or 1-octanol at specified concentrations of 10, 50, 250, 500 or 1,000 mg l^{-1}. OD_{600} readings were taken hourly up to 7 h. Figure 9 shows the final OD_{600} recorded at 7 h.

References

1. USA White House. *National Bioeconomy Blueprint* (The White House, 2012).
2. USDA, US. *Bio-Based Products: Market Potential And Projections Through 2025* (US Department of Agriculture, 2008).
3. Zhang, K., Sawaya, M. R., Eisenberg, D. S. & Liao, J. C. Expanding metabolism for biosynthesis of non-natural alcohols. *Proc. Natl Acad. Sci. USA* **105,** 20653–20658 (2008).
4. Marcheschi, R. J. *et al.* A synthetic recursive "+ 1" pathway for carbon chain elongation. *ACS Chem. Biol.* **7,** 689–697 (2012).
5. Erickson, B. & Winters, P. Perspective on opportunities in industrial biotechnology in renewable chemicals. *Biotechnol. J.* **7,** 176–185 (2012).
6. Plaza, M. *et al.* Biochemical and molecular characterization of α-ketoisovalerate decarboxylase, an enzyme involved in the formation of aldehydes from amino acids by *Lactococcus lactis*. *FEMS Microbiol. Lett.* **238,** 367–374 (2004).
7. Felnagle, E. A. *et al.* Engineering synthetic recursive pathways to generate non-natural small molecules. *Nat. Chem. Biol.* **8,** 518–526 (2012).
8. Bayer, T. S. *et al.* Synthesis of methyl halides from biomass using engineered microbes. *J. Am. Chem. Soc.* **131,** 6508–6515 (2009).
9. Heins, Richard A. *et al.* Phylogenomically guided identification of industrially relevant GH1 β-glucosidases through DNA synthesis and nanostructure-initiator mass spectrometry. *ACS Chem. Biol.* **9,** 2082–2091 (2014).
10. The National Academies. *Industrialization of Biology: A Roadmap to Accelerate Advanced Manufacturing of Chemicals* (The National Academis Press, 2015).
11. Richter, F. *et al.* De novo enzyme design using Rosetta3. *PLoS ONE* **6,** e19230 (2011).
12. Finn, Robert D. *et al.* Pfam: the protein families database. *Nucleic Acids Res.* **42,** D222–D230 (2013).
13. Finn, R. D., Clements, J. & Eddy, S. R. HMMER web server: interactive sequence similarity searching. *Nucleic Acids Res.* **39,** W29–W37 (2011).
14. Song, Yifan *et al.* High-resolution comparative modeling with RosettaCM. *Structure* **21,** 1735–1742 (2013).
15. Huang, Y. *et al.* CD-HIT Suite: a web server for clustering and comparing biological sequences. *Bioinformatics* **26,** 680–682 (2010).
16. Zhang, Y. & Skolnick, J. TM-align: a protein structure alignment algorithm based on the TM-score. *Nucleic Acids Res.* **33,** 2302–2309 (2005).
17. Berthold, C. L. *et al.* Structure of the branched-chain keto acid decarboxylase (KdcA) from Lactococcus lactis provides insights into the structural basis for the chemoselective and enantioselective carboligation reaction. *Acta Crystallogr. D* **63,** 1217–1224 (2007).
18. Bruning, M. *et al.* Structural and kinetic studies on native intermediates and an intermediate analogue in benzoylformate decarboxylase reveal a least motion mechanism with an unprecedented short-lived predecarboxylation intermediate‡. *Biochemistry* **48,** 3258–3268 (2009).
19. Shen, C. R. & Liao, J. C. Metabolic engineering of *Escherichia coli* for 1-butanol and 1-propanol production via the ketoacid pathways. *Metab. Eng.* **10,** 312–320 (2008).
20. Kell, D. B. *et al.* Membrane transporter engineering in industrial biotechnology and whole cell biocatalysis. *Trends Biotechnol.* **33,** 237–246 (2015).
21. Foo, J. L. *et al.* Improving microbial biogasoline production in *Escherichia coli* using tolerance engineering. *mBio* **5,** e01932-14 (2014).
22. Brennan, T. C. R. *et al.* Alleviating monoterpene toxicity using a two-phase extractive fermentation for the bioproduction of jet fuel mixtures in *Saccharomyces cerevisiae*. *Biotechnol. Bioeng.* **109,** 2513–2522 (2012).
23. Schnoes, A. M. *et al.* Annotation error in public databases: misannotation of molecular function in enzyme superfamilies. *PLoS Comput. Biol.* **5,** e1000605 (2009).
24. Parekh, S. Strain improvement. *Desk Encyclopedia of Microbiology* (Academic Press, 2010).
25. Kearse, M. *et al.* Geneious Basic: an integrated and extendable desktop software platform for the organization and analysis of sequence data. *Bioinformatics* **28,** 1647–1649 (2012).
26. Shao, Y. *et al.* Advances in methods and algorithms in a modern quantum chemistry program package. *Phys. Chem. Chem. Phys.* **8,** 3172–3191 (2006).
27. Oechem, T. K. *version 1.7. 7* (OpenEye Scientific Software Inc, 2012).
28. Edgar, R. C. MUSCLE: multiple sequence alignment with high accuracy and high throughput. *Nucleic Acids Res.* **32,** 1792–1797 (2004).
29. Letunic, I. & Bork, P. Interactive Tree Of Life (iTOL): an online tool for phylogenetic tree display and annotation. *Bioinformatics* **23,** 127–128 (2007).
30. Letunic, I. & Bork, P. Interactive Tree Of Life v2: online annotation and display of phylogenetic trees made easy. *Nucleic Acids Res.* **39,** W475–W478 (2011).
31. Gibson, G. D. *et al.* Enzymatic assembly of DNA molecules up to several hundred kilobases. *Nat. Methods* **6,** 343–345 (2009).
32. Datsenko, A. K. & Wanner, B. L. One-step inactivation of chromosomal genes in *Escherichia coli* K-12 using PCR products. *Proc. Natl Acad. Sci. USA* **97,** 6640–6645 (2000).
33. Gasteiger, E. *et al.* *ExPASy*: the proteomics server for in-depth protein knowledge and analysis. *Nucleic Acids Res.* **31,** 3784–3788 (2003).
34. The PyMOL Molecular Graphics System, Version 1.7.4 Schrödinger, LLC.

Acknowledgements

We are thankful for the funding and support from the University of California-Davis, the Joint Genome Institute Community Synthesis Project grant 1109, and the Alfred P. Sloan Foundation grant FG-BR2014-012. R.M. was supported by an NRSA Kirschstein Postdoctoral Fellowship (5F32GM099277-02) during the course of this study. We also thank Alexandre Zanghellini and Elizabeth Kellogg for the many insightful discussions that helped shape many of the ideas for the genomic enzyme mining method implemented here.

Author contributions

J.B.S. and J.C.L. conceived the genomic mining strategy and + 1 pathway, respectively, and designed experiments. W.S.M., S.T. and R.M. designed and performed the *in vivo* and *in vitro* characterization experiments. S.B., J.T. and J.B.S. designed and performed genomic mining and molecular modelling of the GEOs. All authors contributed in writing the manuscript, discussions of data, results as well as implications of the work.

Additional information

Competing financial interests: The authors declare no competing financial interests.

Foldamer-mediated manipulation of a pre-amyloid toxin

Sunil Kumar[1,†,*], Melissa Birol[1,*], Diana E. Schlamadinger[1], Slawomir P. Wojcik[1], Elizabeth Rhoades[1,†] & Andrew D. Miranker[1,2]

Disordered proteins, such as those central to Alzheimer's and Parkinson's, are particularly intractable for structure-targeted therapeutic design. Here we demonstrate the capacity of a synthetic foldamer to capture structure in a disease relevant peptide. Oligoquinoline amides have a defined fold with a solvent-excluded core that is independent of its outwardly projected, derivatizable moieties. Islet amyloid polypeptide (IAPP) is a peptide central to β-cell pathology in type II diabetes. A tetraquinoline is presented that stabilizes a pre-amyloid, α-helical conformation of IAPP. This charged, dianionic compound is readily soluble in aqueous buffer, yet crosses biological membranes without cellular assistance: an unexpected capability that is a consequence of its ability to reversibly fold. The tetraquinoline docks specifically with intracellular IAPP and rescues β-cells from toxicity. Taken together, our work here supports the thesis that stabilizing non-toxic conformers of a plastic protein is a viable strategy for cytotoxic rescue addressable using oligoquinoline amides.

[1] Department of Molecular Biophysics and Biochemistry, Yale University, 260 Whitney Avenue, New Haven, Connecticut 06520-8114, USA. [2] Department of Chemical and Environmental Engineering, Yale University, 260 Whitney Avenue, New Haven, Connecticut 06520-8114, USA. † Present address: Department of Chemistry, New York University, Silver Center for Arts and Science, 100 Washington Square East, 10th Floor, New York, New York 10003, USA (S.K.); Department of Chemistry, University of Pennsylvania, 231 South, 34th Street, Philadelphia, Pennsylvania 19104, USA (E.R.). * These authors contributed equally to this work. Correspondence and requests for materials should be addressed to A.D.M. (email: Andrew.Miranker@yale.edu).

Biopolymers are distinguished from artificial polymers by the presence of a specified sequence of precursors. A discrete sequence affords biopolymers the capacity to fold to a specific structure despite having an astronomically sized conformational landscape. A tiny set of precursors (4 and 20 for RNA and proteins, respectively) create a breadth of folds and functions essential to life. In contrast, synthetic polymers have access to an essentially unlimited array of precursor variants; however, the lack of sequence control and unique conformation results in a breadth of function that is dwarfed by biology.

Synthetic foldamers seek to join the best of these two worlds[1]. New scaffolds with specifiable sequences permit folded and functional structures to be successfully designed. For instance, foldamers based on oligomers of arylamides, β-peptide, α/β-peptide and peptoids have been designed to be an antimicrobial agent[2], agonist of GLP-1 receptor[3], an inhibitor of HIV-cell fusion[4] and an antagonist of vascular endothelial growth factor receptor 2 (ref. 5), respectively. The achievable specificities of these small molecules rival biopolymers. For example, a quinoline-based aromatic foldamer was recently developed using an iterative modification approach to selectively and stereospecifically encapsulate D-fructose[6].

Dynamic binding modes, such as conformational selection, are often observed in protein:ligand interactions[7]. A recent example is the anti-cancer drug Gleevec[8]. Gleevec binds to structurally identical sites in Src and its homologue, Abl. Nevertheless, there is a >1,000-fold difference in binding affinity that can be accounted for by differences in protein dynamics. It stands to reason that the internal degrees of freedom of ligands should be similarly exploitable. However, ligand dynamics are viewed unfavourably as a result of entropy loss on binding. The non-covalent and reversible stabilization of foldamer structure presents an opportunity to limit entropic loss on protein binding without sacrificing opportunities and potential benefits of dynamic sampling.

A novel tetraquinoline amide foldamer, ADM-116, has been developed to counter conformational transitions in islet amyloid polypeptide (IAPP). IAPP is a 37-residue peptide co-packaged with insulin in pancreatic β-cells. In type II diabetes, aggregation of this peptide into amyloid fibres is observed, and pre-amyloid states of IAPP are toxins resulting in β-cell death[9]. ADM-116 displays activity across solution and cell-based assays, as a result of structure-specific binding to pre-amyloid states. Unique to ADM-116 is its capacity to cross the plasma membrane, without assistance from cellular processes, and antagonize toxicity long after the cellular uptake of IAPP is complete. Passive translocation is remarkable because ADM-116 is dianionic and further, can carry an anionic fluorescent cargo across the plasma membrane: properties otherwise observed only for polycations. These gains of function are dependent on the capacity of the foldamer to not only recognize IAPP but also to transiently sample partially folded states.

Results

Time-dependent localization of IAPP. INS-1 cells were incubated with 100 nM IAPP (Fig. 1a) labelled at its N terminus with Alexa-594 ($IAPP_{A594}$). Co-addition of 0 or 13 μM unlabelled IAPP corresponds to non-toxic and toxic conditions, respectively. After 5 h under non-toxic conditions, IAPP is not significantly internalized (Fig. 2). By 12 h, intracellular IAPP is readily observed with maximum extent of internalization apparent at 24 h (Fig. 2). At all time points under this non-toxic condition, IAPP appears as diffuse puncta, possibly a consequence of the energy-dependent cellular uptake under these conditions[10]. Under toxic conditions, slightly elevated uptake of

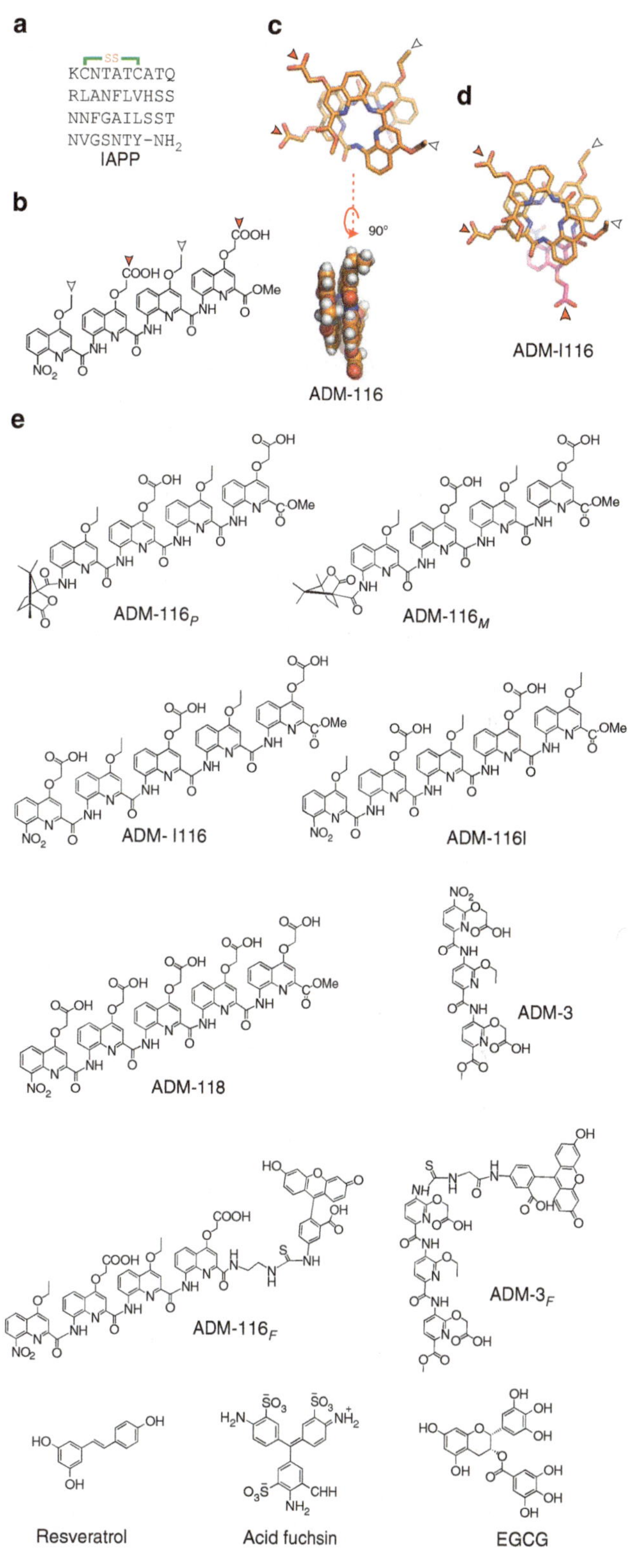

Figure 1 | Structures of compounds. (**a**) Primary sequence of human IAPP. Wild-type IAPP is C-terminally amidated and has an intramolecular disulfide bond between residues 2 and 7. (**b**) Chemical structure of ADM-116. (**c**) Three-dimensional stick (top) and sphere (bottom) model of ADM-116. (**d**) Pentaquinoline, ADM-I116, in which ADM-116 is extended one residue from the N terminus. Arrows in **a**,**b** and **d** indicate surface-exposed functional moieties. Structures in **b** and **d** adapted from a longer oligoquinoline[47]. (**e**) Chemical structures of other molecules used in this study.

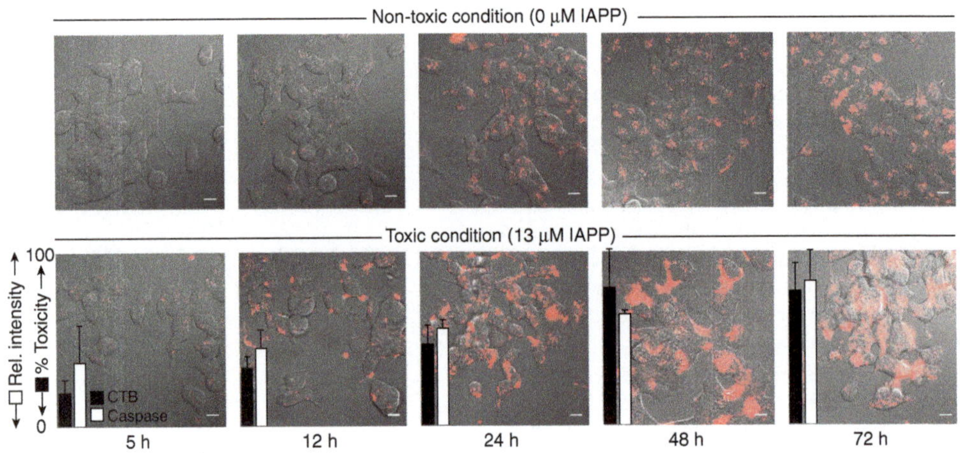

Figure 2 | Time-dependent uptake of IAPP and toxicity. At time zero, 100 nM $IAPP_{A594}$ with (toxic) and without (non-toxic) an additional 13 μM of unlabelled IAPP was added to culture media of INS-1 cells. Confocal images were taken at the indicated time points. Three biological replicates were performed. Scale bar, 10 μm. Inset: colourimetric evaluation of toxicity (CellTiter Blue) and apoptosis (using caspase-3) at the toxic condition relative to vehicle-only controls. Each time point is the average of eight on-plate repeats from each of three independently performed replicates ($n = 24$). rel., relative.

IAPP is apparent by 5 h. By 12 h, contrasting behaviour can be clearly seen with the toxic condition showing external and internalized puncta and larger assemblies[11]. By 24 h, extracellular IAPP is a small fraction of the total IAPP. At 48 h, large intracellular aggregates appear and continue to increase in intensity by 72 h. This progression suggests that under these conditions, self-assembly is mediated by the intracellular environment, consistent with the work showing the culture media and plasma membrane to be inhibitory to amyloid formation[12].

Cytotoxicity is mediated by intracellular IAPP. The time dependence of IAPP-mediated toxicity was measured in parallel to imaging. Using caspase-3/7 activity as an indicator of apoptosis, the effects are apparent as early as 5 h with steady increases over the time course of the observation (Fig. 2). The fraction of cells affected by IAPP can be approximated by monitoring total cytosolic reductase activity relative to IAPP-free controls. The trend clearly parallels the apoptosis marker. Importantly, continued increases in apparent apoptotic activity are evident long after the 24 h time point, where little extracellular IAPP is present (Fig. 2). Taken together, this suggests the site of toxicity is intracellular. *In vivo*, it is debated whether IAPP becomes toxic before or after secretion[9,13]. Certainly, it is possible that IAPP has multiple mechanisms and locations for inducing toxic gains of function. The cell-penetrating peptide quality of IAPP, however, makes such determinations particularly challenging, and possibly irrelevant[14]. Nevertheless, it is clear that the small-molecule modulation of IAPP requires a compound that can be internalized.

Targeting membrane-bound IAPP. We hypothesize that the compounds that target intracellular IAPP are more likely to be found among membrane active inhibitors of IAPP self-assembly[15]. In solution, the assembly of 10 μM IAPP into fibres occurs with a reaction midpoint, t_{50}, of 18 ± 1.4 h (Supplementary Fig. 1). An otherwise identical reaction conducted in the presence of large unilamellar vesicles (LUVs; composed of a 1:1 mixture of DOPG:DOPC) yields a t_{50} of 1.1 ± 0.1 h (Fig. 3a). These LUVs possess a surface anionicity of 50%. Comparable catalysis is apparent at 7:3, DOPC:DOPG, which is closer to the anionic content of cell membrane phospholipids (Supplementary Fig. 2)[16,17]. In previous work, we have successfully used anionicities $\geq 50\%$ to generate insights for

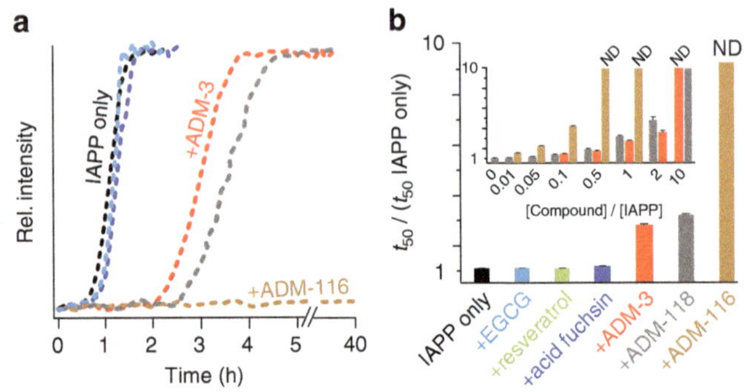

Figure 3 | Effect of ligands on the kinetics of IAPP fibre formation. Representative (**a**) and statistics of reaction midpoints ($t50$, **b**) for 10 μM IAPP fibre formation reactions catalysed by unilamellar vesicles (630 μM monomer units, DOPG:DOPC, 1:1, 100 nm). (**b**) Inhibition with the indicated small molecules is performed at 1:1 IAPP:small molecule. Colours are consistent between the panels. Inset: statistics of dose dependence of inhibition of fibre formation by ADM-116, ADM-3 and ADM-118. An absence of kinetics over the time course of observation (for example, >40 h for ADM-116 at 1:1) are indicated as not detected (ND). All experiments were conducted at least in triplicate with errors reported as ±1 s.d. rel., relative.

IAPP that proved applicable to live cells[18,19]. In a study most relevant to our work here, we used LUVs identical to those presented here to establish cross-correlation of membrane-bound α-helical conformers of IAPP, inhibition of several solution biophysical-based gains of function and cytotoxic rescue[20]. We further note that detergent induced α-helical oligomers of IAPP recently enabled the discovery of conformation-specific antisera in human diabetics[21]. In this work, we use the LUV-based assay as a reference condition under which small molecules can be assessed for activity on IAPP in the context of a lipid bilayer.

Synthetic foldamers based on quinoline have been shown to hold promise in solution biophysical assays of IAPP[22]. Synthesis and screening (Methods; Supplementary Methods) of numerous variants allowed us to identify one derivative, ADM-116 (Fig. 1b,c), with marked activity in LUV-catalysed fibrillation assays. At 1:1 (IAPP:ADM-116), liposome-catalysed fibrillation is undetectable (a t_{50} >40-fold higher than control) with significant inhibition observable even at 1:0.1 (Fig. 3a,b). Amyloid formation is also not observed by electron microscopy (Supplementary Fig. 3) and far-ultraviolet circular dichroism (CD; Fig. 5d). In comparison, our compounds ADM-3 (ref. 20) and OQ5 (referred to in this work as ADM-118) (refs 22,23) inhibit by factors of 2.7 ± 0.1- and 3.3 ± 0.1-fold, respectively. Other compounds such as epigallocatechin gallate (EGCG)[24], acid fuchsin[25] and resveratrol[26], show no detectable effect on liposome-catalysed amyloid assembly (Fig. 3a,b). Clearly, ADM-116 contains functional moieties, the steric and physicochemical properties of which result in exceptional inhibition of lipid-catalysed amyloid assembly.

ADM-116 rescues IAPP-induced toxicity. After 48 h of incubation with 13 μM IAPP, INS-1 viability is decreased 78 ± 8%. Co-addition of ADM-116 at a stoichiometric ratio of 1:1 (IAPP:ligand) wholly restores viability (Fig. 4a,b). The compound ADM-3 (ref. 20), a tripyridylamide, is also active on membrane-bound IAPP and also restores cell viability (Fig. 4a) with a dose dependence comparable to ADM-116 (Fig. 4b). Toxic rescue by ADM-116 and ADM-3 is also reproducible in a non-pancreatic cell line, COS-1 (Supplementary Fig. 4). Our previously published compound, ADM-118, as well as resveratrol[26], EGCG[24] and acid fuchsin[25] were all ineffective in rescuing toxicity under these conditions. Published conditions show that cytotoxic rescue can be achieved following a procedure in which IAPP and small molecule are preincubated for ≥11 h before adding the complexes to cell culture[24,25]. Following this procedure, all of these molecules significantly rescue IAPP-mediated toxicity in INS-1 cells (Supplementary Fig. 5). In contrast, we co-introduce IAPP and small molecule to cell culture. At concentrations investigated in this report, none of the molecules show intrinsic toxicity towards INS-1 cells nor interfered with the colourimetric assay (Supplementary Fig. 6).

ADM-116, and not ADM-3, rescues cells from intracellular IAPP toxicity. In light of the intracellular origins of toxicity indicated above (Fig. 2), cell viability was instead assessed under conditions in which introduction of IAPP was followed by a delay before the addition of small molecule. Remarkably, even after a delay of 24 h, ADM-116 is capable of rescuing IAPP-induced cytotoxicity by 47 ± 4%. In marked contrast, ADM-3 was ineffective in rescuing toxicity when added after 12 h (Fig. 4c). As IAPP is internalized by 24 h of incubation (Fig. 2), this suggests that the mechanism of ADM-116 rescue includes penetration of the plasma membrane.

Molecular specificity of binding. The complex formed by ADM-116 and IAPP is small and monodisperse. A fluorescein derivative of ADM-116, ADM-116$_F$, was prepared by coupling at its C terminus and then monitored by fluorescence correlation spectroscopy. Autocorrelation profiles were readily fit to a single diffusing species with a diffusion time, τD, for ADM-116$_F$ measures 130 ± 10 μs (Fig. 5a) with no apparent presence of soluble aggregates. These observations remained consistent from 25 to 13 μM, the latter corresponding to concentrations used in cell culture (Supplementary Fig. 7a). Self-association and aggregation are similarly absent at 20 μM ADM-116, as measured by nuclear magnetic resonance (NMR; Supplementary Fig. 7b). On titration of a 25 nM solution of ADM-116$_F$ with IAPP, a second component becomes evident with a τD = ~400 μs (Fig. 5a). The diffusion time of this slower component was consistent over the course of the titration, indicative of a discrete complex. Fitting the fractional amplitude of bound to unbound ADM-116$_F$ gives a $K_d = 240 \pm 60$ nM. Isothermal titration calorimetry (ITC) using unlabelled IAPP and ADM-116 yields an exothermic profile that fits a one-site binding model with a $K_a = 2.8 \pm 0.3 \times 10^6\,M^{-1}$ ($K_d = 360 \pm 40$ nM; Fig. 5b; Supplementary Table 1). Plainly, ADM-116 forms a discrete complex with strong affinity to IAPP.

ADM-116 stabilises the α-helical subdomain of IAPP. The far-ultraviolet CD spectrum of membrane-bound IAPP exhibits two minima near 208 and 222 nm, characteristic of α-helical structure[27] (Fig. 5d). This profile transitions to that of β-sheet, characteristic of amyloid, within 1 h. In contrast, IAPP remains predominantly α-helical even after 4 h when incubated with ADM-116 (1:0.5, IAPP:ADM-116). α-Helical structure within the membrane-binding subdomain of IAPP has been mapped to segments containing most of the first 22 residues of IAPP[27,28]. Within this stretch, the sequence variant of IAPP from rat, rIAPP, varies only at position 18 (H18R). Binding of ADM-116 by rIAPP was not detectable (Fig. 5b). An additional five residues differ between hIAPP and rIAPP over residues 23–29. Amyloid nucleation in full-length hIAPP is mediated, in part, by residues 20–29 (ref. 29). The kinetic profile of amyloid formation by 200 μM of the 10-residue sub-peptide, $IAPP_{20-29}$, gives a t_{50} of 6.6 ± 0.3 h (Supplementary Fig. 8), consistent with our earlier work[22]. The presence of 1:1 ADM-116 has little effect on the independent amyloid assembly of $IAPP_{20-29}$. Thus, ADM-116 does not functionally interact with human residues 20-29 nor does it detectably bind to rIAPP. Combined with the observation of membrane-bound α-helical stabilization, this suggests that ADM-116 binds to the α-helical subdomain of IAPP.

Specificity of intracellular binding. To gain mechanistic insight, ADM-116 was applied to INS cells and visualized by confocal microscopy. ADM-116$_F$ at non-rescue (200 nM)

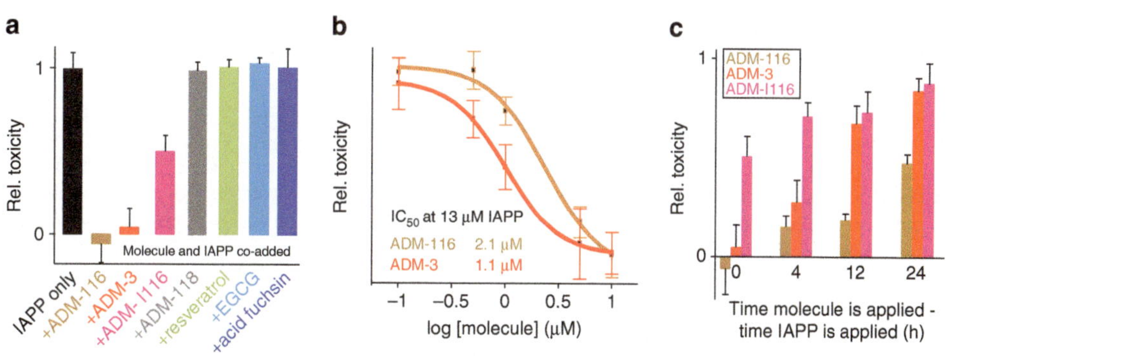

Figure 4 | Rescue of INS-1 cells from IAPP-induced toxicity. (**a**) Statistics of the toxic effect of 13 μM IAPP applied at time zero to INS-1 cells and measured 48 h later using CellTiter Blue. Data are shown for IAPP alone and with equimolar ratio of the indicated small molecule co-added at a 1:1 ratio. Data are renormalized to the toxicity induced only by IAPP. Each condition is the average of eight on-plate repeats from each of three independently performed replicates ($n = 24$). (**b**) Dose dependence of toxic rescue by ADM-3 and ADM-116. (**c**) Relative toxicity of 13 μM IAPP in which a time delay is introduced between the addition of IAPP and addition of molecule. (**a–c**) Error bars are s.d.'s from three sets of experiments conducted on separate occasions. rel., relative.

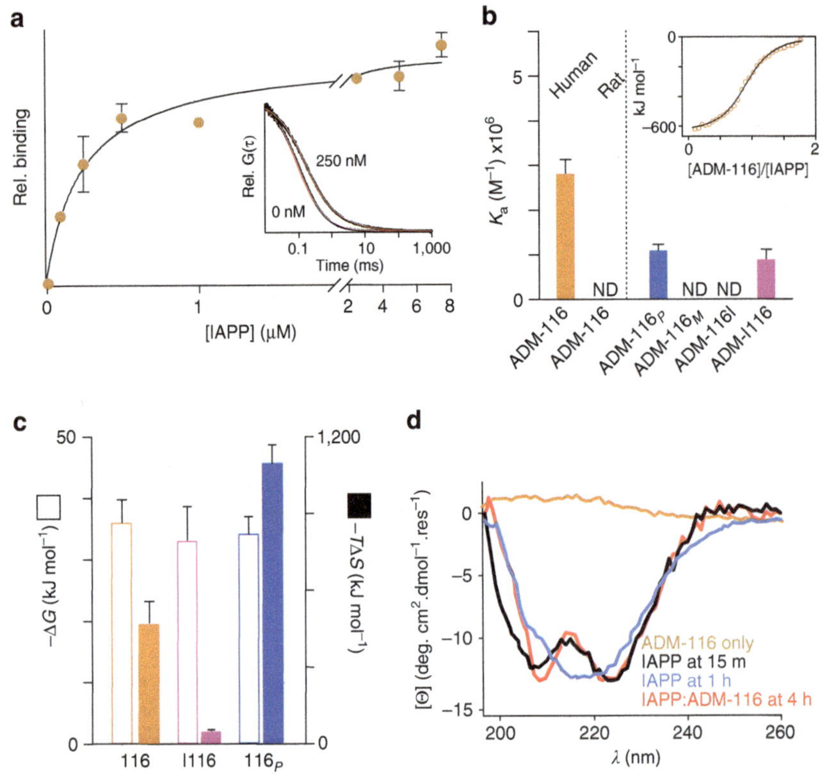

Figure 5 | Small molecule:IAPP binding and structural effects. (**a**) Relative binding of ADM-116 to IAPP using 25 nM ADM-116$_F$, and measuring diffusion by fluorescence correlation spectroscopy as a function of [IAPP]. Inset: representative raw autocorrelation data for ADM-116$_F$ in the presence of 0 nM and 250 nM IAPP. Fits using a single diffusing species for data at 0 nM IAPP and two diffusing species at 250 nM IAPP are shown in red. The data shown are an average of three experimental repeats. (**b**) Association constants for the indicated small molecule with human IAPP measured by ITC. Note, undetectable binding indicated with 'ND'. Inset: representative raw data and fit for ADM-116 binding plotted as a function of stoichiometry. (**c**) Free energy and entropy contributions to binding by ITC for the indicated compounds. (**d**) Far-ultraviolet CD spectra of 25 μM IAPP in the presence of LUVs (1.2 mM in monomeric units, DOPG:DOPC, 1:1, 100 nm) expressed as mean residue ellipticity (MRE). Shown are two time points for IAPP alone (blue, black) and with addition of stoichiometric ADM-116 at $t = 0$. ADM-116 without IAPP is shown as control (orange). rel., relative; ND, not detected.

and rescue (200 nM + 15 μM ADM-116) concentrations is able to penetrate the cell membrane (Fig. 6a). In contrast, fluorescein-labelled ADM-3 (ADM-3$_F$) showed no capacity to populate the cell interior (Fig. 6a). Approximately, 1,800 peptides have been shown to behave as cell-penetrating peptides (CPPs) capable of not only crossing the membranes but facilitating the transport of cargo such as a fluorophore[30]. These peptides tend to be short (10–20 amino acids), strongly cationic (5–10 basic amino acids) and require concentrations > 10 μM for cell penetration to occur without cellular assistance[31]. Cationic oligoquinolines have recently been shown to have properties in keeping with most CPPs[32]. For example, an octaquinoline with eight positive charges, but not a +4 tetraquinoline penetrates cells with energy-independent penetration reported at 100 μM. Here to mimic a cell surface devoid of active endocytosis, giant plasma membrane-derived vesicles (GPMVs) were prepared from live INS-1 cells[12,33]. ADM-116, and not ADM-3, shows a clear capacity to penetrate (Fig. 6b). This is remarkable as ADM-116 is anionic (−2), devoid of positive charges, and yet capable of crossing the bilayer at 200 nM while also carrying a negatively charged fluorophore. Overall, ADM-116 likely enters cells via both active- and energy-independent pathways. The mechanism for the latter, however, appears to be fundamentally different than virtually all CPPs reported to date.

Cellular rescue is associated with co-localization of protein and small molecule. INS-1 cells were treated with toxic concentrations of IAPP (100 nM IAPP$_{A594}$ and 13 μM IAPP). After 20 h, rescuing concentrations of ADM-116 (200 nM ADM-116$_F$ and 15 μM ADM-116) were added. In rescued cells, IAPP co-localizes with ADM-116 and fewer IAPP aggregates are observed (Fig. 7a). An alternative mechanism operates under conditions in which IAPP and ADM-116 are co-introduced. Although IAPP and ADM-116 can separately enter cells (Figs 2 and 6a), when added together, they co-localize on the cell surface (Fig. 7b). A similar observation can be made for ADM-3 (Supplementary Fig. 9). This suggests that the interaction of ADM-116 or ADM-3 with IAPP prohibits the entry of toxic peptide into the cells. In contrast, rescue by delayed addition of ADM-116 appears to disrupt internal toxic aggregate formation.

Intracellular rescue is a consequence of direct interactions between IAPP and ADM-116. Förster resonance energy transfer (FRET) measurements were made in live cells. Here 200 nM ADM-116$_F$ was applied to INS-1 cells 20 h after 100 nM IAPP$_{A594}$. Energy transfer from ADM-116$_F$ to IAPP$_{A594}$ is readily apparent and intracellular (Fig. 7c). Importantly, statistical assessment of the FRET reveals a Gaussian distribution strongly consistent with interactions not only being close (~40 Å) but also discrete and therefore specific (Fig. 7d).

Structure dynamics and function. ADM-116 is a folded molecule with a hydrophobic core. For Food and Drug Administration-approved drugs, measured and calculated octanol-water partition coefficients are closely comparable (Fig. 6c). The predicted and experimental log P for ADM-3 are

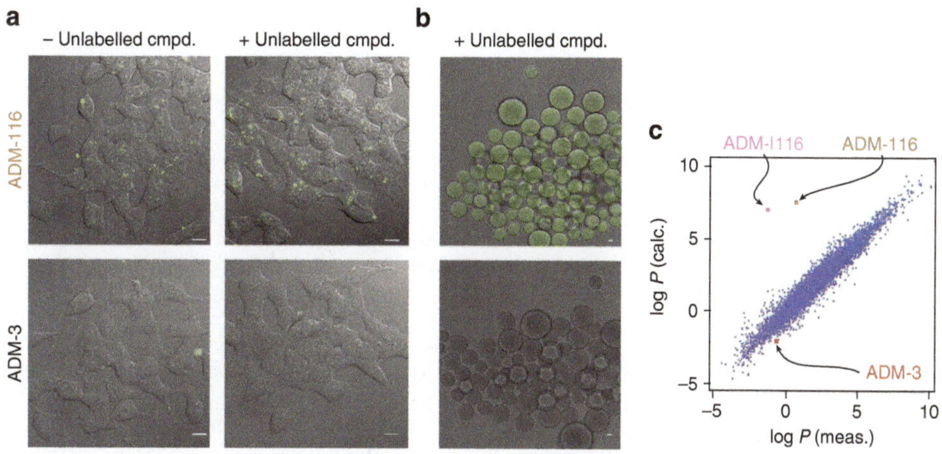

Figure 6 | Membrane translocation and partitioning. (**a**) Confocal fluorescence imaging of INS-1 cells incubated for 12 h with 200 nM ADM-116$_F$ or 200 nM ADM-3$_F$. Conditions shown are with and without addition of a further 15 μM of unlabelled small molecule. (**b**) GPMVs prepared from INS-1 cells incubated for 24 h with 200 nM ADM-116$_F$ or ADM-3$_F$ mixed with 10 μM of unlabelled compound. Scale bars, 5 μm (**a**,**b**). (**c**) Calculated versus measured octanol:water partition coefficients for three molecules used in this work (prepared using MolInspiration, http://www.molinspiration.com). United States approved pharmacopeia shown as blue dots. calc., calculated; cmpd., compound.

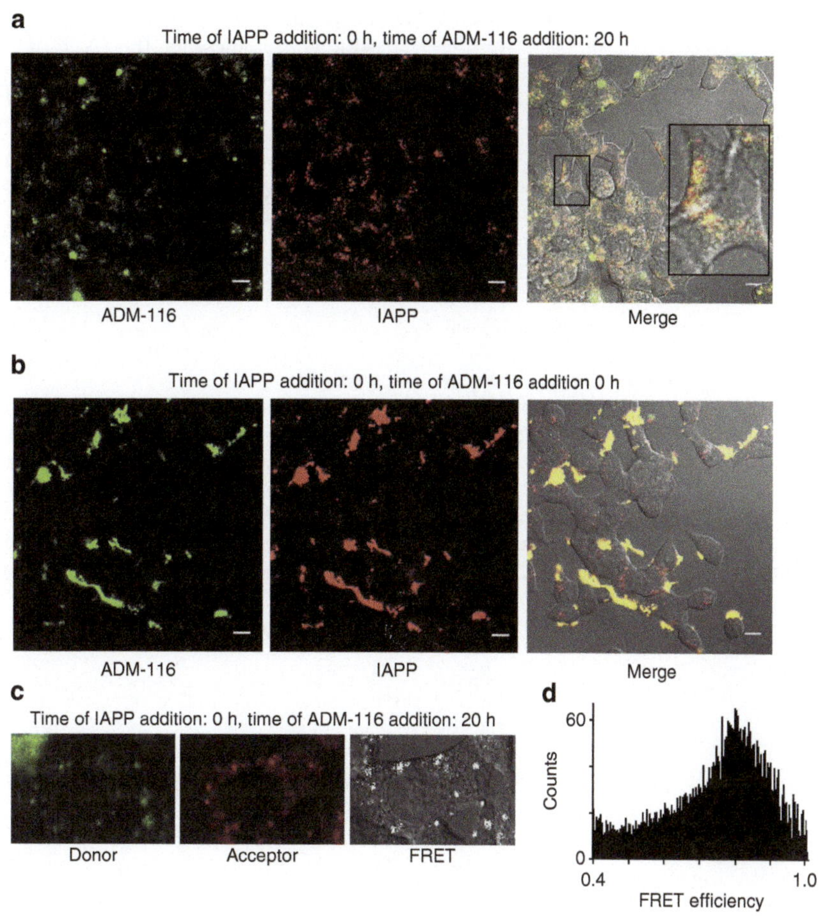

Figure 7 | Co-localization of IAPP and ADM-116. (**a**,**b**) INS-1 cells were exposed to a rescued condition doped with fluorescent variants of protein and ligand (13 μM IAPP, 15 μM ADM-116, 100 nM IAPP$_{A594}$ and 200 nM ADM-116$_F$). In **a**, small molecule and IAPP are co-introduced to the culture media. In **b**, the ligand is added 20 h after protein. Scale bars, 10 μm. (**c**) as in **b**, but without the unlabelled components and initial incubation with IAPP was performed for 18 h. Confocal fluorescence images were collected at the acceptor's emission wavelength using donor (left) or acceptor excitation light (middle). FRET was computed from these channels and is shown as white dots on a DIC image of a representative cell. Only areas showing a FRET efficiency of >0.4 are shown as values below this are indistinguishable from background. Background FRET was determined using parallel experiments that included a further 13 μM of unlabelled IAPP (Supplementary Fig. 11). (**d**) FRET histogram depicting the total counts at the indicated FRET efficiencies across ~50 regions of interest. The data shown have been repeated three times. The FRET efficiency calculated is an average of fifty cells imaged, performed across three independent biological replicates.

− 0.7 and − 2.0, consistent with its practical solubility in buffer (Fig. 6c). For ADM-116, however, there is a seven order of magnitude disparity. The calculated log *P* is 7.6 (comparable to cholesterol), while the measured log *P* is 0.6. This anomaly is a consequence of the folded nature of ADM-116. Overall, ADM-116 is anionic (− 2) and water soluble, yet capable of passive cell membrane penetration. This suggests a paradigm for cell penetration in which dynamic conformational change of the small molecule is central to this process.

The distribution of quinoline conformations is affected by ADM-116:IAPP binding. Conformational dynamics permits compact oligoquinolines to sample an equal mixture of right-(*P*) and left-(*M*) mirror-image helices[34]. The CD spectrum of ADM-116 in aqueous buffer contains equal contributions from these two hands resulting in a flat line at 390 nm (Fig. 8b). In marked contrast, the CD spectra of ADM-116 recorded in the presence of stoichiometric IAPP (Fig. 8a) shows positive ellipticity (Fig. 8a). Binding clearly shifts the tetraquinoline equilibrium between right- and left-handed helical states.

IAPP binds preferentially to the (P) conformer of ADM-116. Following an established method[35], two analogues were synthesized in which the chiral camphanyl groups were coupled with the N terminus of ADM-116 (Fig. 1e). CD spectra confirm the induction of helical bias (Fig. 8b) with ADM-116$_P$ and ADM-116$_M$ defined as the derivatives that give positive and negative profiles at 390 nm, respectively. Importantly, the intensity of ADM-116 in the presence of IAPP is within 90% that of ADM-116$_P$ (Fig. 8a; Supplementary Fig. 10). By ITC, the K_a of IAPP to ADM-116$_P$ is $1.1 \pm 0.1 \times 10^6\,M^{-1}$; within a factor of three of ADM-116. In contrast, binding was not detectable for ADM-116$_M$ (Fig. 5b). This pattern is evident in diverse assays. For lipid-catalysed fibrillation reactions, ADM-116$_P$ is as potent an inhibitor as ADM-116, while ADM-116$_M$ inhibits fibrillation only weakly (Fig. 8c). In cell toxicity assays, ADM-116$_P$ shows ~2.4 times greater inhibition of toxicity than ADM-116$_M$ (Fig. 8c). Thus, in solution biophysical assays and in toxic rescue, IAPP:ADM-116 interactions are stereospecific.

IAPP binding to ADM-116 includes interactions with the C-terminal end of the ADM-116 helix. Enantiomeric interconversion in oligoquinolines is more restricted with increasing length[35,36]. Here two molecules were synthesized in which ADM-116 was lengthened by one subunit either at the N-terminal (ADM-|116) or C-terminal (ADM-116|) end. The lateral surfaces of ADM-|116 and ADM-116| necessarily include the surface of ADM-116 (Fig. 1d,e). If IAPP:ADM-116 binding includes contact with the ends of the quinoline helix, then binding to one of the two variants should be more strongly affected. The K_a of IAPP:ADM-|116 is similar to ADM-116 ($K_a = 0.9 \pm 0.2 \times 10^6\,M^{-1}$). This binding induces a random coil to α-helix transition comparable to ADM-116 (Supplementary Fig. 10b). Note also that the binder, ADM-116$_P$, is also an N-terminal derivative. In marked contrast, no binding is detected for ADM-116|. This indicates that IAPP interacts with the C-terminal end of ADM-116.

Intracellular toxic rescue by ADM-116 is dependent on oligoquinoline length. In toxicity studies, the co-addition of ADM-|116 and IAPP results in diminished toxicity relative to IAPP alone (Fig. 4a). However, if ADM-|116 is added to INS-1 cells after IAPP is internalized, rescue is no longer observed (Fig. 4c). In other words, the addition of a fifth quinoline to ADM-116 results in a molecule that retains activities comparable to ADM-3, that is, a compound that is an effective antagonist provided IAPP and small-molecule encounter each other in the extracellular environment.

Differences in time-dependant rescue by ADM-|116 and ADM-116 can be mapped to the thermodynamic stabilities of the molecules. We observe the $\Delta\Delta G$ of IAPP binding to ADM-|116 versus ADM-116 to be $\sim 3\,kJ\,mol^{-1}$ (Fig. 5c; Supplementary Table 1). Strikingly, changes to entropic contributions ($\Delta T\Delta S$) are $\sim 400\,kJ\,mol^{-1}$ (Fig. 5c). The large difference in entropic contributions resembles a hydrophobic effect similar to observations for globular protein folding. Structural assessment of the IAPP:ADM-|116 complex by CD shows weak ellipticity at 390 nm after only ~2.5 h compared with IAPP:ADM-116. This level of ellipticity persists even after days of incubation (Fig. 8a), indicating that equilibrium has been reached and therefore a reduction in the capacity to shift the enantiomer equilibrium. Note also that as a complex, IAPP:ADM-116 loses cell-penetrating function (Figs 6a and 7b). Taken together, these observations suggest that ADM-116 exists predominantly in its folded state, but is able to sample partially folded states, a subset of which, are relevant to its membrane crossing activities.

Discussion

In this work, we have shown oligoquinoline-mediated perturbation of an intracellular pre-amyloid toxin resulting from stabilization of a weakly sampled conformation. IAPP is not only a target relevant to the pathology of diabetes but also an exemplar of intrinsically disordered proteins (IDPs) and peptide subdomains of larger systems. The widespread importance of IDPs to areas spanning cell signalling and pathological self-assembly make them a key target for basic and applied biomedical research[37,38]. IDPs do not have stable, well-defined

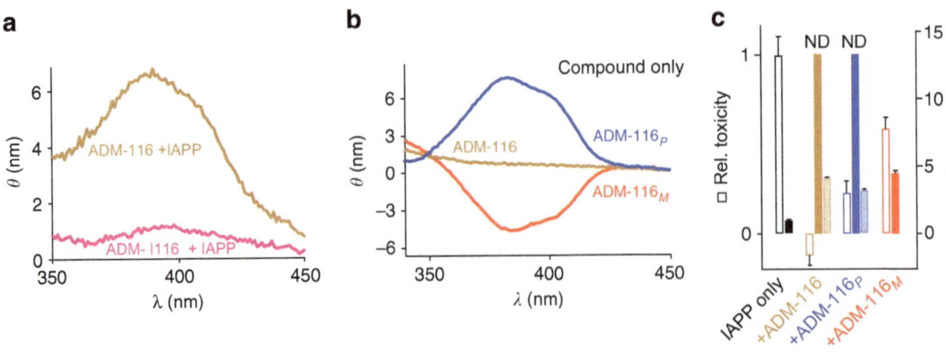

Figure 8 | Selection, perturbation and activity of ADM-116 enantiomers. (**a**) Visible CD of 25 μM ADM-116 or ADM-|116 in the presence of 25 μM IAPP. (**b**) Visible CD of ADM-116 and each of two chiral variants designed to introduce bias into the ADM-116 enantiomer distribution. (**c**) Comparison of the fibre formation inhibition (closed) and toxic rescue (open) activities of enantiomer-biased constructs of ADM-116. Experimental conditions match those used in Fig. 3 and Fig. 4, respectively. Fibre formation inhibition also shown for inhibition of reactions conducted at 10:1 IAPP:small molecule (hash).

structures. Rather, their dynamic nature enables promiscuous yet specific binding to other cellular proteins. The importance of effecting change to an IDP structural ensemble is exemplified by a recent report on 4E-BP2, a functional repressor of translation initiation[39]. In this protein, the conformational ensemble is manipulated into a partially ordered state by phosphorylation. By analogy, the function of any IDP might be controlled using an appropriately derivatized, cell-penetrating tetraquinoline. Once inside the cell, the molecule would regulate the IDP by manipulating its structural transitions.

Oligoquinolines represent a design paradigm that specifically capitalizes on the plastic nature of IDPs. Rather than design a small molecule to complement the surface of a known protein structure, a compact and folded small-molecule structure is used to induce and/or stabilize a subset of peptide conformations. State-of-the-art docking approaches show that it is possible to meaningfully identify fragment moieties suitable for specific docking with computed or measured IDP substates. Recent approaches have targeted α-synuclein from Parkinson's[40], Aβ from Alzheimer's[41] and IAPP[42]. Such approaches are complementary to the presented work as the R-groups of oligoquinolines can be modified to include such identified moieties.

Naturally occurring folded proteins form specific interactions with biotic and abiotic targets in diverse contexts. Each fold represents a topology with tolerance to alterations in its solvent-exposed residues. The immunoglobulin fold alone has a theoretical surface diversity of 10^{12}. The separation, albeit imperfect, of side-chain components relevant to folding from those responsible for molecular recognition contributes to this capacity. The oligoquinoline scaffold adopts a singular fold with as few as four subunits with complete separation of folding and molecular recognition determinants. The significance is that this is a class of compounds that as a result of an exploitable combination of structure and dynamics, can penetrate cellular membranes, transport cargo, and finally localize and perturb intracellular targets.

Methods

Materials. Thioflavin T (ThT) was purchased from Acros Organics (Fair Lawn, NJ, USA), lipids (dioleoylphosphatidylglycerol (DOPG) and dioleoylphosphatidylcholine (DOPC)) from Avanti Polar Lipids, Inc. (Alabaster, AL, USA), 96-well plates (black, w/flat bottom) from Greiner Bio-One (Monroe, NC, USA), silica plates (w/UV254, aluminium backed, 200 μm) and silica gel (standard grade, particle size = 40–63 μm, 230 × 400 mesh) from Sorbent Technologies (Atlanta, GA, USA), dry solvents from Sigma Aldrich (St Louis, MI, USA) or VWR (Bridgeport, NJ, USA), 2,6-dichloro-3-nitropyridine, alkyl iodides, triethylamine (dry), 2-chloro-1-methylpyridinium iodide, *tert*-butyl bromoacetate, trifluoroacetic acid (TFA) and triethylsilane from Sigma Aldrich, and IAPP from Genscript (Piscataway, NJ, USA) and Elim Biopharmaceuticals (Hayward, CA, USA). IAPP was repurified and handled in-house as follows: IAPP (~2 mg) was solubilized in 7 M guanidinium hydrochloride. The solution was filtered (0.2 μm) and transferred to C-18 spin column, washed twice with water (400 μl each), followed by 10% acetonitrile in water and 0.1% formic acid (v/v), and then eluted into 200 μl of 50% acetonitrile in water and 0.1% formic acid (v/v). The concentration of IAPP was calculated at 280 nm ($\varepsilon = 1{,}400\ M^{-1}\,cm^{-1}$). The IAPP solution was divided into several aliquots (20–50 μl, 1–2 mM), lyophilized and stored as a white solid at −80 °C. A fresh stock solution of IAPP was prepared in water from lyophilized aliquots for each experiment. Alexa-594-labelled IAPP was prepared as described previously[10].

Unilameller vesicles. Unless otherwise stated, LUVs used in this work were prepared from a 1:1 mixture of DOPG/DOPC. The lyophilized mixture was hydrated in 100 mM KCl, 50 mM sodium phosphate and pH 7.4 for 20 min. A 20 mg ml^{-1} solution of lipid in buffer was passed 21 times through a polycarbonate membrane (pore diameter = 100 nm). The phospholipid content of the final material was measured using a total phosphorous assay[43].

Kinetic assay. Amyloid reactions were conducted in buffer containing liposome (630 μM lipid) and 20 μM ThT in a black 96-well plate. This was followed by the addition of small molecule dissolved in dimethylsulphoxide (DMSO; final DMSO concentration = 0.5%, v/v). Fibre formation was initiated by the addition of IAPP stock solution. The final volume in each well was 200 μl. Kinetics of fibrillation was monitored by ThT fluorescence (excitation 450 nm and emission 485 nm), using a FluoDia T70 fluorescence plate reader (Photon Technology International, Edison, NJ, USA). The data were blank subtracted and renormalized to the maximum intensity of reactions containing only IAPP. Each kinetic trace was fit to a sigmoidal form:

$$I = \frac{(b_2 + m_2 t) + (b_1 + m_1 t)e^{\frac{t_{50} - t}{\tau}}}{1 + e^{\frac{t_{50} - t}{\tau}}}, \qquad (1)$$

where I is the fluorescence intensity, t is time, and b, m and τ are dependent fitting variables. All samples were run at least in triplicate and error bars shown in the text represent ±1 s.d.

Transmission electron microscopy. IAPP (10 μM) was incubated in buffer (100 mM KCl, 50 mM sodium phosphate, pH 7.4) with 630 μM LUVs in the presence and absence of 1:1 IAPP:small molecule. Aliquots were incubated for 60 s on glow-discharged (25 mA and 30 s) carbon-coated 300-mesh copper grids. After drying, grids were negatively stained for 60 s with uranyl acetate (2%, w/v). Micrographs were taken on a Phillips Tecnai 12 transmission electron microscope (Hillsboro, Oregon, USA) at 120 kV accelerating voltages. All conclusions drawn from images in this work include at least one repeat in which the sample identity was withheld from the investigator preparing and analysing images.

Isothermal calorimetry titration. ITC experiments were conducted in a NANO-ITC (TA instruments, New Castle, DE, USA). Solutions of small molecules (100 μM in 1 mM KCl, 20 mM Tris, pH 7.4) was serially titrated (50 μl injections via rotary syringe) into an isothermal sample cell containing 10 μM IAPP (250 μl in 1 mM KCl, 20 mM Tris, pH 7.4) with a stirring speed of 300 r.p.m. Ten-second injections were spread 240 s apart. The heat associated with each injection (oligoquinoline:IAPP) was extracted by integrating area under each curve using NanoAnalyze software (New Castle, DE, USA). The heat of small molecule binding to IAPP was corrected by subtracting heat of oligoquinoline injections into buffer. The final corrected heats were plotted as a function of molar ratio (oligoquinoline:IAPP) and fitted with a one binding site model.

Circular dichroism. CD spectra were collected on AVIV MODEL 215 (AVIV Instruments, Inc., Lakewood, NJ, USA) for 25 μM IAPP mixed with 1200 μM LUVs in 100 mM KCl and 50 mM sodium phosphate at pH 7.4. The data were collected from 200 to 260 nm at 0.5-nm intervals with 10 s averaging time and an average of four repeats. The CD spectra in the presence of oligoquinolines were recorded as above except collecting from 200 to 500 nm at a stoichiometric ratio of 1:1 (oligoquinoline:IAPP).

Synthesis of ADM-116. To a solution of compound **9** (88 mg, 0.1 mmol, Supplementary Fig. 60) in ethylacetate (10 ml), Pd/C (15 mg) was added and the reaction mixture was started with constant stirring in the atmosphere of H_2 (g, 1 atm.) at room temperature. The progress of the reaction was monitored using thin-layer chromatography. The disappearance of the starting material confirms the completion of the reaction (10 h). The reaction mixture was filtered and the filtrate was dried over rotovap to afford the desired product as a yellow solid, which is used in next step without further characterization.

To a solution of the reduced analogue of compound **9** (51 mg, 0.06 mmol, Supplementary Fig. 60) in dichloromethane (10 ml), triethylamine (20 mg, 0.2 mmol, anhydrous) and 2-chloromethyl-1-methyl pyridinium iodide (15 mg, 0.06 mmol) were added and the reaction was refluxed for 20 min at 50 °C under inert atmosphere. To this solution, compound **3** (17 mg, 0.05 mmol, Supplementary Fig. 59) was added and reaction started with constant stirring at 50 °C under inert atmosphere. The reaction mixture was stirred for 12 h after which the volatiles were removed on rotovap. Flash chromatography (0–35% ethylacetate in hexane, v/v) yielded the desired product (compound **10**) as a yellow solid (54 mg, 83%).

To a solution of compound **10** (44 mg, 0.04 mmol) in dichloromethane (2 ml), triethylsilane (0.1 ml, excess) and trifluoroacetic acid (0.5 ml, excess) were added and the reaction solution was stirred at room temperature for 4 h. The reaction mixture was dried and washed with cold diethyl ether (4 × 3 ml) that resulted in a light brown solid as compound **ADM-116** (35 mg, 89%).

Synthesis of ADM-116$_F$. To a round bottom flask *tert*-butyl ADM-116 (30 mg, 0.027 mmol; Supplementary Fig. 63) was added, followed by the addition of ethylene diamine (2 ml, excess) and 4-dimethylaminopyridine (catalytic amount). The reaction mixture was stirred overnight at 50 °C under inert atmosphere. The reaction mixture was then poured into water (10 ml) that resulted in a white precipitate. The precipitate was dissolved in ethylacetate (10 ml) and washed with 0.1 N HCl (3 × 10 ml) and then dried over Na_2SO_4. The compound was then dissolved in dimethylformamide (3 ml, anhydrous), followed by the addition of 4-dimethylaminopyridine (0.25 ml, excess) and fluorescein-5-isothiocyanate

(12 mg, 0.03 mmol). The reaction ran overnight in dark with constant stirring under inert atmosphere. The solvents were evaporated on rotovap. The orange solid was dissolved in dichloromethane (2 ml), followed by the addition of triethylsilane (0.1 ml, excess) and trifluoroacetic acid (0.4 ml, excess). The solution was stirred in dark for 4 h and then dried on rotovap. The product was washed with cold diethyl ether (3 × 3 ml). The reaction mixture was loaded on a semipreparation reverse-phase high-performance liquid chromatography (Varian ProStar with VYDAC reverse-phase C-18 column) to purify the product. The desired product (**ADM-116$_F$**) was obtained as an orange solid (3.5 mg, 25%, overall yield for three steps).

Synthesis of ADM-3. Synthesis and characterization of ADM-3 are reported elsewhere[20].

Synthesis of ADM-3$_F$. To a solution of *tert*-butyl ADM-3-NCS (20 mg, 0.027 mmol, Supplementary Fig. 64) in pyridine (5 ml, anhydrous), *N*, *N*-diisopropylethylamine (6 µl, 0.06 mmol) was added and the solution was stirred for 10 min. To this solution, 5-(aminoacetamido) fluorescein (22 mg, 0.054 mmol) was added and the reaction was started in dark with continuous stirring under inert atmosphere. The reaction solution was stirred overnight in dark. The product was purified using column chromatography (0–20% methanol in dichloromethane with 0.5 % triethylamine, v/v) as an orange solid (23 mg, 76%).

To a solution of *tert*-butyl ADM-3$_F$ (20 mg, 0.017 mmol, Supplementary Fig. 64) in dichloromethane (4 ml), triethylsilane (0.1 ml, excess) was added, followed by the addition of trifluoroacetic acid (0.4 ml, excess) and the reaction solution was stirred in dark at room temperature for 4 h. The solution was then dried and the orange solid was washed with cold diethyl ether (3 × 5 ml), which afforded the desired product (**ADM-3$_F$**) as an orange solid (13 mg, 72%).

Small-molecule characterization. Final steps of small-molecule purification was conducted by high-performance liquid chromatography using a Varian ProStar with VYDAC reverse-phase columns (4.6 × 100 mm, 1 ml min^{-1}, analytical; 10 × 100 mm, 3 ml min^{-1}, semi-preparation; Supplementary Fig. 12–58; Supplementary Methods). The mobile phase was composed of A: 5% ACN; 95% H_2O; and 0.1% TFA(v/v), and B: 95% ACN; 5% H_2O; and 0.08% TFA (v/v). The solution NMR spectra of small molecules were recorded on 400, 500 and 600 MHz Agilent spectrometers. The deuterated solvents used for O-*tert*-butyl ester protected and deprotected (O-COOH) oligoquinolines were $CDCl_3$ and $(CD_3)_2SO$, respectively. Splitting patterns that were difficult to interpret are indicated as multiplet (m) or broad (b). Mass spectra were obtained using either MALDI-TOF Voyager DE Pro (Yale University, CBIC centre) or University of Illinois Mass Spectrometry Facility. High-resolution electrospray ionization mass spectra were obtained using the Waters Synapt G2-Si ESI MS mass spectrometer (Milford, MA, USA).

Fluorescence correlation spectroscopy. Fluorescence correlation spectroscopy measurements were made on a laboratory-built instrument based around an inverted microscope using an Olympus IX71 microscope (Olmpus, Tokyo, Japan), as described previously[44]. In brief, a continuous-emission 488-nm diode-pumped solid-state 50-mW laser was set to 5-20-mW output power and further adjusted with neutral density filters to 18 µW of power just before entering the microscope. Fluorescence was collected through the objective and separated from the excitation laser using a Z488rdc long-pass dichroic and an HQ600/200m band-pass filter (Chroma, Bellows Falls, VT, USA). Fluorescence was focused onto the aperture of a 50-µm optical fibre coupled with an avalanche photodiode (Perkin Elmer, Waltham, MA, USA). A digital correlator (Flex03LQ-12; Correlator.com, Bridgewater, NJ, USA) was used to generate autocorrelation curves.

Measurements were made in eight-well chambered coverglasses (Nunc, Rochester, NY, USA) that were plasma treated followed by precoating with polylysine-conjugated polyethylene glycol (PEG-PLL) to prevent ADM-116 and/or IAPP from adsorbing to chamber surfaces. Low-density PEG coating was performed by preparing a 100-mg ml^{-1} solution of PEG (MW = 2 kDa, NANOCS, Boston, MA, USA) in poly-L-lysine hydrobromide (Sigma Aldrich). Reaction was performed for 6 h in dark at room temperature, followed by overnight dialysis. Chambers were incubated overnight with PEG-PLL solution, rinsed thoroughly with Millipore water and stored in water before use. All samples were incubated in buffer (20 mM Tris, pH 7.4, 10 mM NaCl) for 1 h before taking measurements. Titration was performed keeping constant concentrations of ADM-116$_F$ (25 nM) and adding increasing concentrations of IAPP. The autocorrelation curves were collected at regular intervals (10 min), and each autocorrelation curve was collected over 10 s and repeated 30 times.

Autocorrelation curves were fitted using Matlab (The MathWorks, Nattick, MA, USA). Global analyses were performed on IgorPro (Wavemetrics, Portland, OR, USA). For ADM-116, the model for a single diffusing species undergoing three-dimensional Brownian diffusion with a triplet state is given by equation (2) (ref. 45).

$$G(\tau) = \frac{1}{N(1-T)} \times \left(1 - T + Te^{\left(\frac{-t}{\tau_{\text{triplet}}}\right)}\right) \times \left[1 + \frac{\tau}{\tau_{d,1}}\right]^{-1} \times \left[1 + \frac{\tau}{s^2\tau_{d,1}}\right]^{-\frac{1}{2}} \quad (2)$$

Here N is the number of ADM-116$_F$ molecules in the detection volume, T is the fraction of molecules in the triplet state and τ_{triplet} is the triplet-state relaxation time. The characteristic translational diffusion time of a diffusing particle is given by $\tau_{d,1}$.

In the presence of IAPP, the model for two-component analysis is given by equation (3) (ref. 45).

$$G(\tau) = \frac{1}{N(1-T)} \times \left(1 - T + Te^{\left(\frac{-t}{\tau_{\text{triplet}}}\right)}\right) \times \left[r \times \left[1 + \frac{\tau}{\tau_{d,1}}\right]^{-1} \times \left[1 + \frac{\tau}{s^2\tau_{d,1}}\right]^{-\frac{1}{2}} + (1-r) \times \left[1 + \frac{\tau}{\tau_{d,2}}\right]^{-1} \times \left[1 + \frac{\tau}{s^2\tau_{d,2}}\right]^{-\frac{1}{2}}\right], \quad (3)$$

where r is the fraction of the fast-diffusing component with a diffusion time $\tau_{d,1}$, whereas $\tau_{d,2}$ is the diffusion time of the slow component. The structure factor, s, was determined as a free parameter for solutions of free Alexa Fluor 488 hydrazide dye and then fixed to the experimentally determined value of 0.17 for all subsequent fittings. For experiments in the presence of IAPP, global analysis was performed by fixing the predetermined values for the diffusion coefficient, triplet diffusion time and amplitude for ADM-116. The triplet state of IAPP-bound and -unbound ADM-116 was considered to be the same.

Confocal microscopy. Images were obtained in eight-well NUNC chambers (Thermo Scientific, Rochester, NY, USA) seeded with 20,000–25,000 cells per well. After culturing for 48 h, the medium was replaced with the medium containing constituents according to the experiment performed. For time-dependent localization experiments of IAPP, the medium contained 100 nM IAPP$_{A594}$, 13 µM unlabelled peptide and incubated for the specified time points. For experiments in the presence of ADM-116$_F$ and ADM-3$_F$, additional fluorescein-labelled and -unlabelled small molecules, 200 and 13 µM, respectively, was introduced in the medium. For delayed addition experiment with small molecules, the medium was removed for the second time, replacing with medium containing the small molecule. Images were acquired after 48-h total incubation time. Imaging was carried out at the Yale Department of Molecular, Cellular and Developmental Biology imaging facility on a Zeiss LSM 510 confocal microscope, using a × 63 Plan-Apo/1.4-NA oil-immersion objective with DIC capability (Carl Zeiss, Oberkochen, Germany). For all experiments reporting on the uptake of labelled IAPP, the gain setting for the red channel was kept constant from sample to sample. Image acquisition and processing were achieved using Zeiss Efficient Navigation and Image J software.

Imaging FRET. The INS-1 growth media was then replaced with media containing 100 nM IAPP$_{A594}$ and incubated for 18 h. Media was then replaced a second time with media containing 200 nM ADM-116$_F$. Images were taken after 5-h incubation. Background FRET was determined using parallel experiments where 100 nM IAPP$_{A594}$ was initially incubated for 18 h in cells in the presence of a further 13 µM of unlabelled IAPP. Media was then replaced with media containing 200 nM ADM-116$_F$. Imaging was carried out at the Yale Department of Molecular, Cellular and Developmental Biology imaging facility on a Zeiss LSM 510 confocal microscope, using a × 100 Plan-Apo/1.4-NA oil-immersion objective with DIC capability (Carl Zeiss). For ADM-116$_F$, fluorescein was excited with a 488-nm Argon2 laser and detected through a 505–550-nm emission filter. IAPP$_{A594}$ was excited with a 561-nm Argon2 laser and detected through a 590–630-nm emission filter. For all experiments, the pinhole was kept constant to the Z-slick thickness of each filter channel. Single-cell images were obtained for donor alone, acceptor alone and donor–acceptor fusion channels. Image acquisition and processing were achieved using Zeiss Efficient Navigation and Image J software[46]. The Image J plugin, RiFRET[44], was used to calculate and remove the bleed through for each channel and to calculate a pixel-based FRET efficiency. The FRET distance was then calculated using:

$$E = \frac{R_0^6}{R_0^6 + r^6} \quad (4)$$

where E is the calculated efficiency of FRET energy transfer, R_0 is the Förster distance (60 Å for fluorescein–Alexa-594 pair) and r is the distance between the donor and the acceptor.

Cell culture. Rat insulinoma INS-1 cells (832/13, Dr Gary W. Cline, Department of Internal Medicine, Yale University) were cultured at 37 °C and 5% CO_2 in phenol red-free RPMI 1640 media supplemented with 10% fetal bovine serum, 1% penicillin/streptomycin (Life Technologies, Carlsbad, CA, USA) and 2% INS-1 stock solution (500 mM HEPES, 100 mM L-glutamine, 100 mM sodium pyruvate and 2.5 mM β-mercaptoethanol). Cells were passaged on reaching ~95% confluence (0.25% Trypsin-EDTA, Life Technologies), propagated and/or used in experiments. Cells used in experiments were pelleted and resuspended in fresh media with no Trypsin-EDTA. COS-1 cells (from immortalized African green monkey kidney; CRL-1650; American Type Culture Collection) were cultured at 37 °C in 5% CO_2 in DMEM (Life Technologies) supplemented with 10% fetal bovine serum and 1% penicillin/streptomycin (Life Technologies). Once the cells reached ~95% confluence, they were passaged (using 0.25% Trypsin-EDTA; Life Technologies) into fractions and propagated or used in experiments.

Giant plasma membrane vesicle isolation. GPMVs were isolated from INS-1 cells as previously described[12,34]. In brief, cells were plated in 35-mm dishes and cultured for 48 h. Cells were washed with a 10 mM HEPES, 150 mM NaCl, 2 mM $CaCl_2$ (pH = 7.4) twice and were then exposed to 10 mM *N*-ethyl maleimide (Sigma Aldrich) for 2 h. Collected samples were then passed over a gravity-flow column (Bio-Rad) containing size-exclusion Sephacryl matrix of pore size 400-HR (Sigma Aldrich), allowing the purification of GPMVs from residual cell debris.

GPMV imaging. Images were obtained in eight-well NUNC chambers (Thermo Scientific) including 250 µl of GMPV stock solution. For experiments in the presence of ADM-116$_F$ or ADM-3$_F$, 200 nM of small molecule was incubated in the GMPV solution for 24 h at room temperature. Imaging was carried out at the Yale Department of Molecular, Cellular and Developmental Biology imaging facility on a Zeiss LSM 510 confocal microscope, using a × 63 Plan-Apo/1.4-NA oil-immersion objective with DIC capability (Carl Zeiss). For all experiments, the gain setting for the green channel was kept constant from sample to sample. Image acquisition and processing were achieved using Zeiss Efficient Navigation and Image J software[46].

Cell viability. Cell viability was measured colourimetrically using the CellTiter Blue (CTB, Promega, Madison, WI, USA) fluorescence-based assay. Cells were plated at a density 5,000 cells per well in 96-well plates (BD Biosciences, San Diego, CA). Peptide was directly introduced to each well after 48 h of culture and then further incubated for an additional 48 h. For time-dependent experiments, cells were incubated with peptide for the specified time points. After the incubation period, 20-µl CellTiter Blue reagent was added to each well and incubated at 37 °C and 5% CO_2 for 2.5–3.5 h. Fluorescence of the resorufin product was measured on a FluoDia T70 fluorescence plate reader (Photon Technology International, Birmingham, NJ, USA). All wells included the same amount of water to account for different concentrations of peptide added to sample wells. Wells that included water vehicle but not peptide served as the negative control (0% toxic), and wells containing 10% DMSO were the positive control (100% toxic). Per cent toxicity was calculated using the following equation:

$$\%\ \text{Toxicity} = 100 - \left[100 \cdot \left(\frac{<S> - <P>}{<N> - <P>}\right)\right] \quad (5)$$

Each independent variable is the average of eight plate replicates from the negative control ($<N>$), positive control ($<P>$) and samples ($<S>$). Results presented for viability experiments are an average of three such experiments conducted independently on different days. Error bars represent the s.e.m.

Apoptosis was measured colourimetrically by detecting caspase-3/7 (Caspase-Glo 3/7 Assay, Promega). Cells were plated at a density 5,000 cells per well in 96-well plates (BD Biosciences). Peptide was directly introduced to each well after 48 h of culture and then further incubated for the times specified in the main text. After the incubation period, 20-µl Caspase-Glo 3/7 reagent (containing a mixture of caspase-3/7 DEVD-aminoluciferin substrate and a proprietary thermostable luciferase in a reagent optimized for caspase-3/7 activity) was added to each well, and incubated at 37 °C and 5% CO_2 for 2 h. Fluorescence of the free aminoluciferin product was measured on a FluoDia T70 fluorescence plate reader (Photon Technology International). For protein concentration dependence measurements, carrier buffer was added as required to ensure identical volumes of protein were added to each well.

References

1. Goodman, C. M., Choi, S., Shandler, S. & DeGrado, W. F. Foldamers as versatile frameworks for the design and evolution of function. *Nat. Chem. Biol.* **3,** 252–262 (2007).
2. Choi, S. *et al.* De novo design and *in vivo* activity of conformationally restrained antimicrobial arylamide foldamers. *Proc. Natl Acad. Sci. USA* **106,** 6968–6973 (2009).
3. Denton, E. V. *et al.* A β-peptide agonist of the GLP-1 receptor, a class B GPCR. *Org. Lett.* **15,** 5318–5321 (2013).
4. Horne, W. S. *et al.* Structural and biological mimicry of protein surface recognition by α/β-peptide foldamers. *Proc. Natl Acad. Sci. USA* **106,** 14751–14756 (2009).
5. Udugamasooriya, D. G., Dineen, S. P., Brekken, R. A. & Kodadek, T. A peptoid "antibody surrogate" that antagonizes VEGF receptor 2 activity. *J. Am. Chem. Soc.* **130,** 5744–5752 (2008).
6. Chandramouli, N. *et al.* Iterative design of a helically folded aromatic oligoamide sequence for the selective encapsulation of fructose. *Nat. Chem.* **7,** 334–341 (2015).
7. Boehr, D. D., Nussinov, R. & Wright, P. E. The role of dynamic conformational ensembles in biomolecular recognition. *Nat. Chem. Biol.* **5,** 789–796 (2009).
8. Wilson, C. *et al.* Using ancient protein kinases to unravel a modern cancer drug's mechanism. *Science* **347,** 882–886 (2015).
9. Mukherjee, A., Morales-Scheihing, D., Butler, P. C. & Soto, C. Type 2 diabetes as a protein misfolding disease. *Trends Mol. Med.* **21,** 439–449 (2015).
10. Magzoub, M. & Miranker, A. D. Concentration-dependent transitions govern the subcellular localization of islet amyloid polypeptide. *FASEB J.* **26,** 1228–1238 (2012).
11. Masters, S. L. *et al.* Activation of the NLRP3 inflammasome by islet amyloid polypeptide provides a mechanism for enhanced IL-1beta in type 2 diabetes. *Nat. Immunol.* **11,** 897–904 (2010).
12. Schlamadinger, D. E. & Miranker, A. D. Fiber-dependent and -independent toxicity of islet amyloid polypeptide. *Biophys. J.* **107,** 2559–2566 (2014).
13. Westermark, P., Andersson, A. & Westermark, G. T. Islet amyloid polypeptide, islet amyloid, and diabetes mellitus. *Phys. Rev.* **91,** 795–826 (2011).
14. Magzoub, M. & Gräslund, A. Cell-penetrating peptides: small from inception to application. *Q. Rev. Biophys.* **37,** 147–195 (2004).
15. Hebda, J. A., Magzoub, M. & Miranker, A. D. Small molecule screening in context: lipid-catalyzed amyloid formation. *Protein Sci.* **23,** 1341–1348 (2014).
16. Van Meer, G., Voelkar, D. R. & Feigenson, G. W. Membrane lipids: where they are now and how they behave. *Nat. Rev. Mol. Cell. Biol.* **9,** 112–124 (2008).
17. MacDonald, M. J., Ade, L., Ntambi, J. M., Ansari, I. U. & Stoker, S. W. Characterization of phospholipids in insulin secretory granules and mitochondria in pancreatic beta cells and their changes with glucose stimulation. *J. Biol. Chem.* **290,** 11075–11092 (2015).
18. Hebda, J. A. *et al.* A peptidomimetic approach to targeting pre-amyloidogenic states in type II diabetes. *Chem. Biol.* **16,** 943–950 (2009).
19. Last, N. B. & Miranker, A. D. Common mechanism unites membrane poration by amyloid and antimicrobial peptides. *Proc. Natl Acad. Sci. U S A.* **110,** 6382–6387 (2013).
20. Kumar, S. *et al.* Islet amyloid-induced cell death and bilayer integrity loss share a molecular origin targetable with oligopyridylamide-based α-helical mimetics. *Chem. Biol.* **22,** 369–378 (2015).
21. Bram, Y. *et al.* Apoptosis induced by islet amyloid polypeptide soluble oligomers is neutralized by diabetes-associated specific antibodies. *Sci. Rep.* **4,** 4267 (2014).
22. Kumar, S., Brown, M. A., Nath, A. & Miranker, A. D. Folded small molecule manipulation of islet amyloid polypeptide. *Chem. Biol.* **21,** 775–781 (2014).
23. Kumar, S. & Miranker, A. D. A foldamer approach to targeting membrane bound helical states of islet amyloid polypeptide. *Chem. Commun.* **49,** 4749–4751 (2013).
24. Meng, F., Abedini, A., Plesner, A., Verchere, C. B. & Raleigh, D. P. The flavanol (-)-epigallocatechin 3-gallate inhibits amyloid formation by islet amyloid polypeptide, disaggregates amyloid fibrils, and protects cultured cells against iapp-induced toxicity. *Biochemistry* **49,** 8127–8133 (2010).
25. Meng, F. *et al.* The sulfated triphenyl methane derivative acid fuchsin is a potent inhibitor of amyloid formation by human islet amyloid polypeptide and protects against the toxic effects of amyloid formation. *J. Mol. Biol.* **400,** 555–566 (2010).
26. Mishra, R., Sellin, D., Radovan, D., Gohlke, A. & Winter, R. Inhibiting islet amyloid polypeptide fibril formation by the red wine compound resveratrol. *ChemBioChem* **10,** 445–449 (2009).
27. Apostolidou, M., Jayasinghe, S. A. & Langen, R. Structure of α-helical membrane-bound human islet amyloid polypeptide and its implications for membrane-mediated misfolding. *J. Biol. Chem.* **283,** 17205–17210 (2008).
28. Williamson, J. A., Loria, J. P. & Miranker, A. D. Helix stabilization precedes aqueous and bilayer-catalyzed fiber formation in islet amyloid polypeptide. *J. Mol. Biol.* **393,** 383–396 (2009).
29. Ruschak, A. M. & Miranker, A. D. The role of prefibrillar structures in the assembly of a peptide amyloid. *J. Mol. Biol.* **393,** 214–226 (2009).
30. Agrawal, P. *et al.* CPPsite 2.0: a repository of experimentally validated cell-penetrating peptides. *Nucleic Acids Res.* **43,** 1–6 (2015).
31. Kauffman, W. B., Fuselier, T., He, J. & Wimley, W. C. Mechanism matters: a taxonomy of cell penetrating peptides. *Trends Biochem Sci.* **40,** 749–764 (2015).
32. Iriondo-Alberdi, J. *et al.* Cellular internalization of water-soluble helical aromatic amide foldamers. *Chembiochem* **11,** 1679–1685 (2010).
33. Sezgin, E. *et al.* Elucidating membrane structure and protein behavior using giant plasma membrane vesicles. *Nat. Protoc.* **7,** 1042–1051 (2012).
34. Delsuc, N. *et al.* Kinetics of helix-handedness inversion: folding and unfolding in aromatic amide oligomers. *Chemphyschem* **9,** 1882–1890 (2013).
35. Kendhale, A. M. *et al.* Absolute control of helical handedness in quinoline oligoamides. *J. Org. Chem.* **76,** 195–200 (2011).

36. Jiang, H., Léger, J., Dolain, C., Guionneau, P. & Huc, I. Aromatic δ-peptides: design, synthesis and structural studies of helical, quinoline-derived oligoamide foldamers. *Tetrahedron* **59,** 8365–8374 (2003).
37. Metallo, S. J. Intrinsically disordered proteins are potential drug targets. *Curr. Opin. Chem. Biol.* **14,** 481–488 (2010).
38. Berlow, R. B., Dyson, H. J. & Wright, P. E. Functional advantages of dynamic protein disorder. *FEBS Lett.* **589,** 2433–2440 (2015).
39. Bah, A. *et al.* Folding of an intrinsically disordered protein by phosphorylation as a regulatory switch. *Nature* **519,** 106–109 (2015).
40. Tóth, G. *et al.* Targeting the intrinsically disordered structural ensemble of α-synuclein by small molecules as a potential therapeutic strategy for Parkinson's disease. *PLOS ONE* **9,** e87133 (2014).
41. Zhu, M. *et al.* Identification of small-molecule binding pockets in the soluble monomeric form of the Aβ42 peptide. *J. Chem. Phys.* **139,** 035101 (2013).
42. Nath, A., Schlamadinger, D. E., Rhoades, E. & Miranker, A. D. Structure-based small molecule modulation of a pre-amyloid state: pharmacological enhancement of iapp membrane-binding and toxicity. *Biochemistry* **54,** 3555–3564 (2015).
43. Chen, P. S., Toribara, T. Y. & Warner, H. Microdetermination of phosphorus. *Anal. Chem.* **28,** 1756–1758 (1956).
44. Middleton, E. R. & Rhoades, E. Effects of curvature and composition on α-synuclein binding to lipid vesicles. *Biophys. J.* **99,** 2279–2288 (2010).
45. Schwille, P., Haupts, U., Maiti, S. & Webb, W. W. Molecular dynamics in living cells observed by fluorescence correlation spectroscopy with one- and two-photon excitation. *Biophys. J.* **77,** 2251–2265 (1999).
46. Schneider, C. A., Rasband, W. S. & Eliceiri, K. W. NIH image to ImageJ: 25 years of image analysis. *Nat. Methods* **9,** 671–675 (2012).
47. Gillies, E. R., Dolain, C., Leger, J. & Huc, I. Amphipathic helices from aromatic amino acid oligomers. *J. Org. Chem.* **71,** 7931–7939 (2006).

Acknowledgements

We thank Professor Abhi Nath for critical reading of the manuscript. This research was supported by NIH GM094693, an American Diabetes Association mentor-based postdoctoral fellowship to D.E.S. and M.B., and an international Fulbright Scholarship to S.P.W.

Author contributions

S.K., M.B. and A.D.M. designed the project and experiments. S.K., M.B., D.E.S. and S.P.W. performed the experiments. S.K., M.B., D.E.S, S.P.W., E.R. and A.D.M analysed the data. S.K., M.B. and A.D.M wrote the manuscript.

Additional information

Competing financial interests: The authors declare no competing financial interests.

10

Generation of a synthetic GlcNAcylated nucleosome reveals regulation of stability by H2A-Thr101 GlcNAcylation

Lukas Lercher[1,*,†], Ritu Raj[1,*], Nisha A. Patel[2], Joshua Price[1], Shabaz Mohammed[1], Carol V. Robinson[2], Christopher J. Schofield[1] & Benjamin G. Davis[1]

O-GlcNAcylation is a newly discovered histone modification implicated in transcriptional regulation, but no structural information on the physical effect of GlcNAcylation on chromatin exists. Here, we generate synthetic, pure GlcNAcylated histones and nucleosomes and reveal that GlcNAcylation can modulate structure through direct destabilization of H2A/H2B dimers in the nucleosome, thus promoting an 'open' chromatin state. The results suggest that a plausible molecular basis for one role of histone *O*-GlcNAcylation in epigenetic regulation is to lower the barrier for RNA polymerase passage and hence increase transcription.

[1] Department of Chemistry, University of Oxford, Chemistry Research Laboratory, Mansfield Road, Oxford OX1 3TA, UK. [2] Department of Chemistry, University of Oxford, Physical and Theoretical Chemistry Laboratory, South Parks Road, Oxford OX1 3QZ, UK. * These authors contributed equally to this work. † present address: EMBL Heidelberg, Meyerhofstraße 1, 69117 Heidelberg, Germany. Correspondence and requests for materials should be addressed to B.G.D. (email: ben.davis@chem.ox.ac.uk).

The post-translational modification (PTM) of histones is a widespread mechanism for the regulation of eukaryotic gene expression[1]. Addition of *O*-linked β-*N*-acetylglucosamine (GlcNAc) to serine- and threonine-residues is an abundant PTM of cytosolic and nuclear proteins[2] and has recently been identified on histones[3,4]. *O*-GlcNAc sites have been detected on all canonical histones, with H2B-Ser112 *O*-GlcNAcylation being linked to transcriptional activation via promotion of H2B ubiquitinylation[5]. More generally, protein *O*-GlcNAcylation functions in nutrient sensing and is essential for development[6].

In mammals, *O*-GlcNAcylation is catalysed by a single enzyme, *O*-linked *N*-acetylglucosamine transferase (OGT). OGT interacts both with transcriptionally repressing[7] and activating[8] histone modifying complexes suggesting a context dependent role for OGT in transcriptional regulation, as occurs with other histone PTMs, for example, acetylation, methylation and phosphorylation. To date, no molecular mechanism for the transcriptional regulation by histone *O*-GlcNAcylation has been reported. To define the molecular roles of OGT in transcriptional regulation, access to homogeneously GlcNAcylated histones is required. Homogeneous *O*-GlcNAcylation is not attainable enzymatically because OGT is promiscuous[9]; *in vitro* reactions on histones yield mixtures of *O*-GlcNAcylated products[5] and attaining complete PTM is difficult. Here, we report the selective synthesis of a nucleosome containing histone H2A bearing a single *O*-GlcNAc-mimic at threonine (Thr) 101, a functionally unassigned *O*-GlcNAcylation site. Nucleosomes contain an octameric core, in which two H2A/B dimers interact on opposite sides of an H3/H4 tetramer. We show here that H2A-Thr101 GlcNAcylation reduces nucleosome stability through destabilization of the H3/H4 tetramer—H2A/B dimer interface. On the basis of these results we suggest that regulation of nucleosome stability by OGT-dependent GlcNAc transfer may contribute to transcriptional regulation.

Results

Generation of homogeneous GlcNAcylated H2A. We have reported a chemical PTM strategy[10,11], that enables generation of homogenously GlcNAcylated proteins including histone H3. Of the few reported *in vivo* *O*-GlcNAcylation sites in nucleosomes, histone H2A-Thr101 (refs 3,5) is at the H2A/B dimer and H3/H4 tetramer interface (Fig. 1a). Transcriptional regulation by PTMs is often achieved through selective binding to PTMs by effector proteins[1]. Since no selective *O*-GlcNAc-histone-binding proteins have been reported, we wanted to explore if GlcNAcylation could function as an epigenetic signal by directly affecting nucleosome structure. Thr101 modification could potentially influence nucleosome stability, by modulating dimer/tetramer association. To date, no functional consequences of H2A-Thr101 modification have been reported.

To generate homogenous GlcNAcylated nucleosomes we produced and purified a recombinant Thr101Cys variant of the *Xenopus laevis* H2A (Supplementary Fig. 1)[12]. H2A-T101C was treated with 2,5-dibromohexanediamide (DBHDA) at 37 °C for 3 h under denaturing conditions (5 M guanidine.HCl) to yield a dehydroalanine (Dha) residue at H2A-101 after di-alkylation and elimination (Fig. 1b and Supplementary Fig. 2). Following removal of excess DBHDA by desalting, GlcNAc-thiol was reacted with the Dha containing protein to yield the first homogenous H2A containing a *O*-GlcNAc-mimic at residue 101 (Fig. 1b and Supplementary Fig. 3), as shown by liquid chromatography-mass spectrometry (LC-MS). We then used the GlcNAcylated H2A to produce modified nucleosomes in a stepwise manner, that is, by initial formation of H2A/B dimers, followed by addition of the H3/H4 tetramer and the high affinity 601 DNA sequence, to give folded nucleosomes (Fig. 1c).

H2A-T101 GlcNAcylation reduces tetramer-dimer association. To investigate the effect of GlcNAcylation we analysed intermediates by size-exclusion chromatography (SEC) and circular dichroism (CD) spectroscopy. SEC gives information on size and homogeneity and is used to purify histone octamers or sub-complexes for nucleosome reconstitution[12], while CD allows comparison of secondary structure content and stability of complexes. H2A/B dimers were constructed using equimolar amounts of histones and then analysed and purified by SEC[12]. Identical SEC traces for (H2A/B)-wt (wild type) and (H2A/B)-GlcNAc indicate that H2A-101 GlcNAcylation does not substantially influence viable dimerization (Fig. 1d). The conservation of dimer structure and stability was further confirmed by variable temperature CD analysis (Supplementary Fig. 4). The obtained melting temperatures were similar with and without GlcNAcylation ((H2A/B)-GlcNAc, 50.7 ± 0.1 °C; (H2A/B)-wt, 52.9 ± 0.1 °C)[13]. Together these results suggest that H2A-101 GlcNAcylation also does not significantly influence H2A/B dimer structure and stability.

Next, synthetic histone octamers were examined. To refold octamers for nucleosome reconstitution, we combined (H2A/B)-wt and (H2A/B)-GlcNAc with a purified wt H3/H4 tetramer (Supplementary Fig. 4d) and analysed the resulting mixture by SEC. Although for the (H2A/B)-wt we observed octamer formation as anticipated, for the (H2A/B)-GlcNAc we observed only separately eluting di- and tetramers, suggesting GlcNAcylation destabilizes the octamer (Fig. 1e,f). The potential exists for the creation of D- and L-epimers at Cα of the GlcNAcylation site. To verify that observed destabilization was not due to an artefact resulting only from differing amino acid stereochemistry, we incubated the H3/H4 tetramers with an excess of modified H2A/B dimers (1.5- and 2-fold) and separated the formed complexes by SEC (Supplementary Fig. 5). Again we observed no octamer formation, but separately eluting di- and tetramers, consistent with destabilization regardless of chirality.

H2A-T101 GlcNAcylation decreases nucleosome stability. In cells histones do not exist as free octamers but are complexed by chaperones or DNA and chromatin-associated proteins[14]. To investigate the effect of GlcNAcylation on complete nucleosome structure, we refolded nucleosomes from separately purified H3/H4 tetramers and the (H2A/B)-GlcNAc dimers using the salt gradient method[12] and a 145 bp DNA construct[15]. We used the selective and tightly binding '601' nucleosome positioning sequence to obtain homogeneous nucleosomes. Using PAGE, we observed similarly migrating bands for the wt nucleosome reconstituted from wt histone octamers and the GlcNAcylated nucleosome, although the band for the GlcNAc-modified nucleosome appeared more diffuse (Fig. 2a). Again, to control for potential artefacts from stereochemistry at the GlcNAcylation site, we reconstituted nucleosomes with different tetramer/dimer ratios (1.5–5 eq dimer). PAGE analysis revealed similarly migrating species for all reconstitutions (Supplementary Fig. 6). To further probe the structure of these nucleosomes we used varying salt concentrations; it has been proposed that when ionic strength is elevated, non-native species dissociate at lower ionic strength than canonical nucleosomes[16,17]. When incubated at intermediate ionic strength (up to 600 mM KCl; Supplementary Fig. 7) both the wt and GlcNAcylated nucleosomes remained stable, indicating a native structure similar to canonical nucleosomes for both species that does not stem from unspecific non-native binding of histones to DNA.

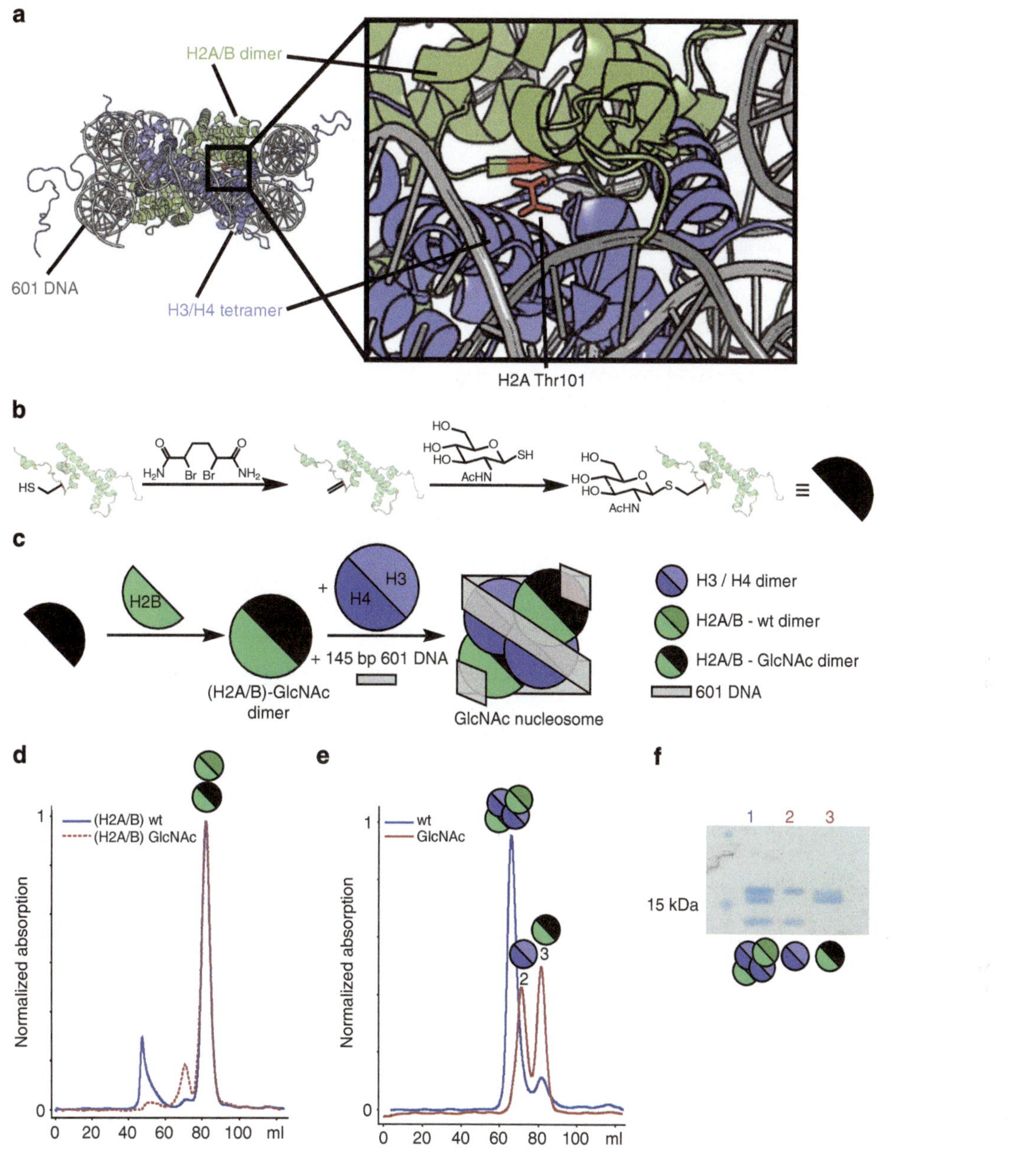

Figure 1 | Modification and refolding of nucleosome complexes with H2A/B-GlcNAc. (**a**) Graphic representation of a nucleosome structure with H2A Thr-101 in red (adapted from 1KX4)[51]. (**b**) Reaction scheme for generation of GlcNAcylated H2A. Cys-101 is converted to dehydroalanine (Dha), which functions as a Michael acceptor for reaction with the GlcNAc-thiol, so replacing the ether linkage with a structurally analogous thioether. (**c**) Schematic representation of H2A/B dimer and nucleosome refolding (wt H2A in dark green and GlcNAcylated H2A in red). (**d**) SEC traces for the refolding of H2A/B dimer from wt (blue) and H2A T101GlcNAc (red) and wt H2B. (**e**) SEC traces for the octamer refolding from wt H3/H4 tetramers and H2A/B-wt (blue) or H2A/B-GlcNAc (red) dimers. (**f**) SDS-PAGE analysis of the main fractions of the marked peaks (original image shown in Supplementary Fig. 16).

To investigate if the H2A-101 GlcNAcylation destabilises nucleosomes, we measured the thermal stability of both wt and GlcNAcylated nucleosome by differential scanning fluorimetry[18] (DSF), CD and electrophoretic analysis (Fig. 2b,c, Supplementary Figs 8–10). The DSF measurements revealed a significant reduction in melting temperature for the GlcNAcylated nucleosomes (59.6 °C (GlcNAc) versus 65.6 °C (wt)). CD analyses also supported the significant destabilization of the H2A-101GlcNAc-containing nucleosome. Although DSF reports on protein melting, by using CD both the protein and/or DNA structural transitions of the nucleosome dissociation process can be measured[19]. The CD spectrum above 250 nm originates almost exclusively from the DNA and is sensitive to structure. The CD spectra of the unmelted wt and the GlcNAcylated nucleosomes (2:1 dimer:tetramer) are near identical (Supplementary Fig. 8); at higher dimer concentrations (4:1 dimer:tetramer) the observed CD is increased above 240 nm suggesting slight changes in DNA structure. Monitoring the thermal dissociation profile for these wt and GlcNAcylated nucleosomes at the higher wavelength 260 nm, revealed a single transition for all complexes (Fig. 2). As for the observations made by DSF, the GlcNAcylated nucleosomes were significantly destabilized (Tm 2:1, 65.5 ± 0.2 °C; 4:1, 67.6 ± 0.4 °C) compared with the wt nucleosome (Tm 71.0 ± 0.3 °C;). When monitoring the thermal dissociation by CD at 220 nm, we observed apparent two-state melting profiles; the first transition for H2A-T101GlcNAc nucleosome was shifted to a lower Tm relative to that of the wt nucleosome (Supplementary Fig. 9, Tm 2:1, 65.8 ± 0.9 °C; 4:1, 66.1 ± 0.4 °C;

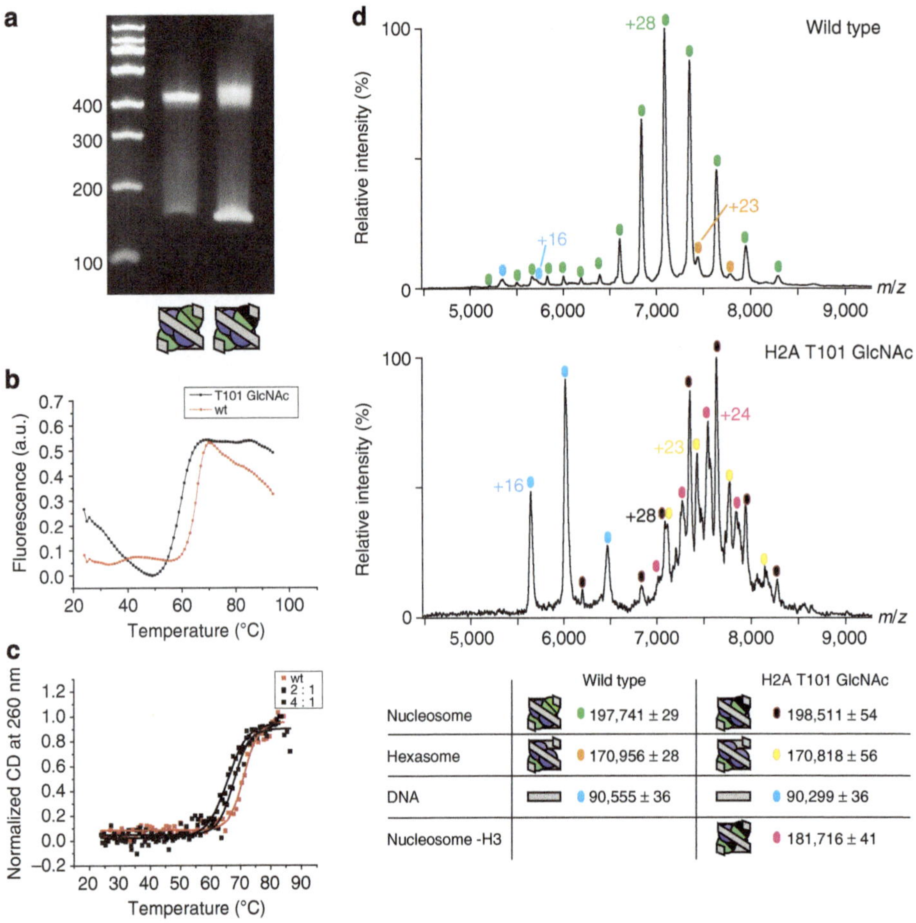

Figure 2 | Comparison of stability and composition of nucleosomes refolded using H2A/H2B-wt or H2A/H2B-GlcNAc dimers and wt H3/H4 tetramers. (**a**) 6% TBE-PAGE gel analysis of salt gradient refolding of wt (left) and GlcNAcylated (right) nucleosomes. (**b**) Melting profile of wild type and H2A Thr-101 GlcNAcylated nucleosome as assayed by DSF. Experiments were performed in duplicates. (**c**) Melting profile of wild type and H2A Thr-101 GlcNAcylated nucleosomes reconstituted with different dimer: tetramer ratios (2:1 and 4:1) as monitored by CD at 260 nm. (**d**) Non-denaturing ESI-MS analysis of wt and GlcNAcylated nucleosomes. Values reported represent the mean value and standard deviation of the observed m/z peaks.

wt, 69.7 ± 0.3 °C). After this first transition both nucleosomes behave similarly with respect to the change in CD with temperature. These observations are consistent with a first transition that corresponds to the dissociation and denaturation of the H2A/B dimer, while the second corresponds to the dissociation and denaturation of the H3/H4 tetramer. To further investigate these dissociation pathways, we incubated wt or GlcNAcylated nucleosomes at different temperatures around the measured Tm and analysed the resulting soluble fraction by SDS–PAGE (Supplementary Fig. 10). In agreement with previous assays, we observed earlier disassembly of the GlcNAcylated nucleosome, indicated by higher migrating bands in the native PAGE (Supplementary Fig. 10a). In the soluble fractions we observed selective loss of H2A and H2B from GlcNAcylated nucleosome at lower temperatures (Supplementary Fig. 10b). These results support the proposed destabilization of the H2A/B dimer in the context of the nucleosome.

GlcNAcylated nucleosomes consist of mixtures of subspecies. We then used nanoelectrospray ionization mass spectrometry (nESI-MS), under conditions that substantially preserve non-covalent interactions[20], to further characterize the GlcNAcylated nucleosome complex. nESI-MS allows accurate analysis and assignment of formed nucleosome complexes. nESI-MS analysis of wt nucleosomes revealed one major set of peaks corresponding to the intact nucleosome, and lower levels of hexasome (~16%) and free DNA (Fig. 2d). In contrast, the GlcNAcylated nucleosome (reconstituted at a 1:2 ratio of H3/H4 tetramer: H2A/H2B dimer) does not constitute a homogenous species; at least three overlapping ion series were observed (Fig. 2d). These were assigned to an intact nucleosome; a species lacking one H3 subunit and a hexasome lacking one H2A/B dimer. Although observation of a hexasome has previously been reported[20], the seven subunit containing species lacking one histone H3 has not been previously described. To test whether the observed species were possible artefacts of nucleosome reconstitution from separated H2A/B dimers and H3/H4 tetramers, we also reconstituted wt nucleosome from mixed wt H2A/B dimers and H3/H4 tetramers (Supplementary Fig. 11). These displayed a near identical behaviour when analysed by non-denaturing gel electrophoresis, CD and nESI-MS. We also analysed GlcNAcylated nucleosomes reconstituted with increasing concentrations of H2A/H2B dimer (up to 4.5 eq with respect to the H3/H4 tetramer) by nESI-MS (Supplementary Fig. 12). Although the relative amount of sub-complexes varied with H2A/H2B equivalents, all samples showed decreased amount of intact nucleosome when compared with the wt sample. In wt nucleosome structures the histone H3 α-3 helix is located close to the modified H2A-101 site. It is possible that a steric clash between this helix and the H2A-T101GlcNAc results in preferred formation of species lacking either one H2A/H2B dimer or one histone H3. Together these results support the proposal that *O*-GlcNAcylation at the H2A-101 site destabilizes nucleosomes.

These combined biophysical studies of GlcNAcylated nucleosomes suggest that PTM mediated by OGT at H2A-T101 may regulate nucleosome stability *in vivo* via modulation of H2A/B dimer/tetramer interactions. This proposal was analysed further by comparison of identified sites of open chromatin and OGT localization in the genome (Fig. 3)[21,22]. DNAse I sensitivity is a measure of chromatin accessibility and thereby nucleosome stability. A strong correlation with almost all identified OGT sites (97%, $P = 0.000999$, Monte Carlo False Discovery rate (MCFDR)) overlapping with reported DNAse I sensitive regions (Fig. 3a) was observed. Transcription start sites (TSS) for active operons contain nucleosome depleted regions, which are DNAse I sensitive; OGT was also associated with protein complexes that regulate transcription at TSS. A strong correlation with 97% of all OGT peaks overlapping with DNAse I sensitive regions at TSS was also observed. These observations are in agreement with the generation of an open chromatin state by OGT recruitment.

GlcNAcylation causes changes in the interactome. Transcriptional regulation by various PTM on the nucleosome involves recruitment of chromatin 'reader' proteins[23]. To gain further insights into the role of H2A-T101 *O*-GlcNAcylation *in vivo*, we probed the effects of nucleosome GlcNAcylation upon nucleosome interaction partners from human cells. We reconstituted nucleosomes using biotinylated DNA containing the 601 sequence[24]; these were then immobilized via biotin onto streptavidin-coated beads and used to interrogate putative nucleosome-binding proteins from HeLa cell nuclear extracts, using synthetically GlcNAcylated and wt nucleosomes ('bait' and control, respectively). The experiments were performed as two independent biological replicates. The enriched proteins were digested and analysed by nLC-MS/MS. MS-based quantitative proteomics has been extensively used to identify reader proteins for various chromatin markers[24,25]. Label-free quantification[26] enabled us to identify 20 proteins, among other identified proteins (Supplementary Data 1), which were over threefold enriched in the GlcNAcylated nucleosome sample (Supplementary Fig. 13); of these 12 (60%) can bind to DNA. Notably, these included USF1 and USF2 that bind an E-Box sequence[27] present in the DNA sequence used here. Inspection of nucleosome crystal structures[15] shows that this region is normally bound by the H2A/B dimer (Supplementary Fig. 14). Additionally, Msh3 (MutS Homolog 3), a DNA mismatch repair protein (MMR), was the most highly enriched H2A T101 GlcNAcylated nucleosome interactor along with Msh2 (MutS Homolog 2), another MMR protein, amongst the top seven enriched interacting partners. The Msh2-Msh6 heterodimer has been implicated in nucleosome remodelling that leads to disassembly of the nucleosome[28]. Together these proteomic results provide evidence of increased accessibility for proteins from a relevant cellular sample to corresponding regions of DNA (for example, USF1 or USF2 to E-Box) facilitated by GlcNAcylation, consistent with the open state resulting from nucleosome destabilization suggested by the *in vitro* biophysical and DNAse sensitivity analyses. They also suggested the putative nucleosome remodellers Msh2 and Msh3 as potential partners that might accelerate nucleosome disassembly upon H2A-T101 *O*-GlcNAcylation.

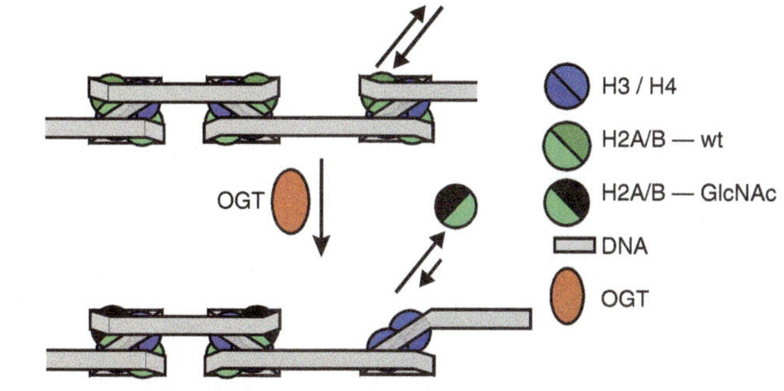

Figure 3 | Analysis of OGT function. (**a**) Genomic snapshots from the UCSC genome browser for OGT ChIP-Seq[21] and DNaseI Hypersensitivity by Digital DNaseI from ENCODE/University of Washington (ES-E14 cells)[22] in mouse embryonic stem cells. (**b**) Model for generation and maintenance of open chromatin states by OGT recruitment. When OGT is present (possibly through OGT recruitment, see Supplementary Fig. 15) H2A Thr-101 can be *O*-GlcNAcylated so causing destabilization of the H2A/H2B dimer in the nucleosome. This promotes hexasome formation and generation of an open chromatin structure.

Discussion

The destabilization of H2A/B dimers in nucleosomes is proposed as an important mechanism to facilitate transcription[29,30]. In one mode, the stability of H2A/B dimers in the nucleosome can be regulated by incorporation of histone H2A variants[31]; for example, labile histone H2A variants (H2A.Z.2.2 and H2A.Bbd) have been reported[16,32,33] that do not form stable octamers without DNA, but do form nucleosome-like species in the presence of DNA, albeit with lower stability. Reduced stability for H2A.Bbd was attributed[33] to changes in the dimer–tetramer docking domain, an interface that notably includes the H2A-T101 site that is the focus of this study. We propose here that GlcNAcylation may regulate nucleosome stability in a post-translational enzyme-mediated manner complementary to co-transcriptional histone variant incorporation, where, in essence, GlcNAcylation generates a labile variant *in situ*. Intriguingly, a third complementary PTM-mediated regulation of the tetramer-dimer interface, which employs a different tetramer-alteration mechanism, has been proposed for H4-K91 acetylation and H4-K91 ubiquitinylation[34,35]. In this third 'H4-alteration model' the same or a similar interface may be modulated but with a vital difference (and hence different mechanistic manifold for exploitation in regulation): in this case the core tetramer would be altered to encourage dissociation of the H2A/H2B dimer (the H4-alteration model), whereas our results suggest a H2A-alteration model that would instead 'trap' an unbound H2A/H2B dimer state. Notably consistent with this, upon *in vitro* OGT treatment, Thr101 modification is observed in isolated H2A, but not when H2A is present in octamers or nucleosomes[5], indicating that the enzymatic *O*-GlcNAcylation occurs selectively on histone H2A in a 'non-bound' state.

Based on these combined observations, we propose the following model connecting OGT recruitment to transcriptional regulation (Fig. 3b). H2A/B dimers are in exchange with a free H2A/B pool, especially in transcriptionally active regions where nucleosomes have to be disassembled every time the RNA polymerase passes through[36]. In the presence of OGT, free

H2A/B dimers could be O-GlcNAcylated. Speculatively, this may be enabled or enhanced by selective OGT recruitment (Supplementary Fig. 15). This would result in decreased nucleosome stability at these sites upon incorporation of O-GlcNAcylated H2A, thus providing an open chromatin state and which could increase binding of additional chromatin modifying complexes and transcription factors (such as those identified in our quantitative proteomic analyses, vide supra) or by reinforcing transcription by lowering the nucleosome barrier for RNA polymerase passage. These proposals are supported by the observation that OGT recruitment to TSS overlaps very well with DNAse I sensitivity at these promoters and by the direct identification of potential partners (for example, USF1 and USF2) from cellular samples. In addition, reduced nucleosome stability might liberate binding sites for other DNA-binding proteins, such as the MMR proteins also identified from cellular samples. It is, of course, important to note that any modification that destabilizes the nucleosome core would be expected to enrich DNA-binding proteins in this way. We also suggest that this mechanism complements both co-transcriptional H2A-variation (H2A.Z.2.2 and H2A.Bbd) and post-translational H4-alteration (H4-K91 acetylation/ubiquitinylation) to provide a third post-translational H2A-alteration/trapping mechanism. In this way, O-GlcNAcylation by OGT presents a potential catalytic mechanism for the amplification of transcriptional activation. In this speculative model, DNA-bound tetramer could remain fixed to a chromatin region thereby allowing such a low abundance modification to play a significant mechanistic role. Additionally, GlcNAcylation could either occur on a transiently 'open' nucleosome (likely under a Curtin–Hammett regime, still DNA-bound), or act on sufficient of the local free dimer pool to maintain the destabilizing effect within a defined local region. The feasibility of such a mechanism (which is speculative) is clearly subject to caveats of suitable kinetics/ lifespan of GlcNAcylation and mobility of any putative free dimer pool. Although the results suggest that OGT may be directly involved in transcriptional regulation, other processes, for example, in histone processing, recycling and localization, could be modulated by O-GlcNAcylation. Indeed, a recent report has revealed that O-GlcNAcylation can occur co-translationally in some proteins[37], although it is currently unknown if this also happens for T101 O-GlcNAcylation. The exact role of OGT in all of these processes necessitates additional studies addressing the *in vivo* function of this and other reported O-GlcNAcylation sites.

Methods

Histone and DNA preparation. Histone plasmids (pET3) encoding for canonical *Xenopus laevis* histones were a kind gift from Dr Rob Klose, Department of Biochemistry, Oxford. The H2A T101C mutation was introduced using the Quik-Change mutagenesis kit (Agilent; #200518) and was verified by sequencing. H2A_T101C_F: ctgctcggaagagtctgtatcgctcagggcgg; H2A_T101C_R: ccgccctgagcga tacagactcttccgagcag.

Histone proteins were produced in *E. coli* BL21 (DE3) pLysS as inclusion bodies and purified by size-exclusion (Superdex S200) and ion-exchange (Hitrap SP HP) chromatography[10,12]. BL21(DE3) pLysS cells transformed with the appropriate plasmid were grown at 37 °C in 2xTY medium containing ampicillin and chloramphenicol at 37 °C until OD600 = 0.4–0.6. Expression was induced by addition of IPTG (0.5 mM) and continued at 37 °C for 2 h (3 h for Histone H4). Cells were harvested by centrifugation and the pellet was re-suspended in lysis buffer (50 mM TRIS pH 7.5, 100 mM NaCl, 1 mM EDTA and 5 mM BME) with protease inhibitor (one tablet per 50 ml, Roche, 11697498001, 40 ml total) and frozen and stored at − 80 °C. Frozen cells were thawed in a water bath, lysed by sonication and inclusion bodies were collected by centrifugation. The supernatant was discarded and the pellet was re-suspended in 40 ml TW buffer (lysis buffer + 1% Triton X) and inclusion bodies were again collected by centrifugation. The pellet was washed two more time with TW buffer and once with wash buffer. 7–15 ml (for 1–2 l culture) unfolding buffer (7 M guanidinum hydrochloride, 10 mM TRIS pH 7.5, 1 mM EDTA, 10 mM DTT, 1 mM benzamidine) was then added, and shaken for 1 h at room temperature (rt), insoluble materials were removed by centrifugation. The solubilized histones were collected and loaded onto an S200 size-exclusion column (equilibrated with SAU-1000 buffer; 7 M Urea, 20 mM NaOAc pH 5.2, 1 M NaCl, 1 mM EDTA and 5 mM BME). The histones were eluted with SAU-1000 buffer, and analysed by SDS–PAGE. Fractions containing histones were pooled dialysed against water containing 2 mM BME, and lyophilized. Histones were further purified by IEX using a HiTrap SP column. The lyophilized histones were dissolved in SAU-200 buffer (like SAU-1000 but with 200 mM NaCl) for 15 min at 4 °C and then loaded onto a HiTrap SP HP (5 ml, GE Healthcare) column equilibrated in SAU-200 buffer. Histones were eluted with a gradient from 0–100% SAU-600 buffer (like SAU-1000 but with 600 mM NaCl) in 20 CV. The fractions containing pure histones were pooled dialysed against water with 2 mM BME, and lyophilized. Lyophilized histones were stored at − 80 °C.

The plasmid encoding for the 145 bp 601 DNA construct[15] was a kind gift from Jinrong Min (SGC Toronto). Large-scale DNA preparation was performed essentially as described by Dyer *et al.*[12] The 601 DNA plasmid (pUC 57) was transformed into HB101 cells, which were plated on ampicillin (100 μg ml^{-1}) containing LB-Agar plates and grown overnight. The next morning, two 15 ml pre-cultures (2xTY containing ampicillin at the same concentration) were inoculated with one colony each and grown at 37 °C until slightly turbid (4–5 h). These where then used to inoculate four flasks (700 ml TB with ampicillin at the same concentration). These were shaken at 37 °C for 22 h. The cells were then harvested (7,000 r.p.m.; 7 min; F10BA-6x500y rotor) and resuspended in a total of 120 ml alkaline lysis buffer 1 (50 mM glucose, 25 mM Tris pH 8.0 and 10 mM EDTA) and split into two 500 ml centrifugation tubes (60 ml per tube). To each tube 120 ml alkaline lysis buffer 2 (0.2 M NaOH, 1% (w/v) SDS) was added and the mixture was shaken and incubated on ice for 10–20 min. Then 210 ml of cold alkaline lysis buffer 3 (4 M sodium acetate, 2 N acetic acid) was added and the mixture was shaken to mix and subsequently incubated on ice for 20 min. The mixture is then centrifuged (1,000*g*, 20 min, 4 °C) and filtered through tissue paper. Then 0.52 v/v isopropanol is added and the mixture is incubated at rt for 15 min and then centrifuged (1000*g*, 30 min, 20 °C; in polypropylene tubes; Beckman Coulter 41121703). The supernatant was then decanted and the pellet transferred to a 50 ml Falcon tube. Then 20 ml TE 10/50 and 21 μl RNase A (100 mg ml^{-1}; Qiagen; 19101) was added and the pellet was dissolved by shaking at 37 °C overnight. Then the mixture was centrifuged (JA25.50; 2000 r.p.m.; 15 min; 4 °C) to remove particulates. To the supernatant one-fifth of 4 M NaCl and two-fifth 40% PEG 6000 was added to precipitate the DNA. The mixture was shaken at 37 °C for 5 min, incubated on ice for 30 min and centrifuged at 3000*g* for 20 min. The supernatant was decanted and the pellet was dissolved in 15 ml TE 10/0.1 and the pellet was dissolved by shaking at 37 °C (2–3 h). The DNA concentration was then determined and 30 units of EcoRV (NEB; R0195) per nmol EcoRV site were added. The reaction was incubated at 37 °C for 16 h and completion was checked by agarose gel. If the reaction is complete, the plasmid is precipitated by addition of 0.18 vol 4 M NaCl and 0.31 vol 40% PEG 6000. The mixture is incubated on ice for 1 h and the centrifuged (JA25.50, 27,000*g*, 20 min, 4 °C). The supernatant is decanted and the insert is recovered by addition of 3 vol EtOH, incubation on ice for 1 h and centrifugation (JA25.50, 27000*g*, 20 min, 4 °C). The insert is dissolved in 3–4 ml TE 10/0.1 and purified by phenol extraction and ethanol precipitation.

Denaturing ESI-MS. The chemical protein modification was monitored using LC-MS. LC-MS was performed using a Micromass LCT (ESI-TOF-MS) and a Shimadzu Prominence HPLC equipped with a Dionex Proswift RP-4H column. The column was run at 0.4 ml min^{-1}. Solvents used were water with 0.1% formic acid and 95:5 Acetonitrile: water with 0.1% formic acid. The proteins were separated using a gradient from 5% B to 95% B in 4 min. After this the column was washed for 2 min with 95% B and then equilibrated with 5% B for 2.5 min. The electrospray source was set to 3200 V capillary voltage and 25 V cone voltage.

Data were analysed using Masslynx and calibrated using myoglobin. Deconvolution was performed using the maximum entropy algorithm.

H2A T101Dha synthesis. The reaction was performed in denaturing reaction buffer (RB; 5 M Gd.HCl, 50 mM TRIS pH 8.0). An amount of H2A T101C (10 mg) was weighed into an Eppendorf tube and dissolved in 500 μl RB. Then 15 mg DTT was added and the protein was shaken at rt for 30 min. Excess DTT was removed by using a PD minitrap column (in RB) giving a final concentration of 10 mg ml^{-1} H2A T101C (Supplementary Fig. 1).

To 900 μl of reduced H2A T101C (10 mg ml^{-1}; 0.74 mM) was added 100 μl of a DBHDA stock solution (0.5 M in DMF; 50 mM final; 90 eq). The reaction was shaken at rt for 30 min and then at 37 °C for 4 h. The reaction was monitored by LC-MS and after completion excess reagent was removed using a PD minitrap column (in RB, loaded with 500 μl). After 4 h at 37 °C full conversion of the cysteine containing starting material to Dha was observed (Supplementary Fig. 2).

H2A T101 GlcNAc synthesis. To 2 ml H2A T101Dha (4 mg ml^{-1}, 270 μM) was added GlcNAc-SH (120 mg, 505 μmol, 250 mM final, 925 eq) and the reaction mixture was shaken at 37 °C for 3.5 h after which time full conversion to H2A T101GlcNAc was observed by LC-MS. Excess GlcNAc-SH was removed by using a PD10 desalting column (in RB) and the resultant product was stored at − 20 °C for dimer reconstitution (Supplementary Fig. 3).

Complex reconstitution. Histone complex reconstitution was performed as described[12,38,39]. For the reconstitution, histones (powder after freeze-drying; 3–4 mg of each histone) were dissolved in 2 ml unfolding buffer (7 M Gd.HCl, 10 mM TRIS pH 7.5, 1 mM EDTA, 1 mM DTT) by rotating at 4 °C for 30 min. Chemically GlcNAcylated H2A was used after the modification reaction in RB. The protein concentration was then measured using a Nanodrop-1000. Wt octamers were reconstituted by combining a 1/1/1.1/1.1 ratio of H3/H4/H2A/H2B. For reconstitution of GlcNAcylated or wt H2A/B dimers equal amounts of each protein were combined. Wt H3/H4 tetramers were refolded by mixing equal amount of each protein. The solution was adjusted to a final concentration of 1–1.5 mg ml^{-1} and dialysed into 3 charges of 2 l refolding buffer (2 M NaCl, 10 mM TRIS, pH 7.5, 1 mM EDTA, 5 mM BME) for at least 4 h each. Precipitated protein was removed by centrifugation and the refolded histones were concentrated to 1–1.5 ml and purified by gel filtration (Superdex S200, 16/60, in refolding buffer). For octamer reconstitution from dimers and tetramers, the indicated dimer and tetramer ratio was mixed in refolding buffer, incubated on ice for 1 h and subsequently purified by gel filtration as above. For the analysis of octamer reconstitutions using different tetramer/dimer ratios (Supplementary Fig. 5) the size-exclusion analysis was performed using a Superdex S200 10/300 GL in refolding buffer.

Nucleosome reconstitution by salt gradient dialysis. Wt nucleosomes were reconstituted by combining wt octamer and DNA in a 1: 0.9 molar ratio. GlcNAcylated nucleosomes were reconstituted by combining dimer: tetramer: DNA in a 2: 1: 0.9 ratio. For dimer titrations the amount of dimer was varied, while the tetramer and DNA amount was kept constant. The mixtures were then transferred into a 6–8 kDa MWCO dialysis tube. The solution in dialysis tubing was dialysed against either a continuous gradient from high-salt buffer (2 M KCl, 10 mM Tris pH 7.5, 1 mM EDTA, 1 mM DTT) to low-salt buffer (250 mM KCl, 10 mM Tris pH 7.5, 1 mM EDTA, 1 mM DTT over 36 h) for large-scale reconstitution or against a series of step gradients for micro-scale reconstitution as described[12]. After completion of the gradient the refolded nucleosomes were dialysed two times against no-salt buffer (10 mM TRIS pH 7.5, 1 mM EDTA, 1 mM DTT)

CD analysis of the H2A/B dimers. CD spectra were recorded using a Chirascan CD-spectrophotometer (Applied Photophysics). H2A/B dimers were buffer exchanged into 20 mM $NaHPO_4$, pH 6.7, 1 mM EDTA, 150 mM NaCl by viva-spin. A final concentration of 10 μM dimer in 500 μl in a micro cuvette (1 mm path length) was used for each measurement. CD spectra were recorded from 200–300 nm with a step size of 1 nm and 0.5 s per point. The initial spectra were recorded in triplicates. The three traces were averaged and smoothed using a moving average (window size = 4) and the background was subtracted. For melting curves the temperature was increased stepwise from 25–90 °C in 1 °C steps with a hold time of 60 s. The temperature of the sample as measured by a sample probe was recorded for each step. The raw data were exported from the Pro-data as CSV and further processed using OriginPro. The melting temperature was extracted by plotting the CD at 220 nm against temperature and fitting using a Boltzmann sigmoidal function. Reported are the fitted parameters including the standard error.

Non-denaturing PAGE analysis. 'Native' PAGE analysis was performed using Novex TBE Gels (6%, 10-well, Invitrogen, EC6265BOX). The gel was prerun in 0.5 × TBE at 150 V for 90 min at 4 °C. Then 1–4 nmol of nucleosome in 5 μl, supplemented with 1 μl 30% sucrose as loading buffer, was added per well. The gel was run at 4 °C at 150 V for 90 min and subsequently stained with SYBR Gold (Invitrogen, S-11494) and visualized using a Bio-Rad Gel Doc XR+ machine.

DSF melting temperature analysis. DSF analysis was performed on a Bio-Rad MJ Mini real-time PCR machine using the Opticon Monitor 3 software. For the analysis 23 μl of 2.5 μM nucleosome in 10 mM $NaHPO_4$, pH 7.5, 1 mM EDTA, 150 mM NaCl and 2 μl of 25 × SYPRO Orange (diluted from 5,000 × Concentrate in DMSO; Life technologies S-6650) were added into a real-time PCR plate and sealed with transparent film. Then the sample was first kept for 1 min at 15 °C and then heated from 15–95 °C in 1 °C steps with one minute at each temperature. Fluorescence was read out for each temperature. The data was analysed using the excel sheets described by Niesen *et al.*[18] Melting temperatures were extracted by fitting using a Boltzmann sigmoidal function in OriginPro. Reported are the fitted parameter including the s.e.m.

CD analysis of the nucleosomes. Nucleosomes (0.62 μM) were buffer exchanged into 10 mM $NaHPO_4$, pH 7.5, 1 mM EDTA, 150 mM NaCl. For the measurement 180 μl of nucleosome solution in a micro cuvette (1 mm path length) was used. CD spectra were recorded from 200–300 nm with a step size of 0.5 nm and 1 s measurement per point. Temperature was ramped stepwise from 25–90 °C in 1 °C steps with 120 s at each temperature. Data was processed as described above for the dimer.

Non-denaturing 'Native' MS. Nucleosomes were buffer exchanged using Micro Bio-Spin chromatography columns (Bio-Rad) equilibrated with 50 mM ammonium acetate pH 7.5. Spectra were acquired on a QToF II MS (Waters) modified for high mass analysis and preservation of non-covalent interactions[40]. Aliquots (3 μl) diluted to 1–2.5 μM were introduced into the mass spectrometer from gold-plated capillary needles (Harvard Apparatus). Capillary voltages were 1.4–1.6 kV, cone voltages were 80–100 V and collision voltage was 10 eV. ToF pressure 2.05×10^{-6} bar, analyser pressure 1.54×10^{-4} bar and backing pressure 4.26×10^{-3} bar. The recorded mass spectra were calibrated externally using caesium iodide in water. Data were acquired and processed with Mass Lynx V4.1 software (Waters) with minimal smoothing and no background subtraction.

ChIP-Seq and DNase I sensitive site analysis. OGT ChIP data published by Vella *et al.*[21] was used for the analysis. DNase I sensitivity data were obtained from data deposited by the ENCODE project (Digital DNaseI from ENCODE/ University of Washington; ES-E14; 129/Ola; E0; hotspot-broadPeak: wgEncodeUwDnaseEse14129olaME0HotspotsRep1)[22,41]. Previously called peaks reported in Vella *et al.*[21] and the DNaseI hotspot-broadPeak were used to analyse overlap between OGT and DNaseI sensitive sites. Data was analysed using Galaxy[42–44] and visualized using the UCSC genome browser[45].

Overlapping regions were found using the 'Operate on Genomic Intervals/ intersect' tool in Galaxy. For 11552 OGT peaks 11233 overlap with DNase-sensitive sites (97%). For DNase-sensitive sites 12257 of 270487 peaks overlap with OGT peaks (4.5%). Further statistical analysis was performed using The Genomic HyperBrowser (https://hyperbrowser.uio.no/)[46]. Using the Statistical analysis tool, we tested for significance of the overlap using following settings: Alternative hypothesis: more, Null model: Preserve segments (T2) and segment length (T1); Randomize Positions (T1) (MC), MCFDR sampling depth: Moderate resolution of global and local *P*-values, Random seed: Random. With these setting we obtained a *P*-value of 0.000999 for the overlap of OGT (T1) and DNase (T2) sensitive sites.

To analyse peaks that are close at TSS, we included all OGT (4,000) and DNase (12,677) peaks that are within + − 500 bp of an annotated TSS. Of these 3,894 OGT peaks (97%) overlap with DNase-sensitive sites and 3,983 DNase-sensitive regions overlap with OGT peaks (31%). Using the same parameters as described above a *P*-value of 0.000999 was obtained for the overlap of OGT (T1) and DNase (T2) sensitive sites.

Quantitative MS with H2A T101 GlcNAc-modified nucleosome. The HeLa nuclear extract was prepared using an adapted protocol: http://www.cyclex.co.jp/resource/protocol/classical_biochemical/preparation_of_nuclear_extracts.pdf] Briefly, HeLa cells were grown to near-confluency in a HYPERFlask cell culture vessel (Corning Life Sciences) containing 500 ml DMEM medium supplemented with 10% FBS and 1% Pen-Strep. Cells were trypsinized and were washed with 10 ml of ice-cold PBS buffer by centrifugation at 300*g* for 5 min at 4 °C. Cells were re-suspended in 10 ml of lysis buffer (10 mM Tris pH 7.5, 10 mM NaCl, 15 mM $MgCl_2$, 250 mM Sucrose, 0.1 mM EGTA, 0.5% NP-40 with complete protease and phosphatase inhibitors) and vortexed for 10 s. Cells were incubated on ice for 15 min followed by being spun through 40 ml of sucrose cushion (30% Sucrose, 10 mM Tris pH 7.5, 10 mM NaCl, 3 mM $MgCl_2$ with complete protease and phosphatase inhibitors) at 1300 g for 10 min at 4 °C. The supernatant was discarded. The isolated nuclei were washed with 10 ml of ice-cold wash buffer (10 mM Tris pH 7.5, 10 mM NaCl with complete protease and phosphatase inhibitors). Nuclei were re-suspended in 1 ml of extraction buffer (50 mM HEPES pH 7.9, 420 mM NaCl, 0.5 mM EDTA, 0.1 mM EGTA, 10% Glycerol, 0.1% NP-40 with complete protease and phosphatase inhibitors) and sonicated for 30 s. The ruptured nuclei were incubated on ice for 30 min followed by centrifugation at 13000 g for 10 min at 4 °C. The supernatant, nuclear extract, was collected. The concentration of nuclear extract was determined by BCA assay. The nuclear extract was diluted using a buffer such that final concentration was 0.5 mg ml^{-1} and buffer conditions were: 20 mM HEPES pH 7.9; 150 mM NaCl; 0.2 mM EDTA; 20% glycerol; and 0.1% DTT with complete protease and phosphatase inhibitors (Binding Buffer). The nuclear extract aliquots were flash-frozen in liquid nitrogen and stored at −80 °C.

The plasmid encoding 186 bp '601' DNA was a kind gift from Dr Till Bartke, Imperial College, London. The nucleosomal DNA was excised and biotinylated as described[12,24] and further used for reconstitution of wt and modified nucleosome as described earlier. Reconstituted nucleosomes were analysed using native PAGE.

Interaction studies between the nucleosome and the nuclear extract were performed as described above. Briefly, 20 μg of each nucleosome were immobilized separately on 100 μl Dynabeads Streptavidin MyOne T1 (Life Technologies) in the reconstitution buffer (10 mM Tris pH 7.5, 250 mM KCl, 1 mM EDTA, 1 mM DTT with 0.1% NP40). The immobilization efficiency was monitored by measuring the absorption value at 260 nm of the supernatant. Immobilized nucleosome was washed with the binding buffer (20 mM HEPES pH 7.9, 150 mM NaCl, 0.2 mM EDTA, 20% Glycerol, 0.1% NP40, 1 mM DTT with complete protease and phosphatase inhibitors) and then incubated with 1 ml of nuclear extract (0.5 mg ml^{-1} in binding buffer) for 4 h at 4 °C. After stringent washes with binding buffer, the interacting partners were eluted using 100 μl of 8 M Urea, 2 M NaCl.

Tryptic digestion of the interacting partners from each nucleosome sample were performed using the filter-aided sample preparation protocol[47]. Digestion was quenched by addition of 1% formic acid solution. Resulting tryptic peptides were separated using an EASY-nLC 1000 UHPLC system (Proxeon) coupled to a Q Exactive mass spectrometer (Thermo Fischer Scientific). Peptides were trapped on an in-house packed trap column (75 μm i.d. × 20 mm, reprosil C18, 3 μm, 120 Å) and separated on an EASY-spray Acclaim PepMap analytical column (75 μm i.d. × 500 mm, RSLC C18, 2 μm, 100 Å). Samples were loaded at a pressure of 500 bar with 100% solvent A (0.1% formic acid, 5% DMSO in water) and the peptides separated by a linear gradient (length: 120 min, 8–28% solvent B (0.1% formic acid, 5% DMSO in acetonitrile), flow rate: 200 nl min^{-1}). The separated peptides were electrosprayed directly into the Q Exactive operated in a data-dependent mode through an EASY-Spray nanoelectrospray ion source. Full-scan MS spectra were acquired in the Orbitrap (scan range 350–1500 *m/z*, resolution 70,000, AGC target 3e6, maximum injection time 100 ms). After the MS scans, the 20 most intense peaks were selected for HCD fragmentation at 30% of normalized collision energy. HCD spectra were also acquired in the Orbitrap (resolution 17,000, AGC target 5e4, maximum injection time 120 ms).

The raw data files generated were processed using MaxQuant software (Version 1.4.3.17), integrated with Andromeda search engine[48,49]. For protein groups identification, peak lists were searched against human database (Swiss-Prot, version 04/13) as well as list of common contaminants by Andromeda. Trypsin with a maximum number of missed cleavages of 2 was chosen. Acetylation (protein N-term) and oxidation (methionine) were used as variable modifications, while carbamidomethylation (Cysteine) was set as a fixed modification. Protein and PSM false discovery rate were set at 0.01. Label-free quantification (LFQ) was performed using the in-built algorithm in the MaxQuant[26]. LFQ intensities were uploaded into Perseus software (Version 1.4.2.27) for further analysis. Contaminant, reverse and protein groups only indentified by site were filtered off. Protein groups identified in both bait and control samples and each replicates were used. Their LFQ intensities were logarithmized and mean value was calculated. The difference between the average LFQ intensities of protein groups between bait and control samples alongside total intensity were evaluated by Significance B to generate a *P*-value.

Salt-based disassembly of the nucleosome. For salt-based disassembly[16,50], to 5 μl of each nucleosome (4 μM in 10 mM Tris pH7.5, 1 mM EDTA) was added 1 μl of a KCl stock solution to give the final salt concentration as indicated. The samples were then incubated at 37 °C for 60 min and analysed by non-denaturing ('native') PAGE.

Thermal disassembly of the nucleosome. 10 μl of each nucleosome (4 μM in 10 mM Tris pH7.5, 20 mM KCl, 1 mM EDTA) were incubated at 63, 66, 69, 72 or 75 °C in a PCR machine for 30 min. The samples were centrifuged and the supernatant was analysed using Native PAGE and SDS-PAGE. SDS-PAGE analysis was performed using 12% Bis-Tris gel (Novex). The gel was run in 1 × MOPS buffer at 200 V for 50 min. The gel was stained with InstantBlue (Expedeon).

References

1. Bannister, A. J. & Kouzarides, T. Regulation of chromatin by histone modifications. *Cell. Res.* **21,** 381–395 (2011).
2. Hart, G. W., Slawson, C., Ramirez-Correa, G. & Lagerlof, O. Cross talk between O-GlcN acylation and phosphorylation: roles in signaling, transcription, and chronic disease. *Annu. Rev. Biochem.* **80,** 825–858 (2011).
3. Sakabe, K., Wang, Z. & Hart, G. W. Beta-N-acetylglucosamine (O-GlcNAc) is part of the histone code. *Proc. Natl Acad. Sci. USA* **107,** 19915–19920 (2010).
4. Fong, J. J. *et al.* β-N-Acetylglucosamine (O-GlcNAc) is a novel regulator of mitosis-specific phosphorylations on histone H3. *J. Biol. Chem.* **287,** 12195–12203 (2012).
5. Fujiki, R. *et al.* GlcNAcylation of histone H2B facilitates its monoubiquitination. *Nature* **480,** 557–560 (2011).
6. Love, D. C., Krause, M. W. & Hanover, J. A. O-GlcNAc cycling: emerging roles in development and epigenetics. *Semin. Cell. Dev. Biol.* **21,** 646–654 (2010).
7. Myers, S. A., Panning, B. & Burlingame, A. L. Polycomb repressive complex 2 is necessary for the normal site-specific O-GlcNAc distribution in mouse embryonic stem cells. *Proc. Natl Acad. Sci. USA* **108,** 9490–9495 (2011).
8. Deplus, R. *et al.* TET2 and TET3 regulate GlcNAcylation and H3K4 methylation through OGT and SET1/COMPASS. *EMBO J.* **32,** 645–655 (2013).
9. Lazarus, M. B., Nam, Y., Jiang, J., Sliz, P. & Walker, S. Structure of human O-GlcNAc transferase and its complex with a peptide substrate. *Nature* **469,** 564–567 (2011).
10. Chalker, J. M., Lercher, L., Rose, N. R., Schofield, C. J. & Davis, B. G. Conversion of cysteine into dehydroalanine enables access to synthetic histones bearing diverse post-translational modifications. *Angew Chem. Int. Ed. Engl.* **51,** 1835–1839 (2012).
11. Chalker, J. M. *et al.* Methods for converting cysteine to dehydroalanine on peptides and proteins. *Chem. Sci.* **2,** 1666–1676 (2011).
12. Dyer, P. N. *et al.* Reconstitution of nucleosome core particles from recombinant histones and DNA. *Methods Enzymol.* **375,** 23–44 (2004).
13. Karantza, V., Baxevanis, A. D., Freire, E. & Moudrianakis, E. N. Thermodynamic studies of the core histones: ionic strength and pH dependence of H2A-H2B dimer stability. *Biochemistry* **34,** 5988–5996 (1995).
14. Burgess, R. J. & Zhang, Z. Histone chaperones in nucleosome assembly and human disease. *Nat. Struct. Mol. Biol.* **20,** 14–22 (2013).
15. Vasudevan, D., Chua, E. Y. D. & Davey, C. A. Crystal structures of nucleosome core particles containing the '601' strong positioning sequence. *J. Mol. Biol.* **403,** 1–10 (2010).
16. Bao, Y. *et al.* Nucleosomes containing the histone variant H2A.Bbd organize only 118 base pairs of DNA. *EMBO J.* **23,** 3314–3324 (2004).
17. Brooks, W. & Jackson, V. The rapid transfer and selective association of histones H2A and H2B onto negatively coiled DNA at physiological ionic strength. *J. Biol. Chem.* **269,** 18155–18166 (1994).
18. Niesen, F. H., Berglund, H. & Vedadi, M. The use of differential scanning fluorimetry to detect ligand interactions that promote protein stability. *Nat. Protoc.* **2,** 2212–2221 (2007).
19. Ausio, J. & Van Holde, K. E. Histone hyperacetylation: its effects on nucleosome conformation and stability. *Biochemistry* **25,** 1421–1428 (1986).
20. Azegami, N. *et al.* Conclusive evidence of the reconstituted hexasome proven by native mass spectrometry. *Biochemistry* **52,** 5155–5157 (2013).
21. Vella, P. *et al.* Tet proteins connect the O-linked N-acetylglucosamine transferase Ogt to chromatin in embryonic stem cells. *Mol. Cell.* **49,** 645–656 (2013).
22. Stamatoyannopoulos, J. A. *et al.* An encyclopedia of mouse DNA elements (Mouse ENCODE). *Genome. Biol.* **13,** 418 (2012).
23. Kouzarides, T. Chromatin modifications and their function. *Cell* **128,** 693–705 (2007).
24. Bartke, T. *et al.* Nucleosome-interacting proteins regulated by DNA and histone methylation. *Cell* **143,** 470–484 (2010).
25. Eberl, H. C., Mann, M. & Vermeulen, M. Quantitative proteomics for epigenetics. *Chembiochem* **12,** 224–234 (2011).
26. Cox, J. *et al.* Accurate proteome-wide label-free quantification by delayed normalization and maximal peptide ratio extraction, termed MaxLFQ. *Mol. Cell. Proteomics* **13,** 2513–2526 (2014).
27. Ferre-D'Amare, A. R., Pognonec, P., Roeder, R. G. & Burley, S. K. Structure and function of the b/HLH/Z domain of USF. *EMBO J.* **13,** 180–189 (1994).
28. Javaid, S. *et al.* Nucleosome remodeling by hMSH2-hMSH6. *Mol. Cell.* **36,** 1086–1094 (2009).
29. Belotserkovskaya, R. *et al.* FACT facilitates transcription-dependent nucleosome alteration. *Science* **301,** 1090–1093 (2003).
30. Pavri, R. *et al.* Histone H2B monoubiquitination functions cooperatively with FACT to regulate elongation by RNA polymerase II. *Cell* **125,** 703–717 (2006).
31. Bönisch, C. & Hake, S. B. Histone H2A variants in nucleosomes and chromatin: more or less stable? *Nucleic Acids Res.* **40,** 10719–10741 (2012).
32. Bönisch, C. *et al.* H2A.Z.2.2 is an alternatively spliced histone H2A.Z variant that causes severe nucleosome destabilization. *Nucleic Acids Res.* **40,** 5951–5964 (2012).
33. Gautier, T. *et al.* Histone variant H2ABbd confers lower stability to the nucleosome. *EMBO Rep.* **5,** 715–720 (2004).
34. Ye, J. *et al.* Histone H4 lysine 91 acetylation: a core domain modification associated with chromatin assembly. *Mol. Cell* **18,** 123–130 (2005).
35. Yan, Q. *et al.* BBAP monoubiquitylates histone H4 at lysine 91 and selectively modulates the DNA damage response. *Mol. Cell* **36,** 110–120 (2009).
36. Kimura, H. & Cook, P. R. Kinetics of core histones in living human cells: little exchange of H3 and H4 and some rapid exchange of H2b. *J. Cell Biol.* **153,** 1341–1354 (2001).
37. Zhu, Y. *et al.* O-GlcNAc occurs cotranslationally to stabilize nascent polypeptide chains. *Nat. Chem. Biol.* **11,** 319–325 (2015).
38. Luger, K., Rechsteiner, T. & Richmond, T. in *Chromatin Protocols.* (ed. Becker, P.) (Humana Press, 1999).
39. Luger, K., Rechsteiner, T. J. & Richmond, T. J. Preparation of nucleosome core particle from recombinant histones. *Methods. Enzymol.* **304,** 3 (1999).
40. Sobott, F., Hernández, H., McCammon, M. G., Tito, M. A. & Robinson, C. V. A tandem mass spectrometer for improved transmission and analysis of large macromolecular assemblies. *Anal. Chem.* **74,** 1402 (2002).
41. Sabo, P. J. *et al.* Genome-scale mapping of DNase I sensitivity *in vivo* using tiling DNA microarrays. *Nat. Methods* **3,** 511 (2006).
42. Goecks, J., Nekrutenko, A., Taylor, J. & Team, T. G. Galaxy: a comprehensive approach for supporting accessible, reproducible, and transparent computational research in the life sciences. *Genome. Biol.* **11,** R86 (2010).
43. Blankenberg, D. *et al.* Galaxy: a web-based genome analysis tool for experimentalists. *Curr. Protoc. Mol. Biol.* Chapter 19 Unit 19.10 1–21 (2010).
44. Giardine, B. *et al.* Galaxy: a platform for interactive large-scale genome analysis. *Genome Res.* **15,** 1451 (2005).
45. Kent, W. J. *et al.* The human genome browser at UCSC. *Genome Res.* **12,** 996 (2002).
46. Sandve, G. *et al.* The Genomic HyperBrowser: inferential genomics at the sequence level. *Genome. Biol.* **11,** R121 (2010).
47. Wiśniewski, J. R., Zougman Nagaraj, N. & Mann, M. Universal sample preparation method for proteome analysis. *Nat. Methods* **6,** 359–362 (2009).
48. Cox, J. & Mann, M. MaxQuant enables high peptide identification rates, individualized p.p.b.-range mass accuracies and proteome-wide protein quantification. *Nat. Biotechnol.* **26,** 1367–1372 (2008).
49. Cox, J. *et al.* Andromeda: a peptide search engine integrated into the MaxQuant environment. *J. Proteome Res.* **10,** 1794–1805 (2011).

50. Muthurajan, U. M. *et al.* Crystal structures of histone Sin mutant nucleosomes reveal altered protein-DNA interactions. *EMBO J.* **23,** 260–271 (2004).
51. Davey, C. A., Sargent, D. F., Luger, K., Maeder, A. W. & Richmond, T. J. Solvent mediated interactions in the structure of the nucleosome core particle at 1.9 Å resolution. *J. Mol. Biol.* **319,** 1097–1113 (2002).

Acknowledgements

We thank Drs Marta Tapia and Thomas Milne for excellent advice and discussions and Dr Skirmantas Kriacionis for assistance with data analysis. We thank the EPSRC and Cancer Research UK (LL, Grant reference NS/A000004/1), Felix Scholarship Foundation (RR), and the Wellcome Trust (NAP, Grant reference 088150/B/09/Z) for financial support. CVR is a Royal Society Professorship recipient and BGD was a Royal Society Wolfson Research Merit Award recipient during the period of this work.

Author contributions

L.L., R.R., S.M., C.V.R., C.J.S., B.G.D. designed the experiments. L.L., R.R., N.A.P., J.P. performed experiments. L.L., R.R., S.M., C.V.R., C.J.S., B.G.D. analysed the data. L.L., R.R., S.M., C.J.S., B.G.D. wrote the paper. All authors read and commented on the paper.

Additional information

Competing financial interests: The authors declare no competing financial interests.

11

Ainsliadimer A selectively inhibits IKKα/β by covalently binding a conserved cysteine

Ting Dong[1,2,*], Chao Li[3,*], Xing Wang[4], Longyang Dian[2], Xiuguo Zhang[2], Lin Li[2], She Chen[2], Ran Cao[2], Li Li[2], Niu Huang[2], Sudan He[4] & Xiaoguang Lei[1,2,3]

Aberrant activation of NF-κB is associated with the development of cancer and autoimmune and inflammatory diseases. IKKs are well recognized as key regulators in the NF-κB pathway and therefore represent attractive targets for intervention with small molecule inhibitors. Herein, we report that a complex natural product ainsliadimer A is a potent inhibitor of the NF-κB pathway. Ainsliadimer A selectively binds to the conserved cysteine 46 residue of IKKα/β and suppresses their activities through an allosteric effect, leading to the inhibition of both canonical and non-canonical NF-κB pathways. Remarkably, ainsliadimer A induces cell death of various cancer cells and represses *in vivo* tumour growth and endotoxin-mediated inflammatory responses. Ainsliadimer A is thus a natural product targeting the cysteine 46 of IKKα/β to block NF-κB signalling. Therefore, it has great potential for use in the development of anticancer and anti-inflammatory therapies.

[1] Graduate School of Peking Union Medical College and Chinese Academy of Medical Sciences, Beijing 100730, China. [2] National Institute of Biological Sciences (NIBS), Beijing 102206, China. [3] Beijing National Laboratory for Molecular Sciences, Department of Chemical Biology, College of Chemistry and Molecular Engineering, Synthetic and Functional Biomolecules Center, and Peking-Tsinghua Center for Life Sciences, Peking University, Beijing 100871, China. [4] Cyrus Tang Hematology Center, Jiangsu Institute of Hematology, First Affiliated Hospital, and Collaborative Innovation Center of Hematology, Soochow University, Suzhou 215123, China. * These authors contributed equally to this work. Correspondence and requests for materials should be addressed to S.H. (email: hesudan@suda.edu.cn) or to X.L. (email: xglei@pku.edu.cn).

The evolutionarily conserved nuclear factor-κB (NF-κB) signalling pathway plays key roles in inflammatory and immune responses and in cell survival by regulating the transcription of numerous target genes[1–4]. The NF-κB family of transcription factors consists of five members, including p50, p52, p65 (RelA), c-Rel and RelB, which form various dimeric complexes. The NF-κB dimers are normally sequestered in the cytoplasm by association with a member of the IκB inhibitory family (for example, IκBα, IκBβ, IκBε) or with the precursor proteins p100 and p105. NF-κB activation typically occurs by nuclear translocation of NF-κB dimers following inducible degradation of IκB, or processing of precursor proteins in response to a variety of stimuli, including the presence of cytokines like TNF-α or IL-1, growth factors, microbial infection and/or chemotherapeutic agents.

Canonical NF-κB activation depends on the degradation of IκB, which is rapidly phosphorylated by an active IκB kinase (IKK) complex. This complex is composed of IKKα and IKKβ catalytic subunits and a regulatory subunit, IKKγ/NEMO (NF-κB essential modulator)[5]. IKKβ is the major subunit responsible for phosphorylation of IκB proteins. For example, IκBα is phosphorylated at Ser-32 and Ser-36 (ref. 6), whereas IκBβ is phosphorylated at Ser-19 and Ser-23 (ref. 7). Phosphorylated IκB subsequently undergoes proteasome-mediated degradation, thereby liberating free NF-κB dimers to translocate to the nucleus that can then promote gene transcription[8]. In addition, an alternative pathway designated as the 'non-canonical NF-κB pathway' relies on the inducible processing of p100 (ref. 9). This pathway mainly activates IKKα, which in turn phosphorylates p100 to trigger its proteolytic processing to p52, leading finally to nuclear translocation of p52-containing NF-κB dimers.

Aberrant activation of the NF-κB signalling pathway is known to be involved in a variety of human diseases including cancer, autoimmune diseases and chronic inflammatory diseases[2,10,11]. The NF-κB pathway is important for cancer development and progression, in that it regulates a wide variety of target genes involved in cell proliferation, cell survival, invasion, angiogenesis and metastasis[12]. Continuous activation of NF-κB is a common feature in the majority of human cancers, including both solid and haematopoietic malignancies[13]. Activated NF-κB induces expression of anti-apoptotic genes, including those of the inhibitor of apoptosis protein family[14], anti-apoptotic Bcl-2 family[15,16] and cellular FLICE-inhibitory protein[17], which is associated with increased resistance of cancer cells to chemotherapy. Moreover, IKKβ has been recently shown to phosphorylate BAD, which results in the blocking of BAD-mediated apoptosis[18]. In addition to its critical role in cancer, enhanced NF-κB activity is a hallmark of various autoimmune and inflammatory diseases. Chronic inflammatory conditions have been shown to drive an increased cancer risk. Examples of this include colitis-associated colon cancer and hepatitis-associated liver cancer[19,20]. Ample evidence suggests that inhibition of NF-κB activity represses cancer cell survival, *in vivo* tumour growth and inflammatory responses. Therefore, strategies focused on reducing NF-κB activity by specific small molecule inhibitors could offer significant therapeutic value for the treatment of these diseases.

Over the past decade, there has been a concerted effort to identify small molecule inhibitors of IKKβ because of its central role in the canonical NF-κB pathway. Some of the small molecule inhibitors that have been identified in these efforts have exerted promising inhibitory effects in various experimental models of tumour and inflammatory diseases[12,21]. However, there is as yet limited clinical experience of the efficacy and safety of such molecules. Therefore, it is of great importance that novel IKKα/β inhibitors with unique binding properties, high efficacy and low toxicity are identified and developed as therapeutic agents to suppress both canonical and non-canonical NF-κB activation. Such efforts also enable the dissection of the mechanisms of IKK regulation of the NF-κB signalling pathway. In the current study, we demonstrate that the natural product ainsliadimer A (**1**) tightly binds IKKα and IKKβ through the conserved cysteine residue 46(C46), leading to the inactivation of both the canonical and the non-canonical NF-κB signalling pathways triggered by multiple stimuli. To our knowledge, ainsliadimer A represents the first small molecule natural product targeting the functional C46 of IKKα/IKKβ. Ainsliadimer A leads to IKKα/IKKβ inactivation through a novel allosteric effect. Moreover, ainsliadimer A powerfully blocks LPS-mediated inflammatory responses and tumour growth *in vivo*. These discoveries not only indicate that ainsliadimer A indeed acts as a potent inhibitor of IKKα/β, but also reveal that ainsliadimer A could be used as a lead compound in the development of new therapeutic agents for the treatment of cancer and inflammatory disorders.

Results

Ainsliadimer A inhibits LPS-induced inflammatory cytokines *in vivo*. Ainsliadimer A (**1**) is a structurally unique and complex sesquiterpene lactone dimer with an unprecedented carbon skeleton (Fig. 1a) that was originally isolated from *Ainsliaea macrocephala*, a plant that has been used in traditional Chinese medicine for the treatment of various diseases, including angina and rheumatoid arthritis[22]. In our systematic chemical genetic studies using bioactive natural products, we have successfully accomplished the first enantioselective total syntheses of ainsliadimer A (**1**)[23], as well as that of other related dimeric and trimeric sesquiterpenoids[24,25]. Such synthetic efforts have provided us with ample material to extensively investigate both the biological activities and potential therapeutic applications of such compounds. Ainsliadimer A (**1**) has been previously shown to exhibit inhibitory activity against the production of nitric oxide (NO) in a mouse macrophage cell line RAW264.7 stimulated by lipopolysaccharide (LPS)[22], a major cell wall component of Gram-negative bacteria. As an endotoxin, LPS is known to cause a systemic inflammatory response and acute tissue injury when injected into mice. Thus, we here attempted to examine the effect of ainsliadimer A (**1**) on LPS-triggered inflammatory responses *in vivo*. Mice were pre-injected with or without Vehicle or ainsliadimer A (**1**), followed by administration of PBS or LPS. As shown in Fig. 1b-d, serum inflammatory cytokines including TNF-α, IL-6 and IL-1β were highly induced in the mice challenged with LPS, as compared with the PBS-treated mice, and the upregulation of these cytokines was greatly attenuated in the mice that had been pretreated with ainsliadimer A (**1**) but not in the Vehicle-pretreated mice. As such, it can be concluded that ainsliadimer A (**1**) has great potential as an anti-inflammatory agent *in vivo*. Furthermore, exogenously administered ainsliadimer A (**1**) ($25\,mg\,kg^{-1}$) was shown to cause a significant decrease in NF-κB activation in the spleen after intravenous injection (as compared with Vehicle-treated mice) following LPS challenge (Supplementary Fig. 1).

Ainsliadimer A acts as a potent inhibitor of NF-κB pathways. The potent anti-inflammatory activity of ainsliadimer A (**1**) prompted us to investigate how it modulates pro-inflammatory cytokine responses. As a well-defined agonist of TLR4 (refs 12,13), LPS specifically activates TLR4 and subsequently results in NF-κB activation and type I IFN responses. We next evaluated the effect of ainsliadimer A (**1**) on LPS-induced NF-κB activation. Phosphorylation of IκB is a key step in the activation of the canonical NF-κB pathway. As expected, phosphorylated IκBα was detected in RAW264.7 cells after exposure to LPS for 10 or 20 min. Strikingly, pretreatment with 8 μM ainsliadimer A (**1**) could effectively block the phosphorylation of IκBα (Fig. 2a). In comparison with the well-known NF-κB inhibitor BMS-345541 [26], which suppressed LPS-induced phosphorylation of IκBα at 20 μM, ainsliadimer A (**1**) exhibited much stronger inhibitory effects against LPS-induced NF-κB activation. Meanwhile, by assessing IFR3 phosphorylation, we found that ainsliadimer A (**1**) had no effect on LPS-induced type I IFN responses, indicating that ainsliadimer A (**1**) specifically targets LPS-trigged NF-κB signalling (Supplementary Fig. 2). Furthermore, ainsliadimer A (**1**) markedly suppressed the phosphorylation of IκBα triggered by synthesized dsRNA poly(I:C), a TLR3 ligand[27] (Fig. 2b). These results demonstrate that ainsliadimer A (**1**) affects a common signalling component shared in both TLR3- and TLR4-mediated NF-κB signalling. In addition to the TLRs, the engagement of death receptors with corresponding ligands including TNF-α is also known to trigger NF-κB activation. We found that ainsliadimer A (**1**) significantly inhibited TNF-α-induced phosphorylation of IκBα in various cell lines (Fig. 2c and Supplementary Fig. 3). Moreover, addition of ainsliadimer A (**1**) abolished TNF-α-induced nuclear translocation of p65 (Fig. 2d). Taken together, these results establish that ainsliadimer A (**1**) acts as a potent inhibitor of the NF-κB pathway induced by diverse stimuli.

TNF receptor-associated factors (TRAFs) are important intermediates in the NF-κB signalling pathway. For example, TRAF2 and TRAF5 regulate TNF-α receptor-triggered IKK activation, and TRAF6 is required for IL-1-induced IKK activation[2]. The latter pathway can be reconstituted in cell-free extracts by adding recombinant TRAF6. As shown in Fig. 2e, phosphorylation of IKKα/β was detected in S100 cell lysates from 293T cells incubated with recombinant TRAF6. Notably, the addition of ainsliadimer A (**1**) completely abolished TRAF6-induced phosphorylation of IKKα/β, indicating that ainsliadimer A (**1**) disrupts the NF-κB pathway by intervening in the function of the IKK complex or that of upstream activators.

We then tested whether ainsliadimer A (**1**) affected the non-canonical NF-κB pathway, which is activated via a subset of TNFR superfamily members, including the receptor activator for nuclear factor κB (RANK), a B-cell-activating factor belonging to the TNF family receptor (BAFFR) and a lymphotoxinβ-receptor (LTβR)[9]. As shown in Fig. 2f, the processing of p100 to p52 triggered by RANKL was suppressed by ainsliadimer A (**1**) in a dose-dependent manner. We further investigated the effect of ainsliadimer A (**1**) on the processing of p100 to p52 triggered by the ectopic expression of NIK. Consistently, addition of ainsliadimer A (**1**) prevented NIK-mediated production of p52 (Supplementary Fig. 4). These results suggest that ainsliadimer A (**1**) also targets the non-canonical NF-κB pathway.

Syntheses and biological evaluations of chemical probes. To explore the functional target(s) of ainsliadimer A (**1**) associated with the NF-κB pathway, we prepared chemical probes for affinity purification. Initial structure and activity relationship studies enabled by total synthesis revealed that the hydroxyl analogue **2** fully retained the biological activity observed with ainsliadimer A (**1**), which suggested that a biotin tag could be attached to this hydroxyl group (Fig. 3a). In addition, the α, β-unsaturated enone moieties were shown to be essential for the biological activity as indicated by the totally inactive analogue **3** (Fig. 3b). This interesting result suggested that the enone moiety might be able to serve as a Michael acceptor to form a covalent bond with the thiol group of the cysteine residue. The inactive analogue **3** could conceivably be used to prepare a negative probe. On the basis of this information, we synthesized both a positive probe (Probe) and a negative probe (NC) from **2** and **3**, respectively, with biotin and a tri(ethyleneglycol) linker (Fig. 3c and Supplementary Information). The biological evaluations of Probe and NC showed that the biotin-tagged positive probe retained the ability to effectively block NF-κB activation at 50 μM, while the biotin-tagged negative probe completely lost the activity at the same concentration (Fig. 3d and Supplementary Fig. 5).

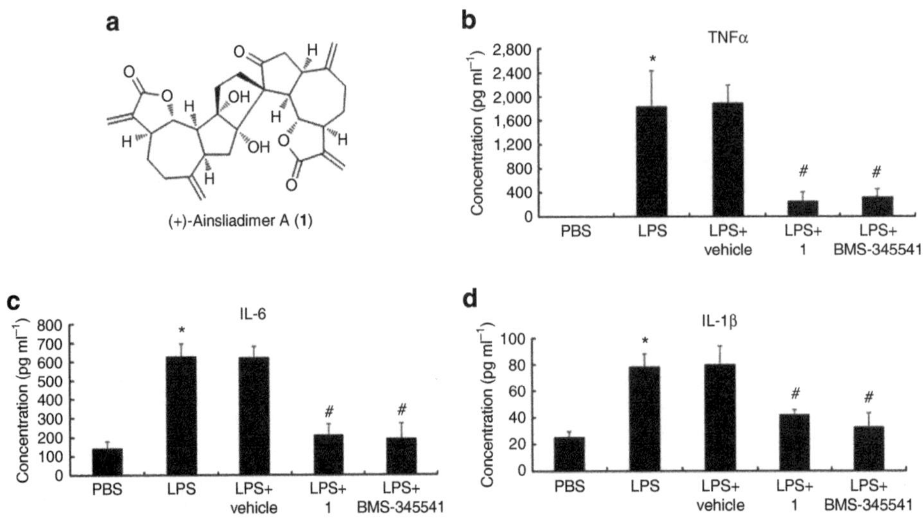

Figure 1 | Ainsliadimer A (1) suppresses the production of inflammatory cytokines *in vivo*. Chemical structure of ainsliadimer A (**1**). (**b-d**) Effects of ainsliadimer A (**1**) on the production of serum TNF, IL-6 and IL-1β induced by LPS. The mice were administered intravenously with ainsliadimer A (1) (25 mg kg^{-1}) (LPS + ainsliadimer A group), BMS-345541(25 mg kg^{-1}) (LPS + BMS-345541group), Vehicle (LPS + Vehicle group), or PBS(PBS group) for 1.5 h before intraperitoneal injection of LPS (1.6 mg kg^{-1}). Serum concentrations of TNF, IL-6 and IL-1β in the four groups were measured after 3.5 h following LPS injection. Mean ± s.e.m. *, $P< 0.05$ versus PBS; #, $P<0.01$ versus LPS. For each group, plasma samples were obtained from at least six mice and were analysed in duplicate. Statistical significance was determined by Student's *t*-test.

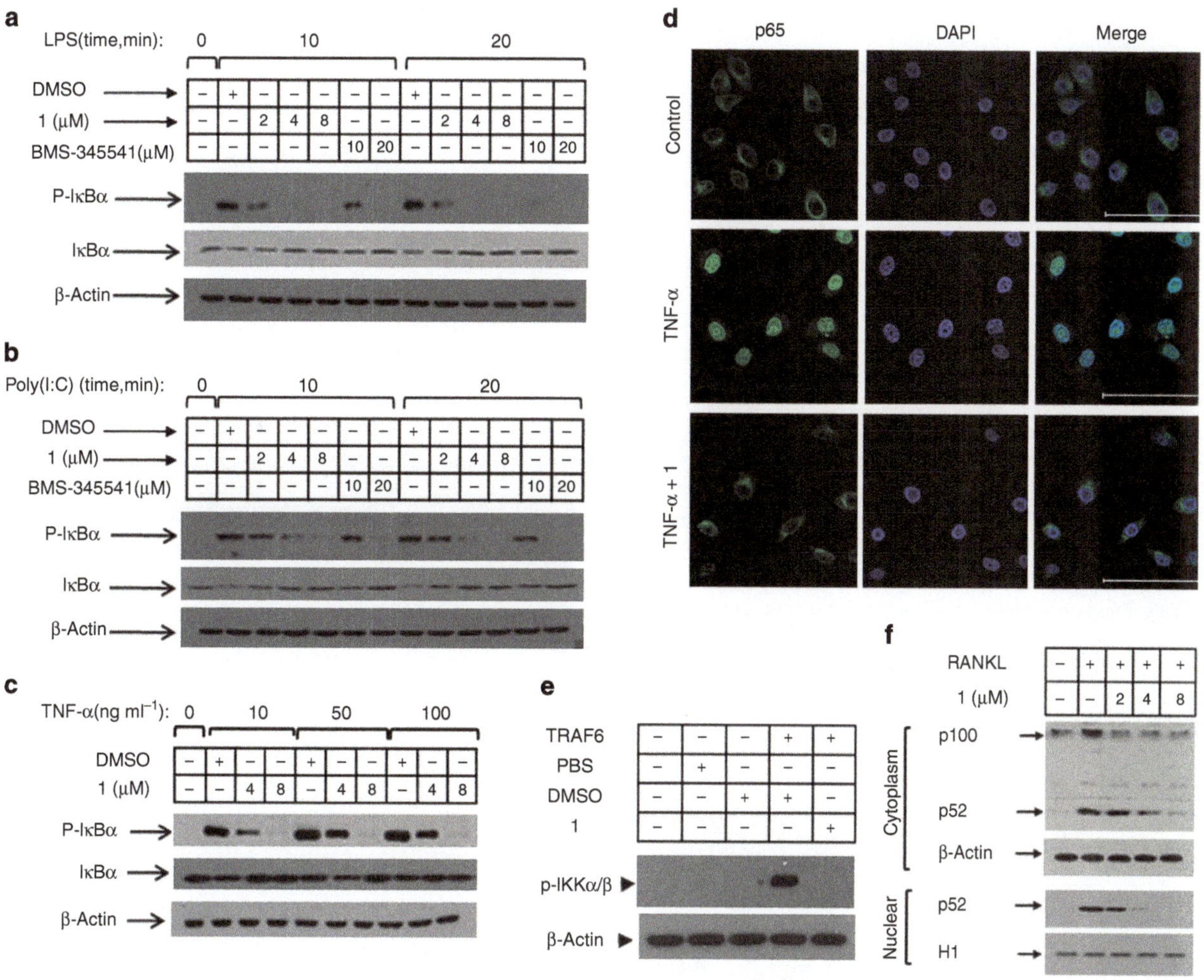

Figure 2 | Ainsliadimer A (1) blocks NF-κB activation induced by diverse stimuli in various cells. Raw 264.7 cells (**a**,**b**), and HeLa cells (**c**) were preincubated with the indicated concentrations of ainsliadimer A (**1**) for 1 h before stimulation with 20 ng ml^{-1} LPS (**a**), 20 μg ml^{-1} Poly(I:C) (**b**) or TNF-α (**c**). Cells were harvested at the indicated time points, and total cell lysates were tested by western blot experiments for the occurrence of IκBα phosphorylation. DMSO and BMS-345541 were used as negative and positive controls, respectively. All experiments were repeated at least three times, and similar results were obtained each time. (**d**) Ainsliadimer A (**1**) inhibited TNF-α-induced NF-κB/P65 nuclear translocation. HeLa cells were pretreated with ainsliadimer A (**1**, 8 μM) for 1.5 h and then treated with TNF-α (20 ng ml^{-1}) for an additional 0.5 h. DMSO was used as a negative control. After treatment, cells were stained with a primary anti-p65 antibody, and then the nucleus was counterstained with DAPI (blue) and examined using fluorescence microscopy. Scar bar, 100 μm. (**e**) Effects of ainsliadimer A (**1**) on TRAF6-induced phosphorylation of IKKα and IKKβ in cell-free extracts. S100 cell lysates from 293T cells were incubated with a PBS control, recombinant TRAF6, DMSO or ainsliadimer A (**1**), and their action was analysed by immunoblotting using antibodies, as indicated. (**f**) Effects of ainsliadimer A (**1**) on the non-canonical NF-κB pathway induced by RANKL. RAW264.7 cells were preincubated with the indicated concentration of ainsliadimer A (**1**) for 1.5 h, and then stimulated with 150 ng ml^{-1} RANKL for 30 min. The cells were then harvested and the cytoplasmic fractions and nuclear fractions were extracted and subjected to immunoblotting with the indicated antibodies.

Ainsliadimer A directly targets IKKα and IKKβ. To identify cellular proteins that interact with ainsliadimer A (**1**), we performed pull-down experiments using Probe and NC as positive and negative probes, respectively. The 293T-cell lysates were incubated with streptavidin agarose beads, which were pre-coupled with Probe or NC. The proteins precipitated by streptavidin agarose beads were resolved by sodium dodecyl sulfate–PAGE (SDS–PAGE) and stained with silver. As shown in (Fig. 4a), two clear bands observed at around 78 and 80 kDa were specifically precipitated by Probe, but not by NC. Peptide mass fingerprinting data analysis revealed that these two proteins were IKKα and IKKβ. We further used immunoblotting to monitor the presence of IKKα and IKKβ in the precipitates. Consistently, both IKKα and IKKβ were specifically pulled down by Probe, but not by NC (Fig. 4b). These data suggest that ainsliadimer A (**1**) targets IKKα and IKKβ or a complex that contains IKKα and IKKβ.

To determine whether ainsliadimer A (**1**) directly binds to IKKα and IKKβ, we generated recombinant IKKα and IKKβ proteins for further analysis. The 293T cells were transfected with Flag-IKKα or Flag-IKKβ plasmids, the recombinant Flag-IKKα or Flag-IKKβ was captured by anti-Flag agarose beads and then eluted off the beads using Flag peptide. The purified Flag-IKKα or Flag-IKKβ was incubated with DMSO or increasing concentrations of Probe, and then the mixtures were resolved by SDS–PAGE, followed by immunoblotting with streptavidin or anti-Flag antibodies. As shown in Fig. 4c, in the presence of Probe, a clear band with a molecular weight of around 80 kDa was detected by a streptavidin antibody. The signal intensity of this band increased with increasing concentrations of Probe, suggesting that there is a strong interaction between Probe and recombinant IKKα. Moreover, we also observed a similar interaction pattern between Probe and recombinant IKKβ

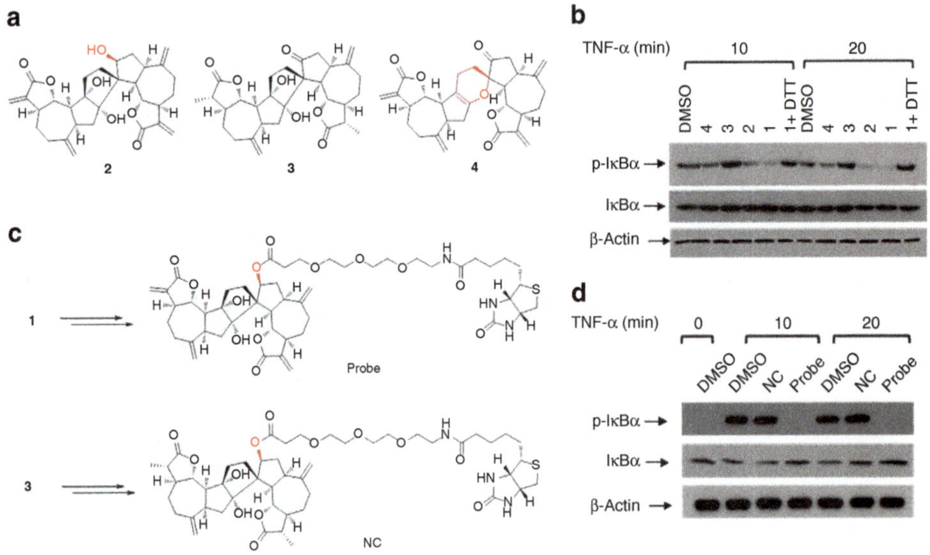

Figure 3 | Syntheses and biological evaluation of chemical probes. (**a**) The chemical structures of analogues of ainsliadimer A (**1**). (**b**) The 293 T cells were preincubated with 8 μM of ainsliadimer A (**1**) or 8 μM of ainsliadimer A analogues for 1 h before stimulation with 10 ng ml^{-1} TNF-α. Cells were harvested at the indicated time points, and the total cell extracts were tested by western blot experiments for the occurrence of IκBα phosphorylation. (**c**) Chemical structures of biotin-labelled ainsliadimer A (positive probe, Probe) and biotin-labelled **3** (negative control, NC). (**d**) The 293 T cells were preincubated with 50 μM Probe or NC for 1.5 h before stimulation with 10 ng ml^{-1} TNF-α. Cells were harvested at the indicated time points, and the total cell extracts were tested by western blot experiments for the occurrence of IκBα phosphorylation. All experiments were repeated at least three times, and similar results were obtained each time.

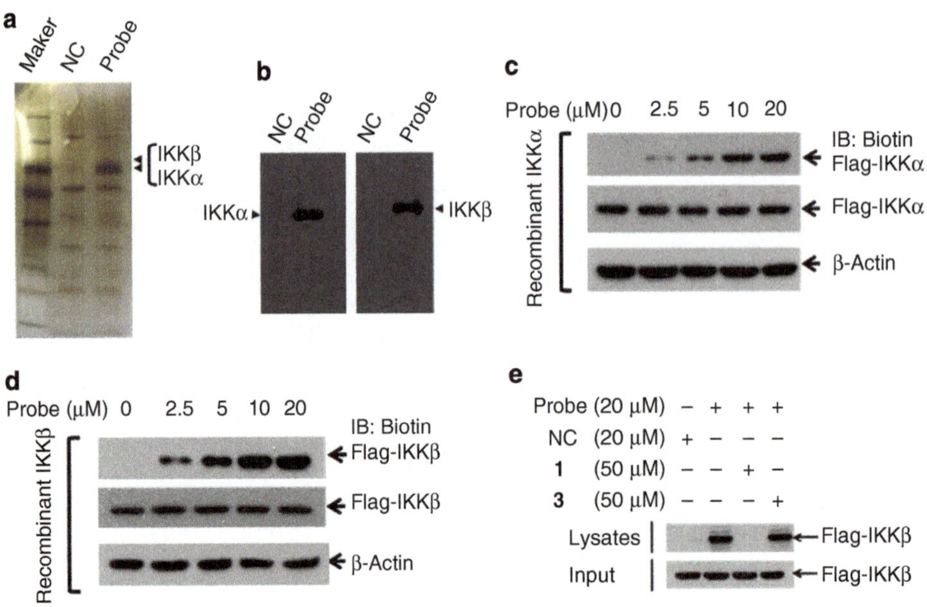

Figure 4 | Ainsliadimer A (1) directly targets IKKα and IKKβ. (**a,b**) The 293T cell lysates were incubated with Probe or NC at 4 °C overnight, the lysates were used for streptavidin-agarose pull-down assays, and the precipitates were resolved by SDS-PAGE, followed by silver staining. The indicated bands were excised and analysed via mass spectrometry (**a**), or detected by western blotting for IKK proteins as indicated (**b**). (**c,d**) The recombinant IKKα and IKKβ proteins were incubated with Probe or NC for 1 h at 37 °C, followed by immunoblotting with biotin (upper band) or flag (lower band). (**e**) The recombinant IKKβ protein was incubated with Probe in the absence or presence of a twofold excess of unlabelled ainsliadimer A (**1**) or **3** for 1.5 h at 37 °C, and the mixtures were blotted for biotin or Flag. All experiments were repeated at least three times, and similar results were obtained each time.

(Fig. 4d). This binding could be out-competed by an excess of unlabelled ainsliadimer A (**1**), but not by **3** (Fig. 4e). Notably, ainsliadimer A (**1**) had no association with IKKγ/NEMO, another subunit of the IKK complex, and had no association with other components in the NF-κB pathway such as TAK1 or TAB1 (Supplementary Fig. 6). Taken together, these results demonstrate that ainsliadimer A (**1**) can efficiently and specifically interact with IKKα and IKKβ, and that the double bond is critical for its binding with IKKα and IKKβ.

C46 of IKKα/β are critical for binding to ainsliadimer A. As ainsliadimer A (**1**) lost its inhibitory activity against NF-κB activation entirely when the enone moieties were saturated or in the presence of excess reducing agent dithiothreitol (DTT) (Fig.3b), the α,β-unsaturated moiety in ainsliadimer A (**1**) is most likely a reactive Michael acceptor, which forms a covalent bond with cysteine residues of its target proteins. Such a bond results in inactivation of the targeting enzymes, as previously demonstrated by natural products such as adenanthin[28]. Therefore, we speculated that the conserved cysteine residues in IKKα and IKKβ might be the binding sites of ainsliadimer A (**1**). We used BLAST analysis with the sequences of IKKα and IKKβ and found nine conserved cysteine residues (Supplementary Fig. 7). To evaluate which of these residues was critical for the binding of ainsliadimer A (**1**), we individually mutated these nine cysteine residues of IKKβ into alanine residues (Fig. 5a) using the primers detailed in Supplementary Table 1. Among these mutants, C46A was the sole mutant that completely lost the ability to interact with ainsliadimer A (**1**), while the other eight mutants retained high-affinity binding for ainsliadimer A (**1**) in a manner similar to wild-type (WT) IKKβ (Fig. 5a). These results indicated that C46 of IKKβ is essential for its binding to ainsliadimer A (**1**). As expected, mutation of C46 into serine in either IKKα or IKKβ abolished its corresponding interaction with ainsliadimer A (Fig. 5b).

To determine the specific residue modified by ainsliadimer A (**1**) in IKKβ, we incubated recombinant IKKβ protein with or without ainsliadimer A (**1**). After the reaction, the IKKβ protein was digested with trypsin and then analysed by LC-MS/MS. Database searching identified a peptide with calculated mass of 2,047.09 Da, which is 506.2 Da larger than the C46-containing

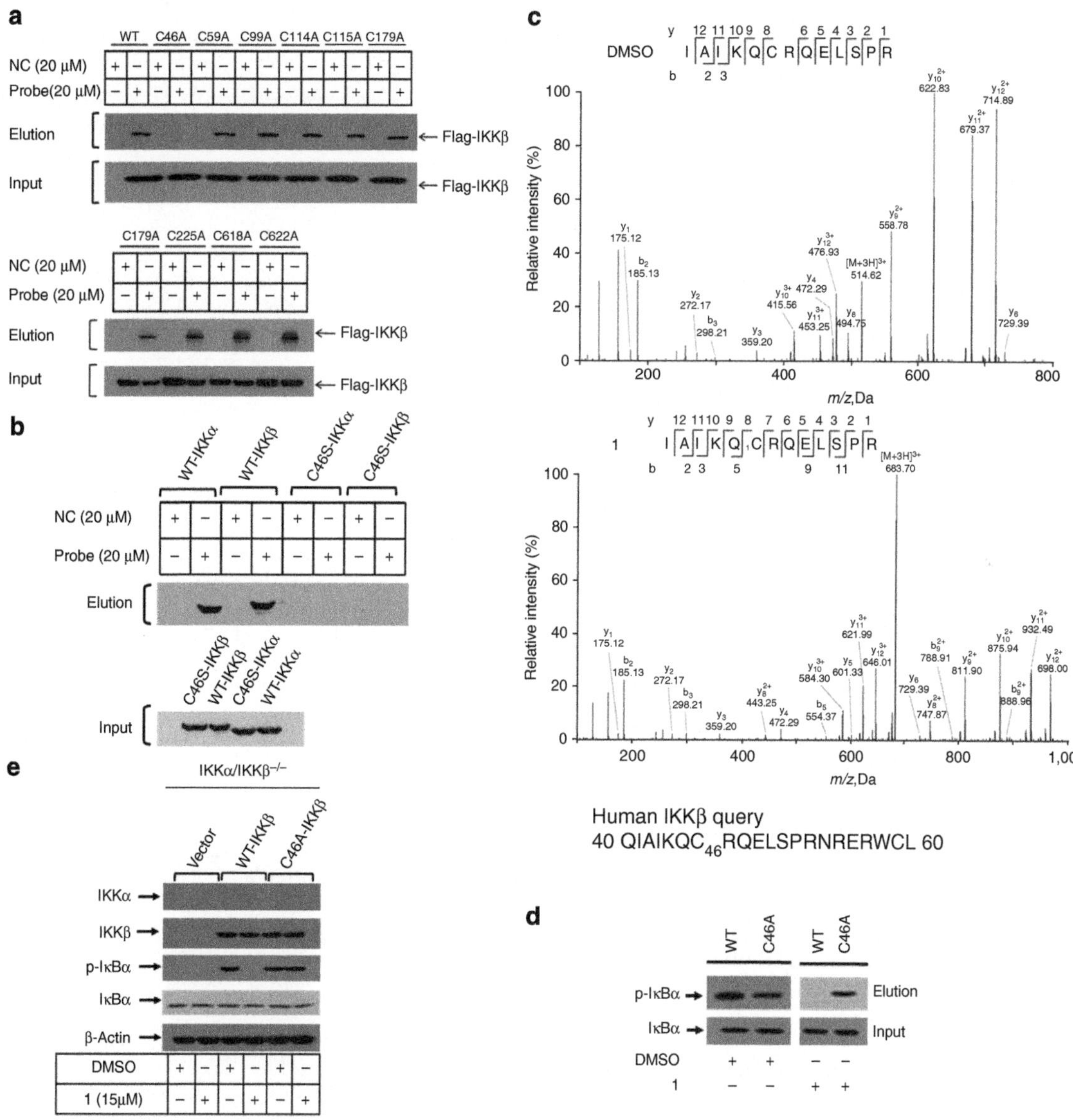

Figure 5 | Cysteine 46 of IKKβ is critical for its binding to ainsliadimer A (1). (**a**) Recombinant WT-IKKβ and IKKβ-mutant proteins were incubated with Probe or NC at 37 °C for 1.5 h, followed by pull-down with streptavidin-agarose; the precipitates were then resolved by SDS-PAGE and blotted for biotin or flag. (**b**) Recombinant WT-IKKα, WT-IKKβ, C46S-IKKα or C46S-IKKβ was incubated with Probe or NC at 37 °C for 1.5 h, followed by pull-down with streptavidin-agarose; the precipitates were then resolved by SDS-PAGE and blotted for biotin or Flag. (**c**) MS/MS analysis of the recombinant IKKβ incubated with or without ainsliadimer A (**1**) for 3.5 h. (**d**) Effects of ainsliadimer A (**1**) on phosphorylation of IκBα *in vitro* using WT-IKKβ or C46A-IKKβ. Purified Flag-IκBα was used as the substrate and the indicated recombinant proteins were used as kinases. The mixtures were incubated with DMSO or ainsliadimer A (10 μM) for 1 h at 37 °C. Immunoblotting using the p-IκBα antibody reflects the kinase activity and the effects of ainsliadimer A (**1**). (**e**) IKKα/IKKβ$^{-/-}$ double-knockout MEF cells were grown to 50% confluence, transfected with 200 ng of the indicated expression plasmids for 8 h, and then treated with ainsliadimer A (**1**) for 1.5 h. Shown are immunoblots of the total cell lysates using the indicated antibodies.

peptide IAIKQCRQELSPR that has a calculated mass of 1,540.85 Da. The mass difference of 506.2 Da matches the molecular weight of a ainsliadimer A (**1**) molecule. Tandem mass spectrometry of this peptide revealed that a 506.2 Da mass shift occurred starting from the y8 to the y12 fragment ions (Fig. 5c), indicating that the C46 residue was covalently modified by ainsliadimer A (**1**). To further confirm this, two synthetic peptides derived from human IKKβ (#1 and #2) containing C46 were incubated with ainsliadimer A (**1**), and the resulting mixture was analysed by MALDI-TOF (Matrix-Assisted Laser Desorption/Ionization Time of Flight) mass spectrometry. As expected (Supplementary Fig. 8), these two synthetic peptides had a mass shift corresponding to a molecular weight of ainsliadimer A (**1**) or Probe after the reaction, indicating that ainsliadimer A (**1**) and Probe could modify C46 *in vitro*. When the C46 in peptide #1 and peptide #2 was replaced by alanine or serine (referred to as peptides 1A, 2A, 1S and 2S), however, the mass of these peptides remained the same after the reaction with either ainsliadimer A (**1**) or Probe (Supplementary Fig. 8). Contrastingly, the synthetic IKKβ C215 containing peptides #3 (AA sequence 202-220) and #4 (AA sequence 212-231) could not be modified by either ainsliadimer A (**1**) or Probe (Supplementary Fig. 8). Together with the mutation binding assays shown in Fig.5b, these results indicate that ainsliadimer A (**1**) binds to IKKβ by covalently modifying its C46 residue.

Ainsliadimer A (**1**) likely modifies IKKβ through Michael addition at C46. To verify the functional significance of this binding event, we performed *in vitro* kinase assays using WT-IKKβ or C46A-IKKβ recombinant proteins as the kinase, and purified Flag-IκBα protein as the substrate. As shown in Fig. 5d, both recombinant WT-IKKβ and C46A-IKKβ phosphorylated IκBα, and the signal intensity of IκBα phosphorylation was similar, suggesting that the replacement of C46 by alanine does not affect the kinase activity of IKKβ. In the presence of ainsliadimer A (**1**), WT-IKKβ-induced phosphorylation of IκBα was significantly suppressed, whereas ainsliadimer A (**1**) did not affect the phosphorylation of IκBα triggered by C46A-IKKβ. Moreover, expression of WT-IKKβ or C46A-IKKβ in IKKα/IKKβ double-knockout MEF cells restored the phosphorylation and degradation of IκBα, and addition of ainsliadimer A (**1**) blocked the phosphorylation and degradation of IκBα induced by WT-IKKβ, but not that induced by C46A-IKKβ (Fig. 5e). Collectively, these results indicate that ainsliadimer A (**1**) inhibits IKKβ by direct modification of its C46 residue. Meanwhile, *in vitro* kinase assays using purified recombinant IKKβ and IκBα showed 2 μM of ainsliadimer A (**1**) could totally inhibit IκBα phosphorylation as BMS-345541 did (Supplementary Fig. 9). We next investigated the ability of ainsliadimer A (**1**) to affect the association of IKKβ with IκBα. As shown in Supplementary Fig. 10, treatment of ainsliadimer A (**1**) reduced the association of IKKβ with IκBα. Thus, ainsliadimer A (**1**) is proven to be able to block the function of IKKβ in the NF-κB pathway.

Ainsliadimer A is a novel allosteric inhibitor of IKKα/β. To further explore how ainsliadimer A (**1**) inactivates IKKα/β, we conducted molecular dynamics (MD) simulations using the first crystal structure of IKKβ (ref. 29) to sample the conformational states of IKKβ on ainsliadimer A (**1**) binding. According to the computational study, the binding mode shows that a stable hydrogen bond forms between the backbone amide group of C46 and the Michael acceptor carbonyl group of ainsliadimer A (**1**) during the simulation; this may be involved in stabilizing the transition state of a Michael addition reaction (Fig. 6a). In addition, ainsliadimer A (**1**) forms favourable hydrophobic interactions with residues including Trp58, Ile62, Val79, Leu91, Pro92 and Leu94 (Fig. 6b), where an allosteric binding pocket was previously identified in 3-phosphoinositide-dependent kinase 1 (PDK1)[30]. In addition, based on our 250 ns MD simulation results, the binding of ainsliadimer A (**1**) introduces significant conformational changes to the ATP binding-site in IKKβ by displacing the conserved structural αC-helix (Supplementary Fig. 11a). As we observed in our simulation of ainsliadimer A (**1**)-bound IKKβ, such an allosteric regulation mechanism was also reported in the crystal structures of PDK-1 kinase, where a covalent inhibitor binds with Cys148 in the same allosteric binding pocket and induces a positional change in the C-helix by moving away from the ATP-binding site (Supplementary Fig. 11b)[30,31].

Notably, ainsliadimer A (**1**) completely blocks solvent access to Trp58, which is located at the bottom of the allosteric pocket and interacts with ainsliadimer A (**1**) favourably throughout the simulation (Fig. 6c). Therefore, binding of ainsliadimer A (**1**) to the allosteric pocket can be experimentally assessed using the tryptophan fluorescence quenching technique, which has been widely used to demonstrate allosteric effects when co-crystal structures are not available[32]. Accordingly, we performed tryptophan fluorescence quenching studies to investigate whether the trypophan (W58) of IKKβ undergoes any changes on binding with ainsliadimer A (**1**). On the basis of the slope of the plot (Fig. 6d), we determined the Stern–Volmer constant for IKKβ quenching by acrylamide with or without the treatment of ainsliadimer A (**1**), which was 7.6709 ± 0.2 and $11.076 \pm 0.2\,M^{-1}$, respectively. This indicates that the binding of ainsliadimer A (**1**) to IKKβ partially protects W58 from the quencher by inducing conformational changes of the target protein. In other words, the process of ainsliadimer A (**1**) binding to IKKβ changes the disposition of W58 from being partially exposed to being in a more completely buried state. As expected, mutation of W58 into alanine in IKKβ abolished its corresponding interaction with ainsliadimer A (Supplementary Fig. 12). We propose that this conformational change induced by ainsliadimer A (**1**) might lead to the inhibition of IKKβ. In addition, it is worth noting that the downward curvature in the Stern–Volmer plot existed in the presence of ainsliadimer A (**1**); this was conceivably caused by the existence of both drug-bound and drug-free receptor populations, whose tryptophan was exposed to differing extents. Together, our data suggest that W58 on the α C-helix is required for the interaction between ainsliadimer A (**1**) and IKKβ. As displacement of α C-helix by a small molecule inhibitor or regulatory protein is able to affect ATP binding[33], we further measured the ATP-binding kinetics of IKKβ in the absence or presence of ainsliadimer A (**1**). The result indicated that ainsliadimer A (**1**) showed ATP-competitive inhibitory effect on IKKβ (Supplementary Fig. 13). Moreover, ainsliadimer A (**1**) significantly suppressed WT-IKKβ-induced phosphorylation of IκBα. Collectively, these data indicate that C46 of IKKα/β likely acts as a novel allosteric regulation site of these enzymes.

Next, we investigated the binding kinetics of ainsliadimer A (**1**), as the conventional IC_{50} value could not well characterize the potency and selectivity for irreversible covalent inhibitors that we observed. We detected the extent of IKKβ binding at several time points for a number of Probe concentrations[28]. The progress of IKKβ binding with ainsliadimer A (**1**) at various concentrations showed time-dependent saturation (Fig. 6e), consistent with an irreversible mechanism of binding. The data were fit to determine the observed rate constants for binding (k_{obs}) at each inhibitor concentration. Plotting these k_{obs} values as a function of Probe concentration revealed a saturation curve (Fig. 6f). These data support a two-step reaction mechanism for the inactivation of IKKβ by ainsliadimer A (**1**), involving the noncovalent reversible

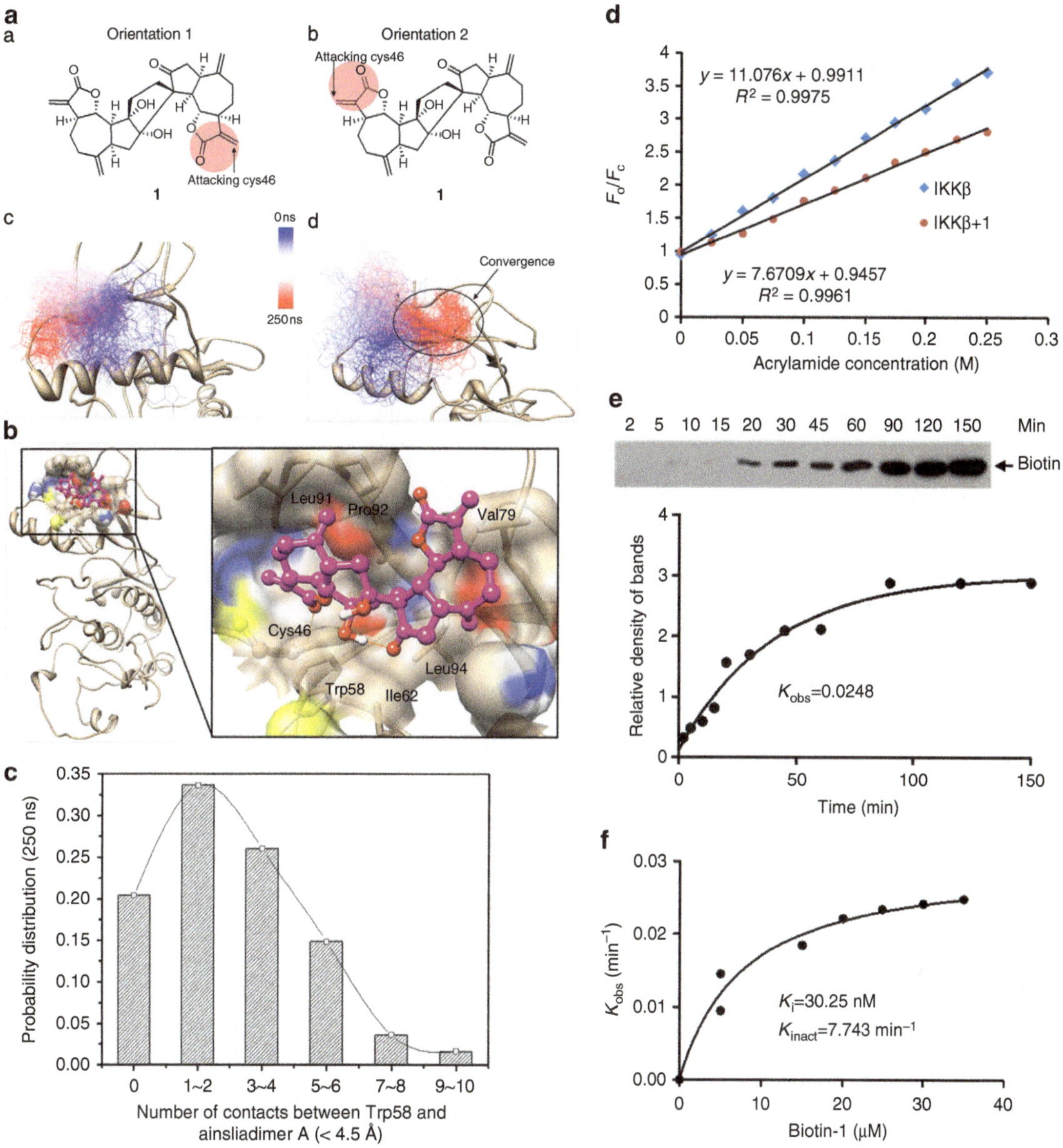

Figure 6 | Ainsliadimer A (1) is a novel allosteric inhibitor of IKKβ. (**a**) (A and B) The two modelled structures used to predict that a Michael acceptor (highlighted using red circle) forms a covalent bond with C46; (C and D) Superimposed structures representing the 250 ns MD simulations in the Orientation-1 and the Orientation-2 systems, respectively. IKKβ is presented in a ribbon model in grey color (sulfur atom of C46 is shown as a yellow sphere), the ligand is shown in a thin line and coloured gradually from deep blue to deep red along the time series. Structural convergence was only observed in the Orientation-2 system, where the compound binds in a relatively stable conformation (highlighted in the black circle) after 180 ns and lasts for the remaining 70 ns (movie S1 records all of the MD simulation trajectories). (**b**) A representative view of a simulated binding complex. (A) The left side shows the global view of the entire IKKβ structure; the right side shows the focused view of ainsliadimer A (1) in the allosteric binding site. Ainsliadimer A (**1**) forms a hydrogen bond with the backbone of C46 and interacts favourably with several hydrophobic residues including Trp58, Ile62, Val79, Leu91, Pro92 and Leu94. (**c**) The number of close contacts (distance <4.5 Å) between any heavy atoms of ainsliadimer A (**1**) and Trp58. (**d**) Binding of ainsliadimer A (**1**) to IKKβ induced conformational change. Stern-Volmer plots for quenching of the intrinsic tryptophan fluorescence of IKKβ by acrylamide are shown. The detailed experimental protocol is described in the Online Methods. (**e**) The determination of k_{obs} for interaction of IKKβ (0.5 μM) with Probe (15 μM) for different times was calculated as described in Online Methods. (**f**) Plotting the k_{obs} value for binding of IKKβ as a function of ainsliadimer A (**1**) concentration.

binding of IKKβ to ainsliadimer A (the initial binding step, K_i), which places the moderately reactive electrophile of ainsliadimer A (**1**) close to a specific nucleophile on the protein, followed by the second chemical step that results in specific covalent linkage (k_{inact}) leading to the inhibited complex[34,35]. On the basis of this model, the k_{inact} value for IKKβ binding with ainsliadimer A (**1**) was determined to be 7.743 min^{-1} (Fig. 6f). The K_i value for IKKβ binding to ainsliadimer A (**1**) was determined to be 30.25 nM (Fig. 6f). Overall, the binding process between ainsliadimer A (**1**) and IKKβ is effective ($k_{inact}/K_i = 4.26 \times 10^6\,M^{-1}s^{-1}$), with good affinity ($K_i = 30.25$ nM) and specific reactivity ($k_{inact} = 7.743\,min^{-1}$) (Fig. 6e,f). We also systematically tested 340 human kinases through a contract service (http://www.reactionbiology.com/) to further confirm this specific binding to IKKβ. As shown in Supplementary Fig. 14, at the concentration of 200 nM, ainsliadimer A (**1**) did not show significant enzyme inhibition effects as compared with the inhibitory effect against IKKβ.

Ainsliadimer A inhibits tumour growth. In addition to its crucial role in controlling immune responses, the NF-κB pathway has been shown to negatively regulate apoptosis and positively regulate the onset and progression of tumours[12]. Aberrant or constitutive NF-κB activation has been observed in many cancer cells, which usually exert increased resistance to chemotherapy. Inhibition of NF-κB results in an increased chemosensitivity of cancer cells to therapeutic drugs as well as cell death. To assess the anticancer effects of ainsliadimer A (**1**), we treated gastric adenocarcinoma BCG-823 cells with ainsliadimer A (**1**) alone, TNF-α alone or with a combination of ainsliadimer A (**1**) and TNF-α for a period of 24 h. As shown in Fig. 7a, pretreatment with TNF-α resulted in a synergistic repression of cell viability. We further examined whether ainsliadimer A (**1**) triggers apoptosis by measuring caspase-3, which is activated via proteolytic cleavage during apoptosis. In BCG-823 cells treated with ainsliadimer A (**1**), we detected cleaved caspase-3 that could be blocked by z-VAD, a pan-caspase inhibitor (Fig. 7b). Furthermore, we found that ainsliadimer A (**1**) suppressed the activation of NF-κB (Supplementary Fig. 3) and downregulated anti-apoptotic genes including c-FLIP, c-IAP and BCL-XL (Fig. 7c). These results demonstrate that the effects of ainsliadimer A (**1**) on the NF-κB pathway facilitate apoptosis of cancer cells. Finally, we evaluated the *in vivo* effects of ainsliadimer A (**1**) in tumour growth. Nude mice bearing a BCG-823 human gastric adenocarcinoma xenograft were treated with Vehicle or ainsliadimer A (**1**) (30 mg kg^{-1}, qdx8) or BMS-345541, a widely used IKKβ inhibitor (30 mg/kg, qdx8). As shown in Fig. 7d, significant xenograft tumour remission was observed after 16 days in all (eight of eight) of the mice treated with ainsliadimer A (**1**), as compared with the Vehicle-treated mice. Notably, ainsliadimer A (**1**) exerted a comparable anti-tumour effect to that of the therapeutic drug BMS-345541. Importantly, the mice did not display any discernible side effects or significant changes in body weight during administration of ainsliadimer A (Fig. 7e). Collectively, these results demonstrate that the inhibition of NF-κB by ainsliadimer A (**1**) resulted in significant death of human cancer cells and suppression of tumour growth with minimal undesired toxicities, thus implicating ainsliadimer A (**1**) as a novel IKK inhibitor with promising potential for anticancer and anti-inflammatory therapeutics.

Discussion

NF-κB transcription factors play a critical function in the regulation of immune responses, inflammation, cell proliferation and survival. Aberrant activity of NF-κB is associated with tumour progression as well as auto-immune and inflammatory diseases. Therefore, crucial regulators in the NF-κB pathway such

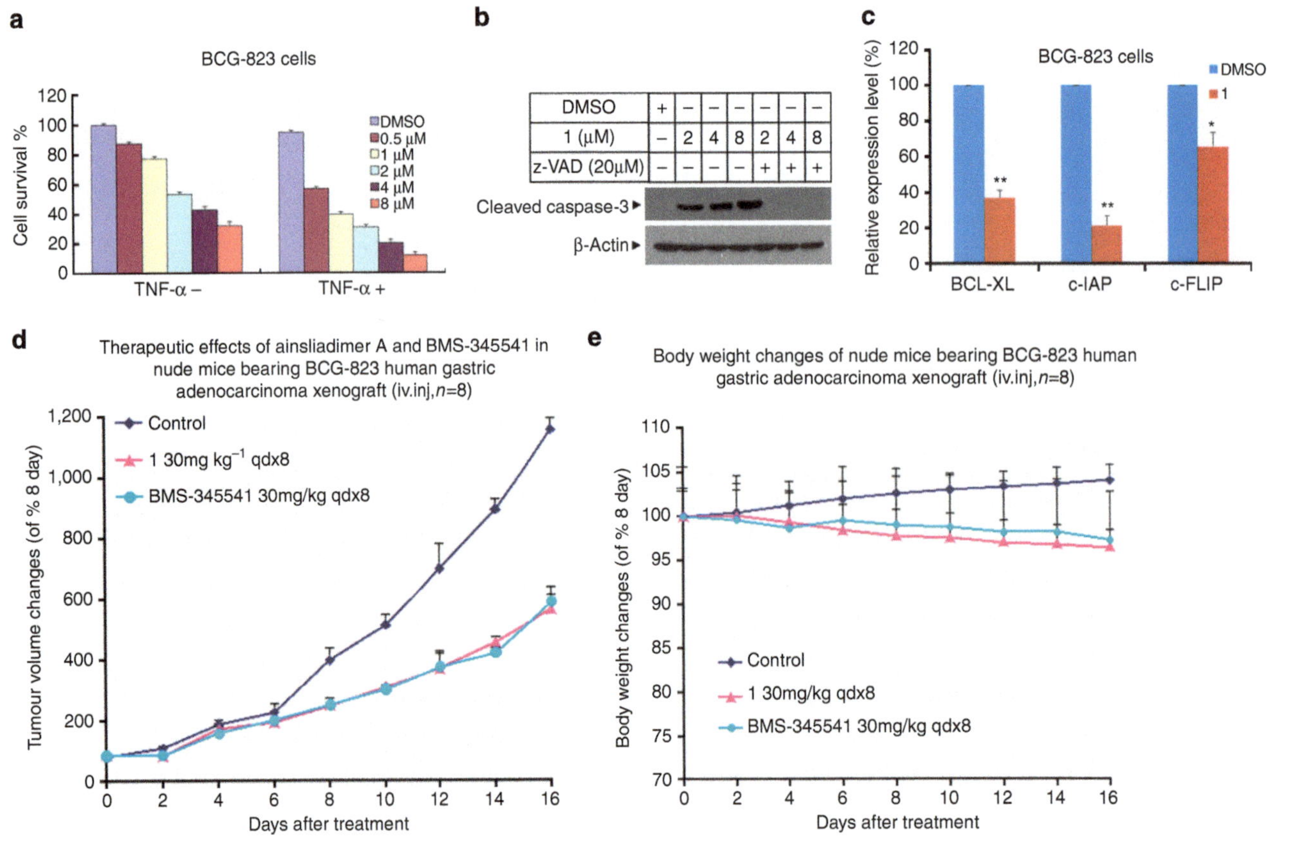

Figure 7 | Ainsliadimer A (1) induces cell death of tumour cells and represses growth of xenograft tumours. (**a**) Ainsliadimer A (**1**) showed anticancer activity in BCG-823 cells. BCG-823 cells were seeded in a 96-well plate and then treated with TNF-α (10 ng ml^{-1}) after pretreatment with the indicated concentration of ainsliadimer A (**1**) for 24 h. Cell viability was assayed by measuring ATP levels. (**b**) Ainsliadimer A (**1**) induced caspase-3 activation. BCG-823 cells were treated with or without the pan-caspase inhibitor z-VAD-FMK (40 μM) 1 h before the addition of 4 μM of ainsliadimer A (**1**), and then caspase-3 activation was measured by western blot using anti-cleaved caspase-3 antibodies. (**c**) Effects of ainsliadimer A (**1**) on NF-κB-regulated genes involved in apoptosis in BCG-823 cells. BCG-823 cells were treated with or without 4 μM ainsliadimer A (**1**) for 16 h, the expression of Bcl-XL, c-IAP and c-FLIP was quantified by real-time PCR. All the experiments were carried out in triplicate. The data are means ± s.d. of the indicated experiments. *$P<0.05$ versus DMSO and **$P<0.01$ versus DMSO. (**d,e**) *In vivo* therapy with ainsliadimer A (**1**) in disseminated BCG-823 human gastric adenocarcinoma xenografts. Data plotted are the representative tumour size (**e**) or mouse body weight (**f**) of two rounds of experiments ($n=8$ for each group).

as IKKα/β are considered as promising therapeutic targets. We show here that a natural product ainsliadimer A (**1**) specifically binds to IKKα/β and subsequently blocks both canonical and non-canonical NF-κB activation.

Ainsliadimer A (**1**) was originally isolated from a Chinese herb[22]. Although ainsliadimer A (**1**) has been shown to suppress LPS-mediated NO production[22], its precise molecular targets and mode of action remained uncharacterized. LPS is known to activate TLR4 and to result in NF-κB activation and type I IFN responses. Interestingly, we found that ainsliadimer A (**1**) specifically blocked LPS-induced NF-κB activation, but not type I IFN responses. Moreover, ainsliadimer A (**1**) exerts potent inhibition of NF-κB activation triggered by different stimuli such as TLR3 ligand poly(I:C) and TNF. Using biotin-labelled **1** as a probe, we found that ainsliadimer A (**1**) directly targeted IKKα and IKKβ, but not other components of the NF-κB pathway such as IKKγ/NEMO, TAK1 and TAB1, suggesting that an exclusive binding mechanism is shared by IKKα and IKKβ.

Ainsliadimer A (**1**) is a structurally unique and complex sesquiterpene lactone dimer with α, β-unsaturated enone moieties that may serve as Michael acceptors. Consistently, our mutagenesis and mass spectrometry results demonstrated that ainsliadimer A (**1**) formed a covalent bond with the conserved C46 of IKKα/β. Through both computational and experimental studies we further uncovered, remarkably, a novel putative allosteric binding pocket around C46 of IKKβ that could explain the high-binding specificity of ainsliadimer A (**1**) bound to IKKβ. Notably, the tryptophan fluorescence quenching results provided direct and strong evidence to support this novel allosteric effect. In addition, ainsliadimer A (**1**) was shown to be an irreversible inhibitor that exhibited specific and potent kinase inhibition activity against IKKβ with k_{inact}/K_i of $4.26 \times 10^6\ M^{-1} s^{-1}$.

Many small molecule inhibitors that interfere with IKKs have been previously described[12,21], but the precise modifications of the target by most of these inhibitors remain obscure. To our knowledge, ainsliadimer A (**1**) represents the first small molecule that blocks NF-κB by directly targeting C46 of IKKα/β. The putative allosteric site around C46, most likely unique to IKKα/β, confers the high selectivity of ainsliadimer A (**1**) for IKKα/β, thereby tending to have low non-specific toxicity and a high therapeutic index. Further, in both *in vitro* and *in vivo* studies, ainsliadimer A (**1**) was shown to be a promising agent against the survival of cancer cells, tumour growth and endotoxin-mediated inflammatory responses. It is tempting to speculate that this unique inhibitory effect of ainsliadimer A (**1**) on IKKα/β should be further exploited for the elucidation of uncharacterized mechanisms of IKKα/β activation, as well as the development of new anticancer and anti-inflammatory therapies.

Methods

Chemical synthesis. The syntheses of **2**, **3**, the positive probe (Probe) and the negative probe (NC) were carried out according to Supplementary Fig 16 and the Supplementary Methods. NMR spectra of the compounds involved in this work were shown in Supplementary Figs 17–28. Ainsliadimer A (**1**) and **4** were synthesized as described in ref. 23.

Reagents. DMEM medium, phosphate buffered saline (PBS), trypsin and penicillin/streptomycin were purchased from GIBCO. ATP, $MgCl_2$, fetal bovine serum (FBS), BMS-345541, DTT, DMSO and LPS were purchased from Sigma. The Cell Titer-Glo Luminescent Cell Viability Assay kit was purchased from Promega. The Protein Assay kit was purchased from Bio-Rad. The HTRF KinEASE kit was purchased from Cisbio. IKKβ kinase protein was purchased from Carna Biosciences. RANKL was purchased from Sino Biological. Human IκBα full length protein was purchased from Abcam (ab59981).

Plasmids. The PCS2-3Flag-IKKα, PCS2-3Flag-IKKβ, PCS2-3Flag-TAK1, PCS2-3Flag-NEMO, PCS2-3Flag-TAB1, PCS2-3Flag-NIK, myc-p100, Flag-IκBα plasmids were kindly provided by Dr Feng Shao and his coworkers (NIBS).

Antibodies. The following antibodies were used for western blotting: p-IκBα (1:1,000 dilution, Cell Signaling, 9246), IκBα (1:1,000 dilution, Cell Signaling, 9242), IKKα (1:1,000 dilution, Cell Signaling, 2682), IKKβ (1:1,000 dilution, Cell Signaling, 2684), p-IKKα/β (1:1,000 dilution, Cell Signaling, 2697), p100 (1:1,000 dilution, Cell Signaling, 4882), H1 (1:1,000 dilution, Santa Cruz Biotechnology, sc-8030), Biotin (1:10,000 dilution, Sigma, A4541), Flag (1:10,000 dilution, Sigma, A8592), myc (1:3,000 dilution, Sigma, C3956), p-IRF3 (1:1,000 dilution, Cell Signaling, 4947) and β-actin (1:3,000 dilution, Cell Signaling, 4967).

Cell cultures. Human cervical cancer HeLa, human glioblastma T98G, human gastric adenocarcinoma BCG-823, human embryonic kidney 293 T and mouse macrophage Raw264.7 cells were obtained from ATCC. These were cultured in DMEM medium supplemented with 10% (v/v) heat-inactivated FBS in a humidified incubator at 37 °C and 5% CO_2/95% air (v/v). All media were supplemented with 10% FBS (Invitrogen) and 100 units ml^{-1} penicillin/streptomycin (Hyclone).

Cell survival assay. The cell survival assay was performed using the Cell Titer-Glo Luminescent Cell Viability Assay kit (Promega) according to the manufacturer's instructions. Luminescence was recorded with a TecanGENios Pro plate reader.

NF-κB activation. Cells were pre-incubated with various concentrations of compound ainsliadimer A (**1**) for 1 h before stimulation with the indicated concentrations of TNF-α, 20 μg ml^{-1} Poly(I:C) or 200 ng ml^{-1} LPS and then harvested at the indicated time points. Total cell extracts were tested by western blotting for the occurrence of IκBα phosphorylation. DMSO and BMS-345541 were chosen as negative and positive controls, respectively.

NF-κB/p65 nuclear translocation immunofluorescence. HeLa Cells were grown to 50–70% confluence in a glass chamber and pretreated with DMSO or 8 μM ainsliadimer A (**1**) for 1.5 h and then stimulated by 20 ng ml^{-1} TNF-α for 30 min. Cell treatment was terminated by washing with PBS, followed by fixation in freshly prepared 4% paraformaldehyde in PBS for 10 min. The fixed cells were washed three times with PBS and then permeabilized in 0.25% Triton X-100 in PBS for 10 min. After blocking with 2 mg ml^{-1} BSA for 1 h at room temperature, we added a P65 antibody at 1:1,000 dilution at 4 °C overnight, washed three times with PBS, then added 1:500 dilution of secondary AlexaFluor488 antibody for 1 h at room temperature in the dark. Finally DAPI was used to stain the nuclei at 37 °C for 30 min in dark. Microscopy was performed using a Nikon eclipse 80i.

Western blot analysis. Cell pellets were collected and resuspended in lysis buffer (20 mM Tris-HCl, pH 7.4,150 mM NaCl, 10% glycerol, 1% Triton X-100, 1 mM Na_3VO_4, 25 mM β-glycerol-phosphate, 0.1 mM PMSF, with a complete Roche protease inhibitor set and a Sigma phosphatase inhibitor set). The resuspended cell pellet or tissue was vortexed for 10 s, incubated on ice for 20 min, and centrifuged at 20,000 × *g* for 20 min. The supernatants were collected for western blot analysis or immunoprecipitation. Images have been cropped for presentation. Full-size images are presented in Supplementary Fig. 15.

Pull-down and MS analysis of 1-bound proteins. The 293 T cells were plated on 10-cm Petri dishes and grown to confluence for 1 day. The cells were harvested and lysed in lysis buffer as described above. Probe or NC was preincubated with streptavidin agarose (Invitrogen) overnight at 4 °C. The beads were then washed three times in lysis buffer. The ainsliadimer A-bound agarose was then incubated with cell lysates overnight at 4 °C. The following day, the beads were washed six times with lysis buffer, then the immunoprecipitations were eluted off the beads using low pH elution buffer at 4 °C for 15 min. Acid elution was neutralized by adding 1/20 volume of 1 M Tris-HCl, pH 9.4, and the preparations were finally boiled in SDS loading buffer. The bead-bound proteins were separated by SDS–PAGE and visualized by silver staining. The protein-containing band in the gel was excised, followed by in-gel digestion and analysis by LC-MS/MS.

Preparation of the wild-type and the site-mutated IKKα or IKKβ. Human IKKα or IKKβ were cloned into a PCS2 vector containing a 3 × Flag tag sequence at the N-terminal region. Site-directed mutagenesis was performed with the QuikChange site-directed mutagenesis kit (Stratagene) using Flag tag-IKKα or Flag tag-IKKβ as a template. These proteins were expressed in 293 T cells and subsequently purified.

MALDI-TOF analysis. The reaction mixtures of peptides or proteins were analysed by matrix-assisted laser desorption/ionization-time of flight mass spectrometry with an Autoflex II MALDI-TOF/TOF mass spectrometer equipped with a nitrogen pulsed laser (Bruker). In brief the peptide mixtures were loaded on C18 Zip Tips separately, washed with 0.1%TFA, eluted with 2, 5-dihydroxybenzoic acid solution (Agilent), then spotted on a Bruker MTP 384 massive stainless steel MALDI target. The matrix spots were allowed to dry at room temperature. Peptide Mass Spectra were acquired in positive reflectron mode with pulsed ion extraction.

Molecular modelling. The crystal structure of IKKβ (PDB entry: 3QA830) was used to generate initial binding complexes. The MD simulation was performed with the Gromacs 4.0.7 package[35,36]. During the simulation, we used the Amber99SB protein force field[37] for proteins and general amber force field[38,39] for small molecules. The complex was placed in a PBC box with at least 12 Å between solvent and solutes. The whole system was neutralized with both sodium and chlorine ions. Distance-dependent harmonic restraint was placed onto the sulfur atom of C46 and the Michael acceptor carbon atom of the ainsliadimer A with the initial distance value set to be 10 Å.

Tryptophan fluorescence quenching studies. Twenty-five micromolars of ainsliadimer A (**1**) or DMSO was preincubated with 5 μM IKKβ for 90 min and titrated with 2.5 μl of 1 M acrylamide; we monitored the tryptophan florescence signal intensity by using 340 nm emission and 295 nm excitation. At the same time, the background signal was corrected by using a paired control that lacked protein. After using the Stern–Volmer equation to analyse F_0/F, we compared the slopes from IKKβ with or without the treatment of ainsliadimer A (**1**). For the Stern–Volmer equation:

$$F_0/F = 1 + K_{S,V}[Q],$$

where F_0 is the emission intensity of the protein without adding quencher, F is the emission intensity of the protein at a given quencher concentration, $K_{S,V}$ is the Stern–Volmer constant for quenching, given by the slope when data are plotted as F_0/F versus $[Q]$.

The ATP binding kinetics of IKKβ. We used an HTRF KinEASE kit (Cisbio Bioassays, Bagnols/Ceze Cedex, France) to measure the ATP binding kinetics of IKKβ. After optimizing the conditions of IKKβ according to the kit protocol, the assays were carried out in the low volume, black 384-well plates (3676 from Corning life Science, MA), with a 10 μl assay volume containing varied concentration of ATP, non-limiting STK Substrate 2-biotin and 1 ng μl^{-1} of IKKβ. Following incubation at 37° for 1.5 h, the reaction was stopped with buffer EDTA, which contained the detection reagents, strepavidin-XL 665 and STK-antibody labelled with Eu3+-cryptate. The resulting TR-FRET signal, calculated as the fluorescence ratio at 665/620 nm, was read on the Envision. The K_m for ATP was determined with the Michaelis–Menten equation fits in GraphPad Prism 5.0.

Kinetic determination of the IKKβ-Ainsliadimer A (1) interaction. The specific interaction between IKKβ and Probe followed two steps: in the initial binding step, ainsliadimer A (**1**) moves close to specific nucleophile residues of IKKβ to form the initial encounter complex IKKβ: ainsliadimer A (**1**), and K_i is the apparent dissociation constant for the initial IKKβ: ainsliadimer A (**1**) complex; in the second chemical step, the formation of specific covalent linkage (k_{inact}) between ainsliadimer A (**1**) and IKKβ resulted in the inhibited complex and k_{inact} the rate constant for IKKβ- ainsliadimer A (**1**) formation at saturating ainsliadimer A (**1**). To get the K_i and K_{inact} value, we incubated IKKβ with a large excess of ainsliadimer A (**1**) for different time periods and then analysed the streptavidin-HRP signal intensity quantified by densitometry. During this process, we performed scanning at the optimal exposure time to ensure that band intensity was proportional to the concentration of protein present. Under the pseudo-first order experimental conditions, the reaction proceeds as in these equations:

$$[\mathrm{IKK}\beta]_t = [\mathrm{IKKb}]_0{}^*E - k_{obs}{}^*t \quad (1)$$

$$[\mathrm{IKK}\beta - 1] = [\mathrm{IKK}\beta]_0 -]\mathrm{IKK}\beta]_t = [\mathrm{IKK}\beta]_0{}^*(1 - E - k_{obs}{}^*t) \quad (2)$$

In equation (1), $[\mathrm{IKK}\beta]_0$ is the total concentration of IKKβ added to the solution and $[\mathrm{IKK}\beta]_t$ is the concentration of IKKβ at time t.

We initially obtained the reaction rate constants of k_{obs} at different ainsliadimer A (**1**) concentrations by fitting to equation (2). We then obtained the value of k_{inact} and K_i by fitting k_{obs} values to equation (3)

$$k_{obs} = k^*_{inact}[1]/([1] + K_i) \quad (3)$$

RNA extraction and real-time PCR analysis. Cells were lysed and total RNA was extracted using the TRIzol Reagent (Invitrogen). Synthesis of cDNA was done with 1 μg of RNA from oligo (dT)-20 primers using the Superscript II first strand synthesis system (Invitrogen). Real-time PCR was performed using specific primers and the SYBR Green ROX Mix (Thermo Scientific). The sequence of the primers is given in the Supplementary Table 2. All experiments were performed in triplicate, and data were normalized to the housekeeping gene *GAPDH*.

Analysis of levels of LPS-induced pro-inflammatory cytokines. Cytokines were assessed by measuring serum concentrations. C57 Male mice at 7–8 weeks of age were randomly assigned into five groups: the PBS, LPS, LPS + Vehicle, LPS + ainsliadimer A (**1**) and LPS + BMS-345541 groups. The mice were administered intravenously with ainsliadimer A (**1**) (25 mg kg^{-1}) (LPS + ainsliadimer A group), BMS-345541(25 mg kg^{-1}) (LPS + BMS-345541group), Vehicle (LPS + Vehicle group) or PBS(PBS group) for 90 min, and then the mice (LPS group, LPS + 1, LPS + Vehicle group) were given intraperitoneal injection of LPS (1.6 mg kg^{-1}). Mice in those groups were killed and their serum was separated from clotted blood at 3.5 h following administration of LPS. Serum was stored at −80 °C, and concentrations of cytokine IL-1β and IL-6 were measured by sandwich ELISA using commercially available reagents according to the manufacturer's instructions (Beijing Gersion Bio-Technology). The concentration of cytokine TNF-α was measured using a Mouse Inflammatory Cytometric Bead Array (BD Biosciences, Palo Alto, CA) according to the manufacturer's protocol. For each group, plasma samples were obtained from at least six mice and analysed in duplicate.

Meantime, we also got the spleen extracts from each mice and then analysed the protein levels of p-IκBα, IκBα and β-Actin by western blot.

Establishment of subcutaneous and disseminated xenografts. Female nude (nu/nu) mice (6–7 weeks old) were purchased from Vital River Company and acclimated for 1 week before the start of the experiment. All animal experiments were performed in accordance with protocols by the Institutional Animal Care and Use Committee at National Institute of Biological Sciences (NIBS). Exponentially growing BCG-823 cells were suspended in PBS and mixed in a 1:1 ratio with Matrigel 0.1 ml suspension containing 2.5–5 × 10^6 cells, and was injected subcutaneously on both the left and the right flank of each mouse above the hind limb. After tumour volumes reached 100 mm^3, the mice were divided into experimental cohorts of eight mice each. The mice were received the intravenous injection of vehicle, ainsliadimer A (30 mg kg^{-1}, qdx8) or BMS-345541, a widely used IKKβ inhibitor (30 mg kg^{-1}, qdx8). Control mice were treated with the same amount of vehicle under the same protocol. Tumour volume was calculated by using the formula mm^3 = $4/3\pi r^3$, where r = (length + width)/4, measuring the two largest perpendicular axes of the tumour.

Statistical analysis. All data are expressed as the mean ± s.e.m. Results are representative examples of three individual experiments. Inferential statistical analyses were performed with a Student's t-test (*$P<0.05$; **$P<0.01$; ***$P<0.001$; NS, not significant). Graph Pad Prism was used for analysis.

References

1. Hayden, M. S. & Ghosh, S. Shared principles in NF-kappaB signaling. *Cell* **132,** 344–362 (2008).
2. Vallabhapurapu, S. & Karin, M. Regulation and function of NF-kappaB transcription factors in the immune system. *Annu. Rev. Immunol.* **27,** 693–733 (2009).
3. Hoffmann, A. & Baltimore, D. Circuitry of nuclear factor kappaB signaling. *Immunol. Rev.* **210,** 171–186 (2006).
4. Hayden, M. S. & Ghosh, S. Signaling to NF-kappaB. *Genes Dev.* **18,** 2195–2224 (2004).
5. Hacker, H. & Karin, M. Regulation and function of IKK and IKK-related kinases. *Science's STKE* **2006,** re13 (2006).
6. Brown, K., Gerstberger, S., Carlson, L., Franzoso, G. & Siebenlist, U. Control of I kappa B-alpha proteolysis by site-specific, signal-induced phosphorylation. *Science* **267,** 1485–1488 (1995).
7. DiDonato, J. *et al.* Mapping of the inducible IkappaB phosphorylation sites that signal its ubiquitination and degradation. *Mol. Cell Biol.* **16,** 1295–1304 (1996).
8. Karin, M. & Ben-Neriah, Y. Phosphorylation meets ubiquitination: the control of NF-[kappa]B activity. *Annu. Rev. Immunol.* **18,** 621–663 (2000).
9. Sun, S. C. Non-canonical NF-kappaB signaling pathway. *Cell Res.* **21,** 71–85 (2011).
10. Ben-Neriah, Y. & Karin, M. Inflammation meets cancer, with NF-kappaB as the matchmaker. *Nat. Immunol.* **12,** 715–723 (2011).
11. Karin, M. Nuclear factor-kappaB in cancer development and progression. *Nature* **441,** 431–436 (2006).
12. DiDonato, J. A., Mercurio, F. & Karin, M. NF-kappaB and the link between inflammation and cancer. *Immunol. Rev.* **246,** 379–400 (2012).
13. Baud, V. & Karin, M. Is NF-kappaB a good target for cancer therapy? Hopes and pitfalls.. *Nat. Rev. Drug Discov.* **8,** 33–40 (2009).
14. Wang, C. Y., Mayo, M. W., Korneluk, R. G., Goeddel, D. V. & Baldwin, Jr. A. S. NF-kappaB antiapoptosis: induction of TRAF1 and TRAF2 and c-IAP1 and c-IAP2 to suppress caspase-8 activation. *Science* **281,** 1680–1683 (1998).
15. Lee, H. H., Dadgostar, H., Cheng, Q., Shu, J. & Cheng, G. NF-kappaB-mediated up-regulation of Bcl-x and Bfl-1/A1 is required for CD40 survival signaling in B lymphocytes. *Proc. Natl Acad. Sci. USA* **96,** 9136–9141 (1999).
16. Tamatani, M. *et al.* Tumor necrosis factor induces Bcl-2 and Bcl-x expression through NFkappaB activation in primary hippocampal neurons. *J. Biol. Chem.* **274,** 8531–8538 (1999).
17. Thome, M. *et al.* Viral FLICE-inhibitory proteins (FLIPs) prevent apoptosis induced by death receptors. *Nature* **386,** 517–521 (1997).
18. Yan, J. *et al.* Inactivation of BAD by IKK inhibits TNFalpha-induced apoptosis independently of NF-kappaB activation. *Cell* **152,** 304–315 (2013).
19. Pikarsky, E. *et al.* NF-kappaB functions as a tumour promoter in inflammation-associated cancer. *Nature* **431,** 461–466 (2004).

20. Greten, F. R. & Karin, M. The IKK/NF-kappaB activation pathway-a target for prevention and treatment of cancer. *Cancer Lett.* **206,** 193–199 (2004).
21. Lee, D. F. & Hung, M. C. Advances in targeting IKK and IKK-related kinases for cancer therapy. *Clin. Cancer Res.* **14,** 5656–5662 (2008).
22. Wu, Z. J. *et al.* Ainsliadimer A, a new sesquiterpene lactone dimer with an unusual carbon skeleton from Ainsliaea macrocephala. *Org. Lett.* **10,** 2397–2400 (2008).
23. Li, C., Yu, X. & Lei, X. A biomimetic total synthesis of (+)-ainsliadimer A. *Org. Lett.* **12,** 4284–4287 (2010).
24. Li, C., Dian, L., Zhang, W. & Lei, X. Biomimetic syntheses of (-)-gochnatiolides A-C and (-)-ainsliadimer B. *J. Am. Chem. Soc.* **134,** 12414–12417 (2012).
25. Li, C., Dong, T., Dian, L., Zhang, W. & Lei, X. Biomimetic syntheses and structural elucidation of the apoptosis-inducing sesquiterpenoid trimers: (−)-ainsliatrimers A and B. *Chem. Sci.* **4,** 1163–1167 (2013).
26. Burke, J. R. *et al.* BMS-345541 is a highly selective inhibitor of I kappa B kinase that binds at an allosteric site of the enzyme and blocks NF-kappa B-dependent transcription in mice. *J. Biol. Chem.* **278,** 1450–1456 (2003).
27. Alexopoulou, L., Holt, A. C., Medzhitov, R. & Flavell, R. A. Recognition of double-stranded RNA and activation of NF-kappaB by Toll-like receptor 3. *Nature* **413,** 732–738 (2001).
28. Liu, C. X. *et al.* Adenanthin targets peroxiredoxin I and II to induce differentiation of leukemic cells. *Nat. Chem. Biol.* **8,** 486–493 (2012).
29. Xu, G. *et al.* Crystal structure of inhibitor ofκB kinase β. *Nature* **472,** 325–330 (2011).
30. Sadowsky, J. D. *et al.* Turning a protein kinase on or off from a single allosteric site via disulfide trapping. *Proc. Natl Acad. Sci. USA* **108,** 6056–6061 (2011).
31. Palmieri, L. *et al.* aC helix displacement as a general approach for allosteric modulation ofprotein kinases. *Drug. Discov. Today* **18,** 408–414 (2013).
32. Murali, R. *et al.* Disabling TNF receptor signaling by inducedconformational perturbation of tryptophan-107. *Proc. Natl Acad. Sci. USA* **102,** 10970–10975 (2005).
33. Huse, M. & Kuriyan, J. The conformational plasticity of protein kinases. *Cell* **109,** 275–282 (2002).
34. Singh, J., Petter, R. C., Baillie, T. A. & Whitty, A. The resurgence of covalent drugs. *Nat. Rev. Drug Discov.* **10,** 307–317 (2011).
35. Ahn, K. *et al.* Discovery and characterization of a highly selective FAAH inhibitor that reduces inflammatory pain. *Chem. Biol.* **16,** 411–420 (2009).
36. Berendsen, H. J., van der Spoel, D. & van Drunen, R. GROMACS: A message-passing parallel molecular dynamics implementation. *Comput. Phys. Commun.* **91,** 43–56 (1995).
37. Hess, B., Kutzner, C., Van Der Spoel, D. & Lindahl, E. GROMACS 4: Algorithms for highly efficient, load-balanced, and scalable molecular simulation. *J. Chem. Theory. Comput.* **4,** 435–447 (2008).
38. Case, D. *et al.* AMBER 12. *University of California, San Francisco* (2012).
39. Wang, J., Wang, W., Kollman, P. A. & Case, D. A. Automatic atom type and bond type perception in molecular mechanical calculations. *J. Mol. Graphics Modell.* **25,** 247–260 (2006).

Acknowledgements

We thank Professor Xiaodong Wang, Professor Feng Shao, Dr Li Zhang, Dr Shan Li and Dr Hao Xu (NIBS) for providing IKKα, IKKβ, TAK1, NEMO, TAB1, NIK and P100 plasmids as well as their helpful suggestions, Ms. Mingyan Zhao (NIBS) for NMR and HPLC-MS analysis, and Dr Jiang Zhou (Peking University) for HRMS analysis. This work was supported by the National Basic Research Program of China (973 Program 2015CB856200, 2013CB910102 and 2012CB837400), the National Natural Science Foundation of China (NNSFC 21222209, 21472010 and 91313303 to X.L., and 31222036 and 31471303 to S.H.), a Project Funded by the Priority Academic Program Development of Jiangsu Higher Education Institutions (PAPD), and a Natural Science Foundation of Jiangsu Province Grant (BK2012004 to S.H.).

Author contributions

X.L. and S.H. designed the study; T.D. and X.W. performed and analysed the biological experiments; C.L. and L.D. performed chemical synthesis; X.Z. performed and analysed the xenograft studies; R.C. performed the docking experiments under the guidance of N.H.; L.L. and S.C. performed MS experiments; X.L., S.H., T.D. and C.L. wrote the manuscript. X.L. and S.H. guided all of the aspects of this study.

Additional information

Competing financial interests: The authors declare no competing financial interests.

12

A method to rapidly create protein aggregates in living cells

Yusuke Miyazaki[1], Kota Mizumoto[2,†], Gautam Dey[1,†], Takamasa Kudo[1], John Perrino[3], Ling-chun Chen[1], Tobias Meyer[1] & Thomas J. Wandless[1]

The accumulation of protein aggregates is a common pathological hallmark of many neurodegenerative diseases. However, we do not fully understand how aggregates are formed or the complex network of chaperones, proteasomes and other regulatory factors involved in their clearance. Here, we report a chemically controllable fluorescent protein that enables us to rapidly produce small aggregates inside living cells on the order of seconds, as well as monitor the movement and coalescence of individual aggregates into larger structures. This method can be applied to diverse experimental systems, including live animals, and may prove valuable for understanding cellular responses and diseases associated with protein aggregates.

[1]Department of Chemical & Systems Biology Stanford University, Stanford, California 94305, USA. [2]Department of Biology Stanford University, Stanford, California 94305, USA. [3]Cell Sciences Imaging Facility Stanford University, Stanford, California 94305, USA. † Present address: Department of Zoology, The University of British Columbia, Vancouver, British Columbia, Canada V6T 1Z3 (K.M.); UCL-MRC Lab for Molecular Cell Biology, London WC1E 6BT, UK (G.D.). Correspondence and requests for materials should be addressed to T.J.W. (email: wandless@stanford.edu).

Proteins are involved in every aspect of cellular function and homoeostasis, playing enzymatic, structural or signalling roles that are typically made possible when the polypeptide folds into a particular three-dimensional conformation. However, proteins do not always achieve their folded conformations during protein synthesis, and are folded; mature proteins may be readily damaged by various types of environmental stresses such as heat and oxidative stress. Misfolded proteins often lose the ability to perform their intended functions, and in some cases misfolded proteins can acquire new pathological functions. Thus, cells have developed a variety of mechanisms to deal with misfolded proteins and utilize a significant amount of energy to maintain cellular protein homoeostasis[1].

Cells attempt to cope with misfolded proteins using two major pathways: the protein refolding and degradation[2,3]. Deficient protein homoeostasis is strongly associated with neurodegenerative diseases including Alzheimer's disease, Huntington's disease, amyotrophic lateral sclerosis and Parkinson's disease[1,4]. It is widely recognized that protein aggregation is often associated with neurodegenerative diseases, and is also strongly associated with aging. For these reasons, one potential therapeutic strategy for treating neurodegenerative and other age-related diseases is to prevent the formation of protein aggregates by modulating protein homoeostasis[4,5].

Although protein aggregates are associated with a wide variety of diseases, we do not yet understand the mechanisms underlying protein aggregation, including how cellular aggregates are formed, how they are cleared, how other cellular proteins are involved in aggregate formation and clearance or why oligomers and aggregates are toxic to cells[6]. Many studies have used forceful perturbations, such as heat shock and proteasome inhibition to generate aggregates[7,8]. However, these stressors are extremely strong and nonspecific, concomitantly inducing global cellular stress responses unrelated to protein aggregation[9]. Others have studied aggregates using genetic perturbations including polyQ-fused mutant Huntingtin (Htt) proteins[10,11]. Despite higher specificity, these systems display relatively little control over the timing of aggregate onset or extent of aggregate formation.

In this study, we describe a novel small molecule-dependent engineered protein that can be used to rapidly create protein aggregates in cells with a high degree of spatial and temporal control. The ability to regulate the formation of aggregates should enable advances in our understanding of how protein aggregation affects protein homoeostasis in cells.

Results

Development of chemically controllable protein aggregates. To develop a conditional system to generate protein aggregates in cells, we utilized destabilizing domain (DD) technology developed by our group[12–14]. This technology was first developed as a method to regulate the metabolic stability of any protein of interest using a cell-permeable ligand in conjunction with a genetically encoded DD domain. We found that these DDs confer instability to any fused-partner proteins and cause proteasome-mediated degradation of the entire fusion protein[12–14]. High-affinity ligands bind to and stabilize these DD-fused proteins in a dose-dependent manner[12–14]. Previous biophysical studies showed that DDs possess a interesting property; the folding state of the DD can be reversibly regulated by the presence or absence of the ligand[15]. We took advantage of this behaviour to interrogate the cellular protein quality-control surveillance mechanisms using DDs as model substrates for unstable proteins[16,17].

To develop a new tool to study the mechanisms of protein aggregation and its effects on cells, we engineered new DDs derived from the human FK506 binding protein 12 (FKBP12) protein (Fig. 1a). The hydrophobicity of N-terminal nuclear export signals (NES) has been previously shown to be a critical factor for aggregation[18]. Thus, we constructed several variations of fluorescently tagged DDs with NES sequences fused to their N termini (Supplementary Fig. 1a). These NES-tagged DDs were stably expressed in HEK293 cells, and we observed that one of these DDs, which we call aggregating destabilizing domain (AgDD), aggregates when the stabilizing ligand (Shield-1 or S1) is withdrawn from the cell-culture media.

Beginning with the N terminus, the AgDD comprised of a 10-amino-acid NES sequence (LALKLAGLDI) derived from a cAMP-dependent protein kinase inhibitor, an FKBP-derived DD (L106P) followed by superfolder green fluorescent protein (GFP). When cells expressing AgDD are cultured in medium containing S1, the specific ligand for FKBP-based DDs, AgDD is distributed uniformly in the cells. However, after removing S1 from the culture medium for 150 min, AgDD coalesced into aggregates in cells (Fig. 1b). This phenomenon of aggregation was also observed in other cell lines, including NIH3T3 and U2OS cells (Supplementary Fig. 1b,c).

To further characterize the time-dependent nature of AgDD aggregation, we collected cells expressing high levels of AgDD by flow cytometry and monitored the behaviour of the AgDD protein using epifluorescence time-lapse microscopy. When cells are cultured with S1, GFP is uniformly distributed throughout these cells with no visible aggregates (Fig. 1c and Supplementary Movie 1). When S1 is withdrawn from the culture media, AgDD rapidly unfolds and forms small aggregates (Fig. 1c and Supplementary Movie 2). On closer examination, the formation of small, mobile aggregates commences within minutes of ligand withdrawal (Fig. 1d and Supplementary Movie 3), and the small aggregates coalesce into larger immobile aggregates over time. Quantification of the aggregates reveals that the cellular area occupied by single aggregates increases in time-dependent manner (Supplementary Fig. 1d), and the total number of aggregates decreases once aggregates appeared (Supplementary Fig. 1e).

Aggregates were allowed to form for 1 h and then S1 was re-introduced into the culture media, leading to the re-appearance of diffuse GFP signal by the 2-h time point (Fig. 1c and Supplementary Movie 4). The addition of cycloheximide along with S1 at the 1-h time point did not prevent or significantly attenuate the growth the diffuse GFP signal (Supplementary Fig. 1f), suggesting that newly synthesized AgDD-GFP is not the primary source of the diffuse GFP signal. These results demonstrate that AgDD is a chemically controllable protein that enables the rapid and pervasive induction of protein aggregation with high-temporal resolution. Another advantage of using the FKBP-based DD is that the S1 ligand does not induce global stress responses produced by stimuli like proteasome inhibitors or heat shock[17,19].

We next monitored the aggregation of AgDD using analytical flow cytometry by collecting AgDD-expressing HEK293 cells at different time points after S1 removal, and subjecting these samples to analysis using the PulSA method[20]. Cells with aggregates displayed narrowly distributed fluorescent signals compared with cells that did not have these punctate structures (Supplementary Fig. 2a). Consistent with the time-lapse microscopy experiments, we found that the population of cells that contained aggregates continuously increased following withdrawal of the ligand (Supplementary Fig. 2b).

AgDD aggregates resemble known aggregates. To further examine the structure of aggregates, we used transmission

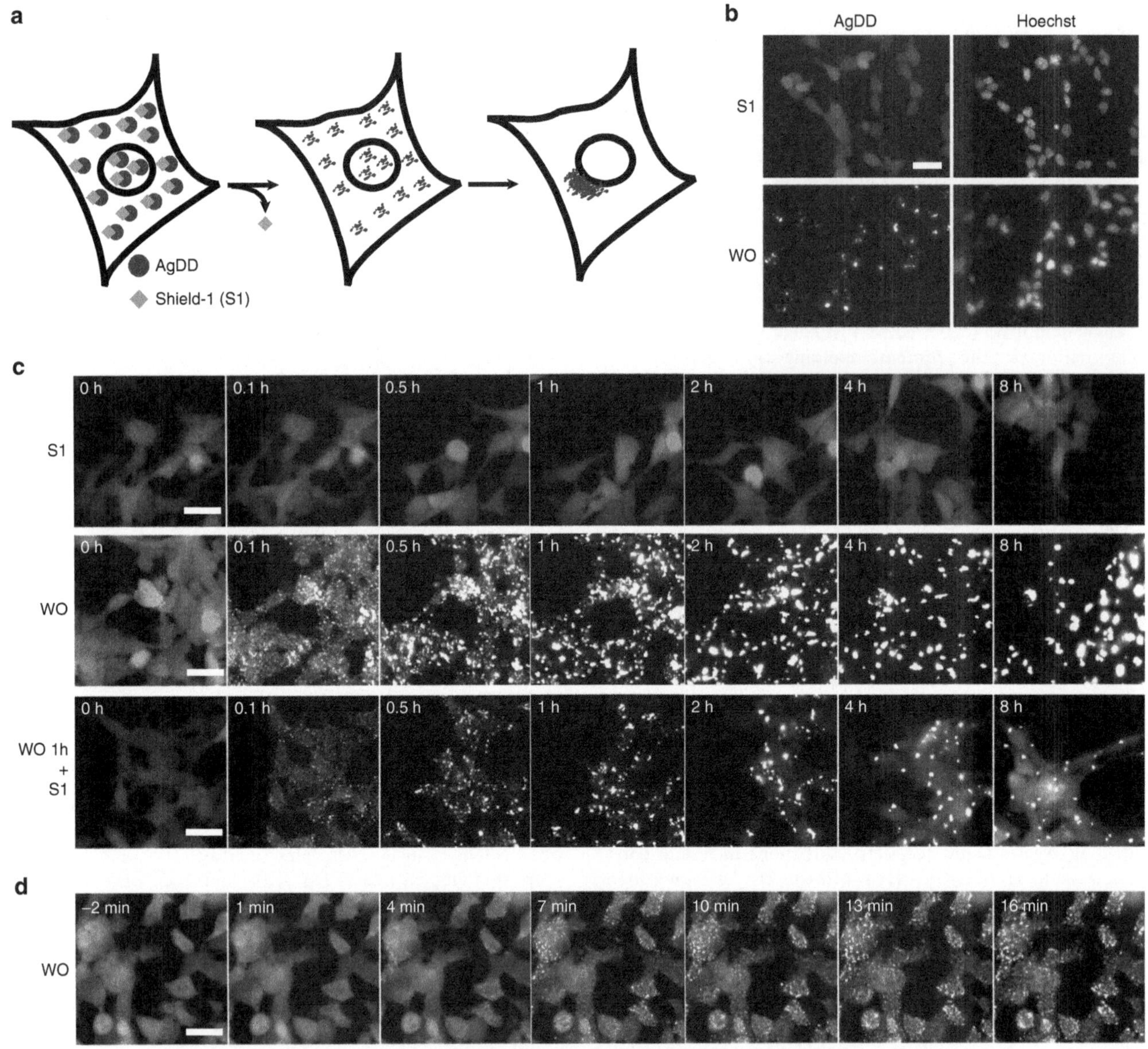

Figure 1 | Development of AgDD to rapidly create protein aggregates in living cells. (**a**) Schematic representation of inducing aggregated proteins inside cells using an AgDD. (**b**) Images of representative HEK cells stably expressing AgDD before and after S1 washout (WO) for 150 min. Scale bar, 10 μm. (**c**) Time-lapse images of representative HEK cells stably expressing AgDD. Top: cells cultured in media with S1 throughout the experiment; middle: cells cultured in media with S1 withdrawn at 0 h and bottom: cells cultured in media with S1 withdrawn from 0-1 h then readministered from 1-8 h. Scale bars, 20 μm. (**d**) Early time-lapse images of representative HEK cells stably expressing AgDD following S1 withdrawal. Scale bar, 20 μm.

electron microscopy (TEM) to visualize negatively stained cells expressing the AgDD-GFP fusion protein. Consistent with the results from fluorescent microscopy, when S1 is withdrawn from the culture media, AgDD forms large, electron-dense aggregates (Fig. 2a,b; Supplementary Fig. 2c,d). The ultrastructural characteristics of the aggregated AgDD-GFP resemble known filamentous aggregated proteins, such as the ΔF508 mutant of cystic fibrosis transmembrane conductance regulator (CFTR)[21].

We also monitored localization of the endogenous p62/SQSTM protein, a well-characterized marker for selective autophagy that is known to colocalize with aggregates[3]. Immunofluorescence staining of AgDD and p62 showed that p62 appears to colocalize with AgDD aggregates, although we observe colocalization only after AgDD has been allowed to aggregate for some time only at later times (5 h following S1 withdrawal, Fig. 2c).

AgDD aggregates affect cellular fitness. Next, we examined the effects of aggregated AgDD on cell growth and viability. For this purpose, we devised a co-culture proliferation experiment in which HEK293 cells expressing AgDD were co-cultured with control HEK293 cells expressing only mCherry in media containing S1. We withdrew S1 to induce AgDD aggregation and analyzed the number of AgDD-positive cells relative to control cells using flow cytometry. Cell expressing the aggregated AgDD proliferated more slowly than control cells but only after a full day of aggregate induction (Fig. 3a and Supplementary Fig. 3a), similar to results from experiments using polyQ aggregates[22].

Aggregates formation depends on the amount of AgDD. We next asked if aggregate formation is sensitive to the intracellular concentration of the AgDD protein. We sorted HEK cells

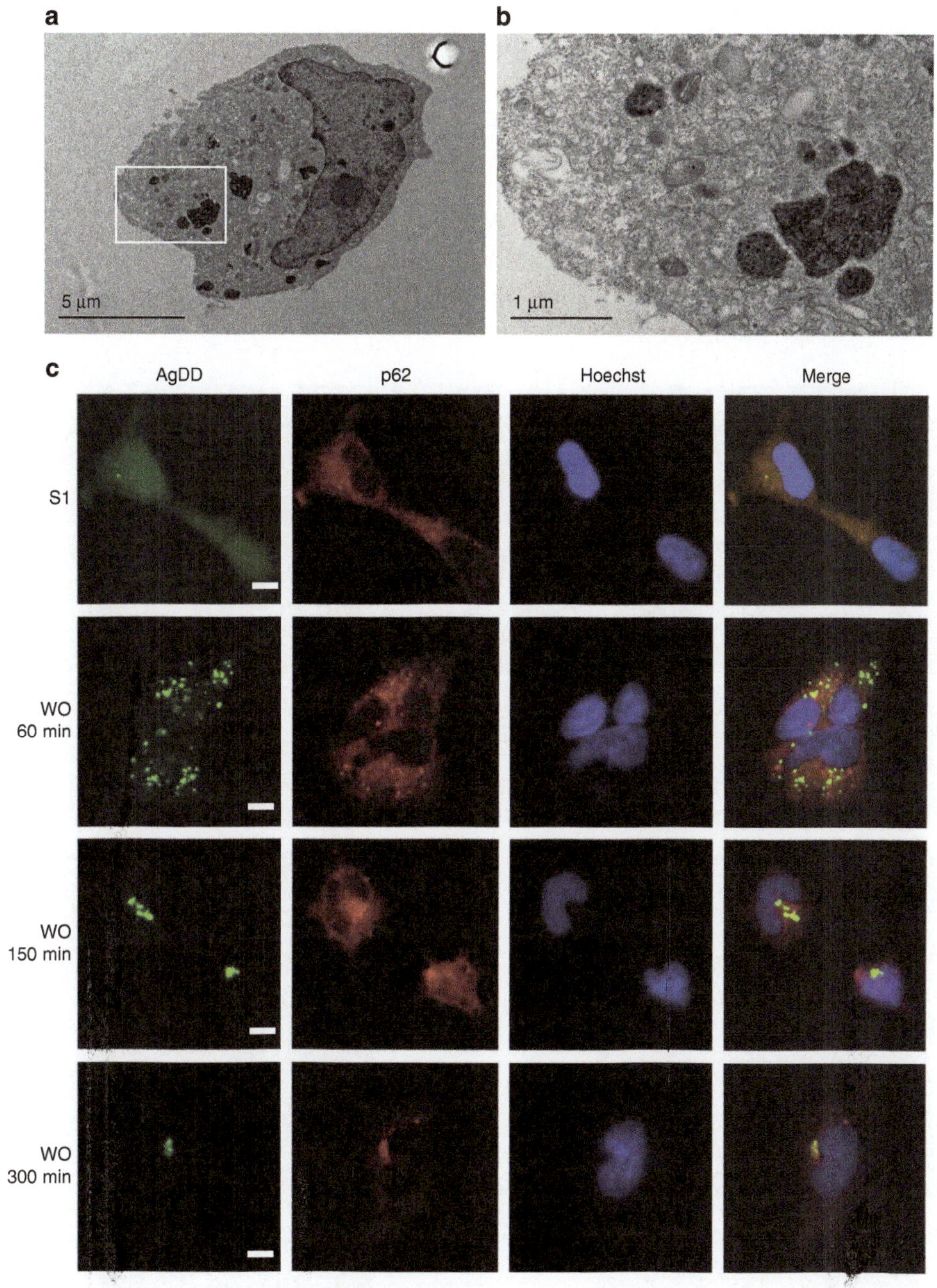

Figure 2 | AgDD aggregates have similar characteristics as filamentous aggregates. (**a**) TEM image of representative HEK cells stably expressing AgDD after S1 washout for 60 min. Scale bar, 5 μm. (**b**) Magnified TEM image of white box inset in **a**. Scale bar, 1 μm. (**c**) Immunofluorescence images of representative cells stably expressing AgDD-GFP before and after S1 removal for the indicated times. Left: AgDD; middle-left: p62; middle-right: Hoechst; right: merge. Scale bars, 10 μm. Texas-Red was used to label the p62 antibody.

stably expressing AgDD into four groups on the basis of their S1-stabilized GFP levels: high, high-med, med-low and low (Supplementary Fig. 3b). By counting the numbers of cells with aggregates 10 min after withdrawal of S1, we observed that cells expressing higher levels of AgDD are more likely to form aggregates (high, 100%; medium-high, 95%; medium-low, 81% and low, 48%, (Supplementary Fig. 3c), suggesting that the formation of AgDD aggregates depends on the concentration of AgDD.

Spatial control using a nuclear-localized AgDD. In order to provide spatial resolution, we next attempted to induce AgDD aggregate formation in specific cellular compartments. Protein aggregates within the nucleus are associated with neurodegenerative diseases, so we first targeted the AgDD-GFP fusion protein to the cell nucleus by fusing a nuclear localization signal to the N terminus of the AgDD construct[23]. When cells expressing nuclear AgDD are cultured in medium with S1, the GFP signal is diffusely localized in the nucleus (Fig. 3b and Supplementary Movie 5). We observed the formation of multiple, small AgDD aggregates immediately after S1 withdrawal (Fig. 3b,c; Supplementary Movies 6 and 7). These small aggregates appeared to be mobile within the cell nucleus and they coalesced somewhat but did not become one large aggregate in contrast to the behaviour of the cytosolic AgDD (Fig. 3b,c; Supplementary Movies 6 and 7). Similar results were obtained when AgDD-GFP was expressed in NIH3T3 cells (Supple-

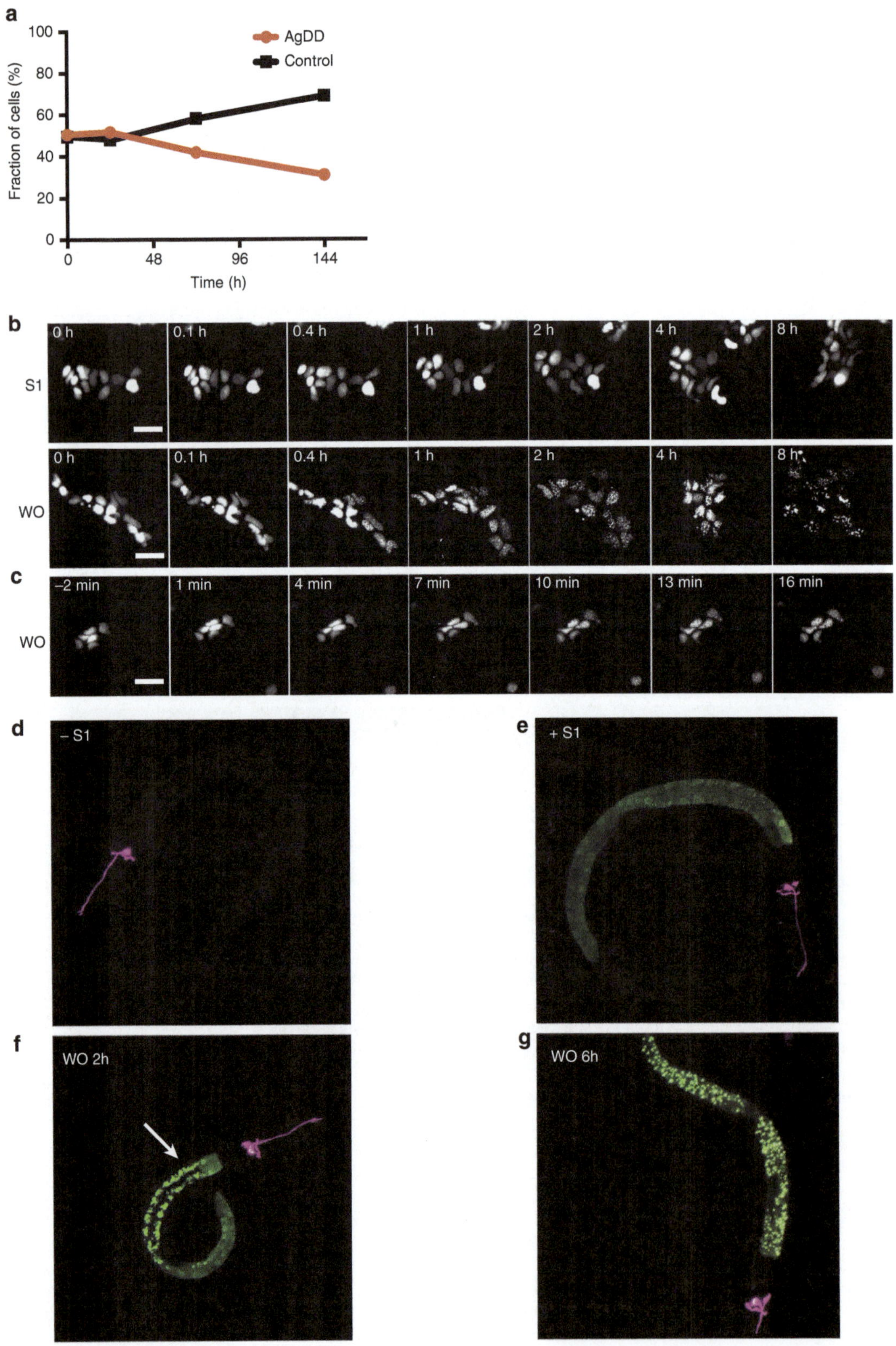

Figure 3 | Applications of AgDD. (**a**) HEK293 cells expressing AgDD and HEK293 cells expressing mCherry were mixed and co-cultured in media containing S1. S1 was withdrawn for the indicated times, and cell populations were quantified using flow cytometry. Replicates of $n = 3$. Error bar is Standard Deviation (STD). (**b**) Time-lapse images of representative HEK293 cells stably expressing Nuclear AgDD. Top: cells cultured in media with S1 throughout the experiment; bottom: cells cultured in media with S1 withdrawn at 0 h. Scale bars, 5 μm. (**c**) Early time-lapse images of representative HEK293 cells stably expressing nuclear AgDD following S1 removal. Scale bar, 5 μm. (**d-g**) Images of representative nematodes expressing AgDD in intestinal cells. Top left: no S1; top right: 50 μM S1, bottom left: S1 removed for 2 h, bottom right: S1 removed for 6 h. Scale bar, 50 μm.

mentary Fig. 3d). These results demonstrate that AgDD can be used to investigate aggregate formation in specific subcellular compartments.

AgDD induces aggregates in live animals. Finally, we used *C. elegans* to test if our AgDD system can be used to conditionally induce aggregate formation *in vivo*. We expressed AgDD in the intestinal cells of *C. elegans* using the *vha-6* promoter. Under normal growth conditions without the stabilizing S1, we did not observe AgDD signal suggesting that the AgDD fusion protein is efficiently degraded as it is when expressed in mammalian cells (Fig. 3d). When worms were grown on plates made with 50 μM S1, the stabilized AgDD-GFP is observed as a diffuse signal in the intestinal cells (Fig. 3e). When worms expressing stabilized AgDD-GFP were transferred to plates lacking S1, the AgDD started to form aggregates (Fig. 3f,g). The onset of the AgDD aggregation was slower than was observed in cultured cells, possibly because of slower rates of S1 clearance in live animals. Formation of the AgDD aggregates typically began in the anterior intestinal cells (arrow in Fig. 3f). Another interesting observation was that the larger aggregates tend to colocalize with the auto-fluorescence signals from gut granules (data not shown), which have been suggested to function as an aggresome-like region in intestinal cells[24].

Discussion

In summary, we developed a system for inducing and monitoring protein aggregates in cells using a small cell-permeable molecule. Compared with previous methods used to study aggregates, AgDD has several advantages: it has fewer off-target effects, it can be precisely regulated both spatially and temporally, and it has low toxicity for cells. Small-protein aggregates are immediately induced following withdrawal of the stabilizing ligand. Thus, this system should enable detailed mechanistic investigations of the mechanisms of the cellular response to protein aggregation. The results also demonstrate that aggregates have similar characteristics as the known filamentous aggregates from TEM and also with p62, a well-known marker for aggregates, appears to associate with the AgDD aggregates at a later stage. One possible explanation for small aggregates could be Q-bodies, which was found in yeast[25].

Furthermore, we demonstrated that the AgDD protein can be used in living organisms, such as the nematode, *C. elegans*, and that aggregate formation could be restricted to specific cellular compartments such as the nucleus. Although further investigation is required, the behaviour of AgDD-derived aggregates in the nucleus appeared to differ from cytosolic aggregates, perhaps reflecting different protein quality-control surveillance mechanisms between these compartments[23]. This system will be extremely useful to investigate nuclear protein-quality control because nuclear protein misfolding and aggregation are strongly associated with the pathology of neurodegenerative diseases[23]. Given the high degree of control inherent in this method, the AgDD constructs may be useful in a discovery screening format to identify potential agents capable of modulating protein aggregate production.

Methods

Cloning. FKBP-derived AgDDs were cloned into Piggybac Transposon System pB vectors as N-terminal fusions to a SuperFolder GFP. For *C. elegans* studies, FKBP-derived AgDDs were fused to the N terminus of sfGFP, and this construct was cloned into a pSM vector where expression of the fusion protein was driven by the a *vha-6* promoter.

Cell culture, transfections and transductions. The NIH3T3 cell lines were cultured in DMEM supplemented with 10% heat-inactivated donor bovine serum (Invitrogen), 2 mM glutamine, 100 U ml^{-1} penicillin and 100 μg ml^{-1} streptomycin. The HEK293 cell lines and U2OS cell lines were cultured with 10% heat-inactivated foetal bovine serum (Invitrogen), 2 mM glutamine, 100 U ml^{-1} penicillin and 100 μg ml^{-1} streptomycin.

To establish AgDD-expressing cell lines, cells were plated at 10×10^4 cells per well of a 6-well plate a day before transfection. The cells were co-transfected with plasmids of a pB vector and PiggyBac transposase vector following standard protocols using TransIT-LT1 (Mirus). After the cells were cultured in growth media for about a week to allow stable integration, the transfected cells were sorted by flow cytometry for high-GFP level cells.

Withdrawal of Shield-1. To remove the Shield-1 (S1) ligand from media containing Shield-1, we directly added purified FKBP(F36V) protein (350–500 μM stock) in cultured media to achieve a final cellular protein concentration of 3.5–5 μM. The presence of extracellular FKBP creates a strong thermodynamic sink facilitating the removal of S1 from stabilizing the intracellular AgDD.

Flow cytometry. Cells were trypsinized and resuspended in culture media before undergoing FACS sorting at the Stanford Shared FACS Facility using BD Aria II. For analytical flow cytometry, cells were trypsinized and resuspended in culture media before analysis using a BD FACS Calibur with no less than 10,000 events represented. Data were analyzed using FlowJo.

Live-cell imaging. Time-lapse imaging was performed with a Nikon Eclipse Ti fluorescence microscope using μManager (https://www.micro-manager.org/). Cells were prepared on an 8-well Nunc Lab-Tek Chambered Coverglass (Thermo Scientific) coated with poly-Lysine (Invitrogen). During time-lapse experiments, the same conditions were maintained including a temperature of 37 °C and a CO_2 concentration of 5%. Other live-cell images were captured on a Zeiss Axioskop 2 epifluorescence microscope equipped with a QICAM FAST 1,394 digital CCD camera.

Immunofluorescence. Following fixation in PBS with 4% paraformaldehyde (Ted Pella) for 20 min, cells were permeabilized using PBS with 0.2% Triton-X (Sigma) for 15 min and incubated in 3% BSA (Sigma-Aldrich) in PBS for 60 min. p62 antibody (mouse, used at 1:400, BD Biosciences 610,832) incubation was carried out at 4 °C overnight followed by Alexa Fluor-conjugated secondary antibodies for 1 h at room temperature (1:1,000; Life Technologies) and Hoechst 33,342 (1:10,000; Life Technologies). Images were acquired on a custom-assembled spinning disk confocal system built around an inverted Zeiss Axiovert 200 M microscope with three lasers (442, 514 and 593.5 nm) and a CoolSNAP CCD camera from Photometrics/Roper Scientific. Imaging was carried out on an integrated ImageXpress (Molecular Devices) high-content analysis system.

***C. elegans* studies.** *C. elegans* strain was maintained as previously described[26]. Transgenic strains were generated by injecting Pvha-6::AgDD-GFP and Podr-1::RFP plasmids into Bristol N2 strain by standard injection methods[27]. Images of fluorescently tagged fusion proteins were captured in live *C. elegans* using a Zeiss LSM710 confocal microscope (Carl Zeiss). Worms were immobilized on a 2% agarose pad with 10 μM levamisole (Sigma-Aldrich). Images were analyzed with Zen software (Carl Zeiss).

Bioinformatic analysis. All statistical analyses were performed in R (www.rproject.org) or Prism 6. All cell image analyses were done using MATLAB and imageJ (http://imagej.nih.gov/ij/).

Quantification of aggregates. The Haar Wavelet transform was applied once to extract the high-frequency image, and this image was resized to its original size with bi-cubic interpolation. It was then subtracted from the original image, which represents the low frequency part of the image. Preliminarily, the illumination of the image was corrected using the similar approach with seven levels of decomposition. These implementations remove noise but retain the aggregates with various size. To identify puncta of various intensity, 2D Gaussian filter was applied to the transformed image, and then if a pixel in the transformed image has 20% larger value than the corresponding pixel in the Gaussian-smoothed image, it was classified as aggregates. This binarized image was further masked by a background region, which was simply determined by a threshold of a constant value.

The average aggregates area and a number of aggregates over time were extracted from the binarized image, and then the mean and standard deviations were calculated from five different fields of views. If a number of aggregates was < 100 in one field of view, the average area was set to 0 in order to suppress an error from segmentation, meaning no aggregates were observed in the image.

References

1. Morimoto, R. I. The Heat shock response: systems biology of proteotoxic stress in aging and disease. *Cold Spring Harb. Symp. Quant. Biol.* **76,** 91–99 (2011).
2. Finley, D. Recognition and processing of ubiquitin-protein conjugates by the proteasome. *Annu. Rev. Biochem.* **78,** 477–513 (2009).
3. Mizushima, N. & Komatsu, M. Autophagy: renovation of cells and tissues. *Cell* **147,** 728–741 (2011).
4. Powers, E. T., Morimoto, R. I., Dillin, A., Kelly, J. W. & Balch, W. E. Biological and chemical approaches to diseases of proteostasis deficiency. *Annu. Rev. Biochem.* **78,** 959–991 (2009).
5. Cohen, E., Bieschke, J., Perciavalle, R. M., Kelly, J. W. & Dillin, A. Opposing activities protect against age-onset proteotoxicity. *Science* **313,** 1604–1610 (2006).
6. Tyedmers, J., Mogk, A. & Bukau, B. Cellular strategies for controlling protein aggregation. *Nat. Rev. Mol. Cell. Biol.* **11,** 777–788 (2010).
7. Kisselev, A. F. & Goldberg, A. L. Proteasome inhibitors: from research tools to drug candidates. *Chem. Biol.* **8,** 739–758 (2001).
8. Miller, S. B. *et al.* Compartment-specific aggregases direct distinct nuclear and cytoplasmic aggregate deposition. *EMBO J.* **34,** 778–797 (2015).
9. Bush, K. T., Goldberg, A. L. & Nigam, S. K. Proteasome inhibition leads to a heat-shock response, induction of endoplasmic reticulum chaperones, and thermotolerance. *J. Biol. Chem.* **272,** 9086-9092 (1997).
10. Kim, S., Nollen, E. A. A., Kitagawa, K., Bindokas, V. P. & Morimoto, R. I. Polyglutamine protein aggregates are dynamic. *Nat. Cell. Biol.* **4,** 826–831 (2002).
11. Kaganovich, D., Kopito, R. & Frydman, J. Misfolded proteins partition between two distinct quality control compartments. *Nature* **454,** 1088–1095 (2008).
12. Banaszynski, L. A., Chen, L.-C., Maynard-Smith, L. A., Ooi, A. G. L. & Wandless, T. J. A rapid, reversible, and tunable method to regulate protein function in living cells using synthetic small molecules. *Cell* **126,** 995–1004 (2006).
13. Iwamoto, M., Björklund, T., Lundberg, C., Kirik, D. & Wandless, T. J. A general chemical method to regulate protein stability in the mammalian central nervous system. *Chem. Biol.* **17,** 981–988 (2010).
14. Miyazaki, Y., Imoto, H., Chen, L.-C. & Wandless, T. J. Destabilizing domains derived from the human estrogen receptor. *J. Am. Chem. Soc.* **134,** 3942–3945 (2012).
15. Egeler, E. L., Urner, L. M., Rakhit, R., Liu, C. W. & Wandless, T. J. Ligand-switchable substrates for a ubiquitin–proteasome system. *J. Biol. Chem.* **286,** 31328–31336 (2011).
16. Chu, B. W. *et al.* The E3 ubiquitin ligase UBE3C enhances proteasome processivity by ubiquitinating partially proteolyzed substrates. *J. Biol. Chem.* **288,** 34575–34587 (2013).
17. Miyazaki, Y., Chen, L.-C., Chu, B. W., Swigut, T. & Wandless, T. J. Distinct transctiptional responses elicited by unfolded nuclear or cytoplasmic protein in mammalian cells. *eLife* doi:10.7554/eLife.07687 (2015).
18. Maynard-Smith, L. A., Chen, L. C., Banaszynski, L. A., Ooi, A. G. L. & Wandless, T. J. A directed approach for engineering conditional protein stability using biologically silent small molecules. *J. Biol. Chem.* **282,** 24866–24872 (2007).
19. Zheng, Z., Li, A., Holmes, B. B., Marasa, J. C. & Diamond, M. I. An N-terminal nuclear export signal regulates trafficking and aggregation of Huntingtin (Htt) protein exon 1. *J. Biol. Chem.* **288,** 6063–6071 (2013).
20. Ramdzan, Y. M. *et al.* Tracking protein aggregation and mislocalization in cells with flow cytometry. *Nat. Methods* **9,** 467–470 (2012).
21. Johnston, J. A., Ward, C. L. & Kopito, R. R. Aggresomes: a cellular response to misfolded proteins. *J. Cell. Biol.* **143,** 1883–1898 (1998).
22. Ripaud, L. *et al.* Overexpression of Q-rich prion-like proteins suppresses polyQ cytotoxicity and alters the polyQ interactome. *Proc. Natl Acad. Sci. USA* **111,** 18219–18224 (2014).
23. Shibata, Y. & Morimoto, R. I. How the nucleus copes with proteotoxic stress review. *Curr. Biol.* **24,** R463–R474 (2014).
24. Hermann, G. J. *et al.* Genetic analysis of lysosomal trafficking in *Caenorhabditis elegans*. *Mol. Biol. Cell.* **16,** 3273–3288 (2005).
25. Escusa-Toret, S., Vonk, W. I. M. & Frydman, J. Spatial sequestration of misfolded proteins by a dynamic chaperone pathway enhances cellular fitness during stress. *Nat. Cell. Biol.* **15,** 1231–1243 (2013).
26. Brenner, S. The genetics of *Caenorhabditis elegans*. *Genetics* **77,** 71–94 (1974).
27. Mello, C. C., Kramer, J. M., Stinchcomb, D. & Ambros, V. Efficient gene transfer in *C. elegans*: extrachromosomal maintenance and integration of transforming sequences. *EMBO J.* **10,** 3959–3970 (1991).

Acknowledgements

We wish to thank members of the Jarosz, Wyoscka, Frydman, Cimprich, Meyer and Covert labs for reagents and helpful advice. We thank the Stanford Shared FACS Facility and the Cell Sciences Imaging Facility. This work was supported by the NIH (GM073046 and P50 GM107615), the Ellison Medical Foundation (AG-SS-2573-10) and the Stanford Discovery Innovation Fund. Y.M. was supported by the Nakajima Foundation.

Author contributions

Y.M. and T.J.W. conceived the project and designed experiments. Y.M. and L-c.C cloned all of the plasmids and established the stable cell lines used in this paper. Y.M., G.D. and T.K. performed cell imaging and image analysis as shown in this paper. Y.M. performed western blot and other biochemical experiments. Y.M. and K.M. performed the experiments in *C. elegans*. L-c.C and J.P. performed the TEM experiments. Y.M., G.D. and T.J.W. wrote and revised the manuscript. All authors edited the manuscript.

Additional information

Competing financial interests: The authors declare no competing financial interests.

13

Atomic basis for therapeutic activation of neuronal potassium channels

Robin Y. Kim[1], Michael C. Yau[1], Jason D. Galpin[2], Guiscard Seebohm[3], Christopher A. Ahern[2], Stephan A. Pless[4] & Harley T. Kurata[1,†]

Retigabine is a recently approved anticonvulsant that acts by potentiating neuronal M-current generated by KCNQ2-5 channels, interacting with a conserved Trp residue in the channel pore domain. Using unnatural amino-acid mutagenesis, we subtly altered the properties of this Trp to reveal specific chemical interactions required for retigabine action. Introduction of a non-natural isosteric H-bond-deficient Trp analogue abolishes channel potentiation, indicating that retigabine effects rely strongly on formation of a H-bond with the conserved pore Trp. Supporting this model, substitution with fluorinated Trp analogues, with increased H-bonding propensity, strengthens retigabine potency. In addition, potency of numerous retigabine analogues correlates with the negative electrostatic surface potential of a carbonyl/carbamate oxygen atom present in most KCNQ activators. These findings functionally pinpoint an atomic-scale interaction essential for effects of retigabine and provide stringent constraints that may guide rational improvement of the emerging drug class of KCNQ channel activators.

[1] Department of Anesthesiology, Pharmacology, and Therapeutics, University of British Columbia, 2176 Health Sciences Mall, Vancouver, British Columbia, Canada V6T 1Z3. [2] Department of Molecular Physiology and Biophysics, University of Iowa, 285 Newton Road, Iowa City, Iowa 52242, USA. [3] Department of Cardiovascular Medicine, University Hospital Münster, Albert-Schweitzer-Campus 1 (Gebäude D3), D-48149 Münster, Germany. [4] Department of Drug Design and Pharmacology (Center for Biopharmaceuticals), University of Copenhagen, Jagtvej 160, DK-2100 Copenhagen, Denmark. † Present address: Department of Pharmacology, University of Alberta, 9-70 Medical Sciences Building, Edmonton, Alberta, Canada T6G 2H7. Correspondence and requests for materials should be addressed to H.T.K. (email: kurata@ualberta.ca).

Epilepsy is a common, heterogeneous and often debilitating disorder, affecting ~1% of the global population. Current clinical management of patients suffering from epilepsy is primarily via medication, although it is generally accepted that ~30% of newly diagnosed epilepsy patients will be resistant to common anticonvulsant drugs[1,2]. Moreover, such resistance is correlated to seizure type—for example, it has been reported that 60% of patients suffering from focal epilepsy will develop and retain resistance to medications[3]. These shortcomings have motivated the search for new antiepileptic drugs, non-conventional treatments for epilepsy (such as the ketogenic diet) and the identification and ongoing validation of new drug targets[4–7].

Pharmacotherapy in epilepsy has been primarily directed towards a small subset of ion channel targets[8]. Overall, the vast majority of drugs developed as anticonvulsants are targeted towards voltage-gated Na^+ channels or GABA-mediated neurotransmission. However, several recently developed drugs have begun to expand the diversity of molecular targets for epilepsy. A good example is the 'first-in-class' Kv channel opener retigabine, presently approved for use as an adjunct therapy to antiepileptic drugs for partial/focal seizures[9]. Among all the antiepileptic drugs in use, retigabine has a unique mechanism—it is the only compound approved for human use that acts by activating/opening Kv channels. The resulting suppression of electrical excitation likely underlies its anticonvulsant and analgesic properties[10,11]. Numerous studies have demonstrated that this drug class is effective and well tolerated over a wide therapeutic window, with comparable effectiveness to many classic anticonvulsants in animal models[7,12,13]. However, despite the therapeutic potential for management of epilepsy and disorders involving membrane excitability, there remains a paucity of molecular understanding of the binding mode(s) and mechanism of action of this new drug class.

Retigabine acts primarily by opening KCNQ2–5 (Kv7.2–7.5) channels[11,14,15]. In the nervous system, KCNQ2–5 channels assemble as heteromers to generate the 'M-current', a subthreshold voltage-gated K^+ conductance that is influenced by muscarinic receptor signalling (primarily via PIP_2) and regulates neuronal excitability[14,16]. Retigabine activates these channels by shifting their voltage dependence to hyperpolarized membrane potentials, effectively causing them to open at voltages where they would normally adopt a closed, non-conducting conformation[17,18]. A conserved tryptophan residue in the pore-forming S5 helix of retigabine-sensitive KCNQ channels (KCNQ2 Trp236, KCNQ3 Trp265) is known to be essential for retigabine effects and is absent in the retigabine-resistant cardiac isoform KCNQ1 (refs 19,20). The identification of retigabine as a potent Kv channel opener has led to numerous compound library screens to generate novel analogues with altered subunit specificity and effectiveness[9,21–25]. However, the molecular details that underlie retigabine–KCNQ channel interactions have remained elusive because of the inherent lack of resolution of conventional site-directed mutagenesis techniques.

We have used unnatural amino-acid mutagenesis to subtly rearrange atoms and electrons of the Trp side chain in the retigabine binding site. Our findings demonstrate that repositioning of the indole nitrogen atom of KCNQ3 Trp265 completely abolishes retigabine effects, suggesting that retigabine interaction requires the formation of a H-bond with Trp265. The importance of this H-bond interaction is further illustrated by introduction of fluorinated Trp analogues (with increased H-bonding propensity) at KCNQ3 Trp265, causing increased drug potency. Using multiple retigabine analogues, we identify an electrostatic fingerprint around a H-bond acceptor that correlates with drug potency. These experimental constraints pinpoint specific atoms and chemical forces involved in retigabine interactions, and highlight approaches that may be used to guide rational improvement of existing drugs.

Results

Retigabine modulation of KCNQ channels through an S5 Trp. Retigabine strongly affects voltage-dependent gating of KCNQ2 and KCNQ3* channels (Fig. 1a,b) by generating a substantial hyperpolarizing shift of the $V_{1/2}$ of activation by ~40–60 mV (KCNQ3* refers to KCNQ3[Ala315Thr]—see Methods). These effects have also been demonstrated in KCNQ4 and KCNQ5, but are absent in KCNQ1 because of the absence of an essential Trp residue (Trp236 in KCNQ2, Fig. 1a; Trp 265 in KCNQ3, Fig. 1b: Leu246 in KCNQ1)[14,26]. This Trp side chain lies in the pore-forming S5 transmembrane segment, near the intracellular voltage-operated gate of the channel. As KCNQ2–5 subunits generally assemble as heteromers in the central nervous system[14,16], we tested the effects of retigabine in oocytes co-injected with KCNQ2 and KCNQ3 and observed large shifts of activation to more negative voltages, although not quite as large as with KCNQ2 or KCNQ3 alone (Fig. 1c). Some KCNQ channel activator molecules also cause a marked increase in peak current; however, the effects of retigabine on KCNQ3* channels are quite modest (rarely greater than a 15% increase in peak current), and throughout this study we have focused on the large gating shifts observed in these channels.

Heteromeric KCNQ2/3 channels were used to alter the number of available retigabine binding sites, by co-injecting KCNQ2[Trp236Phe] with wild-type (WT) KCNQ3, or KCNQ3[Trp265Phe] with WT KCNQ2. Assuming 1:1 stoichiometry of KCNQ2/3 assembly, both conditions are estimated to eliminate roughly half of the retigabine binding sites. Consistent with this, we observed intermediate effects of retigabine with fewer available binding sites, indicating that multiple molecules of the drug may be required to generate the full effect (Fig. 1c). A surprising, currently unexplained observation, was that the retigabine-mediated gating shift in KCNQ2/3 heteromers (~ −40 mV) was noticeably smaller than the shift in KCNQ2 or KCNQ3* homomeric channels. This likely hints that determination of the exact details of stoichiometry and cooperativity of drug binding in heteromeric channels will require alternative experimental approaches. We measured the effects of various mutations at position Trp265 (in KCNQ3*) or Trp236 (in KCNQ2) on retigabine sensitivity (Fig. 1d). In both KCNQ2 and KCNQ3* channels, only a Trp side chain present at this position was sufficient to enable a significant gating shift by retigabine, and even conservative mutations to other aromatic side chains (Phe or Tyr, Fig. 1a,b,d) cannot replicate the retigabine-mediated gating shift, consistent with previous reports[19,20].

We tested whether KCNQ3 position Trp 265 plays a role in regulating channel responses to PIP_2. Using the voltage-sensitive phosphatase ciona intestinalis voltage-sensitive phosphatase (CiVSP) to hydrolyse PIP_2 at depolarized voltages, we pulsed oocytes through a range of prepulse voltages followed by a test pulse to −20 mV to assess channel activity after PIP_2 depletion[27]. We observed similar channel rundown with depolarizing voltage steps in both KCNQ3* and KCNQ3*[Trp265Phe] channels, indicating that Trp265 does not significantly influence the effects of anionic phospholipids on channel function (Fig. 1e,f). Overall, these findings illustrate that KCNQ3 residue Trp265 is essential for the effects of retigabine, but does not play a significant role in regulating channel gating by voltage or PIP_2.

Unnatural amino-acid mutagenesis of KCNQ3 channels. To investigate the underlying mechanism of retigabine interactions with this essential Trp side chain, we employed unnatural amino-acid mutagenesis to introduce subtly altered Trp variants (Fig. 2a). With this method, a stop codon (TAG) is placed in the ion channel gene at a site of interest, and this mRNA is co-injected into *Xenopus laevis* oocytes along with a synthetic amino-acylated tRNA (carrying an unnatural amino acid) that is orthogonal to *Xenopus* tRNA synthetic pathways[28]. When the ribosomal translation machinery encounters the introduced TAG stop codon, the complementary synthetic tRNA (with a CUA anticodon) enables readthrough and incorporation of the appended amino acid into full-length ion channels (Fig. 2a)[29].

Among the KCNQ channels we tested, KCNQ3* channels exhibited large currents when expressed as homomeric channels in *Xenopus* oocytes, with a short latency between injection and expression, and were found to be an effective model system for incorporation of unnatural amino acids. KCNQ3* is also a useful model for retigabine binding because it does not possess a second interaction site that is targeted by some other recently studied KCNQ2 activators (see Discussion)[22,30], and so our results are specifically focused on the retigabine binding site formed by Trp265. Several criteria illustrate the feasibility and fidelity of the nonsense suppression approach in KCNQ3* channels. First, co-injection of KCNQ3*[Trp265TAG] with a full-length tRNA lacking a conjugated amino acid (pdCpA control) resulted in virtually no ionic currents, even several days after injection (Fig. 2b,d). However, when KCNQ3*[Trp265TAG] channels were co-injected with Trp-conjugated tRNA, robust currents with WT-like gating were observed, indicating efficient suppression of the introduced TAG stop codon and synthesis of functional channels with WT-like properties (Fig. 2b,e). Second, although there may be differences in the efficiency of incorporation of unnatural amino acids at this position, we have restricted our investigation to homomeric channels (rather than KCNQ2/KCNQ3 heteromers), and to biophysical parameters ($V_{1/2}$, activation kinetics) that are normally independent of channel expression. Last, we are confident that the synthetic Trp-tRNA was incorporated in a significant fraction of channel subunits because elicited currents responded to retigabine (Fig. 2c,f,g), and only Trp appears to be sufficient for retigabine sensitivity (Fig. 1d).

Loss of retigabine effects after removal of Trp265 H-bonding. Next, we tested retigabine sensitivity of KCNQ3* channels synthesized with the 'Ind' amino acid at position 265. This unnatural amino-acid side chain ablates the hydrogen-bonding capability of Trp by shifting the position of the indole nitrogen atom (Fig. 3a)[31], but retains the steric and chemical compositions of Trp. KCNQ3*[Trp265TAG] currents were efficiently rescued with Ind relative to the pdCpA control, yielding relatively large voltage-activated K^+ currents that closely resembled KCNQ3* (Fig. 3b,c). Remarkably, however, this single atom alteration in each of the KCNQ3* subunits entirely abolishes retigabine activation of KCNQ3* channels (Fig. 3d,e). Notably, Trp265Ind channels displayed WT-like gating, indicating that this residue does not play a significant role in conformational stabilization of channel states in the absence of drug. However, given the potent impact of such a minor manipulation, we examined the effects of the Trp265Ind substitution on channel gating more closely, to rule out the possibility of significant perturbation of channel

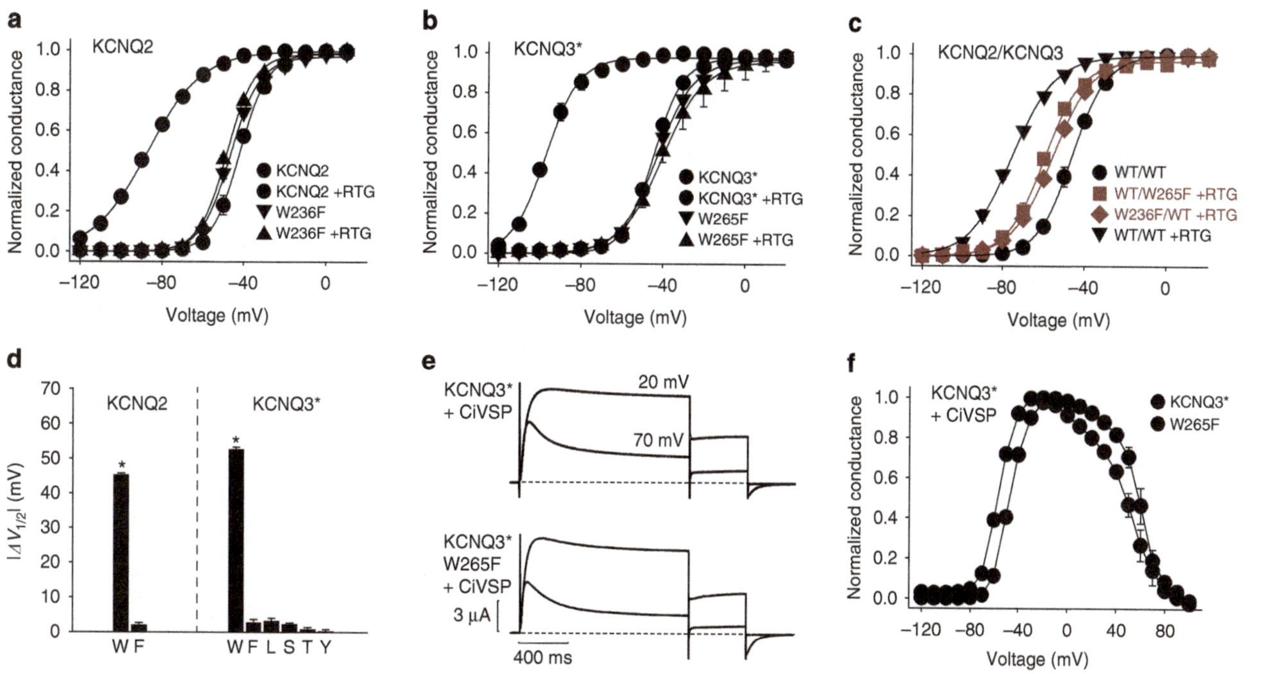

Figure 1 | Multiple retigabine molecules modulate KCNQ2 and KCNQ3 channel subunits via an S5 Trp side chain. (**a**,**b**) Conductance-voltage relationships for (**a**) KCNQ2 ($n=3$) and KCNQ2[Trp236Phe] ($n=6$), and (**b**) KCNQ3* ($n=5$) and KCNQ3*[Trp265Phe] ($n=3$) homomeric channels along with indicated mutants (retigabine concentration of 100 µM). (**c**) Conductance-voltage relationships for heteromeric combinations of KCNQ2 and KCNQ3 (1:1 ratio of injected mRNA, with or without Trp→Phe mutations as indicated, $n=5$ for each combination), used to generate channels with reduced numbers of retigabine binding sites. (**d**) Summary of $V_{1/2}$ shifts in saturating 100 µM retigabine for mutations of KCNQ2 Trp236 and KCNQ3 Trp265 as indicated (*$P<0.05$ in a paired Students t-test comparing control versus 100 µM retigabine in each experimental oocyte, $n=3$-6 per mutant). Only a Trp at either position is sufficient for retigabine sensitivity. (**e**) Exemplar currents of KCNQ3* and KCNQ3*[Trp265Phe] mutant coexpressed with CiVSP, illustrating that the Trp side chain responsible for retigabine sensitivity is not required for PIP2 sensitivity. (**f**) Summary data of tail current magnitude (−20 mV) after prepulses to a range of voltages, in oocytes expressing KCNQ3* ($n=5$) or KCNQ3*[Trp265Phe] ($n=5$) channels, along with CiVSP. In all panels, error bars represent s.e.m.

function. We observed no statistically significant difference in the $V_{1/2}$ or slope of the voltage dependence of activation of KCNQ3* and KCNQ3*[Trp265Ind] (see Table 1 and Fig. 3d,e). Last, as described earlier, the more disruptive Trp265Phe mutation did not alter the inhibition of KCNQ3* currents by PIP_2 depletion (mediated by CiVSP). Taken together, these observations demonstrate a highly specific effect of the Trp265Ind substitution on drug interactions (with little disruption of intrinsic voltage-dependent gating or lipid regulation), indicating that H-bond formation with Trp265 is a crucial step for retigabine action.

Tuning the strength of the Trp265–retigabine interaction. To investigate the possible involvement of other modalities of Trp interactions with the drug, we also examined the consequences of fluorination of Trp265 on retigabine effects. Trp fluorination is typically used to modify the electrostatic surface potential and test for the effects of cation–pi interactions (Fig. 4a)[32]. We were unable to detect functional channels carrying the F_4-Trp side chain (fluorines at ring positions 4,5,6 and 7, as numbered in Fig. 4a); however, robust current rescue was observed for less fluorinated derivatives such as F_3-Trp (Fig. 4b,c). Retigabine potentiation of KCNQ3* channels was retained with F_3-Trp substitution at Trp265 (Fig. 4c–e), a substitution that still potently diminishes the negative electrostatic potential on the face of the side chain (Fig. 4a), indicating that a cation–pi interaction of Trp265 with retigabine or some other entity (perhaps another channel residue) is not required for retigabine effects. It is noteworthy that progressive fluorination of Trp265 caused variable effects on KCNQ3*-gating properties (Table 1), with F_1-Trp causing a modest depolarizing shift of the activation $V_{1/2}$, while F_3-Trp significantly shifted gating in the hyperpolarizing direction. This lack of an obvious trend also suggests that an endogenous cation–pi interaction involving Trp265 does not contribute significantly to the gating process.

The most interesting outcome of fluoro-Trp substitutions further highlights the role of H-bond formation by the indole N–H group. Specifically, fluorination of the Trp side chain significantly increases the polarity of the indole N–H, thereby strengthening the hydrogen-bonding propensity of the indole side chain[33,34]. This is illustrated by the electrostatic surface potentials for Trp and F_3-Trp (note the more polarized electrostatic surface potential of F_3-Trp, highlighted by the blue colour around the N–H bond, in the lower right corner as oriented in Fig. 4a). Consistent with this, retigabine potency was increased with fluoro-Trp analogues at Trp265. For instance, in KCNQ3*[Trp265TAG] channels rescued with Trp (mimicking KCNQ3* channels), 1 μM retigabine elicits very little shift in the voltage dependence of activation (Fig. 4d). However, in channels with F_3-Trp substituted at Trp265, 1 μM retigabine elicits a much larger response (Fig. 4d). Consistent with this effect being attributable to strengthening of the formation of a hydrogen bond by Trp265, we observed a fluorination-dependent increase in retigabine potency, with progressive fluorination shifting the concentration response to lower retigabine concentrations (Fig. 4f). In addition, we observed a small but significant position-dependent effect of mono-fluorination of Trp265, in which fluorination of ring position 7 (closer to the indole N–H) increased retigabine effects slightly more than fluorination of ring position 5 (Fig. 4f). This is consistent with the fluorine substituent at position 7 having a stronger effect on the polarity of the N–H bond.

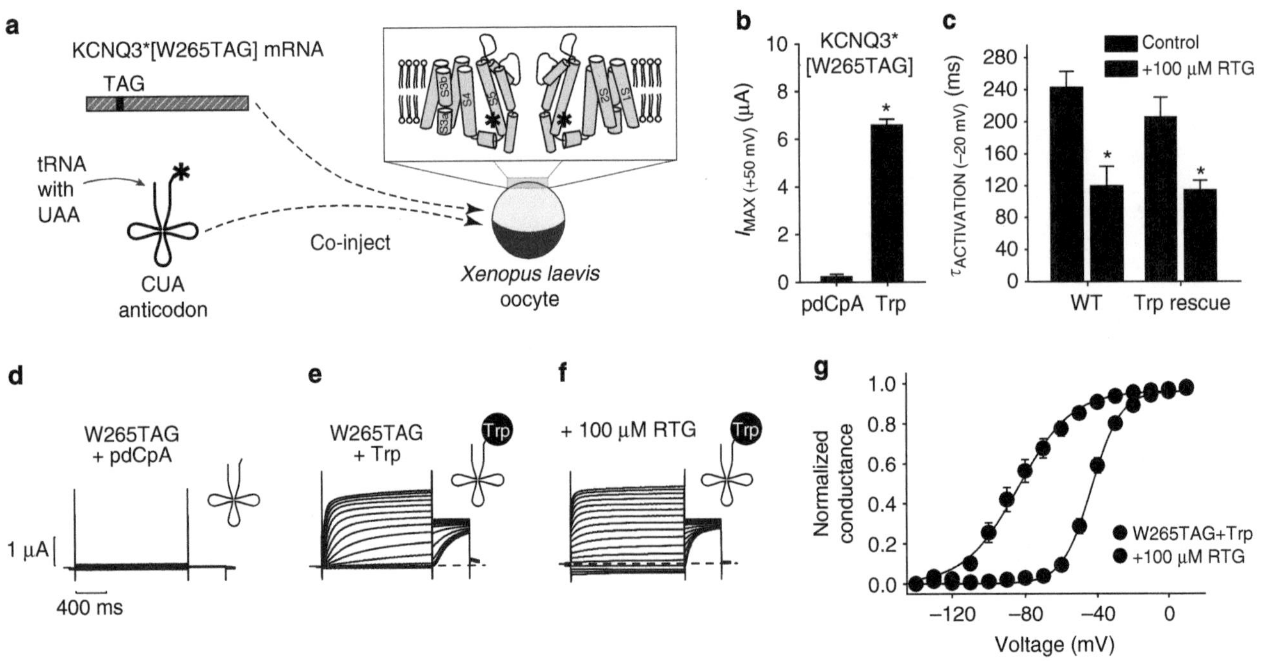

Figure 2 | Nonsense suppression for amino-acid incorporation in KCNQ3* channels. (**a**) Schematic diagram of the nonsense suppression method, in which mRNA (with a stop codon at Trp265) and amino-acylated tRNA are co-injected into *Xenopus* oocytes. Incorporation of the unnatural amino acid enables readthrough of the stop codon and expression of functional channels. (**b**) Current magnitudes in oocytes injected with KCNQ3*[Trp265TAG] mRNA and either an unconjugated tRNA (pdCpA; $n=5$) or tRNA amino-acylated with Trp ($n=8$, $*P<0.05$, Student's t-test). (**c**) Activation kinetics of KCNQ3* ($n=5$) and KCNQ3*[Trp265TAG] ($n=8$) channels rescued with Trp, in the presence or absence of 100 μM retigabine ($*P<0.05$, Student's t-test). (**d**-**f**) Exemplar currents from oocytes injected with KCNQ3*[Trp265TAG] mRNA, and indicated synthetic tRNAs. (**g**) Conductance-voltage relationships for Trp-rescued KCNQ3*[Trp265TAG] ($n=8$), with retigabine response, illustrating faithful incorporation of the desired side chain at position 265. For KCNQ3* channels $V_{1/2}=-44\pm1$ mV, $k=7.5\pm0.5$ mV; for Trp-rescued KCNQ3*[Trp265TAG] $V_{1/2}=-43\pm2$ mV, $k=7.9\pm0.5$ mV (± indicates s.e.). Error bars throughout represent s.e.m.

Since F_3-Trp incorporation itself caused a hyperpolarizing shift of channel activation even in the absence of retigabine (Table 1 and Fig. 4d), a possible alternative explanation was that increased retigabine potency with F_3-Trp at position 265 was a secondary allosteric effect of stabilization of the channel open state, rather than a direct effect on retigabine binding. Thus, as an additional control experiment, we investigated the effects of the Asn220Cys mutation on retigabine sensitivity (in the extracellular S3–S4 linker, distant from the Trp265 putative retigabine binding residue). This mutation shifts channel opening to more negative voltages, even more than F_3-Trp substitution at position 265. In the Asn220Cys mutant, no sensitization to retigabine was observed (Fig. 4g). That is, 1 μM retigabine was unable to generate a marked gating shift, in contrast to the potent retigabine effects with F_3-Trp substitution at Trp265 (curve fits for the F_3-Trp substitution are included for comparison, Fig. 4g). In addition, the retigabine concentration response for Asn220Cys channels was very similar (albeit shallower) to the KCNQ3* background construct, demonstrating that the Asn220Cys mutation *per se* does not inhibit retigabine activation (Fig. 4h). These observations suggest that an indirect effect of open-state stabilization is unlikely to account for increased retigabine potency in F_3-Trp-substituted channels.

Relative contributions of putative drug-binding residues. Given the dramatic effects of the Trp265Ind substitution, we sought to clarify the contributions of other residues reported to play a role in retigabine binding. Using a chimeric approach between KCNQ1 and KCNQ3, previous work identified KCNQ3 residues Thr271 (S5 helix), Leu272 (S5 helix), Leu314 (pore helix) and Leu338 (S6 helix) as important contributors to a putative retigabine binding pocket[19]. We generated several mutations at each of these positions in KCNQ3* and tested retigabine effects over a broad concentration range (Fig. 5a,b). We observed that all functional mutants at these positions retained a large retigabine-mediated gating shift (Fig. 5b), although reduced potency was reflected by a shift to higher drug concentrations in many cases. Moreover, many of these mutations caused intrinsic shifts in channel gating in the absence of retigabine (Fig. 5a). Previous reports have suggested that many of these mutations strongly diminish retigabine action; however, our application of higher concentrations illustrates that, while potency of the drug is weakened, large gating shifts still occur. These data indicate that Thr271, Leu272, Leu314 and L338 make measurable contributions to retigabine binding affinity. However, the Trp265Ind substitution abolishes any response to retigabine at experimentally achievable concentrations. Thus, Trp265Ind remains the most structurally subtle, yet most disruptive, mutation we have identified in terms of retigabine sensitivity, highlighting the potential importance of H-bonding with the Trp265 side chain.

We next used a previously published molecular model of the retigabine binding site to investigate alternative binding modes of the drug[19]. In the published model ('original' in Fig. 5c) there are no obvious H-bond acceptors in the vicinity of Trp265. There are also no apparent H-bond interactions with Thr271. However, this putative binding pocket can accommodate altered orientations (for example, 'flip' in Fig. 5c) in which the drug occupies a similar space (see 'overlay' panel illustrating both binding models in Fig. 5c), but orients the carbamate moiety of retigabine in close proximity to Trp 265. A rotameric shift of Trp265 enables a close interaction between the indole N–H and a carbamate oxygen atom of retigabine. Using molecular dynamics simulations (of 12 retigabine binding sites starting in a 'flip' orientation, for 10 ns each), we measured the mean distance between the retigabine carbamate oxygen and indole N–H group at the end of the simulations as 4.0 ± 0.5 Å (± indicates s.e., four binding sites were measured between 2 and 3 Å, three sites between 3 and 4 Å and five sites were > 5 Å), consistent with the possible formation of a hydrogen bond in this region.

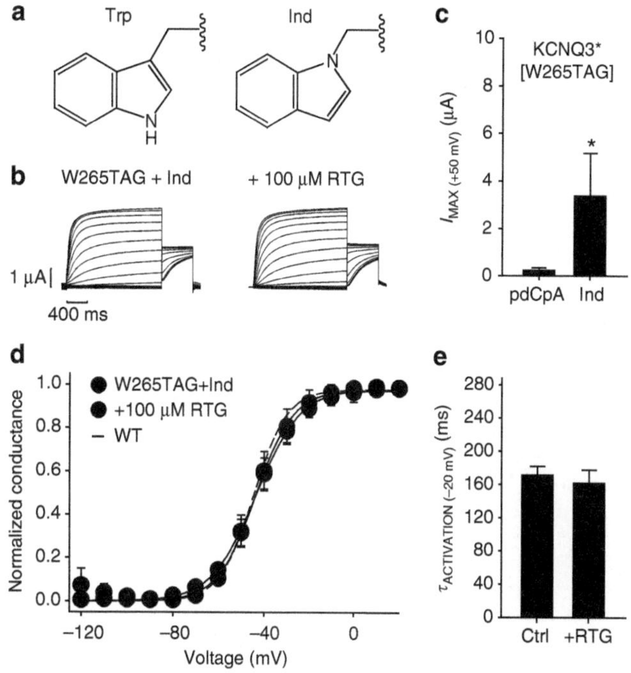

Figure 3 | The position of the Trp 265 indole nitrogen is essential for retigabine sensitivity. (**a**) Chemical structures of Trp and Ind side chains, illustrating the subtle change in the position of the indole nitrogen atom. (**b**) Exemplar currents elicited from a Xenopus oocyte with Ind-rescued KCNQ3*[Trp265TAG] channels illustrating retigabine insensitivity. (**c**) Current magnitudes in oocytes injected with KCNQ3*[Trp265TAG] mRNA and either an unconjugated tRNA (pdCpA; $n = 5$) or tRNA amino-acylated with Ind ($n = 7$, *$P < 0.05$, Student's t-test). (**d**) Conductance-voltage relationships for Ind-rescued KCNQ3*[Trp265TAG], in the presence and absence of retigabine, illustrating the importance of the correct positioning of the N-H group. For KCNQ3*[Trp265Trp], $V_{1/2} = -43 \pm 2$ mV, $k = 7.9 \pm 0.5$ mV; for KCNQ3*[Trp265Ind], $V_{1/2} = -48 \pm 2$ mV, $k = 7.3 \pm 0.6$ mV (no statistical significance, ± indicates s.e.). (**e**) Activation kinetics for KCNQ3*[Trp265Ind] measured at −20 mV in the presence and absence of 100 μM retigabine ($n = 7$, no statistical significance). In all panels, error bars represent s.e.m.

Identification of a likely H-bond donor in retigabine. We further investigated the mechanism of H-bonding with Trp265 by seeking potential H-bond acceptors in the retigabine molecule, a task facilitated by the availability of numerous retigabine analogues. We first considered the analogue ML-213 because it has a simplified chemical scaffold compared with retigabine and a reduced number of possible H-bond acceptors (Fig. 6a,b). We found that ML-213 is a more potent activator of KCNQ3* compared with retigabine ($EC_{50} = 3.6 \pm 0.2$ μM, $n = 5$, versus 11.6 ± 0.4 μM, $n = 5$, respectively, Fig. 6c, ± indicates s.e.). We have also included the calculated electrostatic surface potentials for retigabine and ML-213, illustrating that the negative surface potential around the carbonyl oxygen is more pronounced for ML-213 (red protrusion, Fig. 6a,b).

Expanding on these observations, we examined a spectrum of KCNQ channel activators with varying potency in KCNQ3* channels (Fig. 7). We have depicted all of the drugs tested and colorimetric representations of their calculated electrostatic surface potentials, along with the observed EC_{50} for the shift of activation $V_{1/2}$ of KCNQ3*[Trp265TAG] channels (rescued with Trp, F_3-Trp or Ind). For all drugs tested, effects were abolished by the Trp265Ind substitution, indicating a common mechanism via H-bond formation with Trp265. In addition, F_3-Trp substitution generally resulted in higher drug potency, although this effect was not as pronounced with the less polar analogues ICA-110381 and ICA-069673. Importantly, all of the activator compounds contain a carbonyl oxygen (either in the carbamate moiety of retigabine and flupirtine, or in the amide linker in ML-213 or the ICA compounds), and it is noteworthy that the strength of the negative surface potential correlates well with the potency of the drug for KCNQ3* activation—drugs with a weaker electrostatic surface potential trend towards weaker potency (Fig. 7). The ICA compounds are very weakly potent activators of KCNQ3* channels within our typical experimental concentration range (up to 300 μM), and notably ICA-110381 caused considerably more activation in F_3-Trp-substituted channels, demonstrating that effects of an otherwise weak channel activator can be strengthened by targeting the chemical properties of Trp265. Taken together, these findings illustrate the importance of the H-bonding propensity of Trp265, likely involved in the formation of a H-bond with a carbonyl oxygen present in retigabine and its analogues.

Table 1 | Activation gating parameters for unnatural amino-acid substitutions of KCNQ3*[Trp265TAG] channels.

Substitution	$V_{1/2}$ (mV ± s.e.m.)	Slope factor (mV ± s.e.m.)	n
WT-Trp	− 44 ± 1	7.5 ± 0.5	5
265TAG-Trp	− 43 ± 2	7.9 ± 0.5	12
265TAG-F_1-Trp	− 38 ± 2	9.3 ± 0.5	9
265TAG-F_2-Trp	− 48 ± 2	9.9 ± 0.8	7
265TAG-F_3-Trp	− 62.3 ± 3	10.9 ± 0.8	12
265TAG-Ind	− 48 ± 2	7.3 ± 0.6	7

Fitting of conductance-voltage relationships (see Methods) yielded $V_{1/2}$ (voltage where activation is half-maximal) and the slope factor *k* (voltage range where fractional activation changes e-fold). Data are presented as mean ± s.e.m.

Discussion

Retigabine is currently the only drug widely approved for clinical use that acts via activation of voltage-gated potassium channels[35]. Detailed investigation of this drug prototype may deepen our understanding of how drugs can act as ion channel activators, and accelerate the development of potassium channel activators with increased potency or specificity. Such drugs may have untapped therapeutic potential for management of a variety of 'hyperexcitability' disorders in addition to epilepsy, such as painful syndromes, hypertension and cardiac arrhythmias[9,36]. In addition, recent reports have highlighted retigabine

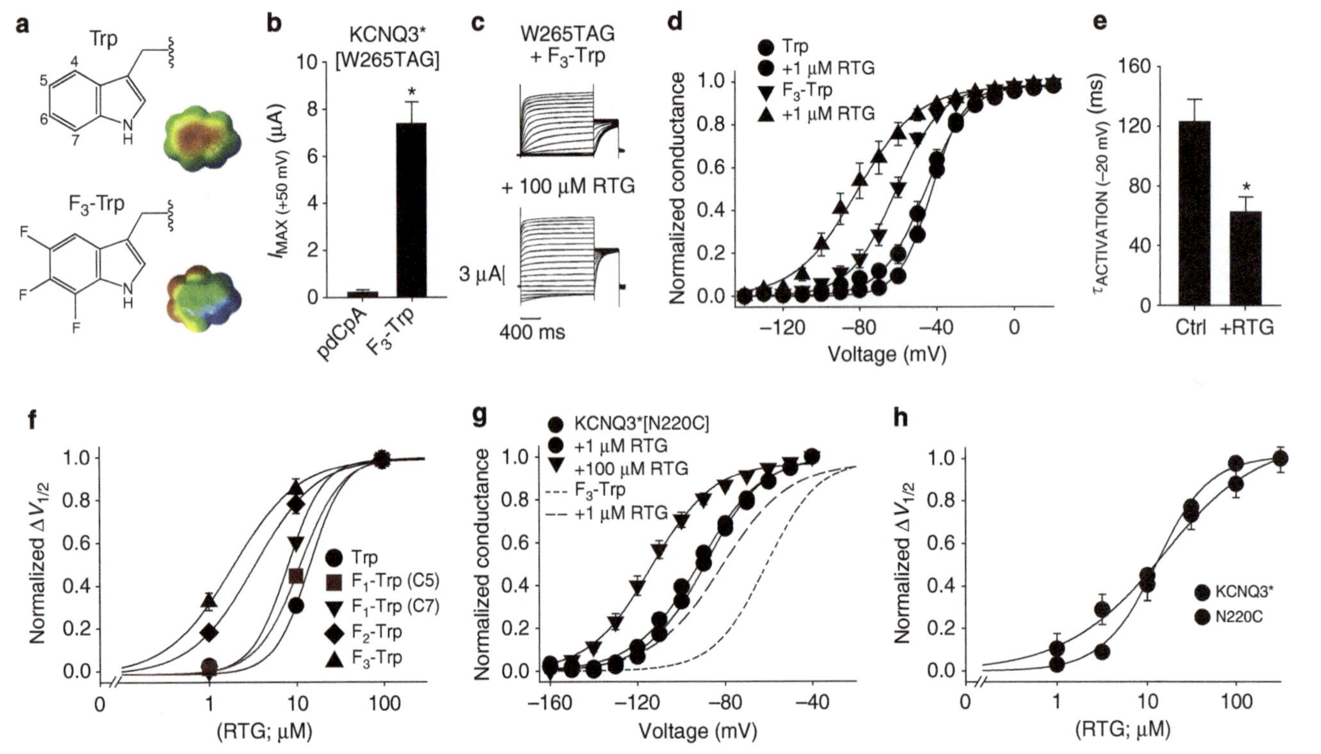

Figure 4 | The polarity of the Trp265 indole nitrogen modulates retigabine sensitivity. (**a**) Chemical structures of Trp (with ring positions labelled) and F_3-Trp, accompanied by colorimetric representations of electrostatic surface potentials. (**b**) Current magnitudes in oocytes injected with KCNQ3*[Trp265TAG] mRNA and either an unconjugated tRNA (pdCpA; $n = 5$) or tRNA amino acylated with F_3-Trp ($n = 12$, *$P < 0.05$, Student's *t*-test). (**c**) Exemplar currents from F_3-Trp-rescued KCNQ3*[Trp265TAG] channels in the presence and absence of retigabine. (**d**) Conductance-voltage relationships illustrating the effects of 1 μM retigabine on Trp-rescued ($n = 12$) and F_3-Trp-rescued KCNQ3*[Trp265TAG] ($n = 12$) channels. (**e**) Activation kinetics (− 20 mV) for F_3-Trp-rescued channels ($n = 3$), in the presence and absence of 100 μM retigabine (*$P < 0.05$, Student's *t*-test). (**f**) Concentration-response curves for retigabine effects on numerous fluoro-Trp analogues ($n = 9$-12 per Trp analogue) substituted at position Trp265, illustrating enhanced retigabine potency with increased fluorination. (**g**) Conductance-voltage relationships with indicated retigabine concentrations on a KCNQ3 mutant (Asn220Cys, $n = 4$) with an intrinsic hyperpolarizing shift in gating. Conductance-voltage relationships for F_3-Trp substituted at Trp265TAG are shown for comparison. (**h**) Concentration-response curves for retigabine effects on KCNQ3* ($n = 5$) and KCNQ3*[Asn220Cys] ($n = 4$) channels. In all panels, error bars represent s.e.m.

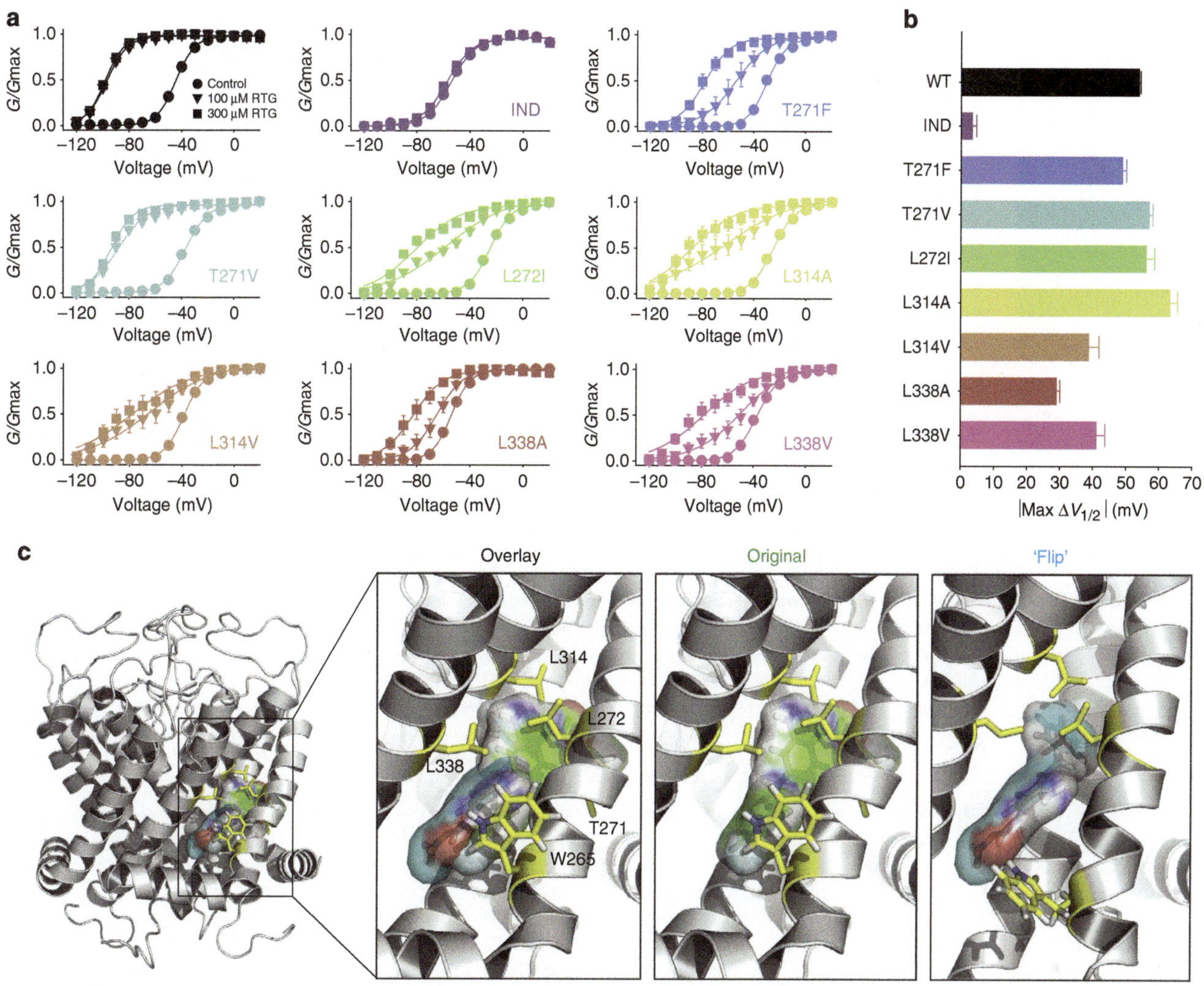

Figure 5 | Detailed characterization of secondary retigabine binding residues and alternative binding site orientations. (**a**) Conductance-voltage relationships were gathered for the indicated KCNQ3* mutant channels ($n = 4$–6 per mutant), in 0, 100 or 300 μM retigabine. (**b**) Maximal $\Delta V_{1/2}$ in 300 μM retigabine measured in each mutant channel. Error bars in **a**,**b** represent s.e.m. (**c**) Retigabine was docked into a molecular model of the pore-forming domain of KCNQ3 (see ref. 19). Two orientations are shown with the carbamate group in either the vicinity of Leu314 ('original' model) or Trp265 ('flip' model). The two binding models are superimposed in the 'overlay', showing the similar space occupied by both drug orientations.

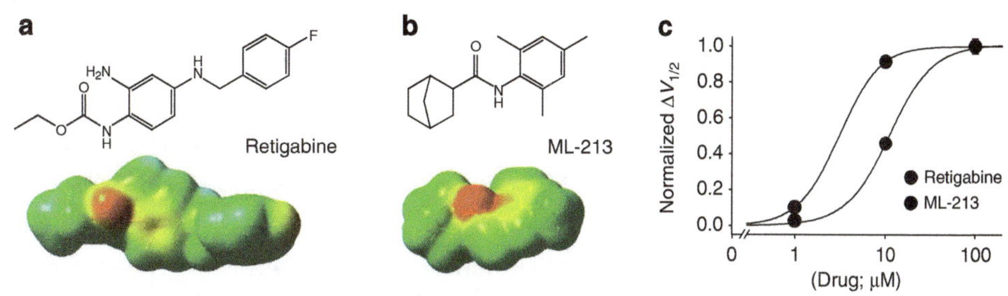

Figure 6 | ML-213 exhibits a stronger electrostatic surface potential and higher potency than retigabine for KCNQ3* activation. (**a**,**b**) Chemical structures and electrostatic surface potentials for retigabine and ML-213. Note the increased negative surface potential in the vicinity of the carbonyl oxygen atom in ML-213. The scale for electrostatic surface potential representation is red: −80 kcal, yellow: 0 kcal, blue: +80 kcal. (**c**) Concentration-response curves for retigabine and ML-213 ($n = 5$) effects on KCNQ3* channels ($n = 5$). Error bars represent s.e.m.

effectiveness in treatment of tinnitus, Parkinson's disease and Huntington's disease[37–39].

At the outset of our study, chimeric studies with the retigabine-insensitive KCNQ1 channel had identified the importance of a Trp residue (KCNQ3 Trp265) in the pore-forming region, near the presumed cytoplasmic channel gate. Retigabine interactions with this site had been rationalized as generic hydrophobic interactions with this conserved Trp residue (also present in KCNQ2, 4 and 5)[19,20].

Although most often considered to be hydrophobic, Trp side chains can deploy diverse chemical forces in ligand interactions[40], and these require specialized chemical-scale approaches to be

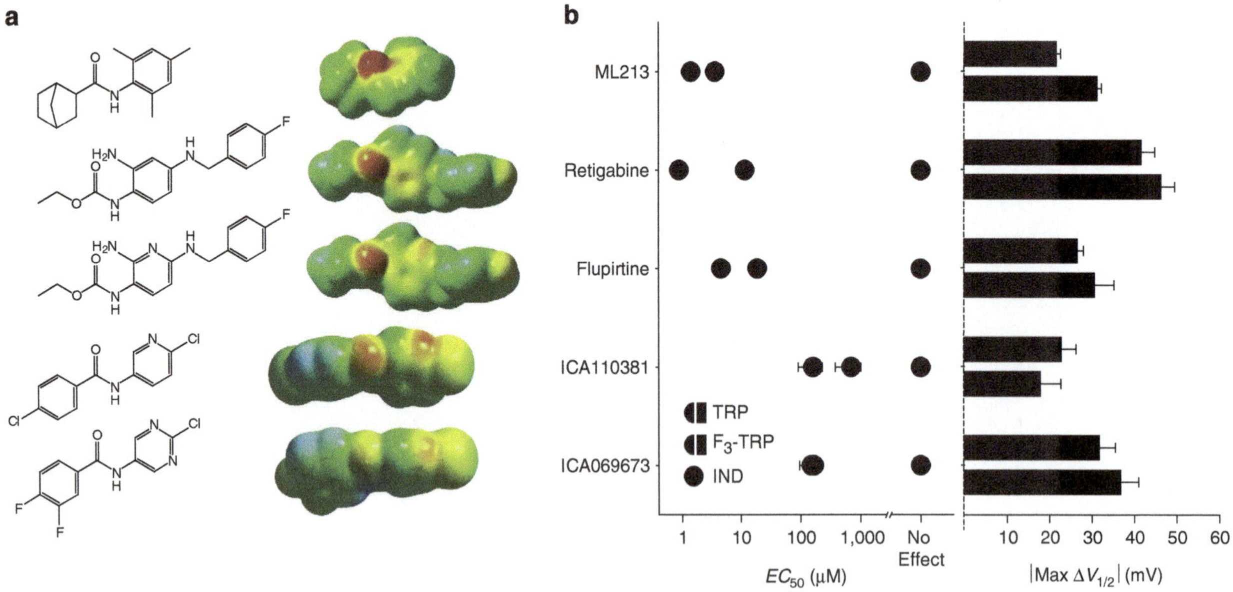

Figure 7 | Effects of retigabine analogues correlate with electrostatic surface potential. (**a**) Chemical structures and electrostatic surface potentials for a series of retigabine analogues. All structures and surface potential maps have been aligned on the basis of the location of the conserved amide-ester bond—note the gradient of the intensity of the negative surface potential around the carbonyl oxygen atom (scaling is the same as in Fig. 6). (**b**) Summary illustrating the EC50 of each drug on Trp-rescued, F_3-Trp-rescued and Ind-rescued KCNQ3*[Trp265TAG] channels ($n = 4$–9 per data point), and the maximal efficacy ($\Delta V_{1/2}$) of each drug in F_3-Trp and Trp-rescued channels (effects on Ind-rescued channels are minimal and thus have been omitted). Error bars represent s.e.m.

distinguished[29]. There is growing recognition of the general importance of electrostatic cation–pi interactions with all aromatic side chains[29,41]. In addition, both Trp and Tyr side chains can act as H-bond donors. Unnatural amino-acid mutagenesis approaches allow us to parse out atomistic modes of interaction, and also hopefully mitigate the possibility of gross structural perturbations that might arise using conventional mutagenesis. By engineering KCNQ3 channels to include subtle Trp variants at position 265, our study identifies the very atoms required for retigabine effects and, further, the chemical forces involved in forming a direct interaction with retigabine. Surprisingly, our findings indicate that the critical property of Trp265 for retigabine activation, rather than hydrophobicity, is the ability to H-bond. Repositioning of the indole nitrogen atom to remove Trp H-bonding abolished the effects of retigabine and all analogues tested (Figs 3 and 7). Moreover, a Trp analogue (F_3-Trp) with altered electrostatics but intact/enhanced H-bonding ability enhances retigabine potency (Figs 4 and 7). A valuable future direction for these experiments will be to develop an independent measure of drug binding, to distinguish whether the H-bond interaction (disrupted by the Trp265Ind mutation) is essential for drug binding, or couples drug binding to a later step involved in channel potentiation. Nevertheless, these observations highlight a rarely investigated mode of drug interactions with Trp, as Trp residues in a binding pocket are often presumed to indicate a hydrophobic site. Our findings illustrate the importance of detailed investigation of the underlying chemistry of these binding sites to better understand mechanisms of drug action.

By demonstrating the effects of H-bond strength on a range of KCNQ3 activators, our findings suggest the importance of a carbonyl oxygen (usually in a carbamate or amide moiety) for the formation of a negative electrostatic surface potential to act as a H-bond acceptor. Previous screens of compound libraries have investigated important physicochemical features important for drug activity on KCNQ2–5 channels. Several studies have reported the importance of an amide bond (or a carbonyl oxygen) as an essential element of the pharmacophore, although the relationship between this functional group and the H-bonding propensity of Trp265 has not been recognized[9,42–44]. Notably, KCNQ activators with marked structural diversity appear to act through Trp265 (several examples are shown in Fig. 8), and these share the common feature of a carbonyl oxygen that we suggest acts as an essential hydrogen bond acceptor. Supporting this notion, our findings illustrate that ML-213 is an effective KCNQ3 activator, correlated with its strong surface potential relative to other drugs tested. In the context of understanding drug interactions with ion channels and other receptors, these observations illustrate the importance of investigating specific chemical forces that enable drug interactions with aromatic side chains known to contribute to binding sites in other ion channels (for example, voltage-gated Na^+ channels, human ether-a-go-go related gene product (hERG))[45–47]. These insights could guide rational 'tuning' of the properties of certain functional groups to alter drug properties in desirable ways.

On the basis of the magnitude of the retigabine-induced shift of the conductance–voltage relationship, saturating effects of retigabine are estimated to stabilize channel opening by ~6 kcal mol^{-1} (calculated as $\Delta G = \Delta(zFV_{1/2})$). As a comparison, formation of a H-bond between a carbonyl and amino groups (N-H: O = C) is generally quite weak, on the order of 2 kcal mol^{-1}. A tetrameric KCNQ channel likely contains four potential retigabine binding sites, and our characterization of heteromeric channels with fewer binding sites (Fig. 1c) indicates that multiple sites may be occupied in saturating retigabine concentrations. Thus, the energetics of the full retigabine-mediated gating shift can be reasonably attributed to multiple retigabine molecules interacting with Trp265 via a H-bond. Although not as prominent as the Trp265 indole nitrogen, other residues also make additional measurable contributions to the overall energy of retigabine binding (Fig. 5 and ref. 19). It should also be noted that cooperativity between retigabine binding sites has not been

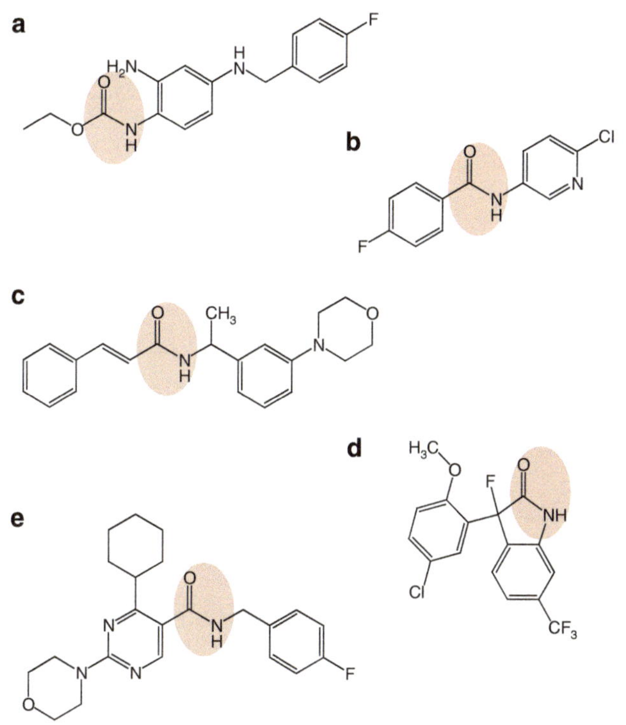

Figure 8 | Diverse structures of KCNQ openers. Multiple structures of KCNQ channel openers are presented to highlight the overall features of an amide group flanked by various ring structures. Our findings highlight the importance of the amide carbonyl for interaction with KCNQ3 Trp 265 and likely equivalent positions in KCNQ2, 4 and 5. Drugs depicted are (**a**) retigabine, (**b**) ztz-240 (described in ref. 24), (**c**) acrylamide (s)-1, (**d**) BMS-204352 and (**e**) an unnamed experimental drug described in ref. 43.

investigated in detail, and so it is not known how multiple retigabine binding events may interact and influence channel function. Investigation of libraries of retigabine analogues has highlighted that variable aromatic ring substituents can alter the apparent affinity, maximal effect and subunit specificity[24,43]. For example, many ring substituents (such as halogenation) alter the polarity of the carbonyl oxygen and thereby alter drug interactions with Trp265. In addition, it seems likely that diverse flanking structures (potentially by interacting with the auxiliary retigabine binding residues investigated in Fig. 5) could influence the orientation of the carbonyl oxygen relative to Trp265.

A final noteworthy detail is that, although our findings delineate an atomic basis for the retigabine interaction with KCNQ3 Trp265 (and analogous residues in KCNQ2,4,5), there is significant evidence for additional interaction sites that enable potentiation of certain KCNQ subunits. For example, the ICA compounds tested (and other related compounds such as ztz-240), along with a family of diclofenac derivatives, have been reported to influence KCNQ2 channels independent of the conserved S5 Trp residue and exhibit dramatically stronger effects in KCNQ2 relative to KCNQ3 (refs 22,24,25,30). In addition, unrelated KCNQ openers such as zinc pyrithione do not appear to depend on the hydrogen bond interaction that we have identified in this study[48]. Therefore, ongoing examination of drug binding to different KCNQ channel subtypes may help to reveal additional sites for KCNQ potentiation.

In conclusion, we have used unnatural amino-acid mutagenesis and available retigabine analogues to localize the effects of retigabine binding to a single H-bond interaction with KCNQ3 Trp265. These findings highlight an unusual mode of drug interaction with a Trp side chain, and demonstrate the importance of careful investigation of the mechanism of drug interactions to guide rational modification of therapeutic compounds.

Methods

Molecular biology and *in vivo* nonsense suppression. KCNQ2 (human) and KCNQ3 (human) channel cDNAs were propagated and manipulated using the pTLN vector (gifts of Dr M. Taglialatela and Dr T. Jentsch). Introduction of point mutations was accomplished using PCR to combine two fragments each of which contained the mutation in an overlapping region of the sequence ('two-step' PCR method). To accomplish this, we used standard PCR approaches to amplify a 5′ fragment and a 3′ fragment (using a 'flanking primer' paired with a mutagenic primer). These fragments were then mixed along with the 'flanking primer' pair for a second round of PCR to combine the overlapping fragments. The resulting fragment was then subcloned into the appropriate parent vector using EcoRI and NotI restriction enzymes, and the sequence was verified by Sanger sequencing approaches (Genewiz). Sequences for the flanking primers and the 'forward' mutagenic primers are provided in Supplementary Table 1 for all mutations that were generated ('reverse' mutagenic primers are simply the reverse complement). In all experiments involving homomeric expression of KCNQ3 channels (including *in vivo* nonsense suppression experiments), the Ala315Thr mutation was introduced to enable efficient trafficking of homomeric KCNQ3. KCNQ3 does not efficiently form functional channels; however, the Ala315Thr mutation enables efficient KCNQ3 functional expression without co-injection of KCNQ2 mRNA (throughout the text we refer to KCNQ3[Ala315Thr] as KCNQ3*)[49]. The plasmid encoding CivSP was kindly provided by Dr Y. Okamura. Complementary RNA was transcribed from the cDNA using the mMessage mMachine Kit (Ambion). Stage V–VI *Xenopus laevis* oocytes were prepared as previously described and were injected with cRNA alone or cRNA plus synthetic tRNA (for nonsense suppression, described below). Oocyte preparation was carried out using a protocol approved by the University of British Columbia Animal Care Committee, in accordance with the Canadian Council for Animal Care guidelines. We used female *Xenopus laevis* frogs 100 g or greater in size. After injection, oocytes were incubated for 12–48 h at 18 °C before recording.

Fluorinated Trp derivatives (5-F_1-Trp; 7-F_1-Trp; 5,7-F_2-Trp (F_2-Trp) and 5,6,7-F_3-Trp (F_3-Trp)) were purchased from Asis Chem (Watertown, MA) and Sigma-Aldrich (St Louis, MO). The 'Ind' Trp variant was synthesized in-house as described previously[31,50]. Incorporation of unnatural amino acids into ion channel proteins was carried out as described extensively in prior publications[29,34,51]. Briefly, unnatural amino acids (aa) were protected with nitroveratryloxycarbonyl and were activated as the cyanomethyl ester, which were then coupled to the dinucleotide pdCpA (GE Healthcare/Dharmacon, Lafayette, CO). This aminoacyl dinucleotide was subsequently ligated to a modified (G73) *Tetrahymena thermophile* tRNA. The amino-acylated tRNA-aa was deprotected by ultraviolet irradiation immediately before oocyte injection. In a typical experiment, 80 ng of tRNA-aa and 40 ng of KCNQ3 cRNA were injected in a 50-nl volume. In control experiments, the cRNA alone or the cRNA together with a tRNA coupled to pdCpA but without an appended amino acid were injected (these parallel control experiments were conducted every experimental day and the results are described throughout the text).

Two-electrode voltage clamp recordings. Voltage-clamped potassium currents were recorded in standard Ringers solution (in mM): 116 NaCl, 2 KCl, 1 $MgCl_2$, 0.5 $CaCl_2$, 5 HEPES (pH 7.4) using an OC-725C voltage clamp (Warner, Hamden, CT). Glass microelectrodes were backfilled with 3 M KCl and had resistances of 0.1–1 MΩ. Data were filtered at 5 kHz and digitized at 10 kHz using a Digidata 1440A (Molecular Devices) controlled by the pClamp 10 software (Molecular Devices). Drugs were purchased from Toronto Research Chemicals (retigabine) or Tocris (ML-213, flupirtine, ICA-069673, ICA-110381), stored as 100 mM stocks in dimethylsulphoxide and diluted to working concentrations each experimental day.

Molecular simulations of retigabine binding. A published model of retigabine docked to the KCNQ3 pore domain was used as the starting point of simulations[19]. In this model the retigabine molecule is positioned with its potential H-bond donor carbonyl oxygen pointed away from Trp265. Therefore, the retigabine was rotated by ~180° ('flipped' upside down). Three further retigabine molecules were docked into a similar position resulting in a model with all four binding sites occupied. This model was energy-minimized using force field AMBER03 and the constraints for the retigabine molecules were assigned with semi-empirical quantum mechanics (QM) using force field NOVA. Subsequently, a short molecular dynamics (MD) simulation was run while constraining the distance between the Trp265 N–H group and the carbonyl oxygen of retigabine to a distance of 4 ± 1.5 Å. The resulting model contained four retigabine molecules in the 'flip' position, and all four Trp265 side chains in a rotated position (see Fig. 5c). Starting with this model, zero, one, two (in opposing or adjacent subunits), three or four retigabines were deleted from the respective model generating all possible combinations of occupied binding-site configurations. Each resulting model was inserted into a virtual membrane and MD simulations were run as described previously[52]. In brief, models were incorporated into membranes, the simulation box was filled with 0.9 mM NaCl/H_2O and, subsequently, all-atom-mobile simulations were performed for 10 ns to reach stable conformations using YASARA Structure version 10.1 (AMBER03, 2 fs time steps; YASARA Biosciences GmbH, Vienna, Austria). The average structures of all models were calculated and the distance between the Trp265 N-H group and the carbonyl oxygen of retigabine was measured individually in each occupied binding site.

Data analysis. Voltage dependence of channel activation was fitted with a standard single component Boltzmann equation of the form $G/Gmax = 1/(1 + e^{-(V - V_{1/2})/k}$ where $V_{1/2}$ is the voltage where channels exhibit half-maximal activation, and k is a slope factor reflecting the voltage range over which an e-fold change in Po is observed. For generation of concentration–response curves to retigabine and derivatives, $\Delta V_{1/2}$ (the shift in half-activation voltage) was normalized to the maximal $\Delta V_{1/2}$ observed, and fit with the following equation: normalized $\Delta V_{1/2} = 1/(1 + EC_{50}/[drug])$. Statistical tests and significance are described in figure legends throughout the text.

References

1. Brodie, M. J. Antiepileptic drug therapy the story so far. *Seizure* **19**, 650–655 (2010).
2. Kwan, P. *et al.* Definition of drug resistant epilepsy: consensus proposal by the *ad hoc* Task Force of the ILAE Commission on Therapeutic Strategies. *Epilepsia* **51**, 1069–1077 (2010).
3. Pati, S. & Alexopoulos, A. V. Pharmacoresistant epilepsy: from pathogenesis to current and emerging therapies. *Cleve. Clin. J. Med.* **77**, 457–467 (2010).
4. Lutas, A. & Yellen, G. The ketogenic diet: metabolic influences on brain excitability and epilepsy. *Trends Neurosci.* **36**, 32–40 (2013).
5. Yellen, G. Ketone bodies, glycolysis, and KATP channels in the mechanism of the ketogenic diet. *Epilepsia* **49**(Suppl 8): 80–82 (2008).
6. Klassen, T. *et al.* Exome sequencing of ion channel genes reveals complex profiles confounding personal risk assessment in epilepsy. *Cell* **145**, 1036–1048 (2011).
7. Luszczki, J. J. Third-generation antiepileptic drugs: mechanisms of action, pharmacokinetics and interactions. *Pharmacol. Rep.* **61**, 197–216 (2009).
8. Wilcox, K. S. *et al.* Issues related to development of new antiseizure treatments. *Epilepsia* **54**(Suppl 4): 24–34 (2013).
9. Miceli, F., Soldovieri, M. V., Martire, M. & Taglialatela, M. Molecular pharmacology and therapeutic potential of neuronal Kv7-modulating drugs. *Curr. Opin. Pharmacol.* **8**, 65–74 (2008).
10. Brodie, M. J. *et al.* Efficacy and safety of adjunctive ezogabine (retigabine) in refractory partial epilepsy. *Neurology* **75**, 1817–1824 (2010).
11. Orhan, G., Wuttke, T. V., Nies, A. T., Schwab, M. & Lerche, H. Retigabine/Ezogabine, a KCNQ/K(V)7 channel opener: pharmacological and clinical data. *Expert. Opin. Pharmacother.* **13**, 1807–1816 (2012).
12. Blackburn-Munro, G., Dalby-Brown, W., Mirza, N. R., Mikkelsen, J. D. & Blackburn-Munro, R. E. Retigabine: chemical synthesis to clinical application. *CNS Drug Rev.* **11**, 1–20 (2005).
13. Porter, R. J., Nohria, V. & Rundfeldt, C. Retigabine. *Neurotherapeutics* **4**, 149–154 (2007).
14. Tatulian, L., Delmas, P., Abogadie, F. C. & Brown, D. A. Activation of expressed KCNQ potassium currents and native neuronal M-type potassium currents by the anti-convulsant drug retigabine. *J. Neurosci.* **21**, 5535–5545 (2001).
15. Otto, J. F., Kimball, M. M. & Wilcox, K. S. Effects of the anticonvulsant retigabine on cultured cortical neurons: changes in electroresponsive properties and synaptic transmission. *Mol. Pharmacol.* **61**, 921–927 (2002).
16. Shah, M., Mistry, M., Marsh, S. J., Brown, D. A. & Delmas, P. Molecular correlates of the M-current in cultured rat hippocampal neurons. *J. Physiol.* **544**, 29–37 (2002).
17. Wickenden, A. D., Yu, W., Zou, A., Jegla, T. & Wagoner, P. K. Retigabine, a novel anti-convulsant, enhances activation of KCNQ2/Q3 potassium channels. *Mol. Pharmacol.* **58**, 591–600 (2000).
18. Main, M. J. *et al.* Modulation of KCNQ2/3 potassium channels by the novel anticonvulsant retigabine. *Mol. Pharmacol.* **58**, 253–262 (2000).
19. Lange, W. *et al.* Refinement of the binding site and mode of action of the anticonvulsant Retigabine on KCNQ K+ channels. *Mol. Pharmacol.* **75**, 272–280 (2009).
20. Schenzer, A. *et al.* Molecular determinants of KCNQ (Kv7) K+ channel sensitivity to the anticonvulsant retigabine. *J. Neurosci.* **25**, 5051–5060 (2005).
21. Xiong, Q., Gao, Z., Wang, W. & Li, M. Activation of Kv7 (KCNQ) voltage-gated potassium channels by synthetic compounds. *Trends Pharmacol. Sci.* **29**, 99–107 (2008).
22. Boehlen, A. *et al.* The new KCNQ2 activator 4-Chlor-N-(6-chlor-pyridin-3-yl)-benzamid displays anticonvulsant potential. *Br. J. Pharmacol.* **168**, 1182–1200 (2013).
23. Yu, H. *et al.* Discovery, synthesis, and structure activity relationship of a series of N-Aryl- bicyclo[2.2.1]heptane-2-carboxamides: characterization of ML213 as a novel KCNQ2 and KCNQ4 potassium channel opener. *ACS Chem. Neurosci.* **2**, 572–577 (2011).
24. Gao, Z. *et al.* Isoform-specific prolongation of Kv7 (KCNQ) potassium channel opening mediated by new molecular determinants for drug-channel interactions. *J. Biol. Chem.* **285**, 28322–28332 (2010).
25. Peretz, A. *et al.* A tale of switched functions: from cyclooxygenase inhibition to M-channel modulation in new diphenylamine derivatives. *PLoS ONE* **2**, e1332 (2007).
26. Wickenden, A. D., Zou, A., Wagoner, P. K. & Jegla, T. Characterization of KCNQ5/Q3 potassium channels expressed in mammalian cells. *Br. J. Pharmacol.* **132**, 381–384 (2001).
27. Murata, Y., Iwasaki, H., Sasaki, M., Inaba, K. & Okamura, Y. Phosphoinositide phosphatase activity coupled to an intrinsic voltage sensor. *Nature* **435**, 1239–1243 (2005).
28. Nowak, M. W. *et al.* *In vivo* incorporation of unnatural amino acids into ion channels in Xenopus oocyte expression system. *Methods Enzymol.* **293**, 504–529 (1998).
29. Beene, D. L., Dougherty, D. A. & Lester, H. A. Unnatural amino acid mutagenesis in mapping ion channel function. *Curr. Opin. Neurobiol.* **13**, 264–270 (2003).
30. Peretz, A. *et al.* Targeting the voltage sensor of Kv7.2 voltage-gated K+ channels with a new gating-modifier. *Proc. Natl Acad. Sci. USA* **107**, 15637–15642 (2010).
31. Lacroix, J. J. *et al.* Intermediate state trapping of a voltage sensor. *J. Gen. Physiol.* **140**, 635–652 (2012).
32. Dougherty, D. A. Cation-pi interactions in chemistry and biology: a new view of benzene, Phe, Tyr, and Trp. *Science* **271**, 163–168 (1996).
33. Deutsch, C. J. & Taylor, J. S. Intracellular pH as measured by 19F NMR. *Ann. N Y Acad. Sci.* **508**, 33–47 (1987).
34. Pless, S. A., Galpin, J. D., Niciforovic, A. P., Kurata, H. T. & Ahern, C. A. Hydrogen bonds as molecular timers for slow inactivation in voltage-gated potassium channels. *Elife* **2**, e01289 (2013).
35. Martyn-St, J. M. *et al.* The efficacy and safety of retigabine and other adjunctive treatments for refractory partial epilepsy: a systematic review and indirect comparison. *Seizure* **21**, 665–678 (2012).
36. Jepps, T. A., Olesen, S. P. & Greenwood, I. A. One man's side effect is another man's therapeutic opportunity: targeting Kv7 channels in smooth muscle disorders. *Br. J. Pharmacol.* **168**, 19–27 (2013).
37. Li, S., Choi, V. & Tzounopoulos, T. Pathogenic plasticity of Kv7.2/3 channel activity is essential for the induction of tinnitus. *Proc. Natl Acad. Sci. USA* **110**, 9980–9985 (2013).
38. Sander, S. E., Lemm, C., Lange, N., Hamann, M. & Richter, A. Retigabine, a K(V)7 (KCNQ) potassium channel opener, attenuates L-DOPA-induced dyskinesias in 6-OHDA-lesioned rats. *Neuropharmacology* **62**, 1052–1061 (2012).
39. Cao, Y. *et al.* Rescue of homeostatic regulation of striatal excitability and locomotor activity in a mouse model of Huntington's disease. *Proc. Natl Acad. Sci. USA* **112**, 2239–2244 (2015).
40. Burley, S. K. & Petsko, G. A. Aromatic-aromatic interaction: a mechanism of protein structure stabilization. *Science* **229**, 23–28 (1985).
41. Gallivan, J. P. & Dougherty, D. A. Cation-pi interactions in structural biology. *Proc. Natl Acad. Sci. USA* **96**, 9459–9464 (1999).
42. Wu, Y. J. *et al.* Synthesis and structure-activity relationship of acrylamides as KCNQ2 potassium channel openers. *J Med. Chem.* **47**, 2887–2896 (2004).
43. Wu, Y. J. & Dworetzky, S. I. Recent developments on KCNQ potassium channel openers. *Curr. Med. Chem.* **12**, 453–460 (2005).
44. Bentzen, B. H. *et al.* The acrylamide (S)-1 differentially affects Kv7 (KCNQ) potassium channels. *Neuropharmacology* **51**, 1068–1077 (2006).
45. Sanguinetti, M. C. *et al.* Physicochemical basis for binding and voltage-dependent block of hERG channels by structurally diverse drugs. *Novartis. Found. Symp.* **266**, 159–166 (2005).
46. Mitcheson, J. S., Chen, J., Lin, M., Culberson, C. & Sanguinetti, M. C. A structural basis for drug-induced long QT syndrome. *Proc. Natl Acad. Sci. USA* **97**, 12329–12333 (2000).
47. Pless, S. A., Galpin, J. D., Frankel, A. & Ahern, C. A. Molecular basis for class Ib anti-arrhythmic inhibition of cardiac sodium channels. *Nat. Commun.* **2**, 351 (2011).
48. Xiong, Q., Sun, H. & Li, M. Zinc pyrithione-mediated activation of voltage-gated KCNQ potassium channels rescues epileptogenic mutants. *Nat. Chem. Biol.* **3**, 287–296 (2007).
49. Zaika, O., Hernandez, C. C., Bal, M., Tolstykh, G. P. & Shapiro, M. S. Determinants within the turret and pore-loop domains of KCNQ3 K+ channels governing functional activity. *Biophys. J.* **95**, 5121–5137 (2008).
50. Belokon, Y. N. *et al.* Nucleophilic addition to an achiral dehydroalanine Schiff base Ni(II) complex as a route to amino acids. A case of stereodetermining asymmetric protonation in the presence of TADDOL. *ARKIVOC.* **3**, 132–150 (2004).

51. Pless, S. A. *et al.* A novel mechanism for fine-tuning open-state stability in a voltage-gated potassium channel. *Nat. Commun.* **4,** 1784 (2013).
52. Strutz-Seebohm, N. *et al.* Structural basis of slow activation gating in the cardiac I Ks channel complex. *Cell. Physiol. Biochem.* **27,** 443–452 (2011).

Acknowledgements

This work has been supported by CIHR MOP-97988 and a HFSP Young Investigator award to H.T.K. (RGY-0081), CIHR CGS-M and CGS-D awards to R.Y.K., a NSERC USRA award to M.C.Y. and CIHR, Banting, and MSFHR awards to S.A.P. C.A.A. is supported by NIH/NIGMS GM106569 and is a member of the Membrane Protein Structural Dynamics Consortium, which is funded by NIH/NIGMS grant no. GM087519. H.T.K. received salary support as a Michael Smith Foundation for Health Research Scholar and as a CIHR New Investigator.

Author contributions

R.Y.K. and M.C.Y. made equal contributions to this study. R.Y.K., M.C.Y., C.A.A., S.A.P., G.S. and H.T.K. all contributed to the design of experiments. J.D.G. and C.A.A. were responsible for chemical synthesis and providing specialized reagents for unnatural amino-acid mutagenesis. R.Y.K., M.C.Y. and S.A.P. performed the majority of experiments and analysed the data. G.S. performed drug-docking simulations. All authors commented on the manuscript. R.Y.K., M.C.Y. and H.T.K. wrote the majority of the manuscript.

Additional information

Competing financial interests: The authors declare no competing financial interests.

14

Inhibition of KRAS codon 12 mutants using a novel DNA-alkylating pyrrole-imidazole polyamide conjugate

Kiriko Hiraoka[1,*], Takahiro Inoue[1,*], Rhys Dylan Taylor[2], Takayoshi Watanabe[1], Nobuko Koshikawa[1], Hiroyuki Yoda[1], Ken-ichi Shinohara[1], Atsushi Takatori[1], Hirokazu Sugimoto[3], Yoshiaki Maru[1], Tadamichi Denda[4], Kyoko Fujiwara[5], Allan Balmain[6], Toshinori Ozaki[3], Toshikazu Bando[2], Hiroshi Sugiyama[2] & Hiroki Nagase[1]

Despite extensive efforts to target mutated RAS proteins, anticancer agents capable of selectively killing tumour cells harbouring KRAS mutations have remained unavailable. Here we demonstrate the direct targeting of KRAS mutant DNA using a synthetic alkylating agent (pyrrole-imidazole polyamide indole-*seco*-CBI conjugate; KR12) that selectively recognizes oncogenic codon 12 KRAS mutations. KR12 alkylates adenine N3 at the target sequence, causing strand cleavage and growth suppression in human colon cancer cells with G12D or G12V mutations, thus inducing senescence and apoptosis. In xenograft models, KR12 infusions induce significant tumour growth suppression, with low host toxicity in KRAS-mutated but not wild-type tumours. This newly developed approach may be applicable to the targeting of other mutant driver oncogenes in human tumours.

[1] Laboratory of Cancer Genetics, Chiba Cancer Center Research Institute, 666-2 Nitona, Chuoh-ku, Chiba 260-8717, Japan. [2] Department of Chemistry, Graduate School of Science, Kyoto University, Kitashirakawa-oiwakecho, Sakyo-ku, Kyoto 606-8502, Japan. [3] Laboratory of DNA Damage Signaling, Chiba Cancer Center Research Institute, 666-2 Nitona, Chuoh-ku, Chiba 260-8717, Japan. [4] Department of Gastroenterology, Chiba Cancer Center, 666-2 Nitona, Chuoh-ku, Chiba 260-8717, Japan. [5] Innovative Therapy Research Group, Nihon University Research Institute of Medical Science, Nihon University School of Medicine, 30-1 Ooyaguchi-kami, Itabashi-ku, Tokyo 173-8610, Japan. [6] Helen Diller Family Comprehensive Cancer Center, University of California, San Francisco, California 94158, USA. * These authors contributed equally to this work. Correspondence and requests for materials should be addressed to H.N. (email: hnagase@chiba-cc.jp).

Recent advances in the next-generation sequencing of human tumours have uncovered hundreds of recurrent oncogenic driver mutations and have identified many novel, actionable therapeutic targets. Oncogenic driver genes, which are often mutated, are continuously expressed and thereby contribute to cancer cell survival, which is a target for clinical therapy. Targeting the RAS–MAPK pathway has led to the development of successful treatments of epidermal growth factor receptor-mutant cancers but has also revealed that unexpected degrees of oncogene addiction and signalling complexity exist[1]. A common outcome of treatment using inhibitors of genes that lie 'downstream' in the RAS pathway is the development of drug resistance[2], and new approaches that target mutant RAS genes may provide a more direct route to inhibit these poor-prognosis tumours.

RAS genes, particularly H-ras, N-ras and K-ras, were the first human oncogenes to be discovered in human tumours more than 30 years ago and are among the most commonly mutated and critical cancer driver genes. The legendary K-Ras protein has been on everyone's 'target' list since its discovery. It has been estimated that ~33,000 newly diagnosed colon cancer patients in the United States during the year 2013 had the KRAS (G12D) mutation, which is considered the most frequent mutation in metastatic colon cancer, or the (G12V) mutation[3,4]. Extensive efforts have therefore been made to develop one or more small chemical compounds that could directly target and inhibit constitutively active KRAS. Unfortunately, direct pharmacological targeting of activated KRAS has, to date, been unsuccessful for clinical use[5]. Recently, a G12C KRAS mutation has been successfully targeted *in vitro* by a small molecule that allosterically controls GTP affinity and effector interactions, representing a promising approach to therapy for this specific subset of KRAS mutant tumours[6]. In the face of the challenges in developing small-molecule drugs that directly target KRAS mutant proteins[7], we have developed a novel approach that directly targets the mutant DNA. Hairpin pyrrole (Py)–imidazole (Im) (PI) polyamides can be designed to bind with high affinity to the minor groove of specific DNA sequences[8]: Py moieties preferentially bind T, A and C bases, but not G, whereas Im is a G-reader. These agents have shown efficacy in preclinical animal models of human diseases, including cancer, with minimal toxicity[9–11]. For example, PI polyamide conjugates with 1,2,9,9a-tetrahydrocyclopropa[1,2-c]benz[1,2-e]indol-4-one (CBI), which is converted to its cyclopropyl form via cyclization of a precursor containing a chloromethyl group to alkylate target DNA at the minor groove of N3 of adenine under neutral pH conditions, is a synthetic analogue of the alkylating moiety of duocarmycin A and can improve the limitations of targeting a coding sequence[12–14]. We therefore synthesized PI polyamide indole-*seco*-CBI conjugates to target common KRAS codon 12 mutations.

Results

KR12 targets KRAS codon 12 (G12D and G12V) mutants. The compounds IPβIIPI-γ-PIβPP-indole-*seco*-CBI (KR12: Fig. 1a) and IIPIβPI-γ-PIPβP-indole-*seco*-CBI (#6)[15] were designed and synthesized to recognize and alkylate adenine residues on the template strand of exon 2 of mutant KRAS, specifically at codon 12 (GTT and GAT) and codon 13 (GAC), respectively, according to the general recognition rule for PI polyamides[8]. The hairpin PI polyamide for KR12 but not for #6 preferentially bound to the mutant KRAS sequences, compared with wild-type, as shown by gel shift and surface plasmon resonance assays (Supplementary Figs 1 and 2, and Supplementary Table 1). Alkylation of theoretical recognition sequences, that is, 5′-WCGCCWWCA-3′ for KR12 and 5′-WCGWCWCCA-3′ for #6 (W indicates A or T), was confirmed using a thermally induced strand cleavage procedure[15]. Dose-dependent, alkylation-mediated selective DNA strand cleavage was detected at the G12D/G12V mutations in the presence of KR12 (Fig. 1b), and #6 showed specific cleavage of the sequence at the codon 13 mutation[15]. Strong, dose-dependent alkylation by KR12 was detected only at the mutated KRAS codon 12 target sites among ~160 base pairs of KRAS EXON2 plus plasmid sequences, which did not have additional full-match recognition sites of KR12.

The KR12-induced cell growth inhibition of a series of KRAS mutated or non-mutated human colon cancer cell lines, that is, HT29 (WT), Caco-2 (WT), SW1463 (12C/12C; GGT-TGT), SW480 (12V/12V; GGT-GTT), SW620 (12V/12V) and LS180 (12D; GGT-GAT/WT), were then analysed using the standard WST (water-soluble tetrazolium salts) assay. The viability of KRAS G12D or G12V mutant cells was significantly lower after treatment compared with wild-type cells or SW1463 cells carrying the G12C mutation, which is not recognized by KR12 (Fig. 2a; see also additional data on cell lines SNU-C2B (12D/12D) and DLD1 (13D/WT) in Supplementary Table. 2). A distinct pattern of cell viability was observed after KR12 treatment. Cell lines with KR12-recognized KRAS mutations (12D or 12V) were more sensitive than cell lines without those sequences, while there was no correlation between cell growth and p53 mutation status.

Mutant KRAS suppression induced by KR12. To analyse the specific alkylation and subsequent downregulation of the KRAS mutant allele after KR12 treatment, human colon cancer-derived, heterozygously mutated LS180 (12V/WT) cells were used to evaluate allele-specific suppression after dimethyl sulfoxide (DMSO), KR12, #6 or KR12 lacking *seco*-CBI (Dp) treatment. Endogenous DNA alkylation at the KRAS codon 12 mutation in LS180 cells after KR12 but not after #6 or DMSO treatment was repeatedly confirmed using ligation-mediated PCR (Supplementary Fig. 3). Unlike DMSO-, #6- or Dp-treated cells, KR12-treated cells showed significant downregulation of total KRAS and mutated KRAS allele expression by quantitative reverse transcription-PCR and colony PCR/sequencing assays, respectively (Fig. 2b,c and Supplementary Table 3). Immunoblot analysis also revealed that KR12, but not #6, induced a significant reduction in KRAS protein levels and in its GST-Raf-bound, active GTP-bound form, whereas KR12 had an undetectable effect on KRAS levels or activity in HT29 cells harbouring wild-type KRAS (Fig. 2d). In contrast to KR12, Dp had an undetectable effect on KRAS codon 12 mutant expression, indicating that the targeted hairpin PI polyamide lacking the CBI alkylating moiety was not sufficient to reduce the expression of the targeted mutant gene (Fig. 2b). Induction of 12V mutation sequence-specific alkylation after KR12 treatment could induce specific reductions in mutated KRAS RNA and in activated KRAS protein levels in heterozygous LS180 cells. We also observed slight reduction of KRAS downstream proteins of phospho-AKT and phospho-ERK expression after KR12 treatment (Supplementary Fig. 4d).

Senescence and DNA damage induced by KR12 in LS180 cells. We next investigated the biological effects of KR12 on KRAS mutant colon carcinoma cells. After 48 h of 50 nM KR12 treatment, a flat cell morphology was displayed in LS180 cells, and SA-β-gal activity was strongly increased compared with DMSO- or #6-treated cells, suggesting that KR12 induced cellular senescence (Fig. 3a). Cellular senescence is a stress response that stably blocks proliferation induced by activation of the RAS pathway, but is dependent on the cellular context[16]. Cell cycle arrest is a critical process in cellular senescence that is induced by the CDK

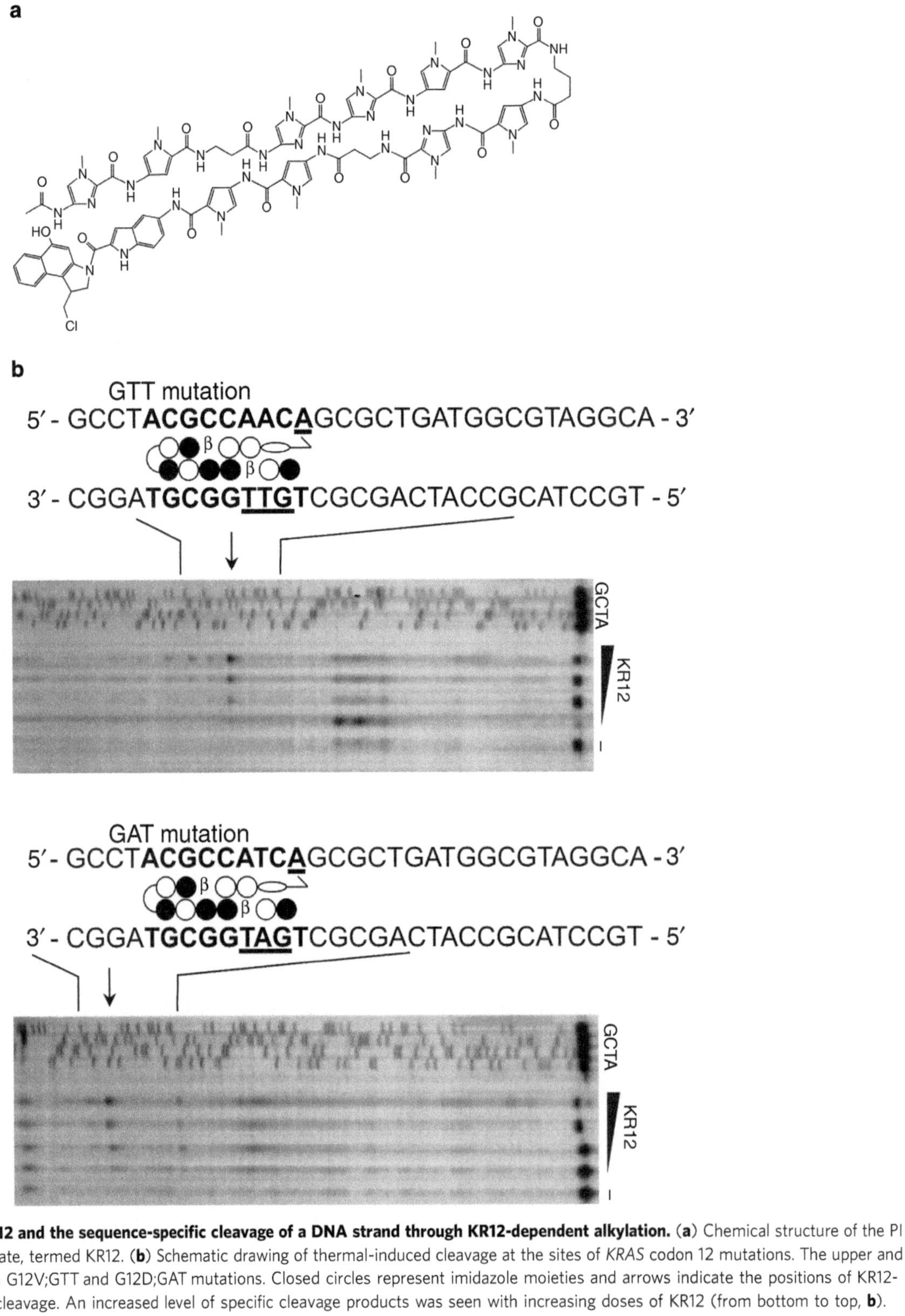

Figure 1 | Structure of KR12 and the sequence-specific cleavage of a DNA strand through KR12-dependent alkylation. (**a**) Chemical structure of the PI polyamide *seco*-CBI conjugate, termed KR12. (**b**) Schematic drawing of thermal-induced cleavage at the sites of *KRAS* codon 12 mutations. The upper and lower panels represent the G12V;GTT and G12D;GAT mutations. Closed circles represent imidazole moieties and arrows indicate the positions of KR12-dependent alkylation and cleavage. An increased level of specific cleavage products was seen with increasing doses of KR12 (from bottom to top, **b**).

inhibitor p21$^{WAF1/CIP1}$ (ref. 17), and 48-h KR12 exposure significantly increased the proportion of G2/M-phase cells (Fig. 3b). In contrast, DMSO and #6 had less of an effect on cell cycle distribution, and none of the three treatments increased the number of sub-G1-phase cells or the level of apoptotic genomic DNA fragmentation (Fig. 3c). KR12 markedly induced phosphorylation of p53 at Ser-15, p21$^{WAF1/CIP1}$ and phosphorylated histone variant H2AX (γH2AX), whereas KR12-mediated cleavage of PARP at the 48-h time point was not observed (Fig. 3d). Phospho-p53 (Ser-15) and γH2AX inductions in KR12-treated cells were also visually confirmed by indirect immunofluorescence staining (Supplementary Fig. 4b). We conclude that KR12 promotes p53-dependent cellular senescence, which is an important tumour suppressor mechanism[16], as well as DNA double-strand breaks, but not apoptosis, after short-term treatment.

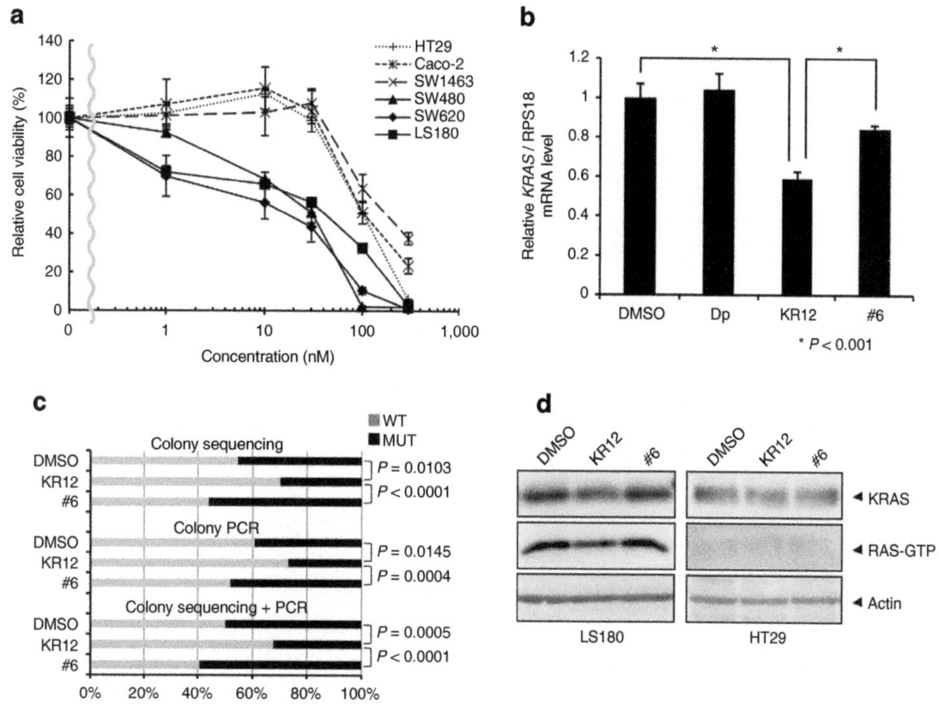

Figure 2 | KR12-mediated specific suppression of KRAS codon 12 mutants in human colon cancer cells. (**a**) WST assay. Human colon cancer cells expressing wild-type KRAS, including HT29 and Caco-2 or human colon cancer cells harbouring *KRAS* codon 12 mutations, including SW1643, SW480, SW620 and LS180, were incubated with the indicated concentrations of KR12. Forty-eight hours after treatment, the percent viable cells were examined by WST assay and depicted in a line graph. Error bars indicate the s.d. of the data from triplicate experiments. (**b**) Quantitative reverse transcription-PCR analysis. Relative KRAS expression 48 h after treatment with DMSO, KR12, #6 or Dp was plotted as a bar graph. Error bars indicate the s.d. of data from triplicate experiments. (**c**) Allele-specific downregulation of *KRAS* codon 12 mutant: cDNA from the treated RNA was subcloned. Transfected antibiotic-resistant/colour-selected colonies were screened by colony PCR and direct colony-sequencing methods. The number of colonies for the wild-type or codon 12 *KRAS* mutant allele of each treated group was counted and reported in Supplementary Table 3. The percentages of wild-type and mutant *KRAS* sequences are also shown in parentheses. (**d**) Immunoblot analysis. Immunoblots for anti-KRAS or anti-actin antibody (top and bottom panels, respectively) for LS180 (12D/WT) and HT29 (WT) cells 48 h after the treatment with either control DMSO solution, KR12 or #6. The GST-Raf-bound proteins from each treated group were pulled down and analysed by immunoblotting with anti-RAS antibody (middle panels).

p53-dependent apoptosis is induced by KR12 in LS180 cells. Strong induction of apoptosis was, however, observed in LS180 cells after 14 days of continuous 50 nM KR12 treatment, as evidenced by the significant increase in sub-G1-phase cells shown by fluorescence-activated cell sorting (FACS) analysis (Fig. 3b and Supplementary Fig. 4a) and typical DNA fragmentation (Fig. 3c), while #6 induced G2/M arrest. LS180 cells, which carry hemizygous wild-type *p53* (ref. 18), showed marked induction of p53 in association with cleaved PARP accumulation and γH2AX, while treatment with #6 resulted in faint p53 activation without PARP cleavage (Fig. 3d), implying that apoptosis induction by long-term KR12 treatment might be p53 dependent. The induction of p53 phosphorylation at Ser-15 and of γH2AX was also visually confirmed by indirect immunofluorescence staining (Supplementary Fig. 4c). On the basis of these data, we propose that treatment of LS180 cells with KR12, but not with #6, introduces DNA breaks at the KRAS locus and induces γH2AX foci and senescence and that after extended exposure to the drug, apoptosis occurs.

KRAS knockdown enhances DNA alkylator-induced cell death. To confirm a possible synergy of DNA alkylator and KRAS suppression in LS180 cells, we assessed LS180 growth inhibition after treatments with short interfering RNA (siRNA) targeting KRAS, control siRNA, #6 (KRAS non-targeting PI polyamide-indole-*seco*-CBI) and combinations of siRNAs and #6. Although single treatments with either KRAS siRNA or #6 with control siRNA resulted in reduced LS180 cell viability, 1 nM of KRAS siRNA induced a partial reduction of cell growth. When treated together with a low dose (0.3 nM) of #6, 1 nM of KRAS siRNA resulted in significantly higher cell growth reduction than samples treated with 1 nM of control siRNA (Supplementary Fig. 5a,b), implying that even partial reduction of KRAS and its downstream signals may sufficiently increase the sensitivity of alkylating agents in LS180 cells. KR12 also caused the partial downregulation of the active KRAS mutant and of DNA alkylation induction; therefore, KR12 increases its own ability to induce DNA damage, which is followed by cell death, in a similar manner.

Preferential *in vivo* tumour growth reduction by KR12. The potential efficacy of KR12 *in vivo* was tested by weekly intravenous injection of DMSO or KR12 (320 μg kg^{-1}) for 5 weeks into nude mice inoculated with HT29 (KRAS WT), LS180 (heterozygous 12D/WT) or SW480 (homozygous 12V) cells. Treatment was started when the tumour size reached 100 mm^3 (Supplementary Table 4 and Supplementary Fig. 6a). KR12 injection had an undetectable effect on the growth of HT29 tumours, whereas a significant reduction of tumour growth, with minimum toxicity, was demonstrated for the SW480 ($P = 0.0009$) and LS180 ($P = 0.0122$) tumours on significantly reducing KRAS expression in those tumours ($P = 0.0388$; Fig. 4 and Supplementary Figs 6b and 7b). Even after cessation of treatment at 5 weeks, KR12-treated SW480 tumours consistently showed

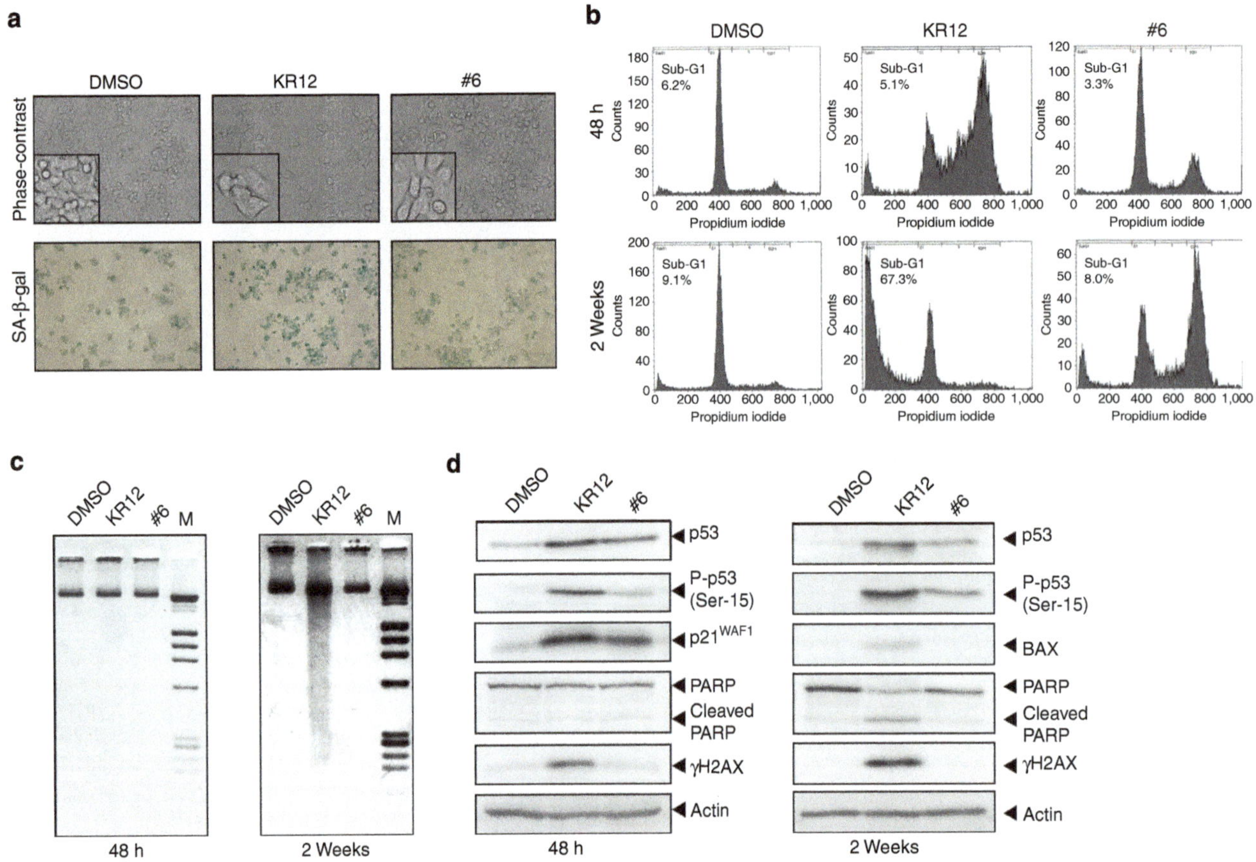

Figure 3 | KR12-dependent induction of cellular senescence and apoptosis. (**a**) SA-β-gal staining. LS180 cells were exposed to DMSO, KR12 or #6 (at a final concentration of 50 nM). Forty-eight hours after treatment, phase-contrast microphotographs were taken (top panels) and the cells were washed in PBS, fixed in 2% formaldehyde plus 0.05% glutaraldehyde and incubated with SA-β-gal staining solution containing X-gal for 24 h at 37 °C (bottom panels). (**b**) Cell cycle distribution analysis. Forty-eight hours (top) and two weeks (bottom) after treatment, cells were subjected to FACS analysis, and the DNA contents of each sample were analysed and depicted following the manufacturer's instructions. (**c**) DNA fragmentation. The attached and floating LS180 cells 48 h (left) and 2 weeks (right) after treatment were collected, and their genomic DNA was prepared and analysed by 1% agarose gel electrophoresis with λ/*Hind*III and ϕ-X174/*Hae*III size markers. (**d**) Immunoblotting. Forty-eight hours (left) and 2 weeks (right) after the treatment, whole-cell lysates were prepared and subjected to immunoblotting with the indicated antibodies. Actin was used as a loading control.

slower growth than the DMSO-treated controls (Supplementary Fig. 7a). Furthermore, the KR12-treated SW480 tumours exhibited movable, hard fibrotic nodules. We also examined the same experiment using an alkylating agent of CBI itself and #6, which should not have performed specific alkylation at the KRAS 12D mutation. Although CBI showed a strong antitumour effect, this tumour growth reduction was slightly weaker than that from KR12 treatment, and #6 did not show any significant antitumour effects in the SW480 xenografted nude mouse model. It was also intriguing that mice treated with KR12, #6 or DMSO remained healthy during the experiment, while CBI-treated mice exhibited weight loss and slight illness (Supplementary Fig. 7a).

Discussion

Although other drugs have been developed against the RAS signalling pathway, they are mainly effective against cancers with wild-type KRAS and are often ineffective in the treatment of KRAS-mutated colorectal cancers[19]. The PI polyamide-based alkylation strategy specifically downregulated constitutively active KRAS and efficiently suppressed the *in vitro* growth of several KRAS mutant cancer cell lines through the induction of DNA damage and cellular senescence followed by apoptosis. The *in vivo* growth of KRAS G12V/G12D mutant tumour xenografts was also efficiently suppressed by weekly injections of KR12 into mice bearing tumours induced by the inoculation of LS180 or SW480 cells. Therefore, KR12 is a promising anticancer drug candidate for unmet KRAS codon 12 mutant cancers. We propose that PI polyamide-mediated-specific silencing of mutant oncogenic driver genes may therefore facilitate the development of a pipeline of next-generation, molecular-targeted therapeutic strategies that could be applied in combination with mutant driver gene identification by sequencing the exomes of human cancers as a route to 'Precision Medicine' that is targeted at the individual patient.

Hairpin *N*-methylpyrrole(Py)-*N*-methylimidazole(Im) polyamide indole-*seco*-CBI conjugates[12,13] can be designed for specific 9-base-pair recognition of the genome following Dervan's recognition rule, as seen in natural minor groove binders, which undergo hydrogen bonding to DNA[8]. Under neutral pH, the open-ringed *seco*-CBI moiety is converted to CBI, which then alkylates the N3 of adenine in the DNA minor groove. When conjugated to PI polyamide indole, the conjugates increase sequence recognition specificity compared with conventional alkylating agents, which mainly recognize the N6 of guanine in the DNA major groove without recognizing the genomic DNA sequence[12–14,20]. Although indole, γ-aminobutyric acid (γ-turn) and beta-alanine linker may require optimization to improve

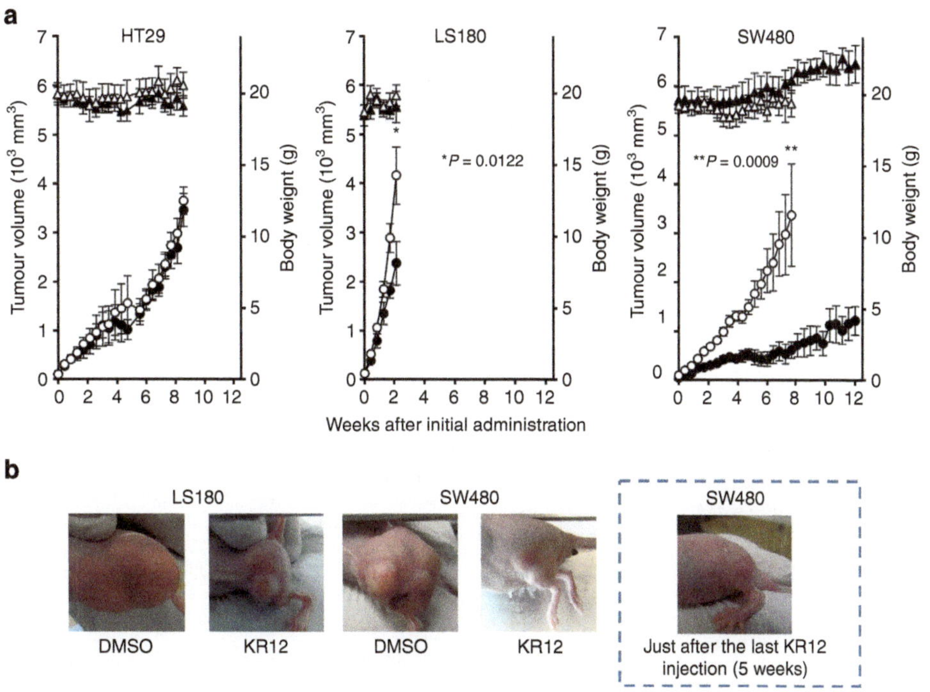

Figure 4 | KR12 suppresses tumour growth *in vivo*. (**a**) The indicated human colon cancer cells, that is, HT29 (WT), LS180 (12D/WT) and SW480 (12V/12V), were injected subcutaneously into BALB/c nude mice. When the tumour volume reached 100 mm^3, DMSO or KR12 (320 μg kg^{-1} body weight) was intravenously injected through the tail vein every 7 days. At the indicated times after administration, the tumour volume was calculated as 1/6 $\pi \times$ the longest diameter × width × height. The mean tumour volume, with s.e. (open circles for DMSO and closed circles for KR12 treatment), and mean body weight, with s.e. (open triangles for DMSO and closed triangles for KR12 treatment), of the KR12-treated group and control group are plotted in a line graph with error bars (s.e.). The numbers of animals used were 6 for HT29 (DMSO and KR12) and LS180 (KR12), 5 for LS180 (DMSO) and SW480 (KR12), and 8 for SW480 (DMSO). (**b**) Images of the euthanized mice of each group are shown. An image of an SW480 xenograft 5 weeks after KR12 treatment, which was the final treatment, is also shown.

their specificity and sensitivity for sequence-specific alkylation, the recognition of specific 9-bp sites in the genome by KR12 allows for multiple recognition sites (~9,121 sites in the human genome (GRCh38/hg38)), where the mutated KRAS site should be included for preferential alkylation.

KR12 treatment resulted in a significant reduction of activated KRAS and a significant induction in DNA damage followed by cell death in KRAS-mutated colon cancer cells. We observed dose-dependent alkylation at the mutated sequences by CBI conjugates, with approximately a 10-fold difference in the dissociation constants of the mutant and wild-type allele sequences, a reduction of mutant KRAS expression in heterozygous LS180 cells, endogenous alkylation of the KRAS mutant genome and *in vitro* and *in vivo* growth reduction of tumour cells with the KRAS mutations. All together, these results suggest that the DNA sequence-specific alkylating activity of hairpin polyamide indole-*seco*-CBI recognizes the match sequence instead of the mismatch site. The number of naked PI polyamides showing biological efficacy as single agents remains limited. Thus, one way to enhance the therapeutic potential of PI polyamides might be to conjugate them to small-molecule drugs, such as alkylating agents[21,22].

In many clinical trials, chemotherapy and targeted therapy combinations result in better outcomes for patients and prolong overall survival. As KR12 is an alkylating agent targeting a restricted number of genomic regions, including the KRAS driver oncogene mutation, it appears suited for combination therapy between molecular-targeted therapy and conventional chemotherapy. Because of the sequence-specific DNA recognition of KR12, the number of genomic regions that are alkylated by KR12 should also be dramatically reduced in comparison with conventional chemotherapy; therefore, the adverse effect of toxicity to normal cells may be remarkably reduced, as demonstrated in animal experiments (Supplementary Fig. 7a). Although potential off-target effects of KR12 should exist among the 9,121 estimated target sites in the reference human genome (GRCh38/hg38), this possibility may not be a major concern for cancer patients with poor prognosis and experience with conventional chemotherapy. Although many issues still remain to be addressed for the clinical development of PI polyamide-indole-*seco*-CBI, as usually seen in cases of drug development, targeting oncogenic driver genes for alkylation may promise a fruitful feature of cancer therapy for untreatable cancer patients with unfavourable prognoses. Extensive genomic sequencing of tumour and normal tissues may provide critical biomarkers to aid in finding appropriate cancer cases that should be treated by KR12 or another specific genome-targeting alkylator.

Tumour selectivity is another important issue for anticancer treatment. In contrast to traditional cytotoxic chemotherapies, which crudely target growth processes shared by normal and cancerous tissues, agents such as imatinib mesylate (Gleevec) for BCR–ABL translocation that target genetic alterations underlying cancer often have mild toxicity because of their greater specificity for cancer cells[23]. In addition to specificity to the cancer genome, PI polyamide has been reported to accumulate specifically in xenografted tumour tissue in some cases[24]. These unique properties of PI polyamide conjugates provide a very efficient capacity for drug delivery, and its alkylating conjugate, which targets oncogenic somatic mutations, could reveal a superb selective advantage by specifically targeting cancer cells with the mutation. This possibility may also support the observed very specific and effective *in vivo* anticancer activity of KR12.

In conclusion, a synthetic PI polyamide alkylator conjugate of KR12 has been designed to target KRAS mutations and induce mutant KRAS suppression followed by tumour cell death *in vitro* and *in vivo* in an oncogenic mutation-specific manner. Thus, KR12 is a possible candidate drug for colorectal patients with either the G12D or G12V mutation.

Methods

Compounds. The compounds ImPyβImImPyIm-γ-PyImβPyPy-indoleCBI (KR12), shown in Fig. 1a, and ImPyβImImPyIm-γ-PyImβPyPy-indoleCBI (#6)[15] were designed and synthesized as described previously[20]. In brief, PI polyamides were synthesized in a stepwise reaction using a previously described Fmoc solid-phase protocol[12] and a peptide synthesizer (PSSM-8, Shimadzu Industry) with a computer-assisted operation system on a 10-μmol scale (9.8 mg of Fmoc-β-alanine Wang resin). After synthesis, 1 ml of 50% 1N LiOH/NMP was mixed with the resin and the mixture was heated at 55 °C for 1 h to detach the PI polyamides from the resin. The PI polyamides were then purified by high-performance liquid chromatography (HPLC) LC-20 (Shimadzu Industry) using a 10 × 150 mm Phenomenex Gemini-NX3u 5-ODS-H reverse-phase column in 0.1% acetic acid in water, with acetonitrile as the eluent, a flow rate of 10 ml min^{-1}, a linear gradient from 30 to 75% acetonitrile over 30 min and detection at 310 nm. Collected KR12-COOH fractions were analysed by LC-MS. DIEA (0.5 μl, 2.86 μmol) and PyBOP (1.0 mg, 1.95 μmol) were added to a solution of Py-Im polyamide carboxylic acid KR12-COOH (1.0 mg, 0.66 μmol) in NMP (100 μl) and the reaction mixture was stirred for 1 h at room temperature. After conversion from KR12-COOH to the activated 1-hydroxy-benzotriazole ester was verified by analytical HPLC (0.1% TFA/CH3CN 30–75% linear gradient, 0–30 min), NH_2-indole-*seco*-CBI (0.6 mg, 1.58 μmol) was added to the reaction vessel and the reaction mixture was stirred for 4 days at room temperature. The reaction mixture was subjected to HPLC purification (0.1% TFA/ CH_3CN 30–75% linear gradient, 0–30 min), and the peak fraction containing the product was collected. After lyophilization, KR12 was produced as a white powder.

High-resolution gel electrophoresis. 5′-Texas Red-labelled DNA fragments containing the indicated KRAS codon 12 mutations were incubated with the indicated concentrations of KR12 for 10 h at room temperature, followed by the addition of calf thymus DNA. The reaction mixtures were then incubated at 90 °C for 5 min to cleave DNA strands at their specific alkylated sites. DNA fragments were then recovered by vacuum centrifugation, dissolved in loading dye, denatured at 95 °C for 20 min and subjected to electrophoresis on a 6% denaturing poly-acrylamide gel.

Cell culture. Human colon carcinoma-derived HT29 (WT), Caco-2 (WT), SW1463 (12C/12C), SW480 (12V/12V) and SW620 (12V/12V) cells were maintained in Dulbecco's modified Eagle's medium supplemented with heat-inactivated 10% fetal bovine serum, 100 U ml^{-1} of penicillin and 100 μg ml^{-1} of streptomycin. Human colon carcinoma-derived LS180 (12D/WT) cells were grown in Eagle's minimum essential medium containing heat-inactivated 10% fetal bovine serum, 100 U ml^{-1} of penicillin and 100 μg of streptomycin. All cell lines were cultured in a 37 °C humidified atmosphere containing 95% air and 5% CO_2.

Cell proliferation analysis. Cells were seeded at a final density of 3×10^3 or 5×10^3 cells per 96-well plate and allowed to attach overnight. The cells were then treated with 0.125% DMSO, 50 nM KR12 or #6. Forty-eight hours after treatment, 10 μl of WST-8 reagent (Dojindo) was added to each culture and the reaction mixtures were incubated at 37 °C for another 2 h. The absorbance readings for each well were performed at 450 nm using a microplate reader (MTP-310; Corona).

Real-time reverse-transcriptase PCR. Total RNA was prepared using the RNAeasy plus mini kit according to the manufacturer's instructions (Qiagen), and 0.5 μg of the RNA was converted to cDNA using the SuperScript VILO cDNA Synthesis System (Invitrogen) according to the manufacturer's protocols. The q-PCR Master Mix and specific primer sets were used for SYBR Green-based quantitative PCR (q-PCR; Applied Biosystems). Three independent measurements were taken and the amounts were estimated by extrapolation from a standard curve. Expression values were normalized against the expression of *RPS18* and used as an endogenous control. The primer sets used in this study were as follows: KRAS, 5′-GGAGAGAGGCCTGCTGAA-3′ (sense) and 5′-TGACCTGCTGTGT CGAGAAT-3′ (antisense); and *RPS18*, 5′-GAGGATGAGGTGGAACGTGT-3′ (sense) and 5′-TCTTCAGTCGCTCCAGGTCT-3′ (antisense).

RAS GTP assay. Cells were washed twice in ice-cold PBS and lysed in Mg^{2+} lysis buffer (MLB; Millipore) and protein concentrations were measured using the Bio-Rad Protein Assay System (Bio-Rad). Equal amounts of cell lysates (500 μg of protein) were mixed with 10 μl of GST-Raf-conjugated agarose beads (Millipore) and incubated for 45 min at 4 °C. After incubation, the agarose beads were extensively washed in lysis buffer and boiled in 40 μl of 2 × SDS-sample buffer for 5 min, and the supernatants were analysed by immunoblotting with anti-KRAS antibody (Abcam, ab55391).

Immunoblot analysis. Equal amounts of cell lysates prepared from LS180 cells exposed to DMSO, KR12 or #6 were separated by 10% SDS–polyacrylamide gel electrophoresis and electrotransferred onto polyvinylidene difluoride membranes (Immobilon-P; Millipore). The membranes were blocked with TBS-T (Tris-buffered saline plus 0.05% Tween 20) containing 5% non-fat dry milk. After blocking, the membranes were incubated with anti-KRAS (ab55391, Abcam, 1:1,000), anti-p21 WAF1 (sc-756, Santa Cruz Biotechnology, 1:1,000), anti-p53 (sc-126, Santa Cruz Biotechnology, 1:4,000), anti-phospho-p53 at Ser-15 (9284, Cell Signaling Technology, 1:1,000), anti-PARP (9542, Cell Signaling Technology, 1:1,000), anti-γH2AX (613402, BioLegend, 1:2,000), anti-Actin antibody (A5060, Sigma, 1:2,000) or anti-BAX (2774, Cell Signaling Technology, 1:1,000) followed by incubation with the appropriate HRP (horseradish peroxidase)-conjugated secondary antibody (rabbit:7074, mouse:7076, Cell Signaling Technology). Following extensive washing in TBS-T, antibodies bound to the indicated proteins were visualized using enhanced chemiluminescence reagents (ECL; GE Life Sciences).

Indirect immunofluorescence assay. LS180 cells treated with DMSO, KR12 or #6 were rinsed twice in ice-cold PBS and then fixed in 3.7% formaldehyde at room temperature for 30 min. After washing in PBS, the cells were permeabilized with 0.1% Triton-X-100 in PBS at room temperature for 5 min and then blocked with 3% bovine serum albumin in PBS at room temperature for 1 h. After washing in PBS, the cells were incubated with anti-phospho-p53 at Ser-15 or with anti-γH2AX at room temperature for 1 h, followed by incubation with FITC-conjugated anti-rabbit IgG or rhodamine-conjugated anti-mouse IgG (Invitrogen), respectively, at room temperature for 1 h. The glass coverslips were washed in PBS and mounted with VectaShield containing DAPI (Vector Laboratories). Fluorescent images were captured using a confocal microscope (Leica).

Flow cytometry. Cell cycle distribution was examined by flow cytometric analysis with a FACScan flow cytometer (BD Biosciences) following the manufacturer's instructions. In brief, the attached and floating cells were collected and fixed in ice-cold 70% ethanol. The cells were then stained with a solution containing 50 μg ml^{-1} of propidium iodide and 100 μg ml^{-1} of RNase A in PBS at 37 °C for 30 min to determine the total DNA content. The cell cycle distribution was then analysed using a FACScan flow cytometer.

Animal tumour models. The animal protocol was approved by Chiba Cancer Center Animal Ethics Committee. HT29 (WT), LS180 (12D/WT) or SW480 (12V/12V) cells were implanted subcutaneously into the dorsal flanks of 6-week-old female athymic mice (Supplementary Fig. 6a). When the tumour volume reached 100 mm^3, administration of a non-toxic dose of DMSO or KR12 (320 μg kg^{-1} body weight) was performed and five subsequent weekly injections with DMSO or KR12 were administered through their tail veins. Tumour-bearing mice were euthanized at the indicated times over 12 weeks after the first administration or when the tumours exceeded 2 cm in diameter.

Statistical evaluation. Statistical analyses of the experimental data were performed using Student's *t*-test. The statistical significance of tumour growth between the treated and control groups was assessed using the statmod software package for R (http://wwww.r-project.org)[25]. The significance level is set at a *P* value of <0.05.

References

1. Perez, R. *et al.* A view on EGFR-targeted therapies from the oncogene-addiction perspective. *Front. Pharmacol.* **4,** 53 (2013).
2. Rokita, M. *et al.* Overexpression of epidermal growth factor receptor as a prognostic factor in colorectal cancer on the basis of the Allred scoring system. *Onco. Targets Ther.* **6,** 967 (2013).
3. Neumann, J. *et al.* Frequency and type of KRAS mutations in routine diagnostic analysis of metastatic colorectal cancer. *Pathol. Res. Pract.* **205,** 858 (2009).
4. The American Cancer Society. *American Cancer Society. Cancer Facts & Figures 2013* (American Cancer Society, 2013).
5. Zimmermann, G. *et al.* Small molecule inhibition of the KRAS-PDEdelta interaction impairs oncogenic KRAS signalling. *Nature* **497,** 631 (2013).
6. Ostrem, J. M. *et al.* K-Ras(G12C) inhibitors allosterically control GTP affinity and effector interactions. *Nature* **503,** 548 (2013).
7. Shima, F. *et al. In silico* discovery of small-molecule Ras inhibitors that display antitumor activity by blocking the Ras-effector interaction. *Proc. Natl Acad. Sci. USA* **110,** 8182 (2013).
8. Dervan, P. B. Molecular recognition of DNA by small molecules. *Bioorg. Med. Chem.* **9,** 2215 (2001).
9. Matsuda, H. *et al.* Transcriptional inhibition of progressive renal disease by gene silencing pyrrole-imidazole polyamide targeting of the transforming growth factor-beta1 promoter. *Kidney Int.* **79,** 46 (2011).

10. Wang, X. *et al.* Inhibition of MMP-9 transcription and suppression of tumor metastasis by pyrrole-imidazole polyamide. *Cancer Sci.* **101,** 759 (2010).
11. Yang, F. *et al.* Antitumor activity of a pyrrole-imidazole polyamide. *Proc. Natl Acad. Sci. USA* **110,** 1863 (2013).
12. Bando, T. & Sugiyama, H. Synthesis and biological properties of sequence-specific DNA-alkylating pyrrole-imidazole polyamides. *Acc. Chem. Res.* **39,** 935 (2006).
13. Wang, Y. D. *et al.* Cell-free and cellular activities of a DNA sequence selective hairpin polyamide-CBI conjugate. *J. Biol. Chem.* **277,** 42431–42437 (2002).
14. Wurtz, N. R. & Dervan., P. B. Sequence specific alkylation of DNA by hairpin pyrrole-imidazole polyamide conjugates. *Chem. Biol.* **7,** 153–161 (2000).
15. Taylor, R. D. *et al.* Sequence-Specific DNA alkylation targeting for Kras codon 13 mutation by pyrrole-imidazole polyamide seco-CBI conjugates. *Eur. J. Chem.* **20,** 1310 (2014).
16. Collado, M. & Serrano, M. Senescence in tumours: evidence from mice and humans. *Nat. Rev. Cancer* **10,** 51 (2010).
17. Campisi, J. Aging, cellular senescence, and cancer. *Annu. Rev. Physiol.* **75,** 685 (2013).
18. Liu, Y. & Bodmer, W. F. Analysis of P53 mutations and their expression in 56 colorectal cancer cell lines. *Proc. Natl Acad. Sci. USA* **103,** 976 (2006).
19. Tortolina, L. *et al.* A multi-scale approach to colorectal cancer: from a biochemical-interaction signaling-network level, to multi-cellular dynamics of malignant transformation. Interplay with mutations and onco-protein inhibitor drugs. *Curr. Cancer Drug Targets* **12,** 339 (2012).
20. Bando, T. *et al.* Efficient DNA alkylation by a pyrrole-imidazole CBI conjugate with an indole linker: sequence-specific alkylation with nine-base-pair recognition. *Bioconjug. Chem.* **17,** 715 (2006).
21. Dickinson, L. A. *et al.* Arresting cancer proliferation by small-molecule gene regulation. *Chem. Biol.* **11,** 1583–1594 (2004).
22. Jespersen, C. *et al.* Chromatin structure determines accessibility of a hairpin polyamide-chlorambucil conjugate at histone H4 genes in pancreatic cancer cells. *Bioorg. Med. Chem. Lett.* **12,** 4068–4071 (2012).
23. Lengauer, C. *et al.* Cancer drug discovery through collaboration. *Nat. Rev. Drug Discov.* **4,** 375–380 (2005).
24. Raskatov, J. A. *et al.* Tumor xenograft uptake of a pyrrole-imidazole (py-im) polyamide varies as a function of cell line grafted. *J. Med. Chem.* **57,** 8471–8476 (2014).
25. Baldwin, T. *et al.* Wound healing response is a major contributor to the severity of cutaneous leishmaniasis in the ear model of infection. *Parasite Immunol.* **29,** 501 (2007).

Acknowledgements

We are grateful to Mr. Y. Hirano and Mr. Y. Suzuki for technical assistance. This work was supported by a Grant-in-Aid for Scientific Research (B) # 23300344 and in Innovative Areas # 25134718 and by a Project for the Development of Innovative Research on Cancer Therapeutics (P-Direct) and Young Scientists (B) #13307606 from the Ministry of Education, Culture, Sports, Science and Technology, Japan.

Author contributions

H.N. planned and oversaw the project. T.W., R.D.T., T.B., H.S. and H.N. designed the chemicals and T.W., R.D.T. and T.B synthesized the chemicals. The bulk of the *in vitro* experiments were performed by K.H. and the bulk of the *in vivo* experiments were performed by T.I., all under the supervision of T.O. and H.N., with technical support from H.Y., H.S., K.F., N.K. and Y.M. K.H. and H.N. performed the statistical analyses. K.H., T.I., T.W., K.S., T.O., T.B., H.S., T.D. and H.N. discussed and interpreted all data. K.H., T.O. and H.N. wrote the manuscript, with input from the other authors. A.B. contributed critical comments to the experimental strategy and editing. A.T., together with K.H., performed the experiments described in Supplementary Figures 3 and 5.

Additional information

Competing financial interests: The authors declare no competing financial interests.

15

Connecting two proteins using a fusion alpha helix stabilized by a chemical cross linker

Woo Hyeon Jeong[1,*], Haerim Lee[2,*], Dong Hyun Song[3], Jae-Hoon Eom[1], Sun Chang Kim[2], Hee-Seung Lee[1], Hayyoung Lee[4] & Jie-Oh Lee[1]

Building a sophisticated protein nano-assembly requires a method for linking protein components in a predictable and stable structure. Most of the cross linkers available have flexible spacers. Because of this, the linked hybrids have significant structural flexibility and the relative structure between their two components is largely unpredictable. Here we describe a method of connecting two proteins via a 'fusion α helix' formed by joining two pre-existing helices into a single extended helix. Because simple ligation of two helices does not guarantee the formation of a continuous helix, we used EY-CBS, a synthetic cross linker that has been shown to react selectively with cysteines in α-helices, to stabilize the connecting helix. Formation and stabilization of the fusion helix was confirmed by determining the crystal structures of the fusion proteins with and without bound EY-CBS. Our method should be widely applicable for linking protein building blocks to generate predictable structures.

[1]Department of Chemistry, KAIST, Daejeon 34141, Korea. [2]Department of Biological Sciences, KAIST, Daejeon 34141, Korea. [3]Agency for Defense Development, Daejeon 34186, Korea. [4]Institute of Biotechnology, Chungnam National University, Daejeon 34134, Korea. * These authors contributed equally to this work. Correspondence and requests for materials should be addressed to J.-O.L. (email: jieoh@kaist.ac.kr).

Linking two protein components to form a predictable and rigid structure is a prerequisite for generating complex protein assemblies in a pre-designed fashion[1]. Most of the chemical cross linkers available have long and flexible spacers to help them approach the reactive side chains of the target proteins. Because of this, the resulting hybrids have significant structural flexibility and the relative orientation and distance between their two components is largely unpredictable. This is the case even when the chemical cross linkers themselves have rigid structures since they are attached to flexible side chains such as cysteines or lysines. Recently several new methods have been proposed to assemble proteins in a pre-designed fashion.

Radford *et al.*[2–4] have shown that two proteins can be linked in a rigid and predictable way using a metal coordination method. Since the bound metal ion has additional coordination sites that are available for binding of another protein molecule, it can induce homodimerization of the protein. Because the metal coordination requires precise positioning of the coordinating amino acid side chains, the resulting protein dimer will have a predictable structure. The authors used a similar metal coordination method to produce a variety of protein assemblies[5–12]. King *et al.*[13] reported a computational method for designing cage-like structures by connecting two protein components. One of them was chosen from among natural proteins that can form stable homodimers and the other from among proteins that can form stable homotrimers. By optimizing the interface between these two protein components in the fusion proteins using computer calculations, they were able to produce several fusion proteins that could assemble into large cage-like structures.

Short alpha-helical linkers joining two protein components have been used to build large and highly symmetric assemblies with rigid and predictable structures. Because an α-helix is under strict structural restraint, the structure of the fusion protein ought to be predictable so long as a connecting helix is formed. Diverse architectures are possible, depending on the specific geometric arrangement between the protein components[14]. Using this method, Lai *et al.*[1,15,16] connected M1 matrix protein, which can form homodimers, and BPO protein, which can form stable homotrimers, with a short linker. By testing various linker sequences they were able to find one that could form an α-helix and supported the formation of cage-like structures. Recently, they applied a similar method to generate large and porous cube-like closed structures with high accuracy[17]. Rigidity of the helical linker is essential for the success of the design because apparently minor deviation of the structure from the intended ideal helix can lead to failure of protein assembly[18].

Here we propose a novel method for connecting the pre-existing α-helices of two proteins into a single extended helix using a chemical cross linker. To test our method, we chose as a model system an ankyrin family protein containing a C-terminal helix and a protein A fragment containing an N-terminal helix. We fused the two proteins by connecting the two terminal α-helices into a single continuous helix. In general, simple genetic ligation of the helices in the proteins does not lead to fusion of the helices. Therefore, we used a synthetic chemical that selectively reacts with α helices and stabilizes them. EY-CBS (3,3′-ethyne-1,2-diylbis-{6-[(chloroacetyl)amino]benzenesulfonic acid}) is a chemical cross linker that can react simultaneously with two cysteines in the i and i + 11 positions of the α-helix[19]. To test whether EY-CBS detects the formation of a helix, we chose two amino acid residues separated by 11 amino acids in the proposed fusion helix and mutated them to cysteines. The successful fusion helices showed high reactivity with EY-CBS, and cross linking with EY-CBS forced the structure of the fusion α-helix to resemble that of an ideal helix. Our method does not require metal ions in the buffer, nor complicated computer simulation in the design process. It can be used to assemble protein components into asymmetrical and non-repeating structures.

Results

Design of the fusion helices. We chose an artificial ankyrin protein, named 'mbp3-16', that has been engineered to bind maltose-binding protein (MBP)[20], and the B4 domain of protein A as a model system to test our fusion helix method. We chose these proteins because both of them are composed entirely of α-helices and lack cysteine residues. In addition to this, they have been proven to be useful as building blocks of artificial protein assemblies because they can be engineered to bind other proteins[21,22]. We produced 17 fusion proteins by connecting the C-terminal helix of the ankyrin domain and the N-terminal helix of the protein A domain (Fig. 1 and Table 1). While designing connections between the two helices, obvious steric collision was avoided by using a molecular graphics program. To generate the EY-CBS reaction site, two amino acid residues separated by 11 amino acids, one from the ankyrin domain and the other from the protein A domain, were chosen and mutated to cysteines. Also, two additional amino acids, in positions i + 4 and i + 7, were mutated to alanines because long side chains may lead to steric collision with the EY-CBS molecules[19]. Using a similar

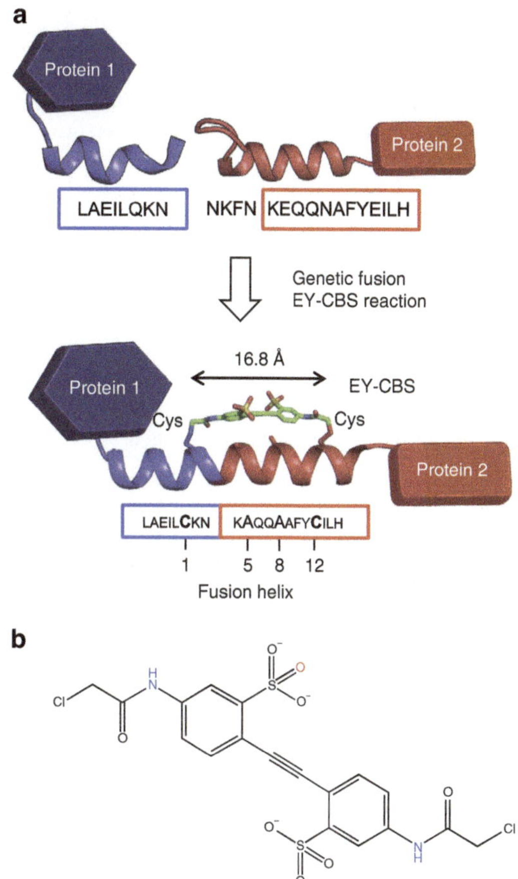

Figure 1 | Design of the fusion proteins. (**a**) The C-terminal helix of protein 1 and the N-terminal helix of protein 2 were genetically connected. Two amino acids separated by 11 amino acids were mutated to cysteines for reaction with EY-CBS. (**b**) Chemical structure of EY-CBS. EY-CBS has a rigid chemical structure with two α-chlorocarbonyl groups reactive with the thiol groups of cysteines.

Table 1 | Design of the fusion proteins.

Name	First protein	Second protein		Sequence of the fusion helix	
3,305	Protein A	mbp3-16	266*	QCAREAAREAAACD	13*
3,306	mbp3-16	Protein A	133	QCAREAAAREAACN	217
3,307	Protein A	mbp3-16	263	NCAQEAAAREAACD	13
3,308	mbp3-16	Protein A	129	ACILQAAAREAACN	217
3,309	Protein A	mbp3-16	266	QCAREAAARDLGCK	17
3,310	mbp3-16	Protein A	133	QCAREAAAREANCE	219
3,311	mbp3-16	Protein A	132	LCKNKAQQAAFYCI	227
3,312	mbp3-16	Protein A	125	NCRLAAILAKNKCQ	220
3,313	Protein A	mbp3-16	259	ACKLNAAQALGRCL	18
3,315	Protein A	mbp3-16	260	ACLNAAQAALGRCL	18
4,254	Protein A	mbp3-16	260	KCLNDALGAKLLCA	21
4,256	Protein A	mbp3-16	260	KCLNDAQAAKLLCA	21
4,257	Protein A	mbp3-16	259	ACKLNAAQALGRCL	18
4,258	Protein A	mbp3-16	259	ACKLNAAQAAGRCL	18
4,260	Protein A	mbp3-16	261	KCNDAAARALLECA	22
4,261	Protein A	mbp3-16	261	KCNDAALGAKLLCA	21
4,262	Protein A	mbp3-16	261	KCNDAAAGAKLLCA	21
6,758	Protein A	Calmodulin	260	KCLNDAEQIAEFCE	15
6,759	Protein A	Calmodulin	260	KCLNDEQIAEFKCA	16
6,760	Protein A	Calmodulin	259	ACKLNAQIAEFKCA	16
6,761	Protein A	Calmodulin	260	KCLNDAQAAAEECI	10

*Residue numbers are taken from the UniProt (protein A and calmodulin code number P38507 and P62158, respectively) and the GenBank (mbp3-16, AY326426) databases.

method, we designed four more fusion proteins with the protein A domain and the calmodulin N-terminal domain[23]. We intentionally positioned positively charged residues such as lysines and arginines in some of the fusion proteins to test whether the negatively charged EY-CBS had higher reactivity with positively charged targets (Fig. 1b). However, we found no obvious correlation between the location of the positively charged residues and the reactivity of the protein towards EY-CBS, and we did not artificially include positively charged amino acids in the later design. When necessary, we inserted alanine residues between the two proteins to control the relative orientation of the two protein components (Table 1). Note that insertion of one amino acid residue in an α-helix will rotate the helix by ~100 degrees. Therefore, one can control the relative orientation of the connected proteins by inserting precise number of amino acids into the fusion helix. Amino acid residues that may have roles in stabilizing the native helices were not changed.

Reaction with EY-CBS. The fusion proteins produced were purified and incubated with 1 mM EY-CBS (Fig. 2 and Supplementary Figs 1–3). Five of the 21 fusion proteins shifted substantially in the SDS–PAGE analysis. The molecular weight of EY-CBS is 521 daltons and if the two cysteines in the fusion helix react with EY-CBS, the shift in molecular weight should be by 521 daltons. In fact the change of molecular weight estimated from the shift of the protein bands was more than ~2 kDa (Fig. 3a). Therefore, we assumed the band shift was not only due to the additional molecular weight of the bound EY-CBS but to a change in the local structure of the fusion helix induced by circularization of the two cysteine side chains by EY-CBS. To analyse the situation further, we chose the 3,311 fusion protein because it had the highest reaction efficiency with EY-CBS (Fig. 3 and Supplementary Figs 4–6). To confirm that the 3,311 fusion indeed reacted with EY-CBS, the absorption spectrum of the reacted protein was measured after removing unreacted EY-CBS by gel filtration chromatography. The isolated EY-CBS molecule has an absorption peak at 340 nm due to the conjugated phenyl rings. As expected, the reacted fusion protein had a strong absorption peak at 340 nm, demonstrating that EY-CBS was covalently linked to the protein (Fig. 3b). The ratio of the absorption peaks at 280 and 340 nm suggests that one EY-CBS molecule is bound per protein molecule.

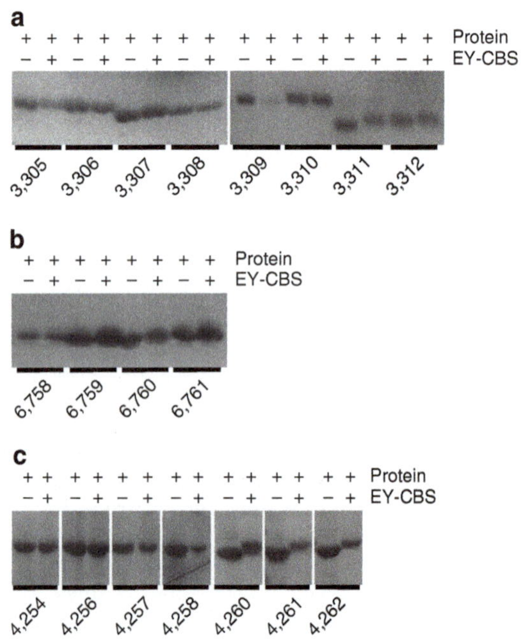

Figure 2 | Reactivity of the designed fusion proteins. The protein A-ankyrin (**a**, **c**) and the protein A-caldmodulin (**b**) fusion proteins were treated with 1mM EY-CBS. Constructs 3,311, 3,312, 4,260, 4,261 and 4,262 are substantially upshifted after the reaction.

To maximize the reaction efficiency of EY-CBS, we measured the reactivity of the 3,311 protein as a function of pH, anticipating that the deprotonated thiol groups of the cysteines would have higher reactivity at higher pH. As expected, the reaction was strongly dependent on pH and proceeded to near completion at pH 8 (Fig. 3c). At pH >9, the protein band and reaction

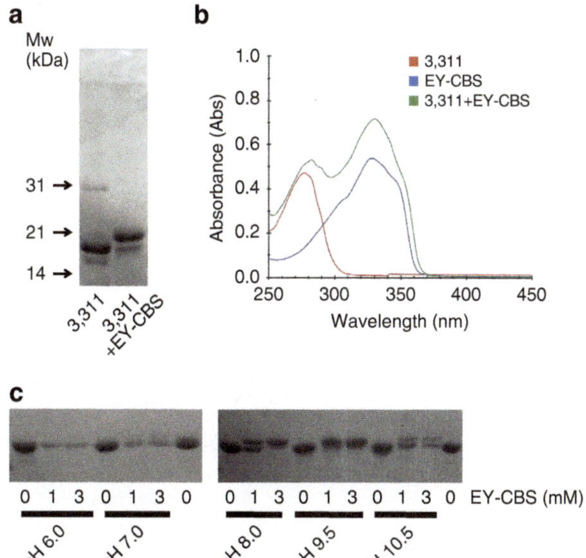

Figure 3 | Reactivity of the 3,311 fusion protein toward EY-CBS. (**a**) SDS-PAGE analysis of the 3,311 protein. Nearly 100% of the 3,311 protein is upshifted by reaction with EY-CBS. (**b**) Absorption spectrum of the 3,311 protein reacted with EY-CBS. Excess EY-CBS after reaction was removed by gel filtration chromatography. (**c**) pH dependence of the EY-CBS reaction. The 3,311 protein was reacted with EY-CBS at different pH and analysed by SDS-PAGE.

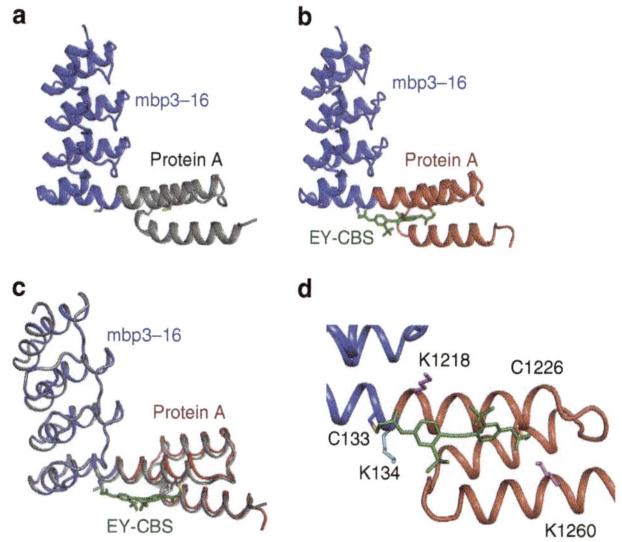

Figure 4 | Crystal structure of the 3,311 fusion protein after reaction with EY-CBS. (**a**) The intended structure of the 3,311 fusion protein. The ankyrin domain of mbp3-16 and the protein A domain are connected by an α-helix. (**b**) Crystal structure of 3,311 EY-CBS. EY-CBS is attached to the fusion helix and shown in green. For clarity the structure of the MBP bound to the ankyrin domain is not shown. (**c**) Superimposition of the intended and crystal structures of the 3,311 fusion. The intended and crystal structures are coloured in grey and blue/red, respectively. (**d**) Close-up view of the fusion helix. Positively charged amino acid side chains that interact with the sulfate groups of EY-CBS are shown.

efficiency declined presumably due to denaturation and precipitation of the fusion protein.

Crystal structure of the 3,311 fusion protein. To verify fusion of the helices, we crystallized the 3,311 fusion protein before and after reaction with EY-CBS and determined the structures of the two products (Figs 4 and 5). Since the ankyrin domain, mbp3–16, had been engineered to bind MBP, we were able to use MBP to facilitate crystallization[20]. In the crystal structure, one MBP molecule interacted exclusively with the mbp3–16 component of the hybrid protein and there was no interaction with either protein A or the fusion helix. The electron density map calculated using the refined protein structure showed a clear density corresponding to the EY-CBS, and one EY-CBS molecule could be modelled into the density (Supplementary Fig. 7). Both reactive ends of the EY-CBS are covalently linked to the cysteines in the fusion helix. In the crystal structure, the fusion helix adopts a nearly ideal α-helical structure and matches closely to the intended structure (Fig. 4c). The Cα atoms of the cysteines are separated by 16.8 angstrom and the mutated alanines are located under the phenyl rings of EY-CBS, as designed. The structures of the ankyrin and protein A domains are easily superimposable on those of the isolated proteins, which demonstrates that the helix fusion has no impact on the structure of the protein other than near the fusion site (Fig. 4b and c). The sulfate moieties of the EY-CBS groups form nonspecific ionic interactions with three lysine groups of the protein A and ankyrin domains (Fig. 4d). Other than that, EY-CBS does not have any noticeable interaction with the protein.

We determined the crystal structure of the unreacted 3,311 fusion protein at 2.8 angstrom resolution to understand why it reacts strongly with EY-CBS (Fig. 5). MBP was not included in the crystallized protein solution. The crystal structure showed that the protein A and ankyrin domains were connected by a slightly bent helix (Fig. 5a). When the structures of the fusion helix with and without the bound EY-CBS were superimposed, the fusion helix was seen to be bent by ~20 degrees with respect to that of the EY-CBS-bound helix (Fig. 5b). The asymmetric unit of the crystal contains 12 fusion proteins that can be divided into two groups, each group containing 6 proteins (Supplementary Fig. 8). The proteins belonging to the same group are in a similar crystal environment and, as expected, have superimposable structures. However, the proteins belonging to different groups are in completely different crystal environments. Despite this, the proteins in the different groups have almost identical structures, with a Cα root mean square difference value of 0.32 angstrom. Furthermore, the simulated annealing omit map calculated using the refined protein structure showed a clear density near the fusion helix (Supplementary Fig. 9). Because of these, we conclude that formation of the fusion helix is unlikely to be a crystallization artifact.

Crystal structure of the unreacted 3,311 clearly explains why the 3,311 fusion protein is strongly reactive with EY-CBS: the fusion helix is already present before EY-CBS is bound; after the covalent reaction with EY-CBS the fusion helix resembles more closely the ideal helix because the distance between the two cysteines is forced to be 16.8 angstrom. In the crystal structure, Asn107 and Lys1218 are separated by 2.9 angstrom and are hydrogen bonded. In addition, Asn135 is connected to Gln1220 via a hydrogen bond and contributes to stability of the fusion helix (Fig. 5c). However, it is not clear whether these relatively weak interactions are enough to stabilize the helical connection between the two protein components. More sophisticated calculations are required to analyse this stabilization.

Crystal structure of a fusion protein that failed the EY-CBS test. Of the 21 fusion proteins, 16 failed to display any significant

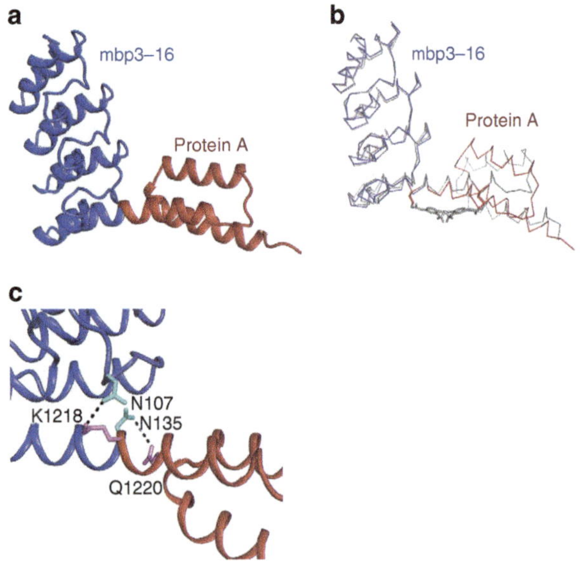

Figure 5 | Crystal structure of the 3,311 fusion protein not reacted with EY-CBS. (**a**) Crystal structure of the fusion protein. The fusion helix is bent by ~20 degrees. (**b**) Superposition of the EY-CBS-bound and -unbound structures of 3,311 fusion protein. The 3,311 EY-CBS and 3,311 structures are coloured in grey and blue/red, respectively. (**c**) Close-up view of the fusion helix. Amino acid residues that may stabilize the fusion helix are labeled.

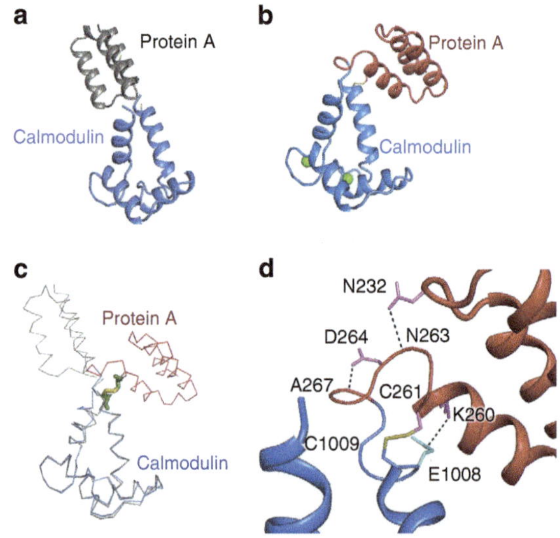

Figure 6 | Crystal structure of the 6,761 fusion protein containing the protein A domain and the calmodulin domain. (**a**) Intended structure of the 6,761 fusion protein. The two protein domains are connected by a continuous α-helix. (**b**) Crystal structure of the 6,761 fusion protein. An unexpected disulfide bridge that connects the mutated cysteines is shown. (**c**) Superposition of the intended and crystal structures of the 6,761 fusion protein. The intended structure is coloured in grey and the crystal structure is coloured in blue and red. (**d**) Close-up view of the fusion site. The cysteine 261 of the protein A domain and cysteine 1,009 of the calmodulin domain form a disulfide bridge. Amino acid residues that stabilize the interface of the two helices are drawn and labeled.

reactivity towards EY-CBS (Fig. 2). We were unable to crystallize any of these proteins except 6,761, presumably due to the structural instability of the interface between the fused protein components.

The crystallized 6,761 fusion protein is composed of an N-terminal protein A and a C-terminal calmodulin domain (Supplementary Fig. 10). Calmodulin is a calcium sensor composed of two homologous calcium-binding domains. Each calmodulin domain contains two EF hand motifs that bind calcium ions[24]. The crystal structure of the 6,761 fusion protein was determined at 2.7 angstrom resolution (Fig. 6). The structures of the two domains of the 6,761 fusion protein were easily superimposable with those of the separate protein A and calmodulin domains, and there were no noticeable structural changes except around the fusion site. In the crystal structure, it could be seen that the two helices designed to fuse had not formed a continuous helix, and the cysteines introduced to bind EY-CBS were close to one another and formed an unexpected disulfide bridge (Fig. 6c). The last and the first turns of the helices of the protein A and calmodulin domains, respectively, formed irregular loops and made sharp ~120 degree turns. The crystal structure of the connecting loop and the disulfide bridge formed a distinct electron density demonstrating that the fusion site was structurally stable (Supplementary Fig. 11). It appears that the 6,761 fusion protein can be crystallized because the interface between the two protein components of the fusion protein is stabilized by the opportunistic disulfide bridge. Several strong interactions can be seen at the fusion site and these account for why the disulfide bridge is formed (Fig. 6d). Lys260 of the protein A domain and Glu1008 of the calmodulin domain are separated by only 3.8 angstrom and form a strong ionic interaction. In addition, hydrogen bonding of the backbone nitrogen of Ala267 and the carboxylic oxygen of Asp264 should also contribute to the stabilization of the disulfide bridge. Furthermore, the side chain of Asn232 and the backbone carbonyl oxygen of Asn263 are separated by 2.9 angstrom and form a strong hydrogen bond.

We conclude that that simple-minded ligation of two helices does not guarantee the formation of a single extended helix, and that high reactivity of the fusion helix with EY-CBS is a reliable indicator of helix formation.

Insertion of a binding adaptor protein into an internal loop. To confirm the validity of our EY-CBS method, we designed several fusion proteins containing the protein A domain inserted into loops preceding internal rather than terminal α-helices of T4 lysozyme. To identify the best insertion site, we tested four loop regions and inserted an artificial protein A analogue, Ztaq (Supplementary Fig. 12)[25]. Three of the four fusion proteins were produced as efficiently as wild-type lysozyme when expressed in *Escherichia coli*. The exception was the 8,133 fusion. All three fusion proteins eluted as monomers in size-exclusion chromatography and were resistant to limited subtilisin digestion, confirming their structural integrity. Of the three possible insertion sites, we chose the loop containing residue 37 and designed three helix-fusion proteins, 8,155, 8,157 and 8,158 containing the EY-CBS-binding sites (Supplementary Figs 13–15). Fusion protein 8,157 has two more amino acids in the fusion helix than protein 8,155 and therefore its protein A domain should be rotated by ~200 degrees and translated by ~7 angstroms. Fusion proteins 8,157 and 8,158 have the same EY-CBS sites except that the positions of the reactive cysteines are shifted by four amino acids.

All three helix-fusion proteins were produced in *E. coli* and purified to homogeneity. Unexpectedly, reaction with EY-CBS did not result in clearly visible upshifts of the protein bands on SDS-PAGE presumably because size of the fusion proteins are too big and resolution of the SDS–PAGE analysis is not good enough to detect small changes in the structure (Fig. 7b and Supplementary Figs 16 and 17). However, we believe that all three

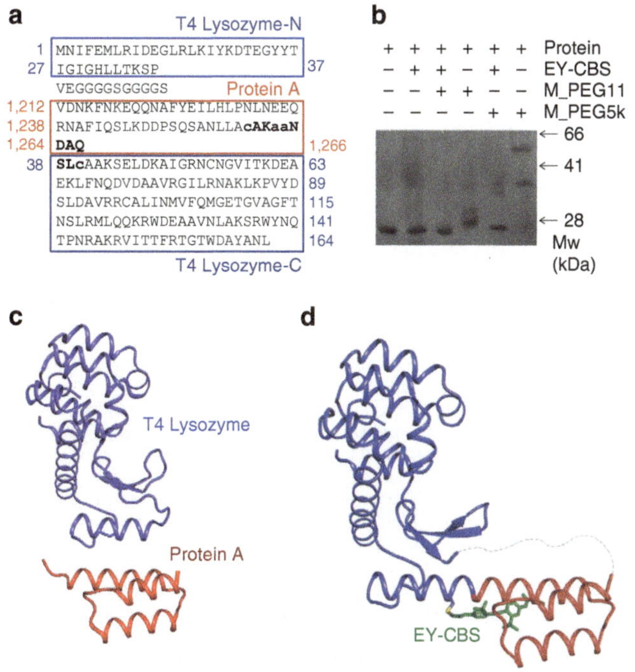

Figure 7 | Insertion of the protein A domain into an internal loop of T4 lysozyme. (**a**) The amino acid sequence of fusion protein 8,155. (**b**) Reactivity of fusion protein 8,155 with EY-CBS. After one hour of the reaction, PEG-Maleimide reagent Maleimide-PEG$_{11}$-Biotin (M_PEG11) or PEG-Maleimide 5000 (M_PEG5k) was added where appropriate. (**c**) Previously reported crystal structures of T4 lysozyme and protein A. (**d**) Crystal structure of fusion protein 8,155. The T4 lysozyme and the protein A parts of the fusion protein are coloured in blue and red, respectively. The 'GGGGS' linker and the N-terminal seven amino acids of protein A that are disordered in the crystal structure are drawn as a broken line.

proteins had reacted with EY-CBS with high efficiency because their cysteines became resistant to two maleimide-containing reagents, Maleimide-PEG$_{11}$-Biotin and PEG-Maleimide 5000. These reagents are highly reactive with the free thiol groups of cysteines, and their reactivity is easier to detect after SDS–PAGE because they have high-molecular weights, 1.1 and 5 kDa, respectively. As shown in Fig. 7b and Supplementary Figs 16 and 17, the fusion proteins were resistant to the PEG-maleimide reagents after EY-CBS treatment, presumably because they had already formed covalent bonds with EY-CBS, whereas the SDS–PAGE bands formed by the same fusion proteins not reacted with EY-CBS were clearly shifted upwards, showing that their cysteines were free to react with the PEG-maleimide reagents.

To confirm fusion of the α-helices connecting protein A and lysozyme, we crystallized 8,155 after reaction with EY-CBS and determined its crystal structure. The purified and reacted 8,157 and 8,158 fusions were also crystallized, but we have not tried to optimize the crystallization conditions nor determined their structures. The 8,155 crystals diffracted X-rays towards 2.7 angstrom resolution. In the crystal structure, one EY-CBS molecule is covalently connected to the two cysteines in the fusion helix as expected (Supplementary Fig. 18). The distance between the Cα atoms of the reacted cysteines is 16.7 angstrom, which is shorter by only 0.1 angstrom than that of 3,311 treated with EY-CBS. The fusion helix thus adopts a nearly ideal α-helical structure and closely matches the intended structure. The lysozyme and protein A parts of the structure can be superimposed with the structures of the individual proteins, demonstrating that fusion of the two helices had little impact on the overall structure of the individual protein components (Supplementary Fig. 19).

We chose the protein A domain as the insertion partner because it can be mutated to bind a variety of target proteins, as shown previously[26]. Because of this, it can be used as a universal adaptor protein mediating dimerization of pairs of target proteins. Provided we identify a suitable connecting helix, we can use the same helix to connect all other mutant protein A's for the following reasons. First, all the mutant protein A proteins adopt an essentially identical conformation, as shown by many crystal and NMR structures. Second, the C-terminal helix where our EY-CBS site is located is not changed in the mutant proteins because the mutations are limited to the first two α-helices. Among the known mutants, the Ztaq and anti-Ztaq proteins were selected for our study because they can form stable heterodimers[25]. To confirm that the mutations in the Ztaq and anti-Ztaq proteins do not affect the EY-CBS reaction, we replaced the protein A regions of fusions 8,155, 8,157 and 8,158 with the Ztaq or anti-Ztaq domain. The substituted fusion proteins retained similar reactivity with EY-CBS, as shown in Supplementary Fig. 16. Because the structure of the Ztaq-anti-Ztaq heterodimer was already known[25], we could predict the structures of the homo- and heterocomplexes 8,155–8,155, 8,157–8,157 and 8,155–8,157 with considerable confidence (Supplementary Fig. 20).

Thus, we have shown that universal adaptor proteins such as protein A or a coiled-coil domain can be inserted into a target protein to generate a predictable homo- or heterodimeric structure using our EY-CBS method.

Discussion

In this research, we have developed a novel method for connecting two proteins using a rigid fusion helix that is stabilized by a chemical cross linker. Using this method, we successfully fused the C-terminal helix of an ankyrin family protein to the N-terminal helix of a protein A fragment. We also succeeded inserting a protein A domain inside T4 lysozyme with predictable structure. Subsequent reaction with EY-CBS stabilized the structure of the fusion helix to that of an ideal helix. Our method, in principle, can be applied to majority of proteins as long as they contain at least one helix in their structure. In our method, EY-CBS reacts with thiol groups of cysteines. Therefore, proteins with natural cysteines may have unwanted reactions with EY-CBS. In such cases all the cysteines exposed to the surface need to be mutated before applying the method. Fortunately, most exposed cysteines found in proteins are not important functionally or structurally and can be converted into other amino acids without deleterious effects. In the PDB database of more than 100,000 protein structures, the majority have at least one α-helix and could therefore be used as building blocks for assembly of complex nanostructures by the helix-fusion method. Furthermore, application area of the helix-fusion method can be expanded by combining it with other methods like the metal chelation or the interface design method.

Several chemicals besides EY-CBS have been reported to stabilize the α-helical content of peptides[27]. Here are some notable examples of these helix stabilizers that can be applied to full size proteins. Two bisarylmethylene bromides named Bph and Bpy can selectively cross-link cysteines at the i and i + 7 positions in the α-helices[28]; the cross-linked peptides show increased helicity and cell permeability. Also spermine contains positively charged amines at 4.6 angstrom and 6.0 angstrom spacings, and has been shown to stabilize a peptide with four negatively charged glutamate or aspartate residues separated by

3–4 residues[29]. The stabilizing effect is highly dependent on pH, suggesting that ionic interactions play the major role in the helix stabilization. Similarly a diguanidium-containing compound has been shown to enhance the helical content of a peptide containing aspartate groups in the i and i + 3 positions[30]. These compounds may be able to replace EY-CBS in the helix-fusion method. Unlike EY-CBS, some of these helix stabilizing compounds do not use cysteines for cross linking. Therefore, they may be useful for connecting proteins with reactive cysteines.

Crystallization chaperones are proteins that can be easily produced and crystallized[31,32]. Fusing them to a target protein has proven successful in improving the crystallization property of the target protein. For example, T4 lysozyme has been inserted into a flexible loop of the β2-adrenergic receptor[33,34]. The authors generated a panel of fusion proteins and chose one where the linker between lysozyme and the receptor was rigid enough for crystallization. A similar method using cytochrome cb562 instead of T4 lysozyme has been essential for crystallizing several G-protein-coupled receptor receptors and determining their structures[35–38]. The success of this method relies on structural rigidity of the fusion site, because proteins with flexible structures cannot readily be crystallized. Since it is generally impossible to predict the structure of fusion sites, a large number of fusion proteins have to be screened and the linker sequence optimized for crystallization. Our method can be used to test whether a fusion protein has a rigid fusion site before extensive and time-consuming crystallization trials. For example, one can insert a helical chaperone protein like cytochrome cb562 into a loop preceding or following an α-helix in the target protein and connect the α-helices of the target and the chaperone protein using EY-CBS as shown in the crystal structure of 8,155 (Fig. 7). The resulting fusion protein will have a rigid helical connection and there will be more chance of successful crystallization.

In this study, we showed that reaction with EY-CBS could force the structure of a fusion helix into that of an ideal helix. The crystal structure demonstrated that the fusion helix was slightly bent at the fusion site before reaction with EY-CBS and was converted to a near-ideal helix after reaction with EY-CBS. For applications that involve crystallization of the fusion protein, the rigidity of the fusion protein is critical and reaction with EY-CBS would be desirable. However, in other applications small deviations from the ideal helical structure and some structural flexibility may be allowed. In such cases the EY-CBS reaction could be useful as a probe to test whether the two helices are indeed fused and a hybrid helix has been formed.

In conclusion, we have developed a new method that can convert two pre-existing α-helices into one long extended helix. This method can be used to link proteins to form a desired structure and may be useful in constructing complex protein nanoassemblies for a variety of purposes.

Methods

Design of fusion proteins. The fusion helices between pairs of proteins were designed using the molecular graphics program, COOT[39], assuming that their structures would not be significantly perturbed by the fusion. The coordinate files of the individual protein components were obtained from the PDB database[20,23,40,41]. The fusion helix was modelled by aligning the helices from the individual proteins with that of an ideal α-helix. Possible clashes between the two proteins were visually checked and avoided by addition or deletion of amino acid residues. Amino acid positions that were open to the solvent for easy approach of EY-CBS were chosen visually and mutated to cysteines for the EY-CBS reaction. Computer programs predicting protein structure were not used in our modelling.

Protein expression and purification. The genes of the fusion proteins were cloned into pET28a vector and over-expressed using *E. coli* strain BL21(DE3). Protein production was induced by adding 0.5 mM IPTG when OD 600 of the culture reached 0.7 and incubating the cells for an additional 20 h at 25 °C.

For proteins 3,305–3,315, glutathione-*S*-transferase was used as the purification tag. The cells were harvested by centrifugation, resuspended in lysis buffer containing 50 mM Tris pH 8.0, 200 mM NaCl and 10 mM β-mercaptoethanol and homogenized using a microfluidizer, model M-110L (Microfluidics). The supernatant after centrifugation was injected into a Glutathione-Sepharose (GE Healthcare) column equilibrated with a buffer containing 20 mM Tris pH 8.0, 200 mM NaCl and 1 mM dithiothreitol (DTT). The glutathione-*S*-transferase-tagged protein was eluted in a buffer containing 50 mM Tris pH 8.0, 100 mM NaCl, 10 mM glutathione and 1 mM DTT. The protein was cleaved overnight by thrombin to remove the purification tag and purified by a Q-Sepharose (GE Healthcare) anion exchange column equilibrated with a buffer containing 20 mM Tris 8.0 and 1 mM DTT. The bound protein was eluted by linear gradient of 1 M NaCl. The fractions containing the target protein were pooled and purified using a Supderdex 200 (GE Healthcare) gel filtration column equilibrated with a buffer containing 20 mM Tris pH 8.0, 200 mM NaCl and 1 mM TCEP.

For proteins 4,254–4,262, chitin-binding domain was used as the purification tag[42]. The harvested cells were homogenized in lysis buffer containing 50 mM Tris pH 8.0, 200 mM NaCl and 1 mM DTT. The supernatant after centrifugation was loaded onto a chitin affinity column (New England Biolab) equilibrated with 20 mM Tris 8.0, 200 mM NaCl and 1 mM DTT. The protein was eluted by thrombin cleavage and purified by a Q-Sepharose anion exchange column equilibrated with a buffer containing 20 mM Tris 8.0 and 1 mM DTT. The bound protein was eluted by linear gradient of 1 M NaCl. The fractions containing the target protein were pooled and purified using a Supderdex 200 gel filtration column equilibrated with a buffer containing 20 mM Tris pH 8.0, 200 mM NaCl and 1 mM TCEP.

Hexa-histidine was used as the purification tag for constructs 6,758–6,761. The harvested cells were homogenized in lysis buffer containing 50 mM Tris pH 8.0, 200 mM NaCl and 1 mM DTT. The supernatant was loaded onto a Ni-NTA affinity column equilibrated with 20 mM Tris 8.0, 200 mM NaCl and 1 mM $CaCl_2$. One millimolar DTT was added to the protein immediately after elution using 500 mM imidazole. The protein was cleaved overnight by thrombin to remove the purification tag. The cleaved protein was purified by Q-Sepharose anion exchange chromatography equilibrated with a buffer containing 20 mM Tris 8.0, 1 mM $CaCl_2$ and 1 mM DTT. The bound protein was eluted by linear gradient of 1 M NaCl. The fractions containing the target protein were pooled and purified using a Supderdex 200 gel filtration column equilibrated with a buffer containing 20 mM Tris pH 8.0, 200 mM NaCl, 1 mM $CaCl_2$ and 1 mM TCEP. The purified proteins were concentrated with an ultrafiltration kit (Amicon) and used for the EY-CBS reaction.

For proteins 8,132–8,136 and 8,155–8,158, hexa-histidine was used as the purification tag. The harvested cells were homogenized in lysis buffer containing 20 mM Tris pH 8.0, 200 mM NaCl, 0.1 mM phenylmethyl sulphonyl fluorideand 10 mM β-mercaptoethanol. The supernatant after centrifugation was loaded onto a Ni-NTA affinity column equilibrated with 20 mM Tris pH 8.0 and 200 mM NaCl. One millimolar TCEP was added immediately to the protein eluted using 300 mM imidazole. The eluted protein was reacted with EY-CBS according to the EY-CBS reaction protocol. The EY-CBS reacted protein was purified using a SP-Sepharose (GE Healthcare) cation exchange column equilibrated with a buffer containing 20 mM MES pH 5.5. The bound protein was eluted by linear gradient of 1 M NaCl. The fractions containing the target protein were pooled and cleaved overnight by thrombin to remove the purification tag. The cleaved protein was purified using a Supderdex 200 gel filtration column equilibrated with a buffer containing 20 mM Tris pH 8.0 and 200 mM NaCl.

EY-CBS reaction. EY-CBS was synthesized as described by Zhang *et al.*[19] The EY-CBS reaction, typically with 1 mg ml^{-1} fusion protein, was carried out at 23 °C for 1 h in 20 mM Tris HCl buffer at pH 8.0 and 200 mM NaCl. One millimolar TCEP was added to the reaction buffer to prevent oxidation of the cysteine during the reaction. Progress of the reaction was monitored by SDS–PAGE. The reaction was stopped and unreacted EY-CBS was removed by Superdex 200 gel filtration or SP-Sepharose ion exchange chromatography. The reacted and purified proteins were concentrated with an ultrafiltration kit and used for crystallization if necessary. The absorption spectrum of the protein after the EY-CBS reaction was measured with a JASCO UV-530 spectrophotometer.

The PEG-Maleimide reaction. For the PEG-Maleimide reaction, 10 mM Maleimide-PEG_{11}-Biotin (Thermo Scientific) or PEG-Maleimide 5000 (Nanocs) was added to fusion proteins that either had, or had not, been reacted with EY-CBS. The progress of the reaction was monitored by SDS–PAGE and the reaction was stopped by adding SDS–PAGE sample buffer containing 570 mM β-mercaptoethanol.

Crystallization. Purified and concentrated proteins were used for crystallization. Initial crystallization conditions were screened using a crystallization robot (Mosquito, TTP Labtech). Unreacted 3,311 fusion protein was crystallized in a buffer containing 0.1 M Bis-Tris pH 6.5, 45% v/v PEG 400 and 10 mM hexamine cobalt (III) chloride. The crystals were frozen in liquid nitrogen. Addition of cryoprotectant was not necessary. After EY-CBS reaction and addition of excess amount of MBP, the 3,311-EY-CBS–MBP complex protein was purified by Superdex 200 gel filtration chromatography and crystallized in a buffer containing 0.1 M MOPS pH 6.5, 0.2 M magnesium chloride and 36% w/v PEG 2000. Crystals were frozen in a buffer

Table 2 | Data collection and refinement statistics.

	3,311 + EY-CBS	3,311	6,761	8,155 + EY-CBS
Data collection				
Space group	$P2_12_12_1$	$P2_1$	$I4_1$	C2
Resolution (Å)	2.30-50	2.80-50	2.70-50	2.60-50
Coordinations::				
a, b, c (Å)	64.3, 75.8, 111.4	85.4, 220.0, 85.4	98.8, 98.8, 31.7	89.5, 149.9, 55.8
α, β, γ (°)	90.0, 90.0, 90.0	90.0, 60.0, 90.0	90.0, 90.0, 90.0	90.0, 122.7, 90.0
R_{sym}	0.063 (0.513)	0.051 (0.152)	0.114 (0.762)	0.079 (0.352)
$I/\sigma I$	31.9	25.3	17.2	18.3
Completeness (%)	99.7 (99.2)	92.9 (80.0)	98.7 (97.0)	99.1 (98.5)
Redundancy	6.8	3.2	6.5	3.5
Search probes (PDB code)	1SVX, 1DEE	1SVX, 1DEE	2SPZ, 1CLL	1LYD, 1DEE
Refinement				
No. of reflections	24,741	62,287	4,460	35,397
R_{work}/R_{free}	0.194/0.262	0.207/0.250	0.206/0.259	0.200/0.259
Twin				
Operator	NA	h, -k, h-l	NA	NA
Fraction	NA	0.46	NA	NA
No. of protein molecules	1 (MBP) + 1 (3311)	12	1	2
No. of atoms	4,330	16,233	978	3,451
Protein	4,188	16,116	966	3,408
Ligands	30	0	0	60
Water	112	117	12	43
Average B factor (Å^2)	51.2	29.6	65.4	57.4
R.m.s. deviations				
Bond length (Å)	0.008	0.010	0.009	0.009
Bond angles (°)	1.080	1.081	1.105	1.451

EY-CBS, 3,3′-ethyne-1,2-diylbis-{6-[(chloroacetyl)amino]benzenesulfonic acid}; NA, not applicable; r.m.s., root mean square.
Highest resolution shell is shown in parenthesis.

containing the crystallization solution plus 14% PEG 2000 as the cryoprotectant. The 6,761 fusion protein was crystallized in a solution containing 10% w/v PEG 1000 and 10% w/v PEG 8000. Crystals were frozen in a buffer containing the crystallization solution plus 30% ethylene glycol as the cryoprotectant. The 8,155 protein reacted with EY-CBS was crystallized in a solution containing 1.84 M NaK phosphate pH 7.5. Crystals harvested with the crystallization buffer were flash frozen in a drop of paraffin oil (Hampton Research) using liquid nitrogen.

X-ray diffraction data and structure refinement. Diffraction data were collected at Pohang Accelerator Laboratory, beam lines 5C and 7A. Diffraction images were integrated and scaled with HKL2000. All the structures were solved by the molecular replacement method using the PHASER software[43]. Initial phasing data were then refined using PHENIX[44]. The program COOT was used for manual correction of the structure. The crystallographic data are summarized in Table 2. The 3,311 crystals were merohedrally twinned with a twin fraction of 0.46. Initially we could not solve the structure by the molecular replacement technique because the diffraction data were incorrectly indexed for the $P6_322$ space group. Later we found that the crystals were almost perfectly twinned (Supplementary Fig. 21). Therefore, the data were re-indexed for space groups with lower symmetry. Promising solutions could be found for both the $P6_3$ and $P2_1$ space groups in molecular replacement calculations. However, refinement could not reduce the free R factor below 40% when the data were indexed for the $P6_3$ space group. Therefore, they were finally indexed for the $P2_1$ space group, and the structure was refined. The free R factor dropped from 0.369 to 0.261 after incorporation of the twin operator in the refinement protocol.

References

1. Lai, Y. T., King, N. P. & Yeates, T. O. Principles for designing ordered protein assemblies. *Trends Cell Biol.* **22,** 653–661 (2012).
2. Radford, R. J., Nguyen, P. C., Ditri, T. B., Figueroa, J. S. & Tezcan, F. A. Controlled protein dimerization through hybrid coordination motifs. *Inorg. Chem.* **49,** 4362–4369 (2010).
3. Radford, R. J., Nguyen, P. C. & Tezcan, F. A. Modular and versatile hybrid coordination motifs on alpha-helical protein surfaces. *Inorg. Chem.* **49,** 7106–7115 (2010).
4. Radford, R. J. & Tezcan, F. A. A superprotein triangle driven by nickel(II) coordination: exploiting non-natural metal ligands in protein self-assembly. *J. Am. Chem. Soc.* **131,** 9136–9137 (2009).
5. Brodin, J. D. *et al.* Metal-directed, chemically tunable assembly of one-, two- and three-dimensional crystalline protein arrays. *Nat. Chem.* **4,** 375–382 (2012).
6. Brodin, J. D. *et al.* Evolution of metal selectivity in templated protein interfaces. *J. Am. Chem. Soc.* **132,** 8610–8617 (2010).
7. Medina-Morales, A., Perez, A., Brodin, J. D. & Tezcan, F. A. *In vitro* and cellular self-assembly of a Zn-binding protein cryptand via templated disulfide bonds. *J. Am. Chem. Soc.* **135,** 12013–12022 (2013).
8. Salgado, E. N. *et al.* Metal templated design of protein interfaces. *Proc. Natl Acad. Sci. USA* **107,** 1827–1832 (2010).
9. Salgado, E. N., Faraone-Mennella, J. & Tezcan, F. A. Controlling protein-protein interactions through metal coordination: assembly of a 16-helix bundle protein. *J. Am. Chem. Soc.* **129,** 13374–13375 (2007).
10. Salgado, E. N., Lewis, R. A., Faraone-Mennella, J. & Tezcan, F. A. Metal-mediated self-assembly of protein superstructures: influence of secondary interactions on protein oligomerization and aggregation. *J. Am. Chem. Soc.* **130,** 6082–6084 (2008).
11. Song, W. J., Sontz, P. A., Ambroggio, X. I. & Tezcan, F. A. Metals in protein-protein interfaces. *Annu. Rev. Biophys.* **43,** 409–431 (2014).
12. Song, W. J. & Tezcan, F. A. A designed supramolecular protein assembly with in vivo enzymatic activity. *Science* **346,** 1525–1528 (2014).
13. King, N. P. *et al.* Computational design of self-assembling protein nanomaterials with atomic level accuracy. *Science* **336,** 1171–1174 (2012).
14. Padilla, J. E., Colovos, C. & Yeates, T. O. Nanohedra: using symmetry to design self assembling protein cages, layers, crystals, and filaments. *Proc. Natl Acad. Sci. USA* **98,** 2217–2221 (2001).
15. Lai, Y. T., Tsai, K. L., Sawaya, M. R., Asturias, F. J. & Yeates, T. O. Structure and flexibility of nanoscale protein cages designed by symmetric self-assembly. *J. Am. Chem. Soc.* **135,** 7738–7743 (2013).
16. Lai, Y. T., Cascio, D. & Yeates, T. O. Structure of a 16-nm cage designed by using protein oligomers. *Science* **336,** 1129 (2012).
17. Lai, Y. T. *et al.* Structure of a designed protein cage that self-assembles into a highly porous cube. *Nat. Chem.* **6,** 1065–1071 (2014).
18. Lai, Y. T., Jiang, L., Chen, W. & Yeates, T. O. On the predictability of the orientation of protein domains joined by a spanning alpha-helical linker. *Protein Eng. Des. Sel.* **28,** 491–500 (2015).
19. Zhang, F., Sadovski, O., Xin, S. J. & Woolley, G. A. Stabilization of folded peptide and protein structures via distance matching with a long, rigid cross-linker. *J. Am. Chem. Soc.* **129,** 14154–14155 (2007).

20. Binz, H. K. *et al.* High-affinity binders selected from designed ankyrin repeat protein libraries. *Nat. Biotech.* **22,** 575–582 (2004).
21. Nygren, P. A. Alternative binding proteins: affibody binding proteins developed from a small three-helix bundle scaffold. *FEBS J.* **275,** 2668–2676 (2008).
22. Pluckthun, A. Designed ankyrin repeat proteins (DARPins): binding proteins for research, diagnostics, and therapy. *Annu. Rev. Pharmacol. Toxicol.* **55,** 489–511 (2015).
23. Chattopadhyaya, R., Meador, W. E., Means, A. R. & Quiocho, F. A. Calmodulin structure refined at 1.7A resolution. *J. Mol. Biol.* **228,** 1177–1192 (1992).
24. Meador, W. E., Means, A. R. & Quiocho, F. A. Target enzyme recognition by calmodulin: 2.4A structure of a calmodulin-peptide complex. *Science* **257,** 1251–1255 (1992).
25. Lendel, C., Dogan, J. & Hard, T. Structural basis for molecular recognition in an affibody:affibody complex. *J. Mol. Biol.* **359,** 1293–1304 (2006).
26. Feldwisch, J. & Tolmachev, V. Engineering of affibody molecules for therapy and diagnostics. *Methods Mol. Biol.* **899,** 103–126 (2012).
27. Henchey, L. K., Jochim, A. L. & Arora, P. S. Contemporary strategies for the stabilization of peptides in the alpha-helical conformation. *Curr. Opin. Chem. Biol.* **12,** 692–697 (2008).
28. Muppidi, A., Wang, Z., Li, X., Chen, J. & Lin, Q. Achieving cell penetration with distance-matching cysteine cross-linkers: a facile route to cell-permeable peptide dual inhibitors of Mdm2/Mdmx. *Chem. Commun.* **47,** 9396–9398 (2011).
29. Tabet, M., Labroo, V., Sheppard, P. & Sasaki, T. Spermine-Induced Conformational-Changes of a Synthetic Peptide. *J. Am. Chem. Soc.* **115,** 3866–3868 (1993).
30. Albert, J. S., Goodman, M. S. & Hamilton, A. D. Molecular recognition of proteins—sequence-selective binding of aspartate pairs in helical peptides. *J. Am. Chem. Soc.* **117,** 1143–1144 (1995).
31. Koide, S. Engineering of recombinant crystallization chaperones. *Curr. Opin. Struct. Biol.* **19,** 449–457 (2009).
32. Lieberman, R. L., Culver, J. A., Entzminger, K. C., Pai, J. C. & Maynard, J. A. Crystallization chaperone strategies for membrane proteins. *Methods* **55,** 293–302 (2011).
33. Cherezov, V. *et al.* High-resolution crystal structure of an engineered human beta2-adrenergic G protein-coupled receptor. *Science* **318,** 1258–1265 (2007).
34. Rosenbaum, D. M. *et al.* GPCR engineering yields high-resolution structural insights into beta2-adrenergic receptor function. *Science* **318,** 1266–1273 (2007).
35. Liu, W. *et al.* Structural basis for allosteric regulation of GPCRs by sodium ions. *Science* **337,** 232–236 (2012).
36. Thompson, A. A. *et al.* Structure of the nociceptin/orphanin FQ receptor in complex with a peptide mimetic. *Nature* **485,** 395–399 (2012).
37. Fenalti, G. *et al.* Molecular control of delta-opioid receptor signalling. *Nature* **506,** 191–196 (2014).
38. Wu, H. *et al.* Structure of a class C GPCR metabotropic glutamate receptor 1 bound to an allosteric modulator. *Science* **344,** 58–64 (2014).
39. Emsley, P., Lohkamp, B., Scott, W. G. & Cowtan, K. Features and development of Coot. *Acta Crystallogr. D Biol. Crystallogr.* **66,** 486–501 (2010).
40. Graille, M. *et al.* Crystal structure of a Staphylococcus aureus protein A domain complexed with the Fab fragment of a human IgM antibody: Structural basis for recognition of B-cell receptors and superantigen activity. *Proc. Natl Acad. Sci. USA* **97,** 5399–5404 (2000).
41. Mooers, B. H. *et al.* Repacking the Core of T4 lysozyme by automated design. *J. Mol. Biol.* **332,** 741–756 (2003).
42. Mitchell, S. F. & Lorsch, J. R. Protein affinity purification using intein/chitin binding protein tags. *Methods Enzymol.* **559,** 111–125 (2015).
43. McCoy, A. J. *et al.* Phaser crystallographic software. *J. Appl. Crystallogr.* **40,** 658–674 (2007).
44. Adams, P. D. *et al.* PHENIX: a comprehensive Python-based system for macromolecular structure solution. *Acta Crystallogr. D Biol. Crystallogr.* **66,** 213–221 (2010).

Acknowledgements

We thank the staff of the beam lines 5C and 7A, Pohang Accelerator Laboratory for help with data collection, and Dr Julian Gross for critical reading of the manuscript. This research was supported by the Intelligent Synthetic Biology Center of Global Frontier Project (2011-0031955) and the National Research Foundation (NRF-2014R1A2A1A10050436) funded by the Ministry of Science, ICT and Future Planning of Korea.

Author contributions

W.H.J., H.L., D.H.S. and J.-H.E. performed experiments and data analysis. W.H.J., H.L., D.H.S., S.C.K., H.-S.L., H.L. and J.-O.L. designed experiments. W.H.J., H.L., S.C.K., H.-S. L., H.L. and J.-O.L. contributed to writing the paper. J.-O.L. supervised the project.

Additional information

Accession codes: Atomic coordinates and diffraction data have been deposited in the Protein Data Bank with code numbers 5CBN, 5CBO, 5COC and 5EWX for the 3311 + EY-CBS, 3311, 6761 and 8155 + EY-CBS structures, respectively.

Competing financial interests: The authors declare no competing financial interests.

16

Carbohydrate scaffolds as glycosyltransferase inhibitors with *in vivo* antibacterial activity

Johannes Zuegg[1,2], Craig Muldoon[2], George Adamson[2], Declan McKeveney[2], Giang Le Thanh[2], Rajaratnam Premraj[2], Bernd Becker[2], Mu Cheng[1], Alysha G. Elliott[1], Johnny X. Huang[1], Mark S. Butler[1], Megha Bajaj[1], Joachim Seifert[2], Latika Singh[2], Nicola F. Galley[3], David I. Roper[3], Adrian J. Lloyd[3], Christopher G. Dowson[3], Ting-Jen Cheng[4], Wei-Chieh Cheng[4], Dieter Demon[5], Evelyne Meyer[5], Wim Meutermans[2] & Matthew A. Cooper[1]

The rapid rise of multi-drug-resistant bacteria is a global healthcare crisis, and new antibiotics are urgently required, especially those with modes of action that have low-resistance potential. One promising lead is the liposaccharide antibiotic moenomycin that inhibits bacterial glycosyltransferases, which are essential for peptidoglycan polymerization, while displaying a low rate of resistance. Unfortunately, the lipophilicity of moenomycin leads to unfavourable pharmacokinetic properties that render it unsuitable for systemic administration. In this study, we show that using moenomycin and other glycosyltransferase inhibitors as templates, we were able to synthesize compound libraries based on novel pyranose scaffold chemistry, with moenomycin-like activity, but with improved drug-like properties. The novel compounds exhibit *in vitro* inhibition comparable to moenomycin, with low toxicity and good efficacy in several *in vivo* models of infection. This approach based on non-planar carbohydrate scaffolds provides a new opportunity to develop new antibiotics with low propensity for resistance induction.

[1] Institute for Molecular Bioscience, The University of Queensland, St Lucia, Queensland 4072, Australia. [2] Alchemia Ltd, PO Box 4851, Eight Mile Plains, Brisbane, Queensland 4113, Australia. [3] School of Life Science, University of Warwick, Gibbet Hill Road, Coventry CV4 7AL, UK. [4] Genomics Research Center, Academia Sinica, 128 Academia Road, Section 2, Taipei 115, Taiwan. [5] Faculty of Veterinary Medicine, Laboratory of Biochemistry, Ghent University, Salisburylaan 133, 9820 Merelbeke, Belgium. Correspondence and requests for materials should be addressed to W.M. (email: wmeutermans@optusnet.com.au) or to M.A.C. (email: m.cooper@uq.edu.au).

Peptidoglycan glycosyltransferases (GT) and transpeptidases (TP) are two key enzymes in the final steps of peptidoglycan (PG) biosynthesis essential for bacterial cell wall integrity and stability. GTs catalyse the polymerization of lipid II disaccharide units, forming a long chain of alternating β-1,4-linked *N*-acetylglucosamines and *N*-acetylmuramic acid, leading to a linear glycan chain and the release of undecaprenyl-pyrophosphate carrier[1–3]. These carbohydrate chains are further crosslinked by TP enzymes, forming linkages between the peptide chain and the D-alanine of a neighboring unit (Fig. 1). GT and TP enzymes are unique to bacteria and are expressed either as individual domains, monofunctional GT (MGT) and penicillin-binding proteins (PBP), respectively or as bifunctional proteins that possess both GT and TP domains (class A PBP)[4,5].

Inhibition of extracellular bacterial cell wall synthesis has been a very successful strategy in the development of many important antibacterial agents, with teixobactin[6], one of the most recently reported. The β-lactam class, which includes cephalosporins, monobactams and carbapenems, inhibit PG crosslinking by covalently binding to the TP enzyme, while glycopeptides such as vancomycin bind directly to the lipid II unit and sterically inhibit further polymerization and crosslinking of PG. Bacteria developed resistance to cell wall inhibitors via β-lactamases, thickened cell walls and modification of the lipid II unit, with extended-spectrum β-lactamases such as NDM-1 and vancomycin-resistant enterococci representing a significant health threat[7]. Glycolipopeptides (for example, ramoplanin), cyclic peptides (for example, AC98-6446) and lantibiotics (for example, nisin or NVB302) also bind to lipid II of Gram-positive bacteria[8]. Only nisin has reached the market, and then only as a food preservative[9].

Antibacterial compounds that bind directly to GT have never been developed for human use. Of the few examples reported in the literature, moenomycin is by far the best described[10]. Moenomycin is produced by various streptomyces species and has a broad-spectrum activity against a range of Gram-positive bacteria. The poor pharmacokinetic properties of moenomycin have prevented further clinical development[10,11], and it has been commercialized only as a 'growth promoter' within animal feed stocks (Flavomycin and Flavophospholipol). Despite evidence that regular application of antibiotics as growth promoters in animals in general leads to increased antibiotic resistance[12,13], remarkably no moenomycin-resistant bacteria in animals have been reported to date[14,15]. *In vitro* resistance induction experiments showed extremely slow development of resistance with low-resistant frequencies[16], as well as no transferable resistance between organisms[17,18], no cross-resistance to other antimicrobials or co-selection of resistant strains[19]. Intrinsic resistance in moenomycin-producing organisms is not associated with the biosynthesis cluster, but is likely to arise from the presence of GT's with low affinity for moenomycin, or some peculiarities of their cell wall organization[10]. Further, moenomycin is primarily accumulated inside of the cells, while its target is located on the cell surface[20,21]. *In vitro*-induced resistance with *S. aureus* showed mutations in the binding site of PBP2 with reduced affinity for moenomycin as well as its ligand, resulting in strains with shorter PG polymers and major cell division defects[16]. The lack of a specific resistance mechanism and the paucity of antibiotics that specifically mimic the carbohydrate portion of bacterial lipid II suggest that direct GT inhibition remains an attractive strategy for the development of novel antibacterial agents with low potential for resistance development.

Moenomycin A is a highly functionalized pentasaccharide attached via a phosphoglycerate linkage to a polyprenyl chain[22] (Fig. 1) that binds competitively to GT enzymes by mimicking the disaccharide–pyrophosphate–prenol linkage of the donor lipid

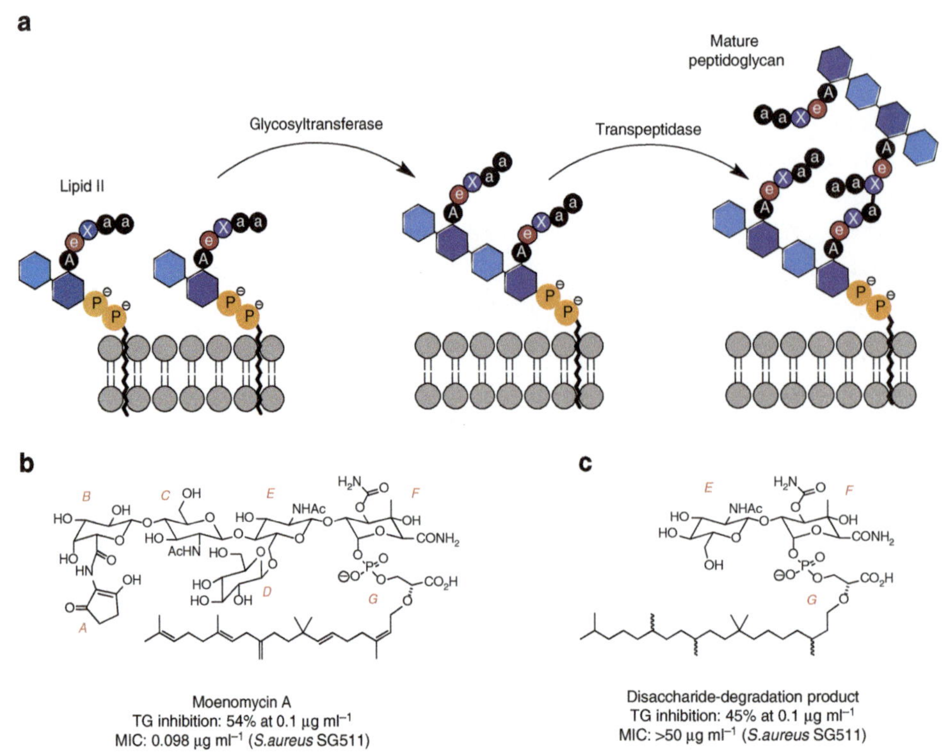

Figure 1 | Overview of PG cell wall synthesis and inhibitors. **(a)** PG synthesis in bacteria from lipid II with subsequent GT and TP catalysis, with A: L-Ala, a: D-Ala, e: D-iGln, X: either D-Lys(Ala_5) in case of *Staphylococcus*, or mDap in case of *Bacillus*. **(b)** Structure and *in vitro* activity of moenomycin A, indicating the different moieties with A to G. **(c)** Structure and *in vitro* activity of moenomycin's disaccharide degradation product.

II[10,23,24]. Although the E and F rings and the phosphoglycerate (G) portion of moenomycin A are important for GT inhibitory activity, analogues of this pharmacophore subunit did not maintain whole cell antibacterial activity[25,26]. Attempts to mimic the EFG structural fragment with disaccharide derivatives[27–30] resulted in compounds (such as TS30153 (ref. 17)) with cell-based activity, that is, minimum inhibitory concentration (MIC) of 3.12 and 12.5 μg ml^{-1} against staphylococci and enterococci, but with no *in vivo* activity. Compound TS30153 (ref. 17) has three hydrophobic binding elements that mimic the acyl and alkyloxy moieties of moenomycin A (Fig. 2b). Attempts to mimic directly lipid II[31], or the ring F of moenomycin[32], with monosaccharide scaffolds gave compounds with only low to medium activity (MIC = 60 μM against *Bacillus cereus*[31]). More recently, *de novo* inhibitors for GT discovered using high-throughput screening[33–35] or *in silico* methods[36], were shown to have improved *in vitro* activity (MIC = 0.25 μM against MRSA[33]), but no *in vivo* activity.

In this study, we explore novel chemistry based on a monosaccharide scaffold[37] to mimic the essential structure features of moenomycin and to improve the drug-like properties, in particular reduced molecular weight and hydrophobicity. Compared with other scaffolds, the monosaccharide scaffold approach provides structural diversity using up to five chiral attachment points within a small volume[38]. This allows for more efficient pharmacophore optimization, while still enabling the generation of a broad structural diversity to scope and improve activity and physicochemical properties. Here we use the approach to produce moenomycin-focused libraries and select compounds with *in vitro* antibacterial activity and *in silico* potential to inhibit the GT enzyme. We demonstrate the strength of this strategy with two of the most promising candidates showing inhibition of GT and PG synthesis in *in vitro* assays, as well as *in vivo* efficacy in eliminating *S. aureus* infection from a mouse mammary gland.

Results

Design and synthesis. We synthesized a small library of compounds by replacing the phosphoglycerate/phosphate moieties (G, Fig. 1) with simpler lipophilic substituents (that is, phenyl, biphenyl or naphthyl groups linked via a urea) and changing the orientation and nature of ring F. This approach yielded compounds like ACL19378 (Fig. 2a, Supplementary Figs 1–8) and ACL19333 (Supplementary Fig. 1), with MICs against Gram-positive bacteria in the range of 2 μg ml^{-1}, but with limited activity in the presence of 50% serum. In the second stage, we used the disaccharide structure–activity relationship information to design more synthetically feasible and smaller monosaccharide molecules. A versatile solid-phase method was developed to rapidly synthesize the representatives of three different core chemotypes M1 to M3, starting from a single monosaccharide building block, that is, 1,5-anhydro-galactitol (see Fig. 3). Chemotype M1 explored the option of using only two of the hydrophobic elements, whereas the other two, M2 and M3, used a benzimidazole moiety as the third hydrophobic group (Fig. 3).

Five hundred compounds were thus synthesized combinatorially on solid-phase resin, purified using high-performance liquid chromatography, and tested for their MIC activity against two Gram-positive staphylococcal strains (methicillin-sensitive (MSSA) and methicillin-resistant *S. aureus* (MRSA)), three enterococcal strains and *Escherichia coli* as a Gram-negative control. Although most compounds were inactive against *E. coli*, many compounds displayed activity against the Gram-positive strains. The derivatives with Gram-positive antibacterial activity generally contained a lipophilic substituent such as an alkyl moiety (minimum length of 10 carbon atoms) or a biaryl, and one or two electron-deficient aryl groups. All active compounds were then tested for haemolytic activity and, after filtering out the haemolytic compounds, a series of compounds of chemotype M3 containing substituted amino-benzimidazoles were selected for further study (Supplementary Table 1). Two compounds (Fig. 4), ACL20215 (Supplementary Figs 10–15) and ACL20964 (Supplementary Figs 16–20), showed broad activity against a range of drug resistant *S. aureus* strains, including MRSA, GISA (glycopeptide-intermediate *S. aureus*), VRSA (vancomycin-resistant *S. aureus*) and DRSA (daptomycin-resistant *S. aureus*) and multi-drug-resistant *S. pneumoniae*, with low haemolytic activity against human red blood cells (Table 1). ACL20215 was assayed for resistance potential and showed a spontaneous mutation frequency against *S. aureus* (ATCC 13709; Smith strain) of less than 2.5×10^{-10} at four times its MIC value.

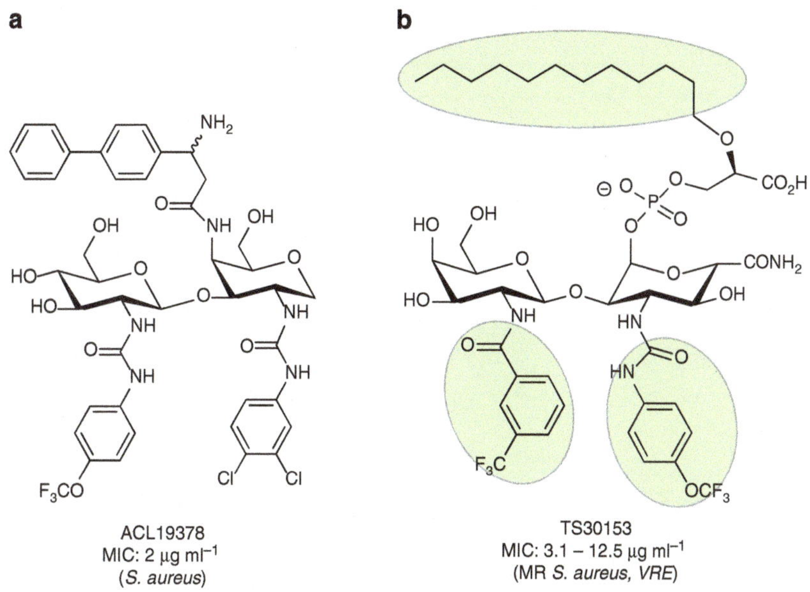

Figure 2 | Disaccharide templates for design of monosaccharides. (**a**) ACL19378, representative compounds from disaccharide library. (**b**) TS30153 (ref. 27) highlighting the three binding elements required for GT inhibition in green.

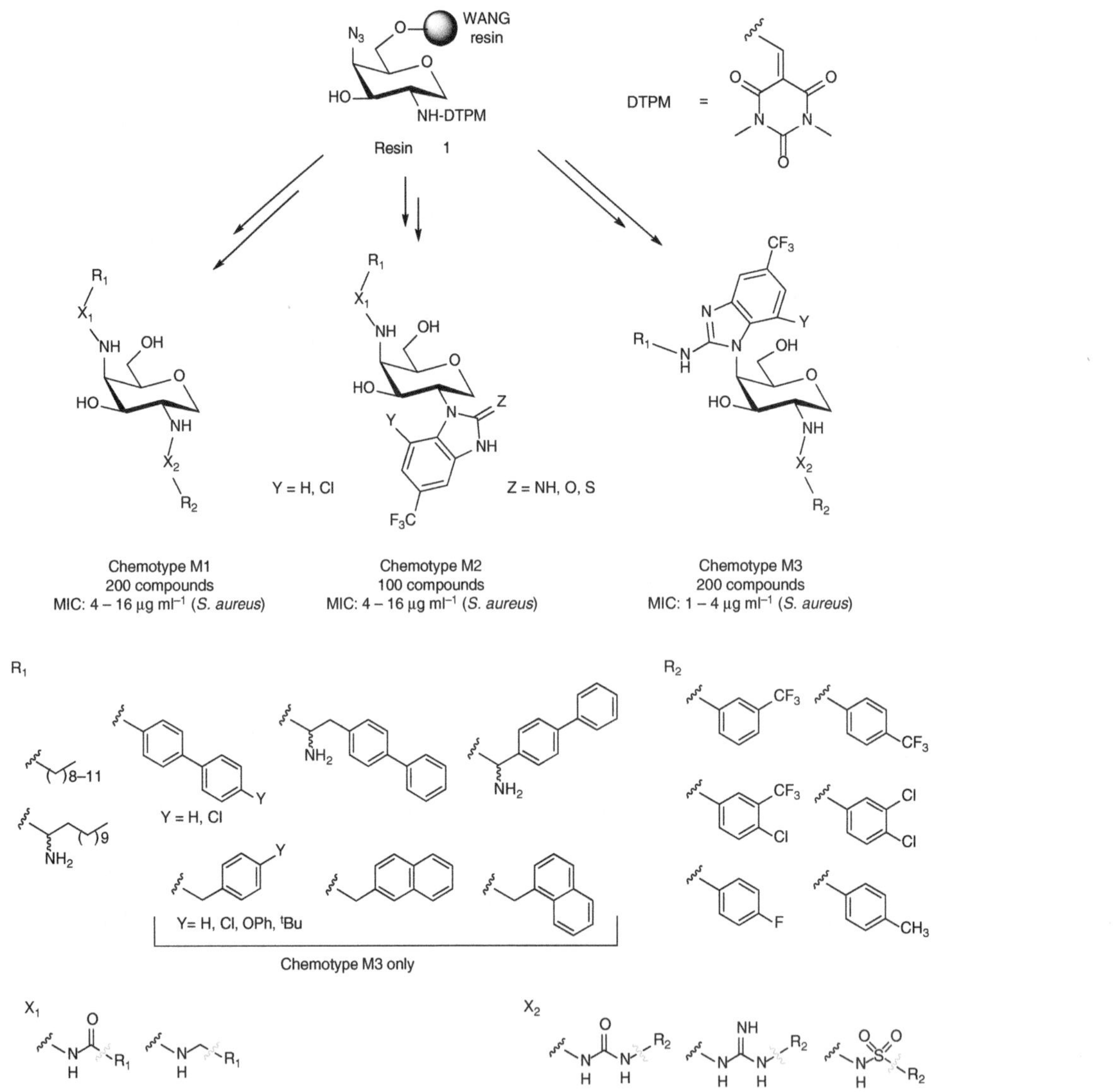

Figure 3 | Design of monosaccharide libraries. The figure illustrates the common starting building block, the three different chemotypes (M1, M2 and M3) and corresponding diversification at each substitution point.

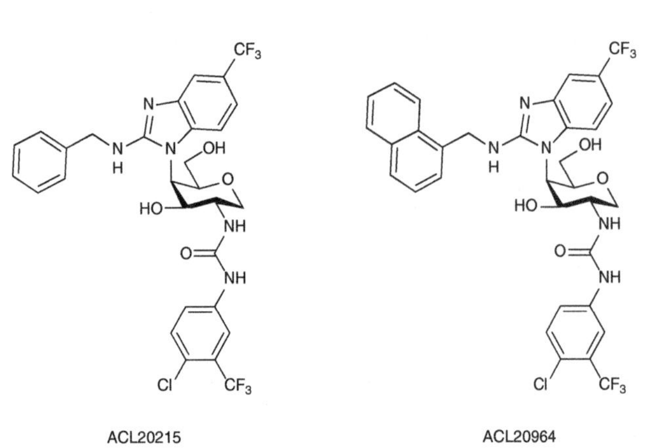

Figure 4 | Structures of ACL20215 and ACL20965. ACL20215 and ACL20965 are two of the most active monosaccharide compounds.

Evaluation of mode of action. To elucidate the mode of action of the inhibitors, we conducted various GT enzyme inhibition assays. We first examined the inhibitory effect of the compounds in a bacterial membrane environment, comparing ACL20215 and ACL20964 in an *in vitro* assay for bacterial PG biosynthesis, using crude *Bacillus megaterium* membrane preparations. This assay monitors [^{14}C]UDP-GlcNAc incorporation into lipid II and mature PG, in the presence of different concentrations of antibiotics[39,40]. Owing to the sequential nature of the glycosyltransfer and transpeptidation, it is not possible to determine at which step PG biosynthesis is blocked in this assay. The low concentration of lipid II can be detected by thin-layer chromatography (TLC) separation and subsequent phosphorimaging of pre-solubilized membranes, which was incubated with the cytoplasmic PG precursors UDP-MurNAc-pentapeptide and UDP-[^{14}C]GlcNAc[41,42]. As shown in Fig. 5 (and Supplementary Table 2 and Fig. 21), both ACL20215 and ACL20964 caused an inhibition of mature PG production at a concentration of 200 μg ml^{-1}, to a similar extent as the controls, vancomycin and moenomycin A. Both inhibitors also caused an accumulation of lipid II, although to a lesser extent compared with vancomycin or moenomycin A.

Membrane-disruption experiments were performed using *S. aureus* (ATCC 25923) in combination with membrane potential-sensitive cyanine dye diSC$_3$5 (ref. 43). Neither compounds showed membrane disruption (Supplementary

Table 1 | *In vitro* activity data of ACL20215 and ACL20964.

Organism	Strain/type	Vancomycin	Moenomycin A	ACL20215	ACL20964
MIC ($\mu g\, ml^{-1}$)					
S. aureus	MSSA, ATCC 25923	1		4	4
	MRSA, ATCC 43300	1	4	4	8
	Newbould 305			2	1
	NRS 17—GISA	8	16-32	8	32
	NRS 1—GISA	4	1	4	16
	VRS 1	>64	8	4	8
	mMRSA, DRSA, *ci*	4	16	4	8
E. faecium	ATCC 35667			16	2
	VanA, ATCC 51559	>64	32	8	>64
E. faecalis	ATCC 29212			4	8
	VanA, *ci*	>64	>64	8	64
S. pneumoniae	MDR, ATCC 700677	2	8-16	4	8-16
E. coli	ATCC 25922	>64	>64	>64	>64
Mutation frequency					
S. aureus (at 4 × MIC)	ATCC 13709			2.5×10^{-10}	
HC_{50} ($\mu g\, ml^{-1}$)					
Human	RBC			74	>100
IC_{50} ($\mu g\, ml^{-1}$)					
S. aureus	MGT			17.1	11.1

ci, clinical isolate; DRSA, daptomycin-resistant *S. aureus*; GISA, glycopeptide-intermediate *S. aureus*; HC_{50}, half maximal haemolytic concentration; IC_{50}, half maximal inhibitory concentration; MDR, multi-drug-resistant; MGT, monofunctional glycosyltranferease; MIC, minimum inhibitory concentration; mMRSA, multi-drug-resistant methicillin-resistant *S. aureus*; Moenomycin A (Sigma, 32404); MRSA, methicillin-resistant *S. aureus*; MSSA, methicillin-sensitive *S. aureus*; RBC, red blood cells; Vancomycin (Sigma, 861987); VRS, vancomycin-resistant *S. aureus*.
All values are $\mu g\, ml^{-1}$.

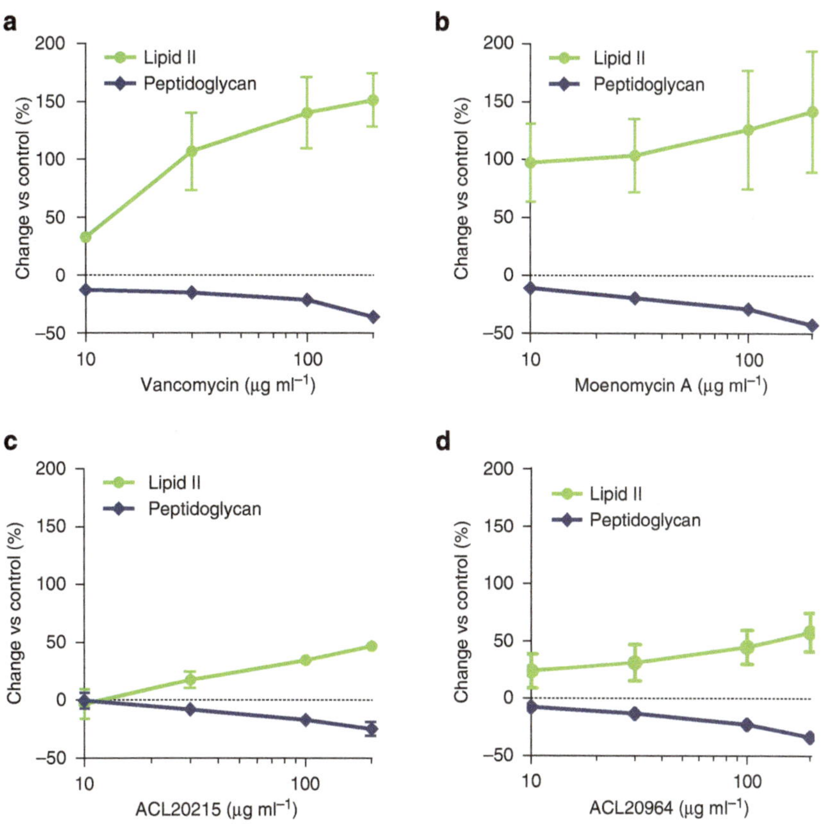

Figure 5 | Inhibition of PG synthesis. The inhibitory effect in bacterial PG biosynthesis within a crude *B. megaterium* membrane is shown for (**a**) vancomycin, (**b**) moenomycin A, (**c**) ACL20215 and (**d**) ACL20964, showing the relative change of lipid II and PG isolated from the crude membrane by TLC, after 3 h, compared with non-antibiotic treatment. Error bars show s.d. for $n = 3$.

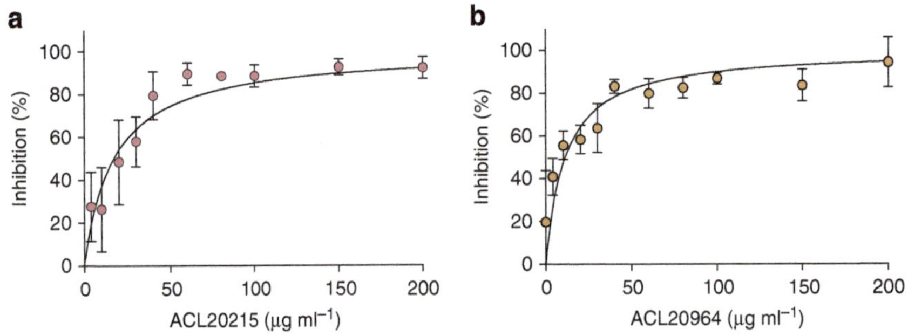

Figure 6 | Inhibition of glycosyltransferase. The inhibition of MGT from *S. aureus* is shown for ACL20215 (**a**) and ACL20964 (**b**), by measuring the transformation rate of fluorescent lipid II analogue and comparing it with the maximum inhibitory effect of moenomycin A at 50 μM or 79.2 μg ml^{-1}. Error bars show s.d. for $n = 4$.

Fig. 22) compared with a positive control Citropin 1.1 (refs 44–46), which suggested that the GT inhibitors disrupt PG biosynthesis without disrupting the cell membrane.

We then monitored the transformation of fluorescent NBD-lipid II by two different GT enzymes, PBP1 from *Clostridium difficile* and MGT from *S. aureus*. The single concentration test at 200 μg ml^{-1} revealed that both inhibitors showed an inhibitory effect against MGT *S. aureus*, while showing only moderate effect against PBP1 *C. difficile* (Supplementary Fig. 23). Confirmative dose–response assays were subsequently conducted with continuous fluorescent activity assay using a dansyl-labelled, lysine-lipid II substrate[47]. The assay revealed that both ACL20215 and ACL20964 were able to inhibit MGT from *S. aureus* with an IC_{50} of 17 and 11 μg ml^{-1}, respectively (Fig. 6). In the same assay, moenomycin A was used as positive control showing an IC_{50} of 5 μM or 8 μg ml^{-1}.

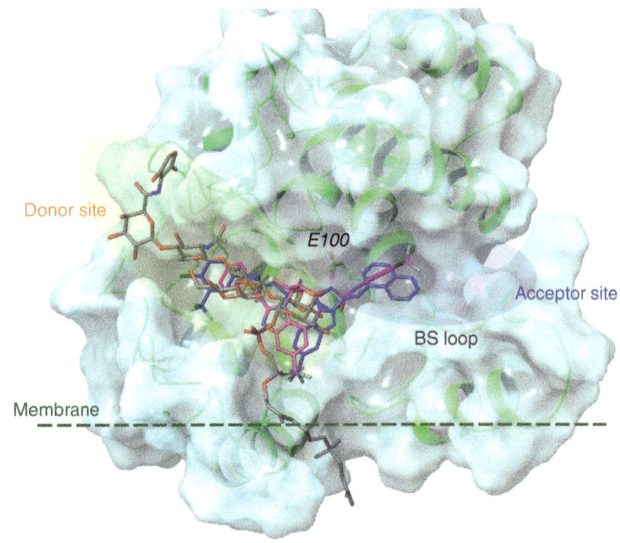

Figure 7 | Virtual docking studies of ACL20215 and ACL20215. The virtual docking poses are shown for ACL20215 (pink) and ACL20964 (blue) within the GT domain of MGT *S. aureus* (pdb: 3HZS[48]) shown as ribbon (green) and surface representation. Binding orientation of moenomycin is shown as found in the corresponding crystal structure (grey). The structure illustrates that both inhibitors are able to occupy the donor, as well as part of the acceptor sites. The dotted line is illustrating the suggested membrane interface[49,50]. BS loop marks the region of the binding site loop with high conformational variation or disorder in the different crystal structure, while E100 marks the active site residue Glu_{100}.

Virtual docking. Several crystal structures of GT domains have been reported for Gram-positive (MGT[48,49] and PBP2 (refs 50,51) from *S. aureus*) and Gram-negative bacteria (PBP1 (ref. 52) from *E. coli* and PGT[11,53] from *Aquifex aeolicus*), showing a high structural similarity between the difference species. One main feature of the structures is a binding site loop (MGT *S. aureus* Phe_{120}–Gly_{130}; PBP2 *S. aureus* Gly_{134}–Gly_{145})[54] located between the donor-binding site, occupied by moenomycin, and the acceptor binding site occupied by the incoming lipid II molecule. This binding site loop is highly flexibly and partly disordered in most of the crystal structures (see Supplementary Table 3). Even when the loop is resolved, it can occupy different conformations, either separating the donor from the acceptor sites or opening a groove between the sites (see Supplementary Fig. 24 and Supplementary Note 1). We have carried out *in silico* virtual docking with the monosaccharides ACL20215 and ACL20964, choosing the receptor model (and loop conformation), which best reproduced the binding orientation of moenomycin. A receptor model based on the crystal structure of MGT from *S. aureus* with a loop conformation blocking the access to the acceptor site, 3HZS[48], was thereby selected (see Supplementary Fig. 25). As shown in Fig. 7, the benzimidazole group of both inhibitors was located similarly to portion G of moenomycin, with one of the other hydrophobic groups located in the donor-binding site (similar to ring E of moenomycin), and the other was located towards the acceptor site not occupied by moenomycin. While the virtual docking experiments were able to reproduce the binding orientation of moenomycin, a degree of uncertainty remained due to the flexibility of the binding site loop (Fig. 7), and its ability to adopt different conformations depending on the ligand [49]. However, the docking experiments clearly indicated the potential of the inhibitors to extend to the acceptor binding site not occupied by moenomycin.

***In vivo* studies**. ACL20215 testing with *in vitro* metabolic stability assays showed no degradation of the compound using both human and mouse liver microsomes. The *in vivo* pharmacokinetic properties of ACL20215 and ACL20964 were investigated using intravenous (i.v.) administration at 3.5 mg kg^{-1} to male rats (Table 2 and Supplementary Table 4). Both compounds had a high apparent half-life ($t_{1/2}$) of 27.2 and 33.8 h, respectively. They also showed a very high volume of distribution (V_D) and a high clearance (Cl_{total}). No urinary excretion was detected for either compound, and no metabolites were observed in plasma. The maximum tolerated dose for both compounds (see Supplementary methods) was determined following intraperitoneal (i.p.) administration of the compounds to mice, and showed good tolerance up to 60 mg kg^{-1}, with no mortality up to 100 mg kg^{-1}. Both compounds induced some minor changes to central/

Table 2 | Pharmacokinetic properties and *in vivo* efficacy of ACL20215 and ACL20964.

	ACL20215	ACL20964
Pharmacokinetic properties		
Metabolic stability (*in vitro*)	No degradation	ND
Rat (i.v.)		
Dose ($mg\,kg^{-1}$)	3.5	3.5
$t_{1/2}$ (h)	27.2	33.8
Plasma Cl_{total} ($ml\,min^{-1}\,kg^{-1}$)	42.1	17.9
Blood Cl_{total} ($ml\,min^{-1}\,kg^{-1}$)	48.9	21.5
V_Z ($l\,kg^{-1}$)	97.2	53.0
Mice (i.p.)		
MTD ($mg\,kg^{-1}$)	100	100
In vivo *efficacy*		
Mouse (i.p.)		
Survival 7 days (%)	100	100
Mouse (i.v.)		
Survival 7 days (%)	10	10
Mouse (mastitis)		
$ED_{2log c.f.u.}$ (µg per gland)	730	510
$ED_{4log c.f.u.}$ (µg per gland)	1,400	770
PD_{50} (µg per gland)	>1,000	800–1,000
PD_{100} (µg per gland)	>1,000	>1,000

Cl, clearance; $ED_{2log c.f.u.}$, effective dose to reduce bacterial load by 2 × log(c.f.u.); $ED_{4log c.f.u.}$, effective dose to reduce bacterial load by 4 × log(c.f.u.); i.p., intraperitoneal injection; i.v., intravenous injection; MTD, maximal tolerated dose; ND, not determined; PD_{50}, 50% protective dose; PD_{100}, 100% protective dose; $t_{1/2}$, half-life; V_Z, volume of distribution.

autonomic responses at the higher dose of $100\,mg\,kg^{-1}$. No adverse effects were observed in a single-dose study ($4\,mg\,kg^{-1}$) following i.v. administration.

ACL20215 and ACL20964 were subsequently tested in a mouse model of septicemia, using 10 male CD-1 (*Crl.*)-derived mice, inoculated i.p. with a LD_{90-100} of *S. aureus* (Smith; 3.7×10^5 c.f.u. per mouse). Both compounds administered i.p. 10 min after inoculation at $50\,mg\,kg^{-1}$ resulted in 100% survival rate after 7 days. However, same studies with i.v. administration of the compounds ($4\,mg\,kg^{-1}$, 10 min after inoculation) showed no antibacterial effect, resulting only in a 10% survival rate, compared with 90% for ampicillin ($0.1\,mg\,kg^{-1}$). The lack of efficacy following i.v. administration is most likely due to a lower dose ($4\,mg\,kg^{-1}$, limited by solubility) combined with the high volume of distribution and serum-binding properties of the compounds, which effectively lowers the free drug concentration at the site of infection. When a higher dose ($50\,mg\,kg^{-1}$ as a suspension) was administered i.p. at the site of infection, the high local concentration of the drug ensures effective clearance of the bacterial infection. Further optimization of compound properties, dose or formulation is required for parenteral administration.

Additional *in vivo* studies were conducted with ACL20215 and ACL20964 using a mouse mammary gland infection (mastitis) model with intraductal inoculation of *S. aureus* (Newbould 305, ATCC 29740). Each compound was instilled at different doses into the teat canal of both contralateral glands from the fourth mammary gland pair of lactating mice at 4 h after bacterial inoculation. Mice were killed at 14 h post infection and both glands were analyzed for c.f.u. counts (Fig. 8, Table 2 and Supplementary Tables 5 and 6). The effective dose to reduce the bacterial load by 2 × log(c.f.u.) (ED_{2log}) were 730 and 510 µg per gland, for ACL20215 and ACL20964, respectively, indicating that both compounds cleared 99% of the staphylococci from the infected mammary gland at a moderate dose. Similarly, the protective doses to clear all bacteria in 50% (PD_{50}) and 100% (PD_{100}) of the glands, respectively, indicated that a high dose of ACL20964 very efficiently cleared *S. aureus* from the infected glands (Table 2). In contrast, no PD values could be determined for ACL20215 as the latter compound was unable to eradicate all bacteria in 50 or 100% of the glands in the analyzed doses to at least the detection limit of the assay.

Discussion

Using moenomycin A and previously reported GT inhibitors[27,28], we designed and synthesized a small library of disaccharide-based compounds with a smaller, more drug-like, hydrophobic tail[29]. These compounds (such as ACL19378) showed good *in vitro* antibacterial activity but had unfavourable physicochemical properties that limited their *in vivo* application. Nevertheless, this set of active compounds gave valuable structure–activity relationship information, which was used to design libraries of compounds based on a smaller monosaccharide scaffold[37]. This strategy also reduced compound hydrophobicity and chemical complexity, enabling the synthesis of the first reported direct GT inhibitors with *in vivo* efficacy against bacteria.

A reductionist approach, moving from disaccharide mimics of the moenomycin EFG fragment to a smaller monosaccharide scaffold, maintains the key pyranose scaffold and the substitution pattern derived from the disaccharide actives. Chemical chirality inherent in the pyranose scaffold ensures a rigid three-dimensional positioning of substituents that is maintained in the series. Second, the solid-phase synthetic method allowed us to make substantial libraries of chemotypes designed to mimic the disaccharide series. In this way, we identified a series of compounds, corresponding to the amino-benzimidazole chemotype, which showed clear antibacterial activity against a range of drug-resistant Gram-positive bacteria. While the cell-based activity suggests a preference for more hydrophobic substituents, some structural variations are not reflected in their activity, such as the difference in activity between a 2- and 1-naphthyl group and the lack of activity of the corresponding biphenyl compound.

The two most promising compounds from this monosaccharide library, that is, ACL20215 and ACL20964, showed good *in vitro* antibacterial activity against a range of Gram-positive bacteria, including those resistant to common antibiotics, that is, MRSA, GISA and VanA enterococci. PG biosynthesis assay data, taken together, suggests that both compounds trigger an accumulation of lipid II and a decrease of mature PG, as is the

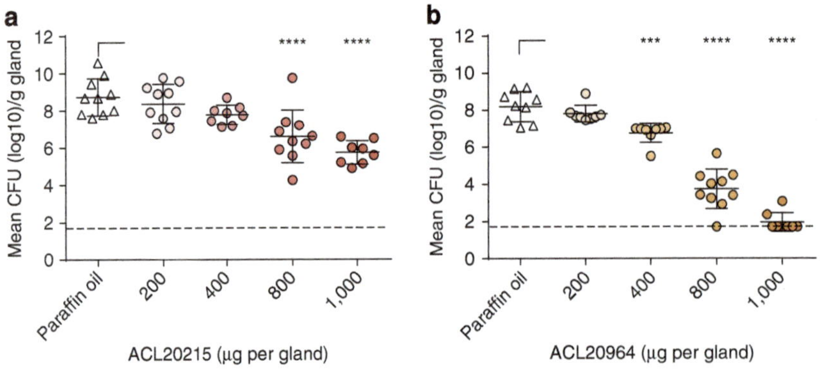

Figure 8 | *In vivo* efficacy in mastitis mouse model. *S. aureus* c.f.u. counts (c.f.u. g^{-1} gland) at 14 h post treatment in infected mouse mammary glands treated with increasing doses of ACL20215 (**a**) and ACL20964 (**b**). Significance compared with control (paraffin oil) are given as $P<0.001$ (***) and $P<0.0001$ (****), calculated by one-way analysis of variance followed by Holm-Sidak *post hoc* test. Dashed line represents the detection limit at 1.7 log_{10}c.f.u. Data values are given in Supplementary Table 5.

case for moenomycin A. Compounds can inhibit the function of GT with IC_{50} values similar to that determined for moenomycin. The virtual docking experiment suggest that the compounds are able to bind in the catalytic site of the GT by occupying part of the donor lipid II-binding site (similar to moenomycin A) as well as part of the acceptor lipid II-binding site (not occupied by moenomycin A).

Both hit compounds can be tolerated in mice up to a dose of 100 mg kg^{-1}, while showing good metabolic stability in rats. Even though the library design aimed to reduce the lipophilicity, it is apparent that GT inhibitory compounds require a certain degree of hydrophobicity to be active *in vitro* and *in vivo*. The monosaccharide scaffold is an excellent scaffold for drug design, as it is able to present various substituents or binding elements (in this case, three hydrophobic elements) in diverse spatial orientation using up to five chiral attachment points[38]. The scaffold is also able to present those substituents in a conformational rigid form, indicated by the fact that both monosaccharide compounds, ACL20215 and ACL20964, existed as two atropisomers[55,56] (see Supplementary Fig. 1 and Supplementary Note 1), conformational restricted isomers or rotamers, which would not occur if the carbohydrate scaffold itself was flexible. Virtual docking experiments show both atropisomers among the top ranked poses. It is reasonable to assume that one isomer will be the preferred binding partner for the GT active site, but our *in silico* and *in vitro* experiments were unable to distinguish them.

The membrane-associated nature of the GT enzyme and the hydrophobicity of its natural substrate lipid II necessitates a certain degree of lipophilicity for a compound with an inhibitory effect. While serum binding could not be eliminated in this pilot series, ACL20215 and ACL20964 showed *in vivo* efficacy without toxicity. GT inhibition hence remains a very attractive drug discovery target[3], as the current inhibitor moenomycin shows extremely low induction of antibiotic resistance[14,15], and also inhibits the conjugative transfer of resistance plasmids[19,57]; significant advantages given the current background of increased antimicrobial resistance.

Methods

Solid-phase synthesis. All monosaccharide compounds were synthesized on solid-phase using an orthogonally protected galactitol-building block attached to WANG resin. The synthesis of ACL20215 and ACL20964 is given in Supplementary Fig. 9 and below as a representative example.

DTPM removal: the resin was treated with a solution of 5% hydrazine hydrate in dimethylformamide (DMF; 10 ml g^{-1} of resin), shaken (1 h, RT), drained and washed (3 × DMF, 3 × DCM, 3 × DMF). Urea formation: the resin was treated with a solution of 4-chloro-3-trifluoromethyl-phenyl isocyanate (0.15 M) in DMF (10 ml g^{-1} of resin), shaken (O/N, RT), drained and washed (3 × DMF, 3 × DCM). The resin was taken up in a solution of sodium methoxide (0.15 M) in MeOH (5 ml g^{-1} of resin) and tetrahydrofuran (20 ml g^{-1} of resin), shaken (3 h, RT), drained and washed (3 × tetrahydrofuran, 3 × MeOH, 3 × DCM, 3 × DMF). Azide reduction: the resin was treated with a solution of lithium *tert*-butoxide (0.2 M) and DL-dithiothreitol (DTT, 0.2 M) in DMF (15 ml g^{-1} of resin), shaken (O/N, RT), drained and washed (3 × DMF, 3 × MeOH, 3 × DCM, 3 × DMF).

Formation of the substituted benzimidazoles: the resin was treated with a solution of 4-fluoro-3-nitro-benzotrifluoride (0.36 M) and DIPEA (0.36 M) in DMF (10 ml g^{-1} of resin), and heated at 50 °C. The resin was drained and washed with (3 × DMF, 3 × DCM, 3 × DMF). Reduction the nitro group: the resin was treated with a solution of $SnCl_2.2H_2O$ in DMF (2.0 M, 10 ml g^{-1} of resin), shaken (O/N, RT), drained and washed (3 × DMF, 3 × DMF/MeOH 1:1, 3 × DCM, 3 × DMF, 3 × DMF/MeOH 1:1, 3 × DCM).

To form the benzimidazole, the resin was treated with DIPEA (0.5 M) in DCM (10 ml g^{-1} of resin), shaken (1 h, RT), drained and washed (3 × DCM), followed by a solution of cyanogen bromide (1.0 M, 10 ml g^{-1} of resin), shaken (O/N, RT), drained and washed (3 × DCM, 3 × MeOH, 3 × DCM). To alkylate the 2-amine, the resin was treated with a solution of benzyl bromide or 1-(bromomethyl) naphthalene (0.4 M) and DIPEA (0.8 M) in DMF (10 ml g^{-1} of resin), shaken (O/N, RT), drained and washed (3 × DCM, 3 × MeOH, 3 × DCM).

Cleavage and purification: each resin was treated with 10% TFA, 20% triethylsilane in dry DCM (1.5 ml), allowed to stand at RT for 3 h, drained into a test tube and washed (3 × DCM). The concentrated samples were treated with a solution of saturated ammonia in methanol (1.0 ml) and left to stand at RT for 2 h, and concentrated by vacuum. Crude samples were purified using preparative high-performance liquid chromatography on a C-18 column (water/acetonitrile gradient).

Analytical data for ACL20215. The analytical data for ACL20125 are given as ^{1}H-NMR, temperature dependent ^{1}H-NMR, ^{13}C-NMR, COSY, edCOSY and HMBC NMR spectra in Supplementary Figs. 10–15, respectively. The structure of ACL20215 exists as two distinctive rotamers or conformational isomers that can be detected in NMR experiments. Transition between the two isomers, or atropisomers, can be achieved by heating the sample to 45 °C (see Supplementary Fig. 11). *In silico* analysis of the structure and conformation of ACL20125 indicate restricted torsional rotation of the C^4–$N^{Benzimidazole}$ bond, due to size of the benzimidazole group. Energy barrier calculation indicate an upper range of 25 kcal/mol for this rotational barrier (see Supplementary Fig. 3), which, in relation to other known atropisomers, corresponds to an interconversion rate from a few hours to a few days[56].

^{1}H-NMR (600 MHz, dimethylsulphoxide (DMSO)-d6): major rotamer δ 9.20 (br s, 1H, 7-NH), 8.10 (d, J = 8.0 Hz, 1H, H-20), 8.02 (d, J = 2.2 Hz, 1H, H-9), 7.50 (d, J = 8.9 Hz, 1H, H-13), 7.48 (dd, J = 2.2, 8.9 Hz, 1H, H-12), 7.42 (s, 1H, H-17), 7.39 (d, J = 7.3 Hz, 2H, H-25, H-29), 7.27 (dd, J = 7.3, 7.3 Hz, 2H, H-26, H-28), 7.22 (d, J = 8.0 Hz, 1H, H-19), 7.19 (dd, J = 7.3, 7.3 Hz, 1H, H-27), 7.14 (t, J = 5.8 Hz, 1H, 15-NH), 6.34 (d, J = 6.0 Hz, 1H, 2-NH), 5.41 (d, J = 5.8 Hz, 1H, 3-OH), 4.91 (m, 1H, H-4), 4.91 (t, J = 5.1 Hz, 1H, 6-OH), 4.63 (m, 2H, H-23), 4.27 (dd, J = 5.2, 11.2 Hz, 1H, H-1β), 4.00 (m, 2H, H-3, H-5), 3.87 (dddd, J = 5.2, 5.8, 11.2, 11.2 Hz, 1H, H-2), 3.40 (m, 1H, H-6a), 3.39 (m, 1H, H-1α), 3.18 (dd, J = 5.8, 5.8, 11.2 Hz, 1H, H-6b); minor rotamer δ 9.25 (br s, 1H, 7-NH), 8.10 (d, J = 2.2 Hz, 1H, H-9), 7.78 (t, J = 5.2 Hz, 1H, 15-NH), 7.54 (m, 2H, H-12, H-13), 7.54 (d, J = 7.3 Hz, 2H, H-25, H-29), 7.45 (s, 1H, H-17), 7.36 (d, J = 8.0 Hz, 1H, H-20), 7.35 (dd, J = 7.3, 7.3 Hz, 2H, H-26, H-28), 7.27 (dd, J = 7.3, 7.3 Hz, 1H, H-27), 7.21 (d, J = 8.0 Hz, 1H, H-19), 6.36 (d, J = 6.0 Hz, 1H, 2-NH), 5.29 (d, J = 6.1 Hz, 1H, 3-OH), 5.00 (m, 1H, H-4), 4.96 (t, J = 5.2 Hz, 1H, 6-OH), 4.65 (m, 1H, H-23a), 4.63 (m, 1H, H-23b), 4.24 (dddd, J = 5.2, 5.8, 11.2, 11.2 Hz, 1H, H-2), 4.16 (dd, J = 5.2, 11.2 Hz, 1H, H-1β), 4.07 (m, 2H, H-3, H-5), 3.48 (ddd, J = 5.2, 5.2, 11.2 Hz, 1H, H-6a), 3.39

(m, 1H, H-1α), 3.27 (dd, $J=5.8$, 5.8, 11.2 Hz, 1H, H-6b); ^{13}C-NMR (150 MHz, DMSO-d6): major rotamer δ 158.9 (C-15), 154.9 (C-7), 143.0 (C-16), 140.1 (C-24), 139.9 (C-8), 137.4 (C-21), 131.8 (C-12), 128.0 (C-26, C-28), 126.8 (C-25, C-29), 126.4 (C-27), 125.5 (q, 1JCF = 271 Hz, C-22), 122.9 (q, 1JCF = 273 Hz, C-14), 122.6 (q, 2JCF = 31 Hz, C-10), 122.3 (C-13), 121.5 (C-11), 121.0 (q, 2JCF = 31 Hz, C-18), 116.1 (C-9), 115.0 (C-19), 112.1 (C-20), 111.4 (C-17), 77.7 (C-5), 70.7 (C-3), 68.8 (C-1), 60.3 (C-6), 55.2 (C-4), 48.8 (C-2), 46.0 (C-23); minor rotamer δ 155.8 (C-15), 155.0 (C-7), 142.1 (C-16), 140.9 (C-21), 139.9 (C-8), 139.0 (C-24), 131.9 (C-12), 128.4 (C-26, C-28), 127.5 (C-25, C-29), 127.0 (C-27), 125.5 (q, 1JCF = 271 Hz, C-22), 122.9 (q, 1JCF = 273 Hz, C-14), 122.6 (q, 2JCF = 31 Hz, C-10), 122.3 (C-13), 121.5 (C-11), 121.0 (q, 2JCF = 31 Hz, C-18), 116.1 (C-9), 115.0 (C-19), 109.4 (C-20), 111.1 (C-17), 77.9 (C-5), 70.8 (C-3), 69.0 (C-1), 59.9 (C-6), 56.4 (C-4), 48.5 (C-2), 46.8 (C-23); HRESIMS (*m/z*): [M + H]+ calcd. for $C_{29}H_{27}Cl_1F_6N_5O_4$, 658.1650; found, 658.1659.

Analytical data for ACL20964. The analytical data for ACL20964 are given as ^{1}H-NMR, ^{13}C-NMR, COSY, edCOSY and HMBC NMR spectra in Supplementary Figs 16–20, respectively. Similar to ACL20215, ACL20964 exists as two conformational isomer, due to rotational restriction of the C^4–$N^{Benzimidazole}$ bond, caused by the large benzimidazole group.

^{1}H-NMR (600 MHz, DMSO-d6): major rotamer δ 9.26 (br s, 1H, 7-NH), 8.14 (d, $J=8.2$ Hz, 1H, H-32), 8.13 (d, $J=8.2$ Hz, 1H, H-20), 8.04 (s, 1H, H-9), 7.94 (d, $J=8.2$ Hz, 1H, H-29), 7.81 (d, $J=8.2$ Hz, 1H, H-27), 7.63 (d, $J=8.2$ Hz, 1H, C-25), 7.54 (m, 2H, H-30, H-31), 7.51 (m, 2H, H-12, H-13), 7.42 (s, 1H, H-17), 7.39 (dd, $J=8.2$, 8.2 Hz, 1H, H-26), 7.22 (d, $J=8.2$ Hz, 1H, H-19), 7.19 (t, $J=$ 5.7 Hz, 1H, 15-NH), 6.34 (br s, 1H, 2-NH), 5.47 (br s, 1H, 3-OH), 5.09 (m, 2H, H-23), 4.97 (m, 1H, H-4), 4.95 (br s, 1H, 6-OH), 4.28 (dd, $J=4.8$, 11.2 Hz, 1H, H-1β), 4.01 (m, 2H, H-3, H-5), 3.92 (m, 1H, H-2), 3.45 (m, 1H, H-6a), 3.39 (m, 1H, H-1α), 3.23 (m, 1H, H-6b); minor rotamer δ 9.26 (br s, 1H, 7-NH), 8.21 (d, $J=8.2$ Hz, 1H, H-32), 8.10 (s, 1H, H-9), 7.96 (d, $J=8.2$ Hz, 1H, H-29), 7.87 (d, $J=8.2$ Hz, 1H, H-27), 7.78 (t, $J=5.1$ Hz, 1H, 15-NH), 7.73 (d, $J=8.2$ Hz, 1H, C-25), 7.60 (dd, $J=8.2$, 8.2 Hz, H-31), 7.55 (m, 2H, H-12, H-13), 7.54 (m, 1H, H-30), 7.47 (s, 1H, H-17), 7.50 (m, 1H, H-26), 7.37 (d, $J=8.2$ Hz, 1H, H-20), 7.22 (d, $J=8.2$ Hz, 1H, H-19), 6.36 (br s, 1H, 2-NH), 5.31 (br s, 1H, 3-OH), 5.16 (dd, $J=5.6$, 15.3 Hz, 1H, H-23a), 5.11 (dd, $J=4.8$, 15.3 Hz, 1H, H-23b), 5.01 (m, 1H, H-4), 4.95 (br s, 1H, 6-OH), 4.24 (m, 1H, H-2), 4.16 (dd, $J=5.2$, 11.2 Hz, 1H, H-1β), 4.05 (m, 1H, H-3), 4.01 (m, 1H, H-5),3.48 (m, 1H, H-6a), 3.39 (m, 1H, H-1α), 3.27 (m, 1H, H-6b); ^{13}C-NMR (150 MHz, DMSO-d6): major rotamer δ 158.7 (C-15), 154.9 (C-7), 143.1 (C-16), 140.0 (C-8), 137.5 (C-21), 134.9 (C-33), 134.2 (C-28), 131.9 (C-12), 130.9 (C-24), 128.5 (C-29), 127.0 (C-27), 126.6 (q, 2JCF = 31 Hz, C-10), 126.1 (C-31), 125.6 (C-30), 125.5 (C-26), 125.4 (q, 1JCF = 271 Hz, C-22), 124.6 (C-25), 123.4 (C-32), 122.8 (q, 1JCF = 273 Hz, C-14), 122.3 (C-13), 121.5 (C-11), 121.0 (q, 2JCF = 31 Hz, C-18), 116.1 (C-9), 115.0 (C-19), 112.1 (C-20), 111.4 (C-17), 77.6 (C-5), 70.7 (C-3), 68.8 (C-1), 60.3 (C-6), 55.1 (C-4), 49.0 (C-2), 44.3 (C-23); minor rotamer δ 155.8 (C-15), 154.9 (C-7), 142.1 (C-16), 140.9 (C-21), 140.0 (C-8), 134.2 (C-33), 133.2 (C-28), 131.8 (C-12), 130.9 (C-24), 128.6 (C-29), 127.7 (C-27), 126.6 (q, 2JCF = 31 Hz, C-10), 126.5 (C-31), 125.8 (C-30), 125.6 (C-26), 125.4 (q, 1JCF = 271 Hz, C-22), 125.4 (C-25), 123.4 (C-32), 122.8 (q, 1JCF = 273 Hz, C-14), 122.3 (C-13), 121.5 (C-11), 121.0 (q, 2JCF = 31 Hz, C-18), 116.1 (C-9), 115.0 (C-19), 111.1 (C-17), 109.5 (C-20), 77.8 (C-5), 70.9 (C-3), 68.5 (C-1), 59.8 (C-6), 56.4 (C-4), 48.5 (C-2), 44.7 (C-23); HRESIMS (*m/z*): [M + H]+ calcd. for $C_{33}H_{29}Cl_1F_6N_5O_4$, 708.1807; found, 708.1797.

Virtual docking. The virtual docking of inhibitor and moenomycin structures into the binding site of MGT from *S. aureus* were done using the software Glide (version 6.5, Schrödinger, LLC, New York, 2014 (ref. 58)) and using several different crystal structures with different binding site loop conformations. In addition, the induced fit protocol in the Schrödinger software package (Induced Fit Docking protocol 2014-4, Glide version 6.5, Prime version 3.7, Schrödinger, LLC[59,60]) was used to allow conformational flexibility of the GT domain. The docking experiments were all done using the standard precision (SP) in Glide and defining a binding site with 13 Å around the crystal structure of moenomycin, large enough to include also the acceptor site which is not occupied by moenomycin. For the structure of moenomycin, a truncated version was used, without the fatty acid chain and only a lactic acid attached to the phosphate group. Models for inhibitors were built in four different conformations, $^4C_1$ and $_1C^4$ chair conformation for the galactose moiety, and two different orientations for the benzimidazole moiety compared with the galactose ring. All structure models were built using Maestro (version 10.0, Schrödinger, LLC, New York, 2014).

The different approaches, that is, crystal structures and protocols, were validated and the best one selected by comparing the docking orientation of moenomycin with the one found in the crystal structures, by visual inspection. The following models for MGT from *S aureus* have been used for the docking experiments: 3HZSm–3HZS[48] changing to wild type, by mutating Asn100 to Asp, 3VMRm–3VMR[49] modelling the missing loop residues (2 residues) with loop search, 3NB6sa-Homology model of MGT *S. aureus* sequence using 3NB6 (ref. 11) (PGT from *A. aeolicus*) with complete loop as template, 3VMSm–3VMS[49] modelling missing loop (7 residues) using the loop in 3NB6 (ref. 11) as template.

From these receptor models, 3HZSm was able to reproduce the binding orientation of moenomycin A as found in the crystal structure (see Supplementary Fig. 25; a, green: crystal structure; grey; docked structure). The 3VMRm model produced similar orientations for moenomycin but with different orientation of the D ring. In 3NB6sa model, moenomycin was binding in the same binding pocket but its orientation was always different, with phosphoglycerate (G) portion of the molecule facing either to the solvent, the donor or acceptor binding site. In the 3VMSm model, which has the most open groove between the donor and acceptor site, moenomycin was actually oriented partly across the acceptor binding site. Induced fit protocol on those models did not produce better moenomycin binding orientations compared with the standard docking protocol.

***In vitro* PG biosynthesis.** The cell-free particulate fraction of *B. megaterium* KM (ATCC13632), capable of catalysing the polymerization of PG from UPD-linked precursors was performed as described previously[61]. *B. megaterium* was grown in standard medium, harvested and washed with Tris-buffer by centrifugation. Resuspended bacteria were subjected to three freeze/thaw cycles (5 min dry ice, followed by 10 min at RT), homogenized by the glass homogenizer and centrifuged all at 4 °C, leaving most of the cell wall in the pellet. Resuspended pellet was combined with UDP-*N*-acetylmuramyl-pentapeptide, [^{14}C]UDP-*N*-acetylglucosamine and individual compounds or antibiotics (that is, vancomycin hydrochloride or moenomycin A), and incubated at for 3 h at RT, placed in a boiling water bath for 3 min to inactivate enzymes and to prevent any further lipid II transformation, and analysed by TLC on silica gel plates. After separation, plates were dried, exposed to phosphorimaging screen (1 week), scanned by Typhoon 8600 calculating the integrated density value of each band on silica gel. Changes of PG or lipid II were calculated as a percentage from negative control (for more details see Supplementary Methods).

Inhibition of glycosyltransferase. Inhibition of glycosyltransferase was measured using a fluorescence detection method[62] by adding 1 μM *S. aureus* MGT to 1.45 μM fluorescent dansyl-Lys Lipid II and different concentration of inhibitors, all in a buffer of 50 mM Tris pH 8 containing 10 mM $MnCl_2$, 0.08% (v/v) decyl PEG, 10% (v/v) DMSO, 100 μg ml^{-1} hen egg-white lysozyme, in 96-well microtiter plates. Initial rates were measured as a decrease in fluorescence (ex/em: 340/521 nm) and calculated as a percentage compared with no inhibitor as negative control (0% of rate inhibition) and moenomycin A (Sigma, Cat. no.: 32404) as positive control (100% of rate inhibition). The data were fitted to a simple saturation model of inhibitor binding to a single site, from which IC_{50} values were extracted. Moenomycin A showed an IC_{50} of 5 μM in this assay.

Minimal inhibitory concentration (MIC) determination. The compounds along with standard antibiotics were serially diluted twofold across the wells of 96-well standard polystyrene non-treated plates (Corning 3370). Compounds and standard antibiotic controls ranged from 1.28 mg ml^{-1} to 0.06 μg ml^{-1} with final volumes of 50 μl per well. Bacteria were cultured in Brain–Heart Infusion (Bacto laboratories, Cat. no. CM1135B) at 37 °C overnight. A sample of each culture was then diluted 40-fold in fresh brain–heart infusion broth and incubated at 37 °C for 2–3 h. The resultant mid-log phase cultures were diluted to 5×10^5 c.f.u. ml^{-1} then 50 μl was added to each well of the compound-containing 96-well plates giving a final compound concentration range of 64 μg ml^{-1} to 0.03 μg ml^{-1} in 2.5×10^5 c.f.u. ml^{-1}. All the plates were covered and incubated at 37 °C for 24 h. MICs were determined visually as the lowest concentration showing no visible growth.

Antibiotic control compound vancomycin (Sigma, Cat. no.: 861987) was prepared as water solution, while moenomycin A (Sigma, Cat. no.: 32404) was dissolved in DMSO and 20 mM ammonium acetate, due to solubility issues. The average MIC for moenomycin for *S. aureus* was, however, within the range of recent literature[35].

***In vivo* mouse mammary gland infection model.** For the infection of the mice, *S. aureus* Newbould 305 (ATCC 29740) isolated from a clinical mastitis cases[63] was used and prepared. The procedure for mouse mammary gland infection has been recently described[64]. CD-1 lactating mice were utilized 12–14 days after giving birth, with pups weaned 1–2 h before bacterial inoculation of the mammary glands. Inoculation of both left (L4) and right (R4) glands of the fourth abdominal mammary gland pair of anesthetized mice with 150 c.f.u. of *S. aureus*, was done using 32-gauge syringes (blunt needle). The antimicrobial formulation was instilled into the mammary gland of anesthetized mice at 4 h after bacterial inoculation, followed by i.p. administration of postoperative analgesic Buprecare. Mice were killed 14 h post treatment, mammary glands (two per mouse) were harvested, weighed and homogenized. Bacterial c.f.u. counts were obtained after quantification of serial logarithmic dilutions of mammary gland homogenates on TSA. The detection limit (DL) was 1.7 $\log_{10}$ c.f.u. g^{-1} gland weight (for more details see Supplementary Methods). The animal experiments were approved by the Ethical Committee of the Faculty of Veterinary Medicine, Ghent University (EC2009/133).

References

1. Derouaux, A., Sauvage, E. & Terrak, M. Peptidoglycan glycosyltransferase substrate mimics as templates for the design of new antibacterial drugs. *Front. Immunol.* **4**, 78 (2013).

2. Bouhss, A., Trunkfield, A. E., Bugg, T. D. & Mengin-Lecreulx, D. The biosynthesis of peptidoglycan lipid-linked intermediates. *FEMS Microbiol. Rev.* **32,** 208–233 (2008).
3. Galley, N. F., O'Reilly, A. M. & Roper, D. I. Prospects for novel inhibitors of peptidoglycan transglycosylases. *Bioorg. Chem.* **55,** 16–26 (2014).
4. Goffin, C. & Ghuysen, J. M. Multimodular penicillin-binding proteins: an enigmatic family of orthologs and paralogs. *Microbiol. Mol. Biol. Rev.* **62,** 1079–1093 (1998).
5. Coutinho, P. M., Deleury, E., Davies, G. J. & Henrissat, B. An evolving hierarchical family classification for glycosyltransferases. *J. Mol. Biol.* **328,** 307–317 (2003).
6. Ling, L. L. *et al.* A new antibiotic kills pathogens without detectable resistance. *Nature* **517,** 455–459 (2015).
7. Arias, C. A. & Murray, B. E. The rise of the *Enterococcus*: beyond vancomycin resistance. *Nat. Rev. Microbiol.* **10,** 266–278 (2012).
8. Gualerzi, C. O., Brandi, L., Fabbretti, A. & Pon, C. L. *Antibiotics: Targets, mechanisms and resistance* (John Wiley & Sons, 2013).
9. Gharsallaoui, A., Oulahal, N., Joly, C. & Degraeve, P. Nisin as a food preservative: part 1: physicochemical properties, antimicrobial activity, and main uses. *Crit. Rev. Food Sci. Nutr.* http://www.tandfonline.com/doi/abs/10.1080/10408398.2013.763766 (2015).
10. Ostash, B. & Walker, S. Moenomycin family antibiotics: chemical synthesis, biosynthesis, and biological activity. *Nat. Prod. Rep.* **27,** 1594–1617 (2010).
11. Fuse, S. *et al.* Functional and structural analysis of a key region of the cell wall inhibitor moenomycin. *ACS Chem. Biol.* **5,** 701–711 (2010).
12. O'Brien, T. F. *et al.* Facts about antibiotics in animals and their impact on resistance. *Clin. Infect. Dis.* **34,** S71–S144 (2002).
13. Silbergeld, E. K., Graham, J. & Price, L. B. Industrial food animal production, antimicrobial resistance, and human health. *Annu. Rev. Public Health* **29,** 151–169 (2008).
14. Butaye, P., Devriese, L. A. & Haesebrouck, F. Differences in antibiotic resistance patterns of *Enterococcus faecalis* and *Enterococcus faecium* strains isolated from farm and pet animals. *Antimicrob. Agents Chemother.* **45,** 1374–1378 (2001).
15. Hentschel, S., Kusch, D. & Sinell, H. J. *Staphylococcus aureus* in poultry—biochemical characteristics, antibiotic resistance and phage pattern. *Zentralbl. Bakteriol. [B]* **168,** 546–561 (1979).
16. Rebets, Y. *et al.* Moenomycin resistance mutations in *Staphylococcus aureus* reduce peptidoglycan chain length and cause aberrant cell division. *ACS Chem. Biol.* **9,** 459–467 (2014).
17. Huber, G. Mechanism of action of antibacterial agents. Moenomycin and related phosphorus-containing antibiotics. *Antibiotics* **5,** 135–153 (1979).
18. Butaye, P., Devriese, L. A. & Haesebrouck, F. Antimicrobial growth promoters used in animal feed: effects of less well known antibiotics on gram-positive bacteria. *Clin. Microbiol. Rev.* **16,** 175–188 (2003).
19. Pfaller, M. A. Flavophospholipol use in animals: positive implications for antimicrobial resistance based on its microbiologic properties. *Diagn. Microbiol. Infect. Dis.* **56,** 115–121 (2006).
20. Makitrynskyy, R. *et al.* Genetic factors that influence moenomycin production in streptomycetes. *J. Ind. Microbiol. Biotechnol.* **37,** 559–566 (2010).
21. Ostash, B. *et al.* Complete characterization of the seventeen step moenomycin biosynthetic pathway. *Biochemistry* **48,** 8830–8841 (2009).
22. Ostash, B., Doud, E. & Fedorenko, V. The molecular biology of moenomycins: towards novel antibiotics based on inhibition of bacterial peptidoglycan glycosyltransferases. *Biol. Chem.* **391,** 499–504 (2010).
23. Van Heijenoort, J. Formation of the glycan chains in the synthesis of bacterial peptidoglycan. *Glycobiology* **11,** 25R–36R (2001).
24. Schneider, T. & Sahl, H. G. An oldie but a goodie - cell wall biosynthesis as antibiotic target pathway. *Int. J. Med. Microbiol.* **300,** 161–169 (2010).
25. Welzel, P. *et al.* Moenomycin A: minimum structural requirements for biological activity. *Tetrahedron* **43,** 585–598 (1987).
26. El-Abadla, N. *et al.* Moenomycin A: the role of the methyl group in the moenuronamide unit and a general discussion of structure-activity relationships. *Tetrahedron* **55,** 699–722 (1999).
27. Sofia, M. J. *et al.* Discovery of novel disaccharide antibacterial agents using a combinatorial library approach. *J. Med. Chem.* **42,** 3193–3198 (1999).
28. Baizman, E. R. *et al.* Antibacterial activity of synthetic analogs based on the disaccharide structure of moenomycin, an inhibitor of bacterial transglycosylase. *Microbiology* **146,** 3129–3140 (2000).
29. Halliday, J., McKeveney, D., Muldoon, C., Rajaratnam, P. & Meutermans, W. Targeting the forgotten transglycosylases. *Biochem. Pharmacol.* **71,** 957–967 (2006).
30. Shih, H. W. *et al.* Effect of the peptide moiety of lipid II on bacterial transglycosylase. *Angew. Chem. Int. Ed.* **51,** 10123–10126 (2012).
31. Kuhn, H. *et al.* Anti-bacterial glycosyl triazoles - Identification of an N-acetylglucosamine derivative with bacteriostatic activity against *Bacillus. Med. Chem. Commun.* **5,** 1213–1217 (2014).
32. Dumbre, S. *et al.* Synthesis of modified peptidoglycan precursor analogues for the inhibition of glycosyltransferase. *J. Am. Chem. Soc.* **134,** 9343–9351 (2012).
33. Cheng, T.-J. R. *et al.* High-throughput identification of antibacterials against methicillin-resistant *Staphylococcus aureus* (MRSA) and the transglycosylase. *Bioorg. Med. Chem.* **18,** 8512–8529 (2010).
34. Huang, S. H. *et al.* New continuous fluorometric assay for bacterial transglycosylase using Forster resonance energy transfer. *J. Am. Chem. Soc.* **135,** 17078–17089 (2013).
35. Gampe, C. M., Tsukamoto, H., Doud, E. H., Walker, S. & Kahne, D. E. Tuning the moenomycin pharmacophore to enable discovery of bacterial cell wall synthesis inhibitors. *J. Am. Chem. Soc.* **135,** 3776–3779 (2013).
36. Wang, Y. *et al.* Structure-based design, synthesis, and biological evaluation of isatin derivatives as potential glycosyltransferase inhibitors. *Chem. Biol. Drug Des.* **84,** 685–696 (2014).
37. Abbenante, G. *et al.* Biological diversity from a structurally diverse library: systematically scanning conformational space using a pyranose scaffold. *J. Med. Chem.* **53,** 5576–5586 (2010).
38. Le, G. T. *et al.* Molecular diversity through sugar scaffolds. *Drug Discov. Today* **8,** 701–709 (2003).
39. Ge, M. *et al.* Vancomycin derivatives that inhibit peptidoglycan biosynthesis without binding D-Ala-D-Ala. *Science* **284,** 507–511 (1999).
40. Lugtenberg, E. J. J., Van Schijndel-Van Dam, A. & Van Bellegem, T. H. M. *In vivo* and *in vitro* action of new antibiotics interfering with the utilization of N-acetylglucosamine-N-acetylmuramyl-pentapeptide. *J. Bacteriol.* **108,** 20–29 (1971).
41. Somner, E. A. & Reynolds, P. E. Inhibition of peptidoglycan biosynthesis by ramoplanin. *Antimicrob. Agents Chemother.* **34,** 413–419 (1990).
42. Ruzin, A. *et al.* Further evidence that a cell wall precursor [C(55)-MurNAc-(peptide)-GlcNAc] serves as an acceptor in a sorting reaction. *J. Bacteriol.* **184,** 2141–2147 (2002).
43. Wu, M., Maier, E., Benz, R. & Hancock, R. E. Mechanism of interaction of different classes of cationic antimicrobial peptides with planar bilayers and with the cytoplasmic membrane of *Escherichia coli. Biochemistry* **38,** 7235–7242 (1999).
44. Chia, C. S., Gong, Y., Bowie, J. H., Zuegg, J. & Cooper, M. A. Membrane binding and perturbation studies of the antimicrobial peptides caerin, citropin, and maculatin. *Biopolymers* **96,** 147–157 (2011).
45. Giacometti, A. *et al. In vitro* activity and killing effect of citropin 1.1 against Gram-positive pathogens causing skin and soft tissue infections. *Antimicrob. Agents Chemother.* **49,** 2507–2509 (2005).
46. Wegener, K. L. *et al.* Host defence peptides from the skin glands of the Australian blue mountains tree-frog *Litoria citropa*. Solution structure of the antibacterial peptide citropin 1.1. *Eur. J. Biochem.* **265,** 627–637 (1999).
47. Lloyd, A. J. *et al.* Characterization of tRNA-dependent peptide bond formation by MurM in the synthesis of *Streptococcus pneumoniae* peptidoglycan. *J. Biol. Chem.* **283,** 6402–6417 (2008).
48. Heaslet, H., Shaw, B., Mistry, A. & Miller, A. A. Characterization of the active site of *S. aureus* monofunctional glycosyltransferase (Mtg) by site-directed mutation and structural analysis of the protein complexed with moenomycin. *J. Struct. Biol.* **167,** 129–135 (2009).
49. Huang, C.-Y. *et al.* Crystal structure of *Staphylococcus aureus* transglycosylase in complex with a lipid II analog and elucidation of peptidoglycan synthesis mechanism. *Proc. Natl Acad. Sci USA* **109,** 6496–6501 (2012).
50. Lovering, A. L., de Castro, L. H., Lim, D. & Strynadka, N. C. J. Structural insight into the transglycosylation step of bacterial cell-wall biosynthesis. *Science* **315,** 1402–1405 (2007).
51. Lovering, A. L., De Castro, L. & Strynadka, N. C. Identification of dynamic structural motifs involved in peptidoglycan glycosyltransfer. *J. Mol. Biol.* **383,** 167–177 (2008).
52. Sung, M.-T. *et al.* Crystal structure of the membrane-bound bifunctional transglycosylase PBP1b from *Escherichia coli. Proc. Natl Acad. Sci. USA* **106,** 8824–8829 (2009).
53. Yuan, Y. *et al.* Crystal structure of a peptidoglycan glycosyltransferase suggests a model for processive glycan chain synthesis. *Proc. Natl Acad. Sci. USA* **104,** 5348–5353 (2007).
54. Zuegg, J. & Meutermans, W. Crystal structures of the PBP2 glycosyltransferase domain: new opportunities for antibacterial drug design. *ChemMedChem* **2,** 1403–1404 (2007).
55. Oki, M. in *Top. Stereochem.* (eds Allinger, N. L., Eliel, N. L. & Wilen, S. H.) Vol. 14, 1–81 (John Wiley & Sons, 1983).
56. Laplante, S. R. *et al.* Assessing atropisomer axial chirality in drug discovery and development. *J. Med. Chem.* **54,** 7005–7022 (2011).
57. Riedl, S., Ohlsen, K., Werner, G., Witte, W. & Hacker, J. Impact of flavophospholipol and vancomycin on conjugational transfer of vancomycin resistance plasmids. *Antimicrob. Agents Chemother.* **44,** 3189–3192 (2000).

58. Friesner, R. A. *et al.* Glide: a new approach for rapid, accurate docking and scoring. 1. Method and assessment of docking accuracy. *J. Med. Chem.* **47,** 1739–1749 (2004).
59. Sherman, W., Beard, H. S. & Farid, R. Use of an induced fit receptor structure in virtual screening. *Chem. Biol. Drug Des.* **67,** 83–84 (2006).
60. Sherman, W., Day, T., Jacobson, M. P., Friesner, R. A. & Farid, R. Novel procedure for modeling ligand/receptor induced fit effects. *J. Med. Chem.* **49,** 534–553 (2006).
61. Reynolds, P. E. Peptidoglycan synthesis in bacilli. I. Effect of temperature on the in vitro system from *Bacillus megaterium* and *Bacillus stearothermophilus. Biochim. Biophys. Acta* **237,** 239–254 (1971).
62. Schwartz, B., Markwalder, J. A., Seitz, S. P., Wang, Y. & Stein, R. L. A kinetic characterization of the glycosyltransferase activity of *Escherichia coli* PBP1b and development of a continuous fluorescence assay. *Biochemistry* **41,** 12552–12561 (2002).
63. Prasad, L. B. & Newbould, F. H. Inoculation of the bovine teat duct with *Staphylococcus aureus*: the relationship of teat duct length, milk yield and milking rate to development of intramammary infection. *Can. Vet. J.* **9,** 107–115 (1968).
64. Demon, D. *et al.* The intramammary efficacy of first generation cephalosporins against *Staphylococcus aureus* mastitis in mice. *Vet. Microbiol.* **160,** 141–150 (2012).

Acknowledgements

This work was supported by Alchemia Ltd, Bayer A.H. (for mouse *S. aureus* mastitis study), NHMRC Australia Fellowship AF 511105 to M.A.C., BBSRC PhD studentship to N.F.G. and MRC research grants G1100127, G500643 and G0600801 as well as Wellcome Trust equipment grants 071998 and 068598 to D.I.R. and C.G.W. A.J.L. was supported by the Birmingham-Warwick Science City initiative, and T.-J.C. and W.-C.C. were supported by Academia Sinica and Ministry of Science and Technology. We thank David L. Paterson (UQCCR, University of Queensland, Brisbane, Australia) for his kind donation of clinical Gram-positive isolates, as well as Professor Joan Faoagali and Dr Narelle George from the Queensland Health Pathology and Scientific Services (QHPS, Queensland Health, Brisbane, Australia) for the MIC testing of actives against a panel of bacterial strains including clinical isolates.

Author contributions

C.M., G.A., D.M., G.L.T., R.P., B.B., L.S. and J.S. developed the chemistry and performed the synthesis, with J.Z. contributing to the design of the library. M.C., A.G.E., J.X.H., M.S.B., T.-J.C., W.-C.C., N.F.G., D.I.R., A.J.L. and C.G.D. developed, performed or supervised the various biochemical assays including analytical methods. D.D. and E.M. performed the *in vivo* assays. J.Z. and M.B. performed the *in silico* experiments. J.Z., C.M. and M.S.B. contributed to the writing of the manuscript. W.M. and M.A.C. supervised the studies, with W.M. establishing the project.

Additional information

Competing financial interests: The authors declare no competing financial interests.

17

Sugary interfaces mitigate contact damage where stiff meets soft

Hee Young Yoo[1,*], Mihaela Iordachescu[1,*], Jun Huang[2], Elise Hennebert[3,4], Sangsik Kim[5], Sangchul Rho[6], Mathias Foo[7], Patrick Flammang[3], Hongbo Zeng[2], Daehee Hwang[6,8], J. Herbert Waite[9] & Dong Soo Hwang[1,5]

The byssal threads of the fan shell *Atrina pectinata* are non-living functional materials intimately associated with living tissue, which provide an intriguing paradigm of bionic interface for robust load-bearing device. An interfacial load-bearing protein (*A. pectinata* foot protein-1, apfp-1) with L-3,4-dihydroxyphenylalanine (DOPA)-containing and mannose-binding domains has been characterized from *Atrina*'s foot. apfp-1 was localized at the interface between stiff byssus and the soft tissue by immunochemical staining and confocal Raman imaging, implying that apfp-1 is an interfacial linker between the byssus and soft tissue, that is, the DOPA-containing domain interacts with itself and other byssal proteins via $Fe3^{+}$-DOPA complexes, and the mannose-binding domain interacts with the soft tissue and cell membranes. Both DOPA- and sugar-mediated bindings are reversible and robust under wet conditions. This work shows the combination of DOPA and sugar chemistry at asymmetric interfaces is unprecedented and highly relevant to bionic interface design for tissue engineering and bionic devices.

[1] Division of Integrative Biosciences and Biotechnology, Pohang University of Science and Technology, Pohang 37673, Republic of Korea. [2] Department of Chemical and Materials Engineering, University of Alberta, Edmonton, Alberta, Canada T6G 2V4. [3] Laboratory of Biology of Marine Organisms and Biomimetics, Research Institute for Biosciences, University of Mons, 7000 Mons, Belgium. [4] Laboratory of Cell Biology, Research Institute for Biosciences, University of Mons, 7000 Mons, Belgium. [5] School of Environmental Science and Engineering, Pohang University of Science and Technology, Pohang 37673, Republic of Korea. [6] Center for Plant Aging Research, Institute for Basic Science (IBS), Daegu 42988, Republic of Korea. [7] School of Engineering, University of Warwick, Coventry CV4 7AL, UK. [8] Department of New Biology, DGIST, Daegu 42988, Republic of Korea. [9] Department of Molecular, Cellular, and Developmental Biology, University of California, Santa Barbara, California 93106, USA. * These authors contributed equally to this work. Correspondence and requests for materials should be addressed to D.S.H. (email: dshwang@postech.ac.kr).

Achieving robust attachment between living tissues and inert materials faces the challenge of mitigating contact damage due to mismatches in critical properties such as stiffness, hardness, strength and Poisson's ratio[1,2]. Contact damage mitigation is especially critical for durable loadbearing prosthetic or implant devices. Prosthetic devices for amputees, for example, are of tremendous benefit to mankind, but contact damage such as inflammation, rashes, blistering and numbness can limit their usefulness to the fullest potential[3]. There is growing belief that the best engineered interfaces are those inspired by nature (bionic), such as between bone, and muscle[4,5], nail and skin[6], and more recently, squid beak and muscle[7]. These natural interfaces are robust, dynamic and durable—all desirable properties for bionic devices—but on their transferability to engineer bionic devices have yet to be fully understood[8]. In particular, the molecular level characterization of the interface remains elusive.

From a mechanical perspective, there appear to be two strategies in biology for mitigating contact damage between mismatched loadbearing materials: (A) using a strong interfacial adhesion over an increased surface area of contact (thereby reducing the load per unit area), or (B) creating an interface that relies on molecular gradients. The former, suitable for bulky materials, is the strategy employed in the tendon–muscle junction[4,5].

In the latter, the stiffer material is gradually ramped down to the more compliant one by decreasing the ratio of hard to soft blocks in naturally occurring block copolymers, such as in *Mytilus* byssus and polychaete jaws[1,2,7,9]. The butt-joint model[10] predicts that, as the stiffness of the two contacting materials begins to coincide, the radial stress goes to zero. Contributions of the Poisson ratio, another factor in this model but difficult to measure accurately in small samples, are usually ignored.

Our work on the fan shell mussel, *Atrina pectinata* L., sheds some light on the molecular interface between living tissue and load-bearing non-living material. *Atrina* uses a byssus to anchor itself to available hard objects in the benthic sediment[11,12]. About a proximal third of each byssal thread is rooted in the soft tissue within the shell (Fig. 1a), while the rest is exposed outside the shell for attachment to pebbles or detritus. The dimensions of the exposed thread are about 25 cm in length and 25 µm in diameter[11]. Interestingly, the rooted thread portion originates in the byssal adductor muscle and proceeds from there along the byssal groove till it emerges from the living tissue and the shell (Fig. 1a,b). Water flow over the exposed shell results in lift and drag forces that tug on the attached threads. The threads transfer and dissipate the loads to the interconnecting tissue without incurring any damage[12–15]. In this work we report insights at a molecular level about the interface between tissue and embedded byssal threads and how this affects tenacity, toughness, and robustness of a bionic holdfast. The characterized *Atrina*'s foot protein has L-3,4-dihydroxyphenylalanine (DOPA) containing domain and sugar-binding domain that is localized at the interface between tissue and embedded byssal thread. Our results show that the combination of this DOPA-containing and sugar-binding domain plays a significant role in achieving robust load-bearing device and this unprecedented finding has high relevance to bionic interface design for tissue engineering and bionic devices.

Results

Mechanical mismatch between living and non-living material. Previous studies on byssal threads in the genus *Mytilus* invoked molecular gradients along the axis of each thread to produce a stiffness gradient to moderate the stiffness mismatch between the threads and where these threads meet in the stem before entering the living tissue[13]. Specifically, two hybrid collagens in each thread, preCOL-D (collagen + silk domains) and preCOL-P (collagen + elastin domains), are self-assembled in such a way that the stiffer preCOL-D predominates distally, whereas the more compliant preCOL-P prevails approaching the living tissue[13–15]. There is no evidence for molecular gradients in *Atrina* threads; each hydrated thread has a uniform stiffness that is ~100 times stiffer than the surrounding tissue (Supplementary Table 1; Supplementary Discussion)[16]. The nano-indentation measurements have also confirmed the absence of this molecular gradient with its cuticle and core having indistinguishable mechanical property (Supplementary Table 2). As such, this

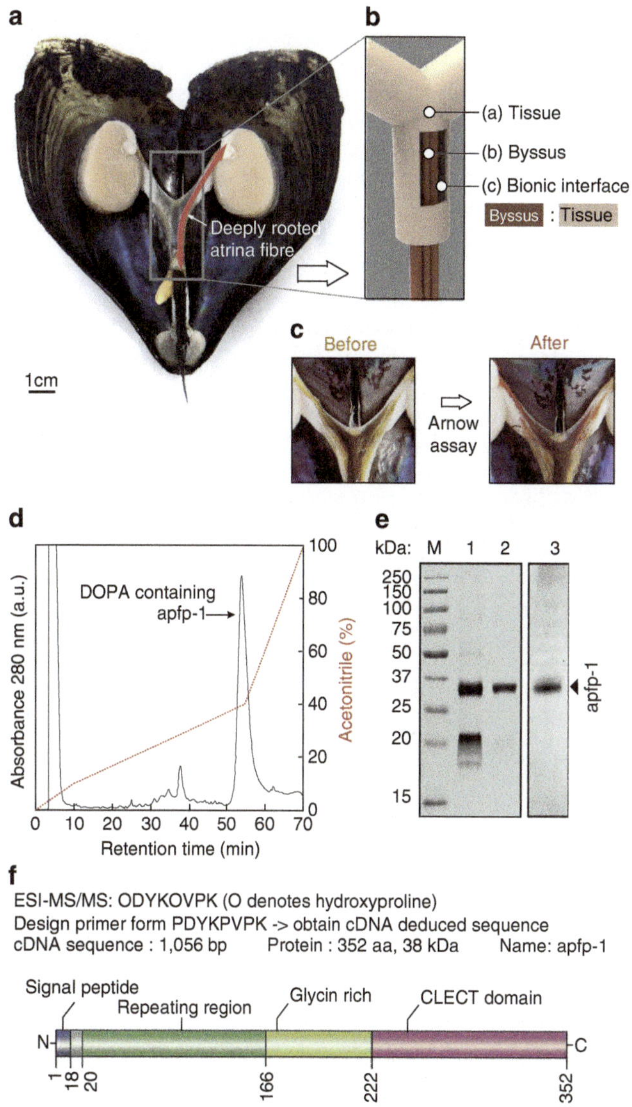

Figure 1 | Characterization of *Atrina pectinata* foot protein-1 (apfp-1). (**a**) Opened fan shell showing the full body after removal of the buccal mass to highlight the deeply rooted byssus through the body. (**b**) Schematic illustration of the dissected byssus wrapping tissue (**c**) Byssus in **b** after staining for DOPA-containing proteins with a catechol-specific reagent. (**d**) C8 HPLC chromatographic separation of apfp-1. (**e**) Gel electrophoresis purification of apfp-1. M: Molecular mass markers, 1: fractions corresponding to GPC major peak; 2: HPLC major peak obtained after injection of the GPC main peak, 3: Catechol-specific staining of apfp-1. (**f**) Schematic representation of apfp-1. Functional domains are drawn to scale at the location within the full protein sequence.

stiffness mismatch almost certainly foreshadows contact damage where thread meets tissue, unless geometry and the large interfacial contact area are designed to dissipate energy efficiently. The *Atrina* byssus meets living tissue in a joint that is very different in geometry from the molecular and mechanical gradients[13–15,17] in *Mytilus* threads. *Atrina* threads exploit the high surface area associated with embedding nearly 10 cm of each thread in the byssal groove. To this, *Atrina* adds numerous sacrificial, strong and reversible lectin-type interactions across the thread–tissue interface to dissipate energy during load transfer.

DOPA as a key adhesive component of the byssus. The chemistry of *Mytilus* byssus has been well studied[18]. A key adhesive signature of the byssus is DOPA, a catecholic amino acid that is post translationally modified from tyrosine[18–20], that is also present in *Atrina* byssus (Supplementary Table 3; Supplementary Discussion). DOPA-containing proteins in *Mytilus* byssus play a key role in both underwater adhesion and load bearing of the byssus[20,21]. Specifically, DOPA in the mussel adhesive proteins forms adhesive bonds with multivalent ions, metal oxides and organic functional groups, which lead to the byssus having robust, stiff and extensible character in underwater adhesion; a vital trait for load bearing by *Mytilus* byssus. Fe^{3+} ions, which are 10^6 fold more enriched in the *Mytilus* byssus than in the ambient seawater, form strong and reversible complexes with DOPA for enhancing the mechanical properties of the *Mytilus* byssus[22–24].

Similar to its *Mytilus* counterpart, the *Atrina* byssus is also predominantly protein (97 wt%, Supplementary Table 4) with abundant metal ions that potentially interacts with DOPA. Additionally, from inductively coupled plasma mass spectrometry (ICP-MS) analyses, we detected a substantial metal ion content, primarily Ca^{2+} and Fe^{3+}, at levels that are 10^3–10^6-fold more enriched in the *Atrina* byssus than in seawater (Supplementary Table 5; Supplementary Discussion). The high metal ion content draws attention to their possible role of mediating protein interactions in the byssus. Given the known interaction between DOPA and Fe^{3+}, we proceeded to look for DOPA-containing proteins in the *Atrina* byssus. The byssus was dissected and, following Arnow staining, DOPA was shown to occur throughout the *Atrina* byssus (Fig. 1c).

Interfacial protein *Atrina pectinata* foot protein-1. Having determined the presence and distribution of DOPA in *Atrina*, we moved on to the purification of the DOPA-containing proteins. The *Atrina* foot proteins were extracted from the foot by using 5% acetic acid, and thereafter, proteins from the *Atrina* foot extract were separated by gel permeation chromatography (GPC) and then subjected to C8 reverse phase high-performance liquid chromatography (HPLC). A major peak was eluted at ~48% (v/v) acetonitrile (Fig. 1d). The major peak was run on SDS–PAGE (polyacrylamide gel electrophoresis) and a dense single band of apparent mass ~36 kDa was observed in the gel (Fig. 1e). Acid urea PAGE (AU-PAGE) of the major peak showed a nitro-blue-tetrazolium-positive band, indicating the presence of DOPA in the purified proteins (Fig. 1e). The major peak was also subjected to matrix-assisted laser desorption and ionization with time-of-flight (MALDI-TOF) mass spectrometry yielding a mass (*m/z*) of 38,705.4 Da (Supplementary Fig. 1) that is consistent with the apparent mass from SDS–PAGE.

The purified DOPA-containing protein (~38 kDa) was transferred to a polyvinylidene difluoride membrane (Immobilion P) and subjected to ESI-MS/MS (electrospray ionization-mass spectrometry) peptide sequencing after *in situ* trypsin digestion. Three peptides were identified following the collision induced dissiciation by ESI-MS/MS: ODYKOVPK (O denotes hydroxylproline) (Supplementary Fig. 2), YELKPGVW, and GLNDVLFPK. In order to obtain a partial cDNA-deduced sequence of the *Atrina* foot protein, 3′ RACE was performed using degenerate oligonucleotides designed from the known amino-acid sequence of the tryptic peptide PDYKPVPK and the Universal Primer Mix. After cloning and sequencing the partial cDNA, a specific primer was designed to amplify the 5′ end of the gene and to retrieve the full-length cDNA. The obtained sequence of the cDNA has a length of 1,071 bp, which translates into a protein of 360 amino acids in length (Supplementary Fig. 3). The predicted molecular mass and theoretical pI of the cDNA-deduced amino-acid sequence without the signal peptide were 38,485.8 Da and 9.47, respectively. The cDNA-deduced amino-acid composition without the signal peptide matched with the amino-acid composition of the protein isolated from the major HPLC peak (~38 kDa, Supplementary Table 3). In addition, the 38 kDa protein from the major peak was subjected to Edman N-terminal sequencing and revealed ASXVPPVD (where X denotes an unknown amino-acid residue) as the starting sequence, which matched that of the cDNA-deduced sequence after the signal peptide. The complete cDNA-deduced protein was named *A. pectinata* foot protein-1 (apfp-1). The mass difference between apfp-1 by MALDI-TOF mass spectrometry and the cDNA-deduced apfp-1 sequence without the signal peptide was around ~220 Da, suggesting apfp-1 has around 13–14 DOPA residues if all the post-translational modification corresponds to DOPA. This, however, is not likely, given that at least two proline residues are also hydroxylated. On the other hand, only trace amounts of hydroxyproline were detected by amino-acid analysis suggesting that Hyp in apfp-1 may be limited to the peptide ODYKOVPK. DOPA content calculated from MALDI-TOF roughly resembled the amino-acid analysis. DOPA content of 38 kDa bands (apfp-1) was 5.2 ± 0.5 mol% based on amino-acid analysis, which suggest 16–19 DOPA residues in native apfp-1 (Supplementary Table 3).

apfp-1 has four distinct regions (Fig. 1f). The first region beginning with the N terminus, is a signal peptide sequence 18 residues long. A domain with degenerate tandem repeats follows including the octapeptide VVPDYKPP, hexapeptide VPKYKS/P along with the heptapeptide PVDYKPP, and a tetrapeptide PVYK that is repeated six times. Similar consensus repeats are also found in other mussel foot protein-1 (mfp-1) analogues[18,20] (Supplementary Table 6). After the repeating region, there is a 55 amino-acid-long glycine-rich region. Within this region, there are nine tyrosine residues, mostly flanked by glycine. This arrangement poses less steric crowding for the DOPA residues and should result in better solvent accessibility, Fe-coordination and catalytic conversion by catechol oxidase of Tyr to DOPA, and later DOPA to Dopaquinone, which is similarly observed in proteins from other organisms including *Aulacomya ater*, *Trichomya hirsute*, *Fasciola hepatica* and *Phragmatopoma californica*[25–27]. The C-terminal domain of the protein is predicted to be a C-type lectin domain, which has a carbohydrate-binding motif[28]. The presence of the C-type lectin domain accounts for the substantial amount of Ca^{2+} detected in our ICP-MS analysis as Ca^{2+} is known to be the metal cofactor in the binding of lectin and sugar[28].

DOPA–Fe^{3+} complexes have a unusual high-stability constant[22,24,29]. Coordination-based crosslinks have been proposed to endow certain biological structures with a number of desirable material properties, including triggered self-assembly, increased toughness, self-repair, adhesion, high hardness in the absence of mineralization and mechanical tunability[18,21–24,29]. The formation of Fe^{3+}–DOPA complexes was detected by resonance Raman spectroscopy[23,29] of purified apfp-1 protein

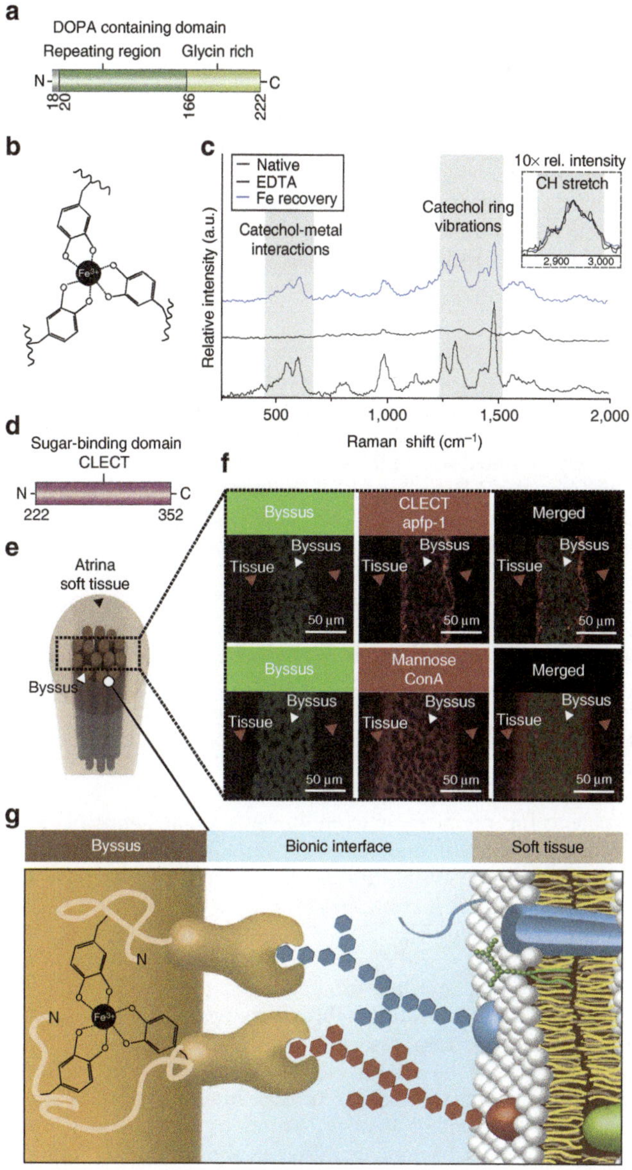

Figure 2 | Role of apfp-1 in *Atrina pectinata*. (**a**) DOPA-containing domain in apfp-1. (**b**) Tris-DOPA-Fe^{3+} complex. (**c**) Resonance Raman spectra from apfp-1 (black), after EDTA treatment (red) and after re-exposure to Fe in apfp-1 (blue). The nearly complete loss of resonance peaks after EDTA treatment is reversed to a nearly complete restoration by means of incubation. in 1 mM $FeCl_3$ (pH 3.2). The three spectra were normalized to the area under the aliphatic CH peak (2,850 to 3,010 cm^{-1}). The inset represents the relative intensity of the non-resonance peak for aliphatic CH stretching from apfp-1 magnified ×10. According to the assignments, the most prominent peaks can be attributed to the interaction of metal with the catecholic oxygens and to the vibrations of the carbon bonds in the catechol ring, respectively. (**d**) Sugar-binding domain in apfp-1. (**e**) Cross-section through the tissue surrounding the *Atrina* byssus. (**f**) Confocal imaging of *Atrina* byssus and the surrounding tissue labelled with anti-apfp-1 antibody to detect apfp-1, and with ConA to detect mannose. Green box represents the auto-fluorescence of the byssus which is the location that shows where the byssus fibre is. Red box represents the apfp-1 antibody to detect apfp-1, and with ConA to detect mannose. (**g**) The byssus is coated with apfp-1. Its DOPA-Fe^{3+} binding property gives stiff and extensible properties to the cuticle region, like fp-1 in mussel byssus. The apfp-1 lectin domain strongly binds to mannose in the cell membrane that surrounds the non-living byssus and acts as the bionic interface between non-living tissue and the cell.

(Fig. 2b,c). Three diagnostic regions of Raman spectra were observed: catechol-Fe^{3+} interaction (550–680 cm^{-1}) and catechol ring vibrations (1,240–1,370 cm^{-1}) are resonance dependent where they are associated with the transfer of an electron from catechol to Fe^{3+}, and CH stretching (2,850–3,010 cm^{-1}), which is not resonance dependent. The Raman spectra acquired from protein purified in the presence of EDTA, which is used to scavenge Fe^{3+} from complexation with catechol showed the loss of Fe^{3+} associated resonance peaks. However, after treatment with $FeCl_3$, the Fe^{3+} associated resonance was partially recovered. Moreover, the EDTA treatment of *Atrina* byssus compromised the mechanical properties of the byssal threads (Supplementary Table 1), suggesting the presence of metal–DOPA interactions. This result offers revealing insights into the Fe^{3+}–DOPA complexes in apfp-1 as it is homologous with the cuticle protein mfp-1, which plays an integral role of metal coordination chemistry in mechanical performance to create a stiff yet extensible coating for *Mytilus* byssus[22,23,30].

Localization of apfp-1 mediates sugar–lectin interaction. To localize apfp-1, a specific polyclonal antibody for apfp-1 was generated. Transmission electron microscopy (Supplementary Fig. 4) and fluorescence imaging of longitudinal and orthogonal cross-sections of the *Atrina* thread (Fig. 2f) revealed that apfp-1 is localized in the cuticle of the byssal threads (Supplementary Discussion). Given the cuticular localization of apfp-1 and colocalization of DOPA–Fe^{3+} complexes (Fig. 2c,f), the similarities between mfp-1 and apfp-1 are impossible to ignore[17,18]. Nevertheless, the lectin domain distinguishes apfp-1 from mfp-1. As lectins are known to bind to sugars surrounding the cell membrane[31], we proceeded to identify the types of sugar specifically bound by apfp-1. Seven biotinylated lectins: (Concanavalin A (ConA): α-D-mannosyl and α-D-glucosyl residues branched α-mannosidic structures (high α-mannose type, or hybrid type and biantennary complex type N-Glycans), Biotinylated Dolichos Biflorus Agglutinin (DBA): α-linked N-acetylgalactosamine Biotinylated Soybean Agglutinin (SBA): α- or β-linked N-acetylgalactosamine, and to a lesser extent, galactose residues, Wheat Germ Agglutinin (WGA): GlcNAcβ1-4GlcNAcβ1-4GlcNAc, Neu5Ac (sialic acid), Ulex europaeus agglutinin (UEA1): Fucα1-2 Gal-R and Ricinus communis (RCA): Galβ1-4GalNAcβ1-R) were applied to sections performed through the adductor muscle and byssus. Only ConA labelling overlapped competitively with apfp-1 localization in *Atrina* tissue and byssus. Since the C-lectin binding domain is structurally predicted to be a mannose-binding lectin, apfp-1 is predicted to bind to mannose, the ligand for ConA. The confocal imaging shows apfp-1 bound to mannose in mucus, the living tissue or cell membranes (Fig. 2f). These findings suggest apfp-1 as an interfacial linker between the byssus and the living tissues via the asymmetric Fe^{3+}–DOPA and the sugar–lectin interactions. We speculated that the linkage to soft tissue is through the membrane bound sugars moieties (Fig. 2g). The Fe^{3+}–DOPA-mediated interactions between the byssus and mfp-1 were shown to be strong and reversible ($W_{ad} \sim -2.0\,mJ\,m^{-2}$ to $-4.0\,mJ\,m^{-2}$ using a surface forces apparatus (SFA))[30]. Lectin–sugar binding is also known to be strong and contributes to a variety of the cellular activities. Previous study using atomic force microscopy and SFA had shown specific and strong binding between lectin and its cognate sugar and a thorough understanding of this interaction holds the key to a better understanding of cell–cell interactions[32,33].

Nano-mechanical quantification of the sugary interfaces. To quantify how strong the sugar–lectin interaction is and how it compares with the Fe^{3+}-DOPA complex, SFA was used to measure the adhesion between the sugar-binding domain of apfp-1 and mannose (Fig. 3a). SFA has been extensively used for measuring various types of biological interactions including specific and non-specific biological interactions and has the unique ability to provide a simultaneous direct measurement of the force F as a function of the absolute surface separation D measured *in situ* and in real time, with a force sensitivity of <10 nN and distance resolution of ~ 1 Å determined by multiple beam interferometry using FECO (Fringes of Equal Chromatic Order)[34].

Mannan is a branched form of mannose derived from yeast cell membranes and was used to measure adhesion between mannose and the lectin domain. There was no adhesion force between two opposing apfp-1 layers at pH 5.5 (Fig. 3b, yellow), however when mannan (average molecular weight 46,000 kDa) was injected (10 μM), the adhesion force increased to $W_{ad} \approx -0.7$ mJ m^{-2} (Fig. 3b, cyan). Notably, when Ca^{2+} cofactor (10 μM) was added, the measured adhesion energy increased to $W_{ad} \approx -1.7$ mJ m^{-2} and was comparable with the Fe^{3+}-DOPA interaction in mfp-1 (ref. 30). Thus, addition of Ca^{2+} cofactor enhanced adhesion (Fig. 3b, red). For a negative control, the adhesion between apfp-1 with Ca^{2+} was measured without mannan, which resulted in $W_{ad} \approx -0.3$ mJ m^{-2} (Fig. 3b, green). Interestingly, the adhesion force of lectin–mannan involving apfp-1, which has only one lectin domain, is comparable with the adhesion involving ~ 100 DOPA residues in mfp-1 protein or the DOPA-containing domain of apfp-1 (refs 29,30). Therefore, a considerable amount of force would be required to pull a lectin-covered thread through a saccharide covered tissue tunnel.

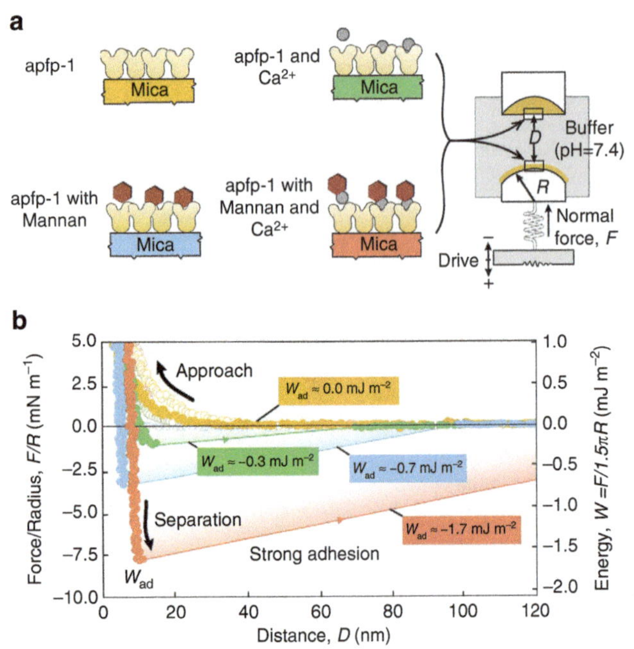

Figure 3 | Biomechanics of apfp-1. (**a**) Different mica surfaces coated to measure adhesion force between apfp-1 and apfp-1 (yellow), apfp-1 and apfp-1 in the presence of mannan (cyan), apfp-1 and apfp-1 in the presence of mannan and calcium (red), apfp-1 and apfp-1 in the presence of calcium (green). (**b**) Force-distance profiles from SFA measurements. The measured force, *F/R* (normalized by the radius of curvature, *R*), is denoted in the ordinate at the left, whereas the corresponding interaction energy per unit area, *W*, between two flat surfaces, defined by $W = F/1.5\pi R$ is on the right. F_{ad} and W_{ad} are the adhesion force and the adhesion energy, respectively.

The finding that sugar–lectin binding is much stronger than Fe^{3+}-DOPA binding within the same byssus system is significant. As already mentioned, strong interfacial adhesion is required to prevent rubbing or sliding friction between the stiff cuticle of byssus and the soft living tissue of *Atrina*. Such friction would occur if the interfacial interactions between cuticle and soft tissue were too weak to prevent extensive slippage during load transfer and result in contact damage. apfp-1 is the ideal linker protein with bi-functionality, which are metal and sugar binding. The binding is weak enough to yield and reform, but not slip.

Discussion

Combining the results of linker protein characterization, confocal imaging and SFA-based nano-mechanics, we have described an intriguing interaction between tissue and non-tissue components in *Atrina* in the absence of gradients. *Atrina* byssal threads are securely mounted within the soft tissue by two strong but yielding interfaces: the C-lectin binding domain is attached to mannose on the glycosylated cell membranes and serves as an interface between the adductor tissues and the embedded portion of byssus thread.

The robust interconnection of non-living tissue and soft tissue is mediated by lectin–sugar interactions. To the best of our knowledge, this study is the first to describe a bionic linker protein, apfp-1, that is involved in interconnecting non-living stiff biomaterial and soft tissue (Fig. 2g). Moreover, this work is also the first to explain the interconnection from the molecular point of view.

What are the implications of this finding? One of the great problems for interfaces between implants and living tissues is the friction between stiff and compliant surfaces that leads to contact damage[1,2,34]. The byssal thread–tissue interface is one in which a stiffness mismatch is mitigated by a large interfacial area that is populated by many sacrificial lectin–sugar bonds that dissipate energy but prevent sliding between surfaces. Specifically, *Atrina* has a two-phase mechanism in their load-bearing mitigation. When the load is transmitted directly to the tissue, *Atrina* exploits its high surface area associated with their embedded thread structure in the byssal groove for mitigation. The mitigation through reversible sacrificial lectin-type interaction would be used when the transmitted load is beyond its first phase handling capability. The advances in this area should inspire improved engineering of bionic implants and devices[35–37]. It is known that sugar-based molecules and polymers have been utilized as conventional adhesives since the beginning of the human history. Thus, a specific and strong adhesion between sugar and lectin in the living organism should be exploited as a key biomedical adhesive for the bionic interface. We envisage that this could shift the research paradigm from conventional research on sugar and lectin that have been focused on cell–cell communications and immune responses towards the focus on bionic implants and adhesives.

Methods

Fan shell mussels. Live fan shell mussels (*A. pectinata*) were collected from a fan mussel farm in Gangjin, South Korea. Before biochemical and biomechanical analyses, the feet and silky byssus were carefully dissected from fresh fan shell mussels. The dissected feet were either used directly or stored at −80 °C until use. For mechanical and biochemical analyses, byssus threads were kept in artificial seawater at 4 °C until use.

Nano-indentation of *Atrina* byssus and surrounding tissues. Hardness value for the *Atrina* byssus was obtained by nano-indentation (Nano indenter G300, Aglient, Santa Clara, CA, USA) with a cube corner diamond tip. Nano-indentation was carried out on microtomed and surface polished surfaces of transverse cross-sections of the byssus and tissues embedded in Epofix. Test specimens hydrated in milliQ water for 24 h before the test. EPOXY used as a standard for the tip area function. Indentation depth 2 μm and all load–displacement curves were analysed using the method described by Oliver and Pharr[38].

Characterization of tensile properties of *Atrina* fibre. The tensile properties of the *A. pectinata* byssus threads were measured under wet conditions using the universal tensile tester (UTS, Instron, USA). The proximal part of the threads was used for the analyses. In the case of wet conditions, threads were kept in stirring artificial seawater at room temperature with or without the additional incubation with 50 mM EDTA in 0.1 M sodium acetate buffer pH 5.5 for 24 h. Byssus threads were loaded to failure at a strain rate of 0.5 mm s^{-1} with the use of a 10 N load cell. The humidity of the chamber was maintained at 99% by gently spraying a fine water mist from a humidifier during the wet condition tests. Strain energy density calculations were performed using SigmaPlot version 12.0 (Systat Software, Inc., San Jose, CA, USA).

Biochemical analysis. Byssus thread samples were incubated in 4 M urea/5% acetic acid solution for 48 h to remove contaminants, washed thoroughly with deionized water, and then freeze-dried, and their dry weight recorded. Dried threads were ground to powder with a mortar and pestle. Extraction of total lipids was performed in a mixture of methanol and chloroform, and the lipid content was determined by a previously reported method[39]. Total protein content was measured by the Lowry method[40] and by amino-acid analysis after sample preparation by acid hydrolysis[12]. Carbohydrates were extracted in 15% trichloroacetic acid from powdered thread samples and their content was determined by the phenol–sulfuric acid method[41]. Five samples were used for each component determination. For the Arnow assay of byssus and foot[42], the tissues were sequentially immersed for 5 min in each of the following solutions: 0.5 N hydrochloric acid, 1.45 M sodium nitrite/0.41 M sodium molybdate and 1 N sodium hydroxide.

Inductively coupled plasma spectroscopy. The byssus thread was dissected from the rest of the thread and rinsed three times with distilled water. Byssal threads from the three sampling dates were pooled for spawned and unspawned mussels to get the 3–4 mg of byssus needed for analysis of metal. Chemical element content of the byssal threads was determined using an IRIS advantage (Thermo, Franklin, MA, USA) ICP in normal mode and with a micro-nebulizer. Analyses were done after acid dissolution of byssal threads with 0.1 ml of concentrated nitric acid (70%) and 0.1 ml of hydrogen peroxide, placed in a water bath for 1 h at 60 °C. Elements detected are Fe, Ca and Zn. To correct for mass bias and instrument drift, a 2% HNO_3 blank solution was run periodically.

Protein purification. For protein extraction, frozen fan shell mussel feet were homogenized in a glass tissue grinder with 5% acetic acid containing two protease inhibitors (1 μg ml^{-1} leupeptin and pepstatin A) and 1 mM EDTA. The homogenate was centrifuged at 20,000*g*, 4 °C for 30 min. Perchloric acid was added drop-wise to the recovered supernatant to a final concentration of 1.4% (w/v). The mixture was further stirred for 10 min, and then centrifuged again at 20,000*g*, 4 °C for 30 min. The collected supernatant was dialyzed against 5% acetic acid overnight, then after addition of 2% acetonitrile, freeze-dried[43] (Supplementary Discussion).

Gel permeation and HPLC. Freeze-dried *Atrina* proteins were resuspended in 500 μl 5% acetic acid, centrifuged at 20,000*g*, 4 °C for 30 min to remove any precipitate, then injected to gel filtration column (Shodex KW 803) and eluted with 5% acetic acid at 0.5 ml min^{-1} at room temperature and monitored at 280 nm. Collected GPC fractions were examined by AU-PAGE for the presence of DOPA-containing proteins[44]. Fractions containing the DOPA-containing proteins were further analysed in a C-8 reverse phase HPLC using an acetonitrile gradient in water with 0.1% trifluoroacetic acid at a flow rate of 1 ml min^{-1}. Separated proteins were freeze-fried and stored at − 80 °C. A small aliquot (100 μl) from each collected peak was freeze-dried separately to be assayed by AU-PAGE for protein separation.

Protein electrophoresis. *Atrina* foot proteins were separated on polyacrylamide gels containing 5% acetic acid and 8 M urea (AU-PAGE), and then transferred to polyvinylidene difluoride (PVDF) membranes (Immobilon P, Millipore) by electrophoresis in 0.7% acetic acid, 200 mA, for 2 h. Proteins were stained with Coomassie Blue R-250 or Coomassie Blue G-250, the latter case when staining PVDF membrane. DOPA-containing proteins were stained with nitro blue tetrazolium redox cycling stain[45]. Protein bands of interest were excised and prepared for amino-acid analysis and peptide sequencing based on electrospray ionization-tandem mass spectrometry (ESI-MS/MS, Thermo, Franklin, MA, USA). *Atrina* foot proteins were also separated on SDS–PAGE, then transferred to PVDF membrane, and sent for N-terminal sequencing by Edman degradation.

Protein and peptide sequencing. N-terminal sequencing of excised PVDF membrane bands containing the protein of interest was done using automated Edman degradation. Also, the excised PVDF membrane bands were transferred into hydrolysis vials containing 100 mM Tris ascorbate pH 7.5, 1 mM $CaCl_2$, 0.1 mg ml^{-1} trypsin. The vials were then flushed with Argon and sealed by flame. Digestion was carried out for 12 h at room temperature. Digestion reaction was terminated by adding glacial acetic acid to a final concentration of 5%. Peptide sequencing was carried out in ESI-MS/MS (Thermo). To confirm the sequence, RT-PCR was run and a cDNA sequence of apfp-1 was obtained. After obtaining the deduced cDNA sequence of apfp-1, additional Edman sequencing of the purified protein was performed twice; the obtained Edman sequencing data was ASPVPPVD. The speculated reason for this discrepancy of the sequence between protein sequence translated from cDNA and Edman sequencing is highly due to the post-translational modification of tyrosine, which results in the difference of Y and P. As such, the following sequence, AS′X′VPPVD is used for the discrepancy part of the protein sequence (Supplementary Discussion).

Mass spectrometry. The masses of the two HPLC fractions analysed were determined by MALDI-TOF mass spectrometry (Applied Biosystems 4700, USA). The MALDI matrix was prepared by dissolving CHCA (a-cyano-4-hydroxycinnamic acid) (7 mg ml^{-1}) in 0.1% trifluoroacetic acid (TFA)/50% acetonitrile (ACN). Purified proteins were dissolved in this matrix solution in a ratio of 10:1 matrix:sample. About 1 μl of sample was analysed (Supplementary Discussion).

Raman spectroscopy. A purified apfp-1 (1 mg ml^{-1}) was added with a solution of $FeCl_3$ in 10 mM Bis-Tris (pH 5.5) and allowed to equilibrate for 10 min. The protein precipitated as the pH was raised to ~8.0 with 0.1 M NaOH. A droplet was allowed to evaporate on a glass slide, and Raman spectra (30 accumulations with 1 s integration time each) were taken from the precipitated protein residue at the edges. Raman spectra were also measured from the protein solution before iron complexation and from solutions of apfp-1 in which Fe^{3+} was not limiting for comparison (no EDTA added in the protein extraction). For Raman spectroscopy, a continuous laser beam was focused on the sample through a confocal Raman microscope (model CRM200, WITec, Ulm, Germany) equipped with a piezo scanner (model P-500, Physik Instrumente, Karlsruhe, Germany). The diode-pumped, 785-nm near-infrared laser excitation (Toptica Photonics AG, Graefelfing, Germany) was used in combination with a 100 × oil immersed (Nikon, numerical aperture (NA) = 1.25) microscope objective. Laser power ranging between 15 and 30 milliwatts was used for all measurements. The spectra were acquired using an air-cooled charge-coupled device (DU401A-DR-DD, Andor, Belfast, North Ireland) behind a grating (300 g mm^{-1}) spectrograph (Acton, Princeton Instruments, Trenton, NJ, USA) with a 6 cm^{-1} spectral resolution. Software ScanCtrlSpectroscopyPlus (version 1.38, Witec) was used for measurement setup. Raman spectra were processed and analysed with Witec Project software (version 2.02). Raman spectra were background subtracted and lightly smoothed using the first order polynomial function and 9-point Savitzky-Golay filter (4th order polynomial), respectively.

Lectin histochemistry. *A. pectinata* adductor muscle, and byssal threads were fixed in 4% paraformaldehyde in sodium phosphate buffer (PBS solution, pH 7.4), rinsed in PBS solution, dehydrated through an ethanol series, embedded in paraffin, and cut into 5 μm-thick sections with a Microm HM 340 E microtome. The sections were subjected to an indirect lectin-labelling method according to the following protocol. Sections were blocked for 30 min in Tris-buffered saline, pH 8.0, containing 0.05% Tween 20 and 3% bovine serum albumin (TBS-T-BSA). The seven biotinylated lectins used for lectin staining (Concanavalin A (ConA): α-D-mannosyl and α-D-glucosyl residues branched α-mannosidic structures (high α-mannose type, or hybrid type and biantennary complex type N-Glycans), Biotinylated DBA: α-linked N-acetylgalactosamine, Biotinylated SBA: α- or β-linked N-acetylgalactosamine, and to a lesser extent, galactose residues, WGA: GlcNAcβ1-4GlcNAcβ1-4GlcNAc, Neu5Ac (sialic acid), Ulex europaeus agglutinin (UEA1): Fucα1-2 Gal-R and Ricinus communis (RCA): Galβ1-4GalNAcβ1-R; Vector Laboratories) were diluted at a concentration of 25 μg ml^{-1} in TBS-T-BSA and applied on the sections for 2 h at room temperature. After three washes in TBS-T, the sections were incubated for 1 h in Texas-Red-conjugated streptavidin (Vector Laboratories) diluted 1:100 in TBS-T-BSA. Following three final washes in TBS-T, they were mounted with Vectashield (Vector Laboratories). Control reactions were performed by substituting the lectins with TBS-T-BSA. For ConA, an additional control reaction was also performed by using the lectin saturated with 0.4 M Methyl α-D-mannopyranoside. Sections were observed using a Zeiss Axioscope A1 microscope equipped with an AxioCam ICc 3 camera.

Transmission electron microscopy. *Atrina* foot samples were frozen using a high pressure freezer (Leica EMPACT2) at a pressure of 2,000–2,050 bar. Freeze substitution was performed in anhydrous acetone (containing 1% OsO4 and 0.1% uranyl acetate) using a Leica EM AFS2 (automatic freeze substitution). The samples were kept at − 85 °C for 3 days, and then at − 60, − 20 and 0 °C for one day at each temperature, and then exposed to room temperature. Spurr's resin was used for infiltration and embedding. The embedded samples were cut to make ultrathin sections at 70–90 nm using a Reichert Ultracut S or Leica EM UC6 (Leica, Vienna, Austria), and the sections were collected using 100 mesh copper grids.

apfp-1 Immunolocalization. A polyclonal anti-apfp-1 antibody was raised against the peptide VLFPKSHKWFWGYGKKQCKWFD (Ontores Biotechnology, China) which is part of the apfp-1 CLECT domain. The specificity of the antibody was established using apfp-1 protein purified from the foot tissue. Antibody labelling was done with apfp-1 antibody. Twenty nanometer gold nano secondary antibody from donkey (Abcam) was used to detect apfp-1 in ultrathin sections of the byssus.

Electron ionization-tandem mass spectrometry. A purified protein (apfp-1) was reduced with 5 mM dithiothreitol for 30 min, and then alkylated with 25 mM iodoacetamide for 30 min at 25 °C in the dark room. The protein was resolved in a digestion solution (6 M urea and 40 mM ammonium bicarbonate) and digested by trypsin (Sequencing-grade, Promega, Madison, WI, USA) for 12 h at 37 °C. The digested peptides were collected, desalted, separated with the C-18 column (Magic C18AQ 5 μ 200 Å particles, Michrom BioResources, Auburn, CA, USA), and analysed on a linear ion trap LTQ XL mass spectrometer (Thermo) interfaced with anelectrospray ion source. The peptides were separated using a linear gradient of ACN/water (2–40% ACN in 90 min), containing 0.1% formic acid, at a flow rate of 260 nl min^{-1}. A full MS scan range was from 300 to 2,000 *m/z*, and MS/MS spectra were collected from the five most intense precursor ions. The collected MS/MS spectra were searched against the *Atrina* protein sequence database using SEQUEST algorithm in the Proteome Discoverer software (Thermo) with the following parameters: semi-trypsin, a mass tolerance of 2.0 and 1.0 Da for precursor and fragment ions, respectively. All peptide identifications by the software were confirmed by manual interpretation of the raw data.

Amino-acid analysis. Protein of interest from excised PVDF membrane were hydrolysed in 6 N HCl with 5% water saturated phenol. The reactions were flushed with Argon and vials sealed by flame. Hydrolysis reaction was carried out for 1 h at 156 °C. Afterwards, the samples were flash evaporated at 60 °C under vacuum and washed till dry twice with 1 ml milliQ water, then twice with 1 ml methanol. The samples were then resuspended in 250–500 μl SYKAM sample dilution buffer. Amino-acid analysis was performed with a ninhydrin-based SYKAM System S4300 Amino Acid Analyzer (SYKAM, Germany).

Molecular cloning. Total RNA was extracted from dissected mussel feet regions that tested positive in the Arnow assay using an RNeasy Mini Kit (QIAGEN, Valencia, CA, USA). Briefly, dissected tissue previously stored in RNAlater solution was disrupted and homogenized in a glass tissue grinder with a Wise Stir homogenizer stirrer. Afterwards, total RNA was extracted according to the protocol specified by the kit. First-strand cDNA synthesis and 3′ and 5′ RACEs were performed by using SMARTer RACE cDNA amplification kit (Clontech, Palo Alto, CA, USA). Degenerate primers were designed based on the sequenced tryptic peptides.

Force versus distance profiles with surface forces apparatus. A surface forces apparatus (SFA) was used to measure the interaction forces of protein films in different aqueous solutions. Detailed setup of the SFA system and its working principles can be found elsewhere[46]. Briefly, two back-silvered thin mica surfaces with thickness ~1 to 5 μm were glued onto cylindrical silica disks with a radius $R \sim 2$ cm. The two disks were mounted in the SFA chamber in a crossed-cylinder configuration, the interaction of which is equivalent to that of a sphere with radius R approaching a flat surface when the separation distance $D \ll R$. Desired buffer solutions were injected between the two surfaces for further force measurement.

The interaction forces of apfp-1 versus apfp-1, in buffer solutions with of Ca^{2+} (10 μM) or mannan (10 μM) or both Ca^{2+} and mannan were measured in PBS buffer (pH 7.4). The film thickness and separation distance can be obtained *in situ* by using an optical technique called multiple beam interferometry using fringes of equal chromatic order (FECO). The adhesion force F_{ad} measured can be correlated to the adhesion energy per unit area of two flat surfaces W_{ad} by:

$$F_{ad} = 1.5 * \pi * R * W_{ad}$$

References

1. Fratzl, P. & Weinkamer, R. Nature's hierarchical materials. *Prog. Mater. Sci.* **52,** 1263–1334 (2007).
2. Bar-On, B., Barth, F. G., Fratzl, P. & Politi, Y. Multiscale structural gradients enhance the biomechanical functionality of the spider fang. *Nat. Commun.* **5,** 3894 (2014).
3. Zettl, J. H., Burgess, E. M. & Romano, R. L. The "Interface" in the immediate postsurgical prosthesis. *Contract* **5261,** 1–6 (1970).
4. Benjamin, M., Evans, E. J. & Copp, L. The histology of tendon attachments to bone in man. *J. Anat.* **149,** 89–100 (1986).
5. Evans, E. J., Benjamin, M. & Pemberton, D. J. Fibrocartilage in the attachment zones of the quadriceps tendon and patellar ligament of man. *J. Anat.* **171,** 155–162 (1990).
6. Kwak, M. K., Jeong, H.-E. & Suh, K. Y. Rational design and enhanced biocompatibility of a dry adhesive medical skin patch. *Adv. Mater.* **23,** 3949–3953 (2011).
7. Miserez, A., Schneberk, T., Sun, C., Zok, F. W. & Waite, J. H. The transition from stiff to compliant materials in squid beaks. *Science* **319,** 1816–1819 (2008).
8. Ono, K. *et al.* Photocrosslinkable chitosan as a biological adhesive. *J. Biomed. Mater. Res.* **49,** 289–295 (2000).
9. Suresh, S. Graded materials for resistance to contact deformation and damage. *Science* **292,** 2447–2451 (2001).
10. Reedy, Jr E. D. Connection between interface corner and interfacial fracture analyses of an adhesively-bonded butt joint. *Int. J. Solids Struct.* **37,** 2429–2442 (2000).
11. Pearce, T. & LaBarbera, M. Biomechanics of byssal threads outside the Mytilidae: *Atrina rigida* and *Ctenoides mitis*. *J. Exp. Biol.* **212,** 1449–1454 (2009).
12. Mascolo, J. M. & Waite, J. H. Protein gradients in byssal threads of some marine bivalve molluscs. *J. Exp. Zool.* **240,** 1–7 (1986).
13. Qin, X.-X. & Waite, J. H. A potential mediator of collagenous block copolymer gradients in mussel byssal threads. *Proc. Natl Acad. Sci. USA* **95,** 10517–10522 (1998).
14. Harrington, M. J. & Waite, J. H. Holdfast heroics: comparing the molecular and mechanical properties of Mytilus californianus byssal threads. *J. Exp. Biol.* **210,** 4307–4318 (2007).
15. Coyne, K. J., Qin, X.-X. & Waite, J. H. Extensible collagen in mussel byssus: a natural block copolymer. *Science* **277,** 1830–1832 (1997).
16. Pearce, T. & LaBarbera, M. A comparative study of the mechanical properties of Mytilid byssal threads. *J. Exp. Biol.* **212,** 1442–1448 (2009).
17. Sun, C. & Waite, J. H. Mapping chemical gradients within and along a fibrous structural tissue, mussel byssal threads. *J. Biol. Chem.* **280,** 39332–39336 (2005).
18. Lee, B. P., Messersmith, P. B., Israelachvili, J. N. & Waite, J. H. Mussel-inspired adhesives and coatings. *Annu. Rev. Mater. Res.* **41,** 99–132 (2011).
19. Waite, J. H. & Tanzer, M. L. Polyphenolic substance of *Mytilus edulis*: novel adhesive containing L-dopa and hydroxyproline. *Science* **212,** 1038–1040 (1981).
20. Silverman, H. & Roberto, F. Understanding marine mussel adhesion. *Mar. Biotechnol.* **9,** 661–681 (2007).
21. Lee, H., Scherer, N. F. & Messersmith, P. B. Single-molecule mechanics of mussel adhesion. *Proc. Natl Acad. Sci. USA* **103,** 12999–13003 (2006).
22. Holten-Andersen, N., Fantner, G. E., Hohlbauch, S., Waite, J. H. & Zok, F. W. Protective coatings on extensible biofibres. *Nat. Mater.* **6,** 669–672 (2007).
23. Harrington, M. J., Masic, A., Holten-Andersen, N., Waite, J. H. & Fratzl, P. Iron-clad fibers: a metal-based biological strategy for hard flexible coatings. *Science* **328,** 216–220 (2010).
24. Wilker, J. J. The iron-fortified adhesive system of marine mussels. *Angew. Chem. Int. Ed.* **49,** 8076–8078 (2010).
25. Waite, J. H. & Rice-Ficht, A. C. Presclerotized eggshell protein from the liver fluke *Fasciola hepatica*. *Biochemistry* **26,** 7819–7825 (1987).
26. Zhao, H., Sun, C., Stewart, R. J. & Waite, J. H. Cement proteins of the tube-building polychaete *Phragmatopoma californica*. *J. Biol. Chem.* **280,** 42938–42944 (2005).
27. Burzio, L. A., Saéz, C., Pardo, J., Waite, J. H. & Burzio, L. O. The adhesive protein of *Choromytilus chorus* (Molina, 1782) and *Aulacomya ater* (Molina, 1782): a proline-rich and a glycine-rich polyphenolic protein. *Biochim. Biophys. Acta* **1479,** 315–320 (2000).
28. Zelensky, A. N. & Gready, J. E. The C-type lectin-like domain superfamily. *FEBS J.* **272,** 6179–6217 (2005).
29. Hwang, D. S. *et al.* Protein- and metal-dependent interactions of a prominent protein in mussel adhesive plaques. *J. Biol. Chem.* **285,** 25850–25858 (2010).
30. Zeng, H., Hwang, D. S., Israelachvili, J. N. & Waite, J. H. Strong reversible $Fe3^{+}$-mediated bridging between dopa-containing protein films in water. *Proc. Natl Acad. Sci. USA* **107,** 12850–12853 (2010).
31. Nicolson, G. L., Yanagimachi, R. & Yanagimachi, H. Ultrastructural localization of lectin-binding sites on the zonae pellucidae and plasma membranes of mammalian eggs. *J. Cell Biol.* **66,** 263–274 (1975).
32. Hench, L. L. Biomaterials: a forecast for the future. *Biomaterials* **19,** 1419–1423 (1998).

33. Menon, S. *et al.* Binding-site geometry and flexibility in DC-SIGN demonstrated with surface force measurements. *Proc. Natl Acad. Sci. USA* **106,** 11524–11529 (2009).
34. Bhushan, B., Israelachvili, J. N. & Landman, U. Nanotribology: friction, wear and lubrication at the atomic scale. *Nature* **374,** 607–616 (1995).
35. Wiegemann, M. Adhesion in blue mussels (*Mytilus edulis*) and barnacles (genus *Balanus*): mechanisms and technical applications. *Aquat. Sci.* **67,** 166–176 (2005).
36. Bandara, N., Zeng, H. & Wu, J. Marine mussel adhesion: biochemistry, mechanisms, and biomimetics. *J. Adhes. Sci. Technol.* **27,** 2139–2162 (2013).
37. Zhou, Y., Snead, M. L. & Tamerler, C. Bio-inspired hard-to-soft interface for implant integration to bone. *Nanomedicine* **11,** 431–434 (2014).
38. Oliver, W. C. & Pharr, G. M. An improved technique for determining hardness and elastic modulus using load and displacement sensing indentation experiments. *J. Mater. Res.* **7,** 1564–1583 (1992).
39. Poulsen, N. *et al.* Isolation and biochemical characterization of underwater adhesives from diatoms. *Biofouling* **30,** 513–523 (2014).
40. Hennebert, E., Wattiez, R. & Flammang, P. Characterisation of the carbohydrate fraction of the temporary adhesive secreted by the tube feet of the sea star *Asterias rubens. Mar. Biotechnol.* **13,** 484–495 (2011).
41. Ohkawa, K., Nagai, T., Nishida, A. & Yamomoto, H. Purification of DOPA-containing foot proteins from green mussel, *Perna viridis*, and adhesive properties of synthetic model copolypeptides. *J. Adhesion* **85,** 770–791 (2009).
42. Smith, A. M., Quick, T. J. & St. Peter, R. L. Differences in the composition of adhesive and non-adhesive mucus from the limpet *Lottia limatula. Biol. Bull.* **196,** 34–44 (1999).
43. Stanley, M. S., Callow, M. E. & Callow, J. A. Monoclonal antibodies to adhesive cell coat glycoproteins secreted by zoospores of the green alga Enteromorpha. *Planta* **210,** 61–71 (1999).
44. Gourdon, D. *et al.* Adhesion and stable low friction provided by a subnanometer-thick monolayer of a natural polysaccharide. *Langmuir* **24,** 1534–1540 (2008).
45. Paz, M. A., Flückiger, R., Boak, A., Kagan, H. M. & Gallop, P. M. Specific detection of quinoproteins by redox-cycling staining. *J. Biol. Chem.* **266,** 689–692 (1991).
46. Israelachvili, J. *et al.* Recent advances in the surface forces apparatus (SFA) technique. *Rep. Prog. Phys.* **73,** 036601 (2010).

Acknowledgements

This work was supported by the Marine Biotechnology program (Marine BioMaterials Research Center) funded by the Ministry of Oceans and Fisheries, Korea (D11013214H480000110, D.S.H.). This research was also supported by the National Research Foundation of Korea Grant funded by the Ministry of Science, ICT and Future Planning (MSIP) (NRF-2014R1A2A2A01006724 and NRF-2011-0029960, D.S.H.), Global PhD Fellowship Program (NRF-2011-0008261), Institute for Basic Science (IBS-R013-G1) through the National Research Foundation of Korea, and a NSERC Discovery Grant and a NSERC RTI Grant (for a surface forces apparatus) from the Natural Sciences and Engineering Research Council of Canada (H. Zeng). P.F. is Research Director of the Fund for Scientific Research of Belgium (F.R.S.-FNRS). We thank Prof. Admir Masic (MIT) for helping with resonance Raman spectroscopy.

Author contributions

H.Y.Y. and M.I. prepared the samples and designed the experiments. J.H. conducted the SFA measurements, E.H. conducted the lectin imaging, S.K. performed the mechanical measurements, S.R. performed the ESI-MS/MS measurements, J.H.W. and D.S.H., conceived the project. M.F., P.F., H.Z., D.H., J.H.W. and D.S.H., advised the experimental design and analysed the results. All authors helped in the preparation of the manuscript.

Additional information

Competing financial interests: The authors declare no competing financial interest.

18

CETSA screening identifies known and novel thymidylate synthase inhibitors and slow intracellular activation of 5-fluorouracil

Helena Almqvist[1,*], Hanna Axelsson[1,*], Rozbeh Jafari[2,†], Chen Dan[3], André Mateus[4], Martin Haraldsson[1], Andreas Larsson[5], Daniel Martinez Molina[2], Per Artursson[4,6,7], Thomas Lundbäck[1] & Pär Nordlund[2,3,8]

Target engagement is a critical factor for therapeutic efficacy. Assessment of compound binding to native target proteins in live cells is therefore highly desirable in all stages of drug discovery. We report here the first compound library screen based on biophysical measurements of intracellular target binding, exemplified by human thymidylate synthase (TS). The screen selected accurately for all the tested known drugs acting on TS. We also identified TS inhibitors with novel chemistry and marketed drugs that were not previously known to target TS, including the DNA methyltransferase inhibitor decitabine. By following the cellular uptake and enzymatic conversion of known drugs we correlated the appearance of active metabolites over time with intracellular target engagement. These data distinguished a much slower activation of 5-fluorouracil when compared with nucleoside-based drugs. The approach establishes efficient means to associate drug uptake and activation with target binding during drug discovery.

[1] Laboratories for Chemical Biology, Karolinska Institutet, Science for Life Laboratory Stockholm, Division of Translational Medicine & Chemical Biology, Department of Medical Biochemistry & Biophysics, Karolinska Institutet, Tomtebodavägen 23A, Solna 171 65, Sweden. [2] Department of Medical Biochemistry & Biophysics, Division of Biophysics, Karolinska Institutet, Scheeles väg 2, Stockholm 171 77, Sweden. [3] School of Biological Sciences, Nanyang Technological University, 61 Biopolis Drive (Proteos), Singapore 138673, Singapore. [4] Department of Pharmacy, Uppsala University, BMC, Box 580, Uppsala SE-751 23, Sweden. [5] School of Biological Sciences, Nanyang Technological University, SBS-04s-45, 60 Nanyang Drive, Singapore 639798, Singapore. [6] Uppsala University Drug Optimization and Pharmaceutical Profiling Platform (UDOPP), Department of Pharmacy, Uppsala University, BMC, Box 580, Uppsala SE-751 23, Sweden. [7] Science for Life Laboratory Drug Discovery and Development platform, Uppsala University, Uppsala SE-751 23, Sweden. [8] Institute of Cellular and Molecular Biology, ASTAR, 61 Biopolis Drive (Proteos), Singapore 138673, Singapore. * These authors contributed equally to this work. † Present address: Clinical Proteomics Mass Spectrometry, Department of Oncology-Pathology, Science for Life Laboratory and Karolinska Institutet, Stockholm, Sweden. Correspondence and requests for materials should be addressed to T.L. (email: thomas.lundback@ki.se) or to P.N. (email: par.nordlund@ki.se).

Therapeutic efficacy is achieved when drugs bind their relevant molecular targets in the physiologically relevant setting. Despite this known fact, insufficient control of target engagement is surprisingly common and contributes to high failure rates in clinical trials[1–3]. Methods that allow for robust measurements of drug target engagement in primary cells, tissues and patient biopsies are thus urgently needed, but have been hard to establish[4,5].

Ligand-induced changes in protein thermal stability are frequently used to monitor binding to isolated proteins in thermal shift assays[6–9]. The recently developed cellular thermal shift assay (CETSA; see Supplementary Note 1 for a list of abbreviations) builds on the discovery that ligand induced thermal shifts can also be measured in the context of cell lysates, whole cells or tissues[10]. This finding effectively allows for biophysical binding studies in native environments—preserving expression levels, posttranslational modifications and the local environment for the endogenous protein. Whereas the original CETSA study included multiple case studies, recent work extends this method to include melting transitions for a significant portion of the proteome, thus expanding the putative use of the methodology to a large number of protein families[11–13]. Of practical importance is that the melting transitions are established for individual proteins by the use of protein affinity reagents[10,14] or quantitative mass spectrometry (MS)[11–13]. As a consequence these measurements are amenable to either high-throughput measurements or proteome-wide multiplexing.

To improve current strategies for drug development, stringent control of target engagement should ideally be established from initial hit identification, through preclinical and clinical development. The same demands apply to the validation of chemical probes discovered in academic settings[2,4,15]. To probe the value of CETSA in earlier stages of the discovery process we applied it for primary screening of thymidylate synthase (TS) in live human myelogenous leukemia cells. TS is a pivotal enzyme in production of thymidine monophosphate and a well validated cancer target[16,17]. Inhibition of TS leads to thymineless death characterized by DNA-damage, chromosomal fragmentation and concomitant induction of apoptosis. Novel classes of TS inhibitors with improved efficacy and resistance profiles could provide important complements to current TS directed drugs, for which there are reports of resistance[18,19].

Here, we show for the first time that a CETSA-based screen for direct physical target engagement constitutes an attractive high throughput screening (HTS) strategy, which allows for the detection of known and novel TS inhibitors with cellular activity. Furthermore, we establish a hit validation strategy, in which time-dependent target engagement is explored in parallel with measurement of intracellular compound concentration. Taken together this provides a sound and efficient strategy to establish control of target engagement from an early stage of the drug discovery process, and which is likely to minimize problems in subsequent stages.

Results

Microplate-based CETSA measuring target engagement of TS. CETSA is based on measurements of remaining soluble target protein against a background of thermally denatured and precipitated proteins following a heat challenge[10,14]. To enable large-scale screening and automation we developed a no-wash immunoassay for TS using AlphaScreen technology in 384-well plates (see Supplementary Figs 1–6 and Supplementary Table 1). As outlined in Fig. 1a the assay workflow starts with a pre-incubation of K562 cells with library compounds or controls to allow cellular uptake, potential compound metabolism and binding to TS. The treated samples in the plates are next transiently heated in a PCR machine, resulting in denaturation and precipitation of intracellular TS unless stabilized by ligand. After cooling to room temperature the cells are lyzed and the remaining (stabilized) levels of TS are measured.

We validated the assay by investigating the response to two drugs of structurally different classes, that is, floxuridine and raltitrexed. Both drugs require intracellular enzymatic conversion prior to high-affinity TS binding[17]. Pyrimidine-based inhibitors, such as floxuridine, bind to TS as the corresponding monophosphate, whereas folate-based drugs, such as raltitrexed, are polyglutamylated and bind TS in a ternary complex with 2′-deoxyuridine 5′-monophosphate (dUMP). Two assay formats were employed for validation. First, the heating was done at a series of different temperatures at a fixed compound concentration to establish aggregation temperature (T_{agg}) curves (Fig. 1b). As expected both drugs resulted in substantial shifts of the thermal stability of TS, thus confirming cellular uptake and intracellular enzymatic conversion to the active forms that bind TS. Based on these curves, 50 °C was selected for further characterization in isothermal dose-response fingerprint ($ITDRF_{CETSA}$) experiments. In these experiments the compound concentration is titrated during the pre-incubation, after which all samples are heated to the same temperature (Fig. 1c,d). Both drugs showed dose-dependent stabilization of TS with half maximal effective concentration (EC_{50}) values in the sub-nM range. Data from parallel experiments using quantitative western blots for assessment of stabilized TS confirmed a significant shift in T_{agg} in the presence of 5 μM of either of these drugs as well as potent dose-dependent stabilization (Supplementary Fig. 7). No change in total TS levels was observable at 5 μM concentration following a 2 h pre-incubation time in the K562 cells, demonstrating that the thermal stabilization data were not influenced by drug-induced changes in total protein levels under these conditions.

Small molecule library screening and hit confirmation. We screened a library of 10,928 compounds at Chemical Biology Consortium Sweden (CBCS; www.cbcs.se) using the TS assay described above. The library includes a structurally-diverse selection of lead-like compounds[20], nucleosides and known drugs (see Methods section for details on the library). These latter subsets include folate and nucleoside-based drugs known to act on TS, suppress thymidine incorporation into DNA and reduce cell proliferation[16,21]. A schematic outline of the screen logistics is available in Supplementary Fig. 8. Screening was performed at a compound concentration of 50 μM and resulted in a reproducible response to the controls and the appearance of several stabilizing compounds (Fig. 2a). Additional graphs illustrating the screen performance and statistics are available in Supplementary Fig. 9 and Supplementary Table 2, respectively. The campaign involved one day of screening, with the AlphaScreen readings done the following morning to ensure equilibration of the antibody recognition.

The threshold for active solutions was calculated at 11.7% stabilization and resulted in 65 hits (Supplementary Table 3). Solutions for 63 of these were available for cherry-picking from our vial-based compound stores, that is, a different source intended for long-term storage. The activities of these solutions were examined in $ITDRF_{CETSA}$ experiments to confirm the screen results (Supplementary Table 3). The majority of compounds with an apparent stabilization above 30% in the primary screen confirmed activity. We also found that 12 out of the 15 vial-based solutions that failed to reproduce activity (highlighted at the bottom of Supplementary Table 3) had been contaminated with

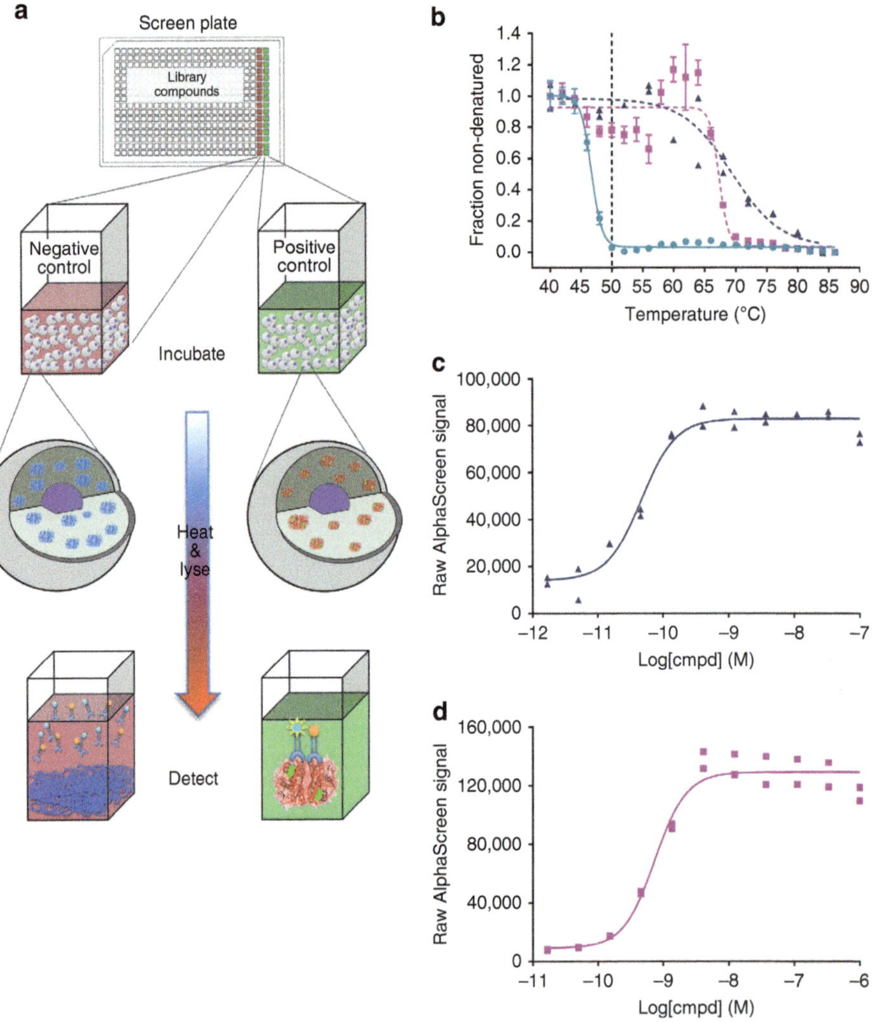

Figure 1 | Development of a no-wash CETSA for human TS. (**a**) Overview of the assay principle with live K562 cells seeded into a 384-well PCR plate. The plate contains controls or library compounds that are taken up by the cells. Following a pre-incubation period the plate is transiently heated for 3 min followed by cooling and cell lysis. Part of the cell lysate is transferred to a detection plate, to which antibodies and AlphaScreen beads are added to allow measurements of remaining soluble TS. (**b**) CETSA derived T_{agg} curves for TS in K562 cells in the presence of DMSO (0.5%) (green circle), 15 μM floxuridine (blue triangle) or 1 μM raltitrexed (magenta square). All data were normalized to the response observed for each treatment condition at the lowest test temperature. The solid line represents the best fit to the Boltzmann sigmoid equation resulting in an apparent T_{agg} of 46.7 ± 0.2 °C for the DMSO control, whereas both floxuridine and raltitrexed stabilized TS above 65 °C (we do not consider higher T_{agg} values reliable as these temperatures influence cell membrane integrity[10]). The vertical dotted line is at 50 °C, the temperature selected for the isothermal screen. Data are provided as the average and standard error of mean (s.e.m.) from two independent experiments performed in duplicate for raltitrexed and as individual data points from one experiment in duplicate for floxuridine. (**c**) $ITDRF_{CETSA}$ of floxuridine (blue triangle) at 50 °C based on raw data from the AlphaScreen readings. The solid line represents the best fit to a saturation binding curve resulting in an EC_{50} of 47 ± 16 pM. Data are provided as two individual data points from one test occasion. (**d**) The corresponding $ITDRF_{CETSA}$ for raltitrexed (magenta square) at 50 °C resulting in an EC_{50} of 0.75 ± 0.2 nM. Data are provided as two individual data points from one test occasion.

highly active compounds because of insufficient tip washing during the transfer from vials to screen library plates (Supplementary Fig. 10). Consequently these hits reproducibly confirmed activity when tested from the contaminated source plates, while the original solutions were inactive. Taken together the confirmation rate was 90% for hits yielding more than 30% apparent thermal stabilization of TS in the original screen.

Fluoropyrimidines, anti-folates and their analogs. The majority of confirmed hits were pyrimidine-based nucleosides and analogs thereof. At the top of the list were substituted 2′-deoxyuridines, including floxuridine (three independent occurrences in the library), 5-trifluoro-2′-deoxythymidine (TFT), and 5-ethynyl-2′-deoxyuridine (EdU) (Fig. 2b and Supplementary Table 3). All of these are known to be taken up and metabolized intracellularly to the active monophosphate forms that interact with TS[16,17]. Novel findings among the nucleosides included the two drugs azacitidine and its deoxyribose analog decitabine, as well as two purine nucleosides (8-bromoadenosine and 8-allyloxy-N2-isobutyryl-2′-deoxyguanosine).

With regards to folate analogs, methotrexate was present at two instances in the library, both as a racemate and as its L form. As expected both appeared as strong hits. The CETSA screen also identified two other marketed drugs, triamterene, a sodium channel inhibitor used to treat hypertension, and pyrimethamine, an inhibitor of dihydrofolate reductase from *Plasmodium falciparum* that is used to treat malaria. They have related

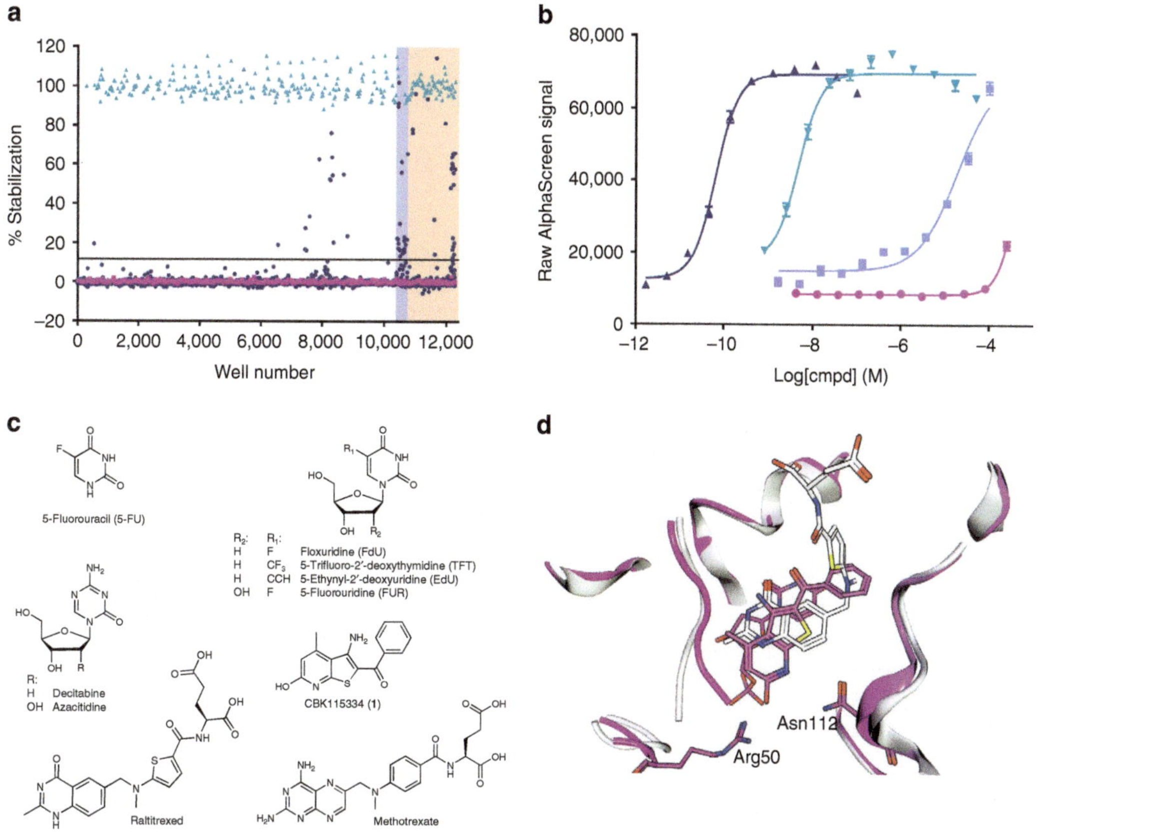

Figure 2 | Primary screen using CETSA to measure target engagement of human thymidylate synthase. (**a**) Scatter plot illustrating normalized screen data, where 0% corresponds to the TS signal observed in the presence of DMSO only (magenta square) and 100% corresponds to the TS signal observed in the presence of 100 nM raltitrexed (green triangle). Data for library compounds at a concentration of 50 μM are shown in blue (blue circle). The hit limit was calculated based on the average plus three standard deviations for the library compounds and is illustrated as a black solid line at 11.7%. The locations of the Prestwick drug set (yellow) and a nucleoside subset (purple) are highlighted. (**b**) $ITDRF_{CETSA}$ data illustrating the ranking of floxuridine (blue upwards triangle), 5-fluorouridine (FUR) (green downwards triangle), and 5-FU (lavender blue square) after 2 h of preincubation time. Data are also included for CBK115334 (magenta circle).The solid lines represent best fits to a saturation binding curve resulting in an apparent EC_{50} of TS at a concentration of 65 ± 9 pM, 47 ± 15 nM, 19 ± 4 μM and 0.46 ± 0.08 mM, respectively. Data are provided as the average and s.e.m. from one independent hit confirmation experiment done in quadruplicate. (**c**) Structures of known drugs and hit compounds discussed in the main text. (**d**) Structure of CBK115334 (magenta) and dUMP bound to TS, shown overlayed on the structure of the complex of raltitrexed (white) and dUMP (PDB 1HVY).

structures and can potentially act as folic acid antagonists[22], but they have not been previously shown to bind TS. Given the scarcity of anti-folates in the hit list we also looked whether there were any obvious false negatives in the screen and confirmed this was not the case (Supplementary Tables 4 and 5).

CBK115334 as a novel TS inhibitor. Besides the pyrimidine- and folate-based inhibitors there were 17 additional weak hits of different chemical classes (Supplementary Table 3). We investigated one of these, CBK115334 or 3-amino-2-benzoyl-4-methylthieno(2,3-b)pyridin-6-ol (**1**), which was chemically distinct from known TS inhibitors (Fig. 2c and Supplementary Fig. 11). It also appeared for the first time as a hit in our screens (Supplementary Table 6). Confirmatory data were obtained using CETSA on K562 cell lysates, which demonstrated a 3.7 °C shift at 200 μM (Supplementary Fig. 12). When applied to isolated recombinant human TS, **1** showed a 2.6 °C shift at 25 μM and a 5.2 °C shift at 100 μM and confirmed binding in the low μM range using surface plasmon resonance (Supplementary Fig. 12). We tested whether binding affected enzymatic activity of TS *in vitro* and observed 60% inhibition at 10 μM concentration and near complete inhibition at 100 μM (Supplementary Fig. 12). A crystal structure of TS with **1** revealed that the compound binds the active site of TS occupying the folate-binding pocket (Fig. 2d). The binding involves π–π stacking interactions with the substrate dUMP and polar interactions with residues lining the catalytic cavity (Asn112 and Arg50 in particular). This constitutes a novel mode of binding as compared with other anti-folates occupying this space[23,24]. Finally, we investigated the impact on cell proliferation in K562 cells. A clear impact was seen with a half-maximal inhibitory concentration (IC_{50}) value just below 100 mM (Supplementary Fig. 12), in line with the weak CETSA response.

Addressing kinetics of compound transport and metabolism. We performed time-traces of the $ITDRF_{CETSA}$ experiments, that is, by varying the time during which cells were exposed to compound prior to the heating step. $ITDRF_{CETSA}$ data obtained after various pre-incubation times are shown in Fig. 3a,b for 5-fluorouracil (5-FU) and floxuridine, demonstrating several orders of magnitude lower potency for 5-FU. The corresponding data on additional nucleosides are available in Supplementary Fig. 13. To examine whether the observed target engagement coincides with appearance of the active forms of these compounds, we monitored levels of compounds and their anticipated active metabolites using liquid chromatography

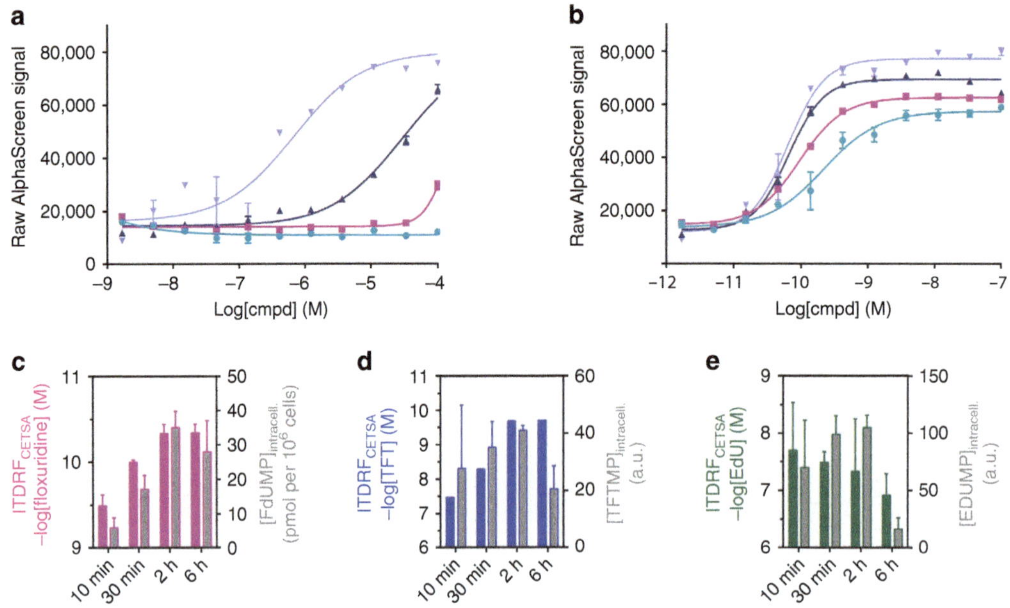

Figure 3 | Time dependence of target engagement and correlation with the appearance of intracellular active metabolites. (**a**) Representative ITDRF$_{CETSA}$ curves for 5-fluorouracil as a function of preincubation time in K562 cells; 10 min (green circle), 30 min (magenta square), 2 h (blue upwards triangle) and 6 h (lavender blue downwards triangle). The solid lines represent best fits to a saturation binding curve function to yield ITDRF$_{CETSA}$ values for half-maximal stabilization of TS. Data are provided as the average and s.e.m. from experiments done in quadruplicate at a single test occasion. (**b**) The corresponding ITDRF$_{CETSA}$ data for floxuridine. (**c**) Half-maximal stabilization of TS (magenta) and intracellular concentration of FdUMP (grey) as a function of preincubation time with floxuridine. The CETSA data are presented as the average and range from two independent experiments. The LC-MS/MS data are provided as the average and s.e.m. from experiments done at three different occasions. (**d**) The corresponding data for TFT (blue) and TFTMP (grey). (**e**) The corresponding data for EdU (green) and EdUMP (grey).

coupled to tandem mass spectrometry (LC–MS/MS)[25,26]. Intracellular and extracellular concentrations as a function of incubation time are shown in Supplementary Fig. 14 for selected nucleosides and their corresponding monophosphate species.

The cellular import and metabolic activation of floxuridine and 5-FU to generate the common active species 5-fluoro-2′-deoxyuridine 5′-monophosphate (FdUMP) require different enzymatic pathways[16,17,27]. CETSA data for floxuridine showed stabilization at low nM concentrations after only 10 min of pre-incubation (Fig. 3b). The potency improved during the first 2 h and persisted throughout the experiment. This time trace was consistent with the intracellular appearance of FdUMP, which was measureable already after 10 min and increased during the first hours of incubation (Fig. 3c). However, for 5-FU the CETSA response increased slowly in the first 6 h (Fig. 3a), with undetectable intracellular levels of FdUMP (Supplementary Fig. 14). Meanwhile the concentration of 5-FU in cells and media remained relatively constant at all time points, thus demonstrating a fast cellular uptake (Supplementary Fig. 14). We hence conclude that the enzymatic conversion of 5-FU to FdUMP is much slower than for floxuridine in K562 cells under these conditions and that the enzymatic conversion to the active species is mirrored by the CETSA responses.

EdU and TFT are structurally related to floxuridine differing only at position 5 of the uracil moiety (Fig. 2c). Although their primary activity on cell viability is believed to result from their misincorporation into DNA, they are also known to inhibit TS following intracellular phosphorylation[28,29]. The uptake and metabolism of TFT, as well as its CETSA time trace, was similar to that observed for floxuridine (Fig. 3d). This was consistent with a build-up of 5-trifluoro-2′-deoxythymidine 5′-monophosphate (TFTMP) and binding to TS in the first hours, generating full target engagement after 2 h of incubation. TFTMP is known to be a tight-binding inhibitor that forms covalent complexes with TS also in absence of the folate-based cofactor[30–32], in line with the observation of a persistent target engagement as measured by CETSA. EdU behaved differently with a more rapid uptake and faster decay of both the extracellular nucleoside and the active form 5-ethynyl-2′-deoxyuridine 5′-monophosphate (EdUMP) (Fig. 3e). The CETSA response was consistent with the fast uptake and activation with an early maximal response that then decayed slightly after the first 2 h, presumably due to the disappearance of the active species.

Phosphorylation and deamination of decitabine. An unexpected hit in the screen was decitabine, which primarily acts as an inhibitor of DNA methyltransferase[33]. The identification of a 2′-deoxycytidine analog as a hit was surprising, but reinforced by the concurrent appearance of the corresponding ribose azacitidine (Fig. 2c). To shed further light on the generation of the active compound, TS stabilization by decitabine itself was first investigated in K562 cell lysates, where activating metabolism is lower because of significant dilution of intracellular enzymes and their substrates. As shown in Supplementary Fig. 15 TS was not stabilized by decitabine in treated lysates. Likewise decitabine did not stabilize recombinant TS in a thermal shift assay, in line with observations for other nucleosides including deoxyuridine, floxuridine and TFT (Supplementary Fig. 15).

The structural analogy to the known nucleoside-based inhibitors of TS triggered the question as to whether decitabine is also phosphorylated, and potentially also deaminated, to generate a TS ligand (Fig. 4a). To investigate the importance of phosphorylation we performed ITDRF$_{CETSA}$ experiments in the presence of DI-82 (ref. 34). This compound is a potent inhibitor of deoxycytidine kinase (DCK) (Fig. 4b), which is required for formation of decitabine monophosphate[35,36]. Dose-dependent stabilization of TS was confirmed in the absence of DI-82,

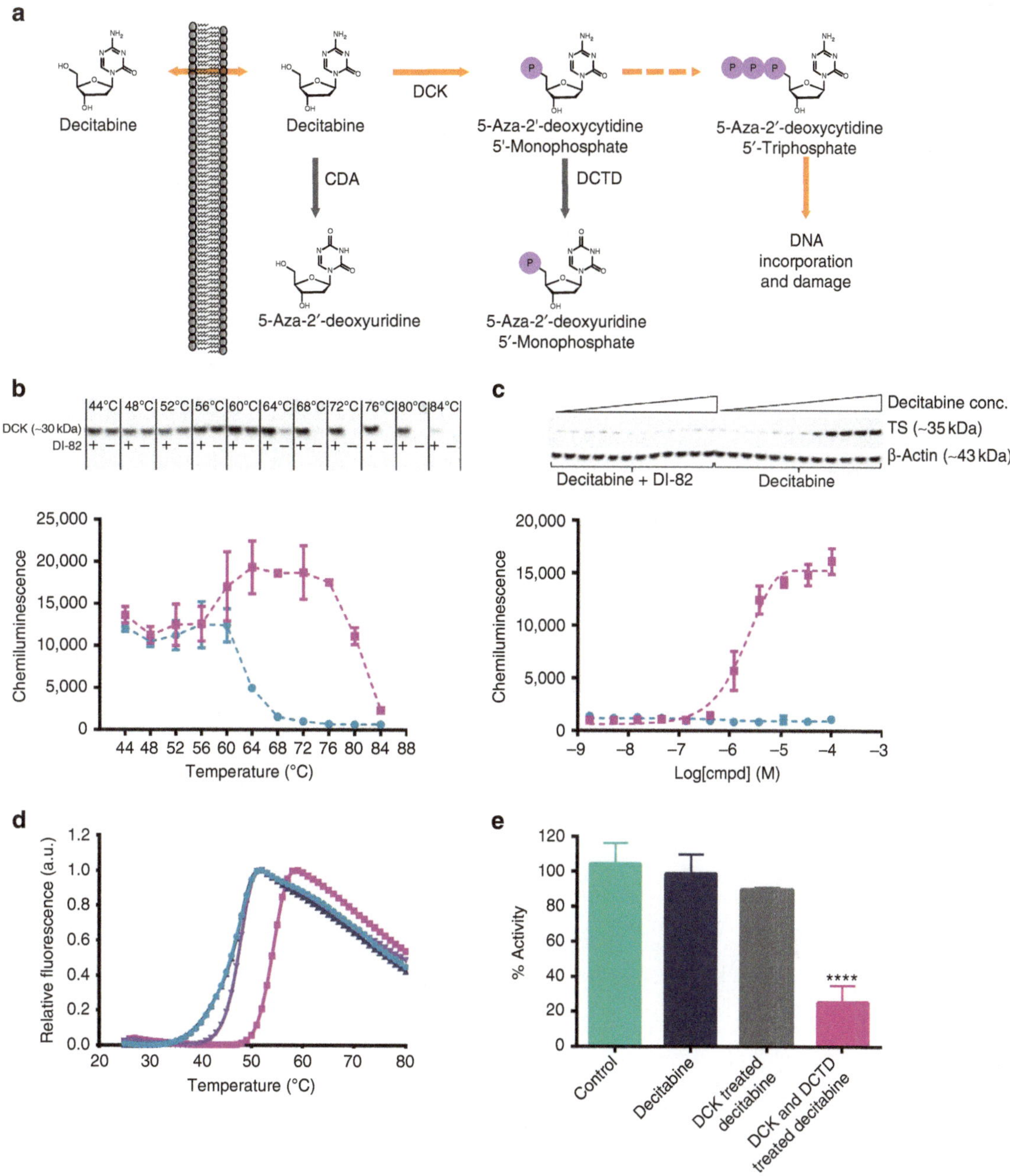

Figure 4 | Target engagement by decitabine is dependent on its metabolic activation. (**a**) Schematic overview of decitabine treatment, cellular uptake and intracellular metabolic conversion. After uptake decitabine is phosphorylated to form 5-aza-2′-deoxycytidine 5′-monophosphate by DCK. This compound is further phosphorylated in two steps to yield the triphosphate that is incorporated into DNA. Cytidine deaminase (CDA) and DCTD are known to be involved in the metabolism and clearance of decitabine[39]. (**b**) T_{agg} experiments for CDK in the absence (green circle) and presence of 200 μM of the DCK inhibitor DI-82 (magenta square). Above the graphs are the chemiluminescence data (full blots are available in Supplementary Fig. 16). The experiments were performed in K562 cells at two independent occasions. (**c**) $ITDRF_{CETSA}$ data for decitabine in the absence (magenta square) and presence of 200 μM of the DCK inhibitor DI-82 (green circle). Full blots are available in Supplementary Fig. 16. The experiments were performed in K562 cells at two independent occasions. (**d**) Normalized thermal shift assay response for recombinant human TS in the absence (magenta square) and presence of 1 mM decitabine without prior enzyme treatment (blue upwards triangle), following DCK treatment (lavender blue downwards triangle) and following treatment with both DCK and DCTD (magenta square). The data are shown as the average and s.e.m. from triplicate samples at one test occasion. (**e**) Enzyme inhibition data for TS in the presence of control and enzymatically treated decitabine samples. The data are shown as the average and s.e.m. from triplicate samples at one test occasion.

whereas its presence at 200 μM completely blunted the ability of decitabine to bind TS (Fig. 4c). However, studies using recombinant DCK to generate decitabine monophosphate resulted in only a marginal stabilization of TS (Fig. 4d). The sample was therefore additionally treated with deoxycytidylate deaminase (DCTD), which is known to deaminate decitabine monophosphate[37], resulting in a thermal shift of nearly 10 °C (Fig. 4d). The inhibitory capacity of these samples mirrored these data, that is, minor inhibition was observed after DCK treatment in the TS enzymatic assay, whereas near full inhibition appeared after treatment with both DCK and DCTD (Fig. 4e). Taken together with the structural analogy to the natural substrate of TS these data strongly infer that the TS ligand is 5-aza-2′-deoxyuridine 5′-monophosphate, that is, the expected product of phosphorylation and deamination of decitabine.

Discussion

Target engagement is essential for efficacy of targeted therapies and validation of new chemical probes[2,4,15]. These validating experiments are ideally performed for many representatives within a chemical series to allow comparisons of structure–activity relationships. To push towards the goal of having a procedure amenable to automation and screening we applied CETSA for assessment of intracellular target engagement at the stage of primary screening. Prior to this work the methodology had been applied on a growing number of drugs and chemical probes[10–14], but it remained challenging to apply to large chemical libraries.

To achieve this evaluation, we developed a homogeneous CETSA and applied it to screening in live, non-engineered cells expressing thymidylate synthase, which is targeted by several different chemical classes of drugs in clinical use[16,17]. Importantly, the screen identified all drugs within the test library that act on TS, as well as novel compounds capable of binding and inhibiting this enzyme. Collectively the known drugs and new inhibitors span over a broad range of affinities. Amongst the new hits was **1**, a μM inhibitor of the purified enzyme with sufficient cell penetration to result in intracellular target engagement and anti-proliferative effects in the high μM range. The binding mode of **1** to TS is partly new and it thus provides a potential starting point for further chemistry optimization.

Several other marketed drugs not previously known to inhibit TS emerged as hits, including triamterene, pyrimethamine and decitabine. As these are clinically-used compounds it is of interest to understand whether there are instances where the interaction with TS plays a role in either efficacy or toxicity. The identification of decitabine illustrates the relevance of monitoring target binding in live cells as this finding was dependent on active cellular metabolism. Decitabine itself was largely inactive on the protein such that, in analogy to the already known uridine-based inhibitors, enzymatic conversion to an active species is a prerequisite for observation of binding. We showed that this conversion does not take place to a significant extent in cell lysates, but can be reproduced with the *in vitro* application of enzymes for phosphorylation and deamination to yield the substrate analog 5-aza-2′-deoxyuridine 5′-monophosphate. Although our data give strong support that a metabolite of decitabine yield significant cellular inhibition of TS, further studies are required to determine whether this is relevant for polypharmacology or toxicity at typical therapeutic doses.

Time-traces of $ITDRF_{CETSA}$ were used to analyze different scenarios for cellular target engagement, thereby integrating aspects of drug transport and metabolism. Combination of these results with measurements of intracellular drug and metabolite concentrations allowed for a comprehensive dissection of cellular drug kinetics. Overall, the intracellular concentrations of the active species of the drugs correlated with the observed target engagement, consistent with the notion that CETSA directly reports on target binding. The combination of high-throughput target engagement studies with LC–MS/MS measurements of intracellular concentrations of drugs and metabolites constitutes a new paradigm for hit validation and optimization in the discovery of chemical probes and drugs. Importantly, CETSA is applicable to studies in native cells and tissue samples[10,14]. Thus the basic scenarios for compound metabolism and target engagement derived from these cell culture studies should be possible to translate towards studies of activities and resistance development to drugs in man. Of particular interest in this regard was the observation that we nearly missed the identification of 5-FU in the screen because of the relatively slow appearance of target engagement. It will be interesting to extend these experiments to patient cells.

The present work demonstrates that CETSA constitutes a robust high-throughput screening strategy that allows for target proteins to be approached in their natural cellular environment. This is in contrast to the majority of targeted cellular HTS assays as these rely on overexpressed and tagged proteins. Since CETSA does not require engineered cells or compounds, it could be particularly attractive for screening in primary cells, tissues or patient-derived material. Our approach can be applied to a large number of different proteins, with the generic assay development path being established in this work. The work also introduces the combination of time-dependent $ITDRF_{CETSA}$ and measurement of intracellular concentrations of metabolites as a stringent approach for hit validation, where the same assay format can be utilized and provide value throughout the drug discovery process.

Methods

Cell culture conditions in AlphaScreen-based experiments. Human myelogenous leukemia cell line K562 (ATCC no. CCL-243) were cultured in RPMI-1640 (SH30027.01, HyClone) supplemented with 10% fetal bovine serum (SV30160.03, HyClone), 0.3 $g\,l^{-1}$ L-glutamine (G7513, Sigma-Aldrich) and 100 units ml^{-1} Penicillin-Streptomycin (P4333, Sigma-Aldrich). The same cell medium composition was used for all experiments unless otherwise stated.

Development of an AlphaScreen-based assay for TS. Measurements of remaining levels of soluble TS in cell lysates were achieved based on an AlphaScreen-based assay. Establishment of this assay required the identification of a pair of antibodies that simultaneously recognize TS (Supplementary Fig. 1). Combinations of four mouse-derived and three rabbit-derived antibodies directed towards different epitopes of TS (see Supplementary Table 1) were tested for this ability. Four different conditions were tested for each pair, with two of those being the absence and presence of target protein. Given our previous experience of ligand induced quenching of protein target recognition by the antibody pair[14] we also included a control containing an excess of dUMP with and without the additional presence of raltitrexed, which are known binders to the active sites of TS. Recombinant TS, diluted in 1 × AlphaLISA buffer (AL000F, PerkinElmer), was preincubated at room temperature in the presence of buffer only, 100 μM dUMP or 100 μM dUMP and 10 μM raltitrexed in a total volume of 4 μl in a ProxiPlate (#6008280, PerkinElmer). After this preincubation all 12 possible combinations of antibody pairs were added to the sample in a volume of 4 μl followed by incubation for 30 min at room temperature. A mix of AlphaScreen acceptor and donor beads was finally added in a volume of 4 μl under subdued light and allowed to incubate at room temperature for 2 h before reading in an Envision plate reader (PerkinElmer). Final concentrations of the reagents in the detection step were 2 nM recombinant TS, 2 nM of each antibody, 40 $\mu g\,ml^{-1}$ AlphaScreen anti-mouse donor beads (#AS104D, PerkinElmer) and 10 $\mu g\,ml^{-1}$ AlphaScreen anti-rabbit acceptor beads (#AL104C, PerkinElmer). The plates were sealed with TopSeal-A PLUS (6050185, PerkinElmer). The data were analyzed using microsoft excel and GraphPad Prism 6.

Four different antibody pair combinations based on sc-376161, WH0007298M1, 15047-1-AP and D5B3 (Supplementary Table 1) were selected from the antibody screen to study the kinetics of their recognition of TS in cell lysate. Two batches of 7.5 million K562 cells per ml in supplemented cell culture medium were prepared to serve as max and min controls. One culture was left at room temperature and the other was heated to 52 °C for 3 min in a PCR machine (TECHNE TC-PLUS thermal cycler). Both batches of cells were then lysed by the addition of an equal volume of 2X AlphaScreen SureFire Lysis Buffer (TGRLB100ML, PerkinElmer). After thorough mixing, 4 μl aliquots of the lysates were transferred to a ProxiPlate and detected and analyzed as described above except that the bead incubation was performed at 2 h, 6 h and overnight.

The sc-376161 and 15047-1-AP antibodies were titrated to match their concentrations with the AlphaScreen bead concentrations, that is, to ensure they do not exceed concentrations where hook effects are observed. Fifteen million K562 cells per ml were prepared in supplemented cell culture medium and lysed as described above. Aliquots of 4 μl were transferred to a ProxiPlate followed by the addition of 4 μl of a mix of different concentrations of the two antibodies (final concentrations of each antibody in the detection varied between 0 and 10 nM). Detection and analysis was done as described above, except the bead incubation was performed overnight.

Optimization of cell numbers was achieved by serial dilution of a cell suspension of K562 cells in supplemented cell culture medium. Each sample was then split into two aliquots, which were either kept at room temperature or heated to 52 °C as described above. Both aliquots were then lysed as described above. After thorough mixing, 3 µl of the lysates were transferred to a ProxiPlate followed by the addition of 6 µl of a mix of antibodies and AlphaScreen acceptor and donor beads in AlphaLISA buffer under subdued light. Detection and analysis was achieved as described above except the antibody concentrations were modified to 1 nM 15047-1-AP and 0.4 nM sc-376161.

The control experiment, in which recombinant TS was seeded to cell lysates was prepared based on a serial dilution of recombinant human TS. Dilutions were done in equal volumes of supplemented cell medium and 2 × AlphaScreen SureFire Lysis Buffer. K562 cells at a cell density of 2 million cells per ml were lysed as described above and split in two samples, which were either kept at room temperature or heated to 52 °C as described above. A 10 µl aliquot of each TS dilution was then added to the same volume of each of the two lysates as well as to the mixture of cell medium and lysis buffer. A 3 µl aliquot of each sample was then transferred to a ProxiPlate and detected and analyzed as described above.

Thermal aggregation experiments using AlphaScreen. Floxuridine and raltitrexed were diluted from dimethyl sulfoxide (DMSO) stock solutions to concentrations of 30 µM and 20 µM respectively in supplemented cell culture medium (final DMSO content 1%). These solutions were transferred in a volume of 10 µl to a skirted Twin.tec PCR 96-wellplate (0030 128 672, Eppendorf). A suspension of K562 cells in a volume of 10 µl and a density of 10 million cells per ml were then added to all wells. The PCR plates containing the compounds and cells were sealed with a breathable plate seal (3345, Corning) and incubated for 2 h in a humidified incubator at 37 °C and 5% CO_2. The cells were then transiently heated to different temperatures ranging from 40 °C to 86 °C for 3 min, followed by a controlled cooling to 20 °C for 1 min using a real-time PCR machine (ProFlex, Applied Biosystems). After the heating step the plate was centrifuged briefly (1,000 × *g* for 1 min) followed by lysis of the heated cells by the addition of 20 µl of 2 × AlphaScreen SureFire Lysis Buffer using a Flexdrop IV (PerkinElmer). To ensure sufficient lysis the cell lysates were mixed by 10 repetitive aspiration and dispensing cycles using a Bravo liquid handling platform (Agilent). The lysates (3 µl) were then transferred to 384-well ProxiPlates followed by the addition of 6 µl of a mix of antibodies and AlphaScreen acceptor and donor beads in AlphaLISA buffer under subdued light. Final concentrations of the assay reagents in the detection step were 1 nM rabbit polyclonal anti-TS IgG (15047-1-AP, Proteintech), 0.4 nM mouse monoclonal anti-TS IgG (sc-376161, Santa Cruz), 40 µg ml^{-1} AlphaScreen anti-mouse donor beads and 10 µg ml^{-1} AlphaScreen anti-rabbit acceptor beads. The plates were sealed with TopSeal-A PLUS and incubated over night at room temperature prior to detection in an Envision plate reader. The data were analyzed using microsoft excel and GraphPad Prism 6.

Composition and storage of the primary screening set. The library of compounds applied in this screening campaign consists of 10,928 compounds and is part of the primary screening set at CBCS. The majority of these compounds was donated by Biovitrum AB and originates from both in-house and commercial sources. Compounds included in the primary screening set were selected to represent a diverse selection of a larger set of 65,000 compounds, while keeping a certain depth to allow crude structure–activity relationship studies. The selection was also biased towards lead-like and drug-like profiles with regards to molecular weight, hydrogen bond donors/acceptors and LogP[20]. The library also includes a nucleoside set from Berry & Associates and a set of approved drugs from Prestwick. Compound stock solutions at 10 mM in DMSO are stored frozen at approximately − 20 °C in individual capped tubes in REMP 96 Storage Tube Racks. The racks are stored in a REMP Small-Size Store, which allows cherrypicking while the solutions are still frozen to minimize repetitive freeze-thaw cycles. For screening purposes the compound solutions have been replicated from the REMP racks to Labcyte 384 LDV plates (LP-0200) and then further into Labcyte 1536 HighBase plates (LP-03730) to enable dispensing using acoustic liquid handling equipment.

Compound handling. Assay ready plates were prepared by transferring 200 nl of the 10 mM DMSO solutions of compounds and controls by means of acoustic dispensing (Echo 550, Labcyte) to 384-well polypropylene plates (784201, Greiner). Compounds were placed in columns 1–22. DMSO controls were placed in column 23 and raltitrexed controls were placed in column 24. The assay ready plates were heat sealed with a Peelable Aluminium seal (24210-001 Agilent) using a thermal microplate sealer (PlateLoc, Agilent) and stored at − 20 °C until use. At the day of the experiment the plates were allowed to thaw for 30 min followed by a brief centrifugation step (1,000 × *g* for 1 min) prior to removal of the seal. The compounds were then diluted with 20 µl supplemented cell culture medium using a Multidrop Combi reagent dispenser (Thermo Scientific). Finally 5 µl of the diluted compounds were transferred to a 384 well hardshell PCR plate (HSR480, BIORAD) using a Bravo liquid handling platform equipped with a 384-well head (Agilent). The final concentrations in the incubation with cells (see below) were 50 µM of test compounds and 100 nM of the positive control raltitrexed. The final concentration of DMSO in the assay was 0.5% in all samples.

For the $ITDRF_{CETSA}$ experiments 11-point dose-response curves with three-fold difference in concentration between wells were generated using the Bravo liquid handling system (all serial dilutions were done in 100% DMSO). The final highest concentrations of the test compounds in the incubation with cells were ranging from 100 nM–50 µM depending on estimated potency. The concentration of DMSO was 0.5% in all samples. The final concentration of the positive control raltitrexed was 100 nM. The assay ready plates were prepared as outlined above for the screen.

Screening and dose-response characterization by AlphaScreen. The screen procedure started with the addition of 5 µl of a suspension of K562 cells at a density of 10 million cells per ml to all wells of a 384 well hardshell PCR plate (HSR480, BIORAD) using an electronic multichannel pipette (Biohit). The plates were then heat sealed with Peelable Aluminium seal in a thermal microplate sealer (PlateLoc, Agilent) and allowed to incubate for 2 h in an incubator at 37 °C and 5% CO_2. For the time-course experiments the incubation times were altered to include also 10 min, 30 min and 6 h. To allow gas exchange during the longer incubation times these plates were instead sealed with at breathable plate seal (3345, Corning). After the incubation step the plates were transiently heated at 50 °C for 3 min followed by a controlled cooling to 20 °C for 1 min using a real-time PCR machine (Light-Cycler480 system, Roche). Plate handling was then as described above for the thermal aggregation experiments. The data were analyzed using microsoft excel and GraphPad Prism 6.

Screen and $ITDRF_{CETSA}$ data analysis. Screen data were imported into microsoft excel and normalized for each compound based on the negative and positive controls on each plate, that is, with the response in the presence of DMSO defining 0% stabilization and the response in the presence of 100 nM raltitrexed defining 100% stabilization. A calculation of the average and standard error of means for each set of controls also allowed an illustration of how these responses varied over the 32 screening plates. The Z' factor[38] is commonly used as a measure of how well the assay separates between the controls and this was calculated as described based on the calculated averages and standard deviations of the controls on a per plate basis. For the T_{agg} shift and the $ITDRF_{CETSA}$ experiments the data were analyzed in GraphPad Prism using the Boltzmann sigmoid equation and the saturation binding curve (rectangular hyperbola; binding isotherm) function, respectively. As already discussed[14] these methods make use of equilibrium models for data analysis although the methodology depends on the irreversible aggregation of denatured material. For this reason we refer to the observed responses as apparent and isothermal dose-response fingerprints and are careful with any quantitative interpretations, being well aware of their dependency on experimental conditions. Experiments are on-going to address the quantitative interpretation of CETSA data.

Measurements of identity and purity of test compound solutions. Assessments of identity and purity of the test solutions that were used for hit confirmation purposes, that is, those being stored in REMP vials, was done by means of high-pressure liquid chromatography coupled to mass spectrometry (HPLC–MS). A small aliquot of each test solution (2 µl of a 10 mM solution) was placed in a 96-well plate (267245, Nunc) and diluted with 20 µl of methanol. The plate was then placed in an Agilent 1,100 HPLC UV/MS with electrospray ionization (ESI +). The HPLC method was based on an ACE C8 3 µm column (3.0 × 50 mm) and a mobile phase (CH_3CN)/(0.1% TFA/H_2O). All solvents were HPLC grade and absorbance was monitored at 220 nm. Compounds that did not give satisfactory data were re-analyzed using a method based on a Waters XBridge C18 3.5 µm column (3.0 × 50 mm), 3.5 min gradient mobile phase (CH_3CN)/(10 mM NH_4HCO_3/H_2O). The instrument software was used to integrate the UV response for each peak and provided a list of the peaks and their associated masses. The estimated purity was calculated based on the integrated area for the expected mass compared with the areas of all other peaks. The result was manually controlled and if there were deviations from the expected outcome a meticulous investigation of the UV-response and MS was performed.

Cell viability assay. A concentration–response curve of CBK115334 was generated using the Bravo liquid handling system (serial dilution of a 50 mM stock solution was done in 100% DMSO). A total of 150 nl of the serially diluted solutions and controls (positive control 0.67 mM staurosporine and negative control DMSO) were transferred to a white 384-well assay plate (3570, Corning) by means of acoustic dispensing (Echo 550, Labcyte). A Multidrop reagent dispenser (Thermo Scientific) was used to dispense 30 µl of a K562 cell suspension at a density of 33 × 10^3 cells per ml in supplemented cell culture medium. The cells were incubated at 37 °C in the presence of 5% CO_2 for 72 h before addition of 30 µl CellTiter-Glo Luminescent Cell Viability Assay reagent (Promega) using a Multidrop Combi reagent dispenser (Thermo Scientific). The plate was placed on a plate shaker for 15 min prior to detection of the luminescence signal in an Envision plate reader (PerkinElmer). The final highest concentration of CBK115334 in the incubation with cells was 250 µM and the final concentration of the positive control staurosporine was 3 µM. All samples contained 0.5% DMSO. The data were analyzed using microsoft excel and GraphPad Prism 6.

Chemicals and buffers in western blot-based experiments. The cell lysis buffer contained 100 mM 4-(2-hydroxyethyl)-1-piperazineethanesulfonic acid (HEPES, pH 7.5), 1 mM Tris-(2-carboxyethyl)phosphine hydrochloride (TCEP) and 10 mM magnesium chloride (Sigma-Aldrich) supplemented with complete (EDTA-free) protease inhibitor cocktail from Roche (Switzerland). Tris-buffered saline with tween (TBST) buffer (150 mM NaCl, 0.05% (v/v) Tween-20, 50 mM Tris-HCl buffer at pH 7.6) was prepared by dissolving TBS-TWEEN tablets obtained from Merck KGaA (Darmstadt, Germany) in ddH_2O. The blocking buffer consisted of 5% (w/v) non-fat milk (Semper AB, Sundbyberg, Sweden) diluted in tris-buffered saline with tween. Hank's Balanced Salt Solution (HBSS) was from Gibco/Life Technologies. Raltitrexed monohydrate, dUMP and decitabine was purchased from Sigma-Aldrich and Selleckchem, respectively. DI-82 was kindly provided by Prof. Caius G. Radu and Raymond M. Gipson at the Department of Molecular and Medical Pharmacology, University of California, Los Angeles.

Cell lines and cultures in western blot-based experiments. Human cell line K562 (ATCC no. CCL-243) was cultured in RPMI-1640 medium (Sigma-Aldrich) supplemented with 0.3 $g l^{-1}$ L-glutamine and 10% fetal bovine serum (FBS, Gibco/Life Technologies, Carlsbad, CA, USA), 100 units per ml penicillin and 100 units per ml streptomycin (Gibco/Life Technologies). Short-term passages (<20) were used for experiments.

Cell lysate thermal shift experiments. For the cell lysate thermal shift experiments, cultured K562 cells were harvested and washed with Hank's Balanced Salt Solution. The cells were diluted in lysis buffer supplemented with complete protease inhibitor cocktail. The cell suspensions were freeze-thawed three times using liquid nitrogen and passed through a 27″ gauge needle five times. The soluble fraction (lysate) was separated from the cell debris by centrifugation at 20,000 × *g* for 20 min at 4 °C. For the thermal aggregation curve experiments cell lysates were diluted with lysis buffer supplemented with 200 μM dUMP and divided into two aliquots, with one aliquot being treated with ligand and the other aliquot with vehicle (control). After 10 min incubation at room temperature the respective lysates were divided into smaller (50 μl) aliquots and heated individually at different temperatures for 3 min in a Veriti thermal cycler (Applied Biosystems/Life Technologies) followed by cooling for 3 min at room temperature. The heated lysates were centrifuged at 20,000 × *g* for 20 min at 4 °C in order to separate the soluble fractions from precipitates. The supernatants containing the remaining soluble proteins were transferred to new 0.2 ml microtubes and analyzed by sodium dodecyl sulfate polyacrylamide gel electrophoresis (SDS-PAGE) followed by western blot analysis.

For the in-cell experiments K562 cells were harvested and resuspended with culture medium to a cell density of 5 million cells per ml and seeded into T25 flasks. Cells were treated with either raltitrexed, floxuridine, DI-82 or vehicle (DMSO) for 2 h in an incubator at 37 °C and 5% CO_2. The cell suspensions were then divided into 100 μl aliquots in 0.2 ml tubes and heated at designated temperatures ranging from 40 to 84 °C for 3 min in a Veriti thermal cycler (Life Technologies) followed by 3 min of cooling at room temperature. The heat-treated cell suspensions were freeze-thawed three times using liquid nitrogen and a heating block set at 25 °C. Tubes were gently vortexed between the freeze-thaw cycles. The resulting cell lysates were centrifuged at 20,000 × *g* for 20 min at 4 °C. The supernatant was removed from the cell debris and aggregates and the remaining soluble TS was analyzed using western blot.

For the $ITDRF_{CETSA}$ in cell experiments, raltitrexed and floxuridine were serially diluted to generate an 11 point dose–response curve with three-fold difference in concentration between each point. K562 cells were treated with each respective compound concentrations and one vehicle as control in 100 μl aliquots in 0.2 ml tubes for 2 h in an incubator at 37 °C and 5% CO_2. The cell aliquots were heated at 50 °C and analyzed with western blot following the procedure described above.

For the decitabine metabolism, decitabine was serially diluted to generate an 11 point dose–response curve and an $ITDRF_{CETSA}$ experiment was performed as described above in absence and presence of 200 μM of the DCK inhibitor DI-82 (referred to as 12R in the main text of the original publication and DI-82 in the Supplementary Material)[34].

SDS-PAGE and western blot. NuPage Novex Bis-Tris 4–12% polyacrylamide gels with NuPAGE MES SDS running buffer (Life Technologies) were used for separation of proteins in the samples. Proteins were transferred to nitrocellulose membranes using the iBlot2 blotting system (Life Technologies). Primary antibodies anti-TS (D5B3) XP (Cell Signaling), anti-dCK (sc-393099), anti-β-actin (sc-69879); secondary goat anti-mouse HRP-IgG (sc-2055) and bovine anti-rabbit HRP-IgG (sc-2374) antibodies (Santa Cruz Biotechnology, Santa Cruz, CA, USA) were used for immunoblotting. All membranes were blocked with blocking buffer; standard transfer and western blot protocols recommended by the manufacturers (listed above) were used. All antibodies were diluted in blocking buffer. The membranes were developed using Clarity Western ECL substrate HRP-Substrate (Bio-Rad) according to the manufacturer's recommendations. Chemiluminescence intensities were detected and quantified using a ChemiDoc XRS + imaging system (Bio-Rad) with Image Lab software (Bio-Rad).

Expression and purification of human thymidylate synthase. The gene encoding human TS (NM_001071.2) was subcloned into the pNIC28-Bsa4 vector and expressed in Rosetta BL21-DE3 Escherichia coli (Novagen) in Terrific Broth media supplemented with 50 $\mu g\, ml^{-1}$ of kanamycin and 34 $\mu g\, ml^{-1}$ chloramphenicol. Cells were grown at 37 °C until $OD_{600\,nm}$ reached about 2.0 and induced with 0.5 mM isopropyl-beta-D-1-thiogalactopyranoside (IPTG) at 18 °C overnight. The cells were harvested by centrifugation at 4,500 × *g* for 15 min at 15 °C. The cell pellet was re-suspended in lysis buffer (100 mM HEPES, 500 mM NaCl, 10 mM imidazole, 10% (v/v) glycerol and 1 mM TCEP at pH 8.0) supplemented with 1:1,000 (v/v) EDTA-free protease inhibitor cocktail (Calbiochem) and 125 $U\, ml^{-1}$ of Benzonase (Merck). Cells were lysed by sonication on ice at 70% amplitude, 3 s on/off for 3 min. The lysate was clarified by centrifugation at 47,000 × *g* for 25 min at 4 °C, and the supernatant was filtered through a 1.3 μm syringe filter to remove cell debris. The cell-free extract was loaded on a pre-equilibrated HisTrapTM HP column (GE Healthcare) in IMAC wash buffer 1 (20 mM HEPES, 500 mM NaCl, 10 mM imidazole, 10% (v/v) glycerol and 1 mM TCEP at pH 7.5) and subsequently washed with 20 column volumes (CVs) of IMAC wash buffer 1 and 15 CVs of IMAC wash buffer 2 (20 mM HEPES, 500 mM NaCl, 25 mM imidazole, 10% (v/v) glycerol and 1 mM TCEP at pH 7.5). Bound protein was eluted with 5 CVs of elution buffer (20 mM HEPES, 500 mM NaCl, 500 mM imidazole, 10% (v/v) glycerol and 1 mM TCEP at pH 7.5) and loaded onto a HiLoad 16/60 Superdex-200 column (GE Healthcare) pre-equilibrated with buffer (20 mM HEPES, 300 mM NaCl, 10% (v/v) glycerol, and 1 mM TCEP at pH 7.5). Based on Nu-PAGE gel results pure protein fractions were pooled and concentrated using a 50 kDa cutoff centrifugal driven filter concentrator (Sartorius Stedium Biotech). The protein concentration was determined by the absorbance at 280 nm using a Nanodrop spectrophotometer (Thermo Scientific).

Thermal shift *in vitro* assay on recombinant protein. The assay was performed on the iCycler iQ Real Time PCR Detection System (Bio-Rad), using the 96-well thin-wall PCR plate (Bio-Rad). The experiment was conducted in a buffer containing 20 mM HEPES at pH 7.5 and 150 mM NaCl. A total volume of 25 μl solution containing 0.2 $mg\, ml^{-1}$ protein, compounds and × 5 Sypro Orange dye (Invitrogen) diluted from 5,000 × stock was dispensed into the 96-well plate. The same amount of DMSO was added in the control wells. The plates were sealed with Microseal B adhesive sealer (Bio-Rad) and heated in iCycler from 25 to 80 °C (56 heating cycles in 28 min). Fluorescent filter used for Sypro Orange measurements was $\lambda_{excitation} = 492$ nm and $\lambda_{emission} = 610$ nm. The calculation of

Table 1 | X-ray diffraction data and refinement statistics.

	TS-dUMP-CBK115334
Data collection	
Space group	$P4_32_12$
Cell dimensions	
a, b, c (Å)	108.2, 108.2, 313.9
α, β, γ (°)	90, 90, 90
Resolution (Å)	30-3.1 (3.2-3.1)*
R_{merge}	0.147 (0.619)
$I/\sigma I$	15.56 (4.98)
Completeness (%)	99.9 (99.9)
Redundancy	11.9 (10.9)
Refinement	
Resolution (Å)	30-3.1 (3.2-3.1)
No. reflections	34,453
R_{work}/R_{free}	18.0/26.8
No. atoms	
Protein	13,596
Ligand dUMP	186
Ligand CBK115334	96
Water	11
B-factors	
Protein	45.3
Ligand dUMP	45.1
Ligand CBK115334	51.9
Water	30.9
R.m.s. deviations	
Bond lengths (Å)	0.014
Bond angles (°)	1.096

*Highest resolution shell is shown in parenthesis.

the midpoint of the curves (T_m) was performed using the software package XLfit from IDBS within microsoft excel.

TS enzyme inhibition assay. Enzymatic activity of recombinant human TS was measured spectrophotometrically at 340 nm by monitoring the absorbance change during the conversion of 5,10-methylenetetrahydrofolate to dihydrofolate using an Infinite M200 spectrometer (Tecan). Measurements were carried out at room temperature and in a buffer of 50 mM Tris at pH 7.5 and 150 mM NaCl. Initial velocities were measured with 250 nM of purified protein, 100 μM 5,10-methylenetetrahydrofolate and 100 μM dUMP in the presence of compounds using the same amount of DMSO as control. Initial rates and activity were analyzed with the software package Prism (GraphPad Software).

Surface plasmon resonance. Recombinant human TS protein (25 μg ml^{-1} in a buffer of 10 mM sodium acetate at pH 5.0, 1 mM dUMP and 200 μM methotrexate) were captured on Sensor Chip S-CM5 via amine coupling to a level of ~5,000 resonance units using Biacore T-200. Raltitrexed was used as a positive control to ensure that the protein remained active after immobilization and during the run. A concentration series (20 nM to 10 μM) of CBK115334 was injected over the prepared surface for 60 s and allowed to dissociate for 60 s with a flow rate of 70 μl min^{-1} at 25 °C. The assay buffer was 20 mM HEPES, 150 mM NaCl, pH 7.5 and 0.005% Tween-20 supplemented with 1% DMSO and 200 μM dUMP. Response data was processed using the BIAevaluation software. Responses were double referenced and solvent-corrected. The data sets were fitted to 1:1 steady-state model for determination of binding constants.

Decitabine treatment with DCK and DCTD. Decitabine at a concentration of 1 mM was incubated with 1 mg ml^{-1} of recombinant DCK and 2.5 mM ATP in a buffer of 50 mM Tris at pH 7.5, 150 mM NaCl and 0.5 mM $MgCl_2$ at room temperature for 60 min to generate the corresponding monophosphates. To probe for nucleotide deamination the samples were further treated with 1 mg ml^{-1} of recombinant DCTD in the presence of 1 mM ATP and 10 μM $ZnCl_2$ for 20 min at room temperature. After incubation, the samples were heated at 95 °C for 10 min and centrifuged for 10 min at the highest speed at 4 °C using a benchtop centrifuge. The supernatant was tested without dilution in thermal shift *in vitro* assays and the TS enzyme inhibition assay. The supernatants from samples without nucleoside were used as the treatment controls.

Crystallization and structure determination. TS protein crystallized in sitting drops comprising equal volume of protein (about 24 mg ml^{-1}) and reservoir solution at 20 °C. The crystallization condition was composed of 0.1 M sodium cacodylate at pH 6.5 and 15% PEG 4,000. Crystals were soaked with 1 mM compound and 2 mM dUMP in cryo-protectant buffer containing 0.1 M sodium cacodylate at pH 6.5 and 25% PEG 4,000 and 10% DMSO for 15 min, followed by flash frozen in liquid nitrogen. Data collection was performed on beamline MX1 at Australian Synchrotron. X-ray diffraction data was collected at 100 K with a wavelength of 0.9537 Å. It should be noted that CBK115334 is subject to keto-enol isomerization. While the resolution is not sufficient to distinguish between these, the included illustrations are based on the enol form (both forms make interactions with Asn112 and Arg50, but with different donor–acceptor pairs). The structure was solved by molecular replacement using Phaser with the hTS-dUMP-raltitrexed structure (PDB code 1HVY) as the search model. Structure was refined with phenix refine. Ligand structures and restraints files were generated using eLBOW. In the TS-dUMP-CBK115334 complex structure, 90.43% of residues were in favoured regions and 7.92% of residues were in allowed regions. The data collection parameters and refinement statistics are summarized in Table 1. An image of the electron density map of the active site in the co-crystal structure is available in Supplementary Fig. 17.

Table 2 | Monitored LC-MS/MS transitions.

Compound/Metabolite	Parent and daughter ions
5-FU	128.8>41.9
FdU	244.9>155.0
FdUMP	324.9>195.0
FUR	260.9>171.0
EdU	250.9>135.8
EdUMP	330.9>195.0
TFT	249.9>179.7
TFTMP	374.9>179.4
Decitabine	227.0>93.8
Decitabine-MP	307.0>195.0
Pyrimethamine*	248.8>176.9
CBK115334*	284.9>188.9

*Analyzed with electrospray ionization in positive mode.

Intracellular compound and metabolite concentrations. Intracellular concentrations of compounds were measured as previously described[25,26]. Briefly, K562 cells were incubated with the compounds for a predefined time at 37 °C in a 5% CO_2 atmosphere. After incubation, the cells were centrifuged (300 × *g* for 5 min) and a medium sample (supernatant) was collected and diluted 1:10 in a 50 nM warfarin solution (internal standard). The cells were washed with phosphate buffered salt solution and lysed with methanol. Methanol was evaporated and the cell samples were reconstituted in 50 nM warfarin. Samples were analyzed with LC–MS/MS with electrospray ionization in negative mode with transitions monitored as listed in Table 2.

References

1. Morgan, P. *et al.* Can the flow of medicines be improved? Fundamental pharmacokinetic and pharmacological principles toward improving Phase II survival. *Drug Discov. Today* **17**, 419–424 (2012).
2. Bunnage, M. E., Chekler, E. L. P. & Jones, L. H. Target validation using chemical probes. *Nat. Chem. Biol.* **9**, 195–199 (2013).
3. Cook, D. *et al.* Lessons learned from the fate of AstraZeneca's drug pipeline: a five-dimensional framework. *Nat. Rev. Drug Discov.* **13**, 419–431 (2014).
4. Simon, G. M., Niphakis, M. J. & Cravatt, B. F. Determining target engagement in living systems. *Nat. Chem. Biol.* **9**, 200–205 (2013).
5. Durham, T. B. & Blanco, M.-J. Target engagement in lead generation. *Bioorg. Med. Chem. Lett.* **25**, 998–1008 (2015).
6. Pantoliano, M. W. *et al.* High-density miniaturized thermal shift assays as a general strategy for drug discovery. *J. Biomol. Screen.* **6**, 429–440 (2001).
7. Ericsson, U. B., Hallberg, B. M., DeTitta, G. T., Dekker, N. & Nordlund, P. Thermofluor-based high-throughput stability optimization of proteins for structural studies. *Anal. Biochem.* **357**, 289–298 (2006).
8. Senisterra, G. A. *et al.* Screening for ligands using a generic and high-throughput light-scattering-based assay. *J. Biomol. Screen* **11**, 940–948 (2006).
9. Niesen, F. H., Berglund, H. & Vedadi, M. The use of differential scanning fluorimetry to detect ligand interactions that promote protein stability. *Nat. Protoc.* **2**, 2212–2221 (2007).
10. Martinez Molina, D. *et al.* Monitoring drug target engagement in cells and tissues using the cellular thermal shift assay. *Science* **341**, 84–87 (2013).
11. Savitski, M. M. *et al.* Tracking cancer drugs in living cells by thermal profiling of the proteome. *Science* **346**, 1255784 (2014).
12. Franken, H. *et al.* Thermal proteome profiling for unbiased identification of direct and indirect drug targets using multiplexed quantitative mass spectrometry. *Nat. Protoc.* **10**, 1567–1593 (2015).
13. Huber, K. V. M. *et al.* Proteome-wide drug and metabolite interaction mapping by thermal-stability profiling. *Nat. Methods* **12**, 1055–1057 (2015).
14. Jafari, R. *et al.* The cellular thermal shift assay for evaluating drug target interactions in cells. *Nat. Protoc.* **9**, 2100–2122 (2014).
15. Arrowsmith, C. H. *et al.* The promise and peril of chemical probes. *Nat. Chem. Biol.* **11**, 536–541 (2015).
16. Longley, D. B., Harkin, D. P. & Johnston, P. G. 5-fluorouracil: mechanisms of action and clinical strategies. *Nat. Rev. Cancer* **3**, 330–338 (2003).
17. Wilson, P. M., Danenberg, P. V, Johnston, P. G., Lenz, H.-J. & Ladner, R. D. Standing the test of time: targeting thymidylate biosynthesis in cancer therapy. *Nat. Rev. Clin. Oncol.* **11**, 282–298 (2014).
18. Assaraf, Y. G. Molecular basis of antifolate resistance. *Cancer Metastasis Rev.* **26**, 153–181 (2007).
19. Gonen, N. & Assaraf, Y. G. Antifolates in cancer therapy: Structure, activity and mechanisms of drug resistance. *Drug Resist Updat.* **15**, 183–210 (2012).
20. Lipinski, C. A. Lead- and drug-like compounds: the rule-of-five revolution. *Drug Discov. Today Technol.* **1**, 337–341 (2004).
21. Houghton, J. A., Tillman, D. M. & Harwood, F. G. Ratio of 2′-deoxyadenosine-5′-triphosphate/thymidine-5′-triphosphate influences the commitment of human colon carcinoma cells to thymineless death. *Clin. Cancer Res.* **1**, 723–730 (1995).
22. Lambie, D. G. & Johnson, R. H. Drugs and folate metabolism. *Drugs* **30**, 145–155 (1985).
23. Almog, R., Waddling, C. A., Maley, F., Maley, G. F. & Van Roey, P. Crystal structure of a deletion mutant of human thymidylate synthase Δ (7-29) and its ternary complex with Tomudex and dUMP. *Protein Sci.* **10**, 988–996 (2001).
24. Sayre, P. H. *et al.* Multi-targeted antifolates aimed at avoiding drug resistance form covalent closed inhibitory complexes with human and Escherichia coli thymidylate synthases. *J. Mol. Biol.* **313**, 813–829 (2001).
25. Mateus, A., Matsson, P. & Artursson, P. Rapid measurement of intracellular unbound drug concentrations. *Mol. Pharm.* **10**, 2467–2478 (2013).
26. Mateus, A., Matsson, P. & Artursson, P. A high-throughput cell-based method to predict the unbound drug fraction in the brain. *J. Med. Chem.* **57**, 3005–3010 (2014).

27. O'Connor, O. A. in *Cancer Drug Discovery and Development: Combination Cancer Therapy: Modulators and Potentiators* (ed. Schwartz, G. K.) 133–174 (Humana Press Inc., 2005).
28. De Clercq, E. *et al.* Thymidylate synthetase as target enzyme for the inhibitory activity of 5-substituted 2′-deoxyuridines on mouse leukemia L1210 cell growth. *Mol. Pharmacol.* **19,** 321–330 (1981).
29. Temmink, O. H., Comijn, E. M., Fukushima, M. & Peters, G. J. Intracellular thymidylate synthase inhibition by trifluorothymidine in FM3A cells. *Nucleosides Nucleotides Nucleic Acids* **23,** 1491–1494 (2004).
30. Danenberg, P. V, Langenbach, R. J. & Heidelberger, C. Structures of reversible and irreversible complexes of thymidylate synthetase and fluorinated pyrimidine nucleotides. *Biochemistry* **13,** 926–933 (1974).
31. Danenberg, P. V & Lockshin, A. Fluorinated pyrimidines as tight-binding inhibitors of thymidylate synthetase. *Pharmac. Ther.* **13,** 69–90 (1981).
32. Eckstein, J. W., Foster, P. G., Finer-Moore, J., Wataya, Y. & Santi, D. V. Mechanism-based inhibition of thymidylate synthase by 5-(trifluoromethyl)-2′-deoxyuridine 5′-monophosphate. *Biochemistry* **33,** 15086–15094 (1994).
33. Stresemann, C. & Lyko, F. Modes of action of the DNA methyltransferase inhibitors azacytidine and decitabine. *Int. J. Cancer* **123,** 8–13 (2008).
34. Nomme, J. *et al.* Structure-Guided Development of Deoxycytidine Kinase Inhibitors with Nanomolar Affinity and Improved Metabolic Stability. *J. Med. Chem.* **57,** 9480–9494 (2014).
35. Momparler, R. L. Molecular, cellular and animal pharmacology of 5-aza-2′-deoxycytidine. *Pharmacol. Ther.* **30,** 287–299 (1985).
36. Momparler, R. L. Pharmacology of 5-aza-2′-deoxycytidine (decitabine). *Semin. Hematol.* **42,** S9–S16 (2005).
37. Momparler, R. *et al.* Kinetic interaction of 5-AZA-2′-deoxycytidine-5′-monophosphate and its 5′-triphosphate with deoxycytidylate deaminase. *Mol. Pharmacol.* **25,** 436–440 (1984).
38. Zhang, J.-H., Chung, T. D. Y. & Oldenburg, K. R. A simple statistical parameter for use in evaluation and validation of high throughput screening assays. *J. Biomol. Screen.* **4,** 67–73 (1999).
39. Chabot, G. G., Bouchard, J. & Momparler, R. L. Kinetics of deamination of 5-aza-2′-deoxycytidine and cytosine arabinoside by human liver cytidine deaminase and its inhibition by 3-deazauridine, thymidine or uracil arabinoside. *Biochem. Pharmacol.* **32,** 1327–1328 (1983).

Acknowledgements

T.L., H.Al., M.H. and H.Ax. acknowledge Karolinska Institutet, SciLifeLab and the Swedish Research Council (Vetenskapsrådet), which funds Chemical Biology Consortium Sweden. P.N. acknowledges funding from Karolinska Institutet (DPA), the Swedish Research Council, the Swedish Cancer Society, the Knut and Alice Wallenberg foundation, as well as a startup grant from Nanyang Technological University. P.A. acknowledges the Swedish Research Council and SciLifeLab. A.M. acknowledges the Portuguese Research Council (Fundação para a Ciência e Tecnologia). R.J. acknowledges Magnus Bergvalls Stiftelse, The Lars Hierta Memorial Foundation and Helge Ax:son Johnsons stiftelse. We are very grateful to Caius G. Radu and Raymond M. Gipson for providing the DCK inhibitor DI-82. We would also like to thank Protein Science Facility (PSF) at Karolinska Institutet for providing DCTD and Staffan Eriksson from the Swedish University of Agricultural Sciences for providing the plasmid for DCK. We are also grateful to Carlo Zaniol at PerkinElmer for valuable support in development of the AlphaScreen assays. Brinton Seashore-Ludlow is acknowledged for careful reviewing and commenting on the manuscript.

Author contributions

H.Al., H.Ax., R.J., D.M.M., T.L. and P.N. conceived the study. H.Al., H.Ax. and T.L. developed the homogenous TS CETSA assay supported by R.J. and D.M.M. and ran the screen. R.J. generated the western blot CETSA data. A.M. and P.A. planned and generated the intracellular concentration data. C.D. generated the *in vitro* thermal shift and surface plasmon resonance data, and solved the crystal structure of CBK115334 together with A.L. M.H. performed the cheminformatics and metabolic pathway analysis. P.N., P.A. and T.L. lead the project. T.L. and P.N. wrote the first draft of the manuscript. All co-authors contributed to the final draft of the manuscript.

Additional information

Accession codes: The crystal structure of CBK115334 and dUMP in complex with TS has been deposited with the RCSB Protein Data Bank under the accession code 5HS3.

Competing financial interests: D.M.M. and P.N. are founders of the company Pelago Bioscience AB. D.C., A.L., A.M., P.A, R.J., T.L., H.Al., M.H. and H.Ax. declare no competing financial interests.

Arginine demethylation is catalysed by a subset of JmjC histone lysine demethylases

Louise J. Walport[1], Richard J. Hopkinson[1], Rasheduzzaman Chowdhury[1], Rachel Schiller[1], Wei Ge[1], Akane Kawamura[1,2] & Christopher J. Schofield[1]

While the oxygen-dependent reversal of lysine N^{ε}-methylation is well established, the existence of bona fide N^{ω}-methylarginine demethylases (RDMs) is controversial. Lysine demethylation, as catalysed by two families of lysine demethylases (the flavin-dependent KDM1 enzymes and the 2-oxoglutarate- and oxygen-dependent JmjC KDMs, respectively), proceeds via oxidation of the *N*-methyl group, resulting in the release of formaldehyde. Here we report detailed biochemical studies clearly demonstrating that, in purified form, a subset of JmjC KDMs can also act as RDMs, both on histone and non-histone fragments, resulting in formaldehyde release. RDM catalysis is studied using peptides of wild-type sequences known to be arginine-methylated and sequences in which the KDM's methylated target lysine is substituted for a methylated arginine. Notably, the preferred sequence requirements for KDM and RDM activity vary even with the same JmjC enzymes. The demonstration of RDM activity by isolated JmjC enzymes will stimulate efforts to detect biologically relevant RDM activity.

[1] Department of Chemistry, Chemistry Research Laboratory, University of Oxford, Mansfield Road, Oxford OX1 3TA, UK. [2] Division of Cardiovascular Medicine, Radcliffe Department of Medicine, Wellcome Trust Centre for Human Genetics, Roosevelt Drive, Oxford OX3 7BN, UK. Correspondence and requests for materials should be addressed to C.J.S. (email: christopher.schofield@chem.ox.ac.uk).

Post-translational modifications, which vastly increase the size of the functional proteome, play central roles in numerous cellular processes, including enzyme catalysis, protein–protein interactions and gene regulation[1]. Modifications to histones, which are especially prevalent on the *N*-terminal tail of histone H3, include N^{ε}-acetylation and ubiquitination of lysine residues, phosphorylation of serine and threonine residues, and $N^{\varepsilon/\omega}$-methylation of lysine and arginine residues[2]. Coupled with other factors, these modifications regulate the sets of genes that are transcribed in a time- and context-dependent manner. While lysine acetylation is usually transcriptionally activating, arginine- and lysine methylation can be either activating or repressive, depending on the site and extent of methylation.

Arginine/lysine methylation is catalysed by *S*-adenosyl-methionine-dependent methyltransferases (protein arginine methyltransferases, PRMTs, and lysine methyltransferases, KMTs, respectively)[3]. Although for many years *N*-methylation was thought to be a stable, irreversible modification, the reversal of lysine methylation is now well established. Two families of histone N^{ε}-methyllysine demethylases (KDMs) have been identified; the flavin-dependent KDM1s (lysine-specific demethylase) and the Fe(II)- and 2-oxoglutarate (2OG)-dependent JmjC-domain-containing enzymes. Both families of KDMs operate via methyl-group oxidizing mechanisms that result in the co-production of formaldehyde[4]. Lysine methylation/demethylation can thus occur on a timescale of only two enzyme-catalysed steps, that is, has the potential to be highly dynamic.

In contrast to lysine methylation, the extent to which arginine methylation is dynamic is much less clear. Arginine residues undergo PRMT-catalysed methylation on their terminal guanidine nitrogens to give monomethyl (MMA/Rme), and symmetric (SDMA/Rme2s) and asymmetric (ADMA/Rme2a) dimethylated forms[5]. The peptidyl-arginine deiminases (PADs) catalyse the hydrolysis of the guanidino arginine side chain to the urea group of citrulline. Both PAD2 and PAD4 catalyse citrulline formation at multiple positions on histone tails[6]. In the case of PAD4, this activity has also been reported on monomethylarginine residues[7], although evidence that this activity occurs in an *in vivo* setting remains limited[8]. In either case, demethylimination is not a 'true' demethylation reaction—it removes but does not reverse N^{ω}-methylation. Neutral citrulline is produced, which has considerably different chemical properties compared to unmethylated arginine. One 2OG oxygenase, JMJD6, has been reported to catalyse methylarginine demethylation of histone H3/H4 residues[9]. Some subsequent reports have not reproduced these observations, and JMJD6 has also been assigned as a lysine C-5 hydroxylase acting on mRNA splicing-regulatory proteins, and potentially histone proteins[10–12]. Other recent reports have identified further arginine demethylation activity by JMJD6 on histone and non-histone substrates[13–16]; however, the biochemical evidence for methylarginine demethylation by JMJD6 is not unequivocal[17].

Despite the controversial history as potential methylarginine demethylases (RDMs), the 2OG oxygenases represent good candidates for such reactions because of the broad spectrum of oxidative reactions that they catalyse. Some JmjC oxygenases catalyse hydroxylation of lysine residues to give stable hydroxylated products[18]. Arginine demethylation could occur via oxidation of the arginine N^{ω}-methyl group and concomitant loss of formaldehyde in a reaction analogous to that of lysine demethylation (Fig. 1a,b). In a recent study we found that the

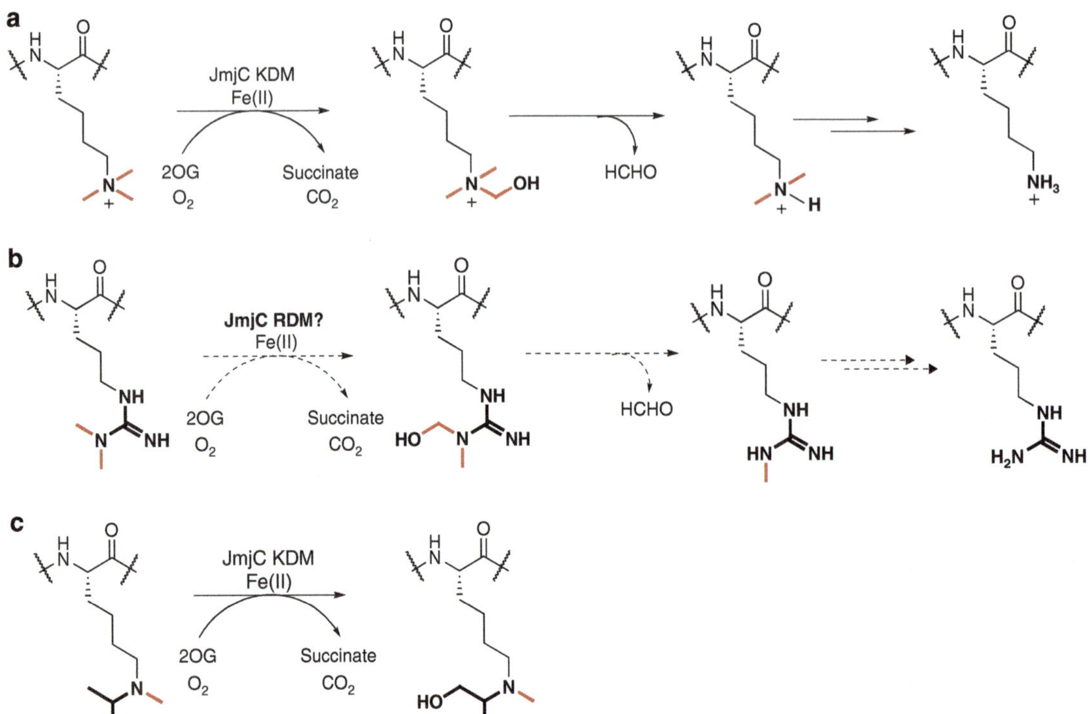

Figure 1 | Mechanism of lysine and arginine demethylation. (**a**) JmjC lysine demethylases (KDM2-7 subfamilies) catalyse oxidative decarboxylation of 2OG to form succinate, carbon dioxide and a reactive iron(IV)-oxo intermediate; this intermediate then facilitates hydroxylation of the lysine N^{ε}-methyl group to form an unstable hemiaminal. Fragmentation of the hemiaminal releases formaldehyde and the unmethylated lysine residue. (**b**) Proposed mechanism for JmjC-catalysed arginine demethylation. (**c**) Hydroxylation of the *N*-methyl,*N*-isopropyllysine derivative is catalysed by various JmjC KDMs at a position analogous to that required for demethylation of asymmetrically dimethylated arginine residues. Methyl groups are bold red lines, dashed arrows represent proposed reactions and HCHO is formaldehyde. Each JmjC KDM/RDM/oxidation reaction is coupled to the conversion of 2OG and oxygen to carbon dioxide and succinate.

JmjC KDMs can catalyse oxidation of N^{ε}-alkyl groups other than methyl groups, including oxidation reactions on isopropyl groups at a position equivalent to that required for oxidation of a N^{ω}-methylarginine group (Fig. 1c)[19]. Here we report systematic studies on the potential of the catalytic domains of representative human JmjC enzymes to catalyse N^{ω}-methylarginine demethylation. The results clearly demonstrate that some, but not all, purified JmjC KDMs can act as RDMs.

Results

JmjC enzymes have RDM activity. To test whether the catalytic domains of the JmjC KDMs can act as RDMs, we produced truncated recombinant proteins containing the catalytic domains of representatives of the six identified human JmjC KDM subfamilies (KDM2A, KDM3A, KDM4E, KDM5C, KDM6B and PHF8 (also known as KDM7B), Fig. 2a)[20,21]. The representative JmjC KDMs were tested for *in vitro* demethylation activity with synthetic fragment sequences of the histone H3 tail with mono- and asymmetric/symmetric dimethylated arginine residues replacing the N^{ε}-methylated lysine residues known to be demethylated in established substrates for each JmjC KDM (Supplementary Table 1). The corresponding 'wild-type' peptides with the naturally positioned N^{ε}-methylated lysine residues were used as positive controls. Reactions were initially analysed using MALDI-TOF mass spectrometry (MS) to test for demethylation (observed as a −14 Da mass shift).

For KDM2A and PHF8, demethylation was observed only for the characterized lysine substrates (Supplementary Figs 1 and 2), that is, no RDM activity was detected. Interestingly, however, in addition to demethylation of the characterized lysine substrates, evidence for demethylation of both asymmetric and symmetric dimethyl- and monomethylarginine residues was accrued by MS for KDM3A, KDM4E, KDM5C and KDM6B (Fig. 2b and Supplementary Figs 3–6), that is, product peptide spectra manifested mass shifts of −14 Da (and in some cases −28 Da, corresponding to the loss of two methyl groups) relative to the methylated substrate residue. Under our incubation conditions, such mass shifts were only observed in the mass spectra of reactions in the presence of enzyme. KDM3A, KDM4E and KDM5C catalysed demethylation of both symmetric and asymmetric dimethylarginine residues to give a mixture of both the monomethylated and unmethylated arginine residues. At least under the tested conditions, however, we observed that KDM6B only catalyses a single demethylation of asymmetric dimethylarginine to the monomethylated species. Within our limits of detection, no activity was observed for KDM6B and the tested peptides containing monomethylarginine or symmetric

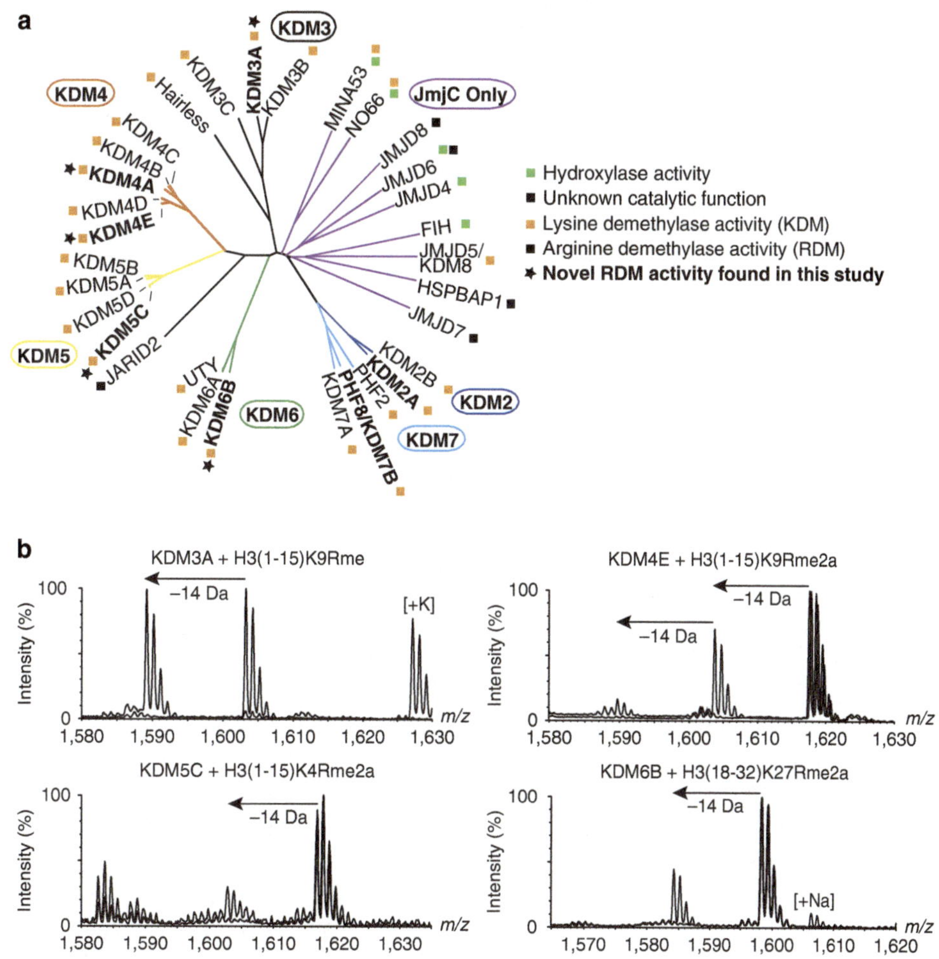

Figure 2 | Some JmjC KDMs catalyse arginine demethylation. (**a**) Human JmjC oxygenases grouped according to the sequence analysis of their catalytic domains showing their assigned/proposed functions and novel functions found in this study (red stars). Some assignments are controversial. Proteins used in this study are in bold. KDM4E is a likely pseudogene. (**b**) MALDI-TOF MS analysis of demethylation of arginine-methylated variant histone peptides by truncated recombinant catalytic domain constructs of KDM3A, KDM4E, KDM5C and KDM6B. Red spectra show reactions including enzyme and black spectra the no enzyme control.

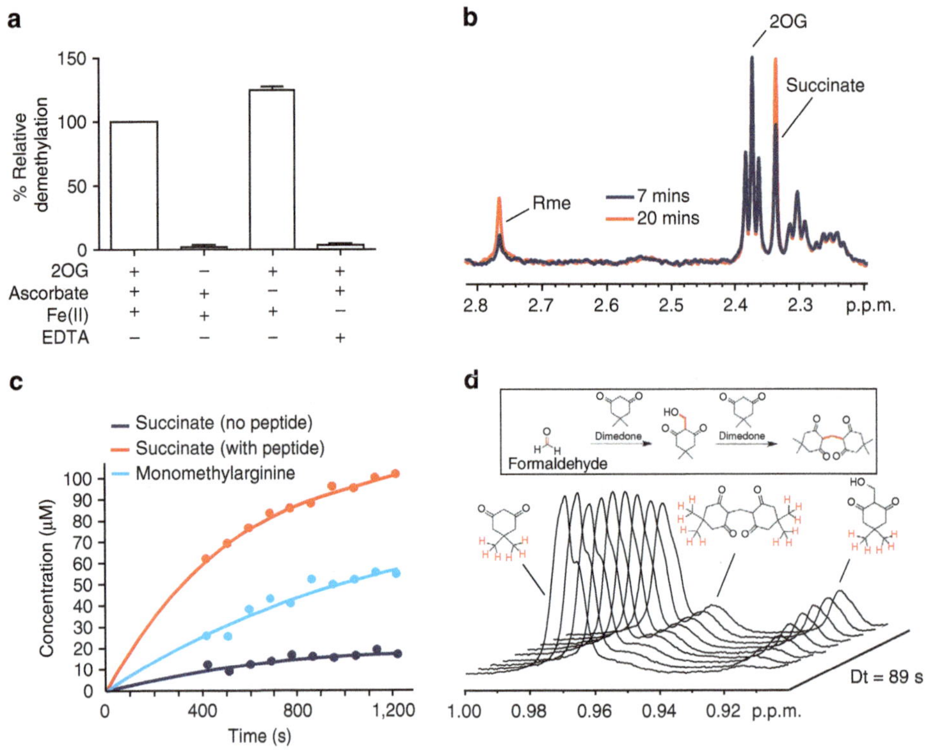

Figure 3 | The mechanisms of arginine and lysine demethylation are similar. (a) The arginine demethylation reaction of KDM6B requires both 2OG and Fe(II) for active demethylation. The degree of arginine demethylation of H3(14–34)K27Rme2a catalysed by KDM6B was quantified by MALDI-TOF mass spectrometry. Data show the mean ± s.e.m. **(b)** ^{1}H NMR analyses of KDM6B-catalysed arginine demethylation. The ^{1}H spectra are of a reaction mixture containing KDM6B (9 μM), H3(14–34)K27Rme2a peptide (1 mM), 2OG (500 μM), ascorbate (1 mM) and iron(II) (100 μM) after 7 min (blue) and 20 min (red) at 298 K. The resonance corresponding to the methyl group of monomethylated arginine (Rme) is highlighted. **(c)** Graphs showing the degree of succinate production and peptide demethylation of H3(14–34)K27Rme2a catalysed by KDM6B as quantified by ^{1}H NMR (700 MHz). **(d)** Detection of formaldehyde release during KDM6B-catalysed arginine demethylation. Dimedone reacts with formaldehyde in aqueous solution to form stable adducts that are detectable using ^{1}H NMR (insert, the formaldehyde-derived carbons in the adducts are highlighted red). Incubation of a reaction mixture containing KDM6B (6.5 μM), H3(18–30)K27Rme2a peptide (1 mM), 2OG (4 mM), ascorbate (1 mM) and iron(II) (100 μM) and dimedone (667 μM) revealed the formation of two dimedone adducts using ^{1}H NMR (700 MHz). Protons responsible for each peak are shown in red.

dimethylarginine (Supplementary Fig. 6a). Interestingly, when the KDM6B substrate was extended from 15 to 21 residues, a double demethylation reaction to the unmethylated arginine residue was observed in the MALDI MS of reactions containing KDM6B (Supplementary Fig. 6b). As expected for a 2OG oxygenase-catalysed reaction, demethylation by KDM6B required the presence of the Fe(II) cofactor and 2OG co-substrate (Fig. 3a).

In order to validate the MS-based results, we used NMR spectroscopy to follow KDM6B-catalysed demethylation of the 15- and 21-residue H3K27Rme2a peptides. Immediately after initiation of the reaction, the samples containing KDM6B, with either the 15- or 21-residue peptide substrate and cofactors/co-substrates, were transferred to NMR tubes and then subjected to ^{1}H NMR (700 MHz) analyses (Fig. 3b and Supplementary Fig. 7a). In both samples, conversion of 2OG to succinate was evidenced by reduction of the ^{1}H resonance at δ_H 2.4 p.p.m. (corresponding to the protons attached to C4 of 2OG) with concomitant emergence of a ^{1}H resonance at δ_H 2.3 p.p.m. (corresponding to the methylene protons of succinate). An additional singlet resonance at δ_H 2.8 p.p.m. was also observed to increase in intensity during the analyses; this was assigned to the *N*-methyl protons of the monomethylarginine product based on ^{1}H and ^{13}C chemical shift analyses (δ_C 27.5 p.p.m., as determined by ^{1}H-^{13}C-HSQC analysis of the sample with the 15-residue peptide, Fig. 3b and Supplementary Fig. 7b). Quantitative analyses on the sample with the 21-residue peptide revealed a faster rate of succinate production relative to the rate of peptide demethylation, indicating some uncoupling of 2OG and peptide oxidation (that is, some conversion of 2OG to succinate without concomitant oxidation of the peptide methyl group). The degree of 2OG oxidation was greater in the presence of peptide than without peptide, suggesting that binding of the peptide to KDM6B stimulates 2OG oxidation but does not always necessarily lead to methyl group oxidation (Fig. 3c and Supplementary Fig. 7c). This 'uncoupling' of 2OG oxidation and peptide demethylation is well precedented for 2OG oxygenases, although it is less pronounced during KDM4E-catalysed demethylation of methyllysine peptides[22]. Addition of the formaldehyde-trapping reagent dimedone to the NMR reaction mixture in the presence of the 15 residue peptide resulted in the emergence of new ^{1}H resonances, which were assigned to two dimedone–formaldehyde adducts (Fig. 3d). These adducts have been previously observed to form upon incubation of dimedone with formaldehyde released during JmjC KDM-catalysed demethylation, and therefore imply that KDM6B-catalysed methylarginine demethylation follows a similar mechanism to that for methyllysine demethylation (that is, methyl group oxidation to release formaldehyde)[22]. The combined MS and NMR results thus reveal that some JmjC oxygenases can catalyse N^{ω}-methyl arginine demethylation.

Preliminary kinetic parameters for selected methylarginine variant peptides were determined using a formaldehyde

Table 1 | Kinetic parameters for arginine demethylation by KDMs (RDMs).

Enzyme	Substrate	K_M/μM	$k_{cat} \times 10^{-3}$ s^{-1}	$k_{cat}/K_M \times 10^{-6}$
KDM3A	**H3(1-15)K9me2**	**44.5 ± 4.8***	**355 ± 95.3***	**7,978.0**
	H3(1-15)K9Rme	12.1 ± 3.1	11.0 ± 1.2	907.6
	H3(1-15)K9Rme2a	11.6 ± 1.4	5.0 ± 0.6	430.7
	H3(1-15)K9Rme2s	23.7 ± 1.9	3.0 ± 0.3	126.5
KDM4E	**H3(1-15)K9me3**	**17.8 ± 4.0**	**170.7 ± 15.2**	**9,575**
	H3(1-15)K9Rme2a	66.0 ± 5.0	4.6 ± 0.7	70.4
	H3(1-15)K9Rme2s	101.1 ± 11.0	8.7 ± 0.3	85.9
	H3(1-15)K9Rme	168.8 ± 35.9	2.7 ± 0.2	15.9
	H3(1-15)R2me2a	*86.9 ± 16.5*	*9.9 ± 1.0*	*114.0*
	H4(1-15)R3me2a	*121.5 ± 10.1*	*2.1 ± 0.1*	*17.7*
KDM5C	**H3(1-15)K4me3**	**7.6 ± 2.9**	**67.1 ± 10.3**	**8,800**
	H3(1-15)K4Rme2a	29.0 ± 0.5	41.2 ± 3.5	1,420.0
	H3(1-15)K4Rme2s	21.1 ± 5.8	4.6 ± 0.7	219.0
	H3(1-15)R2me2a	*19.0 ± 1.9*	*12.0 ± 0.7*	*630.0*
KDM6B	**H3(14-34)K27me3**	**6.7 ± 0.6**	**11.7 ± 0.1**	**1,750.0**
	H3(14-34)K27Rme2a	676.3 ± 118.3	28.8 ± 3.6	42.5

KDM, N^{ε}-methyllysine demethylase; RDM, N^{ω}-methylarginine demethylase.
Apparent kinetic parameters (K_M and k_{cat} values) were determined for each JmjC KDM/RDM using a formaldehyde dehydrogenase-coupled assay. Data show mean ± s.e.m (n = 3). Recognized histone N^{ε}-methylated-lysine peptide substrate for each KDM is marked in bold. 'Natural' histone peptides are shown in italics.
k_{cat}/K_M values are calculated form the average k_{cat} and K_M values for each peptide.
* Kinetic parameters calculated using data points below 100 μM, substrate inhibition observed above 100 μM.

Table 2 | Summary of observed arginine demethylation activity of tested RDMs.

KDM	H3 (1-15) R2X			H3 (1-15) R8X			H3 (10-24) R17X			H3 (18-32) R26X			H4 (1-15) R3X		
	me	me2a	me2s	me	me2a	me2s	me	me2a	me2s	me	me2a	me2s	me	me2a	me2s
KDM3A	−	−	−	−	−	−	−	−	−	−	−	−	−	−	−
KDM4E	+	+	+	+	+	+	−	−	−	+	+	+	−	+	+
KDM5C	+	+	+	−	+	+	−	−	−	−	−	−	−	+	−
KDM6B	−	−	−	−	−	−	−	−	−	−	−	−	−	−	−

+ indicates demethylation activity; − indicates no activity or that below our limits of detection; KDM, N^{ε}-methyllysine demethylase; RDM, N^{ω}-methylarginine demethylase. X corresponds to me/me2a/me2s.

dehydrogenase-coupled demethylation assay (Table 1)[23]. In the case of KDM6B, the methylated arginine peptide probably binds more weakly (as judged by K_M values) in the active site than its methylated lysine analogue (K_M 676.3 ± 118.3 versus 6.7 ± 0.6 μM). Interestingly, however, the k_{cat} value for the arginine peptide was higher than that for the analogous lysine peptide (k_{cat} 28.8 × 10^{-3} ± 3.6 × 10^{-3} versus 11.7 × 10^{-3} ± 0.1 × 10^{-3} s^{-1}), suggesting that in the context of a tighter binding substrate (for example, a protein rather than the peptides tested here) it is possible that arginine demethylation by KDM6B is as efficient as, or favoured relative to, lysine demethylation. For KDM5C, the rate of asymmetric dimethylarginine demethylation was also found to be of similar levels to that of lysine demethylation (k_{cat} 41.2 ± 3.5 versus 67.1 ± 10.3 s^{-1}), while for KDM3A and KDM4E both the rate of demethylation and binding constants were found to be considerably lower for the arginine peptides than the analogous lysine peptides (Table 1). KDM4E appears to bind to the peptide containing symmetric arginine less well than the asymmetric residue (K_M 101.1 ± 11.0 versus 66.0 ± 5.0 μM), but has a faster maximal turnover rate for the symmetric peptide (k_{cat} 8.7 × 10^{-3} ± 0.3 × 10^{-3} versus 4.6 × 10^{-3} ± 0.7 × 10^{-3} s^{-1}), resulting in similar k_{cat}/K_M values. All k_{cat} values were lower and K_M were higher than those for the established H3K9me3 peptide substrate for KDM4E. The monomethylated H3K9Rme peptide had a higher K_M value than both the dimethylated peptides, and displayed the lowest turnover rate (k_{cat} 2.7 × 10^{-3} ± 0.2 × 10^{-3} s^{-1}. Note that monomethylated lysine residues are poor substrates for the KDM4 subfamily[20]). For KDM5C, although both the asymmetric and symmetric dimethylarginine-containing peptides had similar K_M values (K_M 29.0 ± 0.5 versus 21.1 ± 5.8 μM), the turnover rate for the asymmetric peptide was an order of magnitude higher (41.2 × 10^{-3} ± 3.5 × 10^{-3} versus 4.6 × 10^{-3} ± 0.7 × 10^{-3} s^{-1}).

For KDM3A, the monomethylated arginine peptide was a better substrate than the corresponding dimethylated arginine peptides (Table 1). These results reflect the MS screening results, where in the context of overall incomplete substrate peptide demethylation, only conversion to the unmethylated residue was observed, that is, no monomethylated peptide intermediate was detected (Supplementary Fig. 3). Although the monomethylated arginine-methylated substrate showed a lower k_{cat} value than the lysine-methylated substrate (11.0 ± 1.2 versus 355 ± 95.3 s^{-1}), the K_M values for all methylated arginine peptides were found to be smaller than those for H3K9me2 (H3K9Rme = 12.1 ± 3.1 μM, H3K9Rme2a = 11.6 ± 1.4 μM, H3K9Rme2s = 23.7 ± 1.9 μM and H3K9me2 = 44.5 ± 4.8 μM). The k_{cat} value for K9Rme2a and K9Rme2s were lower than that for K9Rme. The combined kinetic results described above for each enzyme imply that the relative efficiency of RDM versus KDM substrate binding and rate of catalysis is sequence and context-dependent.

RDM activity is observed on 'natural' histone peptides. Given the ability of JmjC enzymes to catalyse demethylation of arginine residues in synthetic peptide sequences, we were interested to investigate whether these enzymes could also catalyse arginine demethylation in 'natural' positions in histone peptides. Peptides representing known methylated arginine sites within the canonical histone tail sequences were synthesized and screened as substrates for the four enzymes identified as RDMs (Supplementary Table 1). Interestingly, arginine demethylation activity on the 'natural' histone peptide sequences was observed for some, but not all of the KDMs (Table 2 and Fig. 4). Both KDM4E and KDM5C catalysed demethylation of some of the screened peptides, including peptides methylated at H3R2 and H4R3 (Fig. 4a,b and Supplementary Figs 8 and 9), while KDM3A and KDM6B displayed no activity on any of the new peptides tested (Supplementary Figs 10 and 11). The observation of a lack of RDM activity for KDM6B with these substrates is notable, given that KDM6B displayed such high RDM activity with the unnatural H3K27Rme2a substrate peptide (Table 1). Thus, these results demonstrate that RDM activity is dependent on the sequence context.

Further kinetic characterization was carried out on some of the 'biologically relevant' peptides identified to be substrates for the KDMs (Table 1). Interestingly, despite the different neighbouring peptide sequences to the characterized lysine H3K9me3 substrate of KDM4E, the H3R2me2a peptide was found to be a better substrate for KDM4E than H3K9Rme2a or H3K9Rme2s

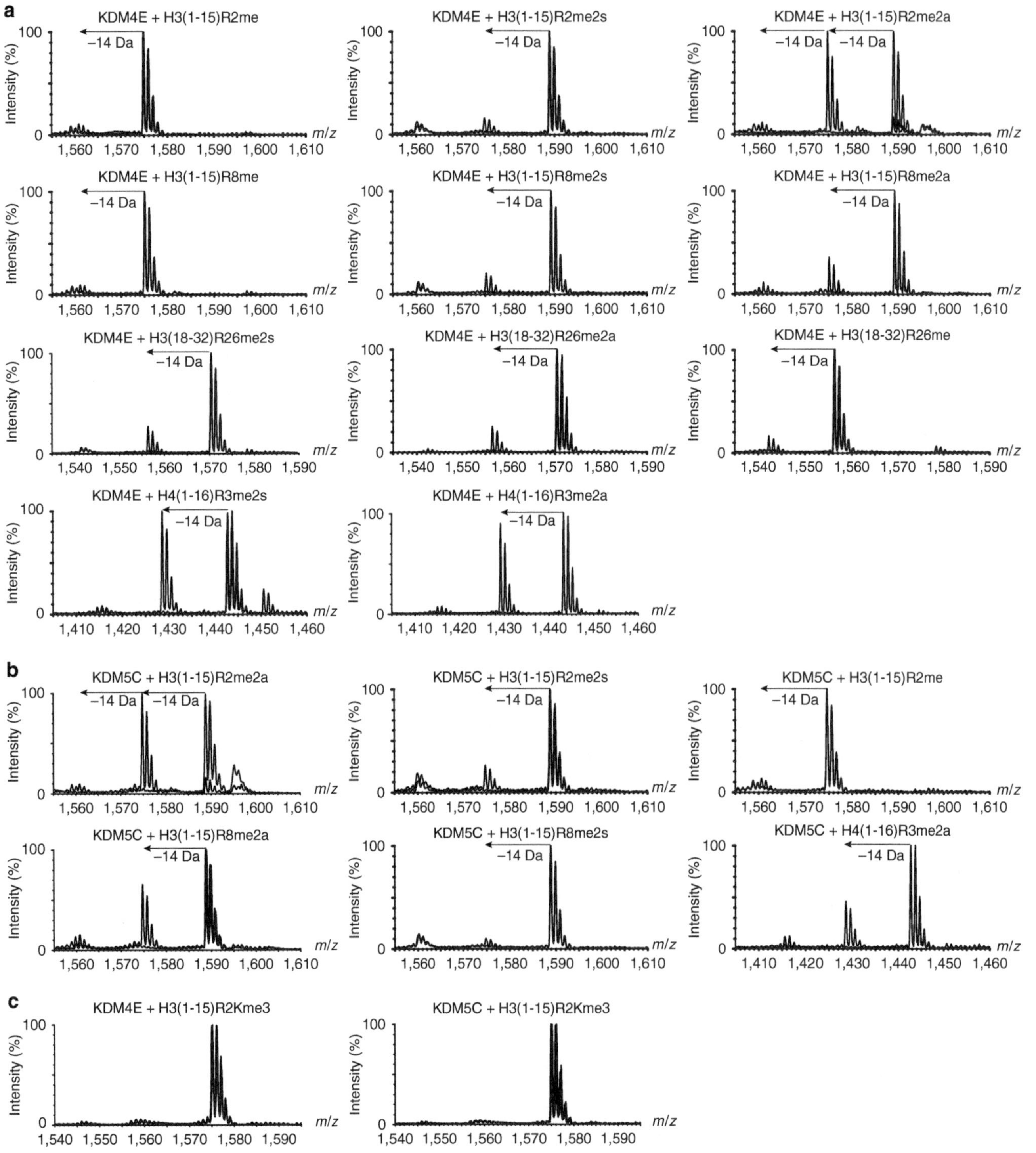

Figure 4 | KDMs catalyse arginine demethylation in 'natural' histone peptides. MALDI-TOF MS of demethylation of arginine methylated 'natural' histone peptides by (**a**) KDM4E and (**b**) KDM5C. (**c**) MALDI-TOF MS revealing no demethylation activity of KMD4E or KDM5C on a H3(1-15)R2Kme3 peptide.

(k_{cat}/K_M 114.0×10^{-6} versus $70.4 \times 10^{-6}\mu M^{-1}s^{-1}/85.9 \times 10^{-6}\mu M^{-1}s^{-1}$). The H4R3me2a peptide was a less efficient substrate for KDM4E, with a turnover rate almost five times lower than that of the H3R2me2a peptide, similar to that observed for H3K9Rme. For KDM5C, the H3R2me2a peptide had a lower K_M value than the H3K4Rme2a peptide, but was an overall slightly less efficient substrate for the enzyme (k_{cat}/K_M $1{,}420 \times 10^{-6}$ versus $630 \times 10^{-6}\mu M^{-1}s^{-1}$). Nevertheless, with a k_{cat}/K_M value only 14 times lower than that for H3K4me3 it appears possible that this reaction could also occur in cells in some contexts, particularly if promoted by auxiliary binding domains or binding partners (as is known to be important for some JmjC KDM activities (for example, PHF8))[24].

Because some JmjC KDMs were found to catalyse demethylation of methylated arginine peptides in sequences substantially different from their characterized lysine substrates (for example, we observed low levels of demethylation of H4R3me2a (SG-Rme2a-GKGGKGLGKGGAK) by both KDM4E and KDM5C, Fig. 4), we were interested to see whether substitution of an arginine residue in the 'natural' histone sequence with a methylated lysine residue would produce a KDM substrate. Interestingly, despite a peptide containing H3R2me2a being a relatively good RDM substrate for both KDM4E and KDM5C, the corresponding H3R2Kme3 peptide (A-Kme3-TKQTARKSTG GKA) was not observed to be a KDM substrate for either enzyme within our limits of detection (Fig. 4c). Overall, the combined results reveal that, at least for some JmjC demethylases, the preferred sequence contexts can differ for RDM and KDM activities.

To investigate whether methylarginine demethylation can be catalysed by full-length JmjC KDMs, as well as by the truncated constructs, full-length proteins were exogenously overexpressed in HEK293T cells either as *N*-terminally Flag-tagged wild-type proteins or active-site variants known to be catalytically inactive[25–27]. The resultant full-length proteins were immunoprecipitated by their Flag-tags and used in MS assays with methylated histone peptides. As with the catalytic domain constructs, arginine demethylation was also observed with all four of the full-length proteins (KDM3A, KDM4A, KDM5C and KDM6B; note KDM4A was used in place of KDM4E as KDM4E is a pseudogene) tested on both 'unnatural' and 'natural' peptide substrates (Fig. 5 and Supplementary Figs 12–16). No RDM activity was observed with the catalytically inactive enzyme variants. Note that, with KDM3A, arginine demethylation activity was observed for the monomethylated peptides, but not for the symmetrically or asymmetrically methylated peptides; this result is consistent with the observed relative reaction efficiencies of these peptides with the truncated construct (Supplementary Fig 13 and Table 1).

JmjC enzymes bind arginine and lysine in a similar manner. Crystallographic studies were then undertaken to investigate the binding modes of methylarginine-containing peptides with JmjC enzymes. KDM4A, a close structural homologue of KDM4E that can also catalyse demethylation of methylarginine peptides (albeit less efficiently in our current assays, Fig. 5), was selected for the crystallographic work because of its relatively robust

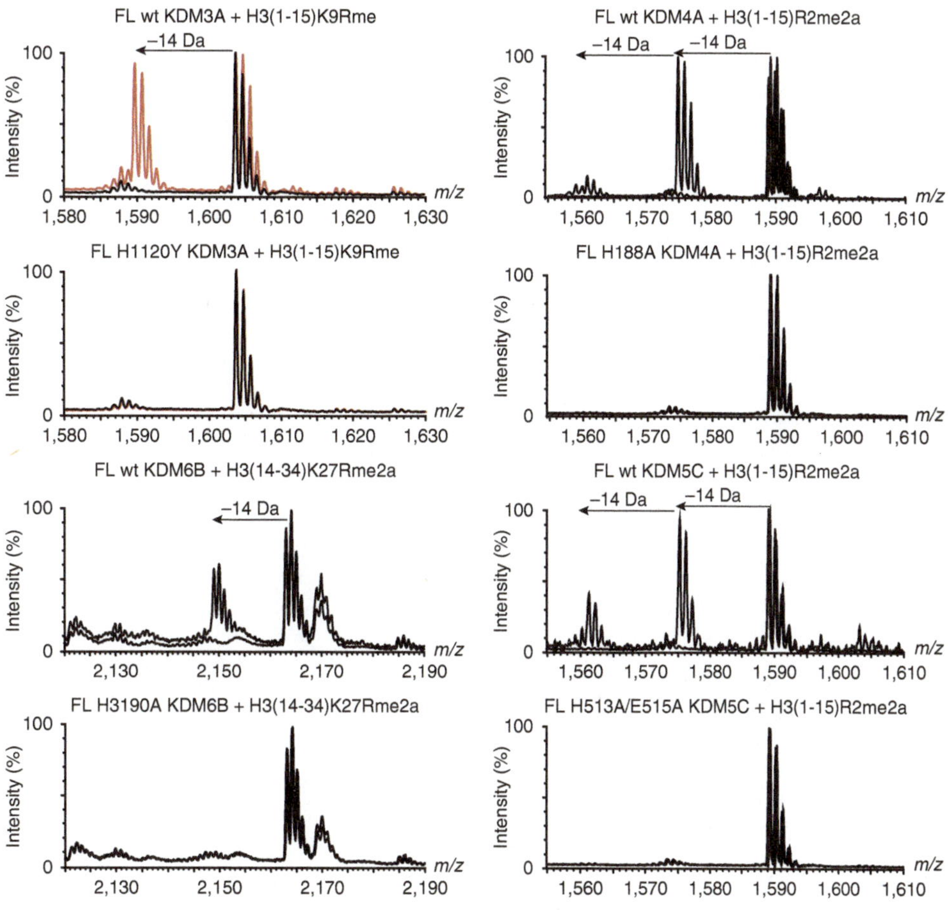

Figure 5 | Full-length (FL) KDMs catalyse demethylation of arginine residues. MALDI-TOF MS of demethylation of arginine methylated peptides by Flag-tagged FL KDMs immunoprecipitated from HEK293T cells.

crystallization properties and because it has previously been crystallized with N^{ε}-methylated lysine peptide substrates[20,28–30]. A suitable crystal was obtained using KDM4A and an H4R3me2s fragment sequence (residues 1–15) with nickel substituting for iron and *N*-oxalylglycine (NOG) acting as a 2OG mimetic). The structure was solved by molecular replacement using PDB ID: 2OX0 as the search model; two molecules of KDM4A (chains A and B) are present in the asymmetric unit. In the initial electron density maps, other than for the methylated arginine itself, only weak difference in electron densities were observed for residues in the substrate peptide, suggesting that the mode of methylated arginine peptide binding could be flexible. We generated an F_o–F_c OMIT map (where different densities were calculated omitting H4R3me2s) as a guide to locate any unmodelled difference peak around the KDM4A (A/B) active sites. This unweighted omit map revealed more than one likely conformation of the methylated arginine side chains and traces of H4R3me2s peptide backbones; only major conformations were modelled and refined in the final structure.

Although care should be taken in over-interpreting 'static' crystal structures of substrate fragments, it is of interest that the arginine side chain was observed to adopt two different extended conformations after refinement. One orientation positions one of the methyl groups of the symmetrically dimethylated arginine residue sufficiently close to the metal centre to enable catalysis (within 4.5 Å; Fig. 6a). The other side chain orientation is a likely catalytically unproductive conformation, wherein the side chain is oriented away from the metal (Fig. 6b; for overlay, see Fig. 6c). The presence of such an unproductive binding conformation might in part account for the higher degree of uncoupled 2OG turnover stimulated by methylated arginine compared with methylated lysine peptide binding, as observed for KDM6B with H3(14–34)K27Rme2a (for overlay of arginine and lysine binding, see Fig. 6d).

Potential for JmjC RDM activity on non-histone substrates. Arginine methylation has been reported for numerous non-histone proteins, as observed by antibody- and/or MS-based approaches, where it is proposed to regulate many cellular processes including alteration of cellular localization, protein stability and protein–protein/DNA interactions[31,32]. Although

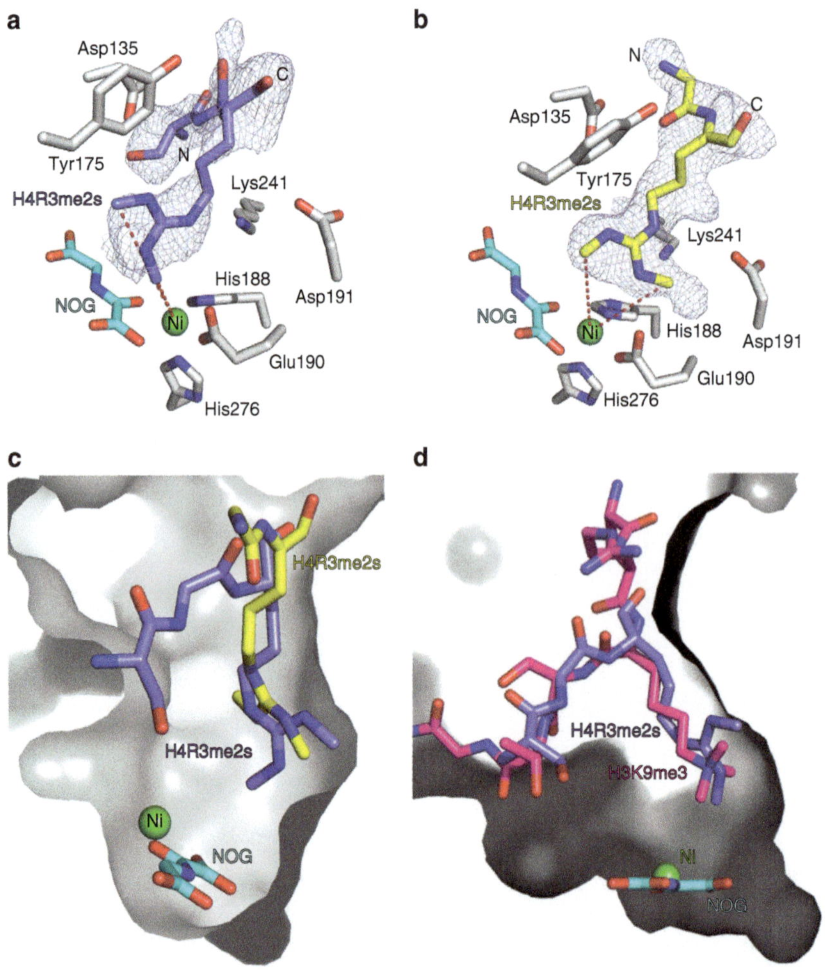

Figure 6 | JmjC KDMs bind methylated arginine peptides in a catalytically-productive binding mode. (**a** and **b**) Views from an X-ray crystal structure of KDM4A in complex with nickel, NOG (a 2OG mimetic) and an H4R3me2s peptide (residues 1–15). Two orientations of peptide binding were refined; one orientation (**a**) positions a methyl group of symmetric dimethylarginine residue sufficiently close to the metal centre to allow catalysis (within 4.5 Å, for crystallographic reasons, nickel was substituted for iron). The other orientation (**b**) is likely not catalytically productive. Fo-Fc OMIT maps contoured to 3σ around the H4R3me2s residues are shown. (**c**) Overlay of the two binding orientations observed for the H4R3me2s peptide in the KDM4A active site. Only one orientation (blue) positions a methyl group close to the catalytic metal centre. (**d**) Overlay of the H4R3me2s peptide (catalytically-productive orientation only) and an H3K9me3 peptide (residues 7–14) bound in the KDM4A active site. The methylated arginine and lysine residues show similar binding modes.

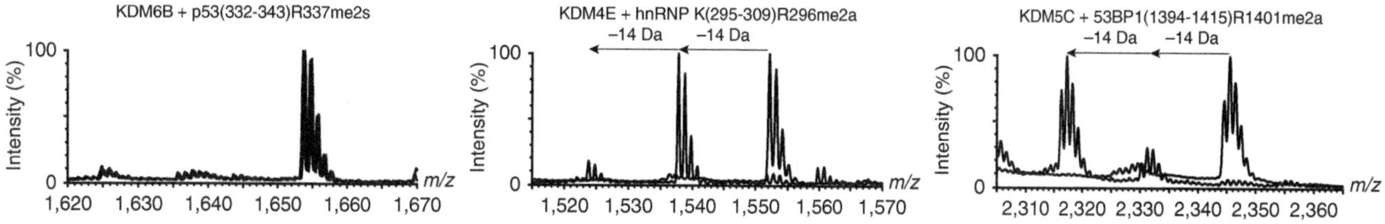

Figure 7 | KDMs catalyse demethylation of arginine residues in non-histone substrates. MALDI-TOF MS assays for demethylation of non-histone arginine methylated peptides by KDMs. Red spectra show reactions including enzyme and black spectra the no-enzyme controls.

limited at present, there are some reports providing at least preliminary evidence for non-histone protein lysine demethylation catalysed by JmjC KDMs[33–35]. Given the ability of the JmjC KDMs to act as RDMs for peptides with sequences substantially different from their characterized lysine sequences, we were interested to investigate whether removal of non-histone arginine methylation could also be catalysed by 2OG oxygenases. As KDM6B is known to interact with tumour protein p53 (p53), peptide sequences containing known arginine methylation sites were synthesized and tested as JmjC RDM substrates[36]; however, no demethylation was observed (Fig. 7 and Supplementary Fig. 17). A selection of peptides from some other nuclear proteins known to be abundantly methylated on arginine residues were also synthesized and tested as RDM substrates by MS. While the majority of these peptides were found not to be substrates, in some cases clear arginine demethylation was observed (Fig. 7 and Supplementary Fig. 17). Demethylation of asymmetric dimethylated arginine residues within peptides from heterogeneous nuclear ribonucleoprotein K (hnRNP K) and tumour suppressor p53-binding protein 1 (53BP1) was observed, as catalysed by KDM4E and KDM5C, respectively. If replicable in cells, these results raise the possibility that arginine demethylation could be a widespread phenomenon extending well beyond the canonical histone tails.

Discussion

Since the observation of N^{ε}-lysine methylation in 1959 (ref. 37), multiple protein residues have been found to undergo *N*- and *O*-methylation[31,38]. Within the context of gene regulation, the *N*-methylation of lysine and arginine residues on histone tails plays critical roles. Multiple methyltransferases have been identified, many of which are apparently highly promiscuous, acting to catalyse methylation on numerous proteins[31,33,39,40]. Despite the pioneering observations in 1973 of the oxidative demethylation of N^{ε}-methylated lysine residues in calf thymus histones[41], the identification of the sequences of N^{ε}-methyl-group demethylases took three decades. Following the identification of KDM1A in 2004 (ref. 42), the subsequent discovery of the JmjC KDMs validated the early studies in identifying oxygenases that catalyse N^{ε}-methyllysine demethylation to release formaldehyde[4,43,44]. Given the plethora of methylated proteins, it appears possible that some KDMs might be found to be equally promiscuous as KMTs.

The possibility of RDM activities related to those of the KDMs is readily envisaged. While no lysine-specific demethylase-type flavin-dependent RDMs have been reported (although are feasible), the JmjC oxygenase JMJD6 has been reported as an RDM acting on histones and other non-histone proteins[9,15,16]. However, in our laboratory's work with JMJD6 we have observed only lysine hydroxylation, not RDM activity[10,45]. Various reports have appeared in the recent literature describing both hydroxylation and RDM activities for JMJD6 (refs 13–17,46). Other JmjC oxygenases have also been the subject of controversial assignments[47–49]. In most cases the cell-based work has principally employed antibody-based methods, in some cases supported by 'proteomic' MS evidence, and at present biochemical evidence for the RDM activity of JMJD6 remains limited. Identification of both RDM and KDM activities in a cellular context using antibodies can be complicated by difficulties in accurately quantifying absolute levels of both methylated substrates and the demethylated products and (along with other factors) by a lack of appropriately specific antibodies for substrates and products[50]. MS-based methods, while rapidly improving, are also not normally sufficiently well optimized for measuring absolute levels in cells, particularly for low abundance post-translational modifications[32,50–52]. Thus, while we appreciate that enzyme activity may differ in the isolated protein form from cellular contexts (for example, as observed for the 2OG-dependent protein hydroxylase factor inhibiting hypoxia-inducible factor[53]), given the complexity and controversy in JmjC assignments, in particular, in the arginine demethylation field, we believe that, where possible, it is important to define the 'biochemical reactions' catalysed by potential RDMs (and KDMs) using isolated enzymes.

The assays applied here directly measure, either by mass shifts or by the appearance of new NMR signals, the product(s) of KDM/RDM activity. It should be noted that in the context of MS analyses we undertook to carry out careful non-enzyme controls to ensure that the conditions of MS analysis did not invoke apparent demethylation, or that we were not observing contamination of the putative product in the starting material, which might not be apparent before incubation.

The results clearly show that, at least under our assay conditions, some, but not all, recombinant forms of JmjC KDMs do have RDM activity. Thus, we observed that KDM3A, KDM4E, KDM5C and KDM6B can catalyse demethylation of methylated arginine residues when substituted for the methylated lysine residue in their characterized peptide substrates. In addition, KDM4E (and full-length KDM4A) and KDM5C were found to catalyse demethylation of histone peptides methylated at H3R2, H3R8, H3R26 (KDM4E, but not KDM5C) and H4R3, and also some non-histone peptide sequences.

One interesting result to emerge from the studies on both histone and non-histone RDM and KDM substrates for the JmjC enzymes is that the apparent sequence selectivities can differ for RDM and KDM activities, even with the same enzyme. Thus, for example, H3R2me2a is a better substrate for KDM4E than H3K9Rme2a (Table 1), and the analogous H3R2Kme3 peptide is not a substrate for KDM4E or KDM5C (Fig. 4c). Although the biological significance of these results is presently unclear, they can be rationalized by the likelihood that in order for productive demethylation to occur, the methylarginine and methyllysine side chains are likely accommodated differently by dual-function KDM/RDM enzymes in order to ensure a catalytically productive relationship between the reactive Fe(IV) = O intermediate and the methyl group at the active site. The differential orientation of the lysine and arginine side chains may in turn have 'knock-on' effects with respect to optimal binding of the surrounding substrate residues. Support for this proposal comes from work revealing that in the case of KDM4A-C, aside from the methylated lysine residues, the H3K9 and H3K36 substrates are accommodated differently at the entrance to the active site[30,54]. Given that for some KDMs post-translational modifications distinct from the residue undergoing demethylation can substantially promote activity, it is possible that RDM activity is similarly promoted. The partial decoupling of 2OG oxidation and demethylation observed for some RDM substrates may reflect non-optimal substrate sequences, or use of a truncated substrate, as precedented with other 2OG-dependent protein hydroxylases[55]. However, it is important to note that, at least in the case of KDM6B, some of the RDM activities we observe are of similar levels to the KDM activity, implying the likelihood of biological relevance of the dual-function KDM/RDM results in some situations (including by competition between substrates).

We appreciate that further work is required on protein substrates in cells for the full biological relevance of these results to be established. Despite considerable efforts, we are not yet able to conclusively determine RDM activity in cells, in part because of the lack of suitably useful antibodies. Nonetheless, the biochemical assignment of RDM activity is interesting and relevant to future work on the functional studies on JmjC enzymes in cells. The JmjC KDMs have been linked to a plethora of human developmental diseases and cancers, and there is growing interest in them as pharmaceutical targets[56–58]. In addition to helping to elucidate the regulatory mechanisms involved in arginine methylation cycles, these results reveal new enzymatic activity and function of the JmjC KDMs, an understanding of which may be crucial to their effective therapeutic targeting. In this regard, it is important to note that, at least some of the roles of JmjC KDMs in disease may involve their production of formaldehyde, which is a product of both KDM and RDM catalysis, and which, at least above threshold levels, is toxic[59]. Now that enzymes that have RDM activity have been conclusively identified, efforts can focus on the demonstration of their biological relevance.

Methods

Materials. Full-length Flag-tagged KDMs were expressed in HEK293T cells (ATCC) using plasmids encoding full-length $KDM4A_{1-1,064}$ (ref. 26), H188A $KDM4A_{1-1,064}$ (ref. 26), $KDM6B_{1-1,636}$ (ref. 27), H1390A $KDM6B_{1-1,636}$ (ref. 27), $KDM3A_{1-1,321}$ (ref. 25) and H1120Y $KDM3A_{1-1,321}$ (ref. 25) as described previously. Plasmids encoding Flag-tagged $KDM5C_{1-4,680}$ and H513A/E515A $KDM5C_{1-4,680}$ were kindly provided by Pavel Savitsky (Structural Genomics Consortium, Oxford). Recombinant constructs of $KDM2A_{1-557}$ (ref. 60), $KDM4A_{1-359}$ (ref. 30), $KDM4E_{1-337}$ (ref. 61), $KDM6B_{1,141-1,590}$ (ref. 62) and $PHF8_{1-447}$ (ref. 63) were produced as *N*-terminally His_6-tagged proteins in *Escherichia coli* cells. $KDM3A_{515-1,317}$ was produced as an *N*-terminally His_6-tagged protein in Sf9 cells[62]. $KDM5C_{1-765}$ was produced as an *N*-terminally His_6-tagged protein in Sf9 cells and was kindly provided by Aleksandra Szykowska (Structural Genomics Consortium). Proteins were purified by Ni-affinity chromatography and size exclusion chromatography as previously described.

Peptide synthesis. Peptides were produced in house using a Liberty Blue automated microwave peptide synthesizer (CEM Corporation) for kinetic and crystallographic studies or either a Multipep RSi (Intavis Bioanalytical Instruments AG) or an S336X peptide synthesizer (CS Bio) for screening purposes. Fluorenylmethyloxycarbonyl-mediated solid-phase chemistry (on MBHA resin) was used to produce the peptides with *C*-terminal amides. After synthesis, peptides were cleaved from the resin by incubation with 92.5% trifluoroacetic acid/2.5% water/2.5% triisopropylsilane/2.5% dimethoxybenzene (3 h) followed by precipitation with ice-cold diethyl ether. Lyophilized peptides produced on the Liberty Blue or S336X synthesizers were purified by reverse-phase high-performance liquid chromatography using a Vydac C18 column (Solvent A = 0.1% trifluoroacetic acid in H_2O, Solvent B = 0.1% trifluoroacetic acid in acetonitrile) to >95% purity as determined by MS. Peptides used for screening were assayed unpurified. Peptide sequences are given in Supplementary Table 1. Methylated arginines suitably protected for solid-phase peptide synthesis were purchased from Novabiochem. In all cases the observed masses for peptides were within 2 Da of the predicted values by MALDI MS.

Activity assays. *MALDI MS activity assays.* Recombinant proteins (2 μM) were incubated with peptides (10 μM, sequences are given in Supplementary Table 1) and cofactors/substrates for 1 h at 37 °C, unless otherwise specified. Reactions were quenched with 1:1 (v/v) methanol and analysed with MALDI MS. Assay conditions for each enzyme are given in Supplementary Table 2. Negative controls without enzyme were included. Note that in some cases small variations ($\leq$2 Da) in the absolute experimentally observed values compared with the calculated values for peptides were observed. However, in all cases where demethylation is assigned the mass shift was 14 Da.

Formaldehyde dehydrogenase-coupled assays. Preliminary kinetic parameters were determined by detection of the formaldehyde by-product of the demethylation reaction using a FDH (formaldehyde dehydrogenase-coupled)/ NAD^+-coupled assay as previously described[61]. Assays were carried out in buffer (enzyme-dependent) with the addition of Fe(II) ammonium sulfate (10 μM), sodium ascorbate (100 μM), 2OG (200 μM), NAD^+ (500 μM), enzyme (0.2–4 μM) and FDH enzyme (0.025 U per assay, Sigma-Aldrich; total volume 30 μl) and were conducted at 37 °C in clear-bottom black 384-well plates. Reactions were monitored over 25 min using a PHERAstar FS (BMG Labtech) plate reader with 355 nm excitation and 460 nm emission. The K_m and V_{max} values were calculated from the reaction rate during the linear phase of formaldehyde production, as determined by the slope of the fluorescence signal recorded.

NMR assays. NMR spectroscopy was carried out as described[22]. NMR spectra were recorded using a Bruker Avance AVIII 700 MHz spectrometer equipped with an inverse TCI cryoprobe, optimized for ^1H observation and installed with the Topspin software. All samples were prepared in Eppendorf tubes (75 μl volume) before being transferred to 2-mm MATCH NMR tubes (Wilgenberg); time course data were then collected over a period of 20 min at 89-s intervals using an automated routine. The solvent deuterium signal was used as an internal lock signal, and the solvent signal was reduced by excitation sculpting[64]. Enzyme stocks were in protiated HEPES/NaCl buffer, pH 7.5, which was diluted with dAFN buffer (prepared by diluting 500 mM ammonium formate, 5 M NaCl in H_2O pH 7.5 to 10% v/v in D_2O) when added to the samples. Chemical shifts are reported relative to the solvent water resonance (δ_H 4.7 p.p.m.).

Cell culture. HEK293T cells (ATCC) were cultured at 37 °C and 5% CO_2 in high-glucose DMEM (Sigma) supplemented with 10% fetal bovine serum (Sigma) and 1% Glutamax (Sigma). Full-length KDMs with Flag-tags were transiently transfected into HEK293T cells using linear polyethylenimine (PEI) (10 μg DNA, 60 μl PEI per T75, 60% confluency, PAA Laboratories) and were harvested after 24–48 h.

On-bead demethylation assays. Cells transiently transfected with the relevant KDM were washed in PBS buffer and lysed in 50 mM HEPES, pH 7.5, 150 mM NaCl, 0.5% NP40, 1% EDTA-free protease inhibitor cocktail and DNAse (10 μg ml^{-1}). Lysate was diluted fivefold with buffer containing 50 mM HEPES, pH 7.5, 150 mM NaCl before being incubated with 20 μl anti-FLAG M2 Magnetic Beads (Sigma-Aldrich) for 3 h at 4 °C. Beads were washed four times with lysis buffer (500 μl) and two times with 50 mM HEPES (500 μl; or 50 mM HEPES, 150 mM NaCl where NaCl was used in the reaction buffer) before MS analysis. Beads were resuspended in a final volume of 30 μl for use in activity assays.

Demethylation reactions were conducted with 2 μl of the immobilized anti-FLAG bead suspension with peptide (10 μM) and cofactors/substrates (conditions given in Supplementary Table 3). Reactions were incubated for 3 h at 37 °C and quenched with 1:1 MeOH by volume. The demethylation products were detected by MALDI MS. Western blotting of lysate samples was carried out to confirm exogenous overexpression of all enzymes using mouse anti-Flag primary (F3165 Sigma, 1:1,000, overnight, 4 °C) and goat anti-mouse-HRP (W4021 Promega, 1:10,000, 1 h, room temperature, Supplementary Fig. 12).

Crystallization. Co-crystals of $KDM4A_{1-359}$.Ni(II).H4(1–15)R3me2s.NOG were obtained in sitting drops grown at 4 °C. Drops contained 10 μg ml^{-1} KDM4A, 5 mM H4(1–15)R3me2s, 5 mM NOG and 4 mM $NiCl_2$. Crystals were cryoprotected with 25% glycerol before being flash-frozen in liquid N_2. Data were collected using a single crystal at 100 K at the Diamond I04-1 MX beam line and were processed with HKL2000 (ref. 65). The structure was solved by molecular replacement using PHASER[66] (search model PDB ID 2OX0) and was refined by alternative cycles of CNS[67] and PHENIX[68], with iterative rebuilding of the refined model using COOT[69]. All residues were in the allowed regions of the Ramachandran plot as calculated by PROCHECK[70]. Data collection and refinement statistics are given in Supplementary Table 4. A stereo image of a portion of the electron density map is shown in Supplementary Fig. 18.

Statistical analysis. Each result shows the mean of three independent experiments with error bars representing the s.e.m. Each experiment was carried out (at least) in technical triplicate.

References

1. Walsh, C. T., Garneau-Tsodikova, S. & Gatto, Jr G. J. Protein posttranslational modifications: the chemistry of proteome diversifications. *Angew. Chem. Int. Ed. Engl.* **44,** 7342–7372 (2005).
2. Rothbart, S. B. & Strahl, B. D. Interpreting the language of histone and DNA modifications. *Biochim. Biophys. Acta* **1839,** 627–643 (2014).
3. Le, D. D. & Fujimori, D. G. Protein and nucleic acid methylating enzymes: mechanisms and regulation. *Curr. Opin. Chem. Biol.* **16,** 507–515 (2012).
4. Walport, L. J., Hopkinson, R. J. & Schofield, C. J. Mechanisms of human histone and nucleic acid demethylases. *Curr. Opin. Chem. Biol.* **16,** 525–534 (2012).
5. Di Lorenzo, A. & Bedford, M. T. Histone arginine methylation. *FEBS Lett.* **585,** 2024–2031 (2011).
6. Zhang, X. *et al.* Peptidylarginine deiminase 2-catalyzed histone H3 arginine 26 citrullination facilitates estrogen receptor alpha target gene activation. *Proc. Natl Acad. Sci. USA* **109,** 13331–13336 (2012).
7. Wang, Y. *et al.* Human PAD4 regulates histone arginine methylation levels via demethylimination. *Science* **306,** 279–283 (2004).
8. Thompson, P. R. & Fast, W. Histone citrullination by protein arginine deiminase: is arginine methylation a green light or a roadblock? *ACS Chem. Biol.* **1,** 433–441 (2006).
9. Chang, B., Chen, Y., Zhao, Y. & Bruick, R. K. JMJD6 is a histone arginine demethylase. *Science* **318,** 444–447 (2007).
10. Webby, C. J. *et al.* Jmjd6 catalyses lysyl-hydroxylation of U2AF65, a protein associated with RNA splicing. *Science* **325,** 90–93 (2009).
11. Han, G. *et al.* The hydroxylation activity of Jmjd6 is required for its homo-oligomerization. *J. Cell. Biochem.* **113,** 1663–1670 (2012).
12. Mantri, M. *et al.* The 2-Oxoglutarate-dependent oxygenase JMJD6 catalyses oxidation of lysine residues to give 5S-hydroxylysine residues. *ChemBioChem* **12,** 531–534 (2011).
13. Gao, W. W. *et al.* Arginine methylation of HSP70 regulates retinoid acid-mediated RARbeta2 gene activation. *Proc. Natl Acad. Sci. USA* **112,** E3327–E3336 (2015).
14. Lawrence, P., Conderino, J. S. & Rieder, E. Redistribution of demethylated RNA helicase A during foot-and-mouth disease virus infection: role of Jumonji C-domain containing protein 6 in RHA demethylation. *Virology* **452-453,** 1–11 (2014).
15. Poulard, C., Rambaud, J., Hussein, N., Corbo, L. & Le Romancer, M. JMJD6 regulates ERalpha methylation on arginine. *PLoS ONE* **9,** e87982 (2014).
16. Liu, W. *et al.* Brd4 and JMJD6-associated anti-pause enhancers in regulation of transcriptional pause release. *Cell* **155,** 1581–1595 (2013).
17. Bottger, A., Islam, M. S., Chowdhury, R., Schofield, C. J. & Wolf, A. The oxygenase Jmjd6--a case study in conflicting assignments. *Biochem. J.* **468,** 191–202 (2015).
18. Markolovic, S., Wilkins, S. E. & Schofield, C. J. Protein hydroxylation catalyzed by 2-oxoglutarate-dependent oxygenases. *J. Biol. Chem.* **290,** 20712–20722 (2015).
19. Hopkinson, R. J. *et al.* Is JmjC oxygenase catalysis limited to demethylation? *Angew. Chem. Int. Ed. Engl.* **52,** 7709–7713 (2013).
20. Williams, S. T. *et al.* Studies on the catalytic domains of multiple JmjC oxygenases using peptide substrates. *Epigenetics* **9,** 1596–1603 (2014).
21. Klose, R. J., Kallin, E. M. & Zhang, Y. JmjC-domain-containing proteins and histone demethylation. *Nat. Rev. Genet.* **7,** 715–727 (2006).
22. Hopkinson, R. J., Hamed, R. B., Rose, N. R., Claridge, T. D. & Schofield, C. J. Monitoring the activity of 2-oxoglutarate dependent histone demethylases by NMR spectroscopy: direct observation of formaldehyde. *ChemBioChem* **11,** 506–510 (2010).
23. Sakurai, M. *et al.* A miniaturized screen for inhibitors of Jumonji histone demethylases. *Mol. Biosyst.* **6,** 357–364 (2010).
24. Feng, W., Yonezawa, M., Jung, Y., Jenuwein, T. & Grummt, I. PHF8 activates transcription of rRNA genes through H3K4me3 binding and H3K9me1/2 demethylation. *Nat. Struct. Mol. Biol.* **17,** 445–450 (2010).
25. Yamane, K. *et al.* JHDM2A, a JmjC-Containing H3K9 Demethylase, FacilitatesTranscription Activation by Androgen Receptor. *Cell* **125,** 483–495 (2006).
26. Hopkinson, R. J. *et al.* 5-Carboxy-8-hydroxyquinoline is a broad spectrum 2-oxoglutarate oxygenase inhibitor which causes iron translocation. *Chemical Science* **4,** 3110–3117 (2013).
27. Kruidenier, L. *et al.* A selective jumonji H3K27 demethylase inhibitor modulates the proinflammatory macrophage response. *Nature* **488,** 404–408 (2012).
28. Couture, J. F., Collazo, E., Ortiz-Tello, P. A., Brunzelle, J. S. & Trievel, R. C. Specificity and mechanism of JMJD2A, a trimethyllysine-specific histone demethylase. *Nat. Struct. Mol. Biol.* **14,** 689–695 (2007).
29. Chen, Z. *et al.* Structural basis of the recognition of a methylated histone tail by JMJD2A. *Proc. Natl Acad. Sci. USA* **104,** 10818–10823 (2007).
30. Ng, S. S. *et al.* Crystal structures of histone demethylase JMJD2A reveal basis for substrate specificity. *Nature* **448,** 87–91 (2007).
31. Bedford, M. T. & Clarke, S. G. Protein arginine methylation in mammals: who, what, and why. *Mol. Cell* **33,** 1–13 (2009).
32. Plank, M. *et al.* Expanding the yeast protein arginine methylome. *Proteomics* **15,** 3232–3243 (2015).
33. Zhang, X., Wen, H. & Shi, X. Lysine methylation: beyond histones. *Acta Biochim. Biophys. Sin.* **44,** 14–27 (2012).
34. Ponnaluri, V. K., Vavilala, D. T., Putty, S., Gutheil, W. G. & Mukherji, M. Identification of non-histone substrates for JMJD2A-C histone demethylases. *Biochem. Biophys. Res. Commun.* **390,** 280–284 (2009).
35. Lu, T. *et al.* Regulation of NF-kappaB by NSD1/FBXL11-dependent reversible lysine methylation of p65. *Proc. Natl Acad. Sci. USA* **107,** 46–51 (2010).
36. Sola, S. *et al.* p53 interaction with JMJD3 results in its nuclear distribution during mouse neural stem cell differentiation. *PLoS ONE* **6,** e18421 (2011).
37. Ambler, R. P. & Rees, M. W. Epsilon-N-Methyl-lysine in bacterial flagellar protein. *Nature* **184,** 56–57 (1959).
38. Lanouette, S., Mongeon, V., Figeys, D. & Couture, J. F. The functional diversity of protein lysine methylation. *Mol. Syst. Biol.* **10,** 724 (2014).
39. Moore, K. E. & Gozani, O. An unexpected journey: lysine methylation across the proteome. *Biochim. Biophys. Acta* **1839,** 1395–1403 (2014).
40. Kudithipudi, S., Kusevic, D., Weirich, S. & Jeltsch, A. Specificity analysis of protein lysine methyltransferases using SPOT peptide arrays. *J. Vis. Exp.* e52203 (2014).
41. Paik, W. K. & Kim, S. Enzymatic demethylation of calf thymus histones. *Biochem. Biophys. Res. Commun.* **51,** 781–788 (1973).
42. Shi, Y. *et al.* Histone demethylation mediated by the nuclear amine oxidase homolog LSD1. *Cell* **119,** 941–953 (2004).
43. Whetstine, J. R. *et al.* Reversal of histone lysine trimethylation by the JMJD2 family of histone demethylases. *Cell* **125,** 467–481 (2006).
44. Tsukada, Y. *et al.* Histone demethylation by a family of JmjC domain-containing proteins. *Nature* **439,** 811–816 (2006).
45. Hahn, P. *et al.* Analysis of Jmjd6 cellular localization and testing for its involvement in histone demethylation. *PLoS ONE* **5,** e13769 (2010).
46. Unoki, M. *et al.* Lysyl 5-hydroxylation, a novel histone modification, by Jumonji domain containing 6 (JMJD6). *J. Biol. Chem.* **288,** 6053–6062 (2013).
47. Sinha, K. M., Yasuda, H., Coombes, M. M., Dent, S. Y. R. & de Crombrugghe, B. Regulation of the osteoblast-specific transcription factor Osterix by NO66, a Jumonji family histone demethylase. *EMBO J.* **29,** 68–79 (2010).
48. Ishimura, A. *et al.* Jmjd5, an H3K36 histone demethylase, modulates embryonic cell proliferation through the regulation of *Cdkn1a* expression. *Development* **139,** 749–759 (2012).
49. Liu, L. *et al.* Hairless is a histone H3K9 demethylase. *FASEB J.* **28,** 1534–1542 (2014).
50. Carlson, S. M. & Gozani, O. Emerging technologies to map the protein methylome. *J. Mol. Biol.* **426,** 3350–3362 (2014).
51. Guo, A. *et al.* Immunoaffinity enrichment and mass spectrometry analysis of protein methylation. *Mol. Cell. Proteomics* **13,** 372–387 (2014).
52. Bremang, M. *et al.* Mass spectrometry-based identification and characterisation of lysine and arginine methylation in the human proteome. *Mol. Biosyst.* **9,** 2231–2247 (2013).
53. Yang, M. *et al.* Asparagine and aspartate hydroxylation of the cytoskeletal ankyrin family is catalyzed by factor-inhibiting hypoxia-inducible factor. *J. Biol. Chem.* **286,** 7648–7660 (2011).
54. Hillringhaus, L. *et al.* Structural and evolutionary basis for the dual substrate electivity of the human KDM4 histone demethylase family. *J. Biol. Chem.* **286,** 41616–41625 (2011).
55. Ehrismann, D. *et al.* Studies on the activity of the hypoxia-inducible-factor hydroxylases using an oxygen consumption assay. *Biochem. J.* **401,** 227–234 (2007).
56. Johansson, C. *et al.* The roles of Jumonji-type oxygenases in human disease. *Epigenomics* **6,** 89–120 (2014).
57. Greer, E. L. & Shi, Y. Histone methylation: a dynamic mark in health, disease and inheritance. *Nat. Rev. Genet.* **13,** 343–357 (2012).
58. Helin, K. & Dhanak, D. Chromatin proteins and modifications as drug targets. *Nature* **502,** 480–488 (2013).
59. Bosetti, C., McLaughlin, C., Tarone, R. E., Pira, E. & LaVecchia, C. Formaldehyde and cancer risk: a quantitative review of cohort studies through 2006. *Ann. Oncol.* **19,** 29–43 (2008).
60. Rotili, D. *et al.* A photoreactive small-molecule probe for 2-oxoglutarate oxygenases. *Chem. Biol.* **18,** 642–654 (2011).
61. Rose, N. R. *et al.* Inhibitor scaffolds for 2-oxoglutarate-dependent histone lysine demethylases. *J. Med. Chem.* **51,** 7053–7056 (2008).
62. Rose, N. R. *et al.* Plant growth regulator daminozide is a selective inhibitor of human KDM2/7 histone demethylases. *J. Med. Chem.* **55,** 6639–6643 (2012).
63. Loenarz, C. *et al.* PHF8, a gene associated with cleft lip/palate and mental retardation, encodes for an Nepsilon-dimethyl lysine demethylase. *Hum. Mol. Genet.* **19,** 217–222 (2010).
64. Hwang, T. L. & Shaka, A. J. Water suppression that works. excitation sculpting using arbitrary wave-forms and pulsed-field gradients. *J. Magn. Reson.* **112,** 275–279 (1995).

65. Otwinowski, Z. & Minor, Jr W. in *Methods in Enzymology.* (eds. Carter Jr, C. W. & Sweet, R. M.) Part A, 307–326 (Academic Press, 1997).
66. McCoy, A. J. *et al.* Phaser crystallographic software. *J. Appl. Crystallogr.* **40,** 658–674 (2007).
67. Brunger, A. T. *et al.* Crystallography & NMR system: a new software suite for macromolecular structure determination. *Acta Crystallogr. D Biol. Crystallogr.* **54,** 905–921 (1998).
68. Adams, P. D. *et al.* PHENIX: a comprehensive Python-based system for macromolecular structure solution. *Acta Crystallogr. D Biol. Crystallogr.* **66,** 213–221 (2010).
69. Emsley, P. & Cowtan, K. Coot: model-building tools for molecular graphics. *Acta Crystallogr. D Biol. Crystallogr.* **60,** 2126–2132 (2004).
70. Laskowski, R. A., MacArthur, M. W., Moss, D. S. & Thornton, J. M. PROCHECK: a program to check the stereochemical quality of protein structures. *J. Appl. Crystallogr.* **26,** 283–291 (1993).

Acknowledgements
This work was supported by grants from Cancer Research UK (C8717/A18245) and the Wellcome Trust (091857/7/10/7), the Biotechnology and Biological Sciences Research Council (studentships to L.J.W. and R.J.H.) and the Engineering and Physical Sciences Research Council (EP/L003376/1). R.J.H. acknowledges a William R. Miller Junior Research Fellowship, St Edmund Hall. A.K. acknowledges a Royal Society Dorothy Hodgkin Fellowship (DH120028) and the BHF Centre of Research Excellence Oxford (RE/13/1/30181). We thank Tristan Smart for purification of KDM2A, the Structural Genomics Consortium (Aleksandra Szykowska) for providing KDM5C protein and the Flag-tagged full-length KDM5C plasmids (Pavel Savitsky), Sarah Madden for KDM4A protein, Matthew Smith for assistance with the NMR experiments and Alexandra Brooks for the H3(14–34)K27Rme2a peptide.

Author contributions
L.J.W. produced proteins and purified peptides, performed the MALDI screening experiments, kinetic experiments and full-length-IPs. R.J.H. produced peptides and carried out the NMR experiments and initial MALDI screening experiments. R.C. carried out crystallographic analyses. R.S. produced and carried out assays with recombinant KDM3A. W.G. produced the peptides for screening. L.J.W., R.J.H., A.K. and C.J.S. conceived and supervised the study and wrote the manuscript.

Additional information

Competing financial interests: The authors declare no competing financial interests.

Caenorhabditis elegans is a useful model for anthelmintic discovery

Andrew R. Burns[1,*], Genna M. Luciani[1,2,*], Gabriel Musso[3,4], Rachel Bagg[1], May Yeo[1], Yuqian Zhang[1], Luckshika Rajendran[1], John Glavin[1], Robert Hunter[1], Elizabeth Redman[5], Susan Stasiuk[5], Michael Schertzberg[1], G. Angus McQuibban[6], Conor R. Caffrey[7], Sean R. Cutler[8], Mike Tyers[9], Guri Giaever[10], Corey Nislow[10], Andy G. Fraser[1,2], Calum A. MacRae[3,4], John Gilleard[5] & Peter J. Roy[1,2,11]

Parasitic nematodes infect one quarter of the world's population and impact all humans through widespread infection of crops and livestock. Resistance to current anthelmintics has prompted the search for new drugs. Traditional screens that rely on parasitic worms are costly and labour intensive and target-based approaches have failed to yield novel anthelmintics. Here, we present our screen of 67,012 compounds to identify those that kill the non-parasitic nematode *Caenorhabditis elegans*. We then rescreen our hits in two parasitic nematode species and two vertebrate models (HEK293 cells and zebrafish), and identify 30 structurally distinct anthelmintic lead molecules. Genetic screens of 19 million *C. elegans* mutants reveal those nematicides for which the generation of resistance is and is not likely. We identify the target of one lead with nematode specificity and nanomolar potency as complex II of the electron transport chain. This work establishes *C. elegans* as an effective and cost-efficient model system for anthelmintic discovery.

[1] The Donnelly Centre for Cellular and Biomolecular Research, University of Toronto, Toronto, Ontario, Canada M5S 3E1. [2] Department of Molecular Genetics, University of Toronto, Toronto, Ontario, Canada M5S 1A8. [3] Cardiovascular Division, Brigham and Women's Hospital, Harvard Medical School, and Harvard Stem Cell Institute, Boston, Massachusetts 02115, USA. [4] Department of Medicine, Harvard Medical School, Boston, Massachusetts 02115, USA. [5] Department of Comparative Biology and Experimental Medicine, Faculty of Veterinary Medicine, University of Calgary, Calgary, Alberta, Canada T2N 4Z6. [6] Department of Biochemistry, University of Toronto, 1 King's College Circle, Toronto, Ontario, Canada M5S 1A8. [7] Center for Discovery and Innovation in Parasitic Diseases and Department of Pathology, University of California, San Francisco, California 94158, USA. [8] Center for Plant Cell Biology, Department of Botany and Plant Sciences, University of California, Riverside, California 92521, USA. [9] Institute for Research in Immunology and Cancer, University of Montreal, Montreal, Quebec, Canada H3T 1J4. [10] Department of Pharmaceutical Sciences, University of British Columbia, Vancouver, British Columbia, Canada V6T 1Z3. [11] Department of Pharmacology and Toxicology, University of Toronto, Toronto, Ontario, Canada M5S 1A8. * These authors contributed equally to this work. Correspondence and requests for materials should be addressed to P.J.R. (email: peter.roy@utoronto.ca).

Parasitic nematodes are estimated to infect about one quarter of all humans and have a dramatic negative impact on human health and productivity in developing nations[1,2]. Nematode infections of agriculturally important plants and animals also result in huge economic losses worldwide[3,4]. Despite this, only a handful of anthelmintic families are currently available. These include the benzimidazoles, macrocyclic lactones (for example, ivermectin), imidazothiazoles (for example, levamisole) and cyclic octadepsipeptides (for example, emodepside), most of which were introduced decades ago. Nematode resistance has been reported for each class of compound, with some natural isolates showing multidrug resistance[3,5]. Anthelmintic resistance is a global issue; although some regions, such as New Zealand, have a particularly high prevalence of resistant parasites[5]. Amino-acetonitrile derivatives (AADs) such as monepantel have recently been introduced to the market; however, resistance to this class of compounds has already been reported[6–8]. While combinatorial strategies may prolong an anthelmitic's utility, growing resistance poses significant challenges for the management of parasitic infections.

One reason for the limited number of available anthelmintics may be related to the difficulty in identifying lead compounds at high throughput. The complex life cycle of parasitic nematodes, which rely on a host for propagation, make it challenging to examine a small molecule's impact on these animals with the throughput required to identify large numbers of candidate molecules for further development. The free-living nematode *Caenorhabditis elegans* may offer a convenient alternative model system to search for new compounds that specifically kill nematodes. *C. elegans*, which is ~1 mm in length as an adult, can be cultured in high-throughput format for multiple generations, allowing the identification of molecules that perturb the worm at any point during its life cycle[9–12].

The majority of marketed anthelmintics are active against *C. elegans*[6,13], and the use of this model system has been instrumental in improving the understanding of the mechanism of action of several anthelmintic compounds, including levamisole, benzimidazole and the amino-acetonitrile derivatives[6,13,14]. Notably, the targets of each of these compounds have been elucidated through forward genetic screens for *C. elegans* mutants that resist their effects. In these screens, *C. elegans* parents are randomly mutagenized and their progeny are subsequently screened for individuals that can resist the effects of a given bioactive molecule. 'Drug'-resistant strains are analysed genetically to identify the resistance-conferring mutant gene. The most frequent resistance-conferring mutant gene within the collection of resistant strains has been shown to encode either the target or the targeted pathway/complex of these bioactive molecules[6,9,14].

Clearly, *C. elegans* is a useful model system to study anthelmintics and offers throughput that is not possible with parasitic species. It therefore follows that it might also be a powerful system with which to screen for anthelmintic lead compounds, as has been suggested over 30 years ago[15]. However, there have only been anecdotal references to the use of *C. elegans* in the anthelmintic screening programs of the pharmaceutical industry[16–20], the details of which have not been publically described. Thus, whether bioactivity in *C. elegans* is generally predictive of bioactivity in parasitic nematode species remains unknown.

Here, we describe our screen for anthelmintic lead compounds using whole *C. elegans* nematodes as the primary model system. We screened 67,012 distinct small molecules for their ability to kill *C. elegans* and re-screened our hits in two widely studied parasitic nematode models: *Cooperia onchophora,* a parasite of cattle, and *Haemonchus contortus*, a parasite of sheep[5,21]. We counter-screened the nematicidal molecules in two vertebrate models of development (HEK293 cells and zebrafish) and identified a set of molecules that kills nematodes but may be inactive in vertebrates. In an effort to identify the protein targets for 39 nematicides, we screened more than 19 million mutant *C. elegans* for resistance. We identified the target of one family of these lethal molecules that is closely related to nematicides that have recently been introduced to the market[22,23], demonstrating the value of *C. elegans* as a model system for the discovery of useful nematicidal molecules.

Results

Molecules that kill *C. elegans* are likely to kill parasites. To identify nematicidal compounds, we screened 67,012 commercially available small drug-like molecules for those that induce obvious phenotypes in *C. elegans* at a concentration of 60 µM or less (see Fig. 1a and Supplementary Data 1). From our preliminary screens, we identified 627 bioactive molecules that we call 'worm actives' or 'wactives'. Rescreening revealed 275 wactives that kill *C. elegans* at a concentration of 60 µM or less (see Supplementary Data 1). By contrast, none of the 182 molecules chosen at random from the set of 67,012 compounds killed *C. elegans* (see Supplementary Data 1).

We next screened the wactive library against the nematode parasites *C. onchophora* and *H. contortus* (Fig. 1a). We chose these two species, both of which are from the same phylogenetic clade as *C. elegans* (clade V), because we could screen them using similar methods that we used to screen *C. elegans* and because many important parasites of humans and domestic livestock are from clade V. We collected nematode eggs from infected animals and tested whether the wactives could kill the eggs or hatched animals (see Methods). Of the 275 wactives that killed *C. elegans,* 129 and 116 killed at least 90% of the *C. onchophora* and *H. contortus* animals, respectively, and 103 killed all three nematode species (Fig. 1b). Of the 182 randomly chosen molecules, none killed *C. onchophora* and five killed *H. contortus.* Hence, molecules that kill *C. elegans* are more than 15 times more likely to kill these parasitic nematodes compared with randomly chosen molecules (Fig. 1c).

We counter-screened the wactive and random control libraries for activity against two vertebrate models: *Danio rerio* (zebrafish) and HEK293 cells (Fig. 1a). Fifty-nine of the 275 wactives that kill *C. elegans* and 28 of the random molecules either kill or cause substantial morbidity in zebrafish (see Methods), representing an enrichment of <1.4-fold (Fig. 1c). Similarly, 76 of the 275 wactives that kill *C. elegans* and 40 of the random molecules caused substantial growth defects in HEK293 cells (see Methods), representing an enrichment of <1.3-fold (Fig. 1c). These results suggest that *C. elegans* is a useful model system with which to identify molecules that are lethal in parasitic nematodes, without generally being cytotoxic in vertebrates.

On analysis of chemical properties, we found that nematicidal compounds had a higher average computed octanol/water partition coefficient (logP; 3.9 versus 3.2, $P<10^{-13}$; Student's *t*-test) and lower average molecular weight (273 versus 328, $P<10^{-20}$; Student's *t*-test) when compared against the complete set of 67,012 compounds (Supplementary Fig. 1). This suggests that molecules that are smaller and with greater lipid-solubility might be more effective nematicides.

Structure analyses reveal 30 classes of anthelmintic leads. To further characterize our 275 *C. elegans*-lethal molecules, we organized them into three separate groups based on their phylogenetic bioactivity profiles (Fig. 2a). Group 1 contains 102 molecules that are lethal to only one or two of the three nematode species tested, but are non-lethal to zebrafish and HEK cells.

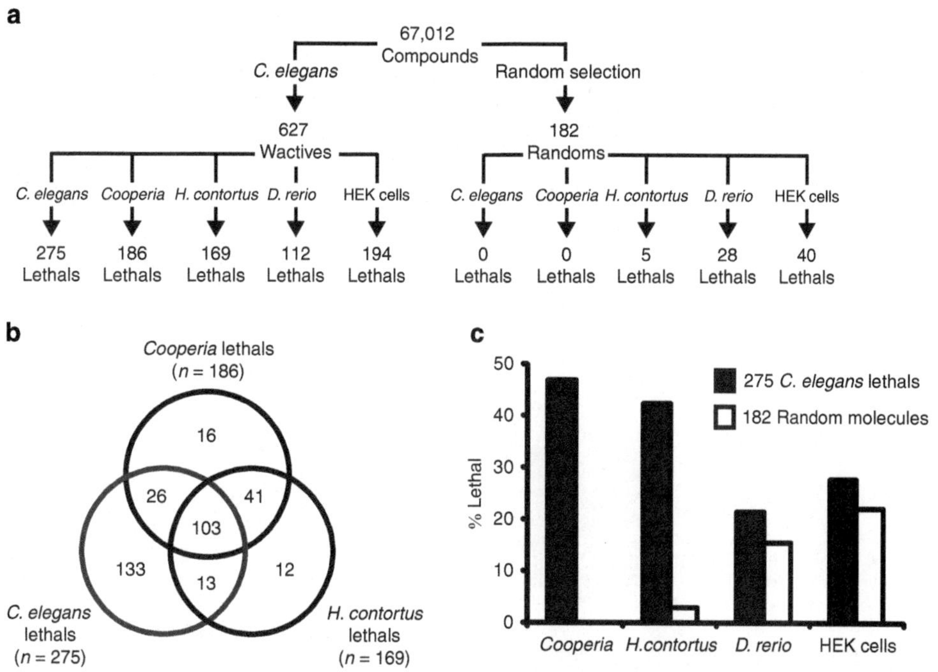

Figure 1 | Molecules that kill *C. elegans* are enriched for those that are lethal to parasitic nematodes. (**a**) Flow chart outlining the multi-organism small-molecule screening pipeline. (**b**) Venn diagram showing the overlap of wactive library molecules that kill *C. elegans*, *Cooperia oncophora* and *H. contortus*. (**c**) Chart showing the enrichment of molecules that kill *Cooperia*, *H. contortus*, zebrafish, and HEK cells in the set of 275 *C. elegans*-lethals, relative to a randomly selected set of 182 compounds.

Group 2 contains 67 compounds that are lethal to all three nematode species, but are non-lethal to zebrafish and human cells. Group 3 contains the remaining 106 compounds that are lethal to fish or HEK cells, and have varied bioactivity in different nematode species. In particular, the 67 molecules in group 2 represent potentially ideal anthelmintic leads; they have activity across multiple nematode species, and appear not to affect vertebrates.

To better understand the structural relationships that exist among our *C. elegans*-lethal compounds, we constructed a structure similarity network that connects molecules if they have a pairwise Tanimoto/FP2 similarity >0.55 (Fig. 2b; see Methods). This network contains 19 isolated clusters composed of three molecules or more, leaving 72 unconnected singletons or pairs. Each cluster represents a unique structural family that could, in principle, target a single protein, although the larger C1 and C2 clusters may contain multiple subfamilies of structures. The 67 group 2 molecules are distributed across 12 clusters, two pairs and 16 singletons, representing 30 structurally unique classes of anthelmintic leads that may target 30 or more distinct protein targets.

To estimate our ability to improve upon the biological activity of the 67 group 2 molecules through medicinal chemistry, we compared the structural similarity of each of these wactives with the 67,012 molecules that we screened (see Methods; Supplementary Data 1). We reasoned that the fraction of molecules within a family of structural analogues that are bioactive may be predictive of the ability to create novel analogues that are bioactive, some of which may have improved bioactivity. We found that 38 (57%) of the 67 wactives are members of a structural analogue family (based on a pairwise Tanimoto/FP2 similarity score of 0.55 or greater) for which more than 10% are lethal to *C. elegans* (Supplementary Fig. 3). Considering only the 18 group 2 wactives that are excluded from the 19 structural clusters in Fig. 2b, we found that 10 (56%) are members of a structural analogue family for which more than 10% are lethal to *C. elegans*. These results suggest that many of the group 2 wactives have the potential for improvement through medicinal chemistry efforts.

Modest structural changes impact phylogenetic specificity. The C3, C13 and C14 clusters in Fig. 2b are composed exclusively of molecules that selectively kill nematodes, suggesting that these families may be targeting nematode-specific proteins. The remaining 16 structural families contain at least one molecule that is lethal to either fish or human cells, suggesting that if the compounds in a cluster target a single protein, the target might be conserved in vertebrates. At the very least, these 16 clusters suggest that relatively modest structural changes can alter the species selectivity of the molecules in a family. To further explore the divergence of species selectivity within these 16 structural families, we focused on pairs of molecules that have a Tanimoto/FP2 pairwise similarity score of 80% or more, and for which one molecule in the pair specifically kills nematodes (group 2 molecules) and the other kills both nematodes and a vertebrate model (group 3 molecules). Seventeen pairs of molecules satisfy these criteria (Supplementary Fig. 2). Inspection of the structural differences of the individual molecules within each pair reveals that very small structural changes can restrict a molecule's bioactivity to nematodes. For example, wact-1, wact-433 and wact-434 in cluster 18 are identical except that an ethyl group in wact-434 replaces the halogen of wact-1 and wact-433. The halogen substitution destroys the core molecule's bioactivity in fish and restricts its activity to nematodes. Medicinal chemistry often yields structural analogues that have reduced or abolished bioactivity. However, our structure–activity analysis has revealed analogues that have not lost bioactivity but have instead become phylogenetically restricted.

Genetic resistance to most nematicides is not easily induced. Forward genetic screens with *C. elegans* have been previously

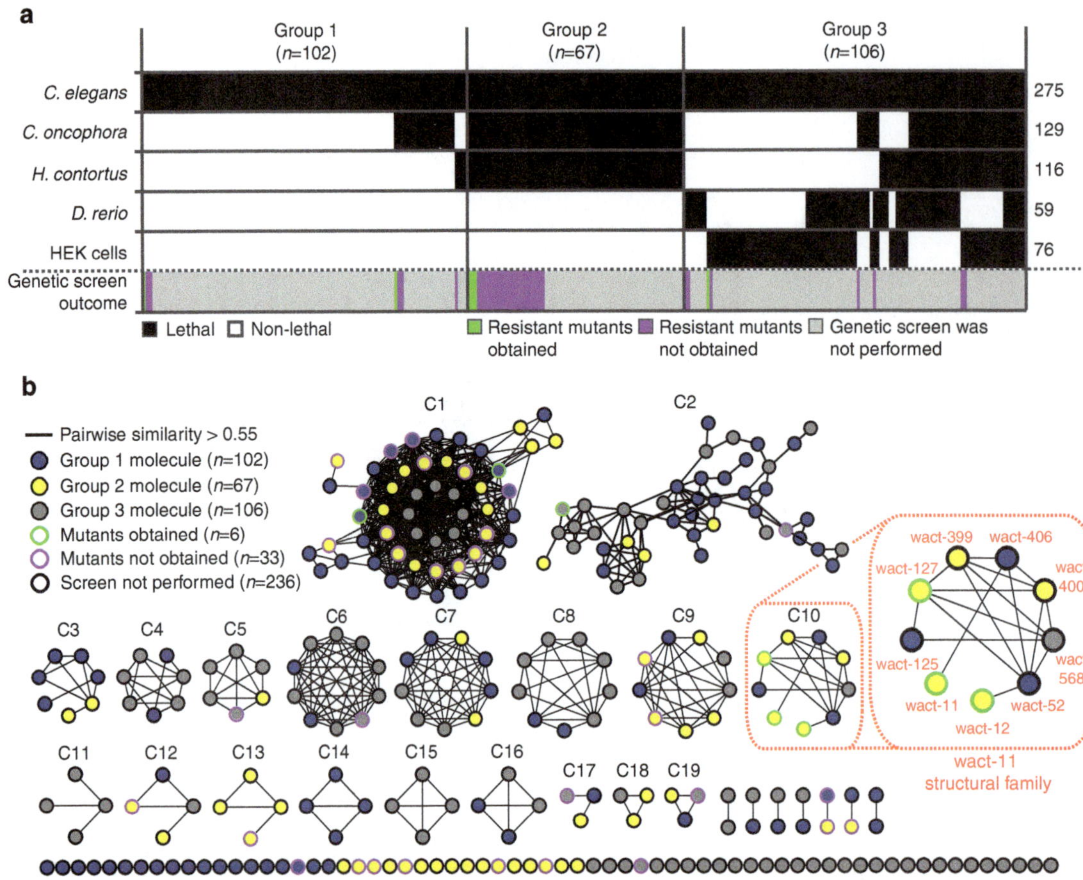

Figure 2 | Nematode selectivity and structural profiling of the 275 *C. elegans*-lethal molecules. (**a**) Heat map indicating the lethality (or lack thereof) induced by each of the 275 *C. elegans*-lethals in two species of parasitic nematode, as well as zebrafish embryos and human embryonic kidney (HEK) cells. For each species, the number of molecules that induce lethality is indicated to the right of the heat map. The molecules segregate into three groups based on their nematode selectivity and cross-species lethality. If a genetic screen for resistant mutants was performed for a given molecule, this is indicated, as well as the outcome of the screen. (**b**) Network based on the structural similarity of the 275 *C. elegans*-lethal molecules. Nodes represent molecules, and edges connect molecules with a pairwise Tanimoto/FP2 score >0.55 (see Methods). The group to which each molecule belongs is indicated by the node fill colour, whereas the genetic screen information is indicated by the node border colour. In the legend, the number of molecules is indicated in parentheses. The 19 clusters containing three or more molecules are named C1 to C19. The wact-11 structural family (cluster C10) is magnified, and the names of each molecule in the family are indicated.

used to identify the targets of bioactive compounds[6,9,14]. Here, we chose 39 compounds from the set of 275 *C. elegans*-lethal wactives to pursue their target genetically (see Methods; Supplementary Data 1; Fig. 2). These 39 molecules span 19 distinct structural classes (Fig. 2b), and the majority of these compounds are found in group 2, which are lethal to all three nematodes tested and non-lethal to zebrafish and HEK293 cells (Fig. 2a).

To identify mutants that are resistant to specific compounds, we randomly mutagenized wild-type *C. elegans* parental (P0) worms using chemical mutagens (see Methods) and screened for animals that resist lethality in either the first (F1) or second (F2) generation. Resistant mutants that arise in F1 screens are dominant and typically encode missense mutations that confer resistance against antagonists, while resistance-conferring mutations in F2 animals are typically recessive reduction-of-function mutations that confer resistance to agonists[24]. Molecules against which F1 screens did not yield resistant mutants were screened again for F2-resistant mutants. We performed F1 screens for all 39 lethal molecules, and F2 screens for 29 of the 39 compounds (Supplementary Table 1). In total, we screened over 19 million mutant genomes and we were able to isolate resistant mutants against only six of the 39 molecules (Supplementary Table 1).

Wact-11-resistant mutants show intra-family cross-resistance. Wact-11, wact-12 and wact-127 are three of the molecules against which we could generate resistant mutants (Supplementary Table 1). These three molecules are part of the C10 cluster (Fig. 2b), which we refer to as the 'wact-11 family', and share an ethyl benzamide moiety (Fig. 3). In total, we isolated 37 mutants that resist the wact-11 family members at a rate of one mutant per 100,000 genomes screened (Supplementary Table 1). Using a representative set of 21 mutants, we performed a detailed dose-response analysis of each mutant against wact-11 and the structurally unrelated nematicide wact-2 in liquid media (Supplementary Fig. 4). All of the tested mutants show at least some resistance to wact-11, but not to wact-2, indicating that they are specifically resistant to wact-11.

Given their structural similarity, we hypothesized that each compound within the wact-11 family may share the same mechanism of action. If true, then mutants that were isolated based on their resistance to one wact-11 family member will resist the lethality that is induced by other family members. To test this, we performed a dose-response analyses of two wact-11-family resistant mutants (isolated based on their respective resistance to wact-11 and wact-12) against all nine wact-11-family members. Both mutants were resistant to all nine wact-11-family members

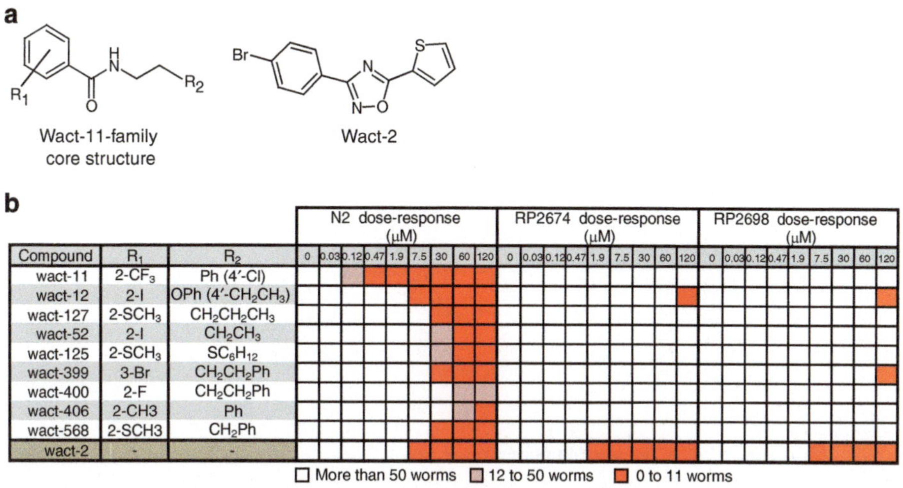

Figure 3 | Wact-11 and wact-12 resistant mutants are cross-resistant to all nine wact-11-family members. (**a**) The wact-11-family core structure and the structure of an unrelated molecule, wact-2, which was used as a negative control throughout this work. (**b**) Heat maps of the wact-11-family dose-response experiments with wild-type worms (N2 strain), as well as two mutant strains, RP2674 and RP2698, isolated as being resistant to wact-12 and wact-11, respectively. The dose-response experiments were carried out using a 96-well plate liquid-based assay (see Methods). White indicates that there were more than 50 worms in three out of four replicate wells. Pink indicates that there were between 12 and 50 worms in three out of four replicate wells. Red indicates that there were between 0 and 11 worms in three out of four replicate wells. In the case of ties, the higher number prevailed (for example, at a given concentration, if two wells had 55 worms, and the other two wells had 20 worms, the chemical would be scored as having more than 50 worms). The R_1 and R_2 groups are indicated for each wact-11-family member. Wact-2 is used here as a negative control.

(Fig. 3), supporting the idea that all nine members of the wact-11 family act by the same mechanism.

The wact-11 family inhibits worm mitochondrial complex II. To identify candidate targets of the wact-11-family, we sequenced the genomes of 33 resistant mutants to identify mutated genes common to multiple strains (Supplementary Data 2 and 3). Ten strains had missense mutations in *sdhb-1*, a different set of 16 strains had a missense mutation in *sdhc-1* (otherwise known as *mev-1*), six other strains had missense mutations in *sdhd-1* and one remaining strain had no commonly mutated gene (Table 1). By contrast, no other gene is represented by distinct mutant alleles in more than four of the 33 strains, and none of the *sdh* genes are mutated in 36 strains that resist the effects of two other unrelated molecules (A.R.B., Houtan Moshiri and P.J.R., unpublished results). Consistent with the isolation of the majority of these mutants in F1 screens, none of the *sdh* mutations are nonsense, frame-shifts or deletions that would be indicative of loss-of-function. Instead, the missense mutations change four unique residues in SDHB-1, seven unique residues in SDHC-1 and three unique residues in SDHD-1 (Table 1).

SDHB-1, SDHC-1 and SDHD-1, along with SDHA-1, are the four protein subunits of *C. elegans* mitochondrial complex II[25] (otherwise known as succinate dehydrogenase or *sdh*), which couples the citric acid cycle to the electron transport chain and is highly conserved among eukaryotes[26]. Complex II couples the oxidation of succinate to fumarate, with the reduction of ubiquinone to ubiquinol[26]. Eukaryotic complex II has at least one ubiquinone-binding site, referred to as the Qp site or Q-site, that exists at the intersection of the SDHB, SDHC and SDHD subunits. In contrast, the succinate-binding site is found exclusively in the SDHA subunit.

A number of Q-site inhibitors are used as fungicides and interest in their use against nematodes is growing[27–29]. For example, flutolanil has been shown to inhibit complex II from the parasitic nematode *Ascaris suum in vitro*, and a co-crystal structure of flutolanil with this complex has been solved[30,31]. We rendered an image of this crystal structure and highlighted the corresponding 14 orthologous residues that are mutated in the wact-11-resistant *C. elegans* mutants (Fig. 4a). Despite residing in three distinct proteins, all the 14 residues cluster around the Q-site where flutolanil is bound. Furthermore, of the 12 residues that are within 4 angstroms of flutolanil's central mass, four are mutated in our wact-11-family resistant mutants (Fig. 4b). Finally, the most frequently mutated residue in our screen is in SDHC-1's R74, which corresponds to *Ascaris*' R89 of SDHC that likely makes electrostatic contacts with the benzene ring of flutolanil's 2-trifluoromethyl-benzamide group[30,31]. Like flutolanil, wact-11 also has a 2-trifluoromethyl-benzamide group, and all wact-11 family members have the benzamide

Table 1 | Complex II residue changes for the wact-11-family-resistant strains.

Number of strains	*C. elegans* mutated gene	*C. elegans* residue change	*Ascaris suum* orthologous residue	Human orthologous residue
1	*sdhb-1*	P145L	P127	P131
2	*sdhb-1*	H146Y	H128	H132
6	*sdhb-1*	P211L	P193	P207
1	*sdhb-1*	I260N	I242	I246
3	*sdhc-1*	T66I	T81	P64
1	*sdhc-1*	G71E	G86	I69
8	*sdhc-1*	R74K	R89	R72
1	*sdhc-1*	G77D	G92	G75
1	*sdhc-1*	C78Y	C93	I76
1	*sdhc-1*	G133E	G148	G131
1	*sdhc-1*	F136S	F151	H134
1	*sdhd-1*	H84Q	H95	H98
4	*sdhd-1*	D95N	D106	D109
1	*sdhd-1*	A97T	G108	V111
1	?			
33 total				

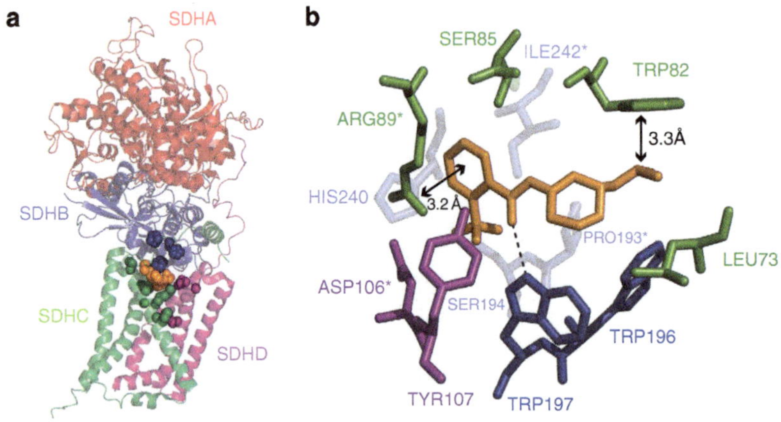

Figure 4 | Complex II residues that are mutated in the wact-11 family resistant mutants cluster near the ubiquinone-binding site (Q-site). (**a**) Rendering of the crystal structure of *Ascaris suum* Complex II bound to the Q-site inhibitor flutolanil (PDB: 3VRB). The side chains of the 14 orthologous residues that are mutated in the wact-11-family resistant mutants are shown as opaque spheres. The atoms of the bound flutolanil molecule are shown as orange-coloured opaque spheres. (**b**) Close-up view of flutolanil bound at the Q-site of Complex II from *Ascaris suum*. The 12 residues shown are no more than 4 Å away from flutolanil, and make up the flutolanil binding pocket. Intermolecular distances are indicated with bidirectional arrows. The dashed line represents a hydrogen bond (H-bond) interaction. Only those H-bonds that occur between Complex II residues and flutolanil are shown; H-bonds that occur between residues of Complex II were omitted for clarity. Bound cofactors, and a bound fumarate molecule, were also omitted for clarity.

moiety. Taken together, our observations suggest that the wact-11-family kills nematodes by binding the Q-site of complex II, and consequently disrupts the interaction of ubiquinone with the complex. Furthermore, the viability of the resistant mutants suggests that the missense mutations alter the Q-site in a way that preserves its function.

We directly tested whether wact-11 and wact-12 can inhibit the enzymatic activity of wild-type *C. elegans* complex II *in vitro* (see Methods). We found that both molecules could inhibit complex II activity with IC50s of 7.4 nM and 5.7 nM, respectively (Table 2, Supplementary Fig. 5). We also tested whether these molecules could inhibit complex II from two independently isolated wact-11-resistant mutants that each harbour the SDHC-1(R74K) missense mutation. We found that complex II from these mutants is insensitive to the molecules up to the highest concentration tested (10 µM; Table 2). As a control, we tested whether wild-type and mutant complex II activity could be inhibited by malonate, which inhibits SDHA-1's succinate-binding activity. We found that malonate inhibits the wild-type complex II to a similar extent as the mutant enzymes (Table 2, Supplementary Fig. 5). These results provide evidence that: (i) mutations in complex II confer resistance to the wact-11-family and (ii) wact-11-family members kill worms by specifically inhibiting complex II at the Q-site.

A SAR analysis reveals potent wact-11 analogues. We performed a focused structure–activity relationship (SAR) analysis with 16 purchasable analogues of the wact-11 family to better understand the structural elements that impact its bioactivity. We first tested the ability of the analogues to inhibit complex II activity *in vitro*. In general, the presence of an electron-withdrawing group at the 2' position of the benzamide benzene ring favours complex II inhibition because the absence of such a group decreases the inhibition by at least 86-fold relative to wact-11 (see wact-11b and wact-11k in Table 2). This observation is consistent with the proposed binding mechanism of flutolanil (see above). The position of the trifluoromethyl group is also important because relocating it to the 4' position (see wact-11 g in Table 2) decreases complex II inhibition by at least 1,300-fold. The R_2 phenyl ring of wact-11 promotes complex II inhibition because removing it (see wact-11i) decreases inhibition 44-fold. Finally, having chloride groups attached to the R_2 phenyl ring promotes complex II inhibition. For example, the two most potent analogues, wact-11 f and wact-11 m, have IC_{50} values of 1 nM, which is over 7-fold lower than that of wact-11, and both of these compounds have electron-withdrawing groups at the $2'$ position of their benzamide benzene rings, as well as chloride groups at the $2'$ and $4'$ positions of their R_2 phenyl rings.

We next tested the *in vivo* potency of the analogues, and found a positive correlation between the *in vitro* IC_{50} and the *in vivo* LD_{100} (dose lethal to 100% of animals tested) values (Table 2; Pearson's correlation coefficient = 0.85), further strengthening the argument that complex II is the *in vivo* target of the wact-11-family. The difference between the *in vitro* and *in vivo* potencies reported in Table 2 is likely due to the resistance of intact *C. elegans* worms to the accumulation of exogenous small molecules[32]. Notably, the two most potent analogues, wact-11 f and wact-11 m, have LD_{100} values in *C. elegans* of 0.469 µM (Table 2). Thus, our SAR analysis revealed nematicides with submicromolar potency.

Few known complex II inhibitors kill *C. elegans*. In addition to flutolanil, we found other commercial complex II Q-site inhibitors that have structural similarity to the wact-11-family (Supplementary Fig. 6). In particular, fluopyram is structurally similar to wact-11 and has recently been developed as part of a crop spray used to kill parasitic nematodes of plants[22,23]. We tested fluopyram and the other structurally related compounds, along with established complex II inhibitors that are structurally unrelated to the wact-11-family (Supplementary Fig. 6), for their ability to kill *C. elegans*. None of these complex II inhibitors are nematicidal up to the highest concentrations tested (120 µM), except for fluopyram and benodanil, which have LD_{100} values of 0.469 µM and 120 µM, respectively (Table 2; Supplementary Fig 7). We found that the wact-11-resistant mutants also resist the lethality induced by fluopyram and benodanil (Supplementary Fig. 7), suggesting that these molecules also target complex II *in vivo*. We also tested the ability of the Q-site inhibitors to inhibit *C. elegans* complex II *in vitro*. Only fluopyram had a potent IC_{50} (1.8 nM), which is almost 2-fold less inhibiting than our two most potent wact-11-family analogues, wact-11 f and wact-11 m.

Table 2 | Complex II IC_{50} and *in vivo* LD_{100} values for wact-11-family molecules and known complex II inhibitors.

Compound	R_1*	R_2*	Species	Strain	SDHC residue change	*In vitro* Complex II IC_{50} (nM)†	*In vivo* LD_{100} (μM)‡
wact-11	2-CF_3	Ph (4′-Cl)	*C. elegans*	N2	None	7.4	7.5
				RP2674	R74K	>10,000	>120
				RP2698	R74K	>10,000	>120
			M. musculus	C57Bl/6	None	>10,000	ND
wact-12	2-I	OPh (4′-CH_2CH_3)	*C. elegans*	N2	None	5.7	7.5
				RP2674	R74K	>10,000	>120
				RP2698	R74K	>10,000	>120
			M. musculus	C57Bl/6	None	>10,000	ND
wact-11f	2-CF_3	Ph (2′-Cl, 4′-Cl)	*C. elegans*	N2	None	1.0	0.469
				RP2674	R74K	>10,000	>120
				RP2698	R74K	>10,000	>120
			M. musculus	C57Bl/6	None	>10,000	ND
wact-11a	2-CF_3	Ph	*C. elegans*	N2	None	26.1	60
wact-11i	2-CF_3	CH_2CH_3	*C. elegans*	N2	None	327.2	>120
wact-11g	4-CF_3	Ph (4′-Cl)	*C. elegans*	N2	None	>10,000	>120
wact-11p	2-I	Ph (4′-Cl)	*C. elegans*	N2	None	5.8	7.5
wact-11e	2-I	Ph (4′-F)	*C. elegans*	N2	None	15.3	60
wact-11d	2-Br	Ph (4′-Cl)	*C. elegans*	N2	None	7.5	1.875
wact-11m	2-Br	Ph (2′-Cl, 4′-Cl)	*C. elegans*	N2	None	1.0	0.469
wact-11j	2-Br	CH_2CH_3	*C. elegans*	N2	None	269.6	120
wact-11c	2-F	Ph (4′-Cl)	*C. elegans*	N2	None	170.1	60
wact-11b	-	Ph (4′-Cl)	*C. elegans*	N2	None	632.9	>120
wact-11k	-	CH_2CH_3	*C. elegans*	N2	None	>10,000	>120
wact-12b	2-I	OPh (4′-CH_3)	*C. elegans*	N2	None	7.7	7.5
wact-12c	2-I	SPh (4′-CH_3)	*C. elegans*	N2	None	132.9	>120
wact-12d	2-I	SPh (4′-Cl)	*C. elegans*	N2	None	128.5	>120
wact-12e	2-Br	OPh (4′-Cl)	*C. elegans*	N2	None	7.8	30
Benodanil			*C. elegans*	N2	None	186.2	120
Boscalid			*C. elegans*	N2	None	549.5	>120
Carboxine			*C. elegans*	N2	None	>10,000	>120
Diazoxide			*C. elegans*	N2	None	>10,000	>120
Fenfuran			*C. elegans*	N2	None	5,279.0	>120
Fluopyram			*C. elegans*	N2	None	1.8	0.469
Flutolanil			*C. elegans*	N2	None	311.2	>120
Harz			*C. elegans*	N2	None	>10,000	>120
Thifluzamide			*C. elegans*	N2	None	3,819	>120
TTFA			*C. elegans*	N2	None	>10,000	>120
Atpenin A5§			*C. elegans*	N2	None	1,678	>120
			M. musculus	C57Bl/6	None	593.3	ND
Malonate‖			*C. elegans*	N2	None	4.9×10^6	>120
				RP2674	R74K	4.2×10^6	ND
				RP2698	R74K	4.2×10^6	ND

Harz, harzianopyridone; ND, not determined; TTFA, thenoyltrifluoroacetone.
*The wact-11-family core structure, with the positions of the R_1 and R_2 groups indicated, is shown in Fig. 3
†The inhibitory curves used to generate the complex II IC_{50} values can be found in Supplementary Fig. 5.
‡The *in vivo* LD_{100} value is defined as the lowest concentration at which no viable animals are visible in four out of four replicate wells, 6 days after 20 first larval-stage worms are deposited. The LD_{100} was determined from a 4-fold dilution series from 120 to 0.00183 μM, and including 60 μM.
§For mouse complex II experiments, atpenin A5 was used as a positive control.
‖For the RP2698 and RP2674 complex II experiments, malonate was used as a positive control.

The wact-11 family fails to inhibit mammalian complex II. The evidence we have presented supports the idea that mitochondrial complex II is the *in vivo* target of the wact-11-family in nematodes. Because the wact-11 family kills nematodes but not human cells, we hypothesized that mammalian complex II is insensitive to these compounds. We tested this hypothesis by assaying the ability of wact-11, wact-12 and wact-11f to inhibit murine complex II *in vitro*. Mouse complex II was insensitive to the highest concentrations tested (10 μM) for all three of these compounds (Table 2; Supplementary Fig. 5).

To better understand the phylogenetic selectivity of the wact-11 family, we inspected the conservation among the SDH subunits in nematodes and vertebrates (Supplementary Fig. 8a). First, we found that 12 of the 14 resistance-conferring residues are conserved in all 10 nematode species examined. Of the two that diverge, SDHC-1's C78 diverges in only one of the 10 species and SDHD-1's A97 diverges in only two species. In each case, the substitutions are conservative. These observations suggest that nematodes may in general be sensitive to the wact-11-family of compounds. Second, we analysed the conservation of the 14 residues among seven vertebrate sequences and found that five of the 14 residues are divergent in all vertebrates examined with four of these substitutions being nonconservative (Supplementary Fig. 8b). Given that mutations of any single one of these residues is sufficient to confer resistance to the wact-11-family, it is reasonable to infer that the vertebrate substitutions at these positions will confer resistance to the wact-11 family of nematicides.

Discussion

Two factors may have discouraged the use of *C. elegans* as a primary high-throughput screening tool to identify novel anthelmintics. First, it is not a parasite and therefore lacks many of the adaptations required for parasitism and the potential anthelmintic targets associated with those processes[16]. Second, the rapid evolution of species within phylum nematoda suggests that *C. elegans* may have essential gene products that may function differently or not exist in some parasitic species[33]. Hence, many of the compounds that kill *C. elegans* may not be effective against parasitic nematodes. However, our work has revealed that molecules that do kill *C. elegans* are more than 15 times more likely to kill parasitic nematodes compared with randomly selected molecules. Given that parasitic nematodes are difficult to screen in high throughput, these results make pre-screening with *C. elegans* an attractive option to increase the throughput of future anthelmintic discovery campaigns.

As expected, not all molecules that kill *C. elegans* are effective against the parasitic models that we tested. In addition, by first screening in *C. elegans*, we have certainly missed molecules that are effective at killing parasitic nematodes but are ineffective in *C. elegans*. However, the speed and ease at which molecules can be screened using *C. elegans* may outweigh the disadvantages it carries as a primary screening system. In principle, *C. elegans* can be used to evaluate hundreds of thousands of molecules at multiple concentrations at a fraction of the cost and time that would be required with most parasitic nematode models.

The extent to which our hits have broad-spectrum activity against distantly-related nematodes is unknown. However, *C. elegans*, *C. oncophora* and *H. contortus* are all clade V nematodes and so molecules that are active against all three of these species are more likely to have broad activity against multiple species in this clade, which include many important parasites of humans and animals[34].

Using forward small molecule screens to identify anthelmintic leads is a powerful approach because it makes no assumptions about what kind of protein makes a good target. Furthermore, these screens have the potential to yield phylum-specific compounds with unexpected and conserved targets that may not have been considered in target-based searches. Our discovery of the activity and target of the wact-11 family provides a good proof-of-principle for the utility of *C. elegans* as a pre-screening model system.

Multiple lines of evidence show that the wact-11 family targets the conserved complex II of the electron transport chain *in vivo*. First, a good correlation exists between complex II *in vitro* inhibition and *in vivo* potency. Second, out of 33 mutants that resist the wact-11-family, 32 have missense mutations in residues that surround the ubiquinone-binding pocket of complex II. By contrast none of the 36 strains that resist unrelated molecules have mutations in complex II. Third, *in vitro* assays show that the wact-11-family can inhibit complex II from wild-type worms but not from worms that have a mutation in complex II that confers resistance to these molecules *in vivo*. Together, these results indicate that the wact-11 family kills *C. elegans* through its inhibition of complex II.

Our screens have revealed that phylogenetically selective bioactivity is highly dependent upon molecular structure. A systematic analysis of close structural analogues revealed 17 pairs of molecules whose phylogenetic bioactivity profile becomes restricted to nematodes with only small alterations in structure. In addition to the 30 groups of anthelmintic lead structures we describe above, the 275 *C. elegans*-lethals comprised 61 structural groups, 33 of which do not contain a single molecule that kills nematodes selectively. However, given the structure–bioactivity analyses presented above, it may be possible to identify structural analogues that specifically kill nematodes for at least some of these structural groups, raising the total number of potential anthelmintic lead structural classes beyond 30.

Of the 30 distinct groups of anthelmintic leads that we have uncovered, we have attempted to screen for resistance against 16 representative molecules and have failed to generate resistance against all but three. The reason behind this is unclear. Typically, we use ethyl methanesulfonate as a mutagen in our genetic screens, which is biased towards inducing G/C to A/T transitions[35] and therefore limits the type of non-synonymous mutations that are induced. This is unlikely to explain the lack of success, however, since we also carried out resistance screens using the *N*-ethyl-*N*-nitrosourea mutagen[35], which produces a variety of nucleotide transitions and transversions, but was equally unsuccessful in yielding resistant mutants (Supplementary Table 1). Instead, we think it more likely that for most lethal molecules, there is no single mutation in the respective target that is capable of conferring resistance. This explanation might be especially true for molecules that inhibit an essential target. It may be that only rare essential targets (like complex II) can be mutated to disrupt the efficacy of an inhibitor without disrupting the target's activity below a viable state. Alternatively, it is possible that many lethal molecules may have multiple essential targets. If true, the generation and isolation of a single mutant animal that has all of the targets mutated to confer resistance would be an exceptionally rare event.

The inability to generate resistant mutants against these molecules has two implications. The first is trivial in that approaches aside from genetics will have to be exploited to identify the target(s) of these molecules. The more important implication is that if the evolution of resistance in the lab can foretell evolution in the field, and there is good evidence for this[3,5–8,13,14], then perhaps the converse is true and that the set of anthelmintic leads for which we are unable to generate resistance should be high priority compounds for further development. Regardless, the evolution of parasitic resistance does not immediately negate the usefulness of an anthelmintic. For example, *C. elegans* mutants that resist the effects of benzimidazoles, imidazothiazoles and cyclooctadepsipeptides can be readily generated in the lab[14,36], yet these molecules have been effective in the field for many years despite the eventual emergence of resistance.

Materials and methods

Chemical sources. The sources for the chemicals and chemical libraries used in our preliminary screens are indicated in Supplementary Data 1. The wact-11-family and structural analogues were purchased from ChemBridge Corporation. The established complex II inhibitors were purchased from Sigma-Aldrich, with the exception of atpenin A5 and harzianopyridone, which were purchased from Enzo Life Sciences, and thifluzamide, which was purchased from AnGene.

***C. elegans* strains and culture methods.** All the animals were cultured using standard methods at 20 °C (ref. 37), unless otherwise indicated. The N2 (wild-type) strain of *Caenorhabditis elegans* was obtained from the *C. elegans* Genetics Center (University of Minnesota).

***C. elegans* liquid-based chemical screening.** The 96-well liquid-based chemical screening assay was adapted from an established RNAi screening protocol[38]. Briefly, saturated HB101 *Escherichia coli* culture was concentrated 2-fold using NGM (nematode growth media) containing 3 mg ml^{-1} NaCl, 2.5 mg ml^{-1} peptone, 5 μg ml^{-1} cholesterol, 1 mM $CaCl_2$, 1 mM $MgSO_4$ and 25 mM KH_2PO_4. A total of 40 μl of NGM + HB101 was dispensed into each well of a 96-well plate, and chemicals were pinned into the wells using a pinning tool with a 300 nl slot volume (V&P Scientific). Approximately 20 synchronized first larval-stage (L1) worms, in 10 μl of M9 buffer (see ref. 24 for the recipe), were then added to each well. The synchronized L1s were obtained from an embryo preparation (see ref. 24 for the protocol) performed the previous day. The final concentration of dimethyl sulfoxide (DMSO) in the wells was 0.6% v/v. A dissection microscope was used to visualize the wells either 5 or 6 days post worm deposition and any obvious chemical-induced phenotypes were noted.

For our preliminary screens, 50,596 out of 67,012 compounds were assayed in liquid at a 60 μM concentration, and the other 16,416 molecules were screened on solid media at 25 μM (see ref. 24 for a description of our solid-based screening method). From these screens, 672 compounds that induced at least a partially penetrant phenotype were re-ordered and arrayed into a 10-plate 'worm active' or 'wactive' library. One hundred and eighty-two molecules chosen at random from the 67,012 compounds were also included in the wactive library, along with many DMSO control wells distributed across the plates.

Using the liquid-based assay, the re-ordered wactive library was re-screened in worms at 7.5, 30 and 60 μM concentrations. Chemicals that induced an obvious phenotype were classified as follows: 'very strong' molecules induced 100% lethality at 7.5 μM, 'strong' molecules induced 100% lethality at 30 μM, 'medium' molecules induced 100% lethality at 60 μM, 'weak' molecules induced incompletely penetrant lethality at 60 μM and 'PEP' molecules induced a non-lethal post-embryonic phenotype such as dumpy (Dpy) or uncoordinated (Unc). The 275 *C. elegans*-lethals that are referred to in the text are made up of the 'very strong', 'strong' and 'medium' molecules.

The 96-well liquid-based assay was also used for all follow-up dose-response experiments in this work. These experiments were carried out in quadruplicate, and the total number of worms in each well was counted 6 days after worm deposition.

***C. oncophora* and *H. contortus* chemical screens.** Fresh cattle and sheep faeces containing eggs of an ivermectin-resistant strain of *C. onchophora*[39] and the MHco3(ISE) strain of *H. contortus*[40,41], respectively, were kindly supplied by Dr Doug Colwell and Dawn Gray (Lethbridge Research Station, Agriculture and Agri-Food Canada). Experimental infections used to generate this material were carried out using established methods[39,41], and were approved by the Lethbridge AAFC Animal Care committee and conducted under animal use license ACC1407. Cattle faeces containing *C. onchophora* eggs were stored anaerobically at 4 °C for a maximum of 3 weeks, whereas sheep faeces containing *H. contortus* eggs were stored at 20 °C for no more than 48 h before harvesting eggs for use. Eggs were isolated from faeces using a standard saturated salt flotation method[42] immediately before each egg hatch assay. Approximately 100 eggs suspended in 100 μl of water were added to each well of a 96-well plate, and the wactive library chemicals were screened at two different concentrations (7.5 and 60 μM, 0.6% DMSO v/v). Baseline egg hatch rates were determined in DMSO control wells ~48 h after the initial set-up of the assay by the addition of iodine tincture to stop development. Plates having DMSO control wells with hatch rates >70% were assayed on a semi-quantitative gradient of '–' to '+ + +', where '–' wells had a hatch rate of <10%, and '+ + +' wells had a hatch rate close to wild type (usually >80%). A dissection microscope was used for visualization purposes. Chemicals were considered bioactive if they consistently had a '–' in more than one trial, with 'very very strong' assigned to compounds that were 90–100% lethal at 7.5 μM in all replicates, 'very strong' if the same was true at 60 μM, 'strong' if replicates had a hatch rate between 10 and 50%, 'medium' if replicates had a 50–80% hatch rate and 'weak' if only one replicate was between 50 and 80% hatch rate. A molecule was considered 'lethal' if it exhibited 'very very strong' or 'very strong' bioactivity.

Zebrafish culture and chemical screening. Wild-type (AB) zebrafish embryos were collected in E3 solution (5 mM NaCl, 0.17 mM KCl, 0.33 mM CaCl2, 0.33 mM MgSO4) immediately after fertilization and arrayed at three embryos per well in 96-well plates, 200 μl per well. Wild-type (AB) zebrafish were originally obtained from Zebrafish International Resource Center (University of Oregon). Chemicals were added to each well at 6 h post fertilization at a final concentration of 10 μM (0.5% v/v DMSO), with each plate containing nine wells of DMSO controls. Embryos were examined for mortality and observable developmental phenotypes at 24 and 48 h post fertilization using an Olympus SZX10 Brightfield microscope. Phenotypes examined were death, developmental delay, reduced pigmentation, cranial oedema or cardiac defects (slow/absent heart rate, abnormal heart size). Each compound was screened in duplicate, with only phenotypes appearing in both replicates associated with a given compound. Any compound producing multiple, distinct phenotypes across replicates (for example, cardiac defects in one replicate, mortality in another) was labelled 'toxic'. A molecule was considered 'lethal' to zebrafish if it induced 'death' or 'toxicity'.

HEK293 cell culture and chemical screening. HEK293T cells (Attisano Lab, University of Toronto) were maintained in Dulbecco's Modified Eagle's Medium (Gibco) supplemented with 10% FBS (DMEM-10% FBS). Cells at a concentration of 50 cells μl^{-1} were seeded into 96-well plates at a final volume of 100 μl (~5,000 cells per well). Chemicals from the wactive library were added to the wells for a final concentration of 60 μM (0.6% DMSO v/v). The wactive library plates were screened at least in triplicate. Cell proliferation was determined using a bromodeoxyuridine (BrdU) assay kit (Exalpha Biologicals Inc.). The BrdU was added 2 days after chemical addition. Fixation and denaturation was performed ~16–18 h later. Anti-BrdU antibody was added, and incorporated BrdU was detected using a horseradish peroxidase-conjugated goat anti-mouse antibody. The BrdU incorporation signal was measured at 450/550 nm for the amount of conversion of tetramethylbenzidine that is proportional to the amount of BrdU incorporated. For each replicate of each plate, the average signal for the DMSO control wells was calculated. For each well in the plate (including the DMSO control wells), the signal was normalized by dividing by the average DMSO signal. The mean and standard deviation for the population of 960 normalized sample signals of the 10-plate wactive library were calculated and found to be 0.6 and 0.44, respectively. A molecule was considered 'lethal' if its normalized signal had a magnitude <0.16 (that is, a value less than the s.d. subtracted from the mean).

Cheminformatics. Chemical structures as supplied by the manufacturers were analysed using the ChemAxon calculator (http://www.chemaxon.com). Specifically, number of hydrogen bond donors, number of hydrogen bond acceptors, mass, atom count, rotatable bond count, logP (a measure of hydrophobicity), Van der Waals surface area, polar surface area, Van der Waals volume and refractivity were computed using default parameters. In the case of salts, all properties were calculated on the largest molecule. Pairwise similarity scores were calculated as the Tanimoto coefficient of shared FP2 fingerprints using OpenBabel (http://openbabel.org). An FP2 fingerprint is a linear fragment of a molecule, containing one to seven atoms. Each pair of compounds to be analysed for similarity were evaluated for presence or absence of any of thousands of possible FP2 fingerprints, and the Tanimoto coefficient represents the number of fingerprints in common between the two compounds divided by the total number of fingerprints present in both compounds. Network visualization for Fig. 2b was performed using Cytoscape[43].

Forward genetic screens for resistant mutants. Our forward genetic screens were carried out as previously described[9,24]. In brief, wild-type parental (P0) worms were mutagenized in 50 mM ethyl methanesulfonate for 4 h, or in 0.5 mM *N*-ethyl-*N*-nitrosourea for 4 h, as previously described. Tens of thousands of synchronized L1s from either the F1 (progeny) or F2 (grand-progeny) generations were dispensed onto 10 cm MYOB agar plates (see ref. 24 for how to prepare MYOB/agar media) containing an ~100% penetrant lethal dose of the nematicide. Candidate-resistant mutants are those worms that can grow in the presence of the chemical. Candidates were picked onto solid MYOB plates without any added chemical, and 12 of their progeny were individually re-tested on a 100%-penetrant lethal dose of the nematicide. Those candidates that re-tested were subsequently homozygosed as previously described[9,24].

Whole-genome sequencing of wact-11-family-resistant mutants. A total 100 μl of packed worms were harvested in a 15 ml conical tube and washed three times with M9 buffer. The worms were then incubated in 6 ml of M9 buffer for 1 h on a nutating shaker at 20 °C. The worms were then washed once with 1 × phosphate-buffered saline (PBS) buffer. The tube was centrifuged and the PBS buffer was aspirated without disturbing the worm pellet. The worms were flash frozen in liquid nitrogen and ground with a pestle until the pellet defrosted. The DNeasy Blood and Tissue Kit (Qiagen) was used to lyse the worm cells and purify the genomic DNA. The Nextera DNA Sample Preparation Kit (Illumina) was used to generate the genomic DNA libraries for sequencing. Individual libraries were quantified with quantitative PCR with reverse transcription using KAPA standards. Multiplexed libraries were sequenced on an Illumina HiSeq2500, paired end reads, 125 bp × 125 bp, using version 4 reagents and flow cells.

Sequence variants were identified using a BWA-GATK pipeline. Briefly, the 125-bp sequencing reads were examined for sequence quality using FastQC (http://www.bioinformatics.babraham.ac.uk/projects/fastqc/) and bases lower than a quality threshold of 30 were trimmed off using Trimmomatic[44]. Reads were aligned to the *C. elegans* N2 reference genome (release W220) using BWA-mem[45]. Alignments were sorted by coordinate order and duplicates removed using Picard (http://picard.sourceforge.net). Before variant calling, reads were processed in Genome Analysis Tool Kit (GATK) v2.5 (ref. 46) for indel realignment and base quality score recalibration, using known *C. elegans* variants from dbSNP build 138 (http://www.ncbi.nlm.nih.gov/SNP/). GATK HaplotypeCaller was used to call variants, and results were filtered for a phred-scaled Q score >30. Finally, called variants were annotated using Annovar[47] to obtain a list of exonic variants for each sample.

***Ascaris suum* complex II structure rendering.** PyMOL[48] was used to generate the images in Fig. 4a,b, using the crystal structure 3VRB (downloaded from the Protein Data Bank). H-bonds were identified by performing the 'find > polar contacts' action in PyMOL[48], using the default settings.

Mitochondria isolation from worms. HB101 *E. coli* cells (*C. elegans* Genetics Center, University of Minnesota) from a 1 l saturated culture were pelleted by centrifugation at 2,500*g* for 10 min, and then resuspended in 50 ml complete S-medium (see ref. 37 for recipe). In all, 450,000 synchronized first larval-stage worms, in M9 buffer, were added to the S-media/HB101 suspension and were grown to adulthood over 3.5 days at 20 °C with shaking at 200 r.p.m. Worms were collected in 6 × 15 ml conical tubes and washed eight times with M9 buffer. The worms from the six tubes were combined into one 15 ml conical tube, and resuspended in 15 ml M9 buffer. One millilitre aliquots of the worm suspension were distributed to each of the 15 × 1.5 ml Sarstedt microcentrifuge tubes. The worms were pelleted by centrifugation at 800*g*, and the M9 buffer was aspirated without disrupting the pellet. A total 600 μl of cold isolation buffer A (250 mM sucrose, 10 mM Tris (pH 7.5), 1 mM EDTA, 1 mM PMSF) was added to each tube. A total 300 μl of cold glass beads was added to each tube. The tubes were cooled on ice for 10 min, and the worms were disrupted by bead beating 6 × 30 s, with 1 min cooling intervals. The tubes were spun down at 1,000*g* for 10 min at 4 °C and the supernatants were transferred to a single 15 ml conical tube on ice. The homogenate was centrifuged at 1,000*g* for 10 min at 4 °C. The supernatant was transferred to a new cold tube, and centrifuged at 16,000*g* for 10 min at 4 °C. The pellet was washed twice by re-suspending gently in 5 ml of cold isolation buffer B (250 mM sucrose, 10 mM Tris (pH 7.5), 1 mM EDTA). After the final wash, 310 μl aliquots of the mitochondrial suspension were distributed across 16 microcentrifuge tubes and the tubes were centrifuged at 16,000*g* for 10 min at 4 °C. The supernatant was aspirated and the pellets were snap frozen in liquid nitrogen and stored at −80 °C until needed.

Mitochondria isolation from mouse liver. The livers from three 8–10-week-old C57Bl/6 female mice (Charles River) were removed, collected in 1.5 ml microcentrifuge tubes, flash frozen in liquid nitrogen and stored at −80 °C until needed. Just before mitochondria isolation, the livers were weighed, placed into a 10-cm petri dish and 10 ml isolation buffer A was added per gram of liver tissue. The tissue was finely minced with a razor blade, transferred to a glass Dounce homogenizer and homogenized by 10 strokes on ice. The homogenate was centrifuged at 1,000*g* for 10 min at 4 °C. The supernatant was collected and centrifuged at 1,000*g* for 10 min at 4 °C. The supernatant was again collected and centrifuged at 16,000*g* for 10 min at 4 °C. The pellet was resuspended in 20 ml of isolation buffer B and centrifuged at 16,000*g* for 10 min at 4 °C. The pellet was resuspended in 20 ml of isolation buffer B, and 1 ml aliquots of the resuspension were distributed into 20 × 1.5 ml microcentrifuge tubes. The aliquoted resuspensions were centrifuged at 16,000*g* for 10 min at 4 °C. The supernatant was aspirated, the pellet was flash frozen in liquid nitrogen and stored frozen at −80 °C until needed. The mice used for experimentation were housed and used in accordance with 'the use and care of experimental animals' guidelines. The animal protocols were reviewed and approved by the Animal Care Committee of the University of Toronto.

Cell-free Complex II functional assay. Mitochondrial pellets were thawed on ice and resuspended in 300 μl of cold isolation buffer B. Using the BCA assay[49], the protein concentration of the mitochondria suspension was determined and was subsequently diluted to a concentration of 0.2 mg ml^{-1} with cold isolation buffer B. Complex II enzymatic assays were carried out in 96-well flat bottom plates. A total 150 μl of complex II assay buffer (1X PBS, 0.35% BSA, 20 mM succinate, 240 μM KCN, 60 μM DCIP, 70 μM decylubiquinone, 25 μM antimycin A, 2 μM rotenone) containing dissolved compounds at the desired concentration, or DMSO (2.4% v/v) alone for control purposes, was added to each well. All of the compounds tested were dissolved in DMSO, except malonate, which was dissolved in water, so DMSO was omitted from the malonate control wells. Five microlitres of the 0.2 mg ml^{-1} mitochondria suspension was added to each well and mixed by pipetting up and down five times. Absorbance for each well was measured at 600 nm using a SpectraMax Plus 384-well Microplate Reader, combined with SoftmaxPro software at 30-s intervals over a 35-min time period. Absorbance versus time was plotted for each well and enzyme activity was calculated as the slope of the line defined by the points ranging from 4 to 10 min. Per cent activity was calculated by dividing the enzyme activity of the chemical-treated wells by that of the DMSO control wells. The per cent activity plotted in Supplementary Fig. 5 is the average of four technical replicates.

References

1. *Parasitic Helminths. Targets, Screens, Drugs and Vaccines* (Wiley-VCH, 2012).
2. Keiser, J. & Utzinger, J. The drugs we have and the drugs we need against major helminth infections. *Adv. Parasitol.* **73,** 197–230 (2010).
3. Besier, B. New anthelmintics for livestock: the time is right. *Trends Parasitol.* **23,** 21–24 (2007).
4. Fuller, V. L., Lilley, C. J. & Urwin, P. E. Nematode resistance. *New Phytol.* **180,** 27–44 (2008).
5. Sutherland, I. A. & Leathwick, D. M. Anthelmintic resistance in nematode parasites of cattle: a global issue? *Trends Parasitol.* **27,** 176–181 (2011).
6. Kaminsky, R. *et al.* A new class of anthelmintics effective against drug-resistant nematodes. *Nature* **452,** 176–180 (2008).
7. Kaminsky, R., Mosimann, D., Sager, H., Stein, P. & Hosking, B. Determination of the effective dose rate for monepantel (AAD 1566) against adult gastro-intestinal nematodes in sheep. *Int. J. Parasitol.* **39,** 443–446 (2009).
8. Scott, I. *et al.* Lack of efficacy of monepantel against *Teladorsagia circumcincta* and *Trichostrongylus colubriformis*. *Vet. Parasitol.* **198,** 166–171 (2013).
9. Kwok, T. C. *et al.* A small-molecule screen in *C. elegans* yields a new calcium channel antagonist. *Nature* **441,** 91–95 (2006).
10. Lemieux, G. A. *et al.* A whole-organism screen identifies new regulators of fat storage. *Nat. Chem. Biol.* **7,** 206–213 (2011).
11. Petrascheck, M., Ye, X. & Buck, L. B. An antidepressant that extends lifespan in adult *Caenorhabditis elegans*. *Nature* **450,** 553–556 (2007).
12. Leung, C. K. *et al.* An ultra high-throughput, whole-animal screen for small molecule modulators of a specific genetic pathway in *Caenorhabditis elegans*. *PLoS ONE* **8,** e62166 (2013).
13. Holden-Dye, L. & Walker, R. J. in *WormBook* 1-13 (The C. elegans Research Community, 2007).
14. Jones, A. K., Buckingham, S. D. & Sattelle, D. B. Chemistry-to-gene screens in *Caenorhabditis elegans*. *Nat. Rev. Drug Discov.* **4,** 321–330 (2005).
15. Simpkin, K. G. & Coles, G. C. The use of *Caenorhabditis elegans* for anthelmintic screening. *J. Chem. Tech Biotechnol* **31,** 66–69 (1981).
16. Geary, T. G. & Thompson, D. P. *Caenorhabditis elegans*: how good a model for veterinary parasites? *Vet. Parasitol.* **101,** 371–386 (2001).
17. Gilleard, J. S. The use of *Caenorhabditis elegans* in parasitic nematode research. *Parasitology* **128**(Suppl 1): S49–S70 (2004).
18. Lee, B. H. *et al.* Marcfortine and paraherquamide class of anthelmintics: discovery of PNU-141962. *Curr. Top. Med. Chem.* **2,** 779–793 (2002).
19. Geary, T. G. in *Parasitic Helminths: Targets, Screens, Drugs and Vaccines (ed. Caffrey, C. R.) 123–134 (Wiley-VCH, 2012).*
20. Woods, D. J. *et al.*in *Parasitic Helminths: Targets, Screens, Drugs and Vaccines.* (ed. Caffrey, C. R.) 297–307 (Wiley-VCH, 2012).
21. Gilleard, J. S. *Haemonchus contortus* as a paradigm and model to study anthelmintic drug resistance. *Parasitology* **140,** 1506–1522 (2013).
22. Hungenberg, H., Fursch, H., Rieck, H. & Hellwege, E. Use of fluopyram for controlling nematodes in crops and for increasing yield. European patent EP2645856 (2013).
23. Broeksma, A., Puetzkuhl, K., Lamprecht, S. & Fuersch, H. VELUM—A evolutionary nematicide for efficient crop production. Presented at the 6th International Congress of NematologyCape Town, South Africa (2014).
24. Burns, A. R. *et al.* High-throughput screening of small molecules for bioactivity and target identification in Caenorhabditis elegans. *Nat. Protoc.* **1,** 1906–1914 (2006).
25. Ichimiya, H. *et al.* Complex II inactivation is lethal in the nematode *Caenorhabditis elegans*. *Mitochondrion* **2,** 191–198 (2002).
26. Sun, F. *et al.* Crystal structure of mitochondrial respiratory membrane protein complex II. *Cell* **121,** 1043–1057 (2005).
27. Sierotzki, H. & Scalliet, G. A review of current knowledge of resistance aspects for the next-generation succinate dehydrogenase inhibitor fungicides. *Phytopathology* **103,** 880–887 (2013).
28. Sakai, C., Tomitsuka, E., Esumi, H., Harada, S. & Kita, K. Mitochondrial fumarate reductase as a target of chemotherapy: from parasites to cancer cells. *Biochim. Biophys. Acta* **1820,** 643–651 (2012).
29. Ishii, H., Miyamoto, T., Ushio, S. & Kakishima, M. Lack of cross-resistance to a novel succinate dehydrogenase inhibitor, fluopyram, in highly boscalid-resistant isolates of *Corynespora cassiicola* and *Podosphaera xanthii*. *Pest Manag. Sci.* **67,** 474–482 (2011).
30. Osanai, A. *et al.* Crystallization of mitochondrial rhodoquinol-fumarate reductase from the parasitic nematode *Ascaris suum* with the specific inhibitor flutolanil. *Acta Crystallogr. Sect. F Struct. Biol. Cryst. Commun.* **65,** 941–944 (2009).
31. Harada, S., Inaoka, D. K., Ohmori, J. & Kita, K. Diversity of parasite complex II. *Biochim. Biophys. Acta* **1827,** 658–667 (2013).
32. Burns, A. R. *et al.* A predictive model for drug bioaccumulation and bioactivity in *Caenorhabditis elegans*. *Nat. Chem. Biol.* **6,** 549–557 (2010).
33. Wasmuth, J., Schmid, R., Hedley, A. & Blaxter, M. On the extent and origins of genic novelty in the phylum Nematoda. *PLoS Negl. Trop. Dis.* **2,** e258 (2008).
34. Holden-Dye, L. & Walker, R. J. in *Parasitic Helminths: Targets, Screens, Drugs and Vaccines*. (ed. Caffrey, C. R.) 23–41 (Wiley-VCH, 2012).
35. Anderson, P. in *Caenorhabditis: Modern Biological Analysis of an Organism.* (eds Epstein, H. F. & Shakes, D. C.) 31–58 (Academic, Inc., 1995).
36. Guest, M. *et al.* The calcium-activated potassium channel, SLO-1, is required for the action of the novel cyclo-octadepsipeptide anthelmintic, emodepside, in *Caenorhabditis elegans*. *Int. J. Parasitol.* **37,** 1577–1588 (2007).
37. Lewis, J. A. & Fleming, J. T. in *Caenorhabditis elegans: Modern Biological Analysis of an Organism*. (eds Epstein, H. F. & Shakes, D. C.) 4–29 (Academic, Inc., 1995).
38. Lehner, B., Tischler, J. & Fraser, A. G. RNAi screens in *Caenorhabditis elegans* in a 96-well liquid format and their application to the systematic identification of genetic interactions. *Nat. Protoc.* **1,** 1617–1620 (2006).
39. Demeler, J., Kuttler, U. & von Samson-Himmelstjerna, G. Adaptation and evaluation of three different *in vitro* tests for the detection of resistance to anthelmintics in gastro intestinal nematodes of cattle. *Vet. Parasitol.* **170,** 61–70 (2010).
40. Laing, R. *et al.* The genome and transcriptome of *Haemonchus contortus*, a key model parasite for drug and vaccine discovery. *Genome Biol.* **14,** R88 (2013).
41. Redman, E. *et al.* Microsatellite analysis reveals marked genetic differentiation between *Haemonchus contortus* laboratory isolates and provides a rapid system of genetic fingerprinting. *Int. J. Parasitol.* **38,** 111–122 (2008).

42. Bartley, D. J. *et al.* A survey of anthelmintic resistant nematode parasites in Scottish sheep flocks. *Vet. Parasitol.* **117,** 61–71 (2003).
43. Cline, M. S. *et al.* Integration of biological networks and gene expression data using Cytoscape. *Nat. Protoc.* **2,** 2366–2382 (2007).
44. Bolger, A. M., Lohse, M. & Usadel, B. Trimmomatic: a flexible trimmer for Illumina sequence data. *Bioinformatics* **30,** 2114–2120 (2014).
45. Li, H. & Durbin, R. Fast and accurate short read alignment with Burrows-Wheeler transform. *Bioinformatics* **25,** 1754–1760 (2009).
46. DePristo, M. A. *et al.* A framework for variation discovery and genotyping using next-generation DNA sequencing data. *Nat. Genet.* **43,** 491–498 (2011).
47. Wang, K., Li, M. & Hakonarson, H. ANNOVAR: functional annotation of genetic variants from high-throughput sequencing data. *Nucleic Acids Res.* **38,** e164 (2010).
48. The PyMOL Molecular Graphics System, Version 1.3r1, (Schrodinger, LLC, 2010).
49. Walker, J. M. The bicinchoninic acid (BCA) assay for protein quantitation. *Methods Mol. Biol.* **32,** 5–8 (1994).

Acknowledgements

We thank Dr Doug Colwell and Dawn Gray at the Lethbridge and Agri-Food Canada Research Centre for supplying faeces from infected sheep and calves for parasite egg collection. We thank Dr Liliana Attisano from the Donnelly CCBR at the University of Toronto for the HEK293T cells. We also thank Dr Derek van der Kooy, Brenda Takabe and Timothy Liao from the Donnelly CCBR at the University of Toronto for the mouse tissue. Whole-genome sequencing was carried out by The Donnelly Sequencing Centre at the University of Toronto. This work was supported by grants to G.M. and C.A.M. from the NIH (HL098938), Burroughs Welcome Foundation and the Harvard Stem Cell Institute; to A.G.F. from the CIHR (488367); to J.G. from the CIHR (230927) and the NSERC-CREATE HPI graduate training programme at the University of Calgary; and to P.J.R. from the CIHR (313296). P.J.R. is a Canada Research Chair in Chemical Genetics.

Author contributions

A.R.B. and G.M.L. performed the primary *C. elegans* chemical screens in P.J.R.'s lab. The chemical re-screens in *C. elegans* were performed by G.M.L. and R.B. in P.J.R.'s lab. G.M.L. performed the *C. oncophora* and *H. contortus* chemical screens, with the help of E.R. and S.S., in J.G.'s lab. G.M. performed the zebrafish chemical screens in C.A.M.'s lab. Y.Z. performed the HEK cell chemical screens in P.J.R.'s lab. A.R.B., G.M.L., G.M. and P.J.R. analysed the chemical screening data. C.R.C. helped perform some chemical screens. G.M. performed all cheminformatic analyses. R.B., L.R., J.G. and R.H. carried out the forward genetic screens in P.J.R.'s lab. Mutant genomic DNA libraries for sequencing were prepared by A.R.B. in P.J.R.'s lab, and sequence analysis was performed by M.S. in A.G.F.'s lab. The complex II functional assays were performed by A.R.B. with input from G.A.M. S.R.C., M.T., G.G., C.N. and P.J.R. contributed chemical libraries for screening. The project was conceived by P.J.R. and the paper was written by A.R.B. and P.J.R.

Additional information

Competing financial interests: The authors declare no competing financial interests.

21

Oscillation of p38 activity controls efficient pro-inflammatory gene expression

Taichiro Tomida[1,†], Mutsuhiro Takekawa[2] & Haruo Saito[1]

The p38 MAP kinase signalling pathway controls inflammatory responses and is an important target of anti-inflammatory drugs. Although pro-inflammatory cytokines such as interleukin-1β (IL-1β) appear to induce only transient activation of p38 (over ~60 min), longer cytokine exposure is necessary to induce p38-dependent effector genes. Here we study the dynamics of p38 activation in individual cells using a Förster resonance energy transfer (FRET)-based p38 activity reporter. We find that, after an initial burst of activity, p38 MAPK activity subsequently oscillates for more than 8 h under continuous IL-1β stimulation. However, as this oscillation is asynchronous, the measured p38 activity population average is only slightly higher than basal level. Mathematical modelling, which we have experimentally verified, indicates that the asynchronous oscillation of p38 is generated through a negative feedback loop involving the dual-specificity phosphatase MKP-1/DUSP1. We find that the oscillatory p38 activity is necessary for efficient expression of pro-inflammatory genes such as *IL-6*, *IL-8* and *COX-2*.

[1] Division of Molecular Cell Signaling, Institute of Medical Science, The University of Tokyo, 4-6-1 Shirokanedai, Minato-ku, Tokyo 108-8639, Japan. [2] Division of Cell Signaling and Molecular Medicine, Institute of Medical Science, The University of Tokyo, 4-6-1 Shirokanedai, Minato-ku, Tokyo 108-8639, Japan. † Present address: Department of Physiology, School of Medicine, Toho University, 5-21-16 Omori-Nishi, Ota-ku, Tokyo 143-8540, Japan. Correspondence and requests for materials should be addressed to T.T. (email: tomida@med.toho-u.ac.jp) or to H.S. (email: h-saito@ims.u-tokyo.ac.jp).

Mitogen-activated protein kinases (MAPKs) constitute major signalling pathways through which eukaryotic cells respond to extracellular stimuli[1,2]. There are several subfamilies of MAPKs in higher eukaryotes, including extracellular signal-regulated kinase (ERK), c-Jun amino-terminal kinase (JNK) and p38 MAPK (p38). All MAPKs are activated through a three-tiered kinase cascade, composed of an MAPK, an MAPK kinase (MAPKK) and a MAPKK kinase (MAPKKK). Distinct MAPKKKs activated by a specific stimulus phosphorylate and thereby activate a cognate MAPKK, which then phosphorylates and activates a downstream MAPK.

The ERK MAPKs are activated by various growth factors and control cell growth and proliferation. In contrast, the p38 and JNK MAPKs are activated by stress conditions, including ultraviolet, oxidative stress and inhibition of protein synthesis by antibiotics such as anisomycin[3]. In general, if the intensity of the stress is moderate, the affected cell will seek to repair the damage. If, however, the stress is too severe for a complete repair, the affected cells are eliminated by apoptosis. For example, sustained activation of p38 and JNK, and concurrent inhibition of ERK induces apoptosis[4]. Thus, the intensity, duration and balance of MAPK activities are important determinants of cell fate.

The p38 MAPK signalling pathway also responds to pro-inflammatory cytokines such as interleukin (IL)-1β and tumour necrosis factor (TNF)-α, and controls immune responses by inducing the expression of effector cytokine genes[5,6]. As a result, the p38 pathway is an important target of anti-inflammatory drugs[7]. Cytokines induce rapid but transient activation of p38 (early phase), which then declines to a sustained low level that is only slightly higher than the pre-stimulation baseline (late phase). Nonetheless, the continued presence of activating cytokines is essential for proper induction of effector cytokines such as IL-6. Although it is believed that the duration and the temporal pattern of MAPK activation affect the quality of output signals[8–10], the dynamics of p38 activity in the seemingly quiet late phase of cytokine activation are unknown. Here we report that, in this late phase, cells go through many cycles of p38 activation and inactivation, and that such p38 activity oscillation is important for efficient induction of its effector functions.

Results

FRET-based monitor of p38 MAPK activity. We first examined the kinetics of IL-1β-induced p38 activation using conventional immunoblotting analyses. As previously shown, p38 is only transiently activated on stimulation of HeLa cells with IL-1β (Fig. 1a; within 1 h, early phase). However, if the p38 inhibitor was added after p38 activity had abated (at 2 h after IL-1β addition, late phase), IL-1β induction of *IL-6* gene expression was severely suppressed (Fig. 1b), indicating that late-phase p38 activity, although very low, must be important for induction of its effector functions. As this low level p38 activity is unlikely to be sufficient to induce expression of p38-dependent pro-inflammatory genes, it is unclear why continuous presence of cytokine is necessary to do so. One possibility is that the p38 response of individual cells might differ from that of the average response of the cell population and thus a small number of cells with prolonged high activity of p38 might be masked by a large number of cells with low p38 activity. We therefore monitored p38 activity in individual living cells following cytokine stimulation. To do so, we developed a p38-specific kinase activity reporter based on the general design of previously reported MAPK reporters[10–14] (Supplementary Fig. 1a). Cells that stably express this p38 reporter (p38 reporter cells) were established using the PiggyBac transposon vector (Supplementary Fig. 1b). p38 reporter cells growing in a multi-well plate were analysed by placing under an automated fluorescence microscope equipped with a media circulation system. As expected, anisomycin, a potent activator of p38, induced strong Förster resonance energy transfer (FRET) signal (yellow fluorescent protein (YFP)/cyan fluorescent protein (CFP) ratio) from the probe-expressing cells (Supplementary Fig. 1c, upper row). More important, the p38 inhibitor SB203580 suppressed the anisomycin-induced FRET signal at concentrations as low as 2 μM, which is unlikely to inhibit other kinases nonspecifically (Supplementary Fig. 1c,d). When p38α expression was inhibited by specific short interfering RNA (siRNA), FRET signal (induced by IL-1β) was suppressed to about 50% of the control cells (Supplementary Fig. 1e). The remaining FRET signal is likely due the uninhibited activity of p38β. Thus, we concluded that the obtained FRET signal was specific to the p38 family kinases.

Long-term oscillation of p38 activity. When p38 reporter cells were stimulated over 60 min with various amounts of anisomycin, the intensity of the FRET signal increased homogenously among the cells in each population (Supplementary Fig. 2a). In contrast, IL-1β induced variable p38 activation within each population, especially at intermediate IL-1β doses ($\sim 10\,ng\,ml^{-1}$) (Supplementary Fig. 2b). When stimulated by higher doses of IL-1β ($> 10\,ng\,ml^{-1}$), p38 activity reached a peak between 20 and 40 min but was on the decline by 60 min, which reflected the dynamics of p38 activation as analysed by immunoblotting analyses (see Fig. 1a). Surprisingly, after p38 activity had completely subsided, it then started to increase again (Fig. 1c).

Quantification of p38 activity in individual cells over time indicated that p38 activity oscillated, and that this oscillation continued with several peaks for at least 8 h (Fig. 1d,e). Mean amplitudes of the second and third peaks were 52% and 61%, respectively, of the initial peak (Fig. 1f). The first peaks displayed similar timing and amplitude for almost all cells, whereas the timing of the second and later peaks was more heterogeneous (Fig. 1g). Mean peak-to-peak intervals became gradually shorter in consecutive rounds: 148 ± 9 min ($n = 56$) for first-to-second and 133 ± 9 min ($n = 46$) for second-to-third intervals. When the responses were averaged over a cell population, p38 oscillatory dynamism was obscured and only the prominent first peak was observed (Supplementary Fig. 3), consistent with immunoblotting analyses (Fig. 1a).

To confirm that the oscillatory FRET signals are due to changing p38 activities in individual cells, we inhibited p38 activity by either 8 μM SB203580 or 1 μM BIRB796 (Supplementary Fig. 4). When SB203580 or BIRB796 was added 30 min before the start of IL-1β stimulation, the FRET signal was completely suppressed (left panels). When SB203580 or BIRB796 was added 120 min after the addition of IL-1β, the first FRET peak was normal, but no subsequent peaks were observed (right panels). Therefore, we conclude that the oscillatory FRET signals in individual cells reliably report changing p38 activities.

Negative feedback by MKP-1 induces oscillatory p38 activity. Biochemical oscillation can be generated by certain network topologies, such as delayed negative-feedback loops[15]. To elucidate the properties of a putative feedback mechanism that might underlie p38 activity oscillation, we examined the input–output relationship following one or two pulsatile (6 min) IL-1β stimulations. On single pulse stimulation, there was only one transient p38 activation peak (Fig. 2a), suggesting that p38 activity oscillation requires the continued stimulation with IL-1β. When two IL-1β pulses were applied with a 60-min interval between them, the p38 activity response to the second stimulation was very weak, indicating that the p38 pathway had entered a

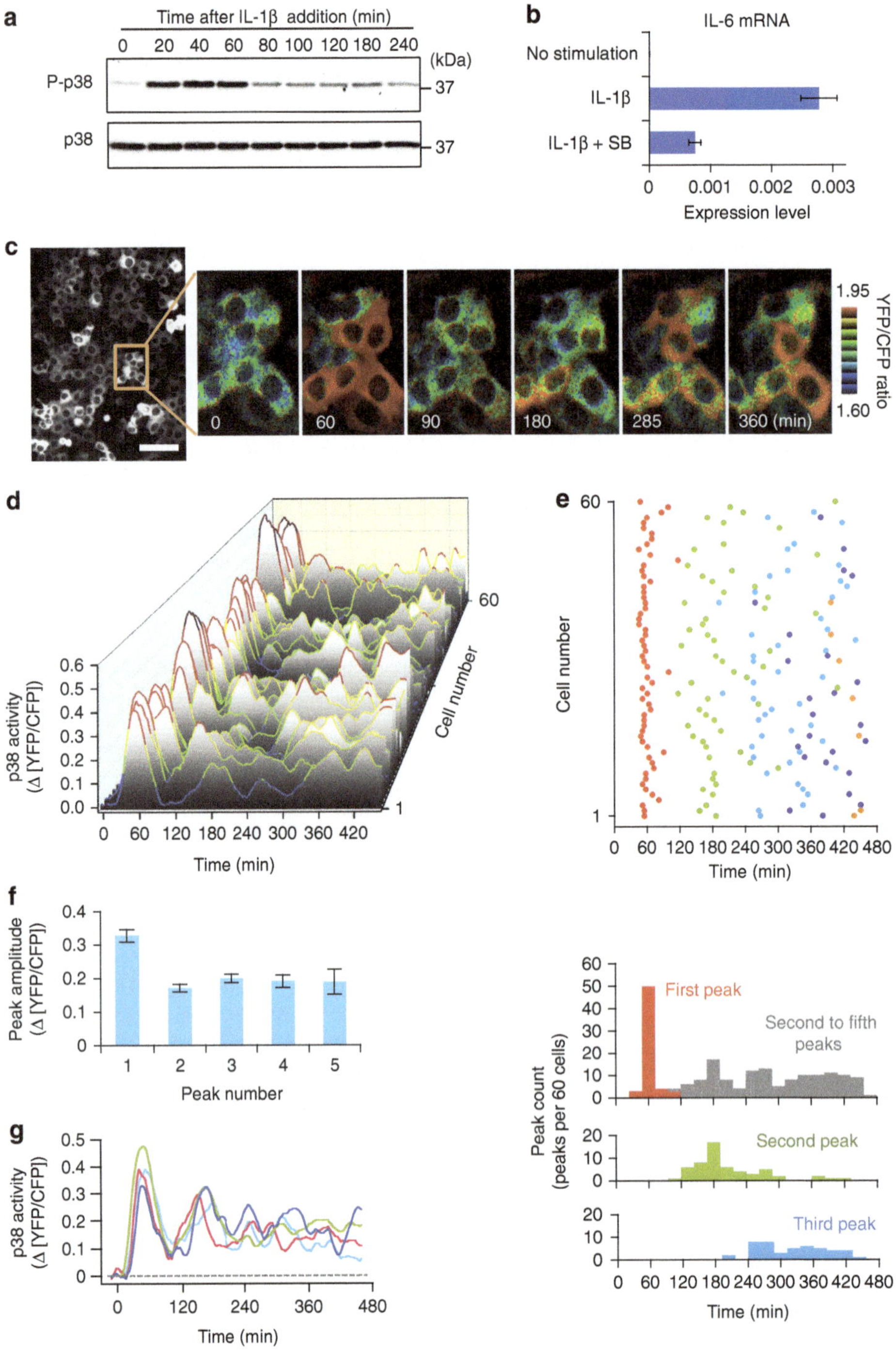

Figure 1 | Dynamics of p38 activation in individual cells. (**a**) Immunoblot analysis of p38 phosphorylation in HeLa cells. IL-1β (30 ng ml^{-1}) was added at time = 0. (**b**) HeLa cells were stimulated with IL-1β (30 ng ml^{-1}) for 360 min and the IL-6 mRNA level was analysed using real-time quantitative PCR. The p38 inhibitor SB203580 (SB; 25 μM) was added at 120 min. Error bars represent s.e.m. ($n = 3$). (**c**) Typical YFP/CFP ratio images showing p38 activation. HeLa cells stably expressing the p38 activity were stimulated continuously with IL-1β (10 ng ml^{-1}). Scale bar, 40 μm. (**d**) Time-course of p38 activation in individual cells. HeLa cells were stimulated continuously with IL-1β (10 ng ml^{-1}). $n = 60$ cells. Colours as per scale bar in **c**. (**e,f**) Quantification of the peaks in **d**. (**e**) Peak times of consecutive rounds of p38 activity are indicated by dots of different colours (top panel). Histograms of peak times are shown below. Grey bars indicate the aggregate counts of the second through the fifth peaks. $n = 60$ cells. (**f**) Mean amplitude of each p38 peak. Error bars represent s.e.m. ($n = 60$, 56, 46, 24 and 7, respectively). (**g**) Overlay plot of typical p38 activity oscillations induced by continuous IL-1β stimulation (10 ng ml^{-1}).

refractory (or inhibited) state after the first strong response (Fig. 2b, top panel). This inhibition was not due to downregulation of IL-1β-specific signalling, as p38 cannot be activated by a subsequent stimulation with TNF-α, which activates p38 via a different receptor (Supplementary Fig. 5). However, the inhibition declined over time, as p38 activity in response to the second stimulation increased as the intervals between the first and second stimuli were lengthened (Fig. 2b, middle and bottom panels). There was high variability in the amplitude of the second peak among individual cells (Fig. 2c).

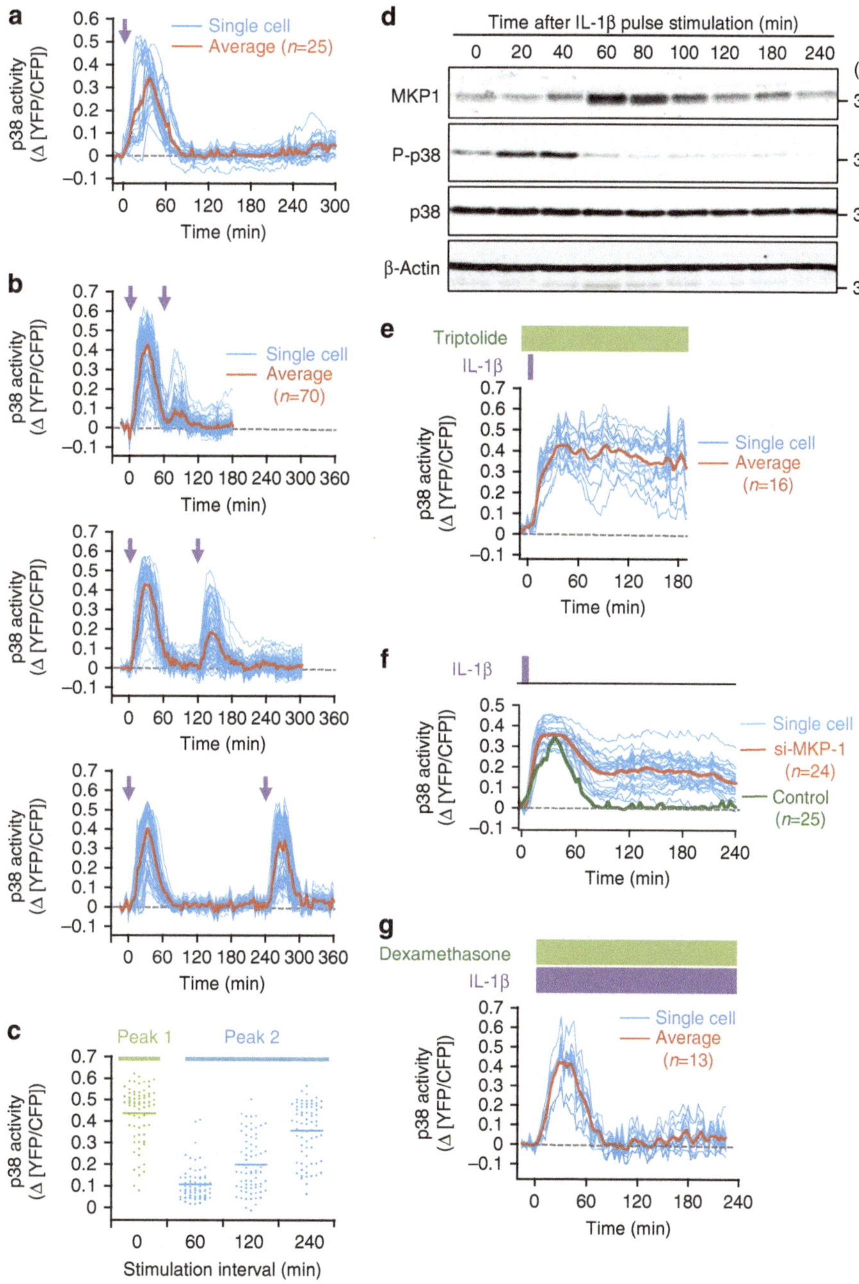

Figure 2 | Variation in the duration of p38 negative feedback regulation in individual cells. p38 responses to single pulse (**a**) or paired-pulse (**b**) IL-1β stimulation, with different intervals between the pulses. Times of the first and second stimulations are indicated by purple arrows. Reporter cells were stimulated with one or two 6-min IL-1β pulses (30 ng ml^{-1}). Responses of individual reporter cells are shown by cyan curves and average traces are in red. $n = 25$ for **a** and 70 for each panel in **b**. (**c**) Peak amplitude of the second peak at the indicated stimulation intervals. $n = 70$ for each group. (**d**) Immunoblot analyses of the time course of p38 phosphorylation and MKP-1 expression in HeLa cells stimulated with a single 6-min IL-1β pulse (30 ng ml^{-1}). (**e-g**) Effects of inhibiting or enhancing MKP-1 expression on p38 activation kinetics. Reporter cells were stimulated as indicated schematically above the graphs. Responses of individual reporter cells are shown by cyan curves and average traces are in red. (**e**) Effect of 30-min pretreatment with the MKP-1 inhibitor triptolide (3 μM) on p38 activation induced by a 6-min IL-1β pulse (30 ng ml^{-1}). (**f**) Effect of inhibiting MKP-1 expression by specific siRNA on p38 activation induced by a 6-min IL-1β pulse (30 ng ml^{-1}). The control curve (green) is the average of 25 siRNA-untreated cells shown in **a**. (**g**) Effect of the presence of the MKP-1 inducer dexamethasone (1 μM) on p38 activation induced by continuous IL-1β (30 ng ml^{-1}) stimulation.

These results indicated that the inhibition of p38 activity lasted as long as 4 h after the initial stimulation, and that there was large variation in the rate of decline of this inhibition among individual cells.

Based on these properties, we suspected involvement of dual-specificity phosphatases, in particular of MKP-1/DUSP1, that can inactivate p38 by dephosphorylation[16]. A 6-min single-pulse of IL-1β stimulation induced MKP-1 expression that reached a maximum level at ~60 min, followed by a gradual decline, whereas p38 phosphorylation decreased rapidly as MKP-1 accumulated (Fig. 2d). As expected, the p38 inhibitors SB203580 (ref. 17) and BIRB796 (ref. 18) suppressed MKP-1 induction at reasonably low concentration: 5 μM SB203580 and 0.1 μM BIRB796, respectively (Supplementary Fig. 6a,b).

When cells were pre-treated with triptolide[19], an inhibitor of MKP-1 expression (Supplementary Fig. 6c), a short pulsatile

IL-1β stimulation induced sustained p38 activation (Fig. 2e). However, as triptolide affects the expression of many other genes as well as *MKP-1*, we also inhibited MKP-1 expression using specific siRNA. When control cells were treated with a short pulsatile IL-1β stimulation, only transient p38 activation was observed (Fig. 2f, green curve). In contrast, when cells transfected with MKP-1 siRNA were similarly treated, sustained p38 activation was observed (Fig. 2f, red curve), indicating that MKP-1 is a major negative regulator of p38 activity.

Dexamethasone is a good inducer of MKP-1 expression[20]. Indeed, simultaneous treatment of cells with IL-1β and dexamethasone induced much higher levels of MKP-1 than treatment with IL-1β alone (Supplementary Fig. 6d). Notably, when dexamethasone was added together with IL-1β, the second and the later oscillating p38 activity peaks were lost (Fig. 2g). Thus, we conclude that the negative feedback cycle between induction of MKP-1 expression by activated p38 and inhibition of p38 activity by MKP-1 governs oscillatory p38 activation.

Mathematical modelling of oscillatory p38 activation. To understand how the p38 activity oscillation arose, we constructed a mathematical model of the p38 feedback regulation (Fig. 3a; see Methods for detail). We estimated the rate constants of the component reactions based on experimental results and published data. When a constant stimulation was applied, the model generated oscillating p38 activation (Fig. 3b). Comparison of p38 activity (i.e., phosphorylated p38) and MKP-1 expression levels indicated that MKP-1 expression and p38 activation peaked alternately, and that new p38 cycles commenced when the MKP-1 concentration dropped to a threshold level. This model also recapitulated the results of paired pulse stimulation (Supplementary Fig. 7a, compare with Fig. 2b) as well as of pharmacological perturbation (Supplementary Fig. 7b, compare with Fig. 2e,g). By this model, it was demonstrated that oscillatory p38 activity arose from the delayed induction of MKP-1 expression by activated p38 as well as a slow degradation of MKP-1.

In the paired-pulse experiments with intervals of 120 or 240 min between pulses, the amplitude of the second response, a measure of the strength of negative feedback, was highly variable (Fig. 2b,c). This finding suggested that the degradation rate of MKP-1 (k_8; see Fig. 3a) might be heterogeneous within a population. The simulation results of paired-pulse stimulations

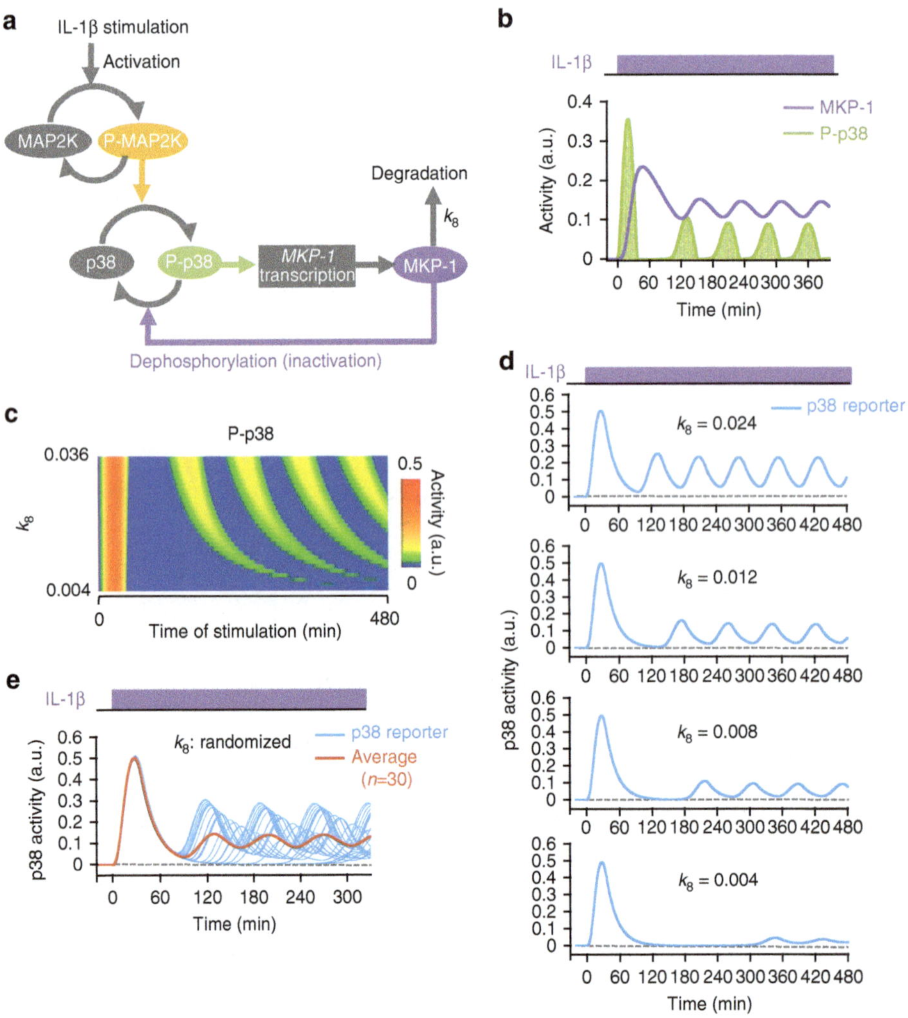

Figure 3 | A mathematical model of p38 activation that generates oscillatory dynamics. (**a**) A schematic diagram of p38 activation. (**b**) Calculated time course of p38 MAPK activity and MKP-1 expression under continuous IL-1β stimulation. a.u., arbitrary units. (**c**) Dependency of p38 activation dynamics on the degradation rate (k_8) of MKP-1. Each horizontal line indicates a calculated time course of p38 activity (phospho-p38) with a different k_8 (vertical axis) under conditions of continuous IL-β stimulation. Activity of p38 is as indicated by the color chart. (**d,e**) Predicted time courses of p38 reporter activity simulated under conditions of continuous IL-1β stimulation. For several fixed k_8 values (**d**) and for 30 randomized k_8 values (**e**). In **e**, individual traces are shown in cyan and the averaged trace is in red.

with varied k_8 values matched well with the experimental results in that the amplitude of the second p38 activity peak was highly sensitive to k_8 change for 120- and 240-min intervals, but not for 60-min intervals (Supplementary Fig. 7c). When the response to a continuous stimulation was simulated using different k_8 values, the second p38 activity peak was significantly delayed and weakened, as k_8 became smaller, whereas the first p38 activity peak was essentially unaffected (Fig. 3c,d). Averaging of simulated p38 traces with varied k_8 values resulted in a robust first peak that was followed by low sustained activity (Fig. 3e), which recreated the actual experimental results. It has been reported that the stability of MKP-1 is variable, owing to posttranslational modifications[21]. Thus, one possible cause of asynchrony in the p38 oscillation is cell-to-cell variability of MKP-1 degradation rate.

p38 oscillation influences pro-inflammatory gene expression. We next investigated whether oscillation of p38 activity plays a role in induction of the pro-inflammatory genes. For that purpose, we applied different temporal patterns of IL-1β stimulation to the HeLa cells (Fig. 4a, protocol numbers 1–6). Robust messenger RNA expression of the *IL-6*, *IL-8* and *COX-2* genes was induced by continuous IL-1β stimulation for 6 h (Fig. 4b; number 2 protocol). As expected, their expression was suppressed by the p38 inhibitor SB203580 (25 μM) (number 3 protocol). The residual gene expression is probably due to non-p38 signalling induced by IL-1β (ref. 22). Addition of 25 μM SB203580 at 2 h, when the initial p38 activation had already subsided (number 4 protocol), did not increase the expression of these genes at 6 h. Lower concentration of SB203580 (8 μM) or another p38-inhibitor BIRB796 (1 μM) had the same effect as 25 μM SB203580 (Supplementary Fig. 8). Consistent with the protocol 4 experiments, a single transient (6 min) IL-1β stimulation, which generated a robust, but single, p38 peak (number 5 protocol), failed to maintain the expression of these genes at 6 h. We found, however, that the continuous presence of IL-1β was unnecessary for efficient gene expression, as repetitive short pulse stimulations (6 min at 60-min intervals) (number 6 protocol) attained gene expression levels equal to or even higher than the continuous IL-1β stimulation.

We thus examined the temporal pattern of p38 activation in cells under the number 6 stimulation protocol. Experimentally, a regimen of five repetitive pulsatile stimulations induced four peaks of p38 activity; the second pulse failed to induce p38 activation (Fig. 4c). The mathematical model very accurately recapitulated these complex dynamics of p38 activation induced by discontinuous IL-1β stimulation (Fig. 4d shows the average response, whereas Supplementary Fig. 9 shows the responses in

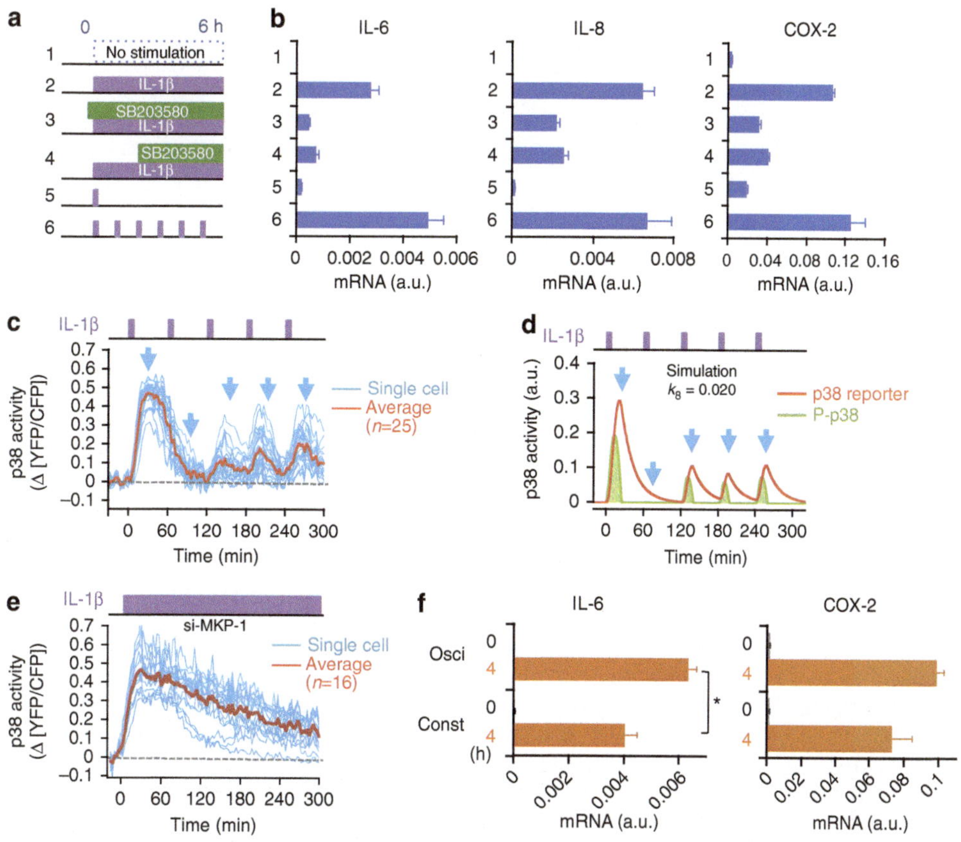

Figure 4 | Oscillation of p38 activity regulates expression of pro-inflammatory genes. (**a**) Diagrams of stimulation/inhibition protocols. HeLa cells were left unstimulated or were stimulated with IL-1β (30 ng ml^{-1}) continuously for 6 h, or with single or multiple 6 min pulses as indicated. SB203580 was added 30 min before (number 3) or 2 h after (number 4) IL-1β addition. (**b**) The mRNA expression of pro-inflammatory genes was determined at the end of the 6-h stimulation protocols using real-time quantitative PCR. Error bars represent s.e.m. ($n = 3$). (**c**) Time course of p38 activation induced by five successive IL-1β pulses (30 ng ml^{-1}, 6 min per pulse with 60-min intervals between pulses) was examined by p38 FRET imaging. Cyan lines indicate individual cell traces and the red line indicates the averaged trace ($n = 25$ cells). (**d**) Simulation of phospho-p38 (green) and p38 reporter activity (red) under the same conditions as in **c**. (**e**) Time course of p38 activation induced by constant IL-1β stimulation (30 ng ml^{-1}) in reporter cells treated with siRNA targeting endogenous MKP-1 (si-MKP-1) was examined by p38 FRET imaging. Cyan and red lines are as in **c** ($n = 16$ cells). (**f**) Expression of pro-inflammatory cytokine mRNA in HeLa cells in which p38 activity oscillates (stimulation conditions as in **c**) or p38 activity is nearly constant (stimulation conditions as in **e**). Amount of mRNA was quantified by quantitative PCR at 4 h. Error bars represent s.e.m. ($n = 3$). *$P = 0.014$, Student's *t*-test.

individual cells). Thus, repetitive pulsatile IL-1β stimulation generated p38 activation time course that was similar to the oscillatory p38 activation generated by the continuous presence of IL-1β, although the former was more synchronous than the latter.

More detailed analyses of gene expression time courses demonstrated that a single p38 activity peak (generated by a single 6 min IL-1β stimulation; number 5 protocol in Fig. 4a) rapidly induced expression of the pro-inflammatory genes that lasted only short period of time (Supplementary Fig. 10a, red curves). In contrast, repetitive activation of p38 (generated by repetitive pulsatile IL-1β stimulations; number 6 protocol in Fig. 4a) maintained high gene expression levels over an extended period (Supplementary Fig. 10a, green curves). These differences were not unexpected, because the cumulative p38 activity was much higher in the second regimen (compare Figs 2a and 4c). We then compared gene expression levels between cells in which p38 was intermittently activated (as in Fig. 4c) and cells in which p38 was continuously activated without oscillation. To create the latter condition, we knocked down MKP-1 expression by specific siRNA and stimulated the cells with continuous IL-1β exposure (Fig. 4e). In these cells, IL-1β stimulation induced sustained (although gradually declining) p38 activity for at least 5 h. By comparing Fig. 4c,e, it is obvious that the cumulative p38 activity is much higher in the latter cells. Nevertheless, mRNA levels of pro-inflammatory genes were similar or even higher in cells in which p38 was intermittently activated than in cells in which p38 was continuously active (Fig. 4f and Supplementary Fig. 10b). Thus, we conclude that oscillatory p38 activity induces pro-inflammatory genes more efficiently than non-oscillatory continuous p38 activity.

Finally, we examined the optimal time interval of p38 oscillation that induces efficient pro-inflammatory gene expression. For this purpose, cells were stimulated with 6 min IL-1β pulses at different (60, 100, 150 and 300 min) intervals (Supplementary Fig. 10c). Most efficient expression of the *IL-6* and *IL-8* genes was observed when cells were repetitively stimulated with either 60- or 100-min interval, whereas the *COX-2* gene was expressed most efficiently when stimulated with 60-min interval (Supplementary Fig. 10d). It is interesting to note that these intervals coincide with those of the intrinsic p38 oscillation induced by constant IL-1β stimulation (see Fig. 1g).

Discussion

We found that constant IL-1β stimulation induces oscillation of p38 MAPK activity. The activity of other MAPKs is also known to oscillate but through different mechanisms. In yeast, the mating pheromone-responsive Fus3 MAPK undergoes sustained oscillation of its phosphorylation and activation, which lasts for at least 8 h with three separate peaks[23]. This oscillation requires the negative regulator of G-protein signalling Sst2, and also partially requires the MAPK phosphatase Msg5. Fus3 oscillation can be detected by immunoblotting analyses, because its oscillation is synchronous. However, in other cases, such as in the present study, oscillation is asynchronous and can be detected only when the response of individual cells are studied. In mammalian cells, rapid nuclear-cytoplasmic shuttling of the ERK MAPK with a rapid periodicity of ~15 min is induced by epidermal growth factor[24]. Much faster oscillation of ERK in the Shankaran study is distinct from that of p38 oscillation, in which periodicity is longer than 2 h. More recently, using a FRET-based ERK activity reporter, Aoki *et al.*[12] reported that ERK activity oscillates with much longer periodicity. This ERK activity oscillation is generated mainly by stochastic and autonomous excitation of the Raf MAPKKK in the absence of external stimulation and is also different from the mechanism of p38 oscillation that depends on external stimulation. In another study, also using a FRET-based ERK activity reporter, Albeck *et al.*[25] reported that, under steady-state conditions, ERK activity oscillates asynchronously with periodicity that is determined by the concentration of external epidermal growth factor. Finally, Regot *et al.*[26] also showed that JNK and ERK (but not p38) activities fluctuate in living cells, by employing multiple kinase activity reporters based on phosphorylation-induced nuclear translocation. Although they did not explore the role of phosphatases in the JNK/ERK activity oscillation, Albeck's and Regot's findings might be mechanistically related to our finding that a negative feedback loop involving the MKP-1 phosphatase is important for the p38 activity oscillation. Thus, oscillation of MAPK activity is more commonplace than previously thought, but in most cases their oscillations are observable only when individual cells are studied.

Oscillatory p38 activation probably satisfies conflicting demands on the p38 pathway. Thus, for proper immune responses, it is important that the target cells continuously respond to the presence of IL-1β or TNF-α through the expression of effector genes. However, although continuous activation of p38 may achieve this goal, it could potentially be deleterious to the cell and may even induce apoptosis[27]. Oscillatory p38 activation therefore may ensure that sufficient expression of effector genes occurs without causing cellular damage. Indeed, we found that oscillatory p38 activation induces the same or even higher levels of the target gene expression than the constant p38 activity induces, in spite of the fact that the cumulative p38 activity is much less in the former case.

An asynchronous p38 oscillation may be important, as a synchronous p38 activity oscillation would lead to periodic and steep rises of effector cytokine expression, which might in turn be damaging to the immune system. Therefore, the asynchronous activation of p38 among cells would ensure that a constant expression level of effector genes is maintained at the population level, without causing damage. In contrast, the synchronous activation of the yeast Fus3 MAPK may be important to ensure that all the cells in a population become mating-competent simultaneously. Thus, although various MAPK pathways employ similar sets of signalling components, they are configured differently to suit specific requirements.

Quantitative analyses of mathematical models are important means to understand the complex interplays among intracellular signalling pathways. To model a signalling pathway mathematically, it is essential to obtain quantitative values for key activities in the pathway. However, the current study as well as others has clearly demonstrated that population level measurements of these values are inadequate in many instances, as the dynamic behaviour of individual cells is totally lost in the population average. Therefore, it is essential to obtain quantitative measurements of these values in individual cells. FRET-based activity reporters are very useful for such measurements.

In conclusion, constant cytokine stimulation generates an oscillatory activation of p38 MAPK that is important for efficient induction of pro-inflammatory gene expression. p38 activity oscillation allows cells to respond optimally to continuous cytokine stimulation such as occurs during infection, while preventing cell damage and apoptosis that can be caused by continuous and excessive p38 activation[4]. These findings should be relevant to the development of safer and more effective anti-inflammatory therapeutics.

Methods

Construction of p38 MAPK activity reporters. p38 MAPK activity reporters were constructed using the pCAGGS-based vector pYL225 (ref. 14), which contains the open reading frames of the enhanced YFP YPet and the super-enhanced CFP. The most efficient reporter so far constructed, termed 'PerKy-38', is composed of the following segments (from the N terminus to the carboxy terminus): YPet (amino acid residues 1–228), Linker 1 (Leu-Glu), the WW domain of Pin1 (amino acid residues 1–54), Linker 2 (Ala-Ser), the p38-docking domain of Mef2A (amino acid residues 266–291), the c-Jun MAPK phosphorylation region excluding the JNK docking site (amino acid residues 54–73), Linker 3 (Gly-Gly-Arg-Gly-Gly) and super-enhanced CFP (amino acid residues 1–237). The nuclear export signal of MEK1 was added at the 3′-end of the reporter, as the reporter without nuclear export signal tended to accumulate in the nucleus.

The DNA segment encoding the Pin1 WW domain was amplified by PCR using a Pin1-encoding plasmid as a template. The DNA segments encoding the Mef2A and c-Jun fragments were prepared by using synthetic oligonucleotides.

Cell lines. Cell lines that constitutively express the PerKy-38 reporter were established using the PiggyBac expression system (System Biosciences, Inc.). The coding sequence of the p38 reporter was inserted between the EcoRI and BamHI cloning sites of the donor plasmid PB510B-1 (PB-CMV-MCS-EF1-Puro). HeLa cells were co-transfected with the donor plasmid for the p38 reporter and the expression plasmid for the PiggyBac transposase. Transfection was carried out using Effectene (Qiagen) in accordance with the manufacturer's standard protocol. Cells that had integrated the p38 reporter into chromosomal locations, and stably express the p38 reporter, were selected using puromycin.

Cell culture and transfection. HeLa cells were maintained in DMEM medium supplemented with 10% fetal bovine serum (FBS), L-glutamine, penicillin and streptomycin. For imaging analyses, cells stably expressing the p38 reporter were plated, a day before imaging analysis, on a 24-well glass-bottom plate (Iwaki, 5826-024) or on a 35-mm-diameter glass-bottom dish (Iwaki, 3910-035) at 30% confluency.

Time-lapse imaging. For time-lapse analyses, cells were starved in Medium 199 (Gibco, 11043-023) supplemented with 0.5% FBS for at least 8 h before imaging. All imaging analyses were carried out at 37°C under 5% CO_2. Fluorescence images of the cells were captured using the Nikon TE-2000E inverted microscope equipped with the CFI PlanApoVC × 20 (numerical aperture 0.75) objective, a CoolSnapHQ CCD (charge-coupled device) camera (Roper), a heat-CO_2 chamber, a mercury lamp (Nikon) and a computer-controlled emission filter changer (Ludl Electronic Products). A FRET filter set (Semrock) was used to excite the fluorescent proteins at 440 nm, and to acquire fluorescence images at 480 nm (for CFP) and at 535 nm (for YFP). Pixel-by-pixel YFP/CFP ratio images were generated using the MetaMorph software (Molecular Device) and were displayed with an intensity modulated display mode based on the YFP fluorescence intensity. Cells were stimulated by rapidly substituting the bathing medium with stimulation medium containing IL-1β or anisomycin using a media circulation pump (ATTO, AC-2120).

Immunoblotting. Cells were plated, a day before analysis, onto a 6-well plate (Falcon, 353046) at 30% confluency. Cells were starved in Medium 199 containing 0.5% FBS for 8 h before stimulation with IL-1β (30 ng ml^{-1}). Immediately after stimulation, the cells were washed twice with ice-cold PBS solution and lysed directly in SDS sample buffer (65 mM Tris-HCl pH 6.8, 5% 2-mercaptoethanol, 3% SDS, 0.0025% bromophenol blue and 10% glycerol). Samples were then boiled for 5 min, separated by SDS–PAGE (10% poly acrylamide gel), transferred onto a nitrocellulose membrane and immunoblotted with antibodies. Anti-p38 (Santa Cruz sc-535 at 1:1,000 dilution) and anti-phospho-p38 (Cell Signaling Technology 9211S at 1:1,000 dilution) were used for detection of the phosphorylation status of endogenous p38 protein in HeLa cells. Anti-MKP-1 (Santa Cruz sc-370 at 1:250 dilution) was used to detect MKP-1 protein expression. Anti-β-actin (Cell Signaling 4967S at 1:1,000 dilution) was used for a loading control. Digital images were captured using the ChemiDoc XRS+ system (Bio-Rad). Full scans of representative immunoblots are presented in Supplementary Fig. 11.

mRNA expression analyses. Endogenous mRNA expression level was analysed using a two-step real-time quantitative PCR method. Cells were starved and stimulated as described for immunoblotting, and cellular RNA was purified using RNeasy Mini (Qiagen) according to the manufacturer's protocol. Reverse transcription of mRNA was performed using PrimeScript reverse transcriptase (Takara), then DNA was analysed using real-time PCR thermal cycler system (Takara, TP800) in the presence of SYBR Green I/ ExTaqII polymerase (Takara, RR820A). The expression level of the target mRNA relative to that of the *GAPDH* mRNA was determined using a standard comparative C_T method. Three independent samples were analysed for each condition. The nucleotide sequences of PCR primers are as follows:

GAPDH forward (Fw) 5′-GCACCGTCAAGGCTGAGAAC-3′,
GAPDH reverse (Rv) 5′-TGGTGAAGACGCCAGTGGA-3′;
IL-6 Fw 5′-GCCAGAGCTGTGCAGATGAG-3′,
IL-6 Rv 5′-TCAGCAGGCTGGCATTTG-3′;
IL-8 Fw 5′-GTGCAGAGGGTTGTGGAGAAGTTT-3′,
IL-8 Rv 5′-ACCAGGAATCTTGTATTGCATCTGG-3′;
COX-2 Fw 5′-TGAAACCCACTCCAAACACAG-3′,
COX-2 Rv 5′-TCAGCATTGTAAGTTGGTGGAC-3′.

Primer sets were obtained from Takara (GAPDH: HA067812, IL-6: HA209655, IL-8: HA217928 and COX-2: HA209441).

siRNA experiments. p38 reporter cells were transfected with siRNAs using Lipofectamine RNAiMAX (Invitrogen) according to the manufacturer's protocol. Expression of p38α/MAPK14 was inhibited by a mixture of two siRNA sequences (catalogue numbers 18464 and 18465, Sigma-Aldrich), whereas the MKP-1-specific siRNA (catalogue number GS1843, Qiagen) contained four siRNA sequences. Negative control siRNA (S5C-0600) was purchased from Cosmo Bio (Tokyo). Transfected cells were subjected to imaging analysis 48 h after transfection.

Mathematical modelling and simulation. The model consists of the following set of differential equations. Implementation of the model and calculation were done using the Mathematica programme (Wolfram Research).

Equations:

(i) Activation of MAP2Ks that activate p38.

$$\mathrm{dMAP2K}[t]/dt = k_0\times \mathrm{S}[t]\times(1-\mathrm{MAP2K}[t]) - k_1\times\mathrm{MAP2K}[t].$$

(ii) Activation and inactivation of p38 MAPK.

$$\mathrm{dMAPK}[t]/dt = k_2\times\mathrm{MAP2K}[t]\times(1-\mathrm{MAPK}[t]) - k_3\times\mathrm{MKP1}[t]\times\mathrm{MAPK}[t]/(k_4+\mathrm{MAPK}[t]).$$

(iii) Transcription of the MKP-1 mRNA.

$$\mathrm{dMKP1RNA}[t]/dt = k_5\times\mathrm{MAPK}[t]\times(1-\mathrm{MKP1RNA}[t]) - k_6\times\mathrm{MKP1RNA}[t]$$

(iv) Protein expression of MKP-1

$$\mathrm{dMKP1}[t]/dt = k_7\times\mathrm{MKP1RNA}[t]\times(1-\mathrm{MKP1}[t]) - k_8\times\mathrm{MKP1}[t].$$

(v) Conversion of p38 activity to the p38 FRET reporter signal.

$$\mathrm{dFRET}[t]/dt = k_9\times\mathrm{MAPK}[t]\times(1-\mathrm{FRET}[t]) - k_{10}\times\mathrm{FRET}[t].$$

Following equations were used for simulating the effect of pharmacological inhibitors.

(vi) Modification of equation (iii) in the presence of dexamethasone.

$$\mathrm{dMKP1RNA}[t]/dt = k_5\times\mathrm{MAPK}[t]\times(1-\mathrm{MKP1RNA}[t]) - k_6\times\mathrm{MKP1RNA}[t] + k_{12}\times(1-\mathrm{MKP1RNA}[t])$$

(vii) Modification of equation (iv) in the presence of triptolide.

$$\mathrm{dMKP1}[t]/dt = k_{11}\times\mathrm{MKP1RNA}[t]\times(1-\mathrm{MKP1}[t]) - k_8\times\mathrm{MKP1}[t]$$

The kinetics of p38 inactivation by MKP-1 was expressed based on Michaelis–Menten equation, which was obtained by means of quasi-steady-state approximation (equation (ii)). In the model, a delay in MKP-1 expression relative to p38 activation, which was evident from our experimental data, was introduced by introduction of a rate-limiting step of MKP-1 transcription in equation (iii).

Variables:

MAP2K[t], the activity of the MAP2Ks that activate p38.
MAPK[t], the activity of the p38 MAPK.
MKP1RNA[t], the *MKP-1* gene transcription level.
MKP1[t], the MKP-1 protein expression level.
S[t], the stimulatory input.
FRET[t], the signal of FRET reporter.
S[t] was set to 0 when there is no stimulation and 1 when stimulated.

Parameters:

k_0, the rate constant for MAP2K activation (0.06 min^{-1}).
k_1, the rate constant for MAP2K deactivation (0.15 min^{-1}).
k_2, the rate constant for p38 MAPK activation (0.15 min^{-1}).
k_3, the maximum reaction rate of p38 MAPK inactivation by MKP-1 (0.16 min^{-1}).
k_4, an equivalent of the Michaelis–Menten constant for the reaction of MKP-1-mediated p38 inactivation (0.0001).
k_5, the rate constant for *MKP-1* gene transcription (0.055 min^{-1}).
k_6, the rate constant for degradation of *MKP-1* gene transcript (0.05 min^{-1}).
k_7, the rate constant for MKP-1 protein expression (0.20 min^{-1}).
k_8, the rate constant for MKP-1 protein degradation (0.02 min^{-1}).
k_9, the rate constant for FRET reporter activation by active p38 MAPK (0.2 min^{-1}).
k_{10}, the rate constant for FRET reporter inactivation (0.05 min^{-1}).
k_{11}, the rate constant for MKP-1 protein expression in the presence of triptolide (0.0002 min^{-1}).
k_{12}, the rate constant for *MKP-1* gene transcription induced by dexamethasone treatment (0.003 min^{-1}).

References

1. Chen, Z. *et al.* MAP kinases. *Chem. Rev.* **101,** 2449–2476 (2001).
2. Raman, M., Chen, W. & Cobb, M. H. Differential regulation and properties of MAPKs. *Oncogene* **26,** 3100–3112 (2007).
3. Kyriakis, J. M. & Avruch, J. Mammalian mitogen-activated protein kinase signal transduction pathways activated by stress and inflammation. *Physiol. Rev.* **81,** 807–869 (2001).
4. Xia, Z., Dickens, M., Raingeaud, J., Davis, R. J. & Greenberg, M. E. Opposing effects of ERK and JNK-p38 MAP kinases on apoptosis. *Science* **270,** 1326–1331 (1995).
5. Zarubin, T. & Han, J. Activation and signaling of the p38 MAP kinase pathway. *Cell Res.* **15,** 11–18 (2005).
6. Cuadrado, A. & Nebreda, A. R. Mechanisms and functions of p38 MAPK signalling. *Biochem. J.* **429,** 403–417 (2010).
7. Kumar, S., Boehm, J. & Lee, J. C. P38 MAP kinases: key signalling molecules as therapeutic targets for inflammatory deiseases. *Nat. Rev. Drug Discov.* **2,** 717–726 (2003).
8. Ashwell, J. D. The many paths to p38 mitogen-activated protein kinase activation in the immune system. *Nat. Rev. Immunol.* **6,** 532–540 (2006).
9. Ventura, J.-J. *et al.* Chemical genetic analysis of the time course of signal transduction by JNK. *Mol. Cell* **21,** 701–710 (2006).
10. Tomida, T., Oda, S., Takekawa, M., Iino, Y. & Saito, H. The temporal pattern of stimulation determines the extent and duration of MAPK activation in a *Caenorhabditis elegans* sensory neuron. *Sci. Signal* **5,** ra76 (2012).
11. Harvey, C. D. *et al.* A genetically encoded fluorescent sensor of ERK activity. *Proc. Natl Acad. Sci. USA* **105,** 19264–19269 (2008).
12. Aoki, K. *et al.* Stochastic ERK activation induced by noise and cell-to-cell propagation regulates cell density-dependent proliferation. *Mol. Cell* **52,** 529–540 (2013).
13. Fosbrink, M., Aye-Han, N.-N., Cheong, R., Levchenko, A. & Zhang, J. Visualization of JNK activity dynamics with a genetically encoded fluorescent biosensor. *Proc. Natl Acad. Sci. USA* **107,** 5459–5464 (2010).
14. Tomida, T., Takekawa, M., O'Grady, P. & Saito, H. Stimulus-specific distinctions in spatial and temporal dynamics of stress-activated protein kinase kinase kinases revealed by a fluorescence resonance energy transfer biosensor. *Mol. Cell Biol.* **29,** 6117–6127 (2009).
15. Ferrell, Jr J. E. Self-perpetuating states in signal transduction: positive feedback, double-negative feedback and bistability. *Curr. Opin. Cell Biol.* **14,** 140–148 (2002).
16. Owens, D. M. & Keyse, S. M. Differential regulation of MAP kinase signalling by dual-specificity protein phosphatases. *Oncogene* **26,** 3203–3213 (2007).
17. Lee, J. C. *et al.* A protein kinase involved in the regulation of inflammatory cytokine biosynthesis. *Nature* **372,** 739–746 (1994).
18. Kuma, Y. *et al.* BIRB796 inhibits all p38 MAPK isoforms *in vitro* and *in vivo*. *J. Biol. Chem.* **280,** 19472–19479 (2005).
19. Chen, P. *et al.* Restraint of proinflammatory cytokine biosynthesis by mitogen-activated protein kinase phosphatase-1 in lipopolysaccharide-stimulated macrophages. *J. Immunol.* **169,** 6408–6416 (2002).
20. Lasa, M., Abraham, S. M., Boucheron, C., Saklatvala, J. & Clark, A. R. Dexamethasone causes sustained expression of mitogen-activated protein kinase (MAPK) phosphatase 1 and phosphatase-mediated inhibition of MAPK p38. *Mol. Cell Biol.* **22,** 7802–7811 (2002).
21. Crowell, S. A. *et al.* Post-translational regulation of mitogen-activated portein kinase phosphatase (MKP)-1 and MKP-2 in macrophages following lipopolysaccharide stimulation - The role of the carboxyl termini of the phosphatases in determining their stability. *J. Biol. Chem.* **289,** 28753–28764 (2014).
22. Molina-Holgado, E., Ortiz, S., Molina-Holgado, F. & Guaza, C. Induction of COX-2 and PGE_2 biosynthesis by IL-1b is mediated by PKC and mitogen-activated protein kinases in murine astrocytes. *Br. J. Pharmacol.* **131,** 152–159 (2000).
23. Hilioti, Z. *et al.* Oscillatory phosphorylation of yeast Fus3 MAP kinase controls periodic gene expression and morphogenesis. *Curr. Biol.* **18,** 1700–1706 (2008).
24. Shankaran, H. *et al.* Rapid and subtained nuclear-cytoplasmic ERK oscillations induced by epidermal growth factor. *Mol. Syst. Biol.* **5,** 332 (2009).
25. Albeck, J. G., Mills, G. B. & Brugge, J. S. Frequency-modulated pulses of ERK activity transmit quantitative proliferation signals. *Mol. Cell* **49,** 249–261 (2013).
26. Regot, S., Hughey, J. J., Bajar, B. T., Carrasco, S. & Covert, M. W. High-sensitivity measurements of multiple kinase activities in live single cells. *Cell* **157,** 1724–1734 (2014).
27. Haq, R. *et al.* Constitutive p38HOG mitogen-activated protein kinase activation induces permanent cell cycle arrest and senescence. *Cancer Res.* **62,** 5076–5082 (2002).

Acknowledgements

We thank Drs P.S. Daugherty, A. Miyawaki and M. Matsuda for providing fluorescent protein vectors; Dr P. O'Grady for her critical reading of the manuscript; and Dr. K. Ichikawa for insightful discussion on mathematical modelling and simulation. Support was provided by several Grants-in-Aid for Scientific Research from the Ministry of Education, Culture, Sports, Science and Technology of Japan to T.T., M.T. and H.S. Materials are available from T.T. or H.S. on request with a material transfer agreement (MTA).

Author contributions

T.T. performed the experiments. T.T., M.T. and H.S. designed the experiments and analysed the data. T.T. and H.S. wrote the paper.

Additional information

Competing financial interests: The authors declare no competing financial interests.

22

Development of background-free tame fluorescent probes for intracellular live cell imaging

Samira Husen Alamudi[1,*], Rudrakanta Satapathy[1,*], Jihyo Kim[2], Dongdong Su[3], Haiyan Ren[4], Rajkumar Das[1], Lingna Hu[1], Enrique Alvarado-Martínez[5], Jung Yeol Lee[1], Christian Hoppmann[4], Eduardo Peña-Cabrera[5], Hyung-Ho Ha[6], Hee-Sung Park[2], Lei Wang[4] & Young-Tae Chang[1,3]

Fluorescence labelling of an intracellular biomolecule in native living cells is a powerful strategy to achieve in-depth understanding of the biomolecule's roles and functions. Besides being nontoxic and specific, desirable labelling probes should be highly cell permeable without nonspecific interactions with other cellular components to warrant high signal-to-noise ratio. While it is critical, rational design for such probes is tricky. Here we report the first predictive model for cell permeable background-free probe development through optimized lipophilicity, water solubility and charged van der Waals surface area. The model was developed by utilizing high-throughput screening in combination with cheminformatics. We demonstrate its reliability by developing CO-1 and AzG-1, a cyclooctyne- and azide-containing BODIPY probe, respectively, which specifically label intracellular target organelles and engineered proteins with minimum background. The results provide an efficient strategy for development of background-free probes, referred to as 'tame' probes, and novel tools for live cell intracellular imaging.

[1] Department of Chemistry & Med Chem Program, National University of Singapore, Singapore 117543, Singapore. [2] Department of Chemistry, Korea Advanced Institute of Science and Technology, Daejeon 305701, Republic of Korea. [3] Laboratory of Bioimaging Probe Development, Singapore Bioimaging Consortium, Agency for Science, Technology and Research, Singapore 138667, Singapore. [4] Department of Pharmaceutical Chemistry and the Cardiovascular Research Institute, University of California San Francisco, California 94158, USA. [5] Departamento de Química, Universidad de Guanajuato, Guanajuato 36050, Mexico. [6] College of Pharmacy and Research Institute of Life and Pharmaceutical Sciences, Sunchon National University, Sunchon 540950, Republic of Korea. * These authors contributed equally to this work. Correspondence and requests for materials should be addressed to H-S.P. (email: hspark@kaist.ac.kr) or to L.W. (email: Lei.Wang2@ucsf.edu) or to Y.-T.C. (email: chmcyt@nus.edu.sg).

Fluorescent probes have emerged as one of the widely-used tools for investigating biological systems, thanks to its sensitivity and technical simplicity. Labelling a biomolecule, particularly protein, with a functional small probe in living cell is a popular approach to obtain greater knowledge of its biological roles and functions. However, labelling a specific intracellular biomolecule is a challenging task because cells are complex, highly organized and composed of various biological events. As a result, most labelling probes reported thus far leave fluorescence background due to nonspecific interactions with other cellular components[1–5]. In addition, probes delivery into live cells can also be demanding as some probes are not able to cross plasma membranes in native condition, thereby limiting their utility for intracellular labelling.

Despite their intricacies, physicochemical properties of probes can be defined numerically through their structural information, so-called descriptors. Descriptors have long been exploited in computational quantitative structure activity relationship study, for instance, to predict aqueous solubility and permeability of drugs[6,7], binding affinities of antitumor drugs[8,9] and also cellular uptake and localization of fluorescent probes[10,11].

Rational design of cell permeable background-free 'tame' probes is tricky and generally involves a long process of trial-and-errors. Here we utilize probes' descriptors for statistics-based quantitative structure activity relationship study to develop a model that can serve as a guideline to design such ideal 'tame' probes by predicting probes' cellular uptake and retention through their structural parameters. The model identifies three features: SlogP (lipophilicity), logS (water solubility) and Q_VSA_FNEG (negatively charged van der Waals surface area), as key factors influencing cell permeability and nonspecific intracellular retention of small fluorescent probes. This new model is then applied to design cyclooctyne- and azide-containing BODIPY 'tame' probes that can be utilized for labelling various intracellular organelles and engineered proteins via copper-free strain-promoted alkyne-azide cycloaddition (SPAAC).

Results

Cellular retention and efflux characteristics of the probes. To expedite an efficient strategy for 'tame' probe development, we set up a system to predict cellular properties of fluorescent probes via phenotypic high-throughput screening (HTS), cheminformatics and statistical analysis (Fig. 1a). As the first step, we conducted phenotypic HTS to investigate cell permeability and nonspecific binding properties of the probes in a training set. We tested about 2,085 probes consisted of a wide range photophysical properties from numerous fluorophore cores such as BODIPY[12–15], xanthone[16,17], rosamine[18], cyanine[19,20] and ACEMAN[21]. Their cellular influx and efflux profiles were examined in two types of mammalian cells, U-2 OS and CHO. These cell lines were chosen due to their flexibility in genetic manipulation, adaptability to various culture conditions and versatility in wide biomedical research such as gene expression, toxicity screening, recombinant protein and live cell imaging. Each of the probes was added into live cells and cell image was taken after 30 min by ImageXpress Micro cellular imaging system (Molecular Device). The cells were then washed with fresh growth media, and after another 10 min, a second image was taken to check the completeness of probe washing out of the cell. The first and second imaging facilitated influx (before washing, BW) and efflux (after washing, AW) measurements, respectively (Supplementary Fig. 1).

Three major phenotypic groups were observed: (i) N-group, non-stain group where probes are unable to pass through the cell membrane, (ii) L-group, low nonspecific binding group where probes enter the cell and leave after washing, and (iii) H-group, high nonspecific binding group where probes pass through cell membrane and retain inside after washing (Fig. 1b). L-group signifies the desired feature of the properly tamed probes as the influx and efflux occur rapidly with minimal background. We observed that probes in L-group came from only the BODIPY family. Probes from the rosamine and cyanine family strongly bound to mitochondria, hence mainly in H-group, whereas most xanthone and ACEMAN probes were not able to pass through cellular membrane. Therefore, we focus only on BODIPY family (920 probes) as a model system for further analysis (Fig. 1c). BODIPY can be an ideal fluorophore as it has an electrically neutral core and superior optical properties in terms of high quantum yield, tunable fluorescence characteristics, high photostability and narrow emission bandwidth[22].

Retaining ratio for automated cellular phenotypic grouping. To measure the influx and efflux of the probes, an automated quantitative intensity-based image analysis was then performed by Multi-Wavelength Cell Scoring analysis using MetaXpress High Content Image software. Retaining ratio (RR, per cent ratio of the average intensity of AW image over BW image) was computed to classify probes into the three phenotypic groups. For each cell in the image, the nuclei and cell region were segmented to provide approximate cellular boundaries. Fluorescence intensity was then quantified on a cell-by-cell basis by the following steps: (i) selecting the image of interest, (2) specifying the width of the stained cell area, (3) adjusting the intensity above local background for detection specificity, and (4) calculating fluorescence intensity with specified reporting parameters. Average fluorescence intensity for each BW and AW images was calculated for each probes and cell lines, and was then used as input in RR calculation (Supplementary Figs 2 and 3). RR heat map showed that about 88% of the probes behaved similarly in both U-2 OS and CHO, implying that these two cell lines are sufficient to serve as a proxy for phenotypic screening (Fig. 2a, Supplementary Fig. 4). From this 88% population, a majority of them (81%) were in H-group, 16% of the probes were observed to be cell-impermeable, and 3% were 'tame' probes (Supplementary Fig. 5).

Molecular descriptors can predict 'tame' probes' properties. Next, we investigated the structural properties of the training set. Active three-dimensional (3D) conformation of the libraries was set through energy minimization by Hamiltonian AM1 method using Molecular Operating Environment (MOE) software. A total of 327 molecular descriptors for each probe were then generated (Supplementary Table 1). Although hundreds of descriptors were generated, not all are relevant to the study; therefore, we cleaned up the data by removing redundant descriptors. Independency between descriptors was evaluated by checking their degree of noncorrelation using SPSS v.13.0.1 software package (SPSS, Inc., Chicago, IL). A number of redundant descriptors were then exempted, such as: highly correlated descriptors (for example, Q_VSA_FHYD and Q_VSA_FPOL, these descriptors have a perfect negative correlation as shown by the independency test, hence one of them can be exempted to eliminate redundancy), duplicated descriptors (for example, PC+ and RPC+; PC− and RPC−), insignificant descriptors (for example, rsynth, which describe the easiness of the synthesis process) and uninterpretable descriptors (for example, BCUT_PEOE_0, BCUT_PEOE_1, BCUT_PEOE_2, BCUT_PEOE_3). This elimination resulted in 39 descriptors for further investigation (Supplementary Table 2).

To identify the important descriptors responsible for probe phenotypic grouping, we carried out significance test-based

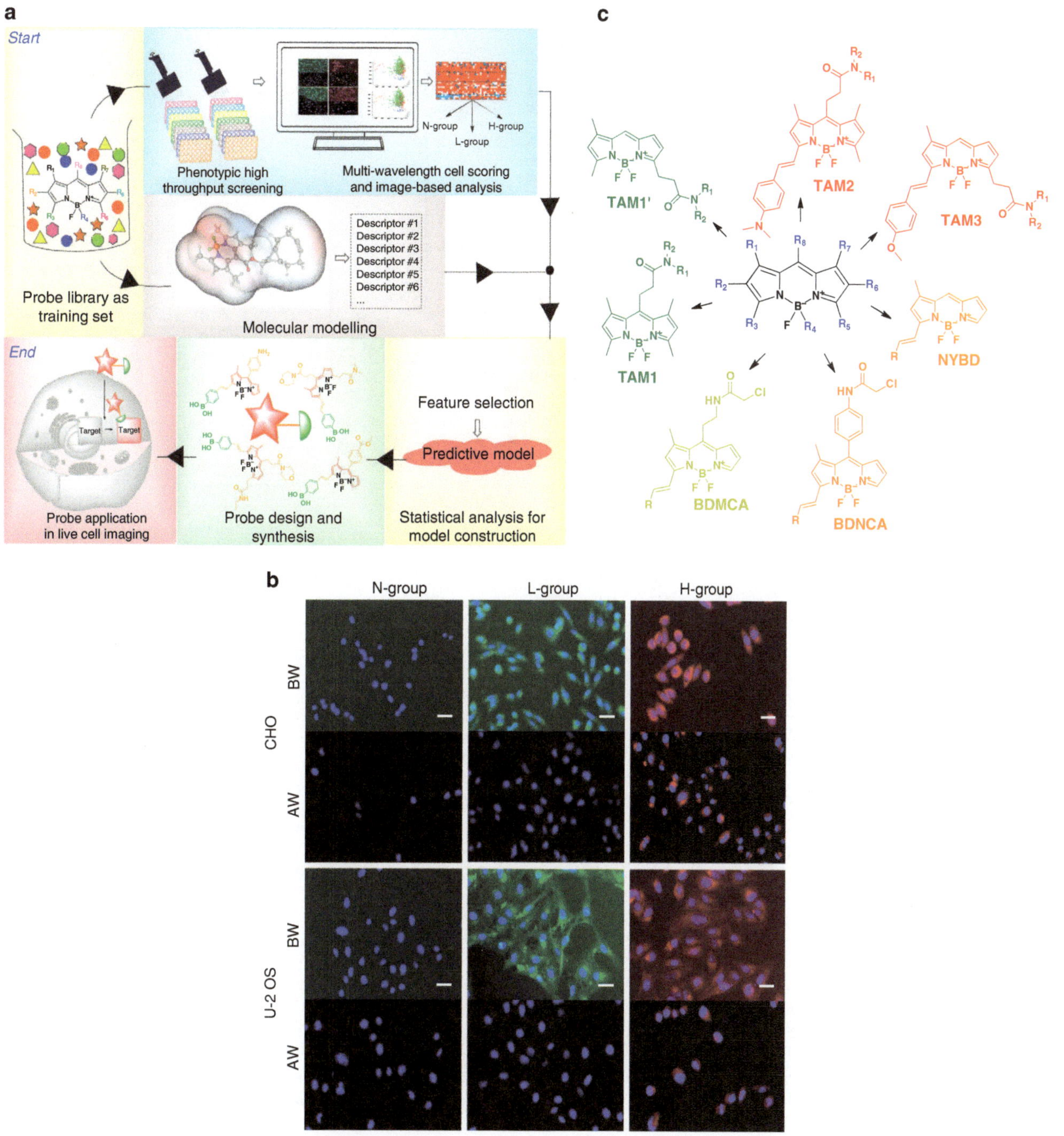

Figure 1 | Experimental set and probes' cellular responses. (**a**) Flowchart of the overall experimental strategy. (**b**) Representative cellular responses in CHO and U-2 OS. Cells were stained with probes at 1 μM final concentration. The overlayed images of Hoechst33324 (blue) and probe signals (according to emission wavelength of the probes) show three phenotypic groups: cell-impermeable group where probes are unable to stain the cell (N-group); cell permeable with low nonspecific binding group (L-group); and cell permeable with high nonspecific binding group (H-group). BW and AW indicate image before and after washing, respectively. Probes signal is significantly decreased after washing in L-group, while is retained in H-group. Scale bar, 10 μm (green, red and blue signals are from FITC, Texas Red and DAPI channel, respectively). (**c**) Core structures of the BODIPY libraries used in training set.

feature selection (STFS). In STFS, the statistical significant differences of descriptors among the three phenotypic groups were calculated to evaluate the pattern recognition ability of individual descriptors. We firstly carried out a normality test to check if our data set follows a normal distribution so as to apply proper statistical methods. Formal statistical Shapiro–Wilk test was used to check data normality. Conceptually, Shapiro–Wilk test involves an arrangement of the sample values and measurement against expected means and/or variances. The measure of central tendency is important: if data distribution is normal, then both mean and median can be measured to be presented; otherwise the median should be more appropriate. Results in Supplementary Table 3 show that nonparametric type STFS analysis was suitable for our database.[23–25]

Next, the nonparametric Kruskal–Wallis test followed by Eta-squared test was conducted to statistically evaluate whether there is a significant difference between descriptors' median in the three phenotypic groups (descriptors were considered to have a

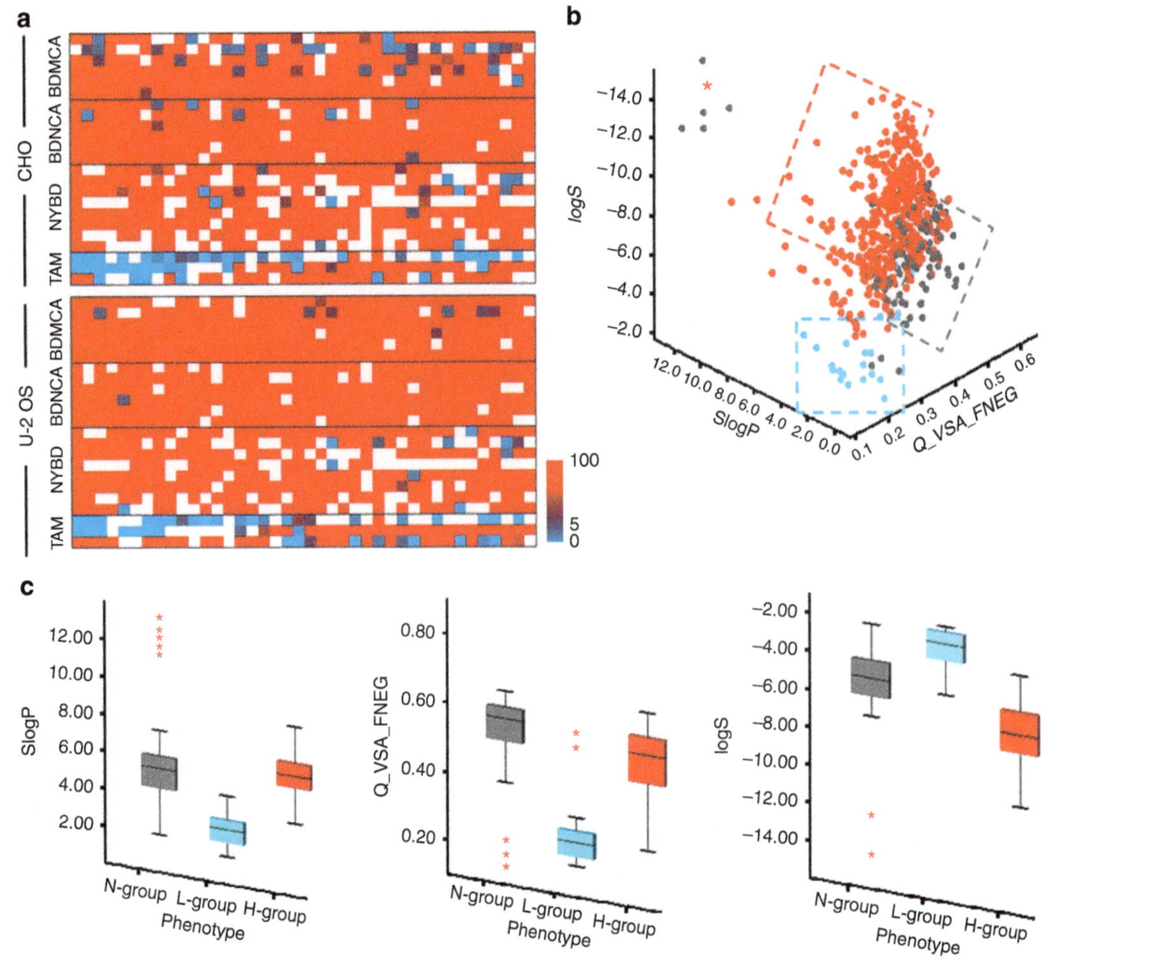

Figure 2 | Predictive model. (**a**) Heat maps of RR for each library in U-2 OS and CHO cell lines. White colour represents probes in N-group. The red and torquise colour represents high and low RR, respectively. (**b**) Three dimensional scatter plot for three key descriptors: SlogP, Q_VSA_FNEG and logS. Grey, turquoise and red dotted box represents area containing mostly probes from N-group, L-group and H-group, respectively. (**c**) Preferred criteria for 'tame' probes. Descriptor values in each group are presented as box plots with medians, quartiles, and interquartile range. The whiskers represent the minimum and maximum scores, and the red asterisks show extreme scores.

statistically significant difference if $P<0.01$). Kruskal–Wallis significance level was estimated on the capability of individual descriptor to form a distinct separable pattern, while Eta-squared test was used to measure the effectiveness of the descriptors intervention relative to phenotypic grouping. Our analysis suggests that lipophilicity, water solubility, surface area properties and size[26] of the probes significantly contributed to their cellular permeability and nonspecific intracellular retention (Supplementary Fig. 6, Supplementary Table 4). To make an interpretable and uncomplicated model, descriptors with large effect in probes' phenotypic grouping ($\eta^2 \geq 0.1$) were selected and analysed by using principal component analysis method to further portray the patterns and highlight the relations of those selected descriptors.

The first three principal components account for 98.0% of the data set variance (Supplementary Table 5). Examination of the component loadings shows that descriptors are arranged into three clusters: lipophilicity (SlogP, logPo/w), water solubility (logS), and charged-surface area properties (Q_VSA_NEG and Q_VSA_FNEG), indicating that these three cluster properties are the major determinant for cellular behaviour of the probes (Supplementary Figs 7 and 8). Interestingly, principal component analysis (PCA) loading plot indicated that three descriptors (Q_VSA_FNEG ($\chi^2(2)=88.823$, $P=0.000$, $\eta^2=0.110$), logS ($\chi^2(2)=110.411$, $P=0.000$, $\eta^2=0.137$), and SlogP ($\chi^2(2)=102.268$, $P=0.000$, $\eta^2=0.1272$)) are sufficient to summarize the data with minimal loss of information (Fig. 2b). Our results thus suggest that criteria for the ideal probes are adequately lipophilic (SlogP $=1$ to 4), moderately negatively charged on its surface (Q_VSA_FNEG $=0.15$ to 0.35) and highly soluble in water (logS $=-2$ to -6; Fig. 2c, Supplementary Fig. 9).

Validation set for the interpretable model. The ultimate goal of a modelling exercise is not only to fit a model into a data set, but importantly to also provide a clear guide to design and synthesize molecules with desired properties. As a proof of concept, we applied the model to an extended set of our in-house BODIPY library (EP library) consists of 107 various structures (Supplementary Fig. 10). On the basis of their descriptors value, the model predicted that nine of the EP probes are in L-group ('tame' probes). Cell experiment showed that eight of the nine probes are indeed in L-group, which is corresponding to a prediction accuracy rate of 89% (Supplementary Table 6).

It is interesting to note that BODIPY is known to be sticky/hydrophobic and usually leaves high background in the cells. To tame BODIPY into background-free probe while keeping its cellular permeability, we synthesized boronic acid-bearing

BODIPY probes, which have a tremendous importance in chemical biology field, by following our model guidelines. Based on their SlogP, Q_VSA_FNEG and logS values, BOR-1H and BOR-2H were predicted to be in H-group while BOR-1 and BOR-2 were in L-group (Supplementary Table 7, Supplementary Fig. 11). We found that the predicted phenotypic groups for these probes are in agreement with the observed cell experimental results, hence validating our predictive model for background-free probes (Supplementary Table 8, Supplementary Fig. 12).

'Tame' probes for intracellular labelling in living cells. To further demonstrate the model and its applications, we synthesized BCN alcohol-containing BODIPY probes, CO-1 and CO-1H, to fit the L-group and H-group, respectively, by varying substituents at different positions of the BODIPY core for SPAAC-based labelling. Their absorbance and emission are similar to GFP, thus enabling their visualization on most standard fluorescence microscope (Supplementary Fig. 13). Fascinatingly, cellular retention and efflux study showed the phenotypic grouping of CO-1 and CO-1H as predicted (Supplementary Fig. 14, Supplementary Table 9).

Next, we investigated intracellular imaging in live cells using CO-1. We synthesized various azide-bearing reporters: TPP-Az, a triphenylphosphonium analogue which accumulates in mitochondria; Sphingo-Az, a ceramide analogue to target golgi apparatus; and Morph-Az, an azide derivative with morpholine moiety as a directing group for lysosome (Fig. 3a). SPAAC covalent labelling reactions between CO-1 and all azide reporters *in vitro* were clean and rapid. We anticipated that the cell permeable CO-1 is capable of labelling any azide reporters effectively, regardless the reporters' cellular localization. We added 2 μM CO-1 to the growth media of U-2 OS cells pre-treated with TPP-Az, Sphingo-Az or Morph-Az separately. Confocal fluorescence images were then taken after probe washing. As shown in Fig. 3b, CO-1 clearly labelled the azide-tagged intracellular organelles in live cells, and CO-1 signal was nicely colocalized with respective organelle probe trackers. Clean fluorescence background was observed due to washable property of the unreacted CO-1. In contrast, when CO-1H was added to the cells pre-treated with TPP-Az, Sphingo-Az or Morph-Az, high-fluorescence background from CO-1H interfered with the observation of the intracellular organelles (Supplementary Fig. 15).

We further demonstrated the utility of CO-1 for visualization of engineered protein with an example of human histone H2B, which localizes at the nucleolus. The red fluorescent protein mKate2 was fused to the C terminus of human histone H2B as a red fluorescent marker. Using the *in vivo* unnatural amino acid (UAA) technology,[27,28] an azide-bearing UAA, *p*-azido-L-phenylalanine (Azi), was site-specifically incorporated at site 16 of mKate2 of the fusion protein H2B-mKate2, which was expressed in live U-2 OS cells. The cells were labelled with CO-1, washed to remove unreacted CO-1, and then imaged with laser scanning confocal microscopy. To our delight, CO-1 was not only able to label cytosolic organelles, but also to pass through nuclear membrane. Clear CO-1 signal was detected in the live cell nucleolus out of low background, and was highly colocalized with mKate2 signals, suggesting specific labelling of histone H2B (Fig. 3c, Supplementary Fig. 16). Moreover, CO-1 exhibited superior photostability under intensive radiation of strong light source in aqueous solution (Supplementary Fig. 17) and was found to be nontoxic and did not interfere with cell proliferation (Supplementary Fig. 18), making CO-1 suitable for prolonged time lapse imaging.

Excited with the results, we further challenged the model and developed another 'tame' azide-bearing probe, AzG-1, to label molecules in mitochondria and the fancy tubulin protein in living cells (Figs 4a,b). To target mitochondria, U-2 OS cells were pre-treated with TPP-BCN (an analogue of triphenylphosphonium) before labelled with 10 μM AzG-1. For tubulin labelling, a cyclooctyne-containing UAA, N-ε-(cyclooct-2-yn-1-yloxy)carbonyl)L-lysine (CoK) was selected for incorporating into tubulin and reacting with AzG-1 via SPAAC.[29] CHO-K1 cells were co-transfected with plasmids pCoKRS-tRNA and pTub-26TAG to incorporate CoK into α-tubulin at position 26. Site-specific incorporation of CoK using an orthogonal $tRNA^{Pyl}$/CoKRS pair was confirmed via western blot analysis (Supplementary Figs. 19 and 20). After incorporation of CoK, cells were then labelled with 10 μM AzG-1 at 37 °C. CHO-K1 cells transfected with plasmid pEGFP-Tubwt expressing EGFP-fused wild-type α-tubulin was used as a control. Confocal fluorescence images were then taken AW of unreacted AzG-1. Cell images clearly show that AzG-1 specifically labelled mitochondria (Fig. 4c) and was also conjugated specifically with CoK-bearing α-tubulin (Fig. 4d, Supplementary Fig. 21).

Discussion

The combination of cheminformatics and HTS of in-house probe library yielded a novel predictive model to guide the development of cell permeable background-free 'tame' probes for intracellular imaging in live cells. Our systematic analysis showed that cellular behaviour of fluorescent probes can be predicted from their photophysical properties; more importantly, we found that lipophilicity, charged van der Waals surface area and water solubility are significant determinants for probes' cellular permeability and nonspecific binding in living cells. These descriptors may account for not only polar noncovalent interactions between probes and phosphate group of the plasma membranes during cellular intake, but also cumulative charge interactions with macromolecules inside the cells.

Our result supports previous studies that hypothesized hydrophobicity and surface charge may correlate with nonspecific binding properties of fluorescent probes[10,30,31]. In addition, water solubility also plays a key factor in slow cellular efflux of the probes. Our observation in cellular retention and efflux study showed that probes were stably solubilized when added to the cells, indicating that extracellular insoluble formation of the probes is unlikely. However, after the probes were incubated in cells for 30 min, the formation of intracellular insoluble or precipitate-like form which is hardly washed out of cells was observed from some of the probes in H-group. It suggests that intracellular precipitation or aggregation due to the low water solubility may be one of the mechanisms behind the slow cellular efflux and high retention of probes in H-group[32–34].

We utilized this model strategy to synthesize probes for copper-free SPAAC in living cells[3,35]. Early study on SPAAC reported nonpermeant cyclooctyne probes for imaging extracellular surface glycans[36–38]. Following that, an attempt was also made to develop cyclooctyne probe for targeting intracellular biomolecules[4]. However, this probe still suffers from high nonspecific binding. Here we show that through structure optimization on the three interpretable descriptors (lipophilicity, charged van der Waals surface area and water solubility), we are able to modulate it to meet the demands of designing smart 'tame' probes.

In summary, our predictive model provides a powerful strategy to efficiently develop background-free BODIPY probes for intracellular labelling in native living cells. We successfully demonstrated the robustness of this strategy by developing

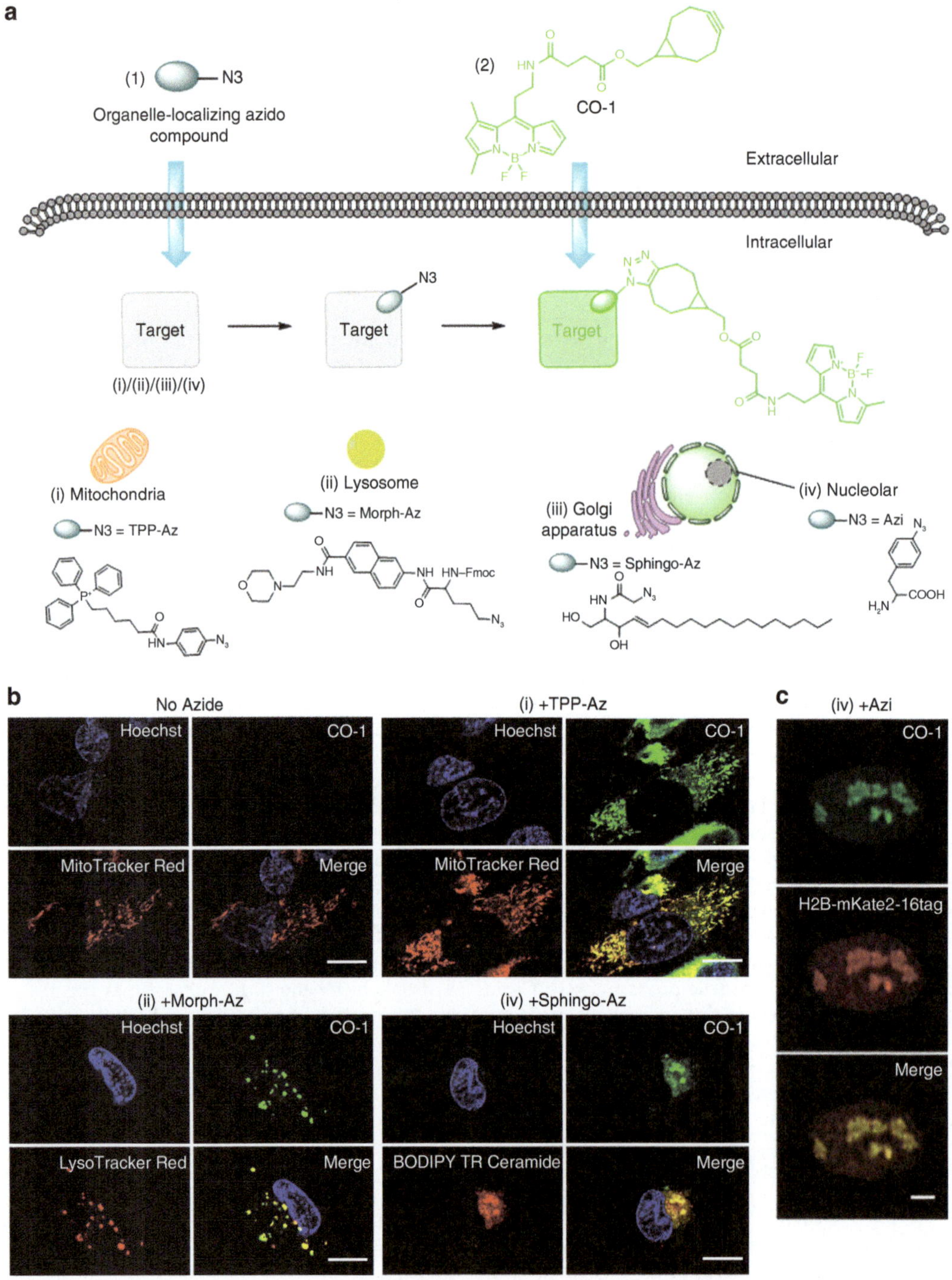

Figure 3 | Live cell imaging with CO-1. (**a**) Chemical structures and schematic illustration of the covalent labelling of azide-tagged organelles using CO-1 in live cells. (**b**) Fluorescence imaging of mitochondria, lysosome and golgi apparatus in U-2 OS cells labelled with CO-1. Cells pre-treated with TPP-Az, Morph-Az or Sphingo-Az were incubated with 2 μM CO-1 in growth media at 37 °C for 1 h and followed by counterstaining with organelle trackers. Scale bar, 15 μm. (**c**) Fluorescence imaging of histone H2B in live U-2 OS. U-2 OS cells were co-transfected with plasmids pIre-Azi3 and pmH2B-6-mKate2-16tag to incorporate UAA Azi into H2B-mKate2 at site 16. After Azi incorporation, cells were labelled with 10 μM CO-1 for 90 min at 37 °C. Cells were then washed and imaged. The mKate2 (red) signals were detected in cell nuclei only, and colocalized with the CO-1 (green) signals in the merged image. Scale bar, 5 μm.

novel labelling tools, CO-1 and AzG-1, which are not only highly cell permeable and washable, but also exhibit a remarkable capability of labelling various intracellular organelles and engineered proteins. The strategy should allow the exploration of dynamic processes (for example, cellular localization, expression level and biological activities) of newly synthesized intracellular biomolecules in their native cellular environment.

Methods

Probes' synthesis and characterization. Details of synthesis and characterization of the probes used in this study can be found in the Supplementary Methods. General synthetic schemes and structures of the BODIPY libraries used in training set can be seen in Supplementary Figs 22 and 23 and Supplementary Tables 10 and 11. For NMR spectra of CO-1, AzG-1 and CO-1H see Supplementary Figs 24–26.

Cellular retention and efflux characteristics test. U-2 OS, a human osteosarcoma cell line, and CHO, a Chinese hamster ovary cell line, (from

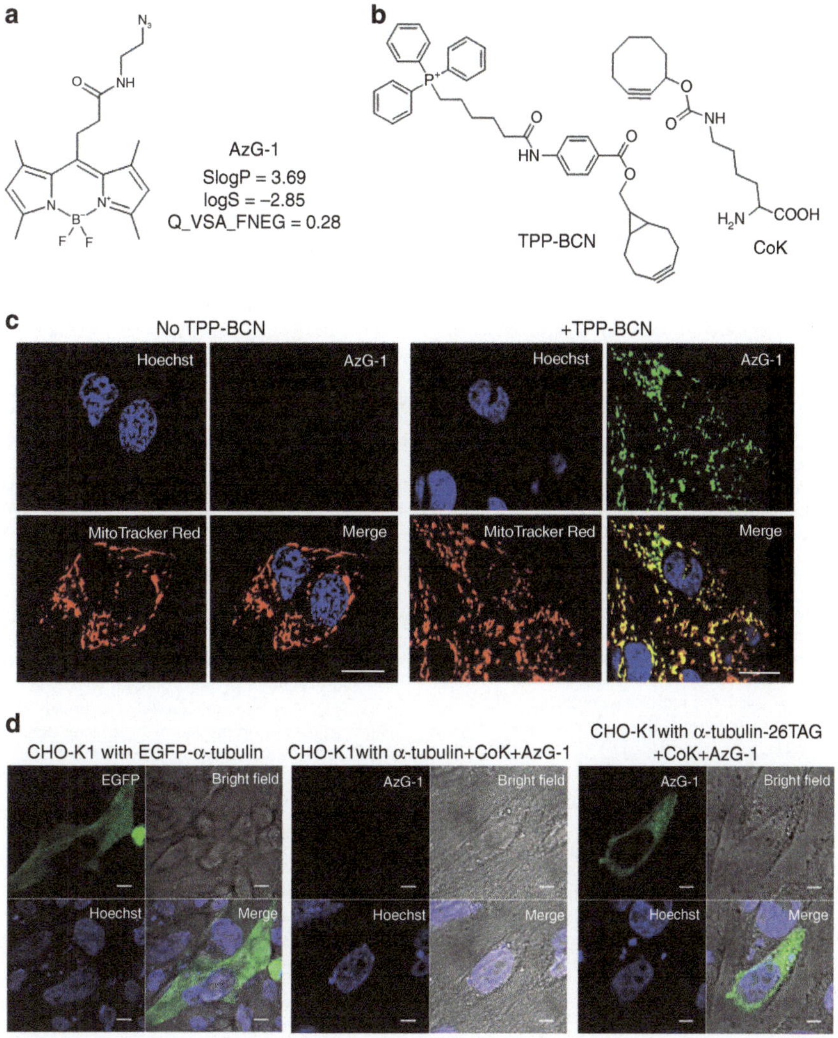

Figure 4 | Live cell imaging with AzG-1 (**a**) Chemical structures of azide-bearing probe AzG-1 and its descriptor values. (**b**) Structure of triphenylphosphonium analogues TPP-BCN and cyclooctyne-containing unnatural amino acid, CoK. (**c**) Fluorescence imaging of mitochondria in U-2 OS cells labelled with AzG-1. Cells were incubated without (left side) and with (right side) TPP-BCN and then labelled with 10 μM AzG-1 for 2 h followed by counterstaining with mitochondria marker. Scale bar, 10 μm. (**d**) Fluorescence imaging of α-tubulin in live CHO-K1 cells. CHO-K1 cells were co-transfected with plasmids pCoKRS-tRNA and pTub-26TAG to incorporate CoK into α-tubulin at position 26. After incorporation of CoK, cells were labelled with 10 μM AzG-1 for 2 h at 37 °C. CHO-K1 cells were also transfected with plasmid pTubwt to express wild type α-tubulin and treated with CoK and AzG-1 in the same way. CHO-K1 cells transfected with plasmid pEGFP-Tubwt expressing EGFP-fused wild type α-tubulin was used as a control. Live cell images show that AzG-1 labelled mitochondria (**c**) and is conjugated specifically with CoK-bearing α-tubulin (**d**). Scale bar, 10 μm.

ATCC) were cultured in Dulbecco's modified eagle's medium (DMEM) (Invitrogen, CA, USA) supplemented with fetal bovine serum (FBS) (10%) and penicillin–streptomycin (1%). Materials used in the cell culture were purchased from Invitrogen. U-2 OS and CHO cells were seeded onto 96-well plate in growth media at 37 °C in the presence of 5% CO_2 and were then allowed to attach and grow to 70–80% confluence. Library probes were dissolved in dimethylsulfoxide (DMSO) to make the 1 mM solution, and stored in −20 °C. Before HTS experiment, the growth media was aspirated and replaced by 200 μl fresh growth media containing probes in final concentration of 1 μM and nuclei dye Hoechst 33342. Plates were incubated for 30 min at 37 °C then were imaged using ImageXpress Micro cellular imaging system (Molecular Device) with × 10 objective lens. Immediately after image acquisition, the cells were washed with fresh growth media, and transferred back to a 37 °C cell incubator for further incubation. After 10 min, cells were again imaged (AW image). The first imaging step allowed influx measurement of BW image , while the second imaging step allowed out-flux measurement of AW image. Images of two regions per well were acquired with DAPI, FITC, TRITC or Texas Red filter sets according to probes' excitation and emission wavelength. Experiment was performed in duplicate. Cellular retention and efflux characteristics test for other synthesized probes were also done by using the same protocol.

RR calculation. RR was computed to quantitatively estimate probes retained inside the cells AW. RR was calculated by:

$$RR(\%)=\left[\frac{FI(AW)}{FI(BW)}\right]\times 100 \qquad (1)$$

$FI_{(AW)}$ is the average fluorescence intensity AW and $FI_{(BW)}$ is the average fluorescence intensity BW. RR is a value ranging from 0 to 100, where higher the value, higher the nonspecific binding of the probes. Background correction was applied to each image to compensate the uneven background intensity, thus improving image segmentation and reducing calculation bias. Probes in which BW images having $FI_{(BW)} \leq 200$ were classified as N-group. Probes ($FI_{(BW)} > 200$) with rapid efflux were observed to have a RR of <5%, and thus were classified into L-group, while those with RR >5% were classified into H-group.

Molecular descriptor for library training set. Chemical Computing Group (CCG) MOE 2011 software package was used for generating 3D structure and molecular descriptors for each probe. Hydrogens and lone-pair electrons were adjusted as required. Partial charge was set and protonation state was corrected for each structure. The active 3D conformation of the libraries was then acquired through energy minimization by Hamiltonian AM1 method. A total of 327 molecular descriptors were generated including topological, molecular connectivity, electrotopological geometric and quantum chemical descriptors.

Significance test-based feature selection. STFS analysis was done using SPSS v.13.0.1 software package (SPSS Inc., Chicago, IL). η^2-test was calculated by:

$$n^2 = \frac{\chi^2}{N-1} \quad (2)$$

where η^2 is eta squared, χ^2 is chi square value and N is sample size. PCA was performed using the same software. In PCA, each of the descriptors and probes were processed and their factors of correlation were extracted.

Cell maintenance and preparation for cell labelling. The reagents Hoechst 33342 (1:5000, H1399), MitoTracker Deep Red FM (referred to as MitoTracker Red), BODIPY TR ceramide and LysoTracker Red were purchased from Life Technologies (Carlsbad, CA, USA). A stock solution of CO-1, CO-1H, TPP-Az, Morph-Az, Hoechst 33342, MitoTracker Red, and LysoTracker Red were first prepared in DMSO. The fluorescence excitation and emission spectra were measured using a SpectraMax M2 plate reader (Molecular devices Corp, USA). U-2 OS cells were cultured at 37 °C with 5% CO_2 in DMEM supplemented with 10% (v/v) FBS, penicillin (100 units per ml) and streptomycin (100 mg ml^{-1}). Cells were passaged two to three times a week, seeded at a density of 75,000 cells per cm into 35-mm glass bottom dishes, and grown in DMEM media overnight before labelling.

Sphingo-Az stock solution preparation. BODIPY TR ceramide were prepared according to manufacturer protocol. Sphingo-Az was prepared as a form of complex with BSA similar to the preparation of BODIPY TR ceramide. Solid Sphingo-Az was dissolved in chloroform:ethanol (19:1 v/v) to give 50 µl of 1 mM stock solution. The stock solution was dried and redissolved in 200 µl 100% ethanol. This solution was then added to 10 ml of HBBS/BSA solution (HBSS + 10 mM HEPES pH 7.4 + 0.34 mg $ml^{-1}$of defatted BSA) on a vortex mixer to give 5 µM Sphingo-Az/5 µM BSA stock solution. This solution can then be stored at − 20 °C.

Mitochondria imaging in live cells. U-2 OS cells were treated with 5 µM TPP-Az, in culture media for 2 h at 37 °C. After incubation, cells were washed and added with 2 µM CO-1 probe. Labelling was allowed to proceed for 1 h at 37 °C in the incubator chamber. Following incubation, cells were treated with nuclei dye Hoechst 33342 (1 µg μl^{-1}) and MitoTracker Red (1 µM). After counterstaining, cells were washed three times with growth media and imaged using Nikon A1R+ confocal laser microscope system.

Lysosome imaging in live cells. U-2 OS cells were treated with 5 µM Morph-Az, in culture media for 1 h at 37 °C. After incubation, cells were washed and added with 2 µM CO-1 probe then proceed for 1 h incubation at 37 °C. Following incubation, cells were treated with nuclei dye Hoechst 33342 (1 µg μl^{-1}) and LysoTracker Red (1 µM). After counterstaining, cells were washed three times with growth media and imaged using Nikon A1R+ confocal laser microscope system.

Golgi apparatus imaging in live cells. U-2 OS cells were treated with 5 µM Sphingo-Az/BSA and 5 µM BODIPY TR ceramide/BSA in HBSS/HEPES for 30 min at 4 °C. Afterwards, cells were washed with cold growth media and further incubated for 30 min at 37 °C. After incubation, cells were washed and incubated with 2 µM CO-1 in culture media for 1 h at 37 °C. Following incubation, cells were treated with nuclei dye Hoechst 33342 (1 µg μl^{-1}). After counterstaining, cells were washed three times with growth media and imaged using Nikon A1R+ confocal laser microscope system.

A1R+ confocal microscopes. Confocal imaging experiments were performed on an inverted Nikon A1R+ confocal laser microscope system using 562/672/405 nm lasers with Plan Apo TIRF 100X DiC oil H H2 objectives (Nikon Instruments, Inc., Japan). Image processing and overlay analysis were performed using NIS Elements 3.10 software (Nikon Instruments, Inc.).

Molecular cloning for histone H2B labelling. Plasmids were cloned in DH10β *Escherichia coli* strain and confirmed by DNA sequencing. PfuTurbo (Stratagene, La Jolla, CA) was used for QuikChange reactions and Phusion (Finnzymes, Finland) for gene amplification. Primers were synthesized by ValueGene (San Diego, CA), and restriction enzymes were purchased from New England Biolabs (Ipswich, MA). Wild-type mKate2 was cloned into pmTagRFP-T-H2B-6 plasmid using the *Age*I and *Not*I sites. The amber stop codon UAG was introduced at site 16 of mKate2 to make the plasmid pmH2B-6-mKate2-16tag.

Newly synthesized histone H2B imaging in live cells. U-2 OS cells were transfected with 1 µg DNA of each plasmid using Lipofectamine2000 in 35 mm glass bottom dishes according to the manufacturer protocol. Azi (purchased from Bachem) was added 1 h before transfection at a final concentration of 0.5 mM. Plasmid pIre-Azi3 (expressing the orthogonal amber suppressor transfer RNA (tRNA) and the E2AziRS specific for Azi) was co-transfected with plasmid pmH2B-6-mKate2-16tag in a 1:1 molar ratio. After 24 h, cells were washed three times with culture medium (DMEM with 10% FBS) (15 min each time). Growth media containing 10 µM CO-1 was then added to the dishes, and cells were incubated at 37 °C for 90 min followed by 4 times washing with growth media (30 min each time). Laser scanning confocal microscopy (Zeiss LSM 700) was used for imaging. The same conditions and settings were used for imaging control and experimental cells.

Plasmid construction for site-specific CoK incorporation. Cyclooctyne-containing UAA, N-ε-(cyclooct-2-yn-1-yloxy)carbonyl)L-lysine (CoK) was selected for copper-free labelling of proteins to avoid the use of cytotoxic metals. To this end, cyclooctynyl lysine, CoK was synthesized according to the previous report[29]. For genetic incorporation of CoK, an engineered orthogonal $tRNA^{Pyl}$/CoKRS pair was constructed as reported before[29]. To express proteins installed with CoK at a designated position in mammalian cells, plasmids expressing the $tRNA^{Pyl}$/CoKRS pair and target protein(s) were constructed in the following way. CoKRS carrying C-terminal haemaglutinin (HA) tag is expressed under the control of EF-1α promoter (derived from pEFIRES vector, Clonetech) and tRNA under the control of human U6 promoter and the CMV enhancer. CoKRS and tRNA expression cassettes were cloned using *KpnI/NotI*, and *BamHI/AscI* sites of pCDNA3 vector (Invitrogen) respectively, generating plasmid pCoKRS-tRNA. For live cell labelling experiments, α-tubulin was used as a model protein. An amber stop codon was inserted into position 26 of α-tubulin and the resulting tubulin-26TAG under the control of EF-1α promoter was cloned between *Kpn*I and *Not*I of pCDNA3, generating plasmid pTub-26TAG. Plasmids pTubwt and pEGFP-Tubwt carrying α-tubulin wild type and EGFP-fused α-tubulin, respectively, were also constructed in a similar way for control experiments. To validate genetic incorporation of CoK using $tRNA^{Pyl}$/CoKRS pair, plasmids pEGFP-Tub-26TAG (carrying EGFP-fused α-tubulin with C-terminal FLAG tag and stop codon TAG at 26) and pEGFP-39TAG (carrying EGFP with stop codon TAG at 39) were constructed for Western blotting and fluorescence microcopy analysis.

Protein labelling on live cells using CoK. CHO-K1 or HeLa cells were grown in DMEM-low glucose 1 g l^{-1} (Gibco) supplemented with 10% FBS (Gibco) in eight-well plate (ibid). At 70 ~ 80% confluency, cells were transfected using 4 µl iN-fect *in vitro* transfection reagent with 100 ng pCoKRS-tRNA and 900 ng pTub-26TAG or with 900 ng pTubwt. After 6 h of incubation, media was replaced with fresh media (DMEM-low glucose medium with 10% FBS, 1% Pen-Strep, 2 Mm L-glutamate) containing 0.25 Mm CoK. After an additional 36 h, media was exchanged with normal growth media without CoK and cells were incubated for 2 h and washed (four times) to remove remaining CoK. Next, growth media containing 1 µM AzG-1 was added to the plates, and cells were incubated for 2 h with four times of washing (30 min each time) using growth media. High-resolution images of live cells were obtained using a Zeiss LSM510 META laser scanning confocal microscope.

Western blotting. To demonstrate site-specific incorporation of CoK into α-tubulin, HEK293T cells were co-transfected with plasmids pCokRS-tRNA and pEGFP-Tub-26TAG at 80–90% confluency and cultured in the presence or absence of 0.5 mM CoK for 36 h. The cells were harvested and subjected to Western blot analysis using anti-HA antibody (Abcam) and anti-FLAG antibody (Agilent) for the detection of HA-tagged CoKRS expression and FLAG-tagged EGFP-α-tubulin expression, respectively.

Cell viability assay. Cell viability was assessed using MTS assay kit (Promega). 1×10^5 U-2 OS cells were seeded on 10 mm culture dishes; cells were stained with or without CO-1 (0–10 M) for 1–16 h. After incubation, the MTS reagent was added and incubated for 4 h at 37 °C. Absorbance was determined at 490 nm using SpectraMax M2 plate reader.

Photostability measurements. 10 µM of CO-1 solution in PBS buffer (pH 7.4) containing 1% DMSO were placed in a 96-well plates. Fluorescence measurement were recorded every 30 s interval for a total period of 12 h (Ex/Em = 490/520) under a xenon flashlamp. For harsh condition experiment, photostability test was done under high intensity UV lamp (Blak Ray, 100 W, 365 nm). Plates were irradiated for 10 min up to 2.5 h at 10 cm distance.

References

1. Jung, D., Min, K., Jung, J., Jang, W. & Kwon, Y. Chemical biology-based approaches on fluorescent labeling of proteins in live cells. *Mol. BioSyst.* **9**, 862–872 (2013).
2. Baskin, J. M. Copper-free click chemistry for dynamic *in vivo* imaging. *Proc. Natl Acad. Sci. USA* **104**, 16793–16797 (2007).
3. Agard, N. J., Prescher, J. A. & Bertozzi, C. R. A strain-promoted [3 + 2] azide-alkyne cycloaddition for covalent modification of biomolecules in living systems. *J. Am. Chem. Soc.* **126**, 15046–15047 (2004).
4. Beatty, K. E., Szychowski, J., Fisk, J. D. & Tirrel, D. A. A BODIPY-cyclooctyne for protein imaging in live cells. *Chembiochem* **12**, 2137–2139 (2011).
5. Fernandez-Suarez, M. *et al.* Redirecting lipoic acid ligase for cell surface protein labeling with small-molecule probes. *Nat. Biotech.* **25**, 1483–1487 (2007).
6. Bergstro, C. A. S. *et al.* Absorption classification of oral drugs based on molecular surface properties. *J. Med. Chem.* **46**, 558–570 (2003).
7. Refsgaard, H. H. F. *et al. In silico* prediction of membrane permeability from calculated molecular parameters. *J. Med. Chem.* **48**, 805–811 (2005).
8. Portugal, J. Evaluation of molecular descriptors for antitumor drugs with respect to noncovalent binding to DNA and antiproliferative activity. *BMC Pharmacol.* **9**, 11–25 (2009).
9. Seedher, N., Bhatia, S. & Singh, B. Quantitative correlation between theoretical molecular descriptors and drug-HSA binding affinities for various COX-2 inhibitors. *Chem. Biol. Drug. Des.* **72**, 297–302 (2008).
10. Horobin, R. W., Rashid-Doubell, F., Pediani, J. D. & Milligan, G. Predicting small molecule fluorescent probe localization in living cells using QSAR modeling. 1. Overview and models for probes of structure, properties and function in single cells. *Biotech. Histochem.* **88**, 440–460 (2013).
11. Horobin, R. W. & Rashid-Doubell, F. Predicting small molecule fluorescent probe localization in living cells using QSAR modeling. 2. Specifying probe, protocol and cell factors; selecting QSAR models; predicting entry and localization. *Biotech. Histochem.* **88**, 461–476 (2013).
12. Vendrell, M. *et al.* Solid-phase synthesis of BODIPY dyes and development of an immunoglobulin fluorescent sensor. *Chem. Commun.* **47**, 8424–8426 (2011).
13. Kang, N. Y. *et al.* Visualization and isolation of Langerhans islets by a fluorescent probe PiY. *Angew. Chem. Int. Ed.* **52**, 8557–8560 (2013).
14. Lee, J. S. *et al.* Synthesis of a bodipy library and its application to the development of live cell glucagon imaging probe. *J. Am. Chem. Soc.* **131**, 10077–10082 (2009).
15. Lee, J. S., Kim, H. K., Feng, H., Vendrell, M. & Chang, Y. T. Accelerating fluorescent sensor discovery: unbiased screening of a diversity-oriented BODIPY library. *Chem. Commun.* **47**, 2339–2341 (2011).
16. Ghosh, K. K., Ha, H. H., Kang, N. Y., Chandran, Y. & Chang, Y. T. Solid phase combinatorial synthesis of a xanthone library using click chemistry and its application to an embryonic stem cell probe. *Chem. Commun.* **47**, 7488–7490 (2011).
17. Ghosh, K. K. *et al.* Development of nucleus staining fluorescent probe for dynamic mitosis imaging in live cells. *Chem. Commun.* **51**, 9336–9338 (2015).
18. Ahn, Y. H., Lee, J. S. & Chang, Y. T. Combinatorial rosamine library and application to *in vivo* glutathione probe. *J. Am. Chem. Soc.* **129**, 4510–4511 (2007).
19. Samanta, A., Vendrell, M., Das, R. & Chang, Y. T. Development of photostable near-IR cyanine dyes. *Chem. Commun.* **46**, 7406–7408 (2010).
20. Zhang, L. *et al.* Development of fluorescent probes specific for parallel-stranded G-quadruplexes by a library approach. *Chem. Commun.* **51**, 7386–7389 (2015).
21. Agrawalla, B. K. *et al.* Glucagon-secreting alpha cell selective two-photon fluorescent probe TP-α: for live pancreatic islet imaging. *J. Am. Chem. Soc.* **137**, 5355–5362 (2015).
22. Lavis, L. D. & Raines, R. T. Bright ideas for chemical biology. *ACS Chem. Biol.* **3**, 142–155 (2008).
23. Guyon, I. & Elisseeff, A. An introduction to variable and feature selection. *J. Mach. Learn. Res.* **3**, 1157–1182 (2003).
24. Levine, T. R. & Hullet, C. R. Eta squared, partial eta squared, and misreporting of effect size in communication research. *Hum. Commun. Res.* **28**, 612–625 (2002).
25. Steiger, J. H. Beyond the F test: effect size confidence intervals and tests of close fit in the analysis of variance and contrast analysis. *Psychol. Methods* **9**, 164–182 (2004).
26. Zheng, N., Tsai, H. N., Zhang, X., Shedden, K. & Rosania, G. R. The subcellular distribution of small molecules: a meta-analysis. *Mol. Pharm.* **8**, 1611–1618 (2011).
27. Coin, I. *et al.* Genetically encoded chemical probes in cells reveal the binding path of Urocortin-i to CRF class B GPCR. *Cell* **155**, 1258–1269 (2013).
28. Takimoto, J. K., Adams, K. L., Xiang, Z. & Wang, L. Improving orthogonal tRNA-synthetase recognition for efficient unnatural amino acid incorporation and application in mammalian cells. *Mol. Biosyst.* **5**, 931–934 (2009).
29. Plass, T., Milles, S., Koehler, C., Schultz, C. & Lemke, E. A. Genetically encoded copper-free click chemistry. *Angew. Chem. Int. Ed.* **123**, 3964–3967 (2011).
30. Nath, S. *et al.* Identification of fluorescent compounds with nonspecific binding property via high throughput live cell microscopy. *PLoS ONE* **7**, e28802 (2012).
31. Zanetti-Domingues, L. C., Tynan, C. J., Rolfe, D. J., Clarke, D. T. & Martin-Fernandez, M. Hydrophobic fluorescent probes introduce artifacts into single molecule tracking experiments due to nonspecific binding. *PLoS One* **8**, e74200 (2013).
32. Shedden, K., Li, Q., Liu, F., Chang, Y. T. & Rosania, G. R. Machine vision-assisted analysis of structure-localization relationships in a combinatorial library of prospective bioimaging probes. *Cytometry* **75**, 482–493 (2009).
33. Min, K. A. *et al.* Massive bioaccumulation and self-assembly of phenazine compounds in live cells. *Adv. Sci. (Weinh)* **2**, pii 1500025 (2015).
34. Fu, D. *et al.* Imaging the intracellular distribution of tyrosine kinase inhibitors in living cells with quantitative hyperspectral stimulated Raman scattering. *Nat. Chem.* **6**, 614–622 (2014).
35. Agard, N. J., Baskin, J. M., Prescher, J. A., Lo, A. & Bertozzi, C. R. A comparative study of bioorthogonal reactions with azides. *ACS Chem. Biol.* **1**, 644–648 (2006).
36. Jewett, J. C., Sletten, E. M. & Bertozzi, C. R. Rapid Cu-free click chemistry with readily synthesized biarylazacyclooctynones. *J. Am. Chem. Soc.* **132**, 3688–3690 (2010).
37. Laughlin, S. T., Baskin, J. M., Amacher, S. L. & Bertozzi, C. R. In vivo imaging of membrane-associated glycans in developing zebrafish. *Science* **320**, 664–667 (2008).
38. Baskin, J. M., Dehnert, K. W., Laughlin, S. T., Amacher, S. L. & Bertozzi, C. R. Visualizing enveloping layer glycans during zebrafish early embryogenesis. *Proc. Natl Acad. Sci. USA* **107**, 10360–10365 (2010).

Acknowledgements

This research was supported by an intramural funding from the A*STAR Biomedical Research Council and Singapore Ministry of Education Academic Research Fund Tier 2 (MOE2010-T2-1-025) and NRF (2014M3A6A4075060, 2011-0020322) to H.P. E.P.-C. thanks CONACyT (grants 129572, 123732) for financial support. L.W. acknowledges support from NIH (1R01GM118384-01). We thank SBIC-Nikon Imaging Centre for confocal microscopy facilities and Prof Go Mei Lin (National University of Singapore) for MOE software.

Author contributions

Y.-T.C. designed the research and analysed the data. S.H.A. performed high-throughput cell screening, cheminformatics, statistical analysis and cellular organelle labelling. R.S., D.S., R.D. and J.Y.L. performed chemical synthesis. L.H. and H.-H.H. synthesized TAM and TAM′ library. H.R., C.H. and L.W. performed histone H2B labelling. E.A.-M. and E.P.-C. synthesized EP library. J.K. and H-S.P. performed tubulin labelling. S.H.A. wrote the manuscript. Y.-T.C., S.H.A., R.S. and L.W. edited the manuscript.

Additional information

Competing financial interests: Y.-T.C., S.H.A. and R.S. are the inventors of CO-1 and AzG-1 for which a patent has been applied The remaining authors declare no competing financial interests.

A broad analysis of resistance development in the malaria parasite

Victoria C. Corey[1], Amanda K. Lukens[2,3], Eva S. Istvan[4], Marcus C.S Lee[5], Virginia Franco[6], Pamela Magistrado[2], Olivia Coburn-Flynn[5], Tomoyo Sakata-Kato[2], Olivia Fuchs[1], Nina F. Gnädig[5], Greg Goldgof[1], Maria Linares[6], Maria G. Gomez-Lorenzo[6], Cristina De Cózar[6], Maria Jose Lafuente-Monasterio[6], Sara Prats[6], Stephan Meister[1], Olga Tanaseichuk[7], Melanie Wree[1], Yingyao Zhou[7], Paul A. Willis[8], Francisco-Javier Gamo[6], Daniel E. Goldberg[4], David A. Fidock[5], Dyann F. Wirth[2,3] & Elizabeth A. Winzeler[1]

Microbial resistance to chemotherapy has caused countless deaths where malaria is endemic. Chemotherapy may fail either due to pre-existing resistance or evolution of drug-resistant parasites. Here we use a diverse set of antimalarial compounds to investigate the acquisition of drug resistance and the degree of cross-resistance against common resistance alleles. We assess cross-resistance using a set of 15 parasite lines carrying resistance-conferring alleles in *pfatp4*, cytochrome bc_1, *pfcarl, pfdhod, pfcrt, pfmdr, pfdhfr,* cytoplasmic prolyl t-RNA synthetase or *hsp90*. Subsequently, we assess whether resistant parasites can be obtained after several rounds of drug selection. Twenty-three of the 48 *in vitro* selections result in resistant parasites, with time to resistance onset ranging from 15 to 300 days. Our data indicate that pre-existing resistance may not be a major hurdle for novel-target antimalarial candidates, and focusing our attention on fast-killing compounds may result in a slower onset of clinical resistance.

[1] Department of Pediatrics, School of Medicine, University of California San Diego, 9500 Gilman Drive 0741, La Jolla, California 92093, USA. [2] Department of Immunology and Infectious Disease, Harvard T.H. Chan School of Public Health, 665 Huntington Avenue, Boston, Massachusetts 02115, USA. [3] Infectious Disease Program, The Broad Institute, 415 Main Street, Cambridge, Massachusetts 02142, USA. [4] Department of Medicine and Microbiology, Washington University School of Medicine, St Louis, Missouri 63110, USA. [5] Department of Microbiology and Immunology, Columbia University College of Physicians and Surgeons, New York, New York 10032, USA. [6] Tres Cantos Medicines Development Campus, Malaria DPU, GlaxoSmithKline, Severo Ochoa 2, Tres Cantos, 28760 Madrid, Spain. [7] The Genomics Institute of the Novartis Research Foundation, 10675 John J Hopkins Drive, San Diego, California 92121, USA. [8] Medicines for Malaria Venture, PO Box 1826, 20 route de Pre-Bois, 1215 Geneva 15, Switzerland. Correspondence and requests for materials should be addressed to E.A.W. (email: ewinzeler@ucsd.edu).

Malaria remains a devastating disease, with an estimated 198 million cases (uncertainty range 124–283 million) and 584,000 deaths (uncertainty range 367,000–755,000) in 2013 alone, affecting primarily children under the age of five[1]. Given that the current vaccine available provides only moderate protection[2], chemotherapeutics constitute the best clinical tools available for the prevention and treatment of malaria. Currently, A number of effective drug therapies exist, most of which target the malaria parasite during its replicative cycle within human erythrocytes, the lifecycle stage that is associated with clinical manifestations of malaria[3]. These consist of 4-aminoquinolines including chloroquine, piperaquine and related compounds; antifolates such as pyrimethamine and cycloguanil; alkanolamines such as halofantrine and lumefantrine; endoperoxides such as artesunate, artemisinin and artemether; and newer synthetic compounds. Antimalarial treatments are typically administered as combination therapies, and artemisinin-based combination therapies (ACTs, such as artemether-lumefantrine) currently represent the therapy class that is most effective and that is the standard of care recommended by the World Health Organization (WHO).

While ACTs have been a highly effective frontline therapy, particularly against multidrug-resistant *Plasmodium falciparum* infections, there is now evidence of resistance emerging to artemisinin and its derivatives[4–7]. Clinical trials with artemisinin mono-therapies have shown that these compounds are taking considerably longer to clear malaria infections in Southeast Asia—typically twice as long as observed a decade ago. Given that parasites have already acquired resistance to several partner drugs, treatment failures are beginning to be observed with combination therapies[8,9]. Thus, new compound classes, ideally with new mechanisms of action, are urgently needed if the gains of the last decade are to be sustained.

In anticipation of eventual widespread ACT failure, there has been a focused and coordinated effort to place new antimalarial drug candidates into the drug development pipeline (see http://www.mmv.org/research-development/rd-portfolio). Leads from phenotypic screens in particular are being progressed into molecules that are suitable for testing in clinical trials. An open question, however, is whether small molecules from phenotypic screens will lead to the identification of new druggable targets and pathways that do not rapidly lose effectiveness in the field because of acquired and pre-existing parasite resistance.

Here we use a set of 50 antimalarial compounds identified in phenotypic screens[10–13] to systematically evaluate whether resistant parasites can be selected and whether or not pre-existing resistance mechanisms confer resistance using a panel of strains containing mutations in a variety of genes, including *P. falciparum* cyclic amine resistance locus (*pfcarl*)[14], cytochrome *bc1* (refs 15–17) and *P. falciparum* ATPase 4 (*pfatp4*)[18–20]. Here we show that pre-existing resistance is less likely to be a problem but that the *de novo* acquisition of resistance occurs rapidly for many compounds. We highlight a set of antimalarial compounds that have thus far defied attempts to create drug-resistant parasites in a variety of different laboratories and identify features that are shared by all, including a rapid rate of killing and lack of pre-existing resistance.

Results

Initial selection of compounds. To systematically investigate both the *de novo* acquisition and pre-existing landscape of drug resistance, we assembled a set of 50 diverse compounds selected from *P. falciparum* asexual phenotypic screens[10–13]. Compounds were initially chosen based on potency, demonstrated by activity against asexual blood stages ranging from 23 nM to 1.67 μM EC_{50} with most compounds having an EC_{50} of $<1\,\mu M$ in the *P. falciparum* strain 3D7 as measured by a hypoxanthine incorporation assay. To minimize non-novel pharmacophores, compounds were compared against the scaffolds of clinical antimalarials, eliminating candidates with similar structures.

Compounds in our set ranged in molecular weight from 261 to 574 $g \cdot mol^{-1}$, with 42 compounds having drug-like properties (compliant with Lipinski's rule of five) and the remaining 8 identified as probe-like compounds. We sought to maximize chemical diversity in our set by eliminating similar compounds as indicated by the Tanimoto coefficient, since compounds with Tanimoto coefficients >0.85, a quantitative measurement of chemical scaffold similarity, are thought to have similar biological activity to one another[21,22]. The resulting compound set displayed an average Tanimoto coefficient of 0.186, ranging from 0.093 to 0.923 (Fig. 1). Although a few compounds were similar to one another (particularly two carbazoles: MMV009063 and MMV665882) a majority of the set was diverse, possessing a variety of functional groups and heterocyclic substructures.

Compound evaluation against a panel of resistant clones. Multidrug resistance alleles, including mutations in *P. falciparum* chloroquine resistance transporter (*pfcrt*) and *P. falciparum* multi-drug resistance transporter 1 (*pfmdr1*) are very common in field isolates. Therefore, we sought to assess the degree to which pre-existing resistance alleles would contribute to a loss of potency for this diverse collection of small molecule compounds. Fifteen clones derived from three main parent lines were chosen to maximize diversity in mutated pathways and genetic backgrounds, thus representing the variety of resistance seen in the field: 3D7 (ref. 23), W2 (ref. 24) or Dd2 (ref. 25). The 3D7 line originates from the clone of a Netherlands clinical isolate strain[23,26] and is generally considered to be drug sensitive, though it does convey resistance to sulfadoxine. Conversely, W2 (ref. 24,27) and Dd2 (ref. 25) are multi-drug-resistant lines originating from the Indochina III/CDC isolate, which contain point mutations in *pfcrt* as well as amplifications in *pfmdr1* and GTP cyclohydrolase[27].

A total of nine known drug-resistance genes were represented by our strain set, containing validated critical single nucleotide variants (SNVs) or copy number variants as well as a handful of additional background mutations. One strain, TM90C2A, was a clinical isolate from Thailand[28], while the remaining 15 lines were created through *in vitro* evolution. Clones contained one or more resistance-conferring alleles in the folate pathway[28], *pfcrt* (ref. 29), *P. falciparum* dihydroorotate dehydrogenase (*pfdhodh*)[30], *pfcarl* (ref. 14), prolyl t-RNA synthetase[31] or heat shock protein 90 (*hsp90*). We also investigated multiple cytochrome bc_1 (refs 15–17) and *pfatp4* (refs 18–20) alleles, including three alleles in the cytochrome bc_1 Q_o site, which confer resistance to atovaquone[15] or a tetracyclic benzothiazepine[16] and one allele in the cytochrome bc_1 Q_i site conferring resistance to a benzylsulfonamide[17]. The five PfATP4 mutant lines possess alleles that map to the transmembrane channel of the sodium-dependent ATP4 transporter, and were acquired by exposing parasites to sub-lethal concentrations of a spiroindolone[18], an aminopyrazole[19], or other scaffolds from the Tres Cantos Antimalarial Set library[20]. These two resistance genes were represented by multiple strains since critical SNV mutations may be located within an active site or result in a change in membrane potential, resulting in each mutation only affecting a subset of PfATP4 or cytochrome *bc1* inhibitors.

To evaluate potential overlapping activity against known antimalarial drug targets, we performed dose-response studies

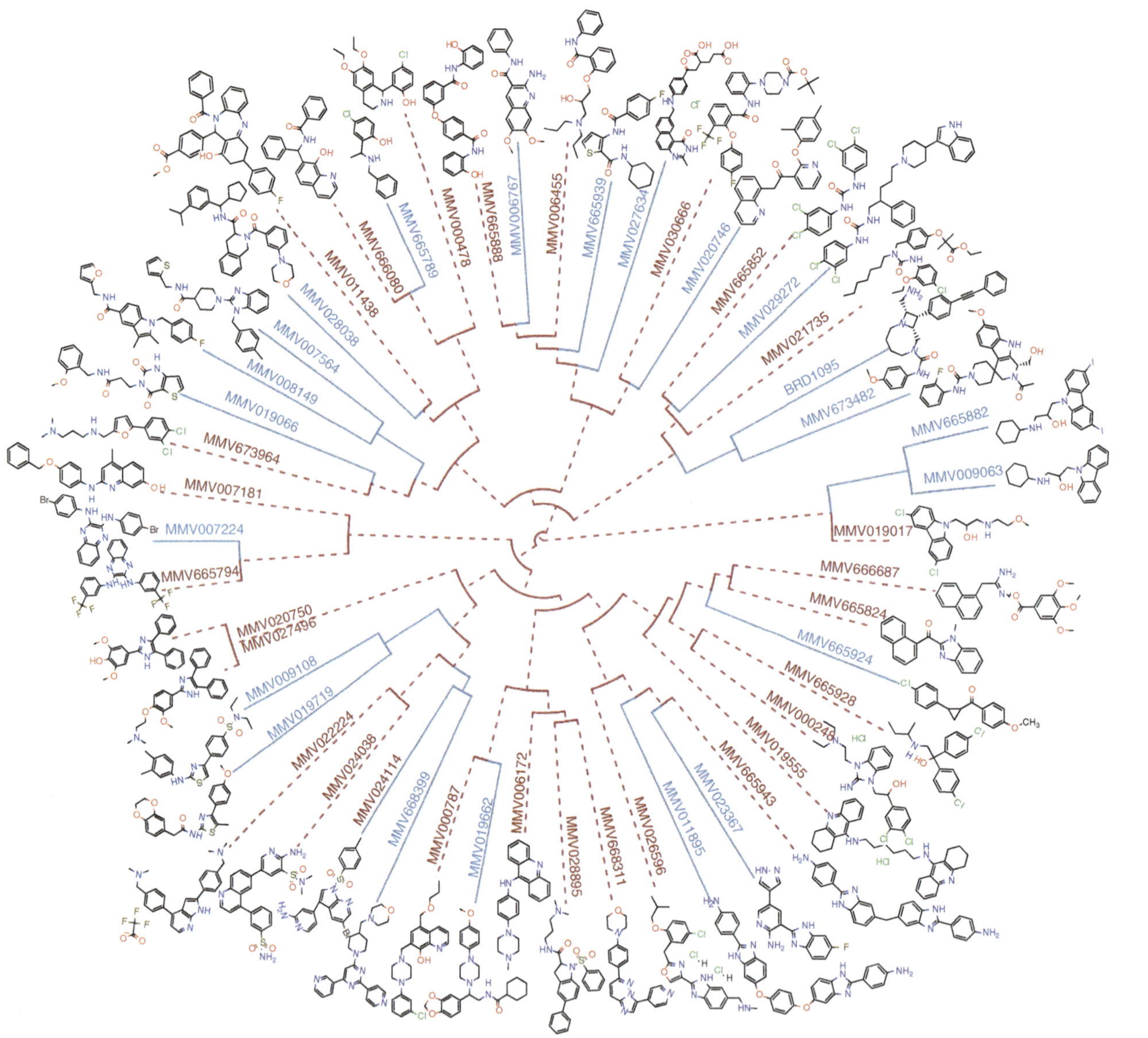

Figure 1 | Chemically diverse compound set. Hierarchical clustering of the 50 compound set. Compounds were clustered by a maximum substructure similarity Tanimoto coefficient. *In vitro* selections that were successful in yielding resistant parasites are highlighted in blue, whereas compounds where resistance development was unsuccessful are highlighted in dashed red.

with each compound against our assembled set of drug-resistant lines and their corresponding parents, in an asexual blood stage proliferation assay (Table 1). Comparisons were made to the drug-sensitive strain 3D7 when a parent was unavailable (TM90C2A, W2 and Dd2). As laboratories conducted EC_{50} assays with different strains and protocols, all comparisons were made between assays from the same laboratory and under the same assay conditions. Compounds displaying >5-fold EC_{50} shifts relative to the parent or sensitive strain were flagged as having potentially non-novel targets. Average fold shifts per strain ranged from 0.82 to 2.34, with a median of 1.23 indicating there was not common resistance to any one resistant clone. No resistance patterns were observed when classifying the compounds based on chemical structure, but this was not surprising given the diversity of compounds chosen. On the other hand, given that these confirmed compounds were derived from larger unconfirmed hit lists to which some filtration criteria may have been applied (for example, elimination of obvious dihydrofolate reductase inhibitors[32]), some[10–13] pre-existing bias may exist in the set.

Alleles in the strain set generally did not confer resistance to the vast majority of the compounds (Fig. 2a), with only two compounds (MMV019066 and MMV008149) losing efficacy in some resistant lines relative to their parent clones (EC_{50} fold-shifts >5-fold). ATQ-R4, bearing a Q_o cytochrome bc_1 allele, was resistant to a propanamide, MMV019066, (P value = 0.0016) with an EC_{50} fold-shift of 16 ×, as determined by a one-way analysis of variance (ANOVA) analysis with a Dunnett's multiple comparison correction. In addition, CYTb-G131S, a cytochrome bc_1 Q_o allele, demonstrated complete resistance to MMV008149, a carboxamide, with a >38-fold change in EC_{50}. Statistically significant resistance was seen against MMV019066 and MMV008149 in other cytochrome bc_1 Q_o mutants (ATQ-R5 and ATQ-R4, respectively), but the EC_{50} fold-shifts seen were under the threshold value set for the study. Neither compound shared any structural features with other cytochrome bc_1 inhibitors, including atovaquone and decoquinate (Fig. 2b).

As had been previously observed[19], mutations in PfATP4 resulted in parasites that were more sensitive to unrelated

Table 1 | Summary of mutated strains testing cross-resistance

Strain	Target/pathway	Parent strain	SNVs	References
ATQ-R4	cytb	3D7	M133I and L144S	15
ATQ-R5	cytb	3D7	F267V	15
CYTb-G131S	cytb	Dd2	G131S	16
CYTbQi	cytb	Dd2	G33A	17
GNF156R	PfCARL	Dd2	L830V, S1076I and M81I	14
NITD609-R2	PfATP4	Dd2	T418N and P990R	18
GNF449-R2	PfATP4	Dd2	I203L and P990R	19
PfATP4-Mut2	PfATP4	W2	P412L	20
PfATP4-Mut1	PfATP4	W2	V178I	Unpublished
PfATP4-Mut3	PfATP4	3D7	F917L	20
TM90C2A	DHFR (folate pathway)	3D7*	Unknown (MR4 origin)	28
PfCRT-Q352R	PfCRT	Dd2	Q352R	29
DHODH	DHODH	3D7	E182D	30
hsp90	cellular chaperone	Dd2	D88Y	Unpublished
ProRS	prolyl t-RNA synthetase	Dd2	L482H	31

*For TM90C2A, there was no official parent strain, so the clone was compared with 3D7 sensitive strain for cross-resistance.
Fifteen mutated strains and their respective parents were used to test for potential cross-resistance present in our compound panel. Mutations (SNVs) responsible for resistance are listed for each strain, along with their corresponding parents and origin.

compounds: ATP4-Mut2 exhibited sensitivity to a sulfonamide (MMV009108—P value = 0.0001; 23 × fold-shift) and a carboxamide (MMV028038—P value = 0.0007; 5 × fold-shift). ATP4-Mut3 additionally conferred sensitivity to MMV009108 (P value = 0.0047; 5.2 × fold-shift). Neither compound was structurally similar to spiroindolones, pyrazoles or dihydroisoquinolones, all previously identified PfATP4 inhibitors (NITD609 (ref. 18), KAE678 (ref. 18), GNF4492 (ref. 19) and (+)-SJ733 (ref. 20)), with a Tanimoto coefficient range of 0.127–0.34 (Fig. 2c).

Finally, two compounds (MMV665939 and MMV028895) lost efficacy in the parent strains W2 or Dd2 when compared with the 3D7 sensitive strain within the same lab. MMV665939 demonstrated fold shifts in W2 compared with 3D7 (EC_{50} fold-shift—9.5 ×), as well as in the other two W2-based clones (PfATP4-Mut1: 16.6 × , PfATP4-Mut2: 18.9 ×) compared with 3D7. Fold shifts in Dd2 compared with 3D7 were also seen in MMV665939, though these shifts were less severe (1.5–3.1 × ; average 2.4 ×). MMV028895 showed reduced efficacy in Dd2 (EC_{50} fold-shift 7.1 ×), which was also seen throughout the other Dd2 clones (2.0–7.5 × ; average—5.5 ×), but unlike MMV665939 maintained potency against W2 strains. These efficacy changes in Dd2 and W2 are most likely due to *pfmdr*, given that W2 and Dd2 contain additional copies of the multi-drug resistance gene (2 and 3–4 copies, respectively) when compared to 3D7.

Selection of resistant parasites. While we found an overall lack of pre-existing resistance, analysing the onset of resistance was critical, as compounds resulting in rapid resistance are not ideal for clinical development. We therefore sought to create drug-resistant parasites using a variety of different selection methods. We implemented either a high-dose method or a ramp-up/pulse method for our *in vitro* selections. Two of the 50 compounds—MMV028895 and MMV665824—were removed from the selection study during experimentation. MMV665824 was removed due to a significant loss of potency, while MMV028895 exhibited cross-resistance to resistant parasites generated by MMV007564. Out of the remaining 48 compounds, resistant parasites were obtained for 23 (Supplementary Data 1). As each selection was performed in triplicate with three independent cultures, we succeeded in generating a total of 66 resistant cultures. Parasites resistant to MMV026596 were not obtained despite 100 days of selection, but the cultures acquired hypersensitivity to mefloquine with a 10-fold reduction in EC_{50} compared with its parental line, 3D7. For the remaining 24 compounds, acquisition of resistant parasites was unsuccessful despite numerous attempts over an extensive period of time.

To determine if the cross-resistance assay was predictive, we sought to determine the target of MMV008149, the compound resulting in the largest EC_{50} fold-shift (> 38 ×) within the cross-resistance assay set by fully sequencing six parasite clones that had acquired resistance to MMV008149. The EC_{50} values for the resistant lines ranged from 1.67 μM to 10.06 μM, a 3–21 × EC_{50}-fold change when compared with the Dd2 parent (EC_{50} = 485 nM).

Genomic DNA (gDNA) was isolated from the Dd2 parent and six clones (two clones per resistant selection flask), and samples were prepared for whole-genome sequencing. Samples were sequenced to > 60 × coverage using paired end reads and aligned to the 3D7 reference genome and variants were called with HaplotypeCaller (GATK). Comparing the nucleotide variation found in the resistant samples to the Dd2 parent clone, which had been isolated immediately before selections, we identified genomic changes that had presumably occurred during selection. Following this comparison, 19 mutations were identified: 11 SNVs and eight insertion/deletions (INDELs) (Table 2, Supplementary Data 2). INDEL mutations were comprised of intergenic (three), codon INDEL (three), intronic (one) and frame-shift (one) mutations. The SNV set showed a mixture of intergenic (five), synonymous (three) and non-synonymous (three) mutations. Comparing all six clones, one gene was mutated across all samples: cytochrome *bc1*. Additionally, variant positions correlated well to EC_{50} fold changes, with the lower fold changes (3.4–5.9 ×) corresponding to the G131S amino acid change, and the higher EC_{50} fold-shifts (18–21 ×) corresponding to the Y126C and V284L amino acid changes. Two of the mutations, G131S and Y126C, were contained in the Q_o site, while the third mutation (V284L) was not in either the Q_o or Q_i binding region[29]. G131S was also the major mutation found in CYTb-G131S, the cross-resistant strain predicting cytochrome *bc1* as a potential target for MMV008149. These results support that the cross-resistance assay was able to successfully identify compounds with overlapping targets, and made us more confident that our compound set consisted of novel targets.

Multi-stage activity profiling. As drugs that eliminate multiple stages of the parasite lifecycle will be critical components in global

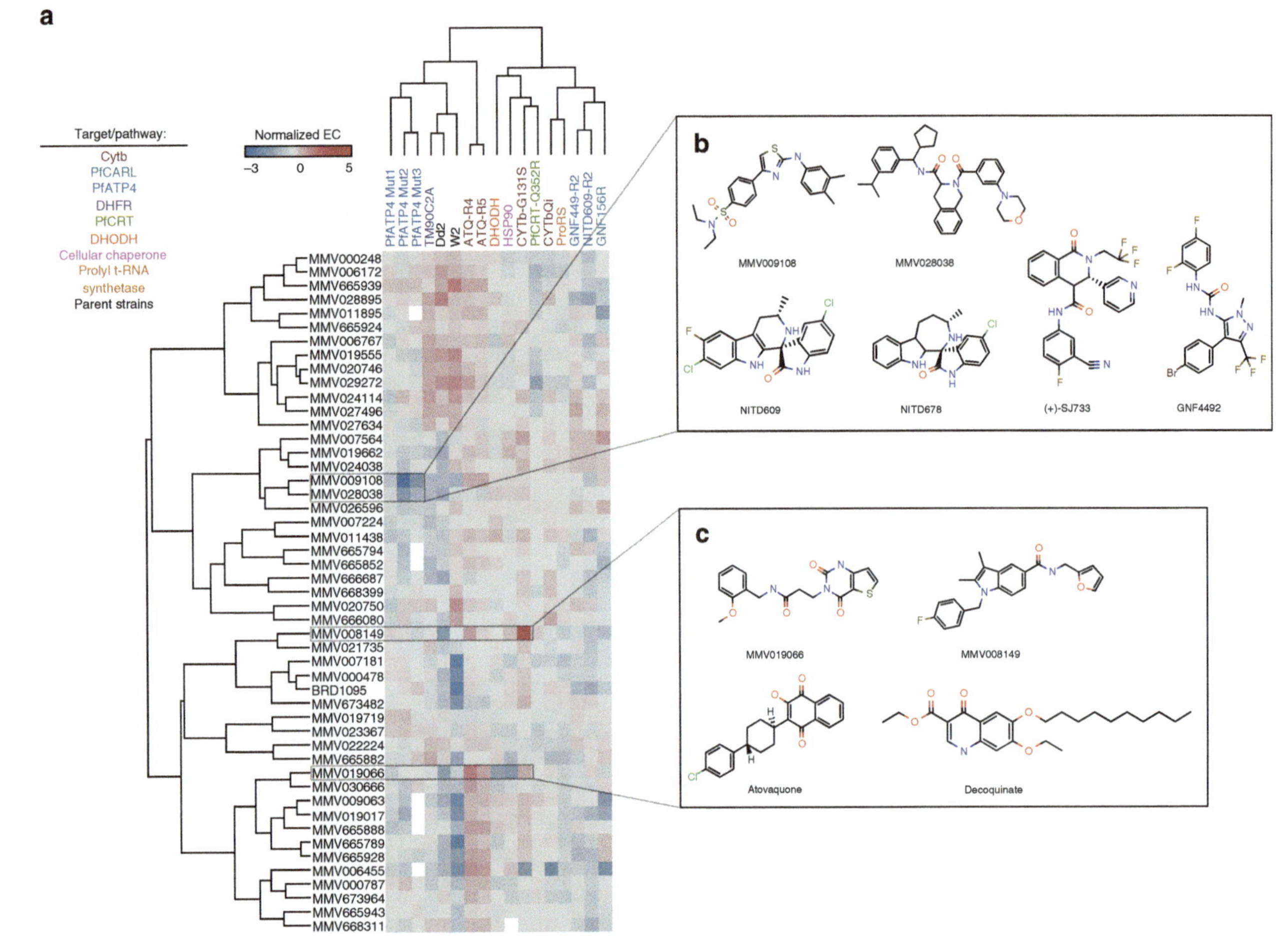

Figure 2 | Cross-resistance fold shifts observed in compound set. (**a**) A total of 15 resistant strains were tested with each MMV compound to identify potential pre-existing cross-resistance. Calculating the fold shifts between each clone and either a corresponding parent or a drug sensitive 3D7 strain generated the heatmap. To normalize conferred resistance and sensitivity, the natural log of each fold shift is displayed. Fold shifts instead of raw data were used as multiple assays were run with varying times and detection indicators. Incomplete cross-resistance assays are depicted in white. All assays were run in triplicate. For one compound (BRD1095), a close analogue (BRD3444) was analysed for TM90C2A and PfATP4-Mut1-3. SMILEs for all compounds are listed in Supplementary Data 1. (**b**) Chemical structures of the two MMV compounds (MMV009108 and MMV028038) with increased efficacy against one or more *pfatp4* mutated clones. Both compounds displayed low structural similarity to a number of other known *pfatp4* inhibitors. (**c**) Chemical structures of the two MMV compounds (MMV019066 and MMV008149) with decreased efficacy against one or more cytochrome *bc1* mutated clones. Atovaquone and decoquinate, two other cytochrome *bc1* inhibitors, were structurally significantly different.

efforts to eliminate malaria[10], we were interested in knowing whether the compounds in our study had additional activities against either the liver and/or transmission stages of the parasite lifecycle. To determine this, we first evaluated the compounds in additional phenotypic assays. The first assay was a *P. berghei* hepatocyte invasion and development assay, which predicts causal prophylactic activity. Hepatocyte toxicity was additionally tested to identify false-positive activity in the liver stage due to host-cell toxicity. Assays were executed in duplicate using a 12-point EC_{50} curve at starting concentrations of 5 μM or 50 μM (Supplementary Table 1). Positive and negative controls testing for hepatocyte toxicity and liver stage activity included atovaquone (PfLuc EC_{50}:0.419 ± 0.044 nM; HepG2 EC_{50}:7720 ± 1370 nM) and puromycin (PfLuc EC_{50}: 32.7 ± 29.7 nM; HepG2 EC_{50}:254 ± 36.6 nM). Of the 47 compounds examined, 15 were found to be active against liver-stage parasites (EC_{50} values < 1 μM (10–880 nM)), nine of which demonstrated a minimal fivefold difference between parasite activity and hepatocyte toxicity (Supplementary Fig. 1). Interestingly, six compounds were more potent in the liver stage than in the asexual stage. The most significant potency change was seen with MMV019066, a propanamide, which had a reduction in EC_{50} from 1.67 μM to 0.31 μM and previously demonstrated cross-resistance with cytochrome bc_1 alleles. This was expected given that cytochrome bc_1 inhibitors are highly active against liver stages. MMV024038, a quinoline sulfonamide, also demonstrated a reduction in EC_{50} from 228 nM to 10 nM. This compound, however, exhibited some hepatocyte toxicity (146 nM); thus the indirect effects of the host-cell environment on parasite growth could not be ruled out.

The second assay was a late-stage (stage V) gametocyte survival assay, which predicts transmission-blocking activity. Assays were executed in duplicate using a 12-point EC_{50} curve at starting concentrations of 1.25 μM or 12.5 μM (Supplementary Table 1). Positive and negative control compounds included puromycin (0.61 ± 0.11 μM) and atovaquone (> 12.50 μM), respectively. For the late-stage gametocyte assay, 12 compounds resulted in EC_{50} values < 1 μM (260–990 nM), 50% of which overlap with the 15 compounds found to have activity in the liver-stage assay. Of the six compounds with complete multi-stage activity, 83% (5/6)

Table 2 | SNV and INDEL mutations in MMV008149

	MMV008149-F1-CIB2	MMV008149-F1-CIB7	MMV008149-F2-CIE2	MMV008149-F2-CIE3	MMV008149-F3-CIB1	MMV008149-F3-CIC10
Genome coverage (x)	84.13	81.56	74.81	90.93	82.73	72.93
% Covered by 15 or more reads	95.6	95.7	95.6	95.7	95.8	95
SNVs identified						
Total mutations	3	3	5	1	4	4
Intergenic	0	2	4	0	1	3
Intronic	0	0	0	0	0	0
Synonymous	2	0	0	0	2	0
Non-synonymous	1	1	1	1	1	1
Genes mutated in all samples (mutation)			*mal_mito_3 (cytochrome bc1)*			
INDELs identified						
Total mutations	6	0	1	2	1	1
Intergenic	1	0	1	2	0	1
Intronic	1	0	0	0	0	0
Frame shift	1	0	0	0	1	0
Codon INDEL	3	0	0	0	0	0
Genes mutated in all samples (mutation)			None			

SNV and INDEL mutations were called and filtered using HaplotypeCaller for six clones isolated from three MMV008149 resistant flasks.

demonstrated hepatocyte toxicity $>10\times$ the liver stage EC_{50}. Unlike the liver stage, none of the compounds had increased potency against sexual stage compared with the asexual parasites, which was expected as previous literature has found that most currently used antimalarials yield higher EC_{50} values against late-stage gametocytes when compared with the asexual stage EC_{50} (ref. 13). Five compounds, however, demonstrated late-stage gametocyte EC_{50} values within a twofold range of the asexual EC_{50} value.

Rate of killing assays. To identify whether selection success could be predicted by compound speed of action, and to gain a greater insight into the potential mechanisms of action, assays were performed to test the killing rate (Supplementary Data 1)[33]. To quantify the killing rate, viability time-course profiles for each compound were compared with antimalarials known to have fast (chloroquine), moderate (pyrimethamine) or slow (atovaquone) rate of action. These rates were then compared with other compound characteristics, including cross-resistance profile, selection success, potency and structural characteristics to identify any possible trends. Overall, the compound set resulted in a fairly even distribution of fast, moderate and slow speeds (Fig. 3a), implying there was no bias in the phenotypic screens for a particular killing rate. One compound's rate of killing, MMV666080, was unidentifiable due to compound availability. No correlation was seen between compound structure and speed of action. Additionally, there was no correlation found between killing rate and compound potency against 3D7, nor between potency and selection success (Supplementary Fig. 2).

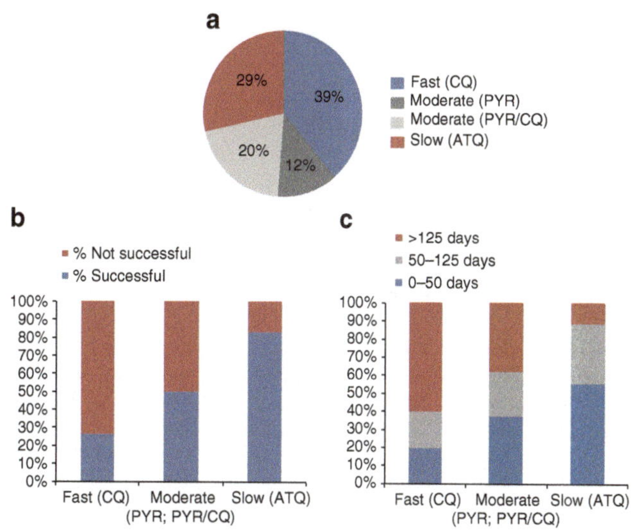

Figure 3 | Killing rate trends. (**a**) Proportion of fast (CQ), moderate (PYR or PYR/CQ), and slow (ATQ) compound killing rates in our compound set (49 compounds in analysis). (**b**) Proportion of failed and successful compound selections sorted by killing rate (47 compounds in analysis). (**c**) Successful selections further proportioned out based on amount of time required to result in resistant parasites. Trend in selection success and killing rate was found to be statistically significant ($P=0.0022$) by a one-tailed Fisher's exact test.

Comparing the selection success rate to compound rate of killing, we saw a significant positive correlation ($P=0.0022$) between speeds of action and selection success, as determined by a one-tailed Fisher's exact test. (Fig. 3b,c). Of the 12 selections with compounds demonstrating a slow speed of action, 83% (10/12) were successful. This was in contrast to compounds with fast-killing rates of which only 26% (5/19) were successful. Additionally, within each set of successful selections, we noticed a trend within the length of time to generate resistant parasites. For the fast-acting compounds, 3/5 successful selections took >125 days, whereas 5/10 resistant parasites were successfully selected for within 50 days for slow-acting compounds. Therefore, slower compounds typically had greater success and required shorter periods of time to develop resistance when compared with fast-acting compounds. This trend remained relatively consistent within individual labs. These results confirm previous findings of faster acting compounds having a lower propensity for developing *de novo* resistance[34]. Fast-acting compounds are already desirable in the clinic due to a quick clearance of parasites and alleviation of symptoms, as well as their propensity towards slower drug resistance development. The inability to develop resistance *in vitro*, however, even when using slow ramp-up selection methods may imply that these benefits come from the target itself. These genes may have minimal mutational flexibility or inhibitors may target several genes, making them ideal for exploitation in antimalarial development. Alternatively, some compounds may have human host targets instead of parasite targets and affect the erythrocyte directly. Further work is required to distinguish between these possibilities.

Discussion

Every therapeutic, whether for the treatment of infectious agents or malignant tumours, is in a race against time; a race against the inevitable development of drug resistance leading to reduced clinical effectiveness. Many factors play a role in the development of resistance, from evolutionary and biochemical constraints of the targeted organism to therapy regimens and the overuse or misuse of therapeutics. Not all therapeutics, however, are created equal. A detailed understanding of a targeted organism's biology and the critical factors required to minimize the emergence of resistance provide opportunities to design more effective and long-lasting drugs and drug combinations. This requires an extensive knowledge of the organism, as well as a comprehensive understanding of how resistance evolves.

This study represents the first systematic analysis of cross-resistance in malaria parasites. We have assessed 50 antimalarial compounds with diverse chemical structures, rates of parasite killing and stage specificity. With few exceptions, the compounds studied did not demonstrate significant cross-resistance to previously identified targets, indicating a large potential to identify additional druggable pathways in the parasite and further our arsenal of antimalarial therapeutics. A lack of cross-resistance to known drug targets does not mean that resistance will not quickly develop, however, and the resistance 'life expectancy' and resulting fitness costs need to be considered for any potential clinical candidate. In selecting for resistant mutants, we found that fast-acting compounds are harder to develop resistance against and generally have a longer onset of resistance when compared with slow-acting compounds. Given that compound killing rates are thought to be determined primarily by their mode-of-action, this resistance feature may be largely due the target or targets themselves. Moving forward, it will be important to verify if this trend holds in compounds with multi-stage activity, as only a small fraction of our compounds (6/50) demonstrated activity against asexual, sexual and liver stage parasites, and multi-stage active antimalarials will be vital in achieving malarial elimination. In addition to killing rate, our compound set provides a list of various physicochemical and structural features that may additionally be predictive of selection success, and the addition of our set to future screens may provide an eventual predictive model, focusing on compounds less likely to be prone to resistance development (Supplementary Figs 3 and 4). Fast-acting compounds have already been a focus in therapeutic development as they rapidly stop disease progression and avoid severe complications, but the additional benefit of reduced evolution of resistance makes them even more attractive candidates for future antimalarial designs.

Methods

Compound origin and computational clustering. Compounds were all publically available and obtained from a variety of sources, including the medicines for malaria venture (MMV) malaria box[10], the GlaxoSmithKline Tres Cantos Antimalarial Set[11], the University of Dundee and the Broad Institute's Diversity Oriented Synthesis libraries, as indicated in Supplementary Data 1. Compound similarity was calculated using the flexible MCS (FMCS) finder package in R. Clustering was conducted using hclust with a ward.D2 method setting. Pairwise distances were calculated between Tanimoto coefficient values from fmcsR.

Strain culture origins and propagation. *P. falciparum* parent strains 3D7 and Dd2 used for selection were obtained from the labs of Dan Goldberg and David Fidock, respectively. The Fidock lab additionally supplied the Dd2Δexo strain for selections with MMV000787 and MMV023367. Functional assay and cross-resistant strains were obtained from the labs of Elizabeth Winzeler, Dyann Wirth, David Fidock and GlaxoSmithKline, as listed in Table 1.

Asexual parasites were grown with 5% haematocrit at 37 °C in RPMI (Rosewell Park Memorial Institute)-1640 medium supplemented with 0.014 mg ml^{-1} hypoxanthine, 38.4 mM HEPES, 0.2% $NaHCO_3$, 0.2% glucose, 3.4 mM NaOH, 0.3 g l^{-1} glutamine and 5% AlbuMAX II. Depending on lab preferences, parasites were grown either in the presence or absence of 0.05 mg ml^{-1} gentamicin and 4.2% human O^+ serum (Interstate Blood Bank, Inc.). Cultures were maintained in a gas mixture of 5% O_2, 5% CO_2 and 90% N_2. When not undergoing selections, cultures were maintained with media changes every other day, keeping parasitemia values at 0.3–4%.

Frozen stocks were prepared using one of two methods: (1) by freezing 100% red blood cells (RBCs) at ~5% parasitemia with equal volumes of a freezing solution composed of 28% glycerol, 3% sorbitol and 0.65% NaCl; (2) freezing 0.2 ml RBCs at ~5% parasitemia with 0.3 ml serum and 0.5 ml of a glycerol solution. Stocks were thawed out by slowly adding 1/5 volume of 12% NaCl, followed by five volumes of 1.6% NaCl. Parasites were then spun down at 800*g* for 5 min at room temperature and washed with supplemented media before standard culturing methods.

Cross-resistance and functional assays. EC_{50} assays were conducted using either a 48 or 72-h assay based on lab preference and specialty (internal assay variance summarized in Supplementary Fig. 5). The 48-h assay was carried out in a 96-well format following standard methods using the ^{3}H-hypoxanthine incorporation assay[35]. The raw EC50 values corresponding to this assay can be found in Supplementary Data 3. Briefly, this assay relies on the parasite incorporation of labelled hypoxanthine that is proportional to *P. falciparum* growth. A culture of parasitized RBCs (0.5% parasitemia with a percentage of ring stage higher than 70% of total parasitemia, 2.0% haematocrit) in RPMI-1640, 5% AlbuMAX and 5 μM hypoxanthine was exposed to drug serial dilutions. Plates were incubated for 24 h at 37 °C, 5% CO_2, 5% O_2 and 90% N_2. After 24 h of incubation, ^{3}H-hypoxanthine was added and plates were incubated for an additional 24-h period. After that, parasites were harvested on a glass fibre filter using a TOMTEC Cell Harvester 96. Filters were dried and melt-on scintillator sheets were used to determine the incorporation of ^{3}H-hypoxanthine. Radioactivity was measured using a microbeta counter. Data were normalized using the incorporation of the positive (parasitized RBCs without drug) and negative (same culture with artesunate at 2 μM) controls. All assays were conducted in triplicate using three independently grown cultures, and EC_{50} values were determined using the XLfit add-in module for Excel (version 2.3.1) from Grafit (version 5.3.1.3), where parameter h is the assay hill slope, x is the compound concentration and y is the inhibition of growth effect. The XLfit equation is:

$$y = [(100)/\left(1+\left(EC_{50}/x\right)^h\right)]$$

Additionally, any EC_{50} outliers were removed using a single iteration of Grubbs' test, eliminating values with z-scores above the critical z-value ($\alpha = 0.05$).

Alternatively, a SYBR Green-based proliferation approach was used for the 72-h assay, as previously described[32]. Briefly, a culture of either synchronized or asynchronous parasitized RBCs (0.3% parasitemia and 4% haematocrit) in screening medium (identical to supplemented medium above except no serum was added) was exposed to serial drug dilutions. Plates were incubated at 37 °C and gassed with 93% N2, 4% CO2 and 3% O2 for 72 h. Following incubation, 10 × SYBR Green I (Invitrogen; supplied in 10,000 × concentration) in lysis buffer (20 nM Tris-HCL, 5 mM EDTA, 0.16% Saponin wt per vol and 1.6% Triton X vol per vol) was added to the wells and the plates were incubated overnight at room temperature before plate reading. EC_{50} assays were repeated three times, taking the average EC_{50} value from each assay set, and removing any outliers using a single iteration of Grubbs' test, eliminating values with z-scores above the critical z-value ($\alpha = 0.05$). The raw EC50 values for synchronized parasites assayed in a 384-well format are listed in Supplementary Data 4. The raw EC50 values for asynchronous parasites assayed in a 96 or 1536-well format are listed in Supplementary Data 5.

EC_{50} values for each strain were compared with those of the corresponding parent strain to determine fold-shift changes and identify resistance. To minimize the cause variance between assays, all comparisons were conducted between strains assayed under identical conditions. A one-way ANOVA followed by a Dunnet's post-test was conducted in GraphPad Prism to identify significant changes between the parent and resistant strain. For the purposes of this study and to minimize false positive, compounds displaying >5-fold EC_{50} shifts and determined to be significantly different from its corresponding parent by a one-way ANOVA analysis were identified as having potentially non-novel targets.

Cross-resistance computational analysis. Computational analysis of cross-resistance was conducted in R. Briefly, EC_{50} fold shift ratios were calculated using log transformed EC_{50} values from the parental and resistant strains. As EC_{50} assays were run across multiple labs, any strain fold-shift calculations were conducted between strains run within the same lab to minimize error due to assay variability.

The natural log of the ratios was loaded into R, and a heatmap analysis was executed using heatmap.2 from the gplots package. Compounds were clustered by structural similarity (fmcsR) and strains were clustered by column mean values.

Evolution of compound-resistant lines. Based on the compound speed of action and lab specialty, selections were conducted using a high-pressure intermittent selection method, a step-wise selection method, or a constant selection method. For high-pressure selections, $\sim 1\text{-}2 \times 10^9$ parasites were treated at a concentration of $3\text{-}10 \times$ EC_{50} until parasites could not be seen by microscopy (4–10 days). Following treatment, compound pressure was removed and cultures were allowed to recover. Once healthy parasites were seen and parasitemia reached $\sim 2\%$, compound pressure was reinstated. For step-wise selections, $\sim 1 \times 10^8$ parasites were treated at a starting concentration resulting in a reduced growth rate of 50%. Cultures were examined daily by microscopy, increasing compound concentration in increments of 5–10% as needed to maintain a 50% reduced growth rate. Selections were carried out until cultures achieved a reproducible EC_{50} fold shift of $> 3 \times$. Finally, constant selections were conducted in a similar manner to the high-pressure intermittent method with the exception that compound pressure was never removed. Following successful selection, cultures were cloned out using a limiting dilution method as previously described[36]. Selections were terminated after 200 days if resistance could not be obtained.

Library preparation and analysis of sequenced samples. gDNA was obtained from parasites by washing infected RBCs with 0.05% saponin and isolating the gDNA using a DNeasy Blood and Tissue Kit (Qiagen), following the standard protocols. To prepare the sequencing libraries, gDNA was tagmented and amplified with the Nextera XT kit (Cat. No FC-131-1024, Illumina) using the standard dual index protocol, and sequenced on the Illumina HiSeq 2500 with a RapidRun mode, sequencing 100 base pairs deep on either end of the fragments. Following sequencing, reads were aligned to the *P. falciparum* 3D7 reference genome (PlasmoDB v. 13.0), following the Platypus pipeline as previously described, with the exception that SNVs and INDELs were called with GATK's HaplotypeCaller[37]. To identify valid variants, mutations were filtered using general recommendations from GATK (Supplementary Table 2). Following the initial filtration, mutations where read coverage were < 5 and/or where mixed read ratios were > 0.2 (reference/total reads) across all samples were removed.

Rate of killing and multi-stage activity assays. Rate of killing was determined for each compound following a previously described methodology, which uses the invasion of fresh erythrocytes as surrogate of parasite viability[33]. Briefly, parasites were treated with compounds for 48 h. Compound was washed out and fresh-labelled erythrocytes added to the treated parasites. Double stained erythrocytes (RBC label plus parasite DNA label) were quantified and per cent of survival determined.

Liver-stage activity was determined by pretreating hepatic human cells (HepG2) for 2 h with compound in 1,536-well plates infected with freshly dissected *P. berghei* sporozoites. After 48 h of incubation, the viability of *P. berghei* exoerythrocytic forms was measured by luminescence reaction light output using BrightGlo (Promega). Varying levels of compound concentration were used (5 μM or 50 μM) due to the stock concentration supplied.

To test sexual-stage activity, compounds were tested against late-stage gametocytes using a MitoTracker fluorescent-based assay[13]. Specifically, synchronized stage V gametocytes were incubated with compound for 72 h. MitoTracker Red CMXRos (Life Technologies) was added to each well (final concentration: 500 nM) together with saponin to lyse the RBCs. Each plate was then imaged with an Operetta High Content Imaging System (Perkin Elmer) for fluorescence (590–640 nm). Varying levels of compound concentration were used (1.25 μM or 12.5 μM) due to the stock concentration supplied.

The gametocytocidal activity was measured using high-content image analysis software (Harmony, Perkin Elmer). The readout was based on number of metabolically active gametocytes per well.

Cheminformatics predictors. We evaluated the association between acquisition of drug resistance and pharmacological, physicochemical and structural characteristics of the 48 compounds for which selection was attempted. Specifically, we considered the following properties: (1) pharmacological properties measured in this study: killing rate, toxicity and EC_{50} at three different stages of parasite development; (2) 51 physicochemical descriptors obtained with QikProp software[38] and 129 descriptors obtained with VolSurf+ (ref. 39); and (3) 2,694 extended-connectivity fingerprints that encode circular substructures with a bond diameter of 10, generated by ChemAxon software (ChemAxon, Kft.)[40]; and (4) 194 hierarchical scaffolds associated with the compounds generated by HierS software[41].

For each feature, we measured statistical significance of its relationship to the selection success, taking into account the nature of the features[42]. In particular, structural features (that is, extended-connectivity fingerprints and scaffolds) are binary and indicate the presence or absence of each structural element in a compound. For these reasons, hypergeometric statistical tests were applied to structural features and *t*-tests were applied to pharmacological and physiochemical features to measure statistical significance.

References

1. World Health Organization. *World Malaria Report* (WHO, Geneva, 2015).
2. RTS,S Clinical Trials Partnership. Efficacy and safety of RTS,S/AS01 malaria vaccine with or without a booster dose in infants and children in Africa: final results of a phase 3, individually randomised, controlled trial. *The Lancet* **386,** 31–45 (2015).
3. Flannery, E. L., Chatterjee, A. K. & Winzeler, E. A. Antimalarial drug discovery - approaches and progress towards new medicines. *Nat. Rev. Microbiol.* **11,** 849–862 (2013).
4. Noedl, H. *et al.* Evidence of artemisinin-resistant malaria in western Cambodia. *N. Engl. J. Med.* **359,** 2619–2620 (2008).
5. Dondorp, A. M. *et al.* Artemisinin resistance in *Plasmodium falciparum* malaria. *N. Engl. J. Med.* **361,** 455–467 (2009).
6. Ashley, E. A. *et al.* Spread of artemisinin resistance in *Plasmodium falciparum* malaria. *N. Engl. J. Med.* **371,** 411–423 (2014).
7. Winzeler, E. A. & Manary, M. J. Drug resistance genomics of the antimalarial drug artemisinin. *Genome Biol.* **15,** 544 (2014).
8. Saunders, D. L., Vanachayangkul, P. & Lon, C. Dihydroartemisinin-piperaquine failure in Cambodia. *N. Engl. J. Med.* **371,** 484–485 (2014).
9. World Health Organization. *Status report on artemisinin resistance. Global Malaria Programme* (2014).
10. Spangenberg, T. *et al.* The open access malaria box: a drug discovery catalyst for neglected diseases. *PLoS ONE* **8,** e62906 (2013).
11. Gamo, F.-J. *et al.* Thousands of chemical starting points for antimalarial lead identification. *Nature* **465,** 305–310 (2010).
12. Guiguemde, W. A. *et al.* Chemical genetics of *Plasmodium falciparum. Nature* **465,** 311–315 (2010).
13. Plouffe, David M. *et al.* High-throughput assay and discovery of small molecules that interrupt malaria transmission. *Cell Host Microbe* **19,** 114–126 (2016).
14. Kuhen, K. L. *et al.* KAF156 is an antimalarial clinical candidate with potential for use in prophylaxis, treatment, and prevention of disease transmission. *Antimicrob. Agents Chemother.* **58,** 5060–5067 (2014).
15. Bopp, S. E. R. *et al.* Mitotic evolution of *Plasmodium falciparum* shows a stable core genome but recombination in antigen families. *PLoS Genet.* **9,** e1003293 (2013).
16. Dong, C. K. *et al.* Identification and validation of tetracyclic benzothiazepines as *Plasmodium falciparum* cytochrome bc1 inhibitors. *Chem. Biol.* **18,** 1602–1610 (2011).
17. Lukens, A. K. *et al.* Diversity-oriented synthesis probe targets *Plasmodium falciparum* cytochrome b ubiquinone reduction site and synergizes with oxidation site inhibitors. *J. Infect. Dis.* **211,** 1097–1103 (2015).
18. Rottmann, M. *et al.* Spiroindolones, a potent compound class for the treatment of malaria. *Science* **329,** 1175–1180 (2010).
19. Flannery, E. L. *et al.* Mutations in the P-type cation-transporter ATPase 4, PfATP4, mediate resistance to both aminopyrazole and spiroindolone antimalarials. *ACS Chem. Biol.* **10,** 413–420 (2015).
20. Jimenez-Diaz, M. B. *et al.* (+)-SJ733, a clinical candidate for malaria that acts through ATP4 to induce rapid host-mediated clearance of *Plasmodium. Proc. Natl Acad. Sci. USA* **111,** E5455–E5462 (2014).
21. Martin, E. J. *et al.* Measuring diversity: experimental design of combinatorial libraries for drug discovery. *J. Med. Chem.* **38,** 1431–1436 (1995).
22. Matter, H. Selecting optimally diverse compounds from structure databases:a validation study of two-dimensional and three-dimensional molecular descriptors. *J. Med. Chem.* **40,** 1219–1229 (1997).
23. Walliker, D. *et al.* Genetic analysis of the human malaria parasite *Plasmodium falciparum. Science* **236,** 1661–1666 (1987).
24. Oduola, A. M., Weatherly, N. F., Bowdre, J. H. & Desjardins, R. E. *Plasmodium falciparum*: cloning by single-erythrocyte micromanipulation and heterogeneity in vitro. *Exp. Parasitol.* **66,** 86–95 (1988).
25. Wellems, T. E. *et al.* Chloroquine resistance not linked to *mdr*-like genes in a *Plasmodium falciparum* cross. *Nature* **345,** 253–255 (1990).
26. Ponnudurai, T., Leeuwenberg, A. D. & Meuwissen, J. H. Chloroquine sensitivity of isolates of *Plasmodium falciparum* adapted to *in vitro* culture. *Trop. Geogr. Med.* **33,** 50–54 (1981).
27. Campbell, C. C., Collins, W. E., Nguyen-Dinh, P., Barber, A. & Broderson, J. R. *Plasmodium falciparum* gametocytes from culture *in vitro* develop to sporozoites that are infectious to primates. *Science* **217,** 1048–1050 (1982).
28. Looareesuwan, S. *et al.* Clinical studies of atovaquone, alone or in combination with other antimalarial drugs, for treatment of acute uncomplicated malaria in Thailand. *Am. J. Trop. Med. Hyg.* **54,** 62–66 (1996).
29. Lukens, A. K. *et al.* Harnessing evolutionary fitness in *Plasmodium falciparum* for drug discovery and suppressing resistance. *Proc. Natl Acad. Sci. USA* **111,** 799–804 (2014).
30. Ross, L. S. *et al. In Vitro* resistance selections for *Plasmodium falciparum* dihydroorotate dehydrogenase inhibitors give mutants with multiple point mutations in the drug-binding site and altered growth. *J. Biol. Chem.* **289,** 17980–17995 (2014).

31. Herman, J. D. *et al.* The cytoplasmic prolyl-tRNA synthetase of the malaria parasite is a dual-stage target of febrifugine and its analogs. *Sci. Transl. Med.* **7,** 288ra77 (2015).
32. Plouffe, D. *et al.* *In silico* activity profiling reveals the mechanism of action of antimalarials discovered in a high-throughput screen. *Proc. Natl Acad. Sci. USA* **105,** 9059–9064 (2008).
33. Linares, M. *et al.* Identifying rapidly parasiticidal anti-malarial drugs using a simple and reliable *in vitro* parasite viability fast assay. *Malar. J.* **14,** 441 (2015).
34. Sanz, L. M. *et al.* *P. falciparum in vitro* killing rates allow to discriminate between different antimalarial mode-of-action. *PLoS ONE* **7,** e30949 (2012).
35. Desjardins, R. E., Canfield, C. J., Haynes, J. D. & Chulay, J. D. Quantitative assessment of antimalarial activity *in vitro* by a semiautomated microdilution technique. *Antimicrob. Agents Chemother.* **16,** 710–718 (1979).
36. Goodyer, I. D. & Taraschi, T. F. *Plasmodium falciparum*: a simple, rapid method for detecting parasite clones in microtiter plates. *Exp. Parasitol.* **86,** 158–160 (1997).
37. Manary, M. J. *et al.* Identification of pathogen genomic variants through an integrated pipeline. *BMC Bioinformatics* **15,** 1–14 (2014).
38. QikProp. version 4.2 Schrödinger, LLC, New York, NY (2014).
39. Cruciani, G., Crivori, P., Carrupt, P. A. & Testa, B. Molecular fields in quantitative structure–permeation relationships: the VolSurf approach. *J. Mol. Struct. THEOCHEM* **503,** 17–30 (2000).
40. Rogers, D. & Hahn, M. Extended-connectivity fingerprints. *J. Chem. Inf. Model* **50,** 742–754 (2010).
41. Wilkens, S. J., Janes, J. & Su, A. I. HierS: hierarchical scaffold clustering using topological chemical graphs. *J. Med. Chem.* **48,** 3182–3193 (2005).
42. Peng, H., Long, F. & Ding, C. Feature selection based on mutual information: criteria of max-dependency, max-relevance, and min-redundancy. *IEEE Trans. Pattern Anal. Mach. Intell.* **27,** 1226–1238 (2005).

Acknowledgements

This work was supported by grants from the Bill and Melinda Gates Foundation (OPP1040406) and the National Institute of Allergy and Infectious Diseases (R01AI103058). Victoria Corey received support from a UCSD Genetics Training Program through an institutional training grant from the National Institute of General Medical Sciences (T32 GM008666).

Author contributions

E.A.W. and V.C.C. wrote the manuscript. Cross-resistance EC_{50} assays were performed and analysed by V.C.C., A.K.L., T.S.K., P.M., M.L., V.F., C.D.C., M.J.L.-M. and S.P. S.M. and M.W. performed multi-stage activity assays. M.G.G.-L., M.L. and V.F. performed killing rate assays. Selections were conducted by V.C.C., O.F., E.S.I., O.C.-F., M.C.S.L., T.S.K., P.M., M.L., M.G.G.-L. and V.F. Prediction model studies were conducted by O.T. and Y.Z. Data was compiled by V.C.C. The manuscript was edited by all authors.

Additional information

Accession codes: All relevant sequence data have been deposited in the National Center for Biotechnology Information (NCBI) Sequence Read Archive database with accession code SRP069308.

Competing financial interests: The authors declare no competing financial interests.

Permissions

All chapters in this book were first published in NC, by Nature Publishing Group; hereby published with permission under the Creative Commons Attribution License or equivalent. Every chapter published in this book has been scrutinized by our experts. Their significance has been extensively debated. The topics covered herein carry significant findings which will fuel the growth of the discipline. They may even be implemented as practical applications or may be referred to as a beginning point for another development.

The contributors of this book come from diverse backgrounds, making this book a truly international effort. This book will bring forth new frontiers with its revolutionizing research information and detailed analysis of the nascent developments around the world.

We would like to thank all the contributing authors for lending their expertise to make the book truly unique. They have played a crucial role in the development of this book. Without their invaluable contributions this book wouldn't have been possible. They have made vital efforts to compile up to date information on the varied aspects of this subject to make this book a valuable addition to the collection of many professionals and students.

This book was conceptualized with the vision of imparting up-to-date information and advanced data in this field. To ensure the same, a matchless editorial board was set up. Every individual on the board went through rigorous rounds of assessment to prove their worth. After which they invested a large part of their time researching and compiling the most relevant data for our readers.

The editorial board has been involved in producing this book since its inception. They have spent rigorous hours researching and exploring the diverse topics which have resulted in the successful publishing of this book. They have passed on their knowledge of decades through this book. To expedite this challenging task, the publisher supported the team at every step. A small team of assistant editors was also appointed to further simplify the editing procedure and attain best results for the readers.

Apart from the editorial board, the designing team has also invested a significant amount of their time in understanding the subject and creating the most relevant covers. They scrutinized every image to scout for the most suitable representation of the subject and create an appropriate cover for the book.

The publishing team has been an ardent support to the editorial, designing and production team. Their endless efforts to recruit the best for this project, has resulted in the accomplishment of this book. They are a veteran in the field of academics and their pool of knowledge is as vast as their experience in printing. Their expertise and guidance has proved useful at every step. Their uncompromising quality standards have made this book an exceptional effort. Their encouragement from time to time has been an inspiration for everyone.

The publisher and the editorial board hope that this book will prove to be a valuable piece of knowledge for researchers, students, practitioners and scholars across the globe.

List of Contributors

Stephanie L. Begg, James C. Paton, Christopher A. McDevitt, Bart A. Eijkelkamp and Jacqueline R. Morey
Research Centre for Infectious Diseases, School of Biological Sciences, The University of Adelaide, Adelaide, South Australia 4072, Australia

Zhenyao Luo, Rafael M. Couñago and Bostjan Kobe
School of Chemistry and Molecular Biosciences, University of Queensland, Brisbane, Queensland 4072, Australia
Australian Infectious Diseases Research Centre, University of Queensland, Brisbane, Queensland 4072, Australia
Institute for Molecular Bioscience, University of Queensland, Brisbane, Queensland 4072, Australia

Cheryl-lynn Y. Ong and Alastair G. McEwan
School of Chemistry and Molecular Biosciences, University of Queensland, Brisbane, Queensland 4072, Australia
Australian Infectious Diseases Research Centre, University of Queensland, Brisbane, Queensland 4072, Australia

Megan J. Maher
La Trobe Institute for Molecular Science, La Trobe University, Melbourne, Victoria 3086, Australia

Megan L. O'Mara
School of Chemistry and Molecular Biosciences, University of Queensland, Brisbane, Queensland 4072, Australia
School of Mathematics and Physics, University of Queensland, Brisbane, Queensland 4072, Australia

Brian J. Cafferty, David M. Fialho, Jaheda Khanam and Nicholas V. Hud
School of Chemistry and Biochemistry, Georgia Institute of Technology, 901 Atlantic Drive, Atlanta, Georgia 30332, USA
NSF-NASA Center for Chemical Evolution, Atlanta, Georgia 30332, USA

Ramanarayanan Krishnamurthy
NSF-NASA Center for Chemical Evolution, Atlanta, Georgia 30332, USA
Department of Chemistry, The Scripps Research Institute, La Jolla, California 92037, USA

Alberto Schena, Rudolf Griss and Kai Johnsson
École Polytechnique Fédérale de Lausanne, Institute of Chemical Sciences and Engineering, Avenue Forel 2, EPFL SB ISIC LIP BCH-4303, CH-1015 Lausanne, Switzerland
École Polytechnique Fédérale de Lausanne, Institute of Bioengineering, CH-1015 Lausanne, Switzerland
National Centre of Competence in Research in Chemical Biology, CH-1015 Lausanne, Switzerland

Sohini Mukherjee, Manjistha Mukherjee and Abhishek Dey
Department of Inorganic Chemistry, Indian Association for the Cultivation of Science, 2A&2B Raja SC Mullick Road, Jadavpur Kolkata 700032, India

Arnab Mukherjee, Ambika Bhagi-Damodaran and Yi Lu
Department of Chemistry, University of Illinois at Urbana-Champaign, Champaign, Illinois 61801, USA

Matthew B. Robers, Melanie L. Dart, Chad A. Zimprich, Thomas Machleidt, Kevin R. Kupcho, James R. Hartnett, Kristopher Zimmerman, Andrew L. Niles, Rachel Friedman Ohana, Danette L. Daniels, Michael Slater, Monika G. Wood, Mei Cong and Keith V. Wood
Promega Corporation, 2800 Woods Hollow Road, Fitchburg, Wisconsin 53711, USA

Thomas A. Kirkland and Sergiy Levin
Promega Biosciences Incorporated, 277 Granada Drive, San Luis Obispo, California 93401, USA

Yi-Qiang Cheng
UNT System College of Pharmacy, Department of Pharmaceutical Sciences, University of North Texas Health Science Center, FortWorth, Texas, USA

Carolyn C. Woodroofe
Promega Biosciences Incorporated, 277 Granada Drive, San Luis Obispo, California 93401, USA
National Institutes of Health, 9800 Medical Center Drive, Rockville MD 20850, USA

Hiroyuki Ohura, Chisato Shirahama, Tomohiro Furuzono, Hiroshi Shimada and Hirotaka Matsuura
Department of Applied Chemistry and Biochemistry, Graduate School of Science and Technology, Kumamoto University, 2-39-1 Kurokami, Chuo-ku, Kumamoto 860-8555, Japan

Toshihiro Ihara and Yusuke Kitamura
Department of Applied Chemistry and Biochemistry, Graduate School of Science and Technology, Kumamoto University, 2-39-1 Kurokami, Chuo-ku, Kumamoto 860-8555, Japan
CREST, Japan Science and Technology Agency, 7 Gobancho, Chiyoda-ku, Tokyo 102-0076, Japan

Rie Sonamoto, Yukiko Okuno and Masatoshi Hagiwara
Department of Anatomy and Developmental Biology, Graduate School of Medicine, Kyoto University, Yoshida-Konoe-cho, Kyoto 606-8501, Japan

Isao Kii and Yuka Koike
Department of Anatomy and Developmental Biology, Graduate School of Medicine, Kyoto University, Yoshida-Konoe-cho, Kyoto 606-8501, Japan
Pathophysiological and Health Science Team, Imaging Application Group, Division of Bio-Function Dynamics Imaging, RIKEN Center for Life Science Technologies, 6-7-3 Minatojima-minamimachi, Chuo-ku, Kobe 650-0047, Japan

Yuto Sumida and Tomoe Kato-Sumida
Laboratory of Chemical Bioscience, Institute of Biomaterials and Bioengineering, Tokyo Medical and Dental University, 2-3-10 Kanda-Surugadai, Tokyo 101-0062, Japan
Chemical Biology Team, Imaging Chemistry Group, Division of Bio-Function Dynamics Imaging, RIKEN Center for Life Science Technologies, 6-7-3 Minatojima-minamimachi, Chuo-ku, Kobe 650-0047, Japan

Suguru Yoshida and Takamitsu Hosoya
Laboratory of Chemical Bioscience, Institute of Biomaterials and Bioengineering, Tokyo Medical and Dental University, 2-3-10 Kanda-Surugadai, Tokyo 101-0062, Japan

Toshiyasu Goto and Hiroshi Shibuya
Department of Molecular Cell Biology, Medical Research Institute, Tokyo Medical and Dental University, 1-5-45, Yushima, Tokyo 113-8510, Japan

Minako Abe, Yosuke Nonaka, Teikichi Ikura and Nobutoshi Ito
Department of Structural Biology, Graduate School of Medical and Dental Sciences, Medical Research Institute, Tokyo Medical and Dental University, 1-5-45, Yushima, Tokyo 113-8510, Japan

Wai Shun Mak and Steve Bertolani
Department of Chemistry, University of California Davis, One Shields Avenue, Davis, California 95616, USA

Stephen Tran
Department of Bioinformatics, University of California, Los Angeles, 172 Boyer Hall Box 951570, 611 Charles E. Young Drive East, Los Angeles, California 90095-1570, USA

Ryan Marcheschi
Department of Chemical and Biomolecular Engineering, University of California Los Angeles, 5531 Boelter Hall, Los Angeles, California 90095-1592, USA

James Thompson and David Baker
Department of Biochemistry, University of Washington, 1705 NE Pacific St Seattle, Washington 98195-7350, USA

James C. Liao
Department of Chemical and Biomolecular Engineering, University of California Los Angeles, 5531 Boelter Hall, Los Angeles, California 90095-1592, USA
Institute for Genomics and Proteomics, University of California Los Angeles, 611 Young Drive East, Los Angeles, California 90095, USA

Justin B. Siegel
Department of Chemistry, University of California Davis, One Shields Avenue, Davis, California 95616, USA
Department of Biochemistry and Molecular Medicine, University of California Davis, 2700 Stockton Boulevard, Suite 2102, Sacramento, California 95817, USA
Genome Center, University of California Davis, 451 Health Sciences Drive, Davis, California 95616, USA

Sunil Kumar
Department of Molecular Biophysics and Biochemistry, Yale University, 260 Whitney Avenue, New Haven, Connecticut 06520-8114, USA
Department of Chemistry, New York University, Silver Center for Arts and Science, 100 Washington Square East, 10th Floor, New York, New York 10003, USA (S.K.)

Elizabeth Rhoades
Department of Molecular Biophysics and Biochemistry, Yale University, 260 Whitney Avenue, New Haven, Connecticut 06520-8114, USA
Department of Chemistry, University of Pennsylvania, 231 South, 34th Street, Philadelphia, Pennsylvania 19104, USA (E.R.)

Melissa Birol, Diana E. Schlamadinger and Slawomir P. Wojcik
Department of Molecular Biophysics and Biochemistry, Yale University, 260 Whitney Avenue, New Haven, Connecticut 06520-8114, USA

Andrew D. Miranker
Department of Molecular Biophysics and Biochemistry, Yale University, 260 Whitney Avenue, New Haven, Connecticut 06520-8114, USA
Department of Chemical and Environmental Engineering, Yale University, 260 Whitney Avenue, New Haven, Connecticut 06520-8114, USA

Lukas Lercher
Department of Chemistry, University of Oxford, Chemistry Research Laboratory, Mansfield Road, Oxford OX1 3TA, UK
EMBL Heidelberg, Meyerhofstraβe 1, 69117 Heidelberg, Germany

Ritu Raj, Joshua Price, Shabaz Mohammed, Christopher J. Schofield and Benjamin G. Davis
Department of Chemistry, University of Oxford, Chemistry Research Laboratory, Mansfield Road, Oxford OX1 3TA, UK

Nisha A. Patel and Carol V. Robinson
Department of Chemistry, University of Oxford, Physical and Theoretical Chemistry Laboratory, South Parks Road, Oxford OX1 3QZ, UK

Ting Dong
Graduate School of Peking Union Medical College and Chinese Academy of Medical Sciences, Beijing 100730, China
National Institute of Biological Sciences (NIBS), Beijing 102206, China

Longyang Dian, Xiuguo Zhang, Lin Li, She Chen, Ran Cao, Li Li and Niu Huang
National Institute of Biological Sciences (NIBS), Beijing 102206, China

Xiaoguang Lei
Graduate School of Peking Union Medical College and Chinese Academy of Medical Sciences, Beijing 100730, China
National Institute of Biological Sciences (NIBS), Beijing 102206, China
Beijing National Laboratory for Molecular Sciences, Department of Chemical Biology, College of Chemistry and Molecular Engineering, Synthetic and Functional Biomolecules Center, and Peking-Tsinghua Center for Life Sciences, Peking University, Beijing 100871, China

Chao Li
Beijing National Laboratory for Molecular Sciences, Department of Chemical Biology, College of Chemistry and Molecular Engineering, Synthetic and Functional Biomolecules Center, and Peking-Tsinghua Center for Life Sciences, Peking University, Beijing 100871, China.

Sudan He and Xing Wang
Cyrus Tang Hematology Center, Jiangsu Institute of Hematology, First Affiliated Hospital, and Collaborative Innovation Center of Hematology, Soochow University, Suzhou 215123, China

Yusuke Miyazaki, Takamasa Kudo, Ling-chun Chen, Tobias Meyer and Thomas J. Wandless
Department of Chemical & Systems Biology Stanford University, Stanford, California 94305, USA

Gautam Dey
Department of Chemical & Systems Biology Stanford University, Stanford, California 94305, USA
UCL-MRC Lab for Molecular Cell Biology, LondonWC1E 6BT, UK (G.D.)

Kota Mizumoto
Department of Biology Stanford University, Stanford, California 94305, USA
Department of Zoology, The University of British Columbia, Vancouver, British Columbia, Canada V6T 1Z3 (K.M.)

John Perrino
Cell Sciences Imaging Facility Stanford University, Stanford, California 94305, USA

Robin Y. Kim and Michael C. Yau
Department of Anesthesiology, Pharmacology, and Therapeutics, University of British Columbia, 2176 Health Sciences Mall, Vancouver, British Columbia, Canada V6T 1Z3
Department of Pharmacology, University of Alberta, 9-70 Medical Sciences Building, Edmonton, Alberta, Canada T6G 2H7

Harley T. Kurata
Department of Anesthesiology, Pharmacology, and Therapeutics, University of British Columbia, 2176 Health Sciences Mall, Vancouver, British Columbia, Canada V6T 1Z3

Jason D. Galpin and Christopher A. Ahern
Department of Molecular Physiology and Biophysics, University of Iowa, 285 Newton Road, Iowa City, Iowa 52242, USA

Guiscard Seebohm
Department of Cardiovascular Medicine, University Hospital Münster, Albert-Schweitzer-Campus 1 (Gebäude D3), D-48149 Münster, Germany

Stephan A. Pless
Department of Drug Design and Pharmacology (Center for Biopharmaceuticals), University of Copenhagen, Jagtvej 160, DK-2100 Copenhagen, Denmark

Kiriko Hiraoka, Takahiro Inoue, Takayoshi Watanabe, Nobuko Koshikawa, Hiroyuki Yoda, Ken-ichi Shinohara, Atsushi Takatori, Yoshiaki Maru and Hiroki Nagase
Laboratory of Cancer Genetics, Chiba Cancer Center Research Institute, 666-2 Nitona, Chuoh-ku, Chiba 260-8717, Japan

Toshikazu Bando, Hiroshi Sugiyama and Rhys Dylan Taylor
Department of Chemistry, Graduate School of Science, Kyoto University, Kitashirakawa-oiwakecho, Sakyo-ku, Kyoto 606-8502, Japan

Hirokazu Sugimoto and Toshinori Ozaki
Laboratory of DNA Damage Signaling, Chiba Cancer Center Research Institute, 666-2 Nitona, Chuoh-ku, Chiba 260-8717, Japan

Tadamichi Denda
Department of Gastroenterology, Chiba Cancer Center, 666-2 Nitona, Chuoh-ku, Chiba 260-8717, Japan

Kyoko Fujiwara
Innovative Therapy Research Group, Nihon University Research Institute of Medical Science, Nihon University School of Medicine, 30-1 Ooyaguchi-kami, Itabashi-ku, Tokyo 173-8610, Japan

Allan Balmain
Helen Diller Family Comprehensive Cancer Center, University of California, San Francisco, California 94158, USA

Woo Hyeon Jeong, Jae-Hoon Eom, Hee-Seung Lee and Jie-Oh Lee
Department of Chemistry, KAIST, Daejeon 34141, Korea

Haerim Lee and Sun Chang Kim
Department of Biological Sciences, KAIST, Daejeon 34141, Korea

Dong Hyun Song
Agency for Defense Development, Daejeon 34186, Korea

Hayyoung Lee
Institute of Biotechnology, Chungnam National University, Daejeon 34134, Korea

Mu Cheng, Alysha G. Elliott, Johnny X. Huang, Mark S. Butler, Megha Bajaj and Matthew A. Cooper
Institute for Molecular Bioscience, The University of Queensland, St Lucia, Queensland 4072, Australia

Johannes Zuegg
Institute for Molecular Bioscience, The University of Queensland, St Lucia, Queensland 4072, Australia
Alchemia Ltd, PO Box 4851, Eight Mile Plains, Brisbane, Queensland 4113, Australia

Craig Muldoon, George Adamson, Declan McKeveney, Giang Le Thanh, Rajaratnam Premraj, Bernd Becker, Joachim Seifert, Latika Singh and Wim Meutermans
Alchemia Ltd, PO Box 4851, Eight Mile Plains, Brisbane, Queensland 4113, Australia

Nicola F. Galley, David I. Roper, Adrian J. Lloyd and Christopher G. Dowson
School of Life Science, University of Warwick, Gibbet Hill Road, Coventry CV4 7AL, UK

Ting-Jen Cheng and Wei-Chieh Cheng
Genomics Research Center, Academia Sinica, 128 Academia Road, Section 2, Taipei 115, Taiwan

Dieter Demon and Evelyne Meyer
Faculty of Veterinary Medicine, Laboratory of Biochemistry, Ghent University, Salisburylaan 133, 9820 Merelbeke, Belgium

Hee Young Yoo and Mihaela Iordachescu
Division of Integrative Biosciences and Biotechnology, Pohang University of Science and Technology, Pohang 37673, Republic of Korea

Jun Huang and Hongbo Zeng
Department of Chemical and Materials Engineering, University of Alberta, Edmonton, Alberta, Canada T6G 2V4

Patrick Flammang
Laboratory of Biology of Marine Organisms and Biomimetics, Research Institute for Biosciences, University of Mons, 7000 Mons, Belgium

Elise Hennebert
Laboratory of Biology of Marine Organisms and Biomimetics, Research Institute for Biosciences, University of Mons, 7000 Mons, Belgium
Laboratory of Cell Biology, Research Institute for Biosciences, University of Mons, 7000 Mons, Belgium

Dong Soo Hwang
Division of Integrative Biosciences and Biotechnology, Pohang University of Science and Technology, Pohang 37673, Republic of Korea
School of Environmental Science and Engineering, Pohang University of Science and Technology, Pohang 37673, Republic of Korea

Sangsik Kim
School of Environmental Science and Engineering, Pohang University of Science and Technology, Pohang 37673, Republic of Korea

Sangchul Rho
Center for Plant Aging Research, Institute for Basic Science (IBS), Daegu 42988, Republic of Korea

Mathias Foo
School of Engineering, University of Warwick, Coventry CV4 7AL, UK

Daehee Hwang
Center for Plant Aging Research, Institute for Basic Science (IBS), Daegu 42988, Republic of Korea
Department of New Biology, DGIST, Daegu 42988, Republic of Korea

J. Herbert Waite
Department of Molecular, Cellular, and Developmental Biology, University of California, Santa Barbara, California 93106, USA

Helena Almqvist, Hanna Axelsson, Martin Haraldsson and Thomas Lundbäck
Laboratories for Chemical Biology, Karolinska Institutet, Science for Life Laboratory Stockholm, Division of Translational Medicine & Chemical Biology, Department of Medical Biochemistry & Biophysics, Karolinska Institutet, Tomtebodavägen 23A, Solna 171 65, Sweden

Rozbeh Jafari
Department of Medical Biochemistry & Biophysics, Division of Biophysics, Karolinska Institutet, Scheeles väg 2, Stockholm 171 77, Sweden
Clinical Proteomics Mass Spectrometry, Department of Oncology-Pathology, Science for Life Laboratory and Karolinska Institutet, Stockholm, Sweden

Daniel Martinez Molina
Department of Medical Biochemistry & Biophysics, Division of Biophysics, Karolinska Institutet, Scheeles väg 2, Stockholm 171 77, Sweden

Chen Dan
School of Biological Sciences, Nanyang Technological University, 61 Biopolis Drive (Proteos), Singapore 138673, Singapore

André Mateus
Department of Pharmacy, Uppsala University, BMC, Box 580, Uppsala SE-751 23, Sweden

Andreas Larsson
School of Biological Sciences, Nanyang Technological University, SBS-04s-45, 60 Nanyang Drive, Singapore 639798, Singapore

Per Artursson
Department of Pharmacy, Uppsala University, BMC, Box 580, Uppsala SE-751 23, Sweden
Uppsala University Drug Optimization and Pharmaceutical Profiling Platform (UDOPP), Department of Pharmacy, Uppsala University, BMC, Box 580, Uppsala SE-751 23, Sweden
Science for Life Laboratory Drug Discovery and Development platform, Uppsala University, Uppsala SE-751 23, Sweden

Pär Nordlund
Department of Medical Biochemistry & Biophysics, Division of Biophysics, Karolinska Institutet, Scheeles väg 2, Stockholm 171 77, Sweden
School of Biological Sciences, Nanyang Technological University, 61 Biopolis Drive (Proteos), Singapore 138673, Singapore
Institute of Cellular and Molecular Biology, ASTAR, 61 Biopolis Drive (Proteos), Singapore 138673, Singapore

Louise J. Walport, Richard J. Hopkinson, Rasheduzzaman Chowdhury, Rachel Schiller, Wei Ge and Christopher J. Schofield
Department of Chemistry, Chemistry Research Laboratory, University of Oxford, Mansfield Road, Oxford OX1 3TA, UK

Akane Kawamura
Department of Chemistry, Chemistry Research Laboratory, University of Oxford, Mansfield Road, Oxford OX1 3TA, UK
Division of Cardiovascular Medicine, Radcliffe Department of Medicine, Wellcome Trust Centre for Human Genetics, Roosevelt Drive, Oxford OX3 7BN, UK

Andrew R. Burns, Rachel Bagg, May Yeo, Yuqian Zhang, Luckshika Rajendran, John Glavin, Robert Hunter and Michael Schertzberg
The Donnelly Centre for Cellular and Biomolecular Research, University of Toronto, Toronto, Ontario, Canada M5S 3E1

Andy G. Fraser and Genna M. Luciani
The Donnelly Centre for Cellular and Biomolecular Research, University of Toronto, Toronto, Ontario, Canada M5S 3E1
Department of Molecular Genetics, University of Toronto, Toronto, Ontario, Canada M5S 1A8

Calum A. MacRae and Gabriel Musso
Cardiovascular Division, Brigham and Women's Hospital, Harvard Medical School, and Harvard Stem Cell Institute, Boston, Massachusetts 02115, USA
Department of Medicine, Harvard Medical School, Boston, Massachusetts 02115, USA

Elizabeth Redman, Susan Stasiuk and John Gilleard
Department of Comparative Biology and Experimental Medicine, Faculty of Veterinary Medicine, University of Calgary, Calgary, Alberta, Canada T2N 4Z6

G. Angus McQuibban
Department of Biochemistry, University of Toronto, 1 King's College Circle, Toronto, Ontario, Canada M5S 1A8

Conor R. Caffrey
Center for Discovery and Innovation in Parasitic Diseases and Department of Pathology, University of California, San Francisco, California 94158, USA

Sean R. Cutler
Center for Plant Cell Biology, Department of Botany and Plant Sciences, University of California, Riverside, California 92521, USA

Mike Tyers
Institute for Research in Immunology and Cancer, University of Montreal, Montreal, Quebec, Canada H3T 1J4

Guri Giaever and Corey Nislow
Department of Pharmaceutical Sciences, University of British Columbia, Vancouver, British Columbia, Canada V6T 1Z3

Peter J. Roy
The Donnelly Centre for Cellular and Biomolecular Research, University of Toronto, Toronto, Ontario, Canada M5S 3E1
Department of Molecular Genetics, University of Toronto, Toronto, Ontario, Canada M5S 1A8
Department of Pharmacology and Toxicology, University of Toronto, Toronto, Ontario, Canada M5S 1A8

Taichiro Tomida
Division of Molecular Cell Signaling, Institute of Medical Science, The University of Tokyo, 4-6-1 Shirokanedai, Minato-ku, Tokyo 108-8639, Japan
Department of Physiology, School of Medicine, Toho University, 5-21-16 Omori-Nishi, Ota-ku, Tokyo 143-8540, Japan

Haruo Saito
Division of Molecular Cell Signaling, Institute of Medical Science, The University of Tokyo, 4-6-1 Shirokanedai, Minato-ku, Tokyo 108-8639, Japan

Mutsuhiro Takekawa
Division of Cell Signaling and Molecular Medicine, Institute of Medical Science, The University of Tokyo, 4-6-1 Shirokanedai, Minato-ku, Tokyo 108-8639, Japan

Samira Husen Alamudi, Rudrakanta Satapathy, Rajkumar Das, Lingna Hu and Jung Yeol Lee
Department of Chemistry & Med Chem Program, National University of Singapore, Singapore 117543, Singapore

Jihyo Kim and Hee-Sung Park
Department of Chemistry, Korea Advanced Institute of Science and Technology, Daejeon 305701, Republic of Korea

Young-Tae Chang
Department of Chemistry & Med Chem Program, National University of Singapore, Singapore 117543, Singapore
Laboratory of Bioimaging Probe Development, Singapore Bioimaging Consortium, Agency for Science, Technology and Research, Singapore 138667, Singapore

Dongdong Su
Laboratory of Bioimaging Probe Development, Singapore Bioimaging Consortium, Agency for Science, Technology and Research, Singapore 138667, Singapore

Haiyan Ren, Christian Hoppmann and Lei Wang
Department of Pharmaceutical Chemistry and the Cardiovascular Research Institute, University of California San Francisco, California 94158, USA

Eduardo Peña-Cabrera and Enrique Alvarado-Martínez
Departamento de Química, Universidad de Guanajuato, Guanajuato 36050, Mexico

Hyung-Ho Ha
College of Pharmacy and Research Institute of Life and Pharmaceutical Sciences, Sunchon National University, Sunchon 540950, Republic of Korea

Victoria C. Corey, Olivia Fuchs, Greg Goldgof, Stephan Meister, Melanie Wree and Elizabeth A. Winzeler
Department of Pediatrics, School of Medicine, University of California San Diego, 9500 Gilman Drive 0741, La Jolla, California 92093, USA

Pamela Magistrado and Tomoyo Sakata-Kato
Department of Immunology and Infectious Disease, Harvard T.H. Chan School of Public Health, 665 Huntington Avenue, Boston, Massachusetts 02115, USA

Dyann F. Wirth and Amanda K. Lukens
Department of Immunology and Infectious Disease, Harvard T.H. Chan School of Public Health, 665 Huntington Avenue, Boston, Massachusetts 02115, USA
Infectious Disease Program, The Broad Institute, 415 Main Street, Cambridge, Massachusetts 02142, USA

Daniel E. Goldberg and Eva S. Istvan
Department of Medicine and Microbiology, Washington University School of Medicine, St Louis, Missouri 63110, USA

David A. Fidock, Olivia Coburn-Flynn, Marcus C.S Lee and Nina F. Gnädig
Department of Microbiology and Immunology, Columbia University College of Physicians and Surgeons, New York, New York 10032, USA

Francisco-Javier Gamo, Maria Linares, Maria G. Gomez-Lorenzo, Cristina De Cózar, Maria Jose Lafuente-Monasterio, Sara Prats and Virginia Franco
Tres Cantos Medicines Development Campus, Malaria DPU, GlaxoSmithKline, Severo Ochoa 2, Tres Cantos, 28760 Madrid, Spain

Yingyao Zhou and Olga Tanaseichuk
The Genomics Institute of the Novartis Research Foundation, 10675 John J Hopkins Drive, San Diego, California 92121, USA

Paul A.Willis
Medicines for Malaria Venture, PO Box 1826, 20 route de Pre-Bois, 1215 Geneva 15, Switzerland

Index